Advances in Intelligent Systems and Computing

Volume 1225

The series "Advances in Intelligent Systems and Computing" contains publications on theory, applications, and design methods of Intelligent Systems and Intelligent Computing. Virtually all disciplines such as engineering, natural sciences, computer and information science, ICT, economics, business, e-commerce, environment, healthcare, life science are covered. The list of topics spans all the areas of modern intelligent systems and computing such as: computational intelligence, soft computing including neural networks, fuzzy systems, evolutionary computing and the fusion of these paradigms, social intelligence, ambient intelligence, computational neuroscience, artificial life, virtual worlds and society, cognitive science and systems, Perception and Vision, DNA and immune based systems, self-organizing and adaptive systems, e-Learning and teaching, human-centered and human-centric computing, recommender systems, intelligent control, robotics and mechatronics including human-machine teaming, knowledge-based paradigms, learning paradigms, machine ethics, intelligent data analysis, knowledge management, intelligent agents, intelligent decision making and support, intelligent network security, trust management, interactive entertainment, Web intelligence and multimedia.

The publications within "Advances in Intelligent Systems and Computing" are primarily proceedings of important conferences, symposia and congresses. They cover significant recent developments in the field, both of a foundational and applicable character. An important characteristic feature of the series is the short publication time and world-wide distribution. This permits a rapid and broad dissemination of research results.

**** Indexing: The books of this series are submitted to ISI Proceedings, EI-Compendex, DBLP, SCOPUS, Google Scholar and Springerlink ****

More information about this series at http://www.springer.com/series/11156

Radek Silhavy
Editor

Artificial Intelligence and Bioinspired Computational Methods

Proceedings of the 9th Computer Science On-line Conference 2020, Vol. 2

Springer

Editor
Radek Silhavy
Tomas Bata University
Zlín, Czech Republic

ISSN 2194-5357 ISSN 2194-5365 (electronic)
Advances in Intelligent Systems and Computing
ISBN 978-3-030-51970-4 ISBN 978-3-030-51971-1 (eBook)
https://doi.org/10.1007/978-3-030-51971-1

This Springer imprint is published by the registered company Springer Nature Switzerland AG
The registered company address is: Gewerbestrasse 11, 6330 Cham, Switzerland

Preface

Artificial Intelligence and Bioinspired Computational Methods papers and topics are presented in this proceedings. This proceedings is a Vol. 2 of the Computer Science On-line Conference. Papers in this part discuss modern hybrid and bioinspired algorithms and their applications.

This book constitutes the refereed proceedings of the Artificial Intelligence and Bioinspired Computational Methods section of the 9th Computer Science On-line Conference 2020 (CSOC 2020), held online in April 2020.

CSOC 2020 has received (all sections) more than 270 submissions from more than 35 countries. More than 65% of accepted submissions were received from Europe, 21% from Asia, 8% from Africa, 4% from America and 2% from Australia.

CSOC 2020 conference intends to provide an international forum for the discussion of the latest high-quality research results in all areas related to computer science.

Computer Science On-line Conference is held online using a modern communication technology. This approach improves the traditional concept of scientific conferences. It brings equal opportunity to participate for all researchers around the world.

I believe that you find the following proceedings exciting and useful for your research work.

April 2020 Radek Silhavy

Organization

Program Committee

Program Committee Chairs

Petr Silhavy	Tomas Bata University in Zlin, Faculty of Applied Informatics, Czech Republic
Radek Silhavy	Tomas Bata University in Zlin, Faculty of Applied Informatics, Czech Republic
Zdenka Prokopova	Tomas Bata University in Zlin, Faculty of Applied Informatics, Czech Republic
Roman Senkerik	Tomas Bata University in Zlin, Faculty of Applied Informatics, Czech Republic
Roman Prokop	Tomas Bata University in Zlin, Faculty of Applied Informatics, Czech Republic
Viacheslav Zelentsov	Doctor of Engineering Sciences, Chief Researcher of St.Petersburg Institute for Informatics and Automation of Russian Academy of Sciences (SPIIRAS), Russia
Roman Tsarev	Department of Informatics, Siberian Federal University, Krasnoyarsk, Russia

Program Committee Members

Boguslaw Cyganek	Department of Computer Science, University of Science and Technology, Krakow, Poland
Krzysztof Okarma	Faculty of Electrical Engineering, West Pomeranian University of Technology, Szczecin, Poland
Monika Bakosova	Institute of Information Engineering, Automation and Mathematics, Slovak University of Technology, Bratislava, Slovak Republic

Pavel Vaclavek	Faculty of Electrical Engineering and Communication, Brno University of Technology, Brno, Czech Republic
Miroslaw Ochodek	Faculty of Computing, Poznan University of Technology, Poznan, Poland
Olga Brovkina	Global Change Research Centre Academy of Science of the Czech Republic, Brno, Czech Republic, and Mendel University of Brno, Czech Republic
Elarbi Badidi	College of Information Technology, United Arab Emirates University, Al Ain, United Arab Emirates
Luis Alberto Morales Rosales	Head of the Master Program in Computer Science, Superior Technological Institute of Misantla, Mexico
Mariana Lobato Baes	Superior Technological of Libres, Mexico
Abdessattar Chaâri	Laboratory of Sciences and Techniques of Automatic Control and Computer Engineering, University of Sfax, Tunisian Republic
Gopal Sakarkar	Shri. Ramdeobaba College of Engineering and Management, Republic of India
V. V. Krishna Maddinala	GD Rungta College of Engineering & Technology, Republic of India
Anand N. Khobragade (Scientist)	Maharashtra Remote Sensing Applications Centre, Republic of India
Abdallah Handoura	Computer and Communication Laboratory, Telecom Bretagne, France

Technical Program Committee Members

Ivo Bukovsky
Maciej Majewski
Miroslaw Ochodek
Bronislav Chramcov
Eric Afful Dazie
Michal Bliznak
Donald Davendra
Radim Farana
Martin Kotyrba
Erik Kral
David Malanik
Michal Pluhacek
Zdenka Prokopova
Martin Sysel
Roman Senkerik
Petr Silhavy
Radek Silhavy
Jiri Vojtesek
Eva Volna
Janez Brest
Ales Zamuda
Roman Prokop
Boguslaw Cyganek
Krzysztof Okarma
Monika Bakosova
Pavel Vaclavek
Olga Brovkina
Elarbi Badidi

Organizing Committee Chair

Radek Silhavy — Tomas Bata University in Zlin, Faculty of Applied Informatics,
email: radek@silhavy.cz

Conference Organizer (Production)

Silhavy s.r.o.
Website: https://www.openpublish.eu
Email: csoc@openpublish.eu

Conference Website, Call for Papers

https://www.openpublish.eu

Contents

Decision Support Model for Assessing Projects by a Group of Investors with Regards of Multi-factors

Valeriy Lakhno[1], Volodymyr Malyukov[1], Berik Akhmetov[2], Nataliia Gerasymchuk[3(✉)], Hennadii Mohylnyi[4], and Petro Kravchuk[5]

[1] Department of Computer Systems and Networks, National University of Life and Environmental Sciences of Ukraine, Kiev, Ukraine
valss21@ukr.net, volod.malyukov@gmail.com
[2] Department of Computer Systems, Yessenov University, Aktau, Kazakhstan
Berik.Akhmetov@yu.edu.kz
[3] Department of Economics, Rzeszow University of Technology, Rzeszów, Poland
90999nag@gmail.com
[4] Department of Computer Systems and Networks, Luhansk Taras Shevchenko National University, Starobilsk, Ukraine
g.mogilniy@gmail.com
[5] Department of Security Management of an Enterprise, Private Higher Educational Establishment, European University", Kiev, Ukraine
p.kr@ukr.net

Abstract. In today's globalized economy, the problems connected with the choice of a rational strategy of investing in advanced information technologies, is still relevant. In order to increase the reliability of solutions in the process of analysis and selection of sound financial strategies for the investor or group of investors (or other decision-makers (DM)) can intellectualized decision support systems. The solution to this problem has its own specific features in each subject area. For example, for tasks related to the assessment of investment projects in the field of enterprises' digitalization, cyber security or Smart City development. A similar problem in the process of decision needs to take into account many factors. This can be done, in particular, with the use of the games theory mathematical apparatus. The development of algorithmic and software component of such expert systems and decision support systems made it necessary to continue the development of models that are based on the quality of bilinear differential game with multiple terminal faces. During the study we considered a new class of bilinear differential game. The article describes a model for the case of the interaction of objects groups in multidimensional space. This approach has enabled to describe adequately the process of finding rational strategies of groups of players (investors). The proposed model was tested in the simulation environment MatLab. The software product being developed will reduce the discrepancies in the data of forecast estimates for investment projects, as well as optimize the choice of strategies for groups of investors.

R. Silhavy (Ed.): CSOC 2020, AISC 1225, pp. 1–10, 2020.
https://doi.org/10.1007/978-3-030-51971-1_1

Keywords: Group of investors · Investment strategy · Multidimensional case · Decision support · Differential game

1 Introduction

The task of analyzing and choosing rational investment strategies has always been quite complicated, since in the course of its solution it is necessary to take into account many very diverse factors. Digital technologies that have emerged over the past decade have aroused great interest among investors, and the development of the digital economy has required a review and, in some cases, a fundamental change in existing investment management models. Many players in the market for investments in technologies related to digitalization of enterprises realize that such financial and investment projects are characterized by a high degree of uncertainty and riskiness [1, 2]. Which, not least, is associated with the multifactorial and multidimensional nature of possible rational investment strategies. Not the least role is played by possible scenarios for the development of the situation in the investment market in related fields, for example, in the field of cybersecurity and the protection of information and communication systems (ICS) of enterprises. Let's note that in some cases of illegitimate interference in the work of ICS can nullify all the efforts of investors. In [2–6], the authors noted that, to improve the efficiency and effectiveness of such projects evaluation (we are talking about investment projects in the field of enterprises' digitalization, cybersecurity, protection ICS, the Smart City technology development, etc.), it is advisable to use the potential of various computerized decision support systems (CDSS). Without a doubt, this statement applies to the major interstate or inter-regional projects of investment in digital technologies and systems [3–5].

All of the above predetermined the research theme relevance, particularly in the aspect of the need to develop new and development of existing models and corresponding software and algorithmic content of decision support systems that will reduce the difference prediction data and the real return on investment in digital technologies and systems.

2 Literature Review

The mathematical and cybernetic aspects of effective financial investment in information technologies and systems (ITS) are considered in publications of many authors [4–10]. At the same time, a number of authors [8, 9], reasonably showed that modern ITS today have become the hallmark of modern enterprises. And not only in the field of high technologies, but also in traditional sectors of the economy. Note that a number of authors [8, 10, 11] showed in their works that in the course of evaluating investment projects in modern ITS, it is no longer sufficient to rely solely on traditional methods and models [10]. For example, in [11, 12] it was shown that, despite the simplicity and logical harmony of the hierarchy analysis method (T. Saati method) [13], it does not always give a reliable result in cases of choosing rational financial strategies of an investor. And although quite a lot of software products are built on this method, including CDSS, their application is often limited by the choice of local criteria and alternatives. It was shown in [9, 10, 14] that the choice of rational strategies by groups of investors and the synthesis of forecasts by investors, which concern the advisability of choosing strategies for investing their financial resources in projects, has not been studied enough. And besides, for a similar and rather complicated statement of the problem, there are no corresponding software implementations in the form of CDSS or expert systems.

In [7, 15, 16], various researchers described models that, in particular, can be used as the basis for CDSS algorithms in the analysis of investor strategies in the context of actions of two parties (players). However, the resulting algorithms did not receive further implementation in the form of complete software applications.

In works [16–18], it was noted that in relation to tasks in which it is necessary to evaluate investor strategies in a multi-factor environment, the most acceptable are models based on game theory. In accordance with this approach [17, 18], the cornerstone is the assumption that one of the parties of the investment process is considered as a certain combination of potential "threats". In this context, a threat is the result of incompetent, uncoordinated actions by investors or a group. Realization of the threat leads to the loss of capital by investors that was spent on the project. And given the multifactorial nature of such tasks, such a statement is more than obvious.

The analysis of publications on the topic of our study, and in particular the works [1, 5, 9, 12–18], revealed the following: (1) many models and algorithms in these publications do not contain the procedure for the investor to receive a clear recommendation on what to do in a certain financial investment strategy; (2) there are practically no models that take into account the multifactorial nature of the task of investing in ITS when it comes to the interaction of a group of investors. Moreover, such an interaction can be conflicting.

A separate area of publications on selected topics are articles and monographs on the application of computer technologies, and, in particular, various intellectualized expert (ES) [19] and CDSS [19, 20] to problems of choosing investment strategies for computer technologies. However, the software products (PP) existing in this segment are still not well understood for the average investor, and are rather aimed at a fairly narrow circle of qualified experts. This is due to the fact that the calculation core mainly contains closed models for evaluating an investment project, the multifactorial nature of the task and behavioral options in the investor group are not taken into account.

The above circumstances ultimately prompted us to focus on the relevance of the problem of choosing a rational strategy by a group of investors. To solve this problem, continued the development of mathematical models for intelligent decision support systems in the search for rational investment strategies in the field of information technology and systems, as well as taking into account the multi-factor nature of the problem.

3 Study Objectives

Objectives – the development of models for intelligent decision support systems in the process of choosing a rational financial strategies by a group of investors.

To achieve this goal, it is necessary to solve the following tasks: develop a model using a toolkit of bilinear multi-step quality games with several terminal surfaces, taking into account the multifactorial nature of the task for finding rational strategies by a group of investors; perform computational experiments to verify the adequacy of the proposed model.

4 Methods and Models

Taking into account publications [12, 14], as well as our previous studies [15, 16, 21], the main areas and, accordingly, strategies of investors in the context of the main task are: investment projects evaluation, for example, in such areas as: digitalization of enterprises; cybersecurity and information security [22]; Smart City [23, 24] and others. The solution of such problems requires taking into account many factors. The latter is not always possible on the basis of classical models used in the process of evaluating an investment project. And even more so, it causes considerable difficulties if you do not use computer decision support systems that can take on a significant part of the routine computing operations to search for options or strategies of a potential investor.

One of the most important tasks facing services providing the development, creation and implementation of advanced technologies in the field of digitalization of enterprises, taking into account multifactorial nature, is the task of financially supporting projects and attracting financial resources ($FinR$) investors. In turn, the decision made on investment in the field of enterprise digitalization should be based on continuous procedures that allow financing take into account all possible factors. Among the factors undoubtedly include the multiplicity of innovative advanced technologies in the field of digitalization, cybersecurity, the development of cloud computing and technology. This is possible if intelligent CDSS or ES are developed and implemented. As the latter, we understand software products for different platforms (Windows, Linux, Android, iOS) that allow making rational decisions on investing financial resources for the development of such technologies.

Our model is based on an analysis of the possibilities of a continuous financing process for groups of players in the field of digitalization of enterprises, taking into account their multifactor nature, which is due to their multiplicity. The model is a continuation of our work [15, 16, 21] and is based on the solution of a bilinear differential quality game of groups of players with several terminal surfaces.

Let's formulate the statement of the problem. There are two groups of investors who seek to invest in projects related to enterprises digitalization (or another similar task mentioned above). Each group of investors acts as a single entity, and it can be considered that each investor in the group is financing a certain digital technology. For example, investors can promote blockchain or Data Mining technologies. Of course, we can talk about other technologies that can both complement and conflict with each other [24]. We believe that the first group is represented by the first player, and the second by the second player. Of course, there may be many investors who act separately, for example, by investing in various technologies as mentioned above. In practice, situations are possible when, for example, some investors are interested in developing e-commerce platforms, some in developing mobile banking, some in developing digital platforms for customer feedback, etc. That is, a different formulation of the problem is possible, for which many investors can act independently from each other, not approved.

So, for our production, players control a dynamic system in multidimensional spaces. This dynamic system sets the change in the financial flows of players. The system is defined by a system of bilinear differential equations with dependent motions. Sets of strategies (U) and (V) of players are defined. Terminal surfaces are also defined S_0,F_0. The goal of the first player (groups of players, hereinafter referred to *Inv*1, as the management of the investee) can bring a dynamic system with its control strategies to the terminal surface S_0. In this case *Inv*1, it does not matter how the second player acts (hereinafter – *Inv*2). The goal of the second player (group of players) *Inv*2 is to bring a dynamic system using his control strategies to the terminal surface F_0. And no matter how it works *Inv*1.

The formulated goal generates two tasks, from the point of view of the first ally player and from the point of view of the second ally player [15, 21].

In our study, we consider the problem from the point of view of the first ally player. This is due to the fact that the problem from the point of view of the second ally player can be solved similarly, due to symmetry. The solution is to find the set of initial states of the players. It is also necessary to find their rational strategies. That is, their strategies that allow objects to bring the system to one or another surface.

In Problem 1 player-ally is treated for *Inv*1, The opposing player is treated for *Inv*2. And vice versa – in task 2 player-ally treated for *Inv*2, and the opposing player – for *Inv*1. Note that the first player is a group of M investors, the second player is a group of K investors. We assume that at the time $t = 0$ at *Inv*1 a set of $h(0) = (h_1(0), \ldots, h_n(0))$ consisting of the vectors $h_i(0)$ which consist of n component. These components characterize the magnitude of the financial resources for the development of j-th new digital technologies (technologies in the enterprise) (j – vector component $h_i(0)$). Accordingly, the same set is available for *Inv*2 – $f(0) = (f_1(0), \ldots, f_n(0))$ ($f_i(0)$ vector, which consists of n component characterizing the magnitude of the financial resources for the development of j-th new digital technologies (technologies in the enterprise) (j – vector component $f_i(0)$).

These sets define forward-looking in the time $t = 0$, the magnitude *FinR* players for each new technology taking into account the multiple factors.

We describe the dynamics of change *FinR* for the players as follows:

$$
\begin{aligned}
& dh_1(t)/dt = -h_1(t) + G_1^1 \cdot h_1(t) - U_1(t) \cdot G_1^1 \cdot h_1(t) - \\
& S_2^{11} \cdot V_1(t) \cdot G_2^1 \cdot f_1(t) - \ldots - S_2^{1K} \cdot V_K \cdot G_2^K \cdot f_K(t); \\
& \ldots\ldots\ldots\ldots\ldots\ldots\ldots\ldots\ldots\ldots\ldots\ldots\ldots\ldots\ldots\ldots\ldots\ldots\ldots\ldots \\
& dh_M(t)/dt = -h_M(t) + G_1^M \cdot h_M(t) - U_M(t) \cdot G_1^M \cdot h_M(t) - \\
& - S_2^{M1} \cdot V_1(t) \cdot G_2^1 \cdot f_1(t) - \ldots - S_2^{MK} \cdot V_K \cdot f_K(t); \\
& df_1(t)/dt = -f_1(t) + G_2^1 \cdot f_1(t) - V_1(t) \cdot G_2^1 \cdot f_1(t) - \\
& S_1^{11} \cdot U_1(t) \cdot G_1^1 \cdot h_1(t) - \ldots - -S_1^{1M} \cdot U_M(t) \cdot G_1^M \cdot h_M(t); \\
& \ldots\ldots\ldots\ldots\ldots\ldots\ldots\ldots\ldots\ldots\ldots\ldots\ldots\ldots\ldots\ldots\ldots\ldots\ldots\ldots \\
& df_K(t)/dt = -f_K(t) + G_2^K \cdot f_K(t) - V_K(t) \cdot G_2^K \cdot f_K(t) - \\
& S_1^{K1} \cdot U_1(t) \cdot h_1(t) - \ldots - S_1^{KM} \cdot U_M(t) \cdot G_1^M \cdot h_M(t);
\end{aligned} \tag{1}
$$

Where h – the magnitude of financial resource ($FinR$) for $Inv1$; f – the magnitude of financial resource ($FinR$) $Inv2$; $h_1(t) \in R^n, \ldots, h_M(t) \in R^n$, $f_1(t) \in R^n, \ldots, f_K(t) \in R^n$; $U_1(t), \ldots, U_M(t), V_1(t), \ldots, V_K(t)$ – square diagonal matrix of order n with positive elements $u_j^i(t) \in [0,1], (i = 1, \ldots, n; j = 1, \ldots, M), v_l^i(t) \in [0,1], \quad i = 1, \ldots, n; l = 1, \ldots, K$; on the main diagonals of the matrices $U_j(t), V_l(t)$; G_1 – financial resource transformation matrix $Inv1$ if successfully implemented, in the field of digitization of enterprises taking into account multiple factors; G_2 – resource transformation matrix $Inv2$; S_1 – matrix, which characterizes the elasticity of the player's actions $Inv2$ with respect to the player 1 (i.e. $Inv1$); S_2 – matrix, which characterizes the elasticity of investment $Inv1$ towards $Inv2$; $G_1^j, G_2^l, S_1^{lj}, S_2^{jl}$ – square matrix of order n with positive elements $g_1^{km}, g_2^{km}, s_1^{km}, s_2^{km}, k = 1, \ldots, n; m = 1, \ldots, n$; Respectively, $t = 0, 1, \ldots$

In Problem 1 player-ally treated for $Inv1$, The opposing player is treated for $Inv2$ And vice versa. We believe that

$$
S_0 = \bigcup_{i=1}^{K \cdot n} \{(h,f) : (h,f) \in R^{2 \cdot n}, h \succ 0, \quad f_i = 0\} \tag{2}
$$

$$
F_0 = \bigcup_{i=1}^{M \cdot n} \{(h,f) : (h,f) \in R^{2 \cdot n}, \quad f \succ 0, h_i = 0\}. \tag{3}
$$

Where S_0, F_0 – terminal surface $Inv1$ and $Inv2$ respectively; $R^{2 \cdot n}$ (2-n-dimensional space).

Interaction is terminated under the conditions:

$$
(h(t), f(t)) \in S_0, \tag{4}
$$

$$(h(t), f(t)) \in F_0. \tag{5}$$

If condition (4) is fulfilled, then we believe that there was not enough financial means to continue the financing procedure. At least one of the promising digital technologies that the investor planned to develop initially. If condition (5) is fulfilled, then we believe that the investment procedure is also completed. That is, it did not have enough financial resources to continue the financing procedure. At least one of his priority technologies in the field of digitalization of the enterprise in which the investor was going to invest resources. If both conditions (4) and (5) are not fulfilled, we believe that the financing procedure continues. The process described by system (1) for the financing procedure is considered in the framework of the positional differential game scheme with full information [21]. The definition of a pure strategy and a set of preferences for the first player was given in [15, 16, 21].

The solution to problem 1 is to find the sets of "preference" *Inv*1 and its optimal strategies. Similarly, the task is posed from the point of view of *Inv*2. As part of the article, we give an example for the case when the first player consists of two investors. The state of these investors is determined by the two-dimensional vector of financial resources $(h_1(t), h_2(t))$. Therefore, the system of differential equations consists of two differential equations. The second player controls the investment process, which is defined by a one-dimensional vector of financial resources $f(t)$. The dynamic system is defined as follows:

$$\begin{aligned} dh_1(t)/dt &= -h_1(t) + (1 - u_1(t)) \cdot h_1(t) - 2 \cdot v(t) \cdot f(t), \\ dh_2(t)/dt &= -h_2(t) + (1 - u_2(t)) \cdot h_2(t) - 2 \cdot v(t) \cdot f(t), \\ df(t)/dt &= -f(t) + (1 - v(t)) \cdot 2 \cdot f(t) - u_1(t) \cdot h_1(t) - u_2(t) \cdot h_2(t); \end{aligned}$$

in this game, condition 4 is satisfied. The set W_1 is written as:

$$\begin{aligned} W_1 = \{(h_1(0), h_2(0), f(0)) : (h_1(0), h_2(0), f(0)) \\ \in \mathrm{int}R^3_+, f(0) \prec (h_1(0) \cdot h_2(0))^{0.5}\}. \end{aligned}$$

The optimal strategy $u^*(.)$ defined as follows:

$$u^*(h_1(0), h_2(0), f(0)) = \begin{cases} 1, (h_1(0), h_2(0), f(0)) \in W_1; \\ 0, (h_1(0), h_2(0), f(0)) \notin W_1, \end{cases}$$

Where R^3_+ – positive orthant in three-dimensional space.

5 Computing Experiment

Computational experiments were performed by MatLab. As initial data were taken on investment projects in the field of digitalization of enterprises in Kiev (Ukraine). The figures show the results of modeling the optimality set of players (investors). Figures 1, 2 and 3 show plurality W_1 of different funding strategies. Offered the first player is a group of two players, and the second player is a group of one player. Then the first Fig. 1 corresponds to the coordinated action of the players in the first group.

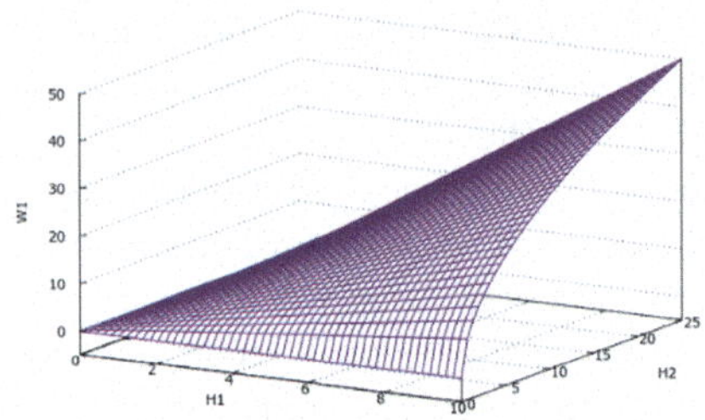

Fig. 1. The result of modeling the coordinated action of players in the first group

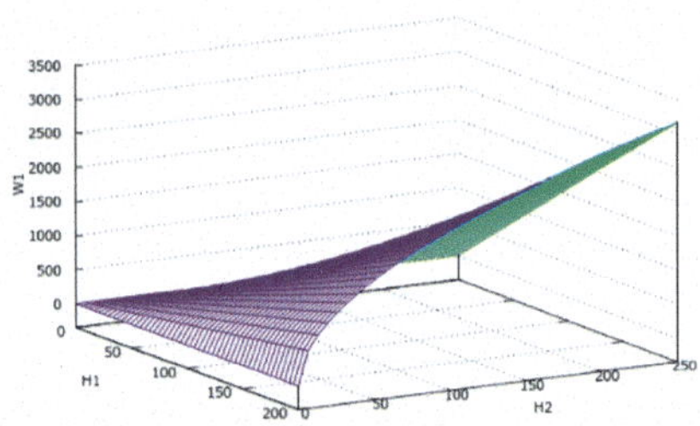

Fig. 2. The result of modeling the optimality set of the first player in the first group with uncoordinated actions of players in the first group

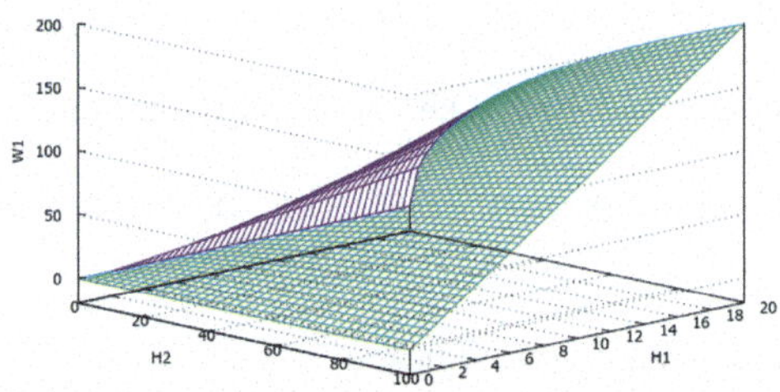

Fig. 3. The result of modeling the optimality set of the second player in the first group in the first group with their uncoordinated actions of the players in the first group

6 Discussion of Computational Experiment Results

In Fig. 1 showed the results of modeling the optimality set of the first player for the case of inconsistency of players' actions in the first group. Figures 2 and 3 correspond to the optimality sets of the first player in the first group and the second player in the first group with their uncoordinated actions. Their union is "less" (by inclusion of sets) of the optimality set of the first group under the coordinated action of the players in the group. This is illustrated in Figs. 2 and 3. In comparison with the available models, which, for example, are described in [8, 12], the proposed model improved the efficiency and predictability for the investor by an average of 7–12%. Further prospects for the development of models and software products described in the article are the transfer of accumulated experience to the actual practice of optimizing strategies for groups of investors. We intend to test the corresponding CDSS module not only on investment projects in the field of enterprise digitalization, but also expand its application to related fields, for example, to select strategies for groups of investors in financing cybersecurity systems, Smart City systems and technologies, etc.

7 Conclusions

A mathematical model is proposed for the selection of rational strategies by a group of investors, taking into account the multifactorial nature of the problem using the example of financing information technologies and systems. The model is focused on the practical application in the program-algorithmic block of decision support systems during the search for rational strategies by a group of investors. The model is based on the toolkit of bilinear multi-step quality games with several terminal surfaces. Unlike existing solutions, a new class of bilinear multi-step games was first considered. The resulting solution made it possible to adequately display graphically the optimality sets of players (investors), for example, in the process of investing in projects in the field of enterprise digitalization, taking into account multifactorial nature. It is shown that such an approach, combined with the use of computer modeling and a decision support system (CDSS), can give an investor more opportunities for analysis and selection of rational financial strategies;

Computer modeling of investor strategies and optimality sets of groups of players was performed based on the developed model in the MatLab system, as well as the prototype of the CDSS module. This CDSS module allows you to reduce the discrepancy between the results of the forecast assessment for groups of investors of the chosen strategy. It is shown that the resulting solution will reduce the discrepancy with the data of forecasting and real return on investment.

References

1. Westerman, G., et al.: The digital advantage: how digital leaders outperform their peers in every industry. MIT Sloan Manag. Capgemini Consult. MA **2**, 2–23 (2012)
2. McArthur, D.: Investing in digital resources. New Direct. Higher Educ. **119**, 77–86 (2002)
3. Andal-Ancion, A., Cartwright, P.A., Yip, G.S.: The digital transformation of traditional businesses. MIT Sloan Manag. Rev. **44**(4), 34–42 (2003)
4. Woodard, C.J., Ramasubbu, N., Tschang, F.T., Sambamurthy, V.: Design capital and design moves: the logic of digital business strategy. MIS Q. **37**(2), 537–564 (2013)
5. Hirt, M., Willmott, P.: Strategic principles for competing in the digital age. McKinsey Q. **5**(1), 1–13 (2014)
6. Zanella, A., Bui, N., Castellani, A., Vangelista, L., Zorzi, M.: Internet of Things for smart cities. IEEE IoT J. **1**(1), 22–32 (2014)
7. Lakhno, V., Malyukov, V., Bochulia, T., et al.: Model of managing of the procedure of mutual financial investing in information technologies and smart city systems. Int. J. Civ. Eng. Technol. (IJCIET) **9**(8), 1802–1812 (2018)

8. Mithas, S., Tafti, A., Mitchell, W.: How a firm's competitive environment and digital strategic posture influence digital business strategy. MIS Q. **37**(2), 511–536 (2013)
9. Tiwana, A., Ramesh, B.: E-services: problems, opportunities, and digital platforms. In: Proceedings of the 34th Annual Hawaii International Conference on System Sciences, p. 8. IEEE, January 2001
10. Mazzarol, T.: SMEs engagement with e-commerce, e-business and e-marketing. Small Enterp. Res. **22**(1), 79–90 (2015)
11. Sedera, D., Lokuge, S., Grover, V., Sarker, S., Sarker, S.: Innovating with enterprise systems and digital platforms: a contingent resource-based theory view. Inf. Manag. **53**(3), 366–379 (2016)
12. Mohammadzadeh, A.K., Ghafoori, S., Mohammadian, A., et al.: A fuzzy analytic network process (FANP) approach for prioritizing Internet of Things challenges in Iran. Technol. Soc. **53**, 124–134 (2018)
13. Selçuk, A.L.P., Özkan, T.K.: Job choice with multi-criteria decision making approach in a fuzzy environment. Int. Rev. Manag. Market. **5**(3), 165–172 (2015)
14. Kache, F., Seuring, S.: Challenges and opportunities of digital information at the intersection of big data analytics and supply chain management. Int. J. Oper. Prod. Manag. **37**(1), 10–36 (2017)
15. Akhmetov, B.B., Lakhno, V.A., Akhmetov, B.S., et al.: The choice of protection strategies during the bilinear quality game on cyber security financing. Bull. Natl. Acad. Sci. Repub. Kazakhstan **3**, 6–14 (2018)
16. Lakhno, V., Malyukov, V., Gerasymchuk, N., et al.: Development of the decision making support system to control a procedure of financial investment. Eastern-Eur. J. Enterp. Technol. **6**(3), 24–41 (2017)
17. Smit, H.T., Trigeorgis, L.: Flexibility and games in strategic investment (2015)
18. Arasteh, A.: Considering the investment decisions with real options games approach. Renew. Sustain. Energy Rev. **72**, 1282–1294 (2017)
19. Gottschlich, J., Hinz, O.: A decision support system for stock investment recommendations using collective wisdom. Decis. Support Syst. **59**, 52–62 (2014)
20. Strantzali, E., Aravossis, K.: Decision making in renewable energy investments: a review. Renew. Sustain. Energy Rev. **55**, 885–898 (2016)
21. Lakhno, V., Malyukov, V., Parkhuts, L., et al.: Funding model for port information system cyber security facilities with incomplete hacker information available. J. Theoret. Appl. Inf. Technol. **96**(13), 4215–4225 (2018)
22. Malyukov, V.P.: Discrete-approximation method for solving a bilinear differential game. Cybern. Syst. Anal. **29**(6), 879–888 (1993)
23. Akhmetov, B., Lakhno, V., Akhmetov, B., Alimseitova, Z.: Development of sectoral intellectualized expert systems and decision making support systems in cybersecurity. Adv. Intell. Syst. Comput. **860**, 162–171 (2019)
24. Akhmetov, B., et al.: Decision support system about investments in smart city in conditions of incomplete information. Int. J. Civ. Eng. Technol. **10**(2), 661–670 (2019)

7-Dimensional Optimization Task: PBO-Nature-Inspired Optimizer Versus 10-Years-Old Differential Evolution Based Optimizer 3rd Generation EPSDE

Jaroslav Moravec(✉)

Faculty of Electrical Engineering and Informatics, University of Pardubice, náměstí Čs. legií 565, 530 02 Pardubice, Czech Republic
jaroslav.moravec@student.upce.cz,
j.moravec.email@seznam.cz

Abstract. Origins of the branch of numerical optimizations with use of evolutionary optimizers date almost 60 years back. It is the area which does not evolve by big jumps and the advancement sometimes hits stagnation periods. At such stagnation times there is a big hungry for the new optimization methods which would fill up the empty space. Many optimizers have appeared in the last decade. These optimizers are more specialized in comparison to optimizers which were proposed dozens of years back. The new optimizers are very often derived from older optimizers. In this presented paper, a 7-dimensional optimization task is solved which is called persons identification using contour of a human hand. The paper is considered as research and comparative study at the same time. An optimizer called EPSDE is compared to the Polar Bear Optimizer. The EPSDE is a deputy of 3rd generation optimizers derived from algorithm differential evolution. The PBO falls into a group of young optimizers marked as "nature inspired". The PBO is three times more time-demanding and primarily significantly worse in solving of given task which is very difficult. A comparison of both optimizers was conducted with use of large comparative database.

Keywords: Persons identification · Differential evolution · Polar Bear Optimizer · Optimization · Evolutionary Algorithms

1 Purpose, Introduction and Related Works

Aim of this paper is to present the comparative study of two different algorithms which can be applied to persons identification using hand contour. The whole algorithmic complex is non-trivial and contains several methods which are interconnected in suitable manners.

For experimental purposes, the Technocampus Hand Image Database was used (THID) [9, 12, 59]. Only the images in the visual spectrum are used.

R. Silhavy (Ed.): CSOC 2020, AISC 1225, pp. 11–25, 2020.
https://doi.org/10.1007/978-3-030-51971-1_2

1.1 Algorithm ICP

An ICP algorithm [3] was presented back in 1992 for the first time. The ICP originates from a publication presented in 1987 [16]. The objective of the ICP is to align/match the two point-clouds in 3-dimensional space with use of operation rotation and translation so as the Euclidean distance of the clouds was as small as possible. The number of points in the clouds can be different. The big advantage of the ICP is that the capability to converge to only one local optimum was proved repeatedly throughout the course of years see [30, 37]. Detailed information can be found in e.g. [25, 42, 44].

1.2 Algorithm EPSDE

A Differential Evolution (DE) algorithm [38, 39] was published by Rainer Storn and Kenth Price in 1996 and falls into area of so-called stochastic non-linear metaheuristic population algorithms. The DE became popular very fast thanks to its big performance, reliability and primarily because it overcomes efficiency of commonly used optimizers in many olden day tasks. The DE overcomes e.g. SGA [15], PSO [27] or ES [40]. After initial enthusiasm came a big disappointment because scientists ascertained that the working parameters of the DE have to be changed in the computational process for many tasks see [8, 50, 51]. The cruel but truthful theorem "No Free Lunch Theorem" [52] which says that in the area of optimization methods nothing comes free holds, sadly, twice. At very end of the 90's arose a second period of development in areas of DE. Many scientists tried to find a solution of this problem in sense of a fully automatic setting of working parameters of the DE algorithm. Many interesting optimizers were proposed which overcame the original DE in many ways. One of them is the algorithm "Differential Evolution Algorithm with Ensemble of Parameters and Mutation and Crossover" (EPSDE) [31]. The EPSDE totally minimizes the number of working parameters. Authors were inspired by previous papers [18, 58].

1.3 Algorithm PBO

The algorithm Polar Bear Optimizer (PBO) [36] arose as a consequence of the permanent effort to extend the areas of Evolutionary Algorithms (EA) of new optimizers, which would be more effective, faster and more powerful in comparison to the optimizers which were described before these days. The PBO is a classic example of a population, meta-heuristic algorithm. The PBO is based on observation of polar bears at the time of seal hunting in arctic lands. The big advantage of the algorithm is that the number of individuals – bears - in the population is variable. The number of individuals can be decreased in accordance with amount of food – seals. The big disadvantage of the PBO is that it has to compute the fitness function several times. Such repetition increases all computational demands which are necessary to reach the optimal solution. The PBO excels in solving of some one-dimensional tasks such as e.g. DeJong's F1 [6]. The PBO is capable to optimize low-dimensional and also high-dimensional tasks type of DeJong's F1. A detailed comparative study can be found in [36].

2 Background

The biometrics and biometrical systems [43] become the center of interest of many research teams and private firms in the last two decades [60]. More or less successfully useful in common life. There is big number of ways how to identify persons with use of biometrics [11, 17, 21]. Modern (new-era) practical applications in biometrics can be found earlier - at the very beginning of the 70's e.g. [10, 41]. Use of biometrics in praxis is really wide today and the market for these technologies grows year by year. The presented paper primarily solves the biometric identification, based on the contour of a human hand see e.g. [9, 19]. The hand capturing can be done in such a way that the hand lies on a pad [20, 46, 56, 57] or the hand is freely inserted under the camera [2, 14, 26]. Another option is to use the pegs [19, 46] which deform hand contour and hence some authors do not use it at all see e.g. [5, 13, 53, 54]. A detailed survey study can be found in [1, 4, 7]. The task can be solved as one-dimensional [45], in 2D space [48, 54, 55], or in 3D space [26, 32]. The disadvantage of the hand image capturing which is not placed onto the pad is an unambiguous pose of whole hand in space. The palm can be fisted a bit. From the general point of view there is a large number of methods. The rational accuracy of all presented works is given by an interval of 87–99% in the time horizon of 6–12 months – see [7]. Advances in last two decades can be found in [1, 7, 34].

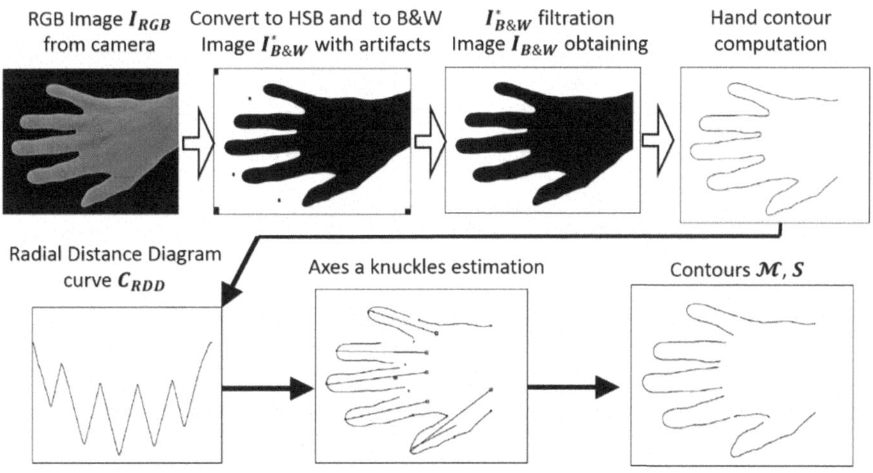

Fig. 1. Scheme of image processing, from RGB to contours $\mathcal{M}$ and S.

Algorithm 1, EPSDE Optimizer

1 Select N_{pop}, L_p^{EPSDE}, T; Given $D_{im}, D_{om}, N_{srt} = 4$
2 Create $\boldsymbol{P_{op}} = (\boldsymbol{X_1}, \dots, \boldsymbol{X_N})$, $\boldsymbol{X_i} = [x_j | \mathrm{j} \in \langle 0, \mathrm{D_{im}}), F]$, $i \in \langle 1, N_{pop})$
3 Init all $\boldsymbol{X_i}$ randomly in D_{om}; $F = rand$; $P_{cr} = rand$
4 $\boldsymbol{CR_{mem}}[i].Strategy(j) = 0.5,\ \forall i, i \in \langle 0, L_p), \forall j, j \in \langle 0, N_{srt})$
5 $\boldsymbol{cr_{mean}}[i] = 0.5;\ \forall i, i \in \langle 0, N_{srt})$
6 $\boldsymbol{cr}[i] = 0.5, \forall i, i \in \langle 0, N_{srt})$
7 Evaluate $\boldsymbol{P_{op}}$, Find the best ind $\boldsymbol{X_{best}}$, $G = 1$
8 **while** (T is not satisfied)
9   **if** $G_{en} > L_p^{EPSDE}$
10   **for** i = 0; i < N_{srt}; i=i+1
11   **for** j = 0; j < $\boldsymbol{CR_{mem}}.count$; j=j+1
12         $R1 = R1 + \boldsymbol{CR_{mem}}[j].Strategy[i]$
13   **endfor**
14       $\boldsymbol{cr_{mean}}[i] = R1/\boldsymbol{CR_{mem}}.count$
15   **endfor**
16   **endif**
17   **for** i = 0; i < N; i=i+1
18   Select randomly $\{\boldsymbol{X_{r1}}, \boldsymbol{X_{r2}}, \dots, \boldsymbol{X_{rk}}\}, rj \neq i, \boldsymbol{X_{rj}} \in \boldsymbol{P_{op}}$
19   Compute trial vector $\boldsymbol{X_t}$ using $\overline{\boldsymbol{V_p}}, \boldsymbol{X_i}, \boldsymbol{X_{rj}}, \boldsymbol{X_{best}}$
20   **if** $\overline{\boldsymbol{V_p}} \neq Current - to - rand/1$ – vector 4., Tab. 1
21   **for** j = 0; j < D_{im}; j=j+1
22         $R1 = U_{niform}\langle 0, \mathrm{D_{im}})$; $R2 = N_{ormal}\langle 0.0, 1.0\rangle$
23   **if** $\neg\left(R2 < \boldsymbol{cr}[\overline{\boldsymbol{V_p}}] \text{ or } j \equiv R1\right)$
24           $\boldsymbol{X_t}.x_j = \boldsymbol{X_i}.x_j$
25   **endif**
26   **endfor**
27   **endif**
28   **if** $fitness(\boldsymbol{X_t})$ is better than $fitness(\boldsymbol{X_i})$
29       $\boldsymbol{X_i}.x_j = \boldsymbol{X_t}.x_j, \forall j, j \in \langle 0, \mathrm{D_{im}})$
30   Remove $\boldsymbol{CR_{mem}}[\boldsymbol{0}]$ from array
31   Add new item at the end of $\boldsymbol{CR_{mem}}$ array
32       $\boldsymbol{CR_{mem}}[L_p - 1].Strategy(j) = \boldsymbol{cr}[j],\ \forall j, j \in \langle 0, \mathrm{D_{im}})$
33   **else**
34       $\overline{\boldsymbol{V_p}} = U_{niform}\langle 0, N_{srt})$; $\boldsymbol{X_i}.F = N_{ormal}(\bar{x} = 0.3, \sigma = 0.5)$
35   **for** j = 0; j < N_{srt}; j=j+1
36   **do** $\boldsymbol{cr}[j] = N_{ormal}(\bar{x} = \boldsymbol{cr_{mean}}[j], \sigma = 0.01)$
37   **while** $(\boldsymbol{cr}[j] < 0 \text{ or } \boldsymbol{cr}[j] > 1.0)$
38   **endfor**
39   **endif**
40   **endfor**
41   Find best ind. $\boldsymbol{X_{best}}$ in $\boldsymbol{P_{op}}$; $G_{en} = G_{en} + 1$
42 **endwhile**
43 $\boldsymbol{X_{best}}$ represents the best estimated solution

$\boldsymbol{P_{op}}$-population of individuals; N_{pop}-the number of individuals; $\overline{\boldsymbol{V_p}}$ -perturbation vector; L_p^{EPSDE} – the learning period; D_{im}-number of dimensions of every individual; D_{om}-domain for every single D_{im}; T-termination condition; $\boldsymbol{X_i}$-chromosome of every individual or one possible solution of the given task; holds that $\boldsymbol{X_{r1}} \neq \boldsymbol{X_{r2}} \neq \cdots \neq \boldsymbol{X_{rk}} \neq \boldsymbol{X_i}$, the number of $\boldsymbol{X_{rk}}$ vectors is given by $\overline{\boldsymbol{V_p}}$. D_{im}, D_{om}-is given by character of the solved task, $\boldsymbol{cr}$ –vector probability of crossover regarding to vector $\overline{\boldsymbol{V_p}}$, $\boldsymbol{cr_{mean}}$–average value of $\boldsymbol{cr}$, N_{srt}–number of used strategies, holds that $N_{srt} = 4$, $\boldsymbol{CR_{mem}}$–memory for individual strategies, the $\boldsymbol{cr_{mean}}$ value of successful strategies is saved in here. End position (row) of $\boldsymbol{CR_{mem}}$ array means $[L_p^{EPSDE} - 1]$ position (row). $N_{ormal}(0.3, 0.5)$ – means random number with normal distribution and with average value $\bar{x} = 0.3$ and standard deviation $\sigma = 0.5$. $U_{niform}\langle 0, N_{srt})$ – a random number with uniform distribution in range of values $\langle 0, N_{srt})$, $[L_p^{EPSDE} - 1]$–means an index in range $\langle 0, L_p^{EPSDE})$ and its value is used on r. 32.

3 Method

Basis of the proposed algorithm for persons identification comes from the EPSDE optimizer, the PBO optimizer and the ICP algorithm. The EPSDE and the PBO make a wrapper of the ICP and define in which way the hand contours will be aligned to each other. Firstly, a database of the model contours is created with which the identified person will be compared. Every person which will have the "granted-access" will have one model-hand-contour stored in the database and this model will be marked as $\mathcal{M}$. A person which asks for free access will be marked as $\boldsymbol{S}$. The result of the identification process is numerically expressed by a measure of correspondence between models $\mathcal{M}$, which are stored in the database with the sample $\boldsymbol{S}$. To obtain the contour $\mathcal{M}$ or $\boldsymbol{S}$, an image capturing is performed with use of a digital camera and a color image (RGB). It is obtained in the required resolution $I_w \times I_h$ – see Fig. 1 and Fig. 2A. The image is first converted to HSB representation – see Fig. 2B, $(H = 0, 360,\ S = 0.0, 1.0,\ B = 0.0, 1.0)$ and then to black and white image B&W – see Fig. 2C. The B&W image is filtered-out and unwilling artifacts are removed and a new filtered image $\boldsymbol{I_{B\&W}}$ is created. Then the algorithm seeks for length of the middle finger. A first, a black pixel from the left side is found in the image $\boldsymbol{I_{B\&W}}$. With use of an algorithm of contour tracing the length of middle finger is then estimated with use of the first black pixel. The length of whole hand from the first pixel to wrist is then calculated as $2.1 \times$ the length of middle finger. To find the hand contour, the Radial Sweep algorithm is used [33, 35] – Fig. 2E and see also [22–24]. An identical algorithm is used to find the length of the middle finger. The start point enabling computation of the contour, lays between the thumb and wrist according to previously determined hand length. The hand contour is used to compute Radial Distance Diagram (RDD) – see Fig. 2D and Fig. 1.

The RDD is segmented in such way that the algorithm seeks for all minima and maxima in XY RDD diagram – see Fig. 2D. The ICP algorithm does not match/compare the whole hand contour but only 4 fingers without thumb, and uses: index finger, middle finger, ring finger and little finger – see Fig. 2G. For every finger the finger-axis is calculated with use of Linear Regression algorithm (LR) [55] – Fig. 2F.

All important points were found and the hand contour is ready to use for classification - see Fig. 2G. The result of the classification process is degree of similarity $\mathcal{E}$ between tested contour $\mathcal{M}$ in the database and a contour $\boldsymbol{S}$ of identified person – see (1). The degree of similarity is given by a single positive value. In the case that the two identical contours $\mathcal{M}$ and $\boldsymbol{S}$ are classified, then the single number representing the similarity is equal to zero. In any other cases the result will be non-zero and a positive number. The degree of similarity is expressed as sum of Euclidean distances of individual points/pixels of the contour $\mathcal{M}$ to closest points of the contour $\boldsymbol{S}$ according to the ICP algorithm.

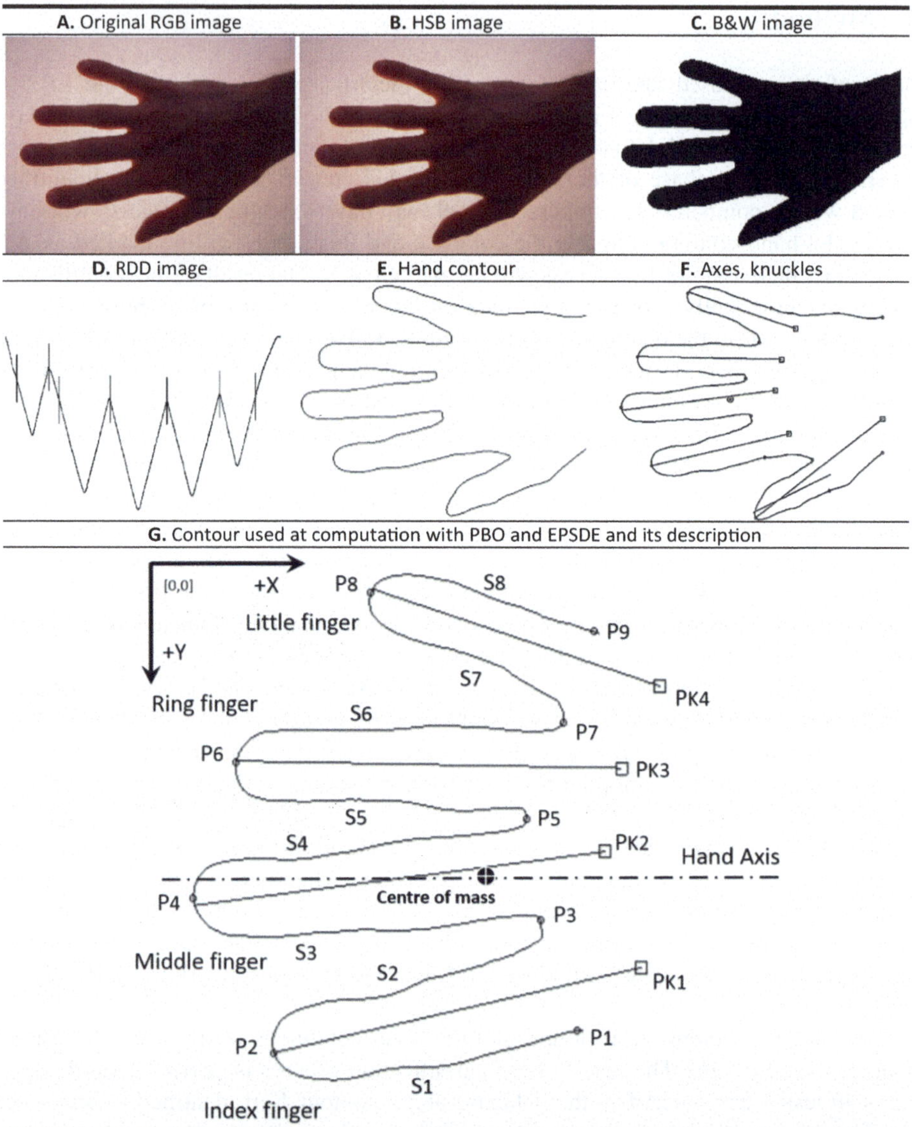

A. – Original RGB image from digital camera. B. – HSB representation of the RGB image. C. – B&W representation prepared to contour calculation. D. – Radial Distance Diagram, vertical lines defines limits of individual fingers, thumb is on the left side. E. – Complete contour created by suitable contour tracking algorithm which was applied to C. F. – calculated axes of individual fingers and positions of knuckles. G. – Description of important parts of fingers, knuckles and parts of hand contour. All fingers are moveable in corresponding knuckles P_K.

Fig. 2. Types of images and detailed description of segmented hand contour.

The evolutionary process with use of selected optimizers PBO or EPSDE can be expressed as follows:

$$fitness = \mathcal{E} = \arg \underset{\mathcal{H} \in \mathbb{R}, S}{\text{opt}} \Im_{\text{EA}}(\mathcal{M}, S, \mathcal{H}) \tag{1}$$

where $\mathcal{H}$ defines a space of possible solutions which is expressed by values X_i. The contour $\mathcal{M}$ is fully static. The contour S is moveable and has all 4 fingers moveable. Angle limits for all fingers is given by (4). All fingers rotate in their knuckles. The function *fitness* is calculated using (2) and in accordance with the operating diagram of the ICP algorithm. Used metric is Euclidean distance.

$$fitness = \sum_{j=1}^{j=n} R_I,\ fitness \in \mathbb{R}^{+,0}, j \in \mathbb{N}^+, \tag{2}$$

where n is a number of points of the contour S. The *fitness* is given in pixels and is defined as the sum of values R_I according to classification criteria $\mathcal{L}_1$ (4). If at time of calculations, a point P_j^S of the contour S is placed out of the area which is given by the image $I_{B\&W}$ a rule is activated $\mathcal{L}_1 = true$. The rule $\mathcal{L}_1 = false$ is used in all other cases. The ICP algorithm seeks for every point P_j^S which is nearest to the point $P_i^{\mathcal{M}}$. Such point is marked as $P_{min}^{\mathcal{M}}$ see (3).

$$R_I = \begin{cases} d\left(P_j^S, P_\Omega\right)^4 & if\ \mathcal{L}_1 = true \\ d\left(P_j^S, P_{min}^{\mathcal{M}}\right) & if\ \mathcal{L}_1 = false \end{cases},\ R_I \in \mathbb{R}^{+,0} \tag{3}$$

$$P_{min}^{\mathcal{M}} : d\left(P_j^S, P_{min}^{\mathcal{M}}\right) = min\, d\left(P_j^S, P_i^{\mathcal{M}}\right),\ i \in \mathbb{N}^+$$

The classification criteria $\mathcal{L}_1$ (4) is strongly bounded with used evolutionary algorithm PBO or EPSDE. The EPSDE algorithm is recorded on Algorithm 1. The PBO algorithm can be found in [36] and its code length corresponds to EPSDE algorithm code length. A chromosome X_i makes identification information of every individual in population. The X_i is expressed as $X_i = (x_0, \ldots, x_6)$. The genes x_0, x_1, x_2 define, in the given order, a change $\Delta X, \Delta Y$ of the position of entre of mass of the contour S in axes X and Y. The x_2 defines a change $\Delta\alpha$ of heading of the hand towards axis X in radians. Value Ω (5) means the space in which at evolution process the center of mass of contour S can moves which is the area of the image $I_{B\&W}$.

$$\mathcal{L}_1 = \begin{cases} true, & if\ x_3 \notin \left\langle -0.40^{RAD}; +0.40^{RAD} \right\rangle \\ true, & if\ x_{4-6} \notin \left\langle -0.30^{RAD}; +0.30^{RAD} \right\rangle \\ true, & if\ x_{0,1} \notin \Omega \\ true, & if\ x_2 \notin \left\langle -0.40^{RAD}; +0.40^{RAD} \right\rangle \\ false, & oth. \end{cases} \tag{4}$$

The space Ω is given as an oblong area with center at position P_Ω as follows:

$$P_\Omega(x_\Omega = 0.5I_w, y_\Omega = 0.5I_h) \tag{5}$$

$$\Omega = [x_\Omega - 0.17I_w, x_\Omega + 0.17I_w, y_\Omega - 0.17I_h, y_\Omega + 0.17I_h]$$

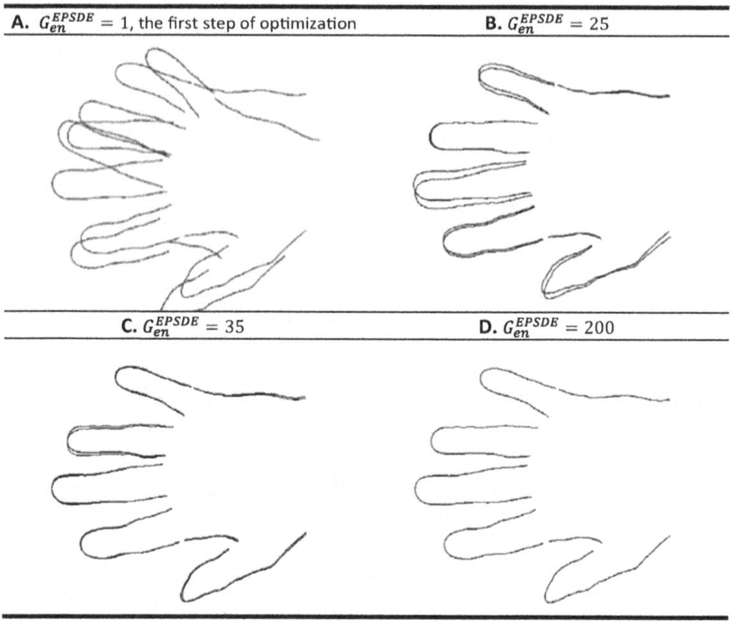

Fig. 3. The alignment process of two identical contours in process of evolution. The EPSDE working parameters: $P_{op} = 15$, $L_p = 10$, $G_{en}^{EPSDE} = 200$, $\kappa_M = \kappa_S = 1$, 640×480 pixels. Sub-optimal solution was found in 82. generation. **B, C, D** – gradual convergence to correct solution. In this case the point-clouds $\boldsymbol{\mathcal{M}}$ and $\boldsymbol{S}$ are identical i.e. $\boldsymbol{\mathcal{M}} \equiv \boldsymbol{S}$, hence $\mathcal{E} = 0$.

In the first step of the evolutionary algorithm (PBO or PESDE) the center of mass of the contour $\boldsymbol{S}$ is placed to point P_Ω – see Fig. 3. The contour $\boldsymbol{\mathcal{M}}$ is for calculation purposes placed to image $\boldsymbol{I_{B\&W}}$ in such way that the center of mass is placed to position P_Ω. Thanks to that, all initial conditions of the ICP algorithm are satisfied. Both the evolutionary optimizers work with seven-dimensional optimizing *fitness* function, $\mathrm{D_{im}} = 7$. The *fitness* function is constrained, non-continuous, non-separable, strongly nonlinear and ill conditioned. The convergence process is recorded in Fig. 3.

Movement of whole hand in plane XY and its rotation towards axis X is given by values of genes x_j of the chromosome $\boldsymbol{X}_i$. All individuals $\boldsymbol{X}_i$ in the population are in randomly deployed the first step of the evolution across whole space of all possible solutions in a certain range of previously defined limits.

4 Results

Section experimental results is divided in to 3 parts: **A.** Election of suitable working parameters of the EPSDE optimizer, **B.** Way of classification of results and experiments arrangements, **C.** Experimental results with PBO and EPSDE optimizers.

4.1 Election of Suitable Working Parameters of the EPSDE Optimizer

The EPSDE optimizer has several working parameters. N_{POP}, N_{GEN} and L_P. The most important is the election of the optimal number of individuals in a population and the number of generations. In Table 1 the result is recorded, which shows suitable values which can be selected so as the final result was as accurate as possible. Identical contours $\mathcal{M}$ and $\boldsymbol{S}$ were compared. The bold area represents suitable values. In experiments this combination of working parameters $N_{POP} = 15$, $N_{GEN} = 200$ and $L_P = 10$ will be used, with regard to time demands. Thanks to the selection of such values the final result does not vary not at 10000 repetitions. The L_P value has only very small effect on the final result. All results in Table 1 were calculated for values $\kappa_M = 1$, $\kappa_S = 1$. If κ_S value is increasing the area of suitable working parameters N_{POP}, N_{GEN} does not change.

Table 1. Effect of N_{POP} and N_{GEN} to final accuracy, *fitness* in pixels

n.	$N_{GEN} \rightarrow$ / $N_{POP} \downarrow$	50	100	150	200	250	300
1	6	645	89	17	16	0	0
2	10	739	89	11	0	0	0
3	15	857	114	24	0	0	0
4	20	741	81	6	0	0	0
5	25	847	88	78	0	0	0
6	30	605	30	2	0	**0**	**0**
7	35	681	47	0	0	**0**	**0**
8	40	769	61	0	0	**0**	**0**

Values are in pixels and recorded values represents average of 50 repetition. $\kappa_M = \kappa_S = 1$.

4.2 Way of Classification and Results Arrangements

A set of standard statistical markers were elected to properly classify all results: *FAR*-false acceptance rate, *FRR*-false rejection rate and calculation of point *EER*-equal error rate. Analogously to [2, 28, 29], the point *EER* is recorded in percent and the corresponding value of *fitness* in pixels.

$$FAR = \frac{\Sigma I_{UL}}{\Sigma I_{TOT}}, \quad FRR = \frac{\Sigma G_{OL}}{\Sigma G_{TOT}} \tag{6}$$

$$\Sigma G_{TOT} = 104 \times 10 = 1040, \quad \Sigma I_{TOT} = 108160 - 1040 = 107120$$

A way to ascertain how robust are all used classifiers, a test which uses a different number of pixels of the contour $\boldsymbol{S}$ is applied. The value κ_S will define what pixel of the contour $\boldsymbol{S}$ is used e.g. every 10th pixel. The $\mathcal{M}$ contour has all pixels in calculation. The selected row of values κ_S is: $\kappa_S \in (1, 5, 10, 15, 20, 25, 30, 35, 40)$. The THID database consists of 1040 images, 104 persons, and 10 images of every person.

4.3 Experimental Results with Optimizers PBO and EPSDE

The results of experiments with use of PBO optimizers and EPSDE optimizers are recorded in Table 2 and 3. Both used optimizers represent a typical populational metaheuristic algorithm. The EPSDE showed better results in comparison to the PBO optimizer. It is well visible that the 3rd generation optimizer EPSDE which is based on 14 years of painstaking research in the given scientific area, provides very good and stable results – see Table 2. For $\kappa_S = 40$, the final accuracy is equal to 13.6 × better than for the PBO for $\kappa_S = 40$. For EPSDE the best reached result is equal to $FAR = 2.11\%$, $FRR = 2.11\%$, $ERR = 2.11\%$. For optimizer PBO the result is equal to $FAR = 8.88\%$, $FRR = 8.88\%$, $ERR = 8.87\%$ - see Table 3. The PBO algorithm is not suitable to solve the tested optimization task.

Table 2. Results for algorithm EPSDE

Values FAR, FRR, EER, 640 × 480, 4 fingers, $D_{im} = 7$							
n.	κ_S	ΣI_{UL}	$FAR\%$	ΣG_{OL}	$FRR\%$	$EER\%$ /pixel	Tot. time
1	**1**	**2269**	**2.11**	**22**	**2.11**	**2.11/1247**	**1.996**
2	5	2462	2.29	23	2.21	2.29/249	0.491
3	10	2620	2.44	23	2.21	2.44/107	0.296
4	15	2712	2.53	24	2.30	2.53/98	0.218
5	20	2911	2.71	25	2.40	2.71/92	0.187
6	25	3078	2.87	26	2.50	2.87/88	0.171
7	30	3102	2.89	28	2.69	2.89/62	0.150
8	35	3378	3.15	28	2.69	2.92/40	0.144
9	40	3517	3.28	29	2.78	3.32/30	0.140

Time demands in [sec.ms] necessary to align/match two samples THID: **P001-03-M** and **P057-08-S**. Total number of evolutions is: 9 rows × 108160 comparisons $\mathcal{M}$ vs $\boldsymbol{S}$ = 973 440 evolutions. $N_{pop}^{EPSDE} = 15$, $G_{en}^{EPSDE} = 200$.

Table 3. Results for algorithm **PBO**

Values *FAR*, *FRR*, *EER*, 640 × 480, 4 fingers, $D_{im} = 7$							
n.	κ_S	ΣI_{UL}	*FAR*%	ΣG_{OL}	*FRR*%	*EER*%/pixel	Tot. time
1	**1**	**9512**	**8.88**	**92**	**8.88**	**8.87/2563**	**11.512**
2	5	9602	8.96	92	8.88	8.96/512	2.215
3	10	9732	9.08	92	8.88	9.08/250	0.920
4	15	9780	9.12	92	8.88	9.12/193	0.873
5	20	9812	9.15	93	8.94	9.15/173	0.780
6	25	9965	9.30	94	9.03	9.30/143	0.633
7	30	10187	9.50	95	9.13	9.50/97	0.468
8	35	10487	9.78	95	9.13	9.78/73	0.452
9	40	10658	9.94	95	9.13	9.94/66	0.421

Time demands in [sec.ms] necessary to align/match two samples THID: **P001-03-M** and **P057-08-S**. Total number of evolutions is: 9 rows × 108160 comparisons $\mathcal{M}$ vs S = 973 440 evolutions.

Table 4. Comparison with other works, our algorithm - see [61, 62]

n.	Author	Method, database	*EER*%
1	[49] (Sharma et al. 2015)	Shape + Geometry, IITD database, EER = 0.52% Shape + Geometry, proprietary db., EER = 0.31%	0.52, 0.31
2	[48] (Santos-Sierra et al. 2011)	SVM, k-NN, HTC db., UST db., IIT db.	2.5,2.0,1.4
3	[47] (Santos-Sierra et al. 2009)	Contour, parametric curve, proprietary db.	3.7
4	[28] (Klonowski et al. 2018)	Set of features, abs. value of difference of features + weigh coefficients, 4 fingers, HGDB db., FAR = 0.00%, FRR = 1.19%	0.59
5	[2] (Barra et al. 2019)	Euclidean distance, THID db.	3.5
6	[2] (Barra et al. 2019)	LDA, hand geometry + subset selection with Decidability, THID db.	0.9
7	[2] (Barra et al. 2019)	Hand geometry, subset selection with Intra/Inter-Class Variability, LDA, 4 fingers, THID db.	0.52
8	**This work, EPSDE**	**Geometry, 4 fingers, 4 knuckles, THID db., 640 × 480**	**2.11**
9	**This work, PBO**	**Geometry, 4 fingers, 4 knuckles, THID db., 640 × 480**	**8.87**

The PBO optimizer is 3 times slower and computational speed is variable according to setting of several internal variables. In comparison to [2] where $ERR = 3.50\%$ is recorded with use of native Euclidean distance, the presented algorithm is better, but it is not capable to reach to Linear Discriminant Analysis (LDA) efficiency [2]. This is because LDA is capable of reaching the result $ERR = 0.52\%$. Results of comparison with several selected authors are recorded in Table 4. Corresponding ROC curves are recorded in Fig. 4.

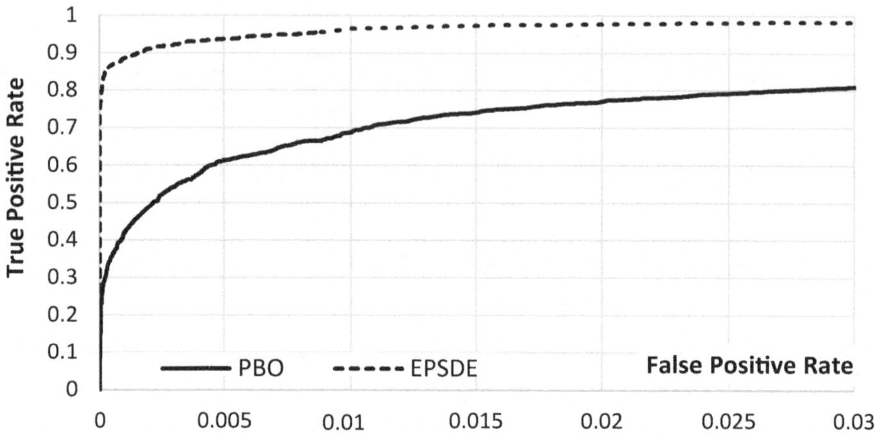

Fig. 4. Receive Operating Characteristics (ROC) curves for the PBO and EPSDE algorithms, $\kappa_M = \kappa_S = 1$.

5 Conclusion

The presented paper shows results of research of hand contour classification. The best possible results were attained with use of the EPSDE optimizer and attained statistical values are $FAR = 2.11\%$, $FRR = 2.11\%$, $ERR = 2.11\%$. The second tested optimizer is PBO and this optimizer recorded significantly worse statistical results $FAR = 8.88\%$, $FRR = 8.88\%$, $ERR = 8.87\%$. Only 4 fingers were used in tests without any other support information. The biggest problem was to correctly estimate the positions of all knuckles because no x-ray apparatus was at our disposal in the preprocessing stage. Sadly, the THID database does not contain information of positions of individual knuckles. The THID database was recorded with very bad ambient lightning which effects the final results as well. A very important factor is the constant press of a hand to the scanner support pad. If the press is changing then the final contour is changing as well. There are a lot of factors which adversely affect the final results. Gradual development and technical advance in future, will ensure better image capturing and hand knuckle position estimating without use of an x-ray apparatus. Such images should assure much better contour classification.

Acknowledgment. The publication was supported by the funds of University of Pardubice, Czech Republic – Student grant competition project (SGS_2020_001).

Author would like to express cordial thanks to Mr. Paul Hooper for his careful English text correction, patience and stamina.

References

1. Bakshe, R.C., Patil, A.M.: Hand geometry techniques: a review. Int. J. Mod. Commun. Technol. Res. **2**(11), 7 (2014)
2. Barra, S., Marsico, M., Nappi, M., Narducci, F., Riccio, D.: A hand-based biometric system in visible mobile environments. Inf. Sci. **479**, 472–485 (2019)
3. Besl, P.J., McKay, H.D.: A method for registration of 3-D shapes. IEEE Trans. Pattrn. Anal. Mach. Intell. **14**(2), 239–256 (1992)
4. Bharathi, S., Sudhakar, R.: Hand biometrics: an overview. Int. J. Autom. Ident. Technol. **3** (2), 101–108 (2011)
5. Cortes, J.M.R., Gil, P.G., Perez, G.S., Castro, P.C.: Shape-based hand recognition approach using the morphological pattern spectrum. J. Electron. Imaging **18**(1), 0130121/1–0130121/6 (2009)
6. Dejong, K.: An analysis of the behavior of a class of genetic adaptive systems. Ph.D. thesis, University of Michigan (1975)
7. Duta, N.: A survey of biometric technology based on hand shape. Pattern Recogn. **42**, 2797–2806 (2009)
8. Eiben, A.E., Hinterding, R., Michalewicz, Z.: Parameter control in evolutionary algorithms. IEEE Trans. Evol. Comput. **3**(2), 124–141 (1999)
9. Faundez-Zanuy, M., Mekyska, J., Font-Aragones, X.: A new hand image database simultaneously acquired in visible, near-infrared and thermal spectrums. Cogn. Comput. **6** (2), 230–240 (2014)
10. Fioretti, W.H., Giordano, A.J., Jacoby, I.H.: Identification corp. (1972)
11. Patent US3648240A. http://www.google.com/patents/US3648240
12. Font-Aragones, X., Faundez-Zanuy, M., Mekyska, J.: Thermal hand image segmentation for biometric recognition. IEEE Aero. Electron. Syst. Mag. **28**(6), 4–14 (2013)
13. Gross, R., Li, Y., Sweeney, L., Jiang, X., Xu, W., Yurovsky, D.: Robust hand geometry measurements for person identification using active appearance models. In: First IEEE International Conference on Biometrics: Theory, Applications, and Systems, vol. 1, no. 6 (2007)
14. Hassanat, A., Al-Awadia, M., Btousha, E., Al-Btousha, A., Alhasanata, E., Altarawnehb, G.: New mobile phone and webcam hand images databases for personal authentication and identification. Procedia Manuf. **3**, 4060–4067 (2015)
15. Holland, J.H.: Outline for a logical theory of adaptive systems. J. ACM **9**(3), 297–314 (1962)
16. Horn, B.K.P.: Closest form solution of absolute orientation using unit quaternions. J. Opt. Soc. Am. **4**(4), 629–642 (1987)
17. Chauhan, S., Arora, A.S., Kaul, A.: A survey of emerging biometric modalities. Procedia Comput. Sci. **2**, 213–218 (2010)
18. Iorio, A., Li, X.: Solving rotated multi-objective optimization problems using differential evolution. In: Australian Conference on Artificial Intelligence, Cairns, Australia, pp. 861–872 (2004)
19. Jain, A.K., Duta, N.: Deformable matching of hand shapes for user verification. In: IEEE International Conference on Image Processing, pp. 857–861 (1999)
20. Jain, A.K., Ross, A., Pankanti, S.: A prototype hand geometry-based verification system. In: 2nd International Conference on Audio- and Video-based Biometric Person Authentication, pp. 166–171 (1999)
21. Jain, A.K., Ross, A., Prabhakar, S.: An introduction to biometric recognition. IEEE Trans. Circ. Syst. Video Technol. **14**(1), 4–20 (2000)

22. Jetenský, P., Marek, J., Rak, J.: Fingers segmentation and its approximation. In: Proceedings of 25th International Conference Radioelektronika (Radioelektronika 2015), pp. 431–434. IEEE (Institute of Electrical and Electronics Engineers) New York (2015). (ISBN 978-1-4799-8117-5)
23. Jetenský, P.: Human hand image analysis extracting finger coordinates using circular scanning. In: Proceedings of 25th International Conference Radioelektronika (Radioelektronika 2015), pp. 427–430. IEEE (Institute of Electrical and Electronics Engineers), New York (2015). (ISBN 978-1-4799-8117-5)
24. Jetenský, P.: Human hand image analysis extracting finger coordinates and axial vectors: finger axis detection using blob extraction and line fitting. In: 2014 24th International Conference Radioelektronika, pp. 1–4. IEEE (Institute of Electrical and Electronics Engineers), New York (2014). (ISBN 978-1-4799-3715-8)
25. Jost, T., Hügli, H.: Fast ICP algorithms for shape registration. In: Joint Pattern Recognition Symposium, pp. 91–99 (2002)
26. Kanhangad, V., Kumar, A., Zhang, D.: Combining 2D and 3D hand geometry features for biometric verification. In: IEEE Computer Society Conference on Computer Vision and Pattern Recognition Workshops (2009)
27. Kennedy, J., Eberhart, R.C.: Particle swarm optimization. In: IEEE International Conference on Neural Networks, vol. 4, pp. 1942–1948 (1995)
28. Klonowski, M., Plata, M., Syga, P.: User authorization based on hand geometry without special equipment. Pattern Recogn. **73**, 189–201 (2018)
29. Luque-Baena, R.M., Elizondo, D., López-Rubio, E., Palomo, E.J., Watson, T.: Assessment of geometric features for individual identification and verification in biometric hand systems. Expert Syst. Appl. **40**(9), 3580–3594 (2013)
30. Maier-Hein, L., et al.: Convergent iterative closest-point algorithm to accommodate anisotropic and inhomogeneous localization error. IEEE Trans. Pattern Anal. Mach. Intell. **34**(8), 1520–1532 (2012)
31. Mallipeddi, R., Suganthan, P.N.: Differential evolution algorithm with ensemble of parameters and mutation and crossover strategies. In: International Conference on Swarm, Evolutionary, and Memetic Computing SEMCCO 2010, pp. 71–78 (2010)
32. Michael, G.K.O., Connie, T., Hoe, L.S., Jin, A.T.B.: Locating geometrical descriptors for hand biometrics in a contactless environment. In: International Symposium in Information Technology, vol. 1, pp. 1–6 (2010)
33. Parker, J.R.: Algorithms for Image Processing and Computer Vision, 2nd edn. Wiley, Hoboken (2010)
34. Pavešic, N., Ribarič, S., Ribarič, D.: Personal authentication using hand-geometry and palmprint features – the state of the art. Technical report (2004)
35. Pavlidis, T.: Algorithms for Graphics and Image Processing. Springer (1982)
36. Polap, D., Wozniak, M.: Polar Bear Optimization Algorithm: Meta-Heuristic with Fast Population Movement and Dynamic Birth and Death Mechanism, Symmetry, pp. 1–20 (2017)
37. Pottmann, H., Huang, Q.X., Yang, Y.L., Hu, S.M.: Geometry and convergence analysis of algorithms for registration of 3D shapes. Int. J. Comput. Vis. **67**(3), 277–296 (2006)
38. Price, K.: Differential evolution: a fast and simple numerical optimizer. In: NAFIPS, pp. 524–527 (1996)
39. Price, K., Storn, R.: Minimizing the real functions of the ICEC contest by differential evolution. In: IEEE International Conference on Evolutionary Computation, pp. 842–844 (1996)
40. Rechenberg, I.: Evolutionsstrategies: Optimierung technischer Systeme nach Prinzipien der biologischen Evolution, Frommann-Holzboog Eds., Stutgart, Germany (1973)

41. Richard, E.H.: Patent US3576537A (1971)
42. Rodrigues, M., Fisher, R., Liu, Y.: Special issue on registration and fusion of range images. Comput. Vis. Image Underst. **87**, 1–131 (2002)
43. Ross, A., Jain, A.K.: Human recognition using biometrics: an overview. Ann. Telecommun. **62**(1–2), 11–35 (2007)
44. Rusinkiewicz, S., Levoy, M.: Efficient variants of the ICP algorithm. In: IEEE Third International Conference on 3-D Digital Imaging and Modeling, vol. 8 (2001)
45. Samiya, S., Abderrahmane, B., Abdelkader, B.: Recognition of individuals based on hand geometry. In: Uncertainty Modeling in Knowledge Engineering and Decision Making, pp. 1023–1029 (2012)
46. Sanches-Reillo, S.R., Sanches-Avila, S.C., Gonzales-Marcos, A.: Biometric identification through hand geometry measurement. IEEE Trans. Pattern Anal. Mach. Intell. **22**(10), 1168–1171 (2000)
47. Santos-Sierra, A., Casanova, J., Avila, C., Vera, V.: Silhouette-based hand recognition on mobile devices. In: International Carnahan Conference on Security Technology, pp. 160–166 (2009)
48. Santos-Sierra, A., Sánchez-Ávila, C., Pozo, G.B., Guerra-Casanova, J.: Unconstrained and contactless hand geometry biometrics. Sensors **11**, 10143–10164 (2011)
49. Sharma, S., Dubey, S.R., Singh, S.K., Saxena, R., Singh, R.K.: Identity verification using shape and geometry of human hands. Expert Syst. Appl. **42**(2), 821–832 (2015)
50. Smit, S.K., Eiben, A.E.: Parameter tuning of evolutionary algorithms: generalist vs. specialist. Appl. Evol. Comput. **6024**, 542–551 (2010)
51. Smit, S.K., Eiben, A.E.: Parameter tuning for configuring and analyzing evolutionary algorithms. Swarm Evol. Comput. **1**(1), 19–31 (2010)
52. Wolpert, D.H., Macready, W.G.: No free-lunch theorems for search. Technical report 95-02-010, SantaFe Institute (1995)
53. Wong, A.L.N., Shi, P.: Peg-free hand geometry recognition. In: Using Hierarchical Geometry and Shape Matching IAPR Workshop on Machine Vision Applications, pp. 281–284 (2002)
54. Xiong, W., Xu, Ch., Ong, S.H.: Peg-free human shape analysis and recognition. In: IEEE International Conference on Acoustics, Speech, and Signal Processing (2005)
55. Yan, X., Su, X.G.: Linear Regression Analysis: Theory and Computing (2009). https://doi.org/10.1142/6986 (2009)
56. Yörük, E., Konukoglu, E., Sankur, B., Darbon, J.: Shape based hand recognition. IEEE Trans. Image Process. **15**(7), 1803–1815 (2006)
57. Yörük, E., Dutagaci, H., Sankur, B.: Hand biometrics. Image Vis. Comput. **24**, 483–497 (2006)
58. Zhang, J., Sanderson, A.C.: JADE: adaptive differential evolution with optional external archive. IEEE Trans. Evol. Comput. **13**(5), 945–958 (2009)
59. Web1. http://splab.cz/en/download/databaze/tecnocampus-hand-image-database. Accessed Mar 2020
60. Web2. www.gemalto.com. Accessed Mar 2020
61. Web3. http://handwork.4fan.cz/. Accessed Mar 2020
62. Web8. http://robomap.4fan.cz/. Accessed Mar 2020

A Hand Contour Classification Using Ensemble of Natural Features: A Large Comparative Study

Jaroslav Moravec(✉)

Faculty of Electrical Engineering and Informatics, University of Pardubice, náměstí Čs. legií 565, 530 02 Pardubice, Czech Republic
jaroslav.moravec@student.upce.cz,
j.moravec.email@seznam.cz

Abstract. Biometrics is a standalone scientific discipline which enjoys more and more attention of many researchers. The provision of the general security plays a key role in many modern branches. In the presented paper a person identification task is solved using the shape of a human hand, and also the hand contour classification algorithm based on an evolutionary estimator is also. The proposed methodology provides the comparison of the identified person with a set of model contours. The examination of the proposed method was performed with use of a database which contains 940 images of the scanned hands from 94 persons, including 10 images from every person. Totally 88360 combinations of the input images. The proposed evolutionary estimator uses an EPSDE algorithm, which is derived from a differential evolution algorithm which was proposed at the end of the 90's. The model of the hand contour of every person is represented by only one image, which has movable finger contours in the classification process regarding the knuckle positions of the hand. Thanks to that, it is not necessary to use the pegs to hold the individual fingers in correct positions. The hand can be both placed on a support desk or can be freely hung in the air. All results obtained at classification time with use of the presented evolutionary estimator provide accuracy of approximately 98%.

Keywords: Biometric identification · Evolutionary algorithms · Hand contour classification · Security

1 Introduction

The biometrics belongs to generally reputable branches today. It has broad use primarily in areas such as security, medical provision or forensic sciences etc. The following biometric systems we can meet in an everyday life: systems which use DNA [52], finger print [6], skin folding on the palm [38], ear auricle structure [26], face structure [15, 54], eye iris structure [9], or hand contour shape [2, 27]. From the general point of view there is no rule which would restrict use of biometric markers. It is possible to encounter methods with different combinations of many biometric features e.g. hand shape and palm print [17] or hand shape and bloodstream [31]. An elementary survey can be found in [1, 18].

R. Silhavy (Ed.): CSOC 2020, AISC 1225, pp. 26–45, 2020.
https://doi.org/10.1007/978-3-030-51971-1_3

There is an uncountable number of methods and instruments enabling classification of the hand contour. Some of proposed methods require to fixate the hand in a special adjusting tool (pegs) [20, 31, 44, 57]. There are other methods which do not require any adjusting tool at all [47, 48, 50, 51]. An adjusting tool can be also hinged [30]. Some methods use a picture of the hand which is freely hung in space [2, 14, 43]. A very popular method is to use a classic office scanner [13, 47]. Some methods are specialized for mobile phones [2, 45]. In practice we can encounter apparatus which perform a person identification without a thumb contour [57]. Some elementary frame rules are, in these days, standardized in the norms ANSI/INCITS 396-2005 - Hand Geometry Format for Data Interchange and corresponding European norm ISO/IEC 19794-10/2007 [58]. The second one was approved early in 2007 and describes only elementary aspects. From the general point of view of the biometric method which use a hand contour, it is always considered less reliable in comparison to e.g. DNA testing. A detailed survey of the actual trends in this area can be found in [1, 5, 11].

For experimental purposes a database GPDS150 was used [13, 47, 55] The "Grupo de Procesado Digital de la señal", GPDS (Digital Signal Processing Group) with DPDS (División de Procesado Digital de la Señal) from "Instituto para el Desarrollo Tecnológico y la Innovación en Comunicaciones IDeTIC", both are active at the "University of Las Palmas de Gran Canaria".

2 Related Works

The person identification development area rapidly grows year by year. Some of the previously proposed methods are commonly accessible e.g. the biometric scanner Allegion company [57], uses only 4 fingers of a hand without thumb and also pegs. The disadvantage of the use of pegs is that the finger contour is deformed by them, see [8, 20] and some parts of individual fingers have to be reconstructed artificially. The advantage of the use of pegs is that they enable fixing of the scanned hand which touches the support pad unambiguously in such way that the stretching fingers is optimal for the next processing step. The hand contour is also deformed by an unsuitable hand pose in 3D space. If the hand in the 3D space is freely hung, it is possible to incline the whole hand in any direction in regards to the camera and thanks to that, the fingers become shorter. Similar information can be found in e.g. [2, 45]. An advantage of the method proposed in [2] is the hygienic cleanliness, because the identified person does not touch the support. The capturing image can be also created in such a manner that the back of the hand is photographed [12, 20]. Some authors use a classic office scanner [50]. [50] published a very interesting paper where they described a method which enables identification of the persons with use hand contour. To obtain the hand images they used a classic office scanner with resolution 45 dpi. They scanned the hand palm and also individual fingers. The authors collected a total of 1374 images of 458 persons for the experimental purposes. The histogram is used in the classification process and also the algorithm Independent Component Analysis (ICA) [3]. All contours of individual fingers are extracted with use a Radial Distance Diagram and in classification are movable in corresponding knuckles. The knuckle positions are only estimated. Hausdorff distance is used as the basic metrics see [10]. All results presented by authors show that the reached accuracy is approximately 98%.

3 Algorithms DE and EPSDE

For experimental purposes the Differential Evolution Algorithm with Ensemble of Parameters and Mutation and Crossover (EPSDE) [29] optimizer was elected. The EPSDE has similar shape as the Differential Evolution (DE) [35, 36, 59, 60], but minimizes the number of working parameters to only three. The EPSDE is derived from DE. The EPSDE differs from the DE by almost 14 years of intensive research in the given area see [7, 19, 53]. No one parameter of the original F, P_C parameters in DE is accessible in the EPSDE. Correct setting of these parameters is driven automatically. The EPSDE algorithm uses a population $\boldsymbol{P_{op}}$ of individuals $\boldsymbol{X_i}|i \in \langle 0, N_{pop}\big)$. Every individual is represented by the vector $\boldsymbol{X_i} = \left[x_j|\mathrm{j} \in \langle 0, \mathrm{D_{im}}), F|\mathrm{j} \in \langle 0.0, 2.0 \rangle\right]$ and this vector represents one possible solution of the given task in a generation G_{en}. Parameter F is marked as mutational (weigh) constant and holds that $0.0 < F \leq 2.0$ [37]. In the area of evolutionary algorithms there are a lot of different interesting algorithms see e.g. [39, 40].

4 Algorithm ICP

An ICP algorithm [4] was presented in 1992 for the first time and originates from a previously published paper [16]. The aim of the ICP is to unambiguously and perfectly align/match the two point-clouds in 3-dimensional space with use of operations rotation and translation so as the Euclidean distance between individual points of the matched clouds was as small as possible. The number of points in both clouds can be different. The original algorithm ICP [4] consists of three steps. The necessary initial condition of the ICP algorithm is that both the clouds have to be aligned to each other with as small a distance as possible. This first step is usually solved by identifying the center of masses of both point clouds. The big advantage of ICP algorithm is that over time, it unambiguously and repeatedly proved its capability to converge to only one local optimum - see [28, 34]. Detailed information and overview study can be found e.g. in [24, 41, 42].

5 Proposed Algorithm - Methodology

The basis of the proposed algorithm for a person identification is made by the EPSDE optimizer and the ICP algorithm. The EPSDE makes a wrapper of the ICP and decides in which way the classified contours will be aligned to each other.

The proposed algorithm for contour classification is recorded in Fig. 1 and can be described in the following steps **1–9**:

1/ In the first step of proposed algorithm a database of model contours of hands is created. Every person which will have a granted access has one hand-model-contour saved in the database. This model will be marked as $\mathcal{M}$. An unknown person which asks for access permission will be marked as $\boldsymbol{S}$. A result of identification process is numerically expressed as a concordance rate between the models $\mathcal{M}$, which are stored

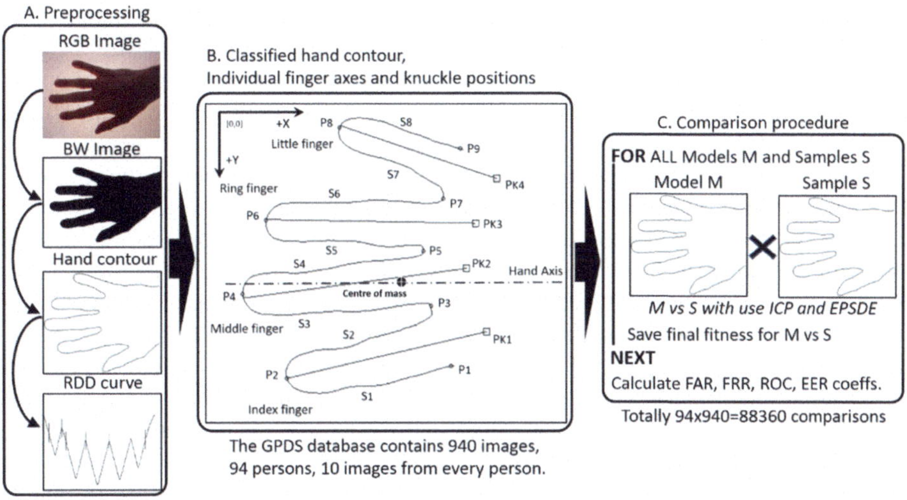

Fig. 1. A chart of the GPDS images processing. (A)-An RGB image is converted to hand contour and Radial Distance Diagram (RDD) conversion. (B)-Fully classified hand contour with finger axes and knuckle positions. (C)-The classification process.

in the database and the data sample $\boldsymbol{S}$. The model contours are created in the same way which is, in a later time, used in person identification, hence both the algorithms will be described together now.

2/ A person which requires the identification or the model contour $\mathcal{M}$ creation inserts his hand into the biometric scanner. The hand image is captured using a digital camera and the color RGB or grayscale image is obtained in suitable resolution $I_w \times I_h$. The hand is inserted into a visual field of the camera in such way that the hand was approximately in the middle of the field and the hand axis and axis X of used coordinate system includes an angle which is as small as possible – see Fig. 1B. For computation purposes it is possible to use both a picture from an office scanner (palm image and fingers) and also image from a digital camera (image of back of the hand or the palm with fingers). The hand may not be overshadowed. A basic assumption is that the hand contour can be extracted easily. The hand does not need to lie on a support pad.

3/ The image is converted from RGB to HSB representation [46] and then to black and white image B&W – see Fig. 2B, C. The image in B&W representation looks as if the image background has white color and the hand is black see Fig. 1C. In the ideal case it is necessary to use ($H = \langle 0, 360 \rangle$, $S = \langle 0.0, 1.0 \rangle$, $B = \langle 0.0, 1.0 \rangle$) only the B channel (intensity) in conversion from HSB to B&W. Only pixels with limit $B \leq 0.4$ are considered. Unfortunately, used database GPDS150 and images in it, requires change of the B value for some groups of images in range $B \leq 0.2 - 0.45$.

4/ Every B&W image is loaded with unwilling artefacts which make the processing more difficult, hence a filtration procedure is used. Every white pixel with its 8-surrounding pixels, which has less than 7 and more than 5 black pixels are colored to black. Every black pixel with its 8-surrounding pixels, which has less than 5 white pixels, are colored to white. All small clusters of black pixels which can be found in an area of size 21 × 21 pixels, which at the same time has zero black pixels in its boundary (2 × 21 + 2 × 19 pixels), are repainted white. The mask of size 21 × 21 pixels is moving across whole the image from the top left corner to bottom right corner. A centering mark of the B&W image is also removed from left middle area. All four margins of width 15 pixels are repainted white.

5/ The filtered B&W image is first used to find a length of the middle finger. A first black pixel in B&W image from the left side is found. For most five-fingered persons the point lies on the top of middle finger. A contour tracing algorithm uses this point to estimate the middle finger length. The length of whole hand is calculated as 2.1x the length of the middle finger. Once the hand length is known, the hand contour is calculated by using the contour tracing algorithm. To find the hand contour a Radial Sweep Algorithm is used (RSA) [32, 33] - see e.g. [21–23]. An identical RSA algorithm is used to find the middle finger length. The starting point for hand contour calculation lies between the wrist and the thumb according to the calculated hand length.

6/ The hand contour calculated in point **5/** uses a Radial Distance Diagram (RDD) for calculation. The RDD is first segmented in such way that all minima and maxima (points of inflexion, a set of important points) are found in the RDD diagram. RDD is considered for such purposes as a one-dimensional curve – see Fig. 2D. Thanks to RDD segmentation it is possible to obtain the set of important points on the top of individual fingers and also all knuckle positions. The RDD is calculated from a point which is situated in the middle of the wrist.

7/ The ICP algorithm does not align/match the whole hand contour but only 4 fingers without thumb – namely index finger, middle finger, ring finger and little finger and several other markers – see Table 1 and Fig. 2A. The thumb contour is not used. The palm contours between wrist and little finger and between thumb and wrist are not used either.

8/ A finger axis is calculated for every finger (4 fingers only – see point **7/**) using a Linear Regression algorithm (LR) [49, 56]. The LR algorithm approximates the set of points which belong to the given finger by line. Only the pixels which belong to every finger are considered + a small correction which is based on excluding several pixels of the contour in a small surrounding of points $P_{1,3,5,7}$. Thanks to this correction it is possible to improve the angle of heading of the regression line. An index finger consists of contours $S_1 + S_2$, a middle finger consists of contours $S_3 + S_4$ etc. Individual finger axes are, in the next processing step, replaced by abscissae. The endpoints of individual abscissae correspond to the limiting points of individual finger contours. All abscissae are extended by several pixels to reach the corresponding finger knuckle. The elongation is performed with the following number of pixels: index finger 25, middle finger 30, index finger 58 and little finger 40. Of course, all the positions are considered as "an approximation" only.

9/ Now, the contour ($\mathcal{M}$ or $\boldsymbol{S}$) is ready and all significant points on the contour were found – see Fig. 2A. If a model contour $\mathcal{M}$ is considered, the contour is inserted into the database. All data samples $\boldsymbol{S}$ are then compared with the model contours $\mathcal{M}$ and it is decided if the contour $\boldsymbol{S}$ belongs to a person which is registered in the database of contours $\mathcal{M}$. The decision, if a contour $\mathcal{M}$ or a contour $\boldsymbol{S}$ will be processed, is driven by a human operator. During classification, all contours $\mathcal{M}$ are gradually selected from the database and are compared (classified) to the hand contour $\boldsymbol{S}$. In the classification process, the model contour $\mathcal{M}$ is fully static and is compared with contour $\boldsymbol{S}$. Individual finger-contours of the contour $\boldsymbol{S}$ rotate around the corresponding knuckles P_{K1}, P_{K2}, P_{K3}, P_{K4} in a maximal angle which is defined in (3). The classification algorithm which uses ICP is trying to correctly align/match the point-clouds of individual finger contours $\mathcal{M}$ and $\boldsymbol{S}$ with use of classification criteria (3,4). The angle of rotation of individual fingers is driven by EPSDE algorithm. Every knuckle position is bound/connected with the center of mass of the contour $\boldsymbol{S}$. In the first step of classification the center of masses of the contours $\mathcal{M}$ and $\boldsymbol{S}$ are mutually identified/matched. The position of center of mass of the contour $\boldsymbol{S}$ is moving according to the requirements of classification algorithm regarding the center of mass of the contour $\mathcal{M}$. The heading of the whole contour $\boldsymbol{S}$ is also changed. The EPSDE algorithm is working until the value *fitness* does not reach an expected value or until an expected number of evolutionary cycles is not reached.

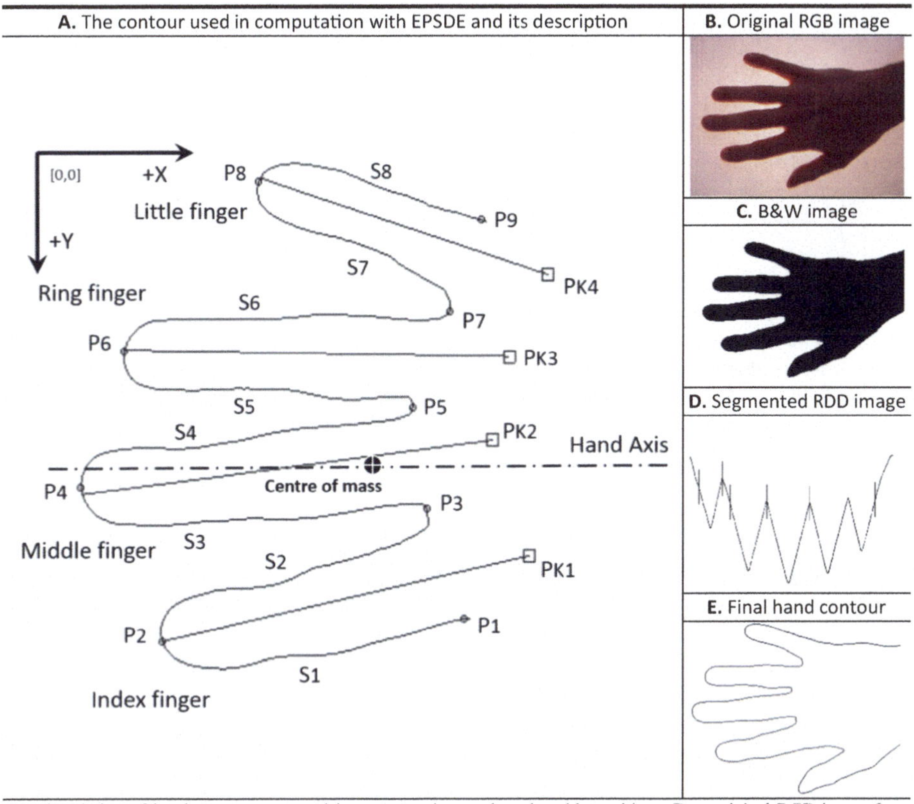

A – description of hand contour, axes and important points such as knuckle positions. **B.** – original RGB image from digital camera. **C.** - B&W representation ready for contour calculation. **D.** – Radial Distance Diagram, vertical lines define the limits of individual fingers, the thumb is on left side. **E.** – Complete contour calculated using Radial Sweep Algorithm.

Fig. 2. A chart displaying the process of hand contour processing – final processed images.

10/ The result of the classification process is expressed as a degree of similarity $\mathcal{E}$ between all model contours $\mathcal{M}$ which are stored in the database and the contour $\boldsymbol{S}$ which represents the identified person – see (3). The degree of similarity is given by a single number. In the case that the contours $\mathcal{M}$ and $\boldsymbol{S}$ are identical, then the single number representing the degree of similarity is equal to zero $\mathcal{E} = 0$. In all other cases the $\mathcal{E}$ is always equal to non-zero value. The degree of similarity which is represented by the single number is given as the sum of Euclidean distances of individual points of the contour $\mathcal{M}$ to the closest points of the contour $\boldsymbol{S}$ according to the methodology of the ICP algorithm and the result of comparison of several other elected markers – see Table 1. For representation of the *fitness* function see (2). The ICP algorithm uses the Euclidean metrics – the distance between two points in 2D.

Table 1. Markers used in classification process

Distances	Contours	Distances	Angles
$f_1^{\cdots} : \overline{P_2P_3}$	$f_7^{\cdots} : S_1 + S_2,\ [P_{K1}],\ \{x_0\}$	$f_{11}^{\cdots} : \overline{P_3P_5}$	$f_{14}^{\cdots} : P_3\widehat{P_5}P_7 = \alpha_1^{\cdots}$
$f_2^{\cdots} : \overline{P_3P_4}$	$f_8^{\cdots} : S_3 + S_4,\ [P_{K2}],\ \{x_1\}$	$f_{12}^{\cdots} : \overline{P_5P_7}$	$f_{15}^{\cdots} : P_5\widehat{P_7}P_3 = \alpha_2^{\cdots}$
$f_3^{\cdots} : \overline{P_4P_5}$	$f_9^{\cdots} : S_5 + S_6,\ [P_{K3}],\ \{x_2\}$	$f_{13}^{\cdots} : \overline{P_3P_7}$	$f_{16}^{\cdots} : P_5\widehat{P_3}P_7 = \alpha_3^{\cdots}$
$f_4^{\cdots} : \overline{P_5P_6}$	$f_{10}^{\cdots} : S_7 + S_8,\ [P_{K4}],\ \{x_3\}$		
$f_5^{\cdots} : \overline{P_6P_7}$			
$f_6^{\cdots} : \overline{P_7P_8}$			

$\overline{P_2P_3}$ - means Euclidean distance of two points. $S_{...} + S_{...}$ – means individual contours – see Fig. 1, which move around corresponding knuckle point $[P_{K...}]$. $P_3\widehat{P_5}P_7$ – means angle $\alpha_1^{\cdots}$, which include lines which led thru the points P_3P_5 and P_5P_7 and which intersect in point P_5. Analogously to other markers $f_{...}$ $[P_{K1}]$ - means a knuckle with which the finger contours S_1, S_2 are connected. All knuckles P_K are then connected together with center of mass of the hand contour. $\{x_{...}\}$ – means corresponding gen of the chromosome. $\alpha_1^{\cdots}, \alpha_2^{\cdots}, \alpha_3^{\cdots}$ - all angles are assumed in radians. $f_1^{\cdots}$ - mean marker $f_1^{\mathcal{M}}$ or f_1^{S}.

In Table 1 are described some measured distances. A relatively big problem is how to correctly estimate the point P_1, which is situated between the index finger and the thumb. The position of this point varies in the range 1–1.5 cm according to the thumb position and how much the thumb is pressed to the index finger. This problem significantly afflicts the whole classification process. On this account only the points $P_{3,5,7}$ are elected and distances are calculated between them.

An evolutionary process using EPSDE optimizer can be expressed as:

$$\mathcal{E} = \arg \underset{\mathcal{H} \in \mathbb{R}, \boldsymbol{S}}{\text{opt}} \ \Im_{\text{EA}}(\mathcal{M}, \boldsymbol{S}, \mathcal{H}). \tag{1}$$

where $\mathcal{H}$ means the space of possible solutions which is defined by variables f_k. The optimization process seeks a correct setting of these variables. The contour $\mathcal{M}$ is fully static and the contour $\boldsymbol{S}$ is moveable.

The *fitness* function (2) is calculated as a sum of real numbers for all individual markers – see Table 1 and (3). The classification criteria use Euclidean metrics for both sets of the markers $f_{1,2,3,4,5,6,11,12,13}$ and $f_{7,8,9,10}$ (points of the contours $\mathcal{M}$ and $\boldsymbol{S}$). An absolute value of difference – see (2), is used to classify the markers $f_{14,15,16}$. The markers mean the angles of heading of individual fingers. The *fitness* function is expressed as follows:

$$fitness = \begin{cases} \left(\sum_{i=1}^{i=16} \wp\right)^6 & if\ \mathcal{L} = 1 \\ \sum_{i=1}^{i=16} \wp & if\ \mathcal{L} = 0 \end{cases}$$

$$fitness, \wp \in \mathbb{R}$$

$$\wp = \begin{cases} 10 \times \left|f_i^{\mathcal{M}} - f_i^{S}\right| & if\ i = 1,2,3,4,5,6 \\ f_{7,8,9,10} & if\ i = 7,8,9,10 \\ 10 \times \left|f_i^{\mathcal{M}} - f_i^{S}\right| & if\ i = 11,12,13 \\ 1000 \times \left|f_i^{\mathcal{M}} - f_i^{S}\right| & if\ i = 14,15,16 \end{cases} \tag{2}$$

$$f_{1,2,3,4,5,6}^{S} : d\left(P_m^S, P_n^S\right), f_{1,2,3,4,5,6}^{\mathcal{M}} : d\left(P_m^{\mathcal{M}}, P_n^{\mathcal{M}}\right)$$

$$f_{7,8,9,10} : d\left(P_j^S, P_{min}^{\mathcal{M}}\right)$$

$$d\left(P_j^S, P_{min}^{\mathcal{M}}\right) = \min d\left(P_j^S, P_i^{\mathcal{M}}\right)\Big|_{i=1}^{i=n}$$

$$f_{11,12,13}^{S} : d\left(P_m^S, P_n^S\right), f_{11,12,13}^{\mathcal{M}} : d\left(P_m^{\mathcal{M}}, P_n^{\mathcal{M}}\right)$$

$$f_{14,15,16}^{S} : \alpha_j^S, f_{14,15,16}^{\mathcal{M}} : \alpha_j^{\mathcal{M}}, j \in (1,2,3)$$

where for distance $\mathrm{d}(A,B)$ of a two points $\mathrm{A}(x_A, y_A)$, $B(x_A, y_A)$ it holds that $\mathrm{d}(A,B) = \sqrt{(x_A - x_B)^2 + (y_A - y_B)^2}$. Indexes m, n for individual points $P_{\ldots}$ see Table 1 and Fig. 1G. $f_i^{\mathcal{M}}$ represents the mark which corresponds to model contour $\mathcal{M}$ and f_i^S represents the mark which corresponds to data sample $\boldsymbol{S}$. It is necessary to use the multipliers 10×, 1000× a because e.g. angle difference is in radians and is very small and all these values would not be useful at time of classification.

Only 4 fingers are used in calculation – see Fig. 1A, 2A. Heading of individual fingers has to be limited by a suitable function which says how much it is possible to rotate every finger in corresponding knuckle. The limiting function (3) will be marked as $\mathcal{L}$ and expressed as follows:

$$\mathcal{L} = \begin{cases} 1 & if\ x_{0,1,2,3} \notin \left\langle -0.30^{RAD}; +0.30^{RAD}\right\rangle \\ 1 & if\ x_{4,5} \notin \boldsymbol{\Psi} \\ 1 & if\ x_6 \notin \left\langle -0.40^{RAD}; +0.40^{RAD}\right\rangle \\ 0, & otherwise \end{cases} \tag{3}$$

where $x_{0,1,2,3,4,5,6}$ means individual genes of the $\boldsymbol{X_i}$ chromosome. Assignment of individual genes of the $\boldsymbol{X_i}$ chromosome to individual markers is given by this scheme: $\{x_0\}$ = index finger, $\{x_1\}$ = middle finger, $\{x_2\}$ = ring finger, $\{x_3\}$ = little finger,

$\{x_4\} = \Delta X$, $\{x_5\} = \mathit{\Delta} Y$, $\{x_6\} = \Delta\alpha$. Genes x_4, x_5, x_6 in given order represent the change of position $\Delta X, \Delta Y$ of the center of mass and the heading $\Delta\alpha$ of the whole contour regarding the center of mass of the contour $\boldsymbol{S}$, as well as to the center of mass of contour $\mathcal{M}$ and axis X of the coordinate system – see Fig. 2A. The contour $\boldsymbol{S}$ is rotated in its center of mass point. The markers $f_{7,8,9,10}$ are connected with the corresponding contour of fingers as follows: f_7-index finger, f_8-middle finger, f_9-ring finger, f_{10}-little finger. In the case that a gen x of the $\boldsymbol{X}_i$ chromosome is set out of the predefined limits (3) the six-power function is used (penalization) to the final function *fitness* – see (2) and (3). The value $\boldsymbol{\Omega}$ means a space in which the center of mass of the contour $\boldsymbol{S}$ can move – the corresponding value is stored in genes x_4, x_5. The space $\boldsymbol{\Omega}$ is given as oblong area with center in the position P_{Ω} as follows:

$$\begin{gathered} P_{\Omega}(x_{\Omega} = 0.5I_w, y_{\Omega} = 0.5I_h) \\ \boldsymbol{\Omega} = [\langle x_{\Omega} - 0.17I_w, x_{\Omega} + 0.17I_w\rangle, \langle y_{\Omega} - 0.17I_h, y_{\Omega} + 0.17I_h\rangle] \end{gathered} \tag{4}$$

I_w means the image width, I_h means the image height. The $\boldsymbol{X}_i$ chromosome effects only the markers f_7, f_8, f_9, f_{10}. Other markers are given by hand contour and participate during computation of the *fitness* function.

Evolutionary optimizer EPSDE works with a 7-dimensional optimization function *fitness*, $\mathrm{D_{im}} = 7$, which is constrained, non-continuous, non-separable, strongly non-linear and ill-conditioned. Part of identification information of every individual $\boldsymbol{X}_i$ is also value F. The correct chromosome should be expressed as $\boldsymbol{X}_i = [x_0, \ldots, x_6, F]$, but value F does not participate while calculation of (2). The value F is elected randomly in the EPSDE and is included into the internal "self-adaptive" mechanism. The genes x_0, x_1, x_2, x_3 contain the values of heading of the individual fingers of the contour $\boldsymbol{S}$ in corresponding knuckles $P_{K\ldots}$. The heading is calculated in radians and is a relative value ($\pm$). All markers $f_{\ldots}$ participate during *fitness* function calculation according to (2) but only markers $f_7, \ldots, f_{10}$ are directly influenced by the EPSDE algorithm. Values for other markers are given as a fixed number from the EPSDE point of view i.e. the EPSDE does not look for their optimal values.

6 Experimental Results

The GPDS150 database [13] contains images from a classic office scanner. The database contains images from 94 persons, 10 images from every person, images resolution is 1521 × 1280 × 8 bit-gray and 1402 × 1018 × 8 bit-gray. There are also different ambient lightning conditions and many images contain different centering marks at the different positions. Filtration of all such unusable and permanently changing parts of images is relatively time demanding. The big disadvantage is that the strong ambient lightning affects many images. One image from every person was selected as the model contour $\mathcal{M}$. Against this reference contour, the comparison was performed with all other samples $\boldsymbol{S}$ in the GPDS150 database. Totally, 94 × 940 = 88360 comparisons of contours were performed for every tested arrangement.

To classify all reached results, several common statistical indicators were selected namely *FAR*-false acceptance rate, *FRR*-false rejection rate and computation of the point *EER*-equal error rate. In order to possibly display the *ROC* curve, several other values were also computed: *FPR*-false positive rate and *TPR*-true positive rate. The algorithm enabling the indicators calculation is recorded in Algorithm 1. It was necessary to experimentally ascertain the upper limit for variable *lim*. If the κ_S value is increasing the value *lim* is also increasing. Usually from value $lim = 1000$ for $\kappa_S = 1$ up to the value $lim = 40000$ for $\kappa_S = 40$. The value *lim* is given in pixels. To estimate how robust are individual tested methods, a various number of pixels of the contour ***S*** were used.

Algorithm 1 The way of computation of the values FAR, FRR, EER

Input:	Output:
lim – max fitness of all genuine	EER, FAR, FRR $TPR_{EER}, FPR_{EER}, TNR_{EER}, FNR_{EER}$ $P_{TP-ERR}, P_{FN-EER}, N_{FP-EER}, N_{TN-EER}$

$\Sigma\, G_{TOT} = 940;\ \Sigma\, I_{TOT} = 87420;\ x \approx$ "∞"
for $\lambda_{EER}=0;\ \lambda_{EER}< lim;\ \lambda_{EER} = \lambda_{EER}+1$
  $P_{TP} = 0;\ P_{FN} = 0;\ N_{FP} = 0;\ N_{TN} = 0$
  for all models $\mathcal{M}$ (94)
    for all samples S (940)
      if $S \cong \mathcal{M} \wedge \varepsilon \le \lambda_{EER}$ **then** $P_{TP} = P_{TP} + 1$ (**SUCCESS**) ☺
      if $S \cong \mathcal{M} \wedge \varepsilon > \lambda_{EER}$ **then** $P_{FN} = P_{FN} + 1$ (**FAILURE**)
      if $S \ncong \mathcal{M} \wedge \varepsilon \le \lambda_{EER}$ **then** $N_{FP} = N_{FP} + 1$ (**FAILURE**)
      if $S \ncong \mathcal{M} \wedge \varepsilon > \lambda_{EER}$ **then** $N_{TN} = N_{TN} + 1$ (**SUCCESS**) ☺
    next
  next
  $TPR = P_{TP}/(P_{TP} + P_{FN});\ FPR = N_{FP}/(N_{FP} + N_{TN})$
  $TNR = N_{TN}/(N_{FP} + N_{TN});\ FNR = P_{FN}/(P_{TP} + P_{FN})$
  $R_1 = P_{FN}/\Sigma\, G_{TOT};\ R_2 = N_{FP}/\Sigma\, I_{TOT}$
  if $x > |R_1 - R_2|$ **then**
    $x = |R_1 - R_2|;\ EER = \lambda_{EER}$
    $P_{TP-ERR} = P_{TP};\ P_{FN-EER} = P_{FN};\ N_{FP-EER} = N_{FP};\ N_{TN-EER} = N_{TN}$
    $TPR_{EER} = TPR;\ FPR_{EER} = FPR$
    $TNR_{EER} = TNR;\ FNR_{EER} = FNR$
    $FAR = R_2;\ FRR = R_1$
    $\Sigma I_{UL} = N_{FP};\ \Sigma G_{OL} = P_{FN}$
  endif
next

$S \cong \mathcal{M}$- means, that investigated contour ***S*** and model contour $\mathcal{M}$ belong to identical person. Detection of correct classification was performed using images name. λ_{EER}-defines range of $fitness$ values in which is seeking point EER. lim-it is necessary to ascertain this value experimentally. x – helper variable. EER-equal error rate point, FAR-false acceptance rate, FRR-false rejection rate, TPR_{EER}-true positive rate at EER point, FPR_{EER}-false positive rate at EER point, TNR_{EER}-true negative rate at EER point, FNR_{EER}-false negative rate at EER point, P_{TP-ERR}-true positive (positive, genuine), P_{FN-EER}-false negative (positive, genuine), N_{FP-EER}-false positive (negative, impostor), N_{TN-EER}-true negative (negative, impostor), FAR-false acceptrance rate, FRR-false rejection rate. TNR-true negative rate, FNR-false negative rate. Values TPR, FPR define Receive operating characteristics (ROC) curve. $FAR = FPR_{EER}$, $FRR = FNR_{EER}$, $\Sigma\, G_{TOT} = (P_{TP} + P_{FN})$, $\Sigma\, I_{TOT} = (N_{FP} + N_{TN})$. $\Sigma\, G_{TOT}$-total number of genuines, $\Sigma\, I_{TOT}$-total number of impostors. ΣI_{UL}-sum of impostors under limit λ_{EER}. ΣG_{OL}-sum of genuines over limit λ_{EER}.

Selected combinations are marked as κ_M, κ_S and represent what number of pixels of the contour ***S*** are used during calculation. The number of pixels of contour $\mathcal{M}$ is not changed i.e. $\kappa_M = 1$ for all experiments. E.g. $\kappa_S = 40$ means that every 40th pixel of contour ***S*** is used in calculation. The set of elected values for κ_S is $\kappa_S \in (1, 5, 10, 15, 20, 25, 30, 35, 40)$. The average number of pixels of a contour in database GPDS150 is 1800. The EPSDE optimizer has only several working parameters: N_{POP}, N_{GEN} and L_P. The most important is selection of the correct number of individuals in the population and the correct number of generations. With regard to time demands, the combination $N_{POP} = 15$, $N_{GEN} = 200$ and $L_P = 10$ will be used in all experiments. The value L_P has only very small effect on the final result. In Fig. 3 are displayed the curves of population convergence towards the correct solution for $N_{POP} = 15$, $N_{GEN} = 200$. The number of iterations is equal to 10. The convergence speed is relatively large from start. Once the number of generations reaches the value 150 then the final *fitness* value changes only minimally. Selected values N_{POP}, N_{GEN} correspond to technical possibilities and time demands of given task. Of course, it is possible to use three times higher values but at the expense of many times higher computational demands.

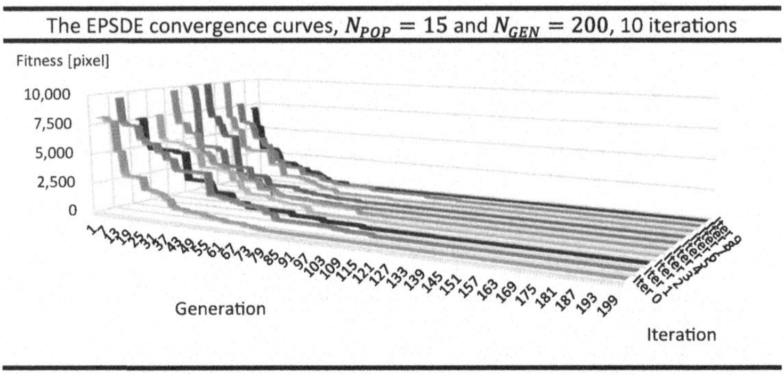

Fig. 3. Convergence curves. Record of 10 iterations.

Totally 5 different combinations of markers were tested - see Table 1, namely E1 – E5 and see [61, 62]. Several tested combinations were elected purely as a matter of interest – E3 and E5. All experiments are marked as **E1, E2, E3, E4, E5** and have the following arrangements and results:

E1/ Used markers $f_{1,..,16}$ where all markers are considered according to Table 1. The contour has 4 fingers – index finger, middle finger, ring finger and little finger. Aligning of two contours $\mathcal{M}$ and $\boldsymbol{S}$ is driven by ICP algorithm; the EPSDE is used as the optimizer. The four fingers without thumb are used in classification. Identical approach is used here e.g. [57].

Results of the classification are recorded in Table 2. The best reached value is $FAR = 1.38\%$, $FRR = 1.38\%$. This is only a slightly worse value in comparison to e.g. [25] where authors reached the value $EER = 1.38\%$. If the value κ_S is increasing, the accuracy decreases relatively faster without regard to the fact that, beside the four-fingered contour, the markers are also used. Differences in results are visible and a proposed arrangement cannot be marked as robust enough, although it provides good results especially for $\kappa_S = 1$.

E2/ This arrangement uses only markers $f_{1,..,6} + f_{11,..,13} + f_{14,..,16}$. No one finger-contour is used nor any other part of the palm-contour. The big advantage of this arrangement is an amazingly fast computation because it is not necessary to use the EPSDE algorithm and the ICP algorithm. If the *fitness* is computed according to (2), then the row with markers $f_{7,8,9,10}$ is missing. The $\wp$ can take only 3 values.

Table 2. Values *FAR*, *FRR*, *EER*

(E1) Markers $f_{1,..,16}$, i.e. 4 fingers + all markers, no Thumb								
n.	κ_M	κ_S	ΣI_{UL}	*FAR*%	ΣG_{OL}	*FRR*%	*EER*/*EER*%	Tot. time
1	**1**	**1**	**1207**	**1.38**	**13**	**1.38**	**2360/1.38**	**4.102**
2	1	5	5254	6.01	57	6.06	1290/6.01	0.655
3	1	10	7174	8.20	77	8.19	1115/8.20	0.639
4	1	15	8090	9.25	87	9.25	1054/9.25	0.561
5	1	20	8840	10.11	95	10.10	1030/10.10	0.452
6	1	25	9210	10.53	99	10.53	1012/10.53	0.390
7	1	30	9436	10.79	102	10.85	998/10.79	0.374
8	1	35	9585	10.96	103	10.95	988/10.96	0.343
9	1	40	9639	11.02	103	10.95	978/11.02	0.327

Tot. time in seconds for one sample alignment. *EER* in pixels and percents. *EER*%-obtained from *ROC* curve. Total number of evolutions is 9 rows × 87420 samples ***S*** = 786780 evolutions.

Results of classification are recorded in Table 3. Value κ_S is not possible to use because the hand contour is not included in the computation process. The reached accuracy values are *FAR* = 2.65%, *FRR* = 2.65%. Such results are relatively bad and this estimator arrangement does not make any sense. Possible improvement can be done using significantly larger number of markers.

Table 3. Values *FAR*, *FRR*, *EER*

(E2) Markers $f_{1,..,6} + f_{11,..,13} + f_{14,..,16}$ i.e. no fingers, no EA is used								
n.	κ_M	κ_S	ΣI_{UL}	*FAR*%	ΣG_{OL}	*FRR*%	*EER*/*EER*%	Tot. time
1	**1**	**1**	**2312**	**2.65**	**25**	**2.65**	**445/2.65**	**0.702**

Tot. time in seconds for one sample calculation.

E3/ This arrangement uses the contour with all five fingers. The thumb and all other parts of the palm are taken into account, according to Fig. 1E. It is all markers $f_{1,..,16}$ + thumb contour + other parts of palm contour. Mutual alignment of the contours $\mathcal{M}$ and ***S*** is driven by the ICP algorithm and the EPSDE optimizer is used.

Table 4. Values *FAR*, *FRR*, *EER*

(E3) All 5 fingers + all markers i.e. $f_{1,...,18}$ + Thumb								
n.	κ_M	κ_S	ΣI_{UL}	$FAR\%$	ΣG_{OL}	$FRR\%$	$EER/EER\%$	Tot. time
1	**1**	**1**	**2230**	**2.55**	**24**	**2.55**	**4497/2.55**	**5.491**
2	1	5	3908	4.47	42	4.46	1727/4.47	1.341
3	1	10	5390	6.16	58	6.17	1353/6.16	0.795
4	1	15	6416	7.33	69	7.34	1224/7.33	0.624
5	1	20	7116	8.14	76	8.08	1160/8.14	0.514
6	1	25	7716	8.82	83	8.82	1122/8.82	0.468
7	1	30	8061	9.22	87	9.25	1093/9.22	0.436
8	1	35	8331	9.52	89	9.46	1073/9.52	0.405
9	1	40	8538	9.76	92	9.78	1058/9.76	0.374

Tot. time in second for one sample alignment. Total number of evolutions is 9 rows × 87420 samples ***S*** = 786780 evolutions. *EER* in pixels and percents.

Results of classification are recorded in Table 4. The best reached values are $FAR = 2.55\%$, $FRR = 2.55\%$. The results are worse in comparison to Table 2. The reason is that the thumb has three knuckles between the wrist and the thumb tip. Accurate positions of individual knuckles of the thumb can be ascertained e.g. using x-ray, but such equipment is not at our disposal. The thumb knuckle on the wrist was assumed only. It is very interesting that the method E1 reaches better results for $\kappa_M = 1, \kappa_S = 1$. If the value κ_S is increasing the final accuracy decreases faster. The method E1 can be, indeed, marked as less robust in comparison to E3, but just a bit.

Table 5. Values *FAR*, *FRR*, *EER*

(E4) 4 fingers and no markers i.e. f_7, f_8, f_9, f_{10} only								
n.	κ_M	κ_S	ΣI_{UL}	$FAR\%$	ΣG_{OL}	$FRR\%$	$EER/EER\%$	Tot. time
1	**1**	**1**	**425**	**0.48**	**4**	**0.42**	**1225/0.48**	**4.710**
2	1	5	751	0.85	8	0.85	261/0.85	1.600
3	1	10	711	0.81	8	0.85	130/0.81	0.670
4	1	15	835	0.95	9	0.95	88/0.95	0.514
5	1	20	1092	1.12	11	1.17	69/1.24	0.452
6	1	25	967	1.10	11	1.17	54/1.10	0.436
7	1	30	1063	1.21	11	1.17	45/1.12	0.374
8	1	35	1157	1.32	14	1.48	39/1.32	0.358
9	1	40	1331	1.52	16	1.70	34/1.52	0.343

Tot. time in seconds for one sample alignment. Total number of evolutions is 9 rows × 87420 samples ***S*** = 786780 evolutions. *EER* in pixels and percents.

E4/ This arrangement uses only 4 fingers without thumb i.e. markers $f_{7,8,9,10}$. The contours $\mathcal{M}$ and S alignment process is driven directly by algorithm ICP and the EPSDE algorithm is used as the optimizer.

Results of experiments with this arrangement are recorded in Table 5. The best reached values are $\boldsymbol{FAR} = \mathbf{0.48\%}$, $\boldsymbol{FRR} = \mathbf{0.42\%}$. These values are the best reached of all selected arrangements E1–E5. The E4 arrangement also enables proposal of a very robust estimator. For $\kappa_M = 1, \kappa_S = 40$ the method provides results $FAR = 1.52\%$, $FRR = 1.70\%$ - see row 9, Table 5. Thanks to the smaller number of pixels of the contour S the calculation is faster. The reason for such very good results is that the finger contours are still identical although the fingers are movable in plane XY in corresponding knuckles $P_{K1,2,3,4}$. This rule is not valid for the thumb because the thumb contour changes its shape during calculation. Several problems were caused by wrong pressing of a hand to the scanner pad and thanks to that, the finger length was shorter for some contours and the contour was deformed.

E5/ This arrangement uses all 5 fingers and no another marker. Alignment of the contours $\mathcal{M}$ and S is driven by the ICP algorithm and EPSDE is used as the optimizer. This combination was elected for comparative purposes because only the combinations mentioned in Table 1 are investigated.

Results of classification are recorded in Table 6. The best reached values are $FAR = 2.12\%$, $FRR = 2.12\%$. The method is relatively robust. If the number of points of the contour is decreasing, the reached accuracy decreases very slowly. For $\kappa_M = 1, \kappa_S = 40$ resulting values are $FAR = 2.98\%$, $FRR = 2.98\%$. Analogously to experiment E4, no other marker $f_{1,..,6}$, $f_{11,..,13}$, $f_{14,..,16}$ is used here. The value *fitness* is calculated purely based on alignment of individual pixels of the contours $\mathcal{M}$ and S.

Table 6. Values *FAR*, *FRR*, *EER*

(E5) 5 fingers and no markers, i.e. f_7, f_8, f_9, f_{10} + Thumb + palm								
n.	κ_M	κ_S	ΣI_{UL}	$FAR\%$	ΣG_{OL}	$FRR\%$	$EER/EER\%$	Tot. time
1	**1**	**1**	**1856**	**2.12**	**20**	**2.12**	**3324/2.12**	**5.319**
2	1	5	1876	2.14	20	2.12	676/2.14	1.294
3	1	10	1894	2.16	21	2.23	345/2.16	0.764
4	1	15	2169	2.48	23	2.44	239/2.48	0.608
5	1	20	1967	2.25	22	2.34	181/2.25	0.483
6	1	25	2218	2.53	25	2.65	150/2.53	0.436
7	1	30	2304	2.63	23	2.44	127/2.63	0.405
8	1	35	2405	2.75	26	2.76	113/2.75	0.374
9	1	40	2613	2.98	28	2.97	100/2.98	0.343

Tot. time in seconds for one sample alignment. Total number of evolutions is 9 rows × 87420 samples S = 786780 evolutions. *EER* in pixels and percents.

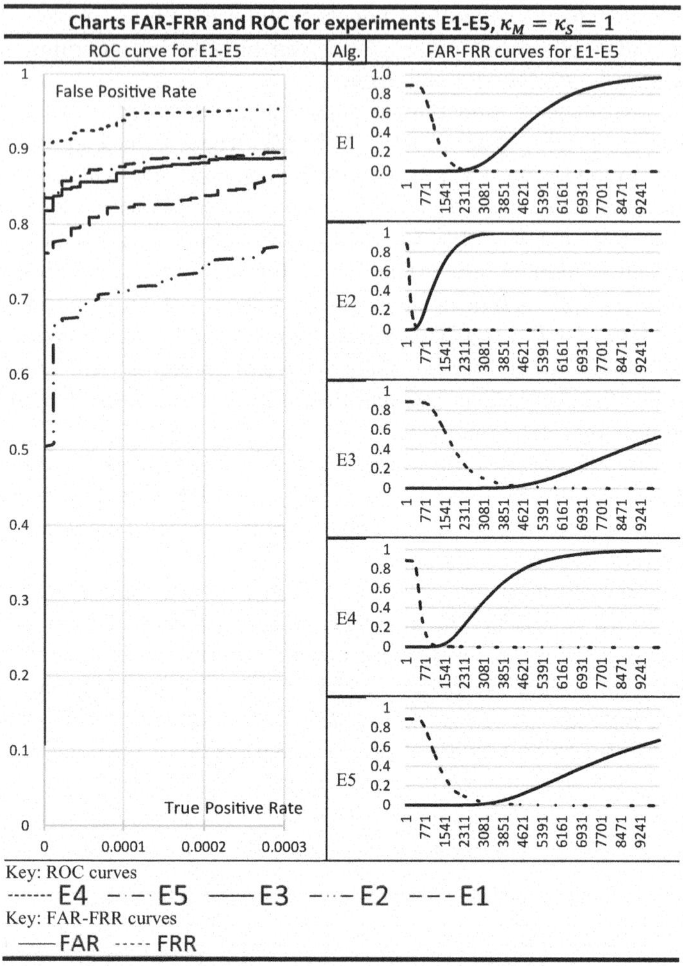

Fig. 4. Curves *FAR*, *FRR* and *ROC* curves for individual experiments.

In Table 7 a summary of the best and the worst reached results is recorded, regarding values κ_M, κ_S. It is clear that in all cases where the markers were used, it is possible to reach better results for $\kappa_M = 1, \kappa_S = 1$; but worse results if the value κ_S is increasing or equal to 40. Any method which includes the set of selected markers is, indeed, less robust.

From individual results in Tables 2, 3, 4, 5 and 6 and also Table 7, it follows that the selected markers do not bring significantly better results. But thanks to the use of all the markers, all results are somewhat worse, but only a bit. Utilizing of standalone markers without the contour brings, according to the Table 1, significantly worse results – see Table 2 and Table 7. The big problem is also that the accurate position of individual knuckles of corresponding fingers is not known. The knuckle positions are estimated only. Unfortunately, in many images of the identical person it is not possible

Table 7. Best and worst attained results *FAR*, *FRR*, *EER*

Recorded only selected values for κ_M, κ_S and individual experiments E1–E5

n.	κ_M, κ_S	Experiment type	b/w	*FAR*%	*FRR*%	*EER*%
1	1, 1	E1 + markers	best	1.38	1.38	1.38
2	1, 1	E2, markers only	best	2.65	2.65	2.65
3	1, 1	E3 + markers	best	2.55	2.55	2.55
4	**1, 1**	**E4, no markers**	**best**	**0.48**	**0.42**	**0.48**
5	1, 1	E5, no markers	best	2.12	2.12	2.12
6	1, 40	E1 + markers	worst	11.02	10.95	11.02
7	1, 40	E2, markers only	worst	2.65	2.65	2.65
8	1, 40	E3 + markers	worst	9.76	9.78	9.76
9	1, 40	E4, no markers	worst	1.52	1.70	1.52
10	1, 40	E5, no markers	worst	2.98	2.97	2.98

b/w-best-best reached results and b/w-worst-worst reached results regarding to values κ_M, κ_S.

to estimate identical knuckle position. Many images in the database have different pre-exposure/ambient lightning and the scanned hand was not inserted to the scanner under the identical angle with regards to ambient light. Also due to the different ambient lighting, the shadow around the hand corresponds to the background color. Under such conditions it is not possible to estimate the hand contour absolutely accurately. One possible way would be to create the hand contour for every image in the GPDS150 database manually, and similarly to estimate the approximate positions of all knuckles.

In Fig. 4 are recorded the resulting curves *FAR*, *FRR* and *ROC* for individual experiments E1–E5. All curves are considered for values $\kappa_M = \kappa_S = 1$. The *X* axis is, for the *ROC* curves, limited to the range $TPR \in \langle 0.0, 0.0003 \rangle$. The best results are provided by the arrangement E4. E4 uses only the finger contours without thumb and without any other makers. Contrary to this, the worst results were reached in experiment E2, where only markers are used. If the whole hand contour is used i.e. thumb taking into account, then the results are very similar to arrangement E4. The *ROC* curves of experiments E1, 3, 4, 5 are very similar. From a practical point of view there is no difference if the attained accuracy was $FAR\% = 0.48\%$ or $FAR\% = 2.65\%$, because no one of these classifiers with such results can be used for sensitive applications in security areas. These classifiers can only be used in everyday situations, where the security does not play a key role.

7 Conclusion

In the presented paper a large comparative study of several different arrangements of evolutionary estimators was described. All tested estimators E1–E5 classify a hand contour. The best reached results are represented by *FAR* values in the range $FAR\% = 0.48\% - 2.55\%$. The classification uses arrangements both with and without use of any other supporting information. The biggest problem was to correctly estimate the knuckle position for individual fingers and also the fact that the database GPDS150 was not acquired with high care. The ambient lightning is varying. Some images contain unwilling objects. A very important factor is the constant pressure of the hand on the scanner support during the hand scanning process. If this pressure is changing then the whole hand contour changes too. There are a lot of factors which afflict the final result. t is no doubt that the gradual development and technical advance in future, will enable catching of better hand images and also the images of bones and corresponding knuckles.

Acknowledgment. The publication was supported by the funds of University of Pardubice, Czech Republic – Student grant competition project (SGS_2020_001).

Author would like to express cordial thanks to Mr. Paul Hooper for his careful English text correction, patience and stamina.

References

1. Bakshe, R.C., Patil, A.M.: Hand geometry techniques: a review. Int. J. Mod. Commun. Technol. Res. **2**(11), 7 (2014)
2. Barra, S., Marsico, M.D., Nappi, M., Narducci, F., Riccio, D.: A hand-based biometric system in visible light for mobile environments. Inf. Sci. **479**, 472–485 (2019)
3. Bartlett, M.S., Lades, H.M., Sejnowski, T.J.: Independent component representations for face recognition. In: Conference on Human Vision and Electronic Imaging III, San Jose, California (1998)
4. Besl, P.J., McKay, H.D.: A method for registration of 3-D shapes. IEEE Trans. Pattern Anal. Mach. Intell. **14**(2), 239–256 (1992)
5. Bharathi, S., Sudhakar, R.: Hand biometrics: an overview. Int. J. Auto. Ident. Technol. **3**(2), 101–108 (2011)
6. Borra, S.R., Reddy, G.J., Reddy, E.S.: A broad survey on fingerprint recognition systems. In: IEEE 2016 International Conference on Wireless Communications, Signal Processing and Networking (WiSPNET) (2016)
7. Brest, J., Boškovič, B., Greiner, S., Žumer, V., Maučec, M.S.: Performance comparison of self-adaptive and adaptive differential evolution algorithms. Soft. Comput. **11**, 617–629 (2007)
8. Covavisaruch, N., Prateepamornkul, P., Ruchikachorn, P., Taksaphan, P.: Personal verification and identification using hand geometry. ECTI Trans. Comput. Inf. Technol. **1** (2), 134–140 (2003)
9. Daugman, J.: How iris recognition works. IEEE Trans. Circ. Syst. Video Technol. **14**(1), 21–30 (2004)
10. Dubuisson, M.P., Jain, A.K.: A modified Hausdorff distance for object matching. In: 12th International Conference on Pattern Recognition, pp. 566–568 (1994)
11. Duta, N.: A survey of biometric technology based on hand shape. Pattern Recogn. **42**, 2797–2806 (2009)
12. Faundez-Zanuy, M., Elizondo, D.A., Ferrer-Ballester, M.A., Travieso-González, C.M.: Authentication of individuals using hand geometry biometrics: a neural network approach. Neural Process. Lett. **26**, 201–216 (2016)

13. Ferrer, M.A., Morales, A., Travieso, C.M., Alonso, J.B.: Low cost multimodal biometric identification system based on hand geometry, palm and finger textures. In: 41st Annual IEEE International Carnahan Conference on Security Technology, pp. 52–58 (2007)
14. Ferrer, M., Vargas-Bonilla, J., Morales, A.: BiSpectral contactless hand based biometric identification device (2011). https://doi.org/10.5772/18096
15. Hemery, B., Mahier, J., Pasquet, M., Rosenberger, C.: Face authentication for banking. In: First International Conference on Advances in Computer-Human Interaction (2008)
16. Horn, B.K.P.: Closest form solution of absolute orientation using unit quaternions. J. Opt. Soc. Am. **4**(4), 629–642 (1987)
17. Charfi, N.: Biometric recognition based on hand shape and palmprint modalities. Image Processing. Ecole nationale supérieure Mines-Télécom Atlantique (2017)
18. Chauhan, S., Arora, A.S., Kaul, A.: A survey of emerging biometric modalities. Procedia Comput. Sci. **2**, 213–218 (2010)
19. Iorio, A., Li, X.: Solving rotated multi-objective optimization problems using differential evolution. In: Australian Conference on Artificial Intelligence, Cairns, Australia, pp. 861–872 (2004)
20. Jain, A.K., Ross, A., Pankanti, S.: A prototype hand geometry-based verification system. In: 2nd International Conference on Audio and Video based Biometric Person Authentication, pp. 166–171 (1999)
21. Jetenský, P., Marek, J., Rak, J.: Fingers segmentation and its approximation. In: Proceedings of 25th International Conference Radioelektronika, RADIOELEKTRONIKA 2015, pp. 431–434. IEEE (Institute of Electrical and Electronics Engineers), New York (2015). ISBN 978-1-4799-8117-5
22. Jetenský, P.: Human hand image analysis extracting finger coordinates using circular scanning. In: Proceedings of 25th International Conference Radioelektronika, Radioelektronika 2015, pp. 427–430. IEEE (Institute of Electrical and Electronics Engineers), New York (2015). ISBN 978-1-4799-8117-5
23. Jetenský, P.: Human hand image analysis extracting finger coordinates and axial vectors: finger axis detection using blob extraction and line fitting. In: 2014 24th International Conference Radioelektronika, pp. 1–4. IEEE (Institute of Electrical and Electronics Engineers), New York (2014). ISBN 978-1-4799-3715-8
24. Jost, T., Hügli, H.: Fast ICP algorithms for shape registration. In: Joint Pattern Recognition Symposium, pp. 91–99 (2002)
25. Kang, W., Wu, Q.: Pose-invariant hand shape recognition based on finger. Geometry **44**(11), 1510–1521 (2014)
26. Kumar, A., Hanmandlu, M., Kuldeep, M., Gupta, H.M.: Automatic ear detection for online biometric applications. In: Third National Conference on Computer Vision, Pattern Recognition, Image Processing and Graphics (2011)
27. Luque-Baena, R.M., Elizondob, D., Lopez-Rubioa, E., Palomoa, E.J., Watsonb, T.: Assessment of geometric features for individual identification and verification in biometric hand systems. Expert Syst. Appl. **40**(9), 3580–3594 (2013)
28. Maier-Hein, L., Franz, A.M., Santos, T.R., Schmidt, M., Fangerau, M., Meinzer, H.P., Fitzpatrick, J.M.: Convergent iterative closest-point algorithm to accommodate anisotropic and inhomogenous localization error. IEEE Trans. Pattern Anal. Mach. Intell. **34**(8), 1520–1532 (2012)
29. Mallipeddi, R., Suganthan, P.N.: Differential evolution algorithm with ensemble of parameters and mutation and crossover strategies. In: International Conference on Swarm, Evolutionary, and Memetic Computing SEMCCO 2010, pp. 71–78 (2010)
30. Moravec, J., Hub, M.: Automatic correction of barrel distorted images using cascaded evolutionary estimator. J. Inf. Sci. **366**, 70–98 (2016)

31. Park, G., Kim, S.: Hand biometric recognition based on fused hand geometry and vascular patterns. Sensors **28**(3), 2895–2910 (2013)
32. Parker, J.R.: Algorithms for Image Processing and Computer Vision, 2nd edn. Wiley, New York (2010)
33. Pavlidis, T.: Algorithms for Graphics and Image Processing, Springer, Heidelberg (1982)
34. Pottmann, H., Huang, Q.X., Yang, Y.L., Hu, S.M.: Geometry and convergence analysis of algorithms for registration of 3D shapes. Int. J. Comput. Vis. **67**(3), 277–296 (2006)
35. Price, K.: Differential evolution: a fast and simple numerical optimizer. In: NAFIPS, pp. 524–527 (1996)
36. Price, K., Storn, R.: Minimizing the real functions of the ICEC contest by differential evolution. In: IEEE International Conference on Evolutionary Computation, pp. 842–844 (1996)
37. Price, K., Storn, R.: Differential evolution – a simple evolution strategy for fast optimization. Dr. Dobb's J. **22**(4), 18–24 and 78 (1997)
38. Ramteke, S.M., Hatkar, S.S.: A survey on security and accuracy in palmprint recognition. Int. J. Eng. Res. Technol. (IJERT) **2**(1), 6 (2013)
39. Rechenberg, I.: Evolutionsstrategies: Optimierung technischer Systeme nach Prinzipien der biologischen Evolution. Frommann-Holzboog Eds., Stutgart, Germany (1973)
40. Rechenberg, I.: Evolutionsstrategie '94. Frommann-Holzboog Ed., Stuttgart (1994)
41. Rodrigues, M., Fisher, R., Liu, Y.: Special issue on registration and fusion of range images. Comput. Vis. Image Underst. **87**, 1–131 (2002)
42. Rusinkiewicz, S., Levoy, M.: Efficient variants of the ICP algorithm. In: IEEE Third International Conference on 3-D Digital Imaging and Modeling, p. 8 (2001)
43. Santos-Sierra, A., Casanova, J.G., Avila, C.S., Vera, J.V.: Silhouette-based hand recognition on mobile devices. In: International Carnahan Conference on Security Technology, pp. 160–166 (2009)
44. Sanches-Reillo, S.R., Sanches-Avila, S.C., Gonzales-Marcos, A.: Biometric identification through hand geometry measurement. IEEE Trans. Pattern Anal. Mach. Intell. **22**(10), 1168–1171 (2000)
45. Santos-Sierra, A., Sánchez-Ávila, C., Pozo, G.B., Guerra-Casanova, J.: Unconstrained and contactless hand geometry biometrics. Sensors **11**, 10143–10164 (2011)
46. Stockman, G., Shapiro, L.: Computer Vision. Prentice Hall, Upper Saddle River (2001)
47. Travieso, C.M., Alonso, J.B., David, S., Ferrer, M.A.: Optimization of a biometric system identification by hand geometry. In: Complex Systems Intelligence and Modern Technological Applications, Cherbourg, France, pp. 581–586 (2004)
48. Xiong, W., Xu, Ch., Ong, S.H.: Peg-free human shape analysis and recognition. In: IEEE International Conference on Acoustics, Speech, and Signal Processing (2005)
49. Yan, X., Su, X.G.: Linear regression analysis: theory and computing. https://doi.org/10.1142/6986
50. Yörük, E., Konukoglu, E., Sankur, B., Darbon, J.: 2006a shape based hand recognition. IEEE Trans. Image Process. **15**(7), 1803–1815 (2009)
51. Yörük, E., Dutagaci, H., Sankur, B.: Hand biometrics. Image Vis. Comput. **24**, 483–497 (2006)
52. Zayaraz, G., Vijayalakshmi, V., Jagadiswary, D.: Securing biometric authentication using DNA sequence and Naccache Stern Knapsack cryptosystem. In: IEEE 2009 International Conference on Control, Automation, Communication and Energy Conservation (2009)
53. Zhang, J., Sanderson, A.C.: JADE: adaptive differential evolution with optional external archive. IEEE Trans. Evol. Comput. **13**(5), 945–958 (2009)
54. Zhi-Peng, F., Yan-Ning, Z., Hai-Yan, H.: Survey of deep learning in face recognition. In: 2014 International Conference on Orange Technologies (2014)

55. Web1. http://www.gpds.ulpgc.es/. Accessed Mar 2020
56. Web2. https://en.wikipedia.org/wiki/Linear_regression. Accessed Mar 2020
57. Web3. https://us.allegion.com. Accessed Mar 2020
58. Web4. http://www.iso.org/iso/home/store/catalogue_tc/catalogue_detail.htm?csnumber=43638. Accessed Mar 2020
59. Web5. http://www1.icsi.berkeley.edu/~storn/code.html. Accessed Mar 2020
60. Web6. https://en.wikipedia.org/wiki/Differential_evolution. Accessed Mar 2020
61. Web7. http://handwork.4fan.cz/. Accessed Mar 2020
62. Web8. http://robomap.4fan.cz/. Accessed Mar 2020

Energy Consumption Reduction in Real Time Multiprocessor Embedded Systems with Uncertain Data

Ridha Mehalaine[1,2] and Fateh Boutekkouk[3(✉)]

[1] ESI: Ecole National Supérieure d'Informatique,
Oued Smar/El Harrach, Algiers 16309, Algeria
r_mahalaine@esi.dz
[2] ICOSI Lab., University of Khenchela,
PoB 1252 El Houria, 40004 Khenchela, Algeria
[3] ReLaCS2 Laboratory, University of Oum El Bouaghi,
04000 Oum El Bouaghi, Algeria
fateh_boutekkouk@yahoo.fr

Abstract. Energy consumption is certainly a major determinant for the success and deployment of embedded systems (ES). Unlike traditional ES, recent ones are more complex, open and interact with dynamic and uncertain environment. In this context, we presented the theoretical results of our flexible scheduling scheme aimed at minimizing energy consumption by using the Dynamic Voltage Scaling (DVS) technique and reusing the time savings that express the difference between the worst case time and the real execution time on a multiprocessor embedded architecture with uncertain data. We performed simulations under Matlab in order to evaluate the behavior of our proposed algorithm. These simulations have in particular confirmed the very good behavior of our proposed algorithm in terms of energy consumption with regard to periodic independent tasks executing on multiprocessor architecture.

Keywords: Multiprocessor embedded systems · Real time scheduling · EDF* · Uncertainty · Fuzzy logic

1 Introduction

An Embedded System (ES) is a specific application-specific computer system that is part of a larger system that is not necessarily computer-based (e.g. electronics, mechanics, etc.) [1]. An ES interacts with the outside world via sensors/actuators and is subject to very strict spatial and energy constraints. Moreover, if an ES must respect the temporal constraints imposed by the external environment, the SE is qualified as real time. In this context, the validity of the system depends not only on the accuracy of the calculations, but also on the moment of production of the results. For a real time system a fair result, but not respecting its temporal constraints is an unusable result that corresponds to a time fault that can have catastrophic consequences in terms of loss of life, ecology, etc. A classification of embedded systems could be carried out in relation to the used source of energy: embedded systems connected to an external source of

R. Silhavy (Ed.): CSOC 2020, AISC 1225, pp. 46–55, 2020.
https://doi.org/10.1007/978-3-030-51971-1_4

energy (ex. the printer, the TV, etc.) and the others, which depend on an autonomous power source with a finite life and often part of the same system (eg the mobile phone, the various components of modern automobiles, etc.). The objective of our work is to find the best scheduling scheme that ensures both the feasibility and the respect of the temporal and energy constraints. Our actual contribution is the application of fuzzy logic to minimize power consumption in real-time multiprocessor embedded systems with independent periodic tasks. The paper is organized as follows: section two is devoted to some recent pertinent related works, some relevant energy consumption estimation and reduction techniques are presented in section three. Our proposed model is detailed in section four and some obtained results from our experimentations are presented in section five before concluding.

2 Some Pertinent Related Works

In [2], a study established a model that predicts and orders the effectiveness of code relocation at the function level. This is based on a complete code profiling that was performed on a real - time system to discover the impact and that was obtained by using a function - level code transfer between the different types of memory, in order to reduce energy consumption. This was accomplished by grouping the assembly instructions to evaluate the distinct power reduction efficiency based on the placement of the function code. In [3], authors examined the challenges faced by battery powered systems, and then explores at more general problems, and several real-world embedded systems. In [4], a new algorithm called Energy Efficient Scheduling Algorithm for the multi-core heterogeneous embedded architectures has been proposed. The algorithm uses the principle of intelligent adaptive mechanism for energy consumption and performance. The algorithm has been tested with different multi-core test benches by integrating the different functionalities of the integrated architectures. The proposed algorithms have been compared to existing algorithms and have been tested in various circumstances. In her PhD thesis, the author developed an application that provides a general pattern of periodic tasks with preceding relationships, having a subset of identical, independent tasks. The deterministic requirements of the application execution and the optimization of the energy consumption converge towards a multiprocessor platform are discussed [5]. In [6], we proposed a fuzzy model to resolve the energy consumption problem in mono-processor embedded systems using fuzzy logic. The work presented in this paper can be considered as an extension of [6] to support shared memory based embedded multiprocessors architectures.

3 Estimation and Reduction of Energy Consumption

3.1 Estimation of Energy Consumption

Energy consumption is a problem that is widely discussed in the field of embedded computing. In this work we are interested in two aspects; the energy estimation and the reduction of the energy consumption at the processor level, as well as proposals to minimize this consumption were presented. We will use the joule, which is the unit of energy in the MKS system (meter-kg-second). A joule (J) is defined as the amount of work done by a force of a Newton moving a mass of one kilogram over a distance of one meter .At the same time a joule is the amount of energy that is dissipated by the power of an electric watt for one second [7]. A battery is an electrical energy storage device .The voltage that exists between the battery terminals can drive an electrical current through a resistor .In resistance, the electrical energy is converted into heat .If an electric current Iflow in a wire with a resistance R(in ohms), then according to Ohm's law: $U = IR$ (1). The dissipated electrical power is given by: $W = IU$ (2) where W represents the dissipated electrical power (in watts), I the current (in amperes). If a constant voltage U is applied to a system with the resistor R over a period of time t, the dissipated energy is equal to: $E = t\ U^2/\ R$ (3). E is the energy dissipated by the Joule effect in t, which denotes the time in second. U represents the voltage and R the resistance respectively [8]. The energy required for a device to execute a program can be expressed as the sum of the following four terms: Etotal = Ecomp + Ecomm + Emem + EIO (4) where Etotal is the total energy needed, Ecomp denotes the energy needed to perform the calculations, Emem represents the energy consumed by the memory subsystem, Ecomm denotes the energy needed for communication, and EIO is the energy consumed by one of the I/O devices. We will examine the term Ecomp in detail by presenting simplified energy models. The numerical values of the parameters depend strongly on the technology used. The most studied energy calculation model in the literature is defined as follows: if the speed of a machine is equal to a value 'v' for a given duration Δ, then the power consumption power is given by $v\alpha$ for a constant $\alpha > 1$, and the total consumed energy is $v^{\alpha} \times \Delta$. If we now assume that the speed of the machine is given by a function of time, v (t), then the total energy consumed during the duration Δ is calculated by integrating the power over the duration Δ: $\int_{\Delta} v(t)^{\alpha}\, dt$ (5) [9]. Each task in the scheduling problem is characterized by an amount of work wi that models the number of processor cycles required to execute it entirely. The total duration of this task depends, therefore on the speed of the system. If this task is executed on a machine running at a speed v, then it will need wi/v units of time to finish and the energy consumed by the machine is $Ei = v^{\alpha}.wi/v$ (6) [9].

3.2 Reduction of Energy Consumption

A first strategy for reducing energy consumption is to work on the technology of hardware components. Thus a decrease in the size of the components made possible by advances in manufacturing techniques allows a lower supply voltage and therefore a lower consumption. A second strategy is to limit the power of a component to the blocks needed for the current processing; for example, a cache memory can be divided into blocks that can be activated independently of each other. Another possibility is to limit the number of state changes in a circuit because each change of state induces an energy cost. Another way that is being explored is to specialize the components for the intended use, for example using reconfigurable FPGA components. The most effective way to lower energy consumption is known as Dynamic Voltage Scaling (DVS). Many modern processors can dynamically lower the voltage to reduce power consumption. The propagation delay limits the clock frequency of the microprocessor: the processor can operate at a lower supply voltage, only if the clock frequency is reduced to support the increase in propagation delay. In most cases, lowering the supply voltage is necessary to lower the operating frequency (for example, microprocessor speed) so all tasks will take longer to run. In real-time systems, if the frequency change is not done correctly, the synchronization requirements of the application may not be respected. Therefore, the advantages of the DVS technique can be exploited in real-time systems only after careful identification of the conditions under which we can safely slow down the processor without missing any deadline (for hard real-time tasks) or missing a limited number of deadlines [10, 11].

4 The Proposed Model

Through this work, we present our new approach which is the formulation of the energy consumption optimization problem under timing constraints using fuzzy logic .We will develop our idea by targeting a multiprocessor architecture with DVS technology. The proposed fuzzy scheduler deals with independent periodic tasks. The proposed algorithm is then validated on an illustrative example. The idea of a dynamic solution is based on the principle of adjusting the speed of future tasks, according to the time gain which is calculated on the basis of tasks worst-case costs and their deadlines. Let Tn = {T1, T2, ..., Tn} be a set of n independent periodic tasks. Each task Ti is characterized by: i: the identity of the task; Pi: the period of the task Ti. Si: the speed of execution of the task Ti; Di: the deadline of the task Ti. ti: the execution time of the task Ti; Ci: the worst case cost for executing the task Ti in processor cycles. gi (S): the power consumed for the execution of the task Ti; Ri: the arrival date of the Ti task. Smin: the minimum speed that ensures system operation. Smax: the maximum speed of the processor (normalized = 1). The execution time is given by: ti = Ci/Si (7). If the execution speed is Si, Si = Smin or Si = S′ then the power consumption is optimal, knowing that S′ is the execution speed which ensures "processors load" equal to 1. In the proposed fuzzy model, the input stage consists of three input variables Ci the worst case execution cost of the task, NBi the battery level and Di the deadline of the task (see Fig. 1). The worst-case execution cost does not depend on the speed of the processor; it is the cost in processor cycles generated by the worst-case execution of the task Ti. The deadline for the task is the deadline which represents the last time before the task Ti must finish .The three input parameters decide the highest task priority from the tasks queue NFAT (Fig. 2). The fuzzy inference rules of our fuzzy model are shown in Fig. 3. All the decisions of the proposed fuzzy system are represented in Fig. 4.

For tasks in different queues, those in the highest priority queue will be scheduled first. Thus, the ready tasks of the high operational queue are given the highest priority. If the high priority sub queue is empty, the pending tasks of the middle priority sub queue are considered. For tasks in the same queue, we adopt the EDF* scheduling which is the task with the shortest distance limit to be set first. This algorithm assigns priorities based on the temporal proximity of each task's deadline, so the task with the closest deadline is assigned the highest priority. At any time, it is the highest priority task that runs (preemptive scheduling). The priority of each task is dynamically recalculated each time the system state changes (arrival of a task, completion of a task).

If there is a ready fuzzy high priority task, then the preemption right is possible. There are so-called hybrid techniques based on collaboration between hardware and software components; for example more or less deep standby strategies of components, or the adaptation of the voltage of the processor. This last class of techniques allows significant reductions in energy consumption. The results obtained with these voltage adaptation techniques become an important research theme in the energy saving community.

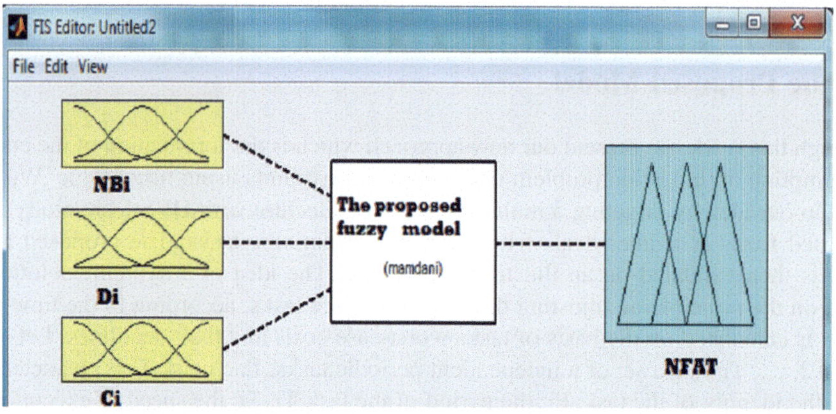

Fig. 1. The proposed fuzzy model

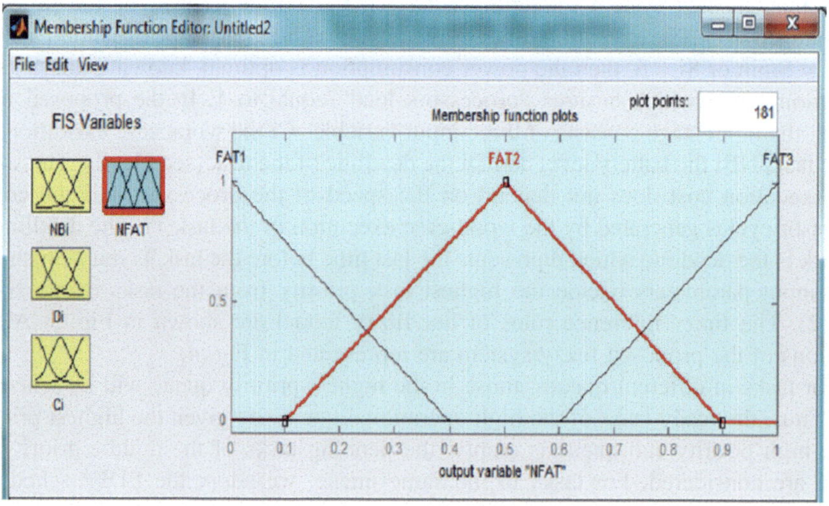

Fig. 2. The output variable NFAT.

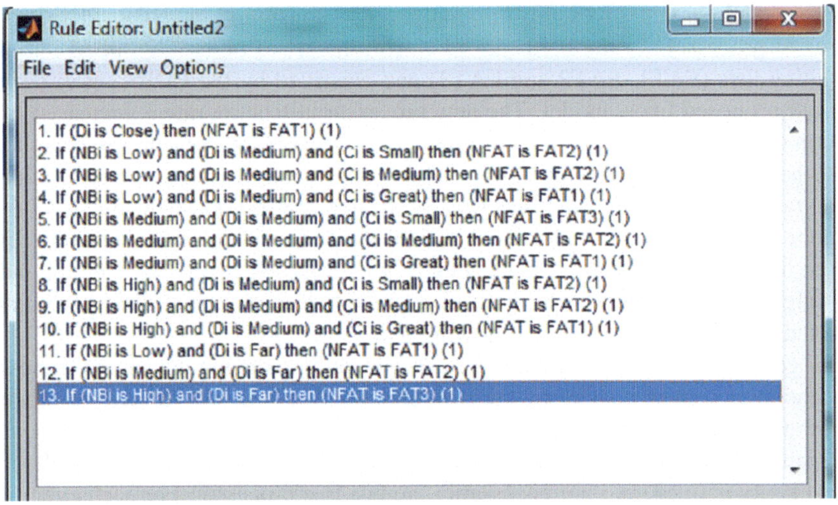

Fig. 3. Fuzzy inference rules

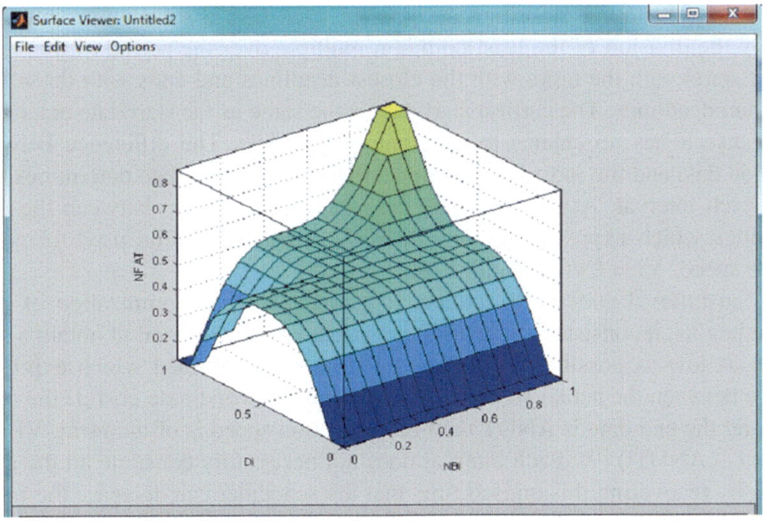

Fig. 4. The decisions of the proposed fuzzy system

Our study is based on several assumptions regarding the execution of tasks on a multiprocessor platform. The execution of the task is not associated with a given processor, and in return for preemption can continue on any processor at any time a task can only run on a single processor. An identical multiprocessor platform (Smin $\leq$ Si $\leq$ Smax) is used; the operation of the processors is based on a common clock. The speed of execution of a task could therefore change with each new allocation of the processor, that is to say the return from preemption. With a known number of processors, there is the proper scheduling that gives the minimal energy consumption. NBR_PROC is the number of processors in the hardware platform; the set of tasks is grouped into three (3) waiting lists (FAT1, FAT2 and FAT3) and classified by the EDF* algorithm after using our proposed fuzzy model. The new processor utilization rate is increased by the insertion of a task and the way the placement of this task. Our algorithm selects the first task of the queue for each free processor or becomes free; for each execution termination of an iteration Tij, it reinserts this task (that is, the iteration Tik, knowing that k = j + 1) in the queue calculated by the model fuzzy if j > (P/Pi). In the case of arrival of a new task, it must be inserted in the position calculated by the proposed fuzzy logic model and the EDF* algorithm; no new jobs inserted in a queue before its arrival date, but it can be reinserted in the case of iteration j, knowing that j > 1.

Our goal is to maximize the CPU utilization rate by reducing the speeds in possible cases; the migration is ensured by the assignment of the first task to the first processor becoming free. In the case of preemption (no free processor and the arrival of a task with higher priority than running tasks) the processor that executes the task of the greatest laxity is freed, and not necessarily the last free processor. For each event if the number of tasks is less than or equals to the number of processors then we execute each task with the minimum speed that allows to meet its deadline without taking into consideration other tasks. No preemption before the processor cycle termination is running. Our scheduling algorithm assigns for each task a set of time intervals calculated by the division of the least common multiple over the period of this task. The scheduler starts with the tasks with the closest deadlines and ends with those with the most distant deadlines. The earliest start date is the same as the start date entered by the user, because it has no chance to start before that date. The difference between the termination date and the most late and the earliest termination date determines the time gain for each interval. At this point we calculate the minimum between the different time savings which expresses the length of the time that can be used to reduce the processor speed. Vi = Ci/(Ci + gain) (8)

Once an interval ends its execution, a second phase of optimization of the consumption begins; it consists in adjusting the speeds of each interval to obtain a speed of operation as low as possible, by the use of time not consumed which expresses the difference between the actual cost of execution and the worst case cost .If the start date is BGN and the end date is ANDT then the execution speed is of the form: Vi = Ci/(Ci + BGNi+1 − ANDTi) (9). Each interval does not necessarily consume all the available time, it is by recovering this unused time that the scheduler can decrease the frequency of the processor by decreasing the speed of execution of the following intervals in order to obtain a better scheduling.

5 Experimental Study

In our case study, we assume that our system includes five (05) periodic independent tasks T1, T2, T3, T4 and T5 and two (02) processors such as: T1(R1 = 0, C1 = 2, Tex1 = 1, D1 = 6, P1 = 20), T2(R2 = 1, C2 = 2, Tex2 = 2, D2 = 4, P2 = 10), T3(R3 = 1, C3 = 3, Tex3 = 3, D3 = 10, P3 = 20), T4(R4 = 3, C4 = 2, Tex4 = 1, D4 = 7, P4 = 10), T5(R5 = 11, C5 = 2, Tex5 = 2, D5 = 19, P5 = 20). Vmax = 1, Vmin = 1/2, NBR_PROC = 2.

At the instant t = 0, the number of tasks is less than the number of processors then we execute these tasks with the minimum speed to meet their deadlines.

The study period is P = ppcm (20) = 20. After using our fuzzy model and the EDF* algorithm we obtain the following interval: [0, T1, 2] with the maximum speed Vmax.

For each interval [x, y, z], x: represents the start date, y: represents the task Ti, z: represents the end date. The date of termination is calculated at the latest [0, T1, 6] then the time gain between the two intervals equals4 and for the regions we have a single region so the time saving of this region is 4. The execution speed is: V11 = 2/(2 + 4) = 2/6 = 1/3, but Vmin = 1/2 then V11 = Max {1/3, 1/2} = 1/2.

Task T11 executes its first processor cycle in 02 units of time by processor 01.

tex11 = 1 so task T1 completes its first (and last) iteration on date 02 without pre-emption. At time t = 1, processor 01 is busy (executes T11) and has not completed execution for the first processor cycle and tasks T21, T31 have already arrived then the number of tasks is greater than the number of processors. The study period is P = ppcm {20,10,20} = 20. After using our fuzzy model and the EDF * algorithm we obtain the following intervals: [1, T2, 3], [2, T1, 3], [1, T3, 4] using the maximum speed Vmax. Termination dates are calculated at the latest: [1, T2, 4], [2, T1, 6], [1, T3, 10] and the time savings between each two intervals are 1, 3, 6 and for the regions there is only one region: {2, 1, 3} and the time saving of this region is 1. We have a single free processor P2 so T2 is executed; the execution speed for the task V21 = 2/(2 + 1) = 2/3 then the task T21 executes its first processor cycle in 1.5 unit of time. At time t = 2, T1 ends its execution by the termination of its first processor cycle; the number of tasks equals the number of processors then we execute each task with the minimum speed that allows to meet its deadline without taking into account other tasks. The intervals are [2.5, T2, 3.5], [2, T3, 5] with the maximum speed Vmax. The completion dates are calculated at the latest [2.5, T2, 4], [2, T3, 10] and the time savings between each two intervals are 0.5 and 5.

At time t = 2 task, T2 is running and has not completed the execution of its first processor cycle so the execution speed for task T21remains unchanged V21 = 2/3, and the execution speed for task V31 = 3 / (3 + 5) = 3/8, then V31 = 1/2. Task T31 executes its first processor cycle in 02 units of time.

At time t = 4, T2 ends its execution of the first iteration by the termination of its second processor cycle; T3 completes the execution of its first processor cycle. The number of tasks is equal to the number of processors then each task is executed with the minimum speed that allows meeting its deadline without taking into consideration other tasks. The intervals are [4, T4, 6], [4, T3, 6] with the maximum speed Vmax. The completion dates are calculated at the latest [4, T4, 7], [4, T3, 10]. The time savings between the two intervals is 14 then the execution speed for task T41 is V41 = 2/(2 + 1) = 2/3, and the execution speed for task T31 is V31 = 2/(2 + 4) = 1/3 then V31 = 1/2 . Task T41 executes its first processor cycle in1.5 unit of time.

At time t = 5.5, T1 completes its execution of the first iteration by the termination of its first processor cycle (tex4 = 1); no task in the queue until time t = 8; T3 completes its execution of the first iteration by the termination of its third processor cycle. No task in the queue until the moment t = 11, the arrival of the task T5; the number of tasks equals the number of processors then we execute each task with the minimum speed that allows to meet its deadline without taking into account other tasks. The intervals are [11, T2, 13], [11, T5, 13] with the maximum speed Vmax.

We calculate the termination dates at the latest [11, T2, 14], [11, T5, 19] so the time savings between each of the two intervals are 1 and6, then the execution speed for the Task T22 is: V22 = 2/(2 + 1) = 2/3, and the execution speed for task T51 is V51 = 2/(2 + 6) = 1/4, V51 = 1/2. Task T22 executes its first and second processor cycles in 03 time units. At time t = 13, a preemption is occurred because of the arrival of task T4 for its second iteration. The processor P1 is occupied by the last processor cycle of the task T2. The intervals are [13, T4, 15], [13, T5,15] with the maximum speed Vmax .

Termination dates are calculated at the latest [13, T4, 17], [13, T5, 19] and the time savings between each of the two intervals are 2 and4. The execution speed for task T42 is V42 = 2/(2 + 2) = 1/2, and the execution speed for task T51 is V51 = 2/(2 + 2) = 1/2 so task T42 executes its first processor cycle in 02 units of time.

At the instant t = 14, task T2 ends its execution and the task T5 is the only task in the queue and the processor P1 is free then the execution speed for task T51 is: V51 = 1/2. Task T51 executes its second processor cycle in 02 units of time. At time t = 16, task T5 completes its execution (Fig. 5).

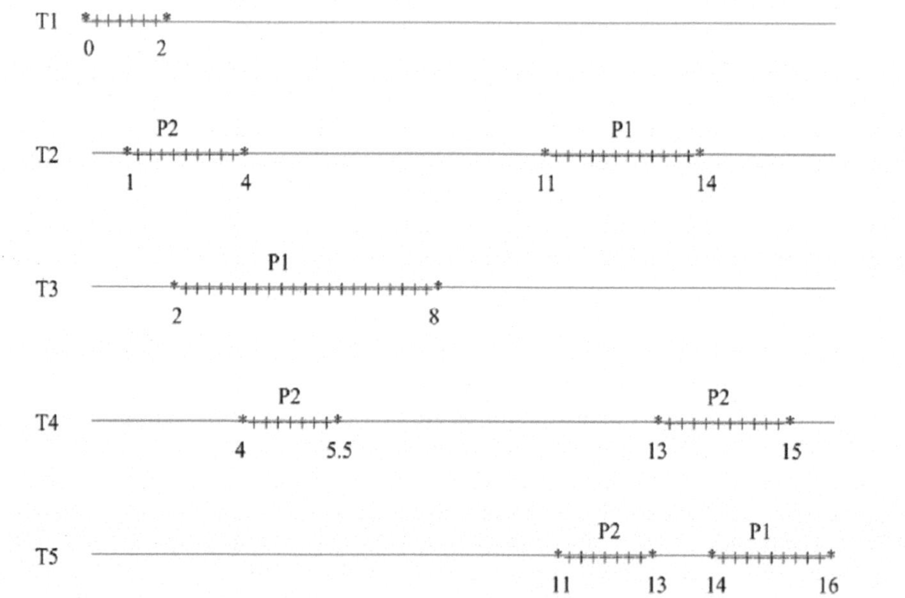

Fig. 5. Tasks scheduling results

The total energy consumed over a period of time is calculated by the integration of power, such as the power dissipation function gi(S) = $a_i S^r$, at $a_i > 0$ and $r \geq 2$ ($a_i = 1$, $r = 2$) is: Etot = (1/2)2 * 2 + (2/3)2 * 3 + (2/3)2 * 3 + (1/2)2 * 6 + (2/3)2 * 1.5 + (1/2) 2 * 2 + (1/2)2 * 4 = 6.83 joules.

6 Conclusion

An analysis of real time multiprocessor scheduling problem that takes into account the minimization of energy consumption leads us to note that this problem could be solved effectively using fuzzy logic. This is justified by the fact that we are dealing with inaccurate and uncertain information. This information includes, for example, the arrival times of tasks, the actual execution times of tasks which are generally very far from their execution times at the worst case, the best clock frequency of the processor which leads to the minimal energy consumption, the right time to migrate a task to another processor, etc. Intrinsic uncertainty in real-time systems and in particular dynamic systems increases the difficulties of conventional scheduling algorithms to optimize energy consumption. By integrating fuzzy logic into the real-time scheduling problem, the scheduler's decisions regarding choice of the best processor clock rate, priorities, and task migration dates can be improved considerably.

References

1. Wolf, W.: Computers and Components Principles of Embedded Computing System Design. Morgan Kaufman Publishers, Burlington (2000)
2. Choi, H., Koo, Y., Park, S.: Modeling the power consumption of function-level code relocation for low-power embedded systems. Appl. Sci. **9**(11), 2354 (2019)
3. Malewski, M., Cowell, D.M.J., Freear, S.: Reviewof battery powered embedded systems design for mission-critical low-power applications. Int. J. Electron. **105**(6), 893–909 (2017)
4. Anuradha, P., Rallapalli, H., Narsimha, G.: Energy efficient scheduling algorithm for the multicore heterogeneous embedded architectures. Des. Autom. Embed. Syst. **22**(1–2), 1–12 (2018)
5. Rohárik Vîlcu, D.M.: Optimal scheduling of tasks for CPU power consumption. Ph.D. thesis, Université Paris XII – Val de Marne (2004)
6. Mehalaine, R., Boutekkouk, F.: Fuzzy energy aware real time scheduling targeting mono-processor embedded architectures. In: CSOC 2016: 5th Computer Science On-line Conference 2016. Springer Series: Advances in Intelligent Systems and Computing - ISSN 2194, vol. 5357. pp. 81–91 (2016)
7. Smith, J.S.: Application Specific Integrated Circuits. Addision Wesley, Boston (1997)
8. Yao, F., Demers, A., Shenker, S.: A scheduling model for reduced CPU energy. In: Proceedings of the 36th Annual Symposium on Foundations of Computer Science, FOCS, Washington, DC, USA, pp. 374–382 (1995)
9. Kacem, F.: Algorithmes Exacts et Approchés pour des problèmes d'Ordonnancement et de Placement. Ph.D. thesis (2012)
10. Buttazzo, G.C., Lipari, G., Abeni, L., Caccamo, M.: Soft RealTime Systems: Predictability vs. Efficiency. Springer, Heidelberg (2005)
11. Pedram, M., Nazarian, S.: Thermal modeling, analysis and management in VLSI cicuits: principles and methods. Proc. IEEE **94**(8), 1487–1501 (2006)

Developing an Efficient Method for Automatic Threshold Detection Based on Hybrid Feature Selection Approach

Heba Mamdouh Farghaly[1(✉)], Abdelmgeid A. Ali[1], and Tarek Abd El-Hafeez[1,2]

[1] Department of Computer Science, Faculty of Science, Minia University, El-Minia, Egypt
heba.mamdouh@mu.edu.eg
[2] Computer Science Unit, Deraya University, El-Minia, Egypt

Abstract. Dimensionality reduction is an interesting area of research in data mining. An effective way to reduce dimensions is feature selection that removes irrelevant information meanwhile helping to understand the learning model better and improving prediction accuracy. In this paper, we face a challenge of filter methods to determine number of significant features that achieves better performance since filters don't evaluate performance based on accuracy but use certain criteria to rank features based on some scores. To handle this challenge, we proposed an effective hybrid approach for feature selection that is a filter-based method inspired by concepts of chi-square, Relief-F, and mutual information. It provides a score for each feature then specifies threshold value automatically based on dataset in use to select important subset of features used to build model, which reduces required execution time and amount of memory. Our proposed approach was analyzed empirically and theoretically to demonstrate its efficiency.

Keywords: Feature selection · Relief-F · Chi-square · Mutual information · Classification · Data mining

1 Introduction

Data mining is the knowledge discovery process. Data mining approaches developed to discover knowledge automatically and identify patterns from a large amount of data [1]. In the past years, the dimensions of data used for data mining and machine learning tasks have increased dramatically. Feature selection is used to reduce data dimensionality in a simplified way that provides a better description of the dataset with fewer features compared to the original feature set. To achieve this purpose, it removes irrelevant or redundant features from the original dataset [2] that results in saving

R. Silhavy (Ed.): CSOC 2020, AISC 1225, pp. 56–72, 2020.
https://doi.org/10.1007/978-3-030-51971-1_5

computing time, accelerating predictive performance, enhancing data comprehensibility, and providing better visibility of data.

Feature selection methods are divided into three categories based on different research strategies which are embedded, filter, and wrapper methods [3]. Filter methods select best features based on data characteristics. It evaluates each feature without utilizing any classification algorithms [4]. In wrappers, the quality of the selected subset of feature is evaluated using the feedback from classification accuracy. Although the wrappers are one of the most preferred methods of feature selection, it is expensive and slower than the filters [5]. The embedded methods search for an optimal feature subset during the construction of classifier. Although embedded methods are less in intensive computationally than wrappers, they select a subset of features based on the learning algorithm and have high computational complexity [6]. Due to these limitations, in this study, we focus on filters methods specifically filter approaches employ independent evaluation measures which include information theory, dependency, and consistency measures [7]. Thus they are fast, can be scaled easily to a very high-dimensional dataset and it is able to omit the irrelevant and redundant features effectively [8]. Filter models consist of the following steps: first, the features are ranked according to certain criteria. Second, the highest ranked features are selected to induce models of classification [9].

Nowadays, researchers are interested in developing new hybrid feature selection methods since they speed up the process of eliminating irrelevant features and improve the accuracy of classification compared to other methods [10, 11]. In this work, a hybrid feature selection approach is proposed for classification by aggregating mutual information, chi-square, and Relief-F methods. The combination of filter-based feature selection approaches is performed in order to discover feature interactions. Theoretically, the mutual information tries to solve the problem of redundancy where it is used to estimate the relevance of a subset of features [12], the chi uses mathematical statistics to measure the independence of two features [13], and the Relief-F method concerns with evaluating features according to the similarity of the neighboring samples in a set of instances that are being analyzed [14]. The performance of the proposed approach has been evaluated using well-known datasets.

Problem Statement: mining of a large amount of data can be infeasible and may take a long time therefore when applying feature selection techniques; it should reduce dataset without losing any valuable data. The classifier efficiency is also another problem of classification where it depends not only on the classification technique but also on the method of feature selection. The selection of irrelevant features increases complexity therefore, the method of feature selection plays a vital role as it increases the efficiency of the classifier.

The main problem of using filter feature selection methods is how to determine the number of significant features that achieve better performance. Filter methods, first, score and rank features based on their relevance for the class label and then select them based on a threshold value that is predefined by the user.

Goals: The essential goal of this research is to develop a method for detecting threshold value automatically using a hybrid filter-based feature selection approach based on mutual information, chi-square and Relief-F methods to identify an optimal value for threshold automatically based on a dataset in use to select features subset that yielding a fewer number of features and a better or even similar classification performance than using a set of all features.

Contributions: main contributions of this work can be summarized as follows:

The proposed method, develop a hybrid feature selection method based on the filters that combine mutual information, chi-Square, and Relief-F which can remove irrelevant features effectively, achieve better performance for classification, and reduce the required amount of memory and execution time.

- The proposed method solves the problem of how to find an optimal threshold value for retaining important features used to achieve desirable performance.
- The proposed method can be applied and integrated into real-life applications which can help experts in decision making.

The rest of this paper is organized as follows: Sect. 2 presents the details of related work. Section 3 explains our proposed hybrid feature selection method. Section 4 demonstrates experimental results. Finally, Sect. 5 presents the conclusion and future work.

2 Related Work

Identifying useful features from thousands of related features is a challenging task. Feature selection techniques are applied for classification problems in order to select the most relevant features in the problem domain, thus improving prediction accuracy and computational speed. Table 1 briefly summarizes several existing studies associated with the feature selection methodologies to remove irrelevant features.

Table 1. Summarizes several existing studies associated with the feature selection methodologies

Authors	Objective	Feature selection methods	Techniques	Dataset	Evaluation metrics
Sulaiman M. A. et al. [15]	Proposed a method for feature selection based on the algorithm of greedy feed forward for selecting a subset of feature from the set of all features	Mutual Information (MI)	• Support Vector Machines (SVM) • Artificial neural networks (ANN)	Well log data from wells of a Middle Eastern region	Accuracy
Jabbar, M. A. et al. [16]	Proposed a classification model to predict heart disease that used chi-square and random forest	Chi-square	Random Forest (RF)	Coronary heart disease	• Confusion matrix • Specificity • Sensitivity • Accuracy • Positive and Negative predictive value
H. Djellali et al. [17]	Proposed an FS_MRMR_SVM feature selection method to classify the medical data	• Fisher score • SVM-RFE • mRMR	SVM	• Wisconsin Breast cancer • Hepatitis	Accuracy
Huda, S. et al. [18]	Address imbalanced medical data challenges	• GANNIGMA • MRMR	• SVM • Bagging • Decision Tree (DT)	Brain tumor	• True and false Positive Rate • Precision • F-measure • ROC curve
Liu, X. et al. [19]	Proposed a hybrid classification system that aiding to diagnosing heart disease	• Relief-F • Rough Set (RFRS)	Ensemble classifier with the C4.5 decision tree	Statlog (Heart) dataset	• Confusion Matrix • Sensitivity • Specificity • Accuracy
Qin, C. J. et al. [10]	Proposed a novel algorithm that integrating multiple feature selection methods into the ensemble algorithm	• MI • Chi-square	• Logistic Regression (LR) • RF • SVM • Gradient • Boosting Decision Tree (GBDT) • Multi-Layer perceptron (MLP) • K-Nearest Neighbor (KNN) • Adaboost	Coronary heart disease	• Recall • Accuracy • F-measure • ANOVA
Haq, A. U. et al. [11]	Proposed a hybrid intelligent machine-learning based system used for heart disease diagnosis	• Relief • MRMR • LASSO	• LR • KNN • ANN • SVM • DT • Naive Bayes (NB)	Cleveland heart disease	• Accuracy • Specificity Sensitivity • Matthews' correlation coefficient • Execution time

Although the selection of a value of the threshold is critical in getting the optimal subset of features, it is clear from the above studies that emphasis has been placed on proposing improved methodologies for feature selection rather than determining a minimum value for the threshold to retain important features. The strong point of this paper relies on developing a method to automatically detect threshold value for selecting the most significant features for building the classifier.

3 Methodology

Unfortunately, most of the real-life data used to build model may contain redundant information so extracting the most meaningful information is a challenging task. Feature selection is an important phase in the classification process; it removes insignificant features from datasets. Feature selection improves both the quality of the model and the efficiency of the modeling process. Therefore, this research aims at proposing a hybrid feature selection technique where three well–known filter feature selection methods were combined in a specific way and were applied to an original dataset that provides for each feature a degree of importance. Then, we introduce a method for automatically determining the optimal value of the threshold used for selecting a subset of important features for building the classifier. The proposed system consists of four main steps: data preprocessing, feature selection, a classification algorithm that was used to classify selected features into different classes and finally evaluate the performance of the proposed technique. Figure 1 and Fig. 2 shows the framework and main steps of the algorithm of the proposed technique.

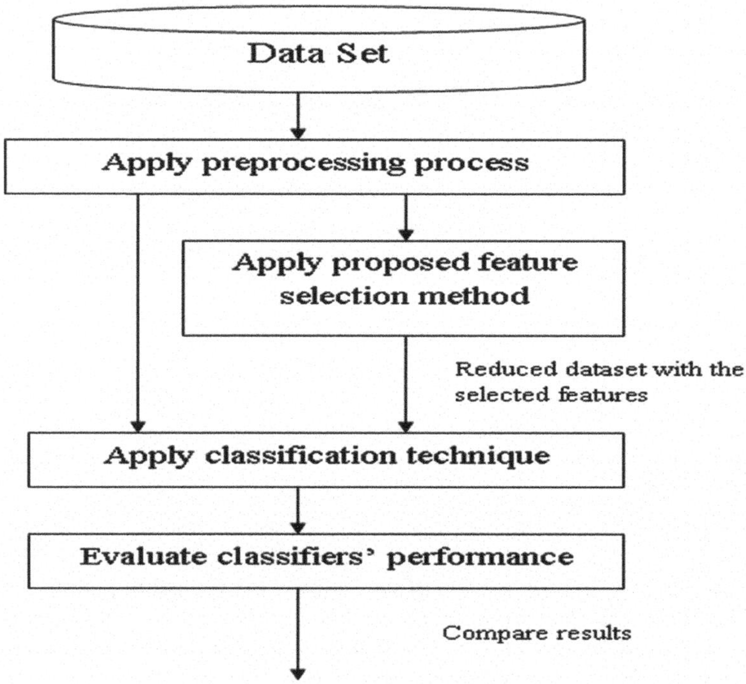

Fig. 1. Proposed system framework.

Algorithm 1:

Input: dataset D, feature set F = {f1, f2, f3,…, fn} with n features (attributes).

Output: classification accuracy rate.

Begin:

```
 1. Load dataset D
 2. For each attribute Ai in D where i = 1: n
 3.      Replace null values by the mean value of
    that attribute.
 4. Apply feature selection in D to select best set
    of features :
 5.     Scores = Apply_FS (F)
 6.     Select best features according to threshold
    ∂:
 7.                y= Best_Features(Scores)
 8. Reshape dataset D to the selected features Y as
    D1.
 9. Split to D1 training set T and test set S.
10. Train classifier using T.
11. Make a prediction on S.
12. Measure the classifiers' performance.
13. Return classification accuracy rate.
```

End

Fig. 2. Proposed algorithm.

3.1 Data Preprocessing

During preprocessing step missing values of some attributes are replaced and the noise values are removed. The datasets used in experiments contain some missing records; data cleaning replacing any missing value with the mean value of that attribute. Then, the dataset was randomly divided into 80% as a training set and 20% as the testing set.

3.2 Feature Selection

The essential goal of feature selection process is to remove the redundant and irrelevant features from the original dataset to improve the classification accuracy. In this paper, as shown in Fig. 3, a hybrid filter feature selection technique that uses mutual information, chi-square, and Relief-F, was proposed.

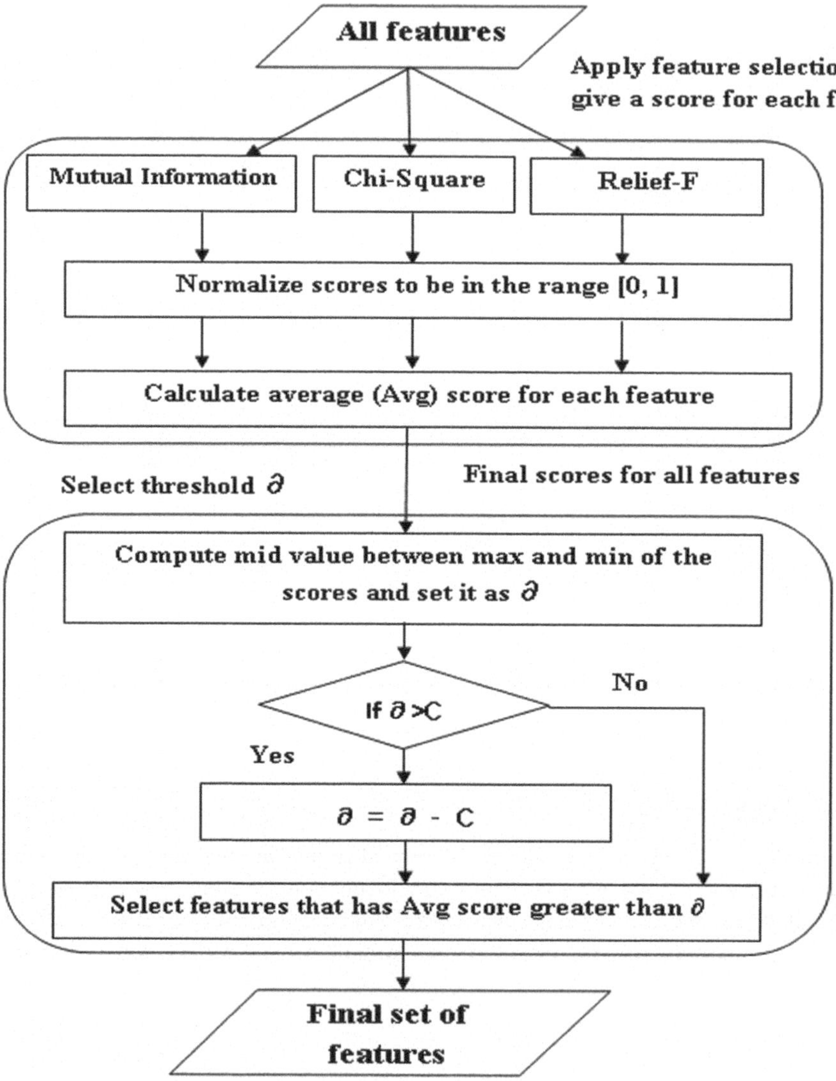

Fig. 3. The proposed hybrid feature selection method.

In the proposed method as shown in Fig. 4 and Fig. 5, each filter method was applied to the original dataset providing a score for each feature. To automatically specify an optimal threshold value for retaining an important subset of features used to build the classifier, scores for each filter method were first normalized to obtain values in the range [0, 1]. Second; they were combined by taking the average value of the three scores calculated for each feature. And finally, a threshold was set to the middle value (mid) between the maximum and minimum values of those scores in order to keep only features whose score is greater than the specified threshold. The reduced dataset was subjected to the classifier.

Algorithm 2: Apply_FS

```
Input: original feature set F = {f1, f2, f3, …,
fn}   where n is the number of features .

Output: average score for all features.
Begin:

1. Let FS={"Mutual information", "Chi-square", "Re-
   lief-F"}
2. Let m=  m = |FS|
3. For i=1... n
4.   For j =1... m
5.        Compute  score[i] [j]
6.   End for
7. End for
8. //    Normalize scores for each feature:
9. For j =1... m |
10.      Nrom_score[i] =  score [i]/ ∑_{i=1}^{n} score [i]
11. End for
12. //   Compute average score between sores of all
    feature selection methods:
13. For i=1... n
14.      Avg_score[i]=∑_{j=1}^{m} Norm_score [j] / m
15. End for
16. Return   average   score   for   all   features
    (Avg_score).

End.
```

Fig. 4. Algorithm of applying feature selection methods

```
Algorithm 3: Best_Features

Input:
  • set of original features F = {f1,f2,f3,…fn}
    where n is the number of features
  • set of average scores for all features
    (Avg_score)
  • C : user defined parameter

Output:     set of selected features F^1
Begin
 1. // compute threshold ∂|
 2. Min _score= min (Avg_score[i])     where i =1:n
 3. Max _score= max (Avg_score[i])     where i =1:n
 4. mid =( Max _score+ Min _score)/2.0
 5. If (mid > C)
 6.      mid = mid -C
 7. End if
 8. ∂ = mid
 9. Let F^1 = [ ]
10. For i=1 … n
11.      If (Avg_score[i] > ∂)
12.           Append f_i to F^1
13.      End if
14. End for
15. Return F^1
End.
```

Fig. 5. Algorithm of selecting best features upon optimal threshold that automatically selected.

Mutual Information

The mutual information is a ranking criterion for information theory [12, 20] that measures dependency between two features. Therefore, it is used for evaluating the relevance of a subset of features. For these reasons, mutual information considered to be the most widely investigated and most preferred for filters: 1) Measuring different types of relationship among random variables [21]. 2) Can successfully reduce data dimensions and improve or maintain the accuracy of classification over using a set of all features [22, 23].

Given two variables X and Y, the information obtained from Y about the variable X is the mutual information that is denoted by I(Y; X).

$$I(Y;X) = H(Y) + H(X) - H(Y,X) = H(Y) - H(Y|X) = H(X) - (X|Y) \quad (1)$$

Where H(Y) is the entropy of y and H(Y | X) is the entropy of y after observing x:

$$H(Y) = -\sum\nolimits_{y \in Y} P(y)\log(P(y)) \quad (2)$$

$$H(Y|X) = -\sum\nolimits_{x \in X} \sum\nolimits_{y \in Y} P(y,x) \log P(y|x) \quad (3)$$

Or can be written as follow:

$$I(X;Y) = -\sum\nolimits_{x \in X} \sum\nolimits_{y \in Y} P(y,x) \log \frac{P(y,x)}{P(y)P(x)} \quad (4)$$

Where p(y, x) is the joint probability of random variables y and x and p(y), p(x) are the probability density functions of variable Y and X respectively. The zero value of mutual information signifies that the two variables are not correlated while a large value indicates a high correlation between two variables.

Chi-Square

Chi-square test [13, 24] is one of the filter methods used for feature selection which is a statistical test measures divergence from the distribution of the occurrence of a feature that is independent of the value of the class. It tests whether the feature distribution differs among groups. The chi-square score calculated by using the summation of the squared difference among expected and observed values divided by expected values. Using Eq. 3 is estimated as follow:

$$\chi^2 = \sum\nolimits_j \frac{(O_j - e_j)^2}{e_j} \quad (5)$$

Where e and O are expected and observed frequency of cases in the category j, respectively.

Relief-F Algorithm

The main principle of the algorithm of Relief-F [14], which is the extension of Relief [25], is to randomly choose feature instances, calculate its nearest neighbors, and feature weighting vector is optimized to give more weight (importance) to features that distinguish the instance from its neighbors of different categories. Relief-F algorithm has high efficiency, is fast, not limited by data types, deal with datasets with discrete or continuous, not limited to problems of two classes, can deal with data that is noisy and incomplete, and is more robust. The formula of Relief-F algorithm to update the weight value of the feature is:

$$W_f^{i+1} = W_f^i + \sum_{c \notin class(y)} \left(\frac{\frac{P(y)}{1-p(class(y))} \sum_{j=1}^{k} diff_f(X, M_j(y))}{m*k} \right) - \sum_{j=1}^{k} \left(\frac{diff_f(X, H_j(y))}{m*k} \right) \quad (6)$$

Where:

On the feature f, $diff_f()$ Is the distance of two samples, $H_j(y)$ is the neighbor sample from the sample of kind y, $M_j(y)$ is the neighbor samples from neighbors of different classes, and P(y) is the class probability.

3.3 Classification Process

Classification algorithms used to assign a class for an unseen record accurately. Performing classification on the features selected by the feature selection process leads to higher prediction accuracy and more efficient time. In this study, well-known classifiers, Random Forest (RF) [26], Naïve Bayes (NB) [27], Decision Tree (DT) [28], and Logistic Regression (LR) [29], were applied over the reduced dataset with the selected features.

3.4 Evaluation Metrics

The performance of the proposed method can be measured using well-known evaluation metrics - the accuracy of the classification [30] and the Balanced Classification Rate BCR [31] that can be calculated using Eqs. (7) and (8) respectively.

$$\text{Accuracy} = \frac{TP + TN}{TP + FP + TN + FN} \quad (7)$$

$$BCR = 0.5 * \left(\frac{TP}{TP + FN} + \frac{TN}{TN + FP} \right) = 0.5(\text{Sensitivity} + \text{specificity}) \quad (8)$$

4 Experimental Results

The performance of the proposed feature selection method, developed in python, was examined by four standard datasets selected from the UCI data mining repository [32]. Table 2 provides a brief description of these datasets.

Table 2. Datasets

Dataset	Data type	Null values	Number of instances	Number of attributes	Number of classes
Diabetes	Continuous	Yes	768	8	2
Breast	Continuous	Yes	699	9	2
Heart	Discrete, continuous	Yes	303	13	2
Hepatitis	Discrete, continuous	Yes	155	19	2

To examine the effect of feature selection on the efficiency of classifier, for each dataset we conducted a well-known classification techniques, namely, DT, NB, RF and LR, for prediction in two scenarios: 1) without applying the proposed feature selection method and 2) with applying the proposed feature selection method then evaluation metrics listed in Sect. 3.4 were used to measure the performance of classification techniques. The default parameters for each classification techniques were used.

At the feature selection stage, mutual information, chi-square, and Relief-F techniques were applied to the original dataset to provide a score for each feature. To overcome the problem of filter feature selection methods in how to select an optimal set of features, the proposed feature selection method automatically determines an optimal threshold value based on the dataset in use. Table 3 shows the value of the threshold and the number of features was selected for each dataset. The user defined the value of C to be 0.1.

Table 3. Threshold values and number of the selected features for each dataset

Dataset	Threshold	Number of features
Diabetes	0.093	4
Heart	0.039	10
Hepatitis	0.035	9
Breast	0.027	8

As shown in Table 4, the performance of various classification techniques was compared when not using or using the proposed feature selection method in terms of BCR and classification accuracy.

Table 4. Comparison of different classifiers with and without using the proposed feature selection method in terms of BCR and accuracy

		Without feature selection				With the proposed feature selection			
		NB	DT	RF	LR	NB	DT	RF	LR
Diabetes	Accuracy	75.324	70.779	77.272	77.272	79.870	72.727	**79.870**	79.870
	BCR	72.323	68.787	73.030	71.818	73.583	70.829	**74.776**	72.390
Heart	Accuracy	77.049	68.852	75.409	75.409	81.967	80.327	**85.245**	85.245
	BCR	76.989	68.870	75.430	75.268	81.154	80.065	**85.620**	84.858
Hepatitis	Accuracy	62.068	72.413	79.310	86.206	**89.655**	79.310	**89.655**	86.206
	BCR	61.250	59.583	71.666	67.916	**83.500**	67.00	**83.500**	81.500
Breast	Accuracy	96.428	92.857	94.285	96.428	**97.142**	94.285	97.142	96.428
	BCR	96.784	91.077	92.164	95.289	**97.326**	93.368	97.005	96.417

From Table 4, we noticed that when applying classification techniques with the proposed approach to select features, it performed better for all datasets where the highest values achieved for accuracy and BCR are highlighted in bold format. For a diabetes dataset, RF, NB, and LR algorithms achieved higher performance with an accuracy of 79.87% but the RF algorithm achieved higher BCR of 74.78%. BCR is considered to be a good alternative to the accuracy of the classifier that evaluates the classifier effectiveness by calculating the probability of the correctly classified samples without taking the class into account while the BCR considers the balance between two classes. For this reason, we can say that RF achieved the best performance for the diabetes dataset. Similarly, the performance of using RF can generate a high result with 85.25% accuracy and BCR of 85.62% for heart dataset. In the hepatitis dataset, RF and NB achieved higher accuracy and BCR with 89.66% and 83.50 respectively. Although applying LR algorithm with and without the proposed feature selection method has the same accuracy, the LR algorithm using the second scenario has achieved the same accuracy with a smaller set of features when compared to using the original set of features. Moreover, it has a higher BCR value (81.50%). For the breast dataset, NB achieved the best accuracy (97.14%) and BCR (97.33%).

Figure 6 summarizes the performance rate for all classification techniques with and without using the proposed feature selection techniques for all datasets. It shows that when applying classification technique using the second scenario, it performs much better than applying the algorithms using the first scenario. In addition, we can be observed that RF classifier using the second scenario has the best performance with the highest values in average for accuracy (87.98%) and BCR (85.23).

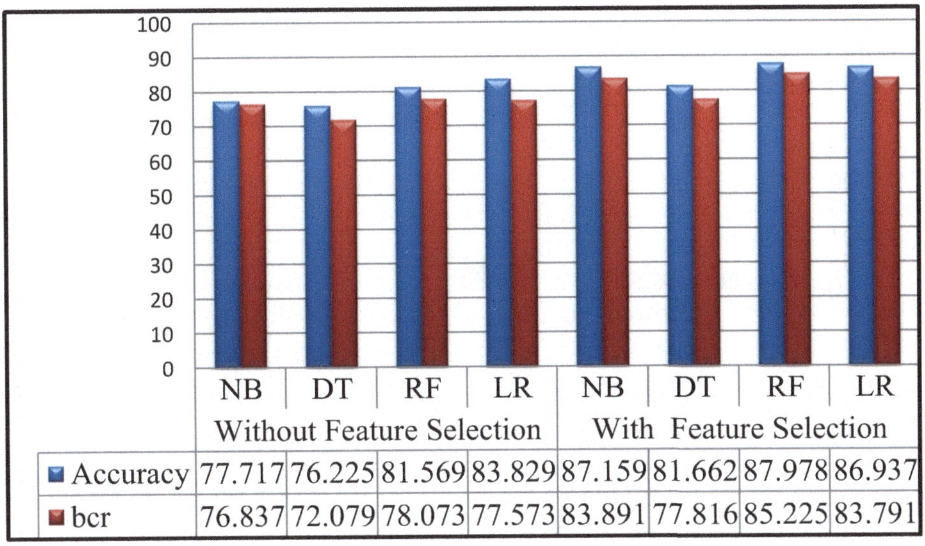

Fig. 6. Comparison of the performance rate for classifiers with and without using the proposed feature selection techniques for all datasets

Table 5 and Fig. 7 indicate the learning time of the model in seconds when using or not using the proposed feature selection techniques for all datasets.

Table 5. Learning time in seconds

	Without feature selection				With the proposed feature selection			
	NB	DT	RF	LR	NB	DT	RF	LR
Diabetes	0.003349	0.006847	0.048613	0.012317	**0.002121**	0.003349	0.047735	0.006746
Heart	0.003064	0.003582	0.043558	0.004871	**0.002000**	0.002321	0.043198	0.003674
Hepatitis	0.002681	0.003328	0.041099	0.004725	**0.001175**	0.001635	0.040361	0.002690
Breast	0.003142	0.003232	0.043854	0.004807	**0.001812**	0.001878	0.043057	0.003562

From Table 5 we observed that classifiers using the second scenario consumed less time to build the model than using the first scenario. Further, the results show that the NB algorithm that used the proposed approach for feature selection is much more effective than all considered algorithms. In fact, only 0.002121 s, 0.002000 s, 0.001175 s, and 0.001812 s were required to build their model for diabetes, heart, and hepatitis and breast datasets, respectively.

As shown from Fig. 7, RF classifier uses an average longer time to create a model while NB uses less time for all datasets.

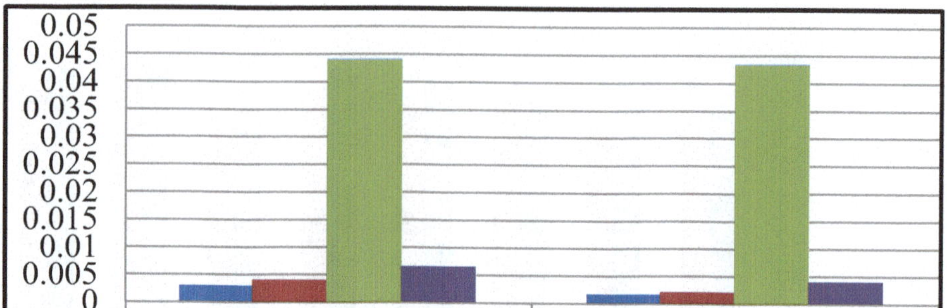

	Without Feature Selection	With the proposed Feature Selection
NB	0.003059	0.001777
DT	0.00424725	0.00229575
RF	0.044281	0.04358775
LR	0.00668	0.004168

Fig. 7. Average learning time for all datasets

5 Conclusions and Future Works

A new hybrid filter feature selection method was proposed in this study. This study illustrated the main problem of the filter feature selection methods that is how to determine the number of features selected to improve the classification accuracy, decrease execution time and the amount of memory that required for constructing the classification model. To solve this problem, the proposed feature selection approach selected the final set of features obtained through the application of mutual information, chi-square, and Relief-F techniques on the original dataset to provide a score for each feature. Then, an optimal threshold value was automatically determined based on the dataset in use in order to keep only those features whose average score was greater than the selected threshold value. The efficiency of the proposed feature selection method was investigated through two experiments on four standard datasets from UCI. For each dataset, well-known classification techniques were used for prediction in two scenarios: the dataset with all features assigned as a first scenario and dataset with a reduced set of features selected by the proposed feature selection method as a second scenario.

The results of the experiments that presented in this study are summarized as follows:

- When applying classification techniques using the second scenario, the results were improved compared to the first scenario in terms of BCR and classification accuracy.

- The threshold was automatically selected based on the dataset in use in order to improve the accuracy of the classification technique where the selected value for the threshold is different for each database.
- The time required to build a classifier using a reduced dataset that was selected using the proposed feature selection method is lower than using the original dataset.
- Although RF classifier using a reduced dataset had the best performance for all datasets with the highest average values for accuracy and BCR, it took more time to build the model, but it still less than the time required to build a model using the original dataset.
- NB algorithm using the reduced dataset took less time to build the model.

As concluding observations, the work presented in this research has proved the objectives of this study where the results obtained demonstrate its effectiveness. In the future, we will focus on applying our proposed method to other purposes such as multi-class classification, or clustering. Moreover will apply it on large datasets and will provide a comprehensive analysis for the feature selection methods and classification techniques.

References

1. Tang, J., Alelyani, S., Liu, H.: Feature selection for classification: a review. Data Classif. Algorithms Appl. **37**, (2014)
2. Jain, D., Singh, V.: An efficient hybrid feature selection model for dimensionality reduction. Proc. Comput. Sci. **132**, 333–341 (2018)
3. Xue, B., Zhang, M., Browne, W.N., Yao, X.: A survey on evolutionary computation approaches to feature selection. IEEE Trans. Evol. Comput. **20**(4), 606–626 (2016)
4. Miao, J., Niu, L.: A survey on feature selection. Proc. Comput. Sci. **91**, 919–926 (2016)
5. Nguyen, H.B., Xue, B., Liu, I., Zhang, M.: Filter based backward elimination in wrapper based PSO for feature selection in classification. In: 2014 IEEE Congress on Evolutionary Computation (CEC), pp. 3111–3118. IEEE (2014)
6. Guyon, I., Elisseeff, A.: An introduction to variable and feature selection. J. Mach. Learn. Res. **3**(3), 1157–1182 (2003)
7. Dash, M., Liu, H.: Feature selection for classification. Intell. Data Anal. **1**(1–4), 131–156 (1997)
8. Yu, L., Liu, H.: Feature selection for high-dimensional data: a fast correlation-based filter solution. In: Proceedings of the 20th International Conference on Machine Learning (ICML), pp. 856–863 (2003)
9. Kumar, V., Minz, S.: Feature Selection. SmartCR **4**(3), 211–229 (2014)
10. Qin, C.J., Guan, Q., Wang, X.P.: Application of ensemble algorithm integrating multiple criteria feature selection in coronary heart disease detection. Biomed. Eng. Appl. Basis Commun. **29**(06), 1750043 (2017)
11. Haq, A.U., Li, J.P., Memon, M.H., Nazir, S., Sun, R.: A hybrid intelligent system framework for the prediction of heart disease using machine learning algorithms. Mob. Inf. Syst. (2018)
12. Steuer, R., Kurths, J., Daub, C.O., Weise, J., Selbig, J.: The mutual information: detecting and evaluating dependencies between variables. Bioinformatics **18**(2), S231–S240 (2002)

13. Khalid, S., Khalil, T., Nasreen, S.: A survey of feature selection and feature extraction techniques in machine learning. In: 2014 Science and Information Conference, pp. 372–378. IEEE (2014)
14. Kononenko, I.: Estimating attributes: analysis and extensions of RELIEF. In: European conference on machine learning, pp. 171–182. Springer, Heidelberg (1994)
15. Sulaiman, M.A., Labadin, J.: Feature selection based on mutual information. In: 2015 9th International Conference on IT in Asia (CITA), pp. 1–6. IEEE (2015)
16. Jabbar, M.A., Deekshatulu, B.L., Chandra, P.: Prediction of heart disease using random forest and feature subset selection. In: Innovations in Bio-Inspired Computing and Applications, pp. 187–196. Springer, Cham (2016)
17. Djellali, H., Zine, N.G., Azizi, N.: Two stages feature selection based on filter ranking methods and SVMRFE on medical applications. In: Modelling and Implementation of Complex Systems, pp. 281–293. Springer, Cham (2016)
18. Huda, S., Yearwood, J., Jelinek, H.F., Hassan, M.M., Fortino, G., Buckland, M.: A hybrid feature selection with ensemble classification for imbalanced healthcare data: a case study for brain tumor diagnosis. IEEE Access **4**, 9145–9154 (2016)
19. Liu, X., Wang, X., Su, Q., Zhang, M., Zhu, Y., Wang, Q., Wang, Q.: A hybrid classification system for heart disease diagnosis based on the RFRS method. Comput. Math. Methods Med. (2017)
20. Guyon, I., Elisseeff, A.: An introduction to variable and feature selection. J. Mach. Learn. Res. **3**(3), 1157–1182 (2003)
21. Tsanas, A., Little, M.A., McSharry, P.E.: A simple filter benchmark for feature selection. J. Mach. Learn. Res. **1**, 1–24 (2010)
22. Peng, H., Long, F., Ding, C.: Feature selection based on mutual information: criteria of max-dependency, max-relevance, and min-redundancy. IEEE Trans. Pattern Anal. Mach. Intell. **8**, 1226–1238 (2005)
23. Cervante, L., Xue, B., Zhang, M., Shang, L.: Binary particle swarm optimisation for feature selection: A filter based approach. In: 2012 IEEE Congress on Evolutionary Computation, pp. 1–8. IEEE (2012)
24. Liu, H., Setiono, R.: Chi2: feature selection and discretization of numeric attributes. In: Proceedings of 7th IEEE International Conference on Tools with Artificial Intelligence, pp. 388–391. IEEE (1995)
25. Kira, K., Rendell, L.A.: The feature selection problem: traditional methods and a new algorithm. In: AAAI, vol. 2, pp. 129–134, July 1992
26. Breiman, L.: Random forests. Mach. Learn. **45**(1), 5–32 (2001)
27. Friedman, N., Geiger, D., Goldszmidt, M.: Bayesian network classifiers. Mach. Learn. **29**(2–3), 131–163 (1997)
28. Wagacha, P.W.: Induction of decision trees. Found. Learn. Adapt. Syst. **12**, 1–14 (2003)
29. Kleinbaum, D.G., Dietz, K., Gail, M., Klein, M., Klein, M.: Logistic regression. Springer, New York (2002)
30. Sokolova, M., Japkowicz, N., Szpakowicz, S.: Beyond accuracy, F-score and ROC: a family of discriminant measures for performance evaluation. In: Australasian Joint Conference on Artificial Intelligence, pp. 1015–1021. Springer, Heidelberg (2006)
31. Tharwat, A.: Classification assessment methods. Appl. Comput. Inf. (2018)
32. UCI Machine Learning Repository. http://archive.ics.uci.edu/ml

Comparison of Hybrid ACO-k-Means Algorithm and Graph Cut for MRI Images Segmentation

Samer El-Khatib[1], Yuri Skobtsov[2(✉)], Sergey Rodzin[1], and Semyon Potryasaev[3]

[1] Southern Federal University, Rostov-on-Don, Russia
samer_elkhatib@mail.ru, srodzin@yandex.ru
[2] St. Petersburg State University of Aerospace Instrumentation, Saint Petersburg, Russia
ya_skobtsov@list.ru
[3] St. Petersburg Institute of Informatics and Automation, Russian Academy of Sciences (SPIIRAS), Saint Petersburg, Russia
spotryasaev@gmail.com

Abstract. Image segmentation is the process of dividing an image into consistent and homogeneous in some characteristics regions. Image segmentation widely used in medical diagnostics. Segmentation algorithms are used for anatomical features ex-traction from medical images. The Hybrid Ant Colony Optimization (ACO) – k-means and Graph Cut image segmentation algorithms for MRI images segmentation are considered in this paper. The proposed algorithms and sub-system for the medical image segmentation have been implemented. There is no universal algorithm for medical image segmentation. In image processing and computer vision, segmentation is still a challenging problem in many real time applications and hence more research work is required. The experimental results show that the proposed algorithm has good accuracy in comparison to Graph cut.

Keywords: MRI images segmentation · Ant Colony Optimization · K-means algorithm · Swarm intelligence · Graph Cut

1 Introduction

1.1 A Subsection Sample

The development of image recognition methods is one of the relevant and difficult tasks in artificial intelligence. When creating recognition systems, which contain high requirements for accuracy and performance, occurs needless to apply new methods to automate the procedure of image recognition. Despite the fact that the task of developing image recognition methods is well researched theoretically, however, there is no universal method for solving it, and the practical solution seems to be very difficult.

At computer processing and recognition of images the wide range of problems is solved. One of the main stages of recognition is the process of dividing the image into

R. Silhavy (Ed.): CSOC 2020, AISC 1225, pp. 73–80, 2020.
https://doi.org/10.1007/978-3-030-51971-1_6

non-overlapping areas (segments) that cover the entire image and are homogeneous by some criteria. Segmentation simplifies the analysis of homogeneous areas of the image, as well as brightness and geometric characteristics. The segmentation is implemented using special methods. Their goal is to separate the analyzed object, structure or area of interest from the surrounding background. This is a difficult task, the performance quality of which significantly affects the accuracy and the possibility of subsequent computer analysis of images, since there are difficulties associated with noise, blurring of images, etc.

To solve the image segmentation problem, there have been developed many methods based on the luminance, gradient and texture information of the image [1]. Results of image segmentation and recognition methods research are set forth in works of R. Woods, R. Gonsalez, D. Canny, U. Pratt, D. Prewitt, L. Roberts, I. Sobel, R. Haralick.

A suitable way to effectively solve the problem of image segmentation is to use mathematical transformations describing the collective behavior of a decentralized self-organized system that consists of a multiple agents interacting locally with each other and with the environment to achieve a predetermined goal. In nature, examples of such systems are swarm systems [2, 7]. Each agent functions autonomously, using own rules. At the same time, the behavior of the entire system is surprisingly daunting [3, 8–10]. This article is dedicated to development and evaluation for Hybrid ACO-k-means in compare to Graph Cut segmentation method for MRI images.

2 Hybrid Ant Colony Optimization k-Means Method for Image Segmentation

To obtain an efficient algorithm for image segmentation we have developed a method in which all the advantages of k-means and ACO algorithms are used. The first step is to set the number of clusters and initialize their centers. Then, according to the clustering algorithm k-means, we need to determine the belonging of each image pixel to a particular cluster. At this stage, the most important role is played by the ACO algorithm. It defines the relationship of each pixel with clusters of the image. This is done according to the probability, which is inversely proportional to the distance between the pixel, cluster center and the variable, which represents the pheromone level. The pheromone level is determined in proportion to the minimal distance between each pair of cluster centers and inversely proportional to the distance between each pixel and its center. Thus, the pheromone level value increases with increase of a distance between the centers of clusters, as well as with increase in compactness of pixels in the cluster. Under the same conditions, the probability of the pixel attachment to the cluster increases.

The evaporation of the pheromone is calculated in order to reduce the impact of the previously made choices which are of lower priority. Similarly to k-means algorithm, in the distributed state the cluster centers are updated by recalculation of the average value of pixels in each cluster. It lasts as long as the change in value of the cluster center does not vary substantially. In contrast to the k-means algorithm, the developed method does not stop at this stage. The process of clustering continues to be performed by m ants, each of which ultimately finds its solution. The criteria for finding the best

solutions and the updated pheromone level are prior for the next group of m ants respectively. When the stopping criterion is reached, the clustering is completed and the best solution is found.
Hybrid ACO algorithm for image segmentation consists of the following steps:

Algorithm 1. Hybrid ACO-k-means segmentation algorithm

```
Begin
  Initialize([number clusters], [number ants]);
  Repeat
  For each ant do
    M: For each pixel do
        Calc(probability belonging pixel to cluster) (1);
    End
    Update (cluster center);
    If (NewCenter <> OldCenter) then
        goto M;
    Else
        Save(current solution);
  End
  Select Best Solution From All Ants (5);
  Update(for each pixel) (6, 7);
  Correct(common solution);
  Until criteria not reached
End
```

Software implementation of the algorithm starts with determination of the level of pheromone τ and assignment of heuristic information η for each pixel. Then, each ant determines pixel's belonging to the cluster with probability P, which can be calculated in the following way (1):

$$P_i(X_n) = \frac{[\tau_i(X_n)]^{\alpha}[\eta_i(X_n)]^{\beta}}{\sum_{j=0}^{K}[\tau_j(X_n)]^{\alpha}[\eta_j(X_n)]^{\beta}} \quad (1)$$

where $\tau_i(X_n)$ and $\eta_i(X_n)$ - information about pheromone and heuristic variable values of belonging of a pixel to the cluster i respectively, α and β - heuristic coefficients of ACO algorithm, k – number of clusters. Heuristic information is obtained as follows:

$$\eta_i(X_n) = \frac{\mathrm{k}}{CDist(X_n, CC_i) * PDist(X_n, PC_i)} \quad (2)$$

Where X_n n[th] pixel, CC_i - i[th] spectral cluster center, PC_i - i[th] spatial cluster center, $CDist(X_n, CC_i)$ - distance between (X_n, CC_i) according to color characteristics – (3), $PDist(X_n, PC_i)$ - the geometrical distance between (X_n, PC_i) - (4), k – constant value.

$$CDist(X_n, CC_i) = |Int(X_n) - Int(CC_i)| \tag{3}$$

where $Int(X_n)$ is the intensity of the pixel X_n.

$$PDist(X_n, PC_i) = \sqrt{(X_n.x - PC_i.x)^2 + (X_n.y - PC_i.y)^2} \tag{4}$$

where $X_n.x$ and $X_n.y$ are x and y coordinates respectively for the pixel X_n.

It is clear that the color distance between the of different cluster centers should be maximum, and the color and geometric distance between the cluster center of and the pixels of own cluster should be minimal. So in the given modification we suggest to use the following set of simple rules as the target function (optimality criterion of a better solution):

- $\max_{i=1..m}\left(\sum_{k=1..K-1}\sum_{j=k+1..K} CDist(C_k, C_j)\right)$ - the maximum sum value of the color distance between the cluster centers for all ants (the distance between the clusters, in terms of color characteristics, should be at a maximum, then clusters will be different from each other), where CDist – the color distance between two pixels, C_k - center of the cluster k.
- $\min_{i=1..m}\left(\sum_{k=1..K}\sum_{p=1..S_k} PDist(C_k, X_p)\right)$ - the minimum value of the sums of geometric distances between the centers of the clusters and the pixels belonging to the cluster (the sum of the Euclidean distances between the center of the cluster and each of its pixels should be minimal, according to the spatial characteristics, respectively, the cluster will be more homogeneous), where S_k - the number of pixels in the cluster k, PDist – Euclidian distance between two pixels, C_k- center of the cluster k.
- $\min_{i=1..m}\left(\sum_{k=1..K}\sum_{p=1..S_k} CDist(C_k, X_p)\right)$ - the minimum sum value of the color distances between the centers of the clusters and the pixels belonging to the cluster (sum of the Euclidean distances between the center of the cluster and its each pixel, according to color characteristics, should be minimal, then, the cluster will be more compact), where S_k - the number of pixels in the cluster k, CDist – the color distance between two pixels, C_k - center of the cluster k.

The fitness function of the ant colony is 3-criterial and can be obtained as follows:

$$f(m_i) = \begin{cases} \sum_{k=1..K-1}\sum_{j=k+1..K} CDist(C_{m_ik}, C_{m_ij}) \\ \sum_{k=1..K}\sum_{p=1..S_{m_ik}} PDist(C_{m_ik}, X_{m_ip}) \\ \sum_{k=1..K}\sum_{p=1..S_{m_ik}} CDist(C_{m_ik}, X_{m_ip}) \end{cases} \tag{5}$$

Choice of the best solution can be presented as:

$$f(best) = \begin{cases} \max(f_1) \\ \min(f_2) \\ \min(f_3) \end{cases} \tag{6}$$

Pheromone updating is calculated as in standard ACO algorithm according to:

$$\tau_i(X_n) \leftarrow (1 - \rho)\tau_i(X_n) + \sum_i \Delta\tau_i(X_n) \tag{7}$$

and pheromone evaporation is calculated as:

$$\Delta\tau_i(X_n) = \begin{cases} \frac{Q*Min(k')}{AvgCDist(k',i)*AvgPDist(k',i)}, & if\ X_n \subset cluster\ i \\ 0, & otherwise \end{cases} \tag{8}$$

where Q is a constant value, $Min(k')$ - the minimum color outer cluster distance found by the most successful ant, $AvgCDist(k',i)$ and $AvgPDist(k',i)$- the average value of the spatial Euclidean and color distances between each pixel and cluster centers for the most successful ant.

2.1 Graph Cut Image Segmentation Algorithm

The Graph Cuts algorithm has become the basis of many modern interactive segmentation algorithms [5]. Graph Cut presents a graph image. A lot of vertices consists of image pixels and two additional, artificially added vertices - source and drain.

For two adjacent vertices i and j, the edge weight is given by the expression

$$B(i,j) = e^{-\left(\frac{|C_i - C_j|}{2\delta^2}\right)} \frac{1}{d(i,j)} \tag{9}$$

where Ci and Cj are the colors of the pixels, δ is a user-defined parameter, d(i, j) is the Euclidean distance between the pixels.

The user interactively indicates several pixels that belong to the object, and several pixels that relate to the background. The vertices belonging to the object are connected to the source by edges having infinite weight. The vertices belonging to the background are associated with the drain by ribs having infinite weight. For the obtained graph with drain and source, a minimal cut is sought, dividing the graph into 2 parts [4]. Pixels that are in the same subgraph as the source refer to the object, the rest to the background. The infinite weight of the edges connecting the pixels with the source and drain is necessary so that all the pixels of the object are in the object, and all the pixels of the background are assigned to the background. The formula for the edge weight is chosen in such a way that pixels with the most different colors are connected by an edge of minimum weight, which leads to a section of the graph along the most contrasting border. An increase in the speed of this algorithm in comparison with other algorithms has not been proven, but on the benchmarks of the reference problems of image segmentation, this algorithm works faster than its counterparts.

3 Testing Hybrid ACO-k-Means and Graph Cut Segmentation Methods

Developed Hybrid ACO-k-means algorithm for image segmentation was investigated. Investigation parameters are: Ossirix benchmark [6], set of 150 images with various initial conditions - good quality (no noise and other artifacts) and contrast images. ACO-k-means method was compared with Graph Cut segmentation algorithm.

Results are presented by processing images with various initial conditions (Figs. 1 and 2). Segmentation accuracy is presented on Figs. 3 and 4. As we can observe ACO-k-means algorithm outperforms Graph cut's accuracy.

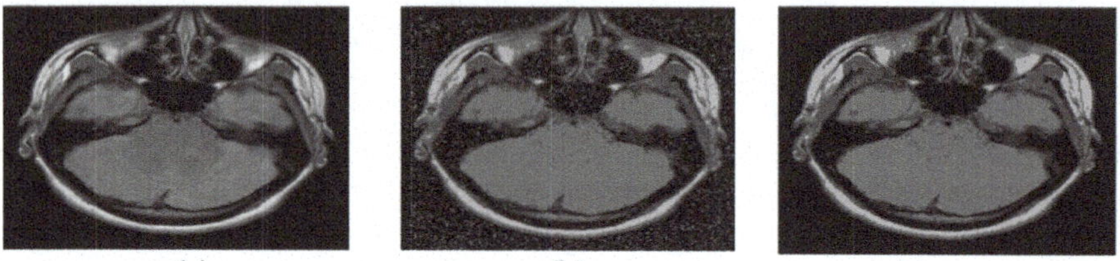

(a) (b) (c)

Fig. 1. (a) Initial image; (b) Graph Cut segmentation result; (c) Hybrid ACO-k-means segmentation result

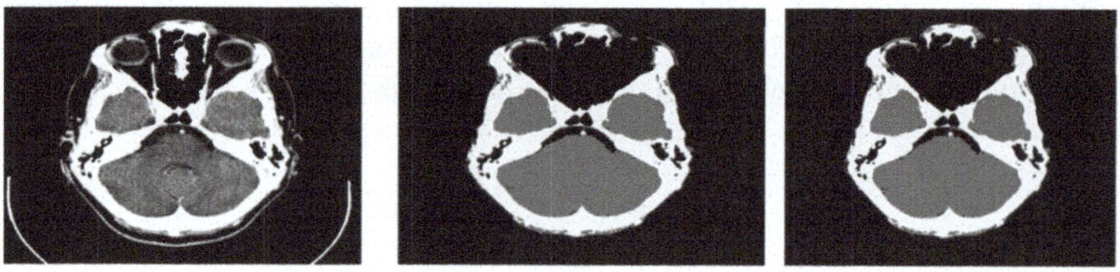

(a) (b) (c)

Fig. 2. (a) Initial image; (b) Graph Cut segmentation result; (c) Hybrid ACO-k-means segmentation result

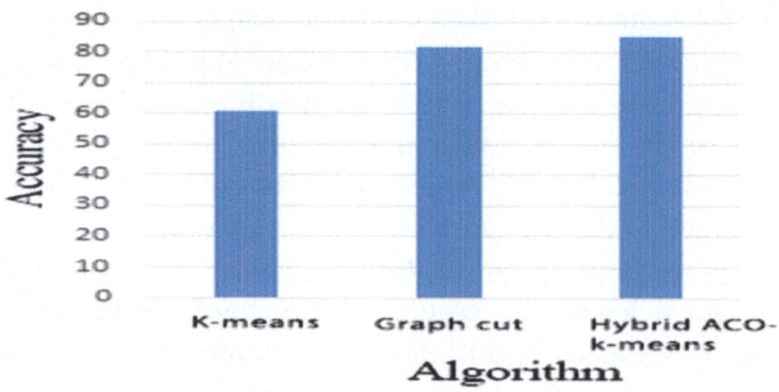

Fig. 3. Evaluation of segmentation accuracy for good quality images

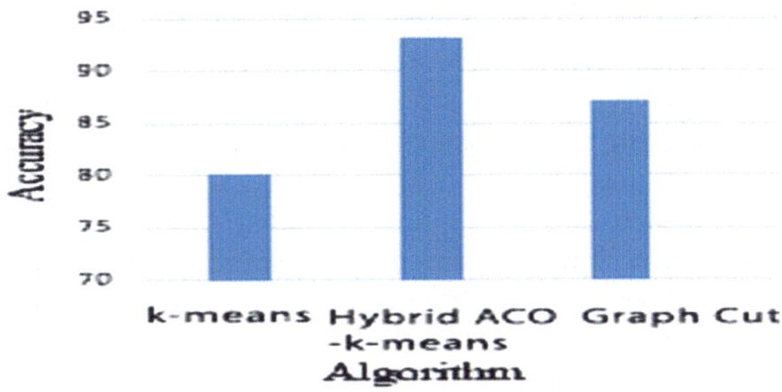

Fig. 4. Evaluation of segmentation accuracy for contrast images

4 Conclusion

In the article was presented comparison of Hybrid ACO-k-means and Graph Cut methods for MRI images segmentation. There were obtained comparative evaluations for the quality for the ACO-k-means segmentation method on different quality images from Ossirix benchmark. Experimentally obtained scientific data indicates the following: ACO-k-means method outperforms Graph cut segmentation quality at average 5%.

Acknowledgements. The reported study was funded by Russian Foundation for Basic Research according to the research project №. 19-07-00570 "Bio-inspired models of problem-oriented systems and methods of their application for clustering, classification, filtering and optimization problems, including big data".

The research described in this paper is partially supported by the Russian Foundation for Basic Research (grants ***17-29-07073-ofi-i, 18-07-01272, 18-08-01505, 19–08–00989****), state research* ***0073–2019–0004****.*

References

1. Gonzalez, R.C., Woods, R.E.: Digital Image Processing Using MATLAB. Prentice-Hall, Upper Saddle River (2008)
2. Kennedy, J., Eberhart, R.C.: Particle swarm intelligence. In: Proceedings of the IEEE International Joint Conference on Neural Networks, pp. 1942–1948 (1995)
3. El-Khatib, S., Rodzin, S., Skobtcov, Y.: Investigation of optimal heuristical parameters for mixed aco-k-means segmentation algorithm for MRI images. In: Proceedings of III International Scientific Conference on Information Technologies in Science, Management, Social Sphere and Medicine (ITSMSSM). Part of series Advances in Computer Science Research, vol. 51, pp. 216–221. Atlantis Press (2016). ISBN (on-line): 978–94-6252-196-4. https://doi.org/10.2991/itsmssm-16.2016.72
4. Saatchi, S., Hung, C.C.: Swarm intelligence and image segmentation swarm intelligence. ARS J **1**, 163–178 (2007)
5. Boykov, Y., Veksler, O., Zabih, R.: approximate energy minimization via graph cuts. IEEE Trans. Pattern Anal. Mach. Intell. **23**(11), 1222–1239 (2001)
6. Ossirix image dataset. http://www.osirix-viewer.com/

7. Skobtsov, Y.A., Speransky, D.V.: Evolutionary computation: hand book, 331p. The National Open University "INTUIT", Moscow (2015). (in Russian)
8. El-Khatib, S., Skobtsov, Y., Rodzin, S.: Improved particle swarm medical image segmentation algorithm for decision making. In: Kotenko, I., Badica, C., Desnitsky, V., El Baz, D., Ivanovic, M. (eds.) Intelligent Distributed Computing XIII. IDC 2019. Studies in Computational Intelligence, vol. 868. Springer, Cham (2020)
9. El-Khatib, S.A., Skobtsov, Y.A., Rodzin, S.I.: Theoretical and experimental evaluation of hybrid ACO-k-means image segmentation algorithm for MRI images using drift-analysis. Proc. Comput. Sci. **150**, 324 (2019)
10. El-Khatib, S., Skobtsov, Y., Rodzin, S., Potryasaev, S.: Theoretical and experimental evaluation of PSO-K-means algorithm for MRI images segmentation using drift theorem. In: Silhavy, R. (ed.) Artificial Intelligence Methods in Intelligent Algorithms. CSOC 2019. Advances in Intelligent Systems and Computing, vol. 985. Springer, Cham (2019)

Parallel Deep Neural Network for Motor Imagery EEG Recognition with Spatiotemporal Features

Desong Kong[1(✉)] and Wenbo Wei[2]

[1] School of Computer Science, Chongqing University of Posts and Telecommunications, Chongqing 400065, China
874148831@qq.com

[2] School of Computer Science, The University of Birmingham, Edgbaston, Birmingham B15 2TT, UK

Abstract. In emerging research field of interdisciplinary studies, EEG plays an important role in brain-computer interface due to the good portability, low cost and high temporal resolution of EEG devices. In this paper, a new neural network model called parallel deep neural network is proposed to extract the spatiotemporal features of the motor imagery EEG signal. Unlike traditional EEG classification algorithms, which often discard the EEG spatial feature, Fast Fourier Transform is performed on the EEG time series for each trial to construct 2-D EEG maps. The convolutional neural network is used in training the 2-D EEG maps to extract EEG spatial features. In addition, the original time series channel signals are trained in parallel based on long short-term memory to extract the EEG time series features. Finally, the spatial and temporal features are fused and classified using feature mosaicing. The experimental results show that the parallel deep neural network has good recognition accuracy and is superior to other latest recognition algorithms.

Keywords: Convolutional neural network · EEG recognition · EEG feature map · Long Short-term memory

1 Introduction

Electroencephalogram (EEG) [1, 2] is a comprehensive reflection of the physiological activity of cerebral cortex and scalp brain cells, which contains a large amount of physiological and disease information. As an extension of human-computer interaction, the brain-computer interface (BCI) [3] has been widely concerned by scholars and researchers in the scientific community, such as emotion recognition based on single-channel EEG signals [4], and epilepsy research based on EEG [5]. Motor imagery (MI) is a physical activity that is subjectively imagined by the human brain, such as imagining a left-hand shake, imagining a right-hand shake, and imagining leg flexion and extension. Through the analysis of MI EEG signal, the intention of human brain can be identified to achieve brain-computer control. Therefore, the study of motor imagery EEG signal processing can accelerate the cognition of brain nerves, the rehabilitation of cerebral diseases, and the exploration of cerebral cortex signals.

R. Silhavy (Ed.): CSOC 2020, AISC 1225, pp. 81–92, 2020.
https://doi.org/10.1007/978-3-030-51971-1_7

There are two parts in the EEG signal analysis including feature extraction and feature classification. The common feature extraction methods include Fast Fourier Transform (FFT) [6], common spatial pattern (CSP) [7], and wavelet transform (WT) [8]. These feature extraction methods not only require a large amount of artificial data processing, but also are sensitive to noise, which will cause feature confusion easily. The common feature classification methods include artificial neural network (ANN) [9], support vector machine (SVM) [10], etc. Because of the complex generation mechanism of EEG signal, these feature classification methods have the problems of shallow iteration level and insufficient feature extraction.

In recent years, deep learning is applied to EEG data analysis due to the ability of processing non-linear and high dimensional data. Langkvist et al. [11] use deep learning algorithm to strengthen the EEG time series feature to solve the problem that the mean, variance, and frequency change with time. Tang et al. [12] train a hidden-layer visible deep stacking network to solve the problem of insufficient training of hidden layer EEG data. Soleymano et al. [13] find that in emotional tasks, using visual imagination and EEG signals can provide continuous and temporal description of emotions, which also provides new ideas for other research directions of deep learning methods in EEG signal analysis.

The EEG signal contains the spatial information represented by the electrode position as well as the inherent temporal information. Since the EEG acquisition device only visualizes the EEG time series data, the previous EEG classification algorithms mainly extract EEG features in time series. In order to make full use of spatiotemporal information to strengthen EEG feature extraction, the parallel deep neural network is proposed to mine EEG temporal features and spatial information. FFT is performed on the time series of each experiment to extract Theta (4–8 Hz), alpha (8–12), and beta (12–36 Hz). These bands are mapped to the 2-D map of electrode positions. Furthermore, convolutional neural network (CNN) is used in training the 2-D EEG maps. At the same time, the EEG time series data is trained based on long short-term memory (LSTM). Finally, the feature mosaicing will be used to fuse the spatial and temporal features. The proposed method is used to analyze the motor imagery EEG data collected by Emotiv and compare with the main EEG analysis methods.

2 Spatiotemporal Feature Extraction of Motion Imagination Based on Parallel Deep Neural Network

2.1 Convolutional Neural Network

Local receptive field, weight sharing, and pooling operation can realize deformations such as translation, scaling, and distortion invariance of image, making CNN have strong anti-noise capability and can reduce the dimensionality of data features.

The convolutional layer is used for feature extraction. The mathematical expression of the convolutional layer is:

$$X_j^l = f(\sum_{i \in m_j^l} X \times w + b) \tag{1}$$

where f is the nonlinear function, m_j^l is the index vector of feature mapping i in layer l, w is the filter, b is the bias.

The pooling layer is used for dimensionality reduction. The downsampling formula of the pooling layer is:

$$X_j^l = down(X_j^{l-1}, N^l) \tag{2}$$

where $down()$ is the sampling function,N^l is the size of window boundary required by the sub-sampling layer l.

2.2 Long Short-Term Memory

LSTM is a special recurrent neural network (RNN), which overcomes the defect of vanishing gradient by introducing gate mechanism and memory unit. It is suitable for extracting time series features. Weight update formula is:

$$i_t = \varepsilon(W_{xi}x_t + W_{hi}h_{t-1} + b_i) \tag{3}$$

$$f_t = \varepsilon(W_{xf}x_t + W_{hf}h_{t-1} + b_f) \tag{4}$$

$$o_t = \varepsilon(W_{xo}x_t + W_{ho}h_{t-1} + b_o) \tag{5}$$

$$\tilde{c}_t = \tanh(W_{xc}x_t + W_{hc}h_{t-1} + b_c) \tag{6}$$

$$c_t = f_t * c_{t-1} + i_t * \tilde{c}_t \tag{7}$$

$$h_t = o_t * \tanh(c_t) \tag{8}$$

where c_t is the status information of memory unit, $\tilde{c}$ is accumulated information of historical moments, w is the weight matrix, b is the bias, ε, tanh are sigmoid activation function and hyperbolic tangent activation function.

2.3 2-D EEG map

MI EEG has strong time series and spatial features. Brain nerves in different locations produce different signal responses during a time period. Due to non-intrusive EEG equipment which uses copper foil sensor and human brain response, there will be a certain lag during collecting data. In this paper, the overlap method is adopted to deal with the EEG data to achieve the purpose of discarding useless data, and closer to the actual scene where the human brain reacts to signals. First, the original EEG data are chopped up into overlapping a time window. Furthermore, the Hanning window is applied. The formula is defined as

$$\begin{cases} x_i = x_{i-1} + f - o * f & i \neq 0 \\ x_i = 0 & i = 0 \end{cases} \quad (9)$$

where x is the starting point of segmentation, i is the number of the sample, f is the frequency, o is overlap factor.

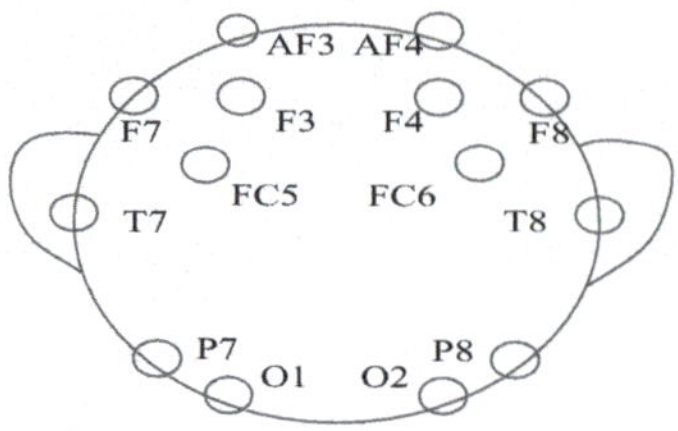

Fig. 1. The placement position of Emotiv

In the actual scene, EEG electrodes existing in 3-D space are distributed at various positions in the human cerebral cortex. In order to obtain the 2-D EEG map, the 3-D EEG electrodes need to be projected onto a 2-D plane. In this paper, Emotiv containing 14 EEG electrodes and 2 positioning electrodes is used for collecting data. The 2-D distribution map of 14 electrodes is shown in Fig. 1.

MI EEG contains multiple time series channels, which are located at various positions on the cerebral cortex and collect the response signals of the brain nerves around the channel area. According to the research of [14], oscillatory cortical activity contributing to working memory primarily exists in three frequency bands, which are respectively theta, alpha, and beta. Therefore, we perform FFT on each channel signal to estimate the power spectrum of the signal. Furthermore, on one side of the frequency range, the scalar values of the three frequency bands are obtained. Finally, the obtained three frequency band data matrices are interpreted as RGB channels and projected onto Fig. 1 to construct 2-D EEG maps. The generated 2-D EEG maps effectively retains the inherent features in EEG space, frequency and time. The process of generating an EEG map is shown in Fig. 2.

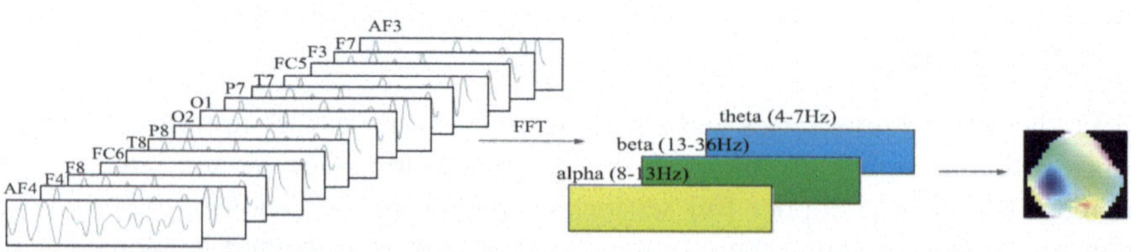

Fig. 2. Generating 2-D EEG map: (1) EEG time series data from 14 channels are acquired; (2) The theta, alpha, and beta are extracted by applying FFT on EEG time series data; (3) three frequency band data are used to form a 2-D EEG map

2.4 Parallel Deep Neural Network

EEG signals triggered by the imagination of brain or environmental stimulation include temporal information, spatial feature, and the inherent noise invariance. Traditional EEG algorithms often use EEG time series data for network training, which cannot provide an effective feature representation for classification tasks. In this paper, a parallel deep neural network (PDNN) is proposed in order to make full use of the temporal and spatial features of EEG signals. PDNN is designed for parallel training. Convolutional networks benefit from a series of operations such as convolution and pooling that can efficiently handle spatial and frequency transformations. Therefore, CNN is used in PDNN to extract spatial features of 2-D EEG maps reconstructed from the original channel signals. LSTM trains feature data by inputting weight parameter cyclically, which can accurately extract the time series features of the original signals. The method of constructing 2-D EEG map proposed in Sect. 2.3 can represent the spatial features of EEG data. However, there is a defect of losing the temporal features in the reconstruction process. Therefore, LSTM is used in PDNN for training the EEG time series data in parallel to extract the time series features. Figure 3 shows the topology of PDNN.

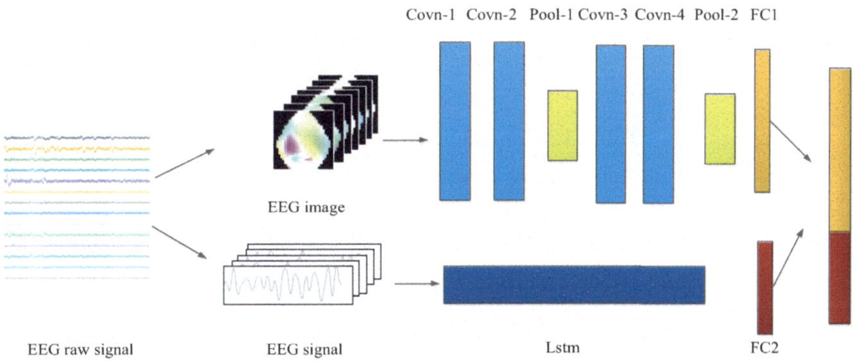

Fig. 3. The topology of PDNN. It consists of four parts: (1) spectral power within three prominent frequency bands is extracted for each location and used to form 2-D EEG maps which are fed into a CNN. (2)The EEG time series data are fed into a LSTM. (3) Based on the feature mosaicing method, the fully connected layer of CNN and LSTM is stitched as a new fully connected layer. (4)The Softmax is defined as a classifier of the fully connected layer.

For extracting spatial features, we construct a CNN to train 2-D EEG maps in PDNN. It consists of convolutional layers and pooling layers. The network structure is shown in Table 1. Convolutional layer 1 involves two convolutional layers stacked together, followed by a pooling layer. convolutional layer 2 adds on top of Convolution layer 1 two more convolutional layers, and followed by another pooling layer. Finally, a fully connected layer is constructed. A dropout of 0.5 will be applied to the fully connected layer.

For extracting temporal features, we construct a LSTM to perform feature extraction on the original time series data in PDNN. The network structure is shown in Table 2. In order to further extract the EEG temporal features, attention mechanism is added to the LSTM. First, take the EEG data matrix corresponding to each moment as input. Then, the LSTM-attention model extracts the corresponding time series features $\{A_{t-T}, \ldots, A_{t-1}\}$ from the structural sequence data $\{h_{t-T}, \ldots, h_{t-1}\}$, and pays attention to the hidden layer vector h through the attention mechanism. Finally, the fully connected layer is used as the output.

The fully connected layer of PDNN is based on feature mosaicing. The process of feature fusion is as follows: By stacking a new fully connected layer on the last layer of CNN and LSTM, PDNN realize the fusion of temporal and spatial features. The dimension of the new fully connected layer is the sum of the fully connected layer of CNN and the fully connected layer of LSTM. The EEG features obtained from the fully connected layer of CNN is denoted as $FC1$. The EEG features obtained from the fully connected layer of LSTM is denoted as $FC2$. The EEG features based on feature mosaicing is denoted as

$$FC = FC1 \oplus FC2 \tag{10}$$

Where, FC is new fully connected layer, $\oplus$ is mosaicing operation. The new fully connected layer is used as the input of the Softmax to perform classification task.

Table 1. Convolutional network structure

Input (28 * 28)	
Convolutional layer 1	Covn3-32
	Covn3-32
Max pooling layer	
Convolutional layer 2	Covn3-64
	Covn3-64
Max pooling layer	
Fully connected layer	

Table 2. LSTM network structure

Input (1 * 1792)
LSTM layer
Fully connected layer

3 Simulation Experiment and Result Analysis

3.1 Experimental Data

In order to verify the superior classification performance of PDNN proposed in this paper, EEG data is obtained through Emotiv EEG signal acquisition experiment. The proposed algorithm is compared with other superior performance EEG classification algorithms. Before describing the data set in detail, we will outline the Emotiv acquisition channel data in this paper.

EEG signal Acquisition Based on Emotiv
EEG channel signal acquisition is one of the important components of EEG data processing and application. The EEG acquisition device used in this paper is Emotiv *c*, which produced by American Emotiv System Company. As shown in Fig. 4, Emotiv *c* is composed of electrode cap, eletrode cao caditive, liquid, USB receiver and tampon.

In data collection, we recruited 3 subjects with no brain disease and right-handed users to perform the motor imagery EEG experiment. This experiment was conducted in a relatively quiet and temperature-friendly room. Before the experiment, the subjects will relax for 5 min. The experimental process is shown in Fig. 5. At the beginning of the recording (t = 0), the subjects were quiet and in a relaxed state. When t = 2, the subjects performed left-hand fist or right-hand fist imaginary task for 2 s according to the on-screen instructions. When t = 4, stop imaginary task according to the instructions on the screen. Repeat the trial process to reach the required size of data set. The experimental data set contains two categories (left fist movements and right fist movements) and 3 subjects. Each subject contains *900* samples, of which left motor imagination and right motor imagination each account for half. The sample size of each subject is *900 * 14 * 128*. In order to further verify the superior classification performance and good generalization ability of the proposed method, we combined the data of three subjects into one data. The size of the merged data set is *3 * 900 * 14 * 128*. Figure 6 shows the original channel data for two channels in a time window.

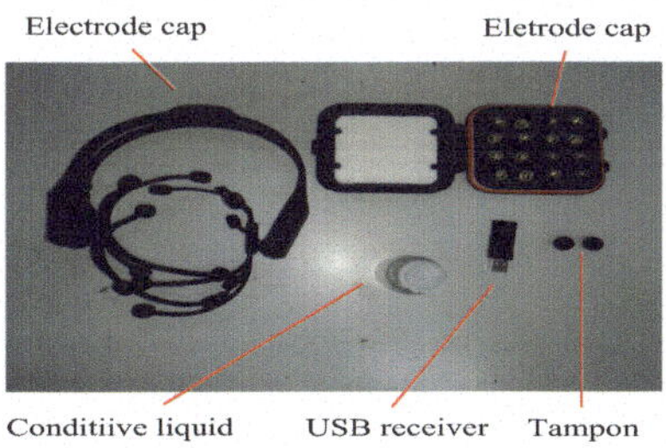

Fig. 4. Emotiv EEG signal acquisition instrument

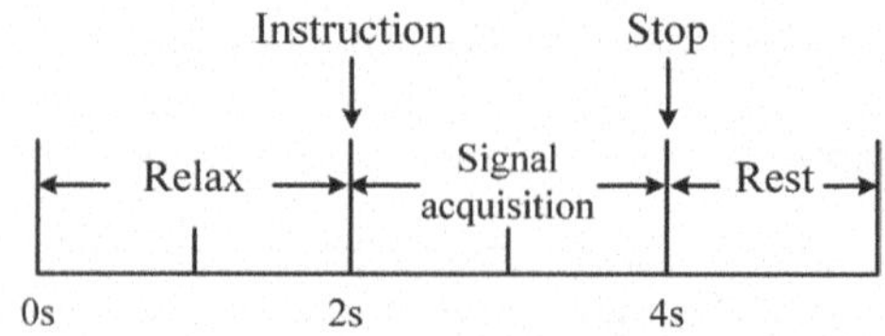

Fig. 5. Experimental procedure for one trial

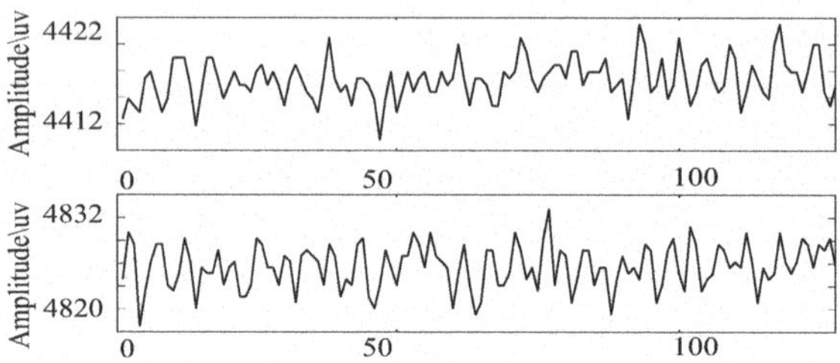

Fig. 6. Original EEG channel signal

Data Processing

Generally, the original EEG channel data obtained from experiments contains noise such as electromyography and electrooculogram, which is not suitable for network training. Therefore, before performing feature extraction, researchers will do a series of data processing to improve the signal-to-noise ratio (SNR), such as high-pass filtering, normalization, and the removal of the artifact. In the paper, we preprocessed the experimental data in following steps. Step 1: Remove the average. In order to avoid the influence of values that differ greatly in the experiment, the mean value of the data is subtracted from the amplitude. Step 2: Normalization. The original data are performed by linear transformation to make values map between [0, 1]. Figure 7 shows the FC5 channel of three subjects after data processing.

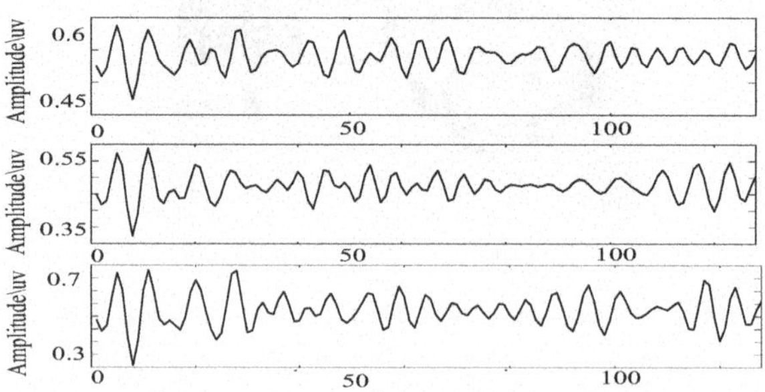

Fig. 7. FC5 channel of three subjects

EEG data contains rich temporal and spatial features. We process the data according to Sect. 2.3 to construct the required 2-D EEG maps. In order to get theta, alpha, and beta frequency band, FFT is performed on 14 channels. Further, the frequency band data matrix is mapped to 2-D map of head. We give 2-D EEG maps of three groups of three subjects, where each group includes two categories (right-hand fist and left-hand fist), as shown in Fig. 8. From Fig. 8, we can find that the generated 2-D EEG maps have obvious location features and the degree of activation of different types of samples has its own characteristics. These characteristics show that the complexity of EEG data features and designing effective EEG feature extraction methods are the keys to achieving high-precision EEG classification.

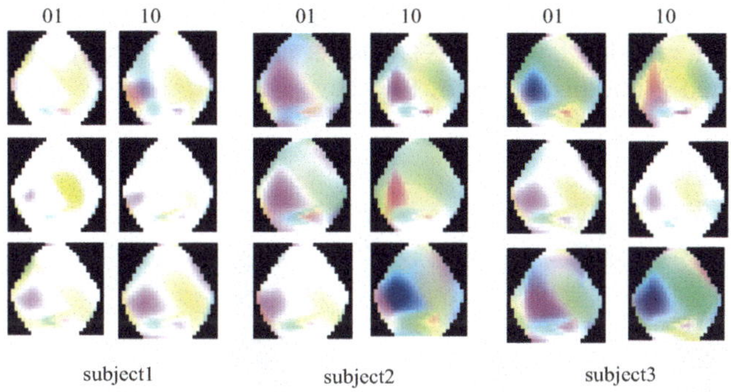

Fig. 8. 2-D EEG map of three subjects

3.2 Analysis of Simulation Result

As can be seen from Sect. 2.3, in processing original EEG channel data with overlap, the choice of overlap factor o has a great influence on the performance of the model. In this experiment, we observe the recognition accuracy of the corresponding model by setting different o values to prove that the processing of EEG data with overlap is effective. All accuracy rates are the highest accuracy rates achieved by experiments. Figure 9 shows the results of classification performance based on the proposed methods with different o values (o =0, o =0.4, o =0.6, o =0.8), where the x-axis represents the values of o and the y-axis represents the accuracy rate.

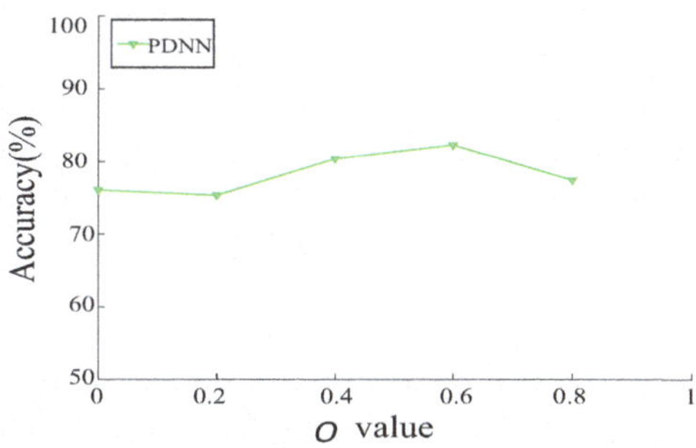

Fig. 9. The recognition accuracy corresponding to different o values

As can be seen from Fig. 9, the performance of model classification can be improved by using the proposed method ($o \neq 0$). It shows that the proposed method of processing EEG data based on overlap can reduce the delayed effect caused by the device and brain reaction, and has a positive impact on the training of the network. On the other hand, we can see that if the o is large than 0.6, causing too much duplicate data, it is easy to lead to the decline of model performance. Further, it can be seen from Fig. 9 that better results are achieved between 0.4 and 0.6. Therefore, in the subsequent experiments, the model in this paper uses o =0.6.

In order to verify the superior performance of the proposed algorithm in extracting the spatiotemporal features of EEG data, we applied multiple classification algorithms to the MI EEG data collected from the experiment. CNN using the original time series data as the training set is called ACNN. CNN using 2-D EEG maps as the training set is called BCNN. The hidden layer of LSTM is set at 200. The test results of these four methods are listed in the histogram of Fig. 10.

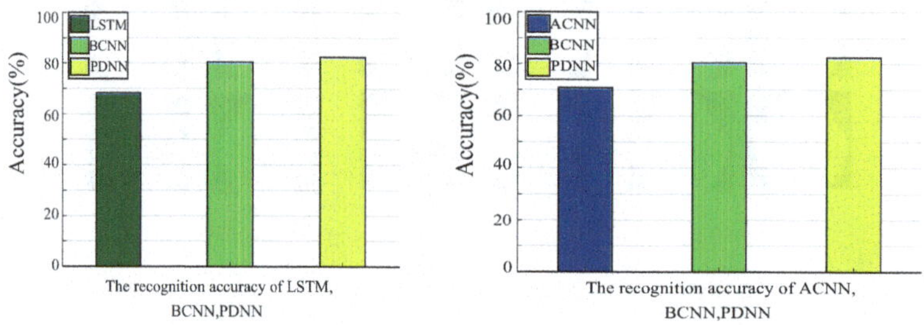

Fig. 10. The recognition accuracy of four algorithms

It can be seen from Fig. 10 that the recognition accuracy of EEG data obtained by PDNN is over 80%, which is higher than that of the other three algorithms. The comparison of ACNN, BCNN, and PDNN shows the accuracy rate of ACNN is the lowest. ACNN uses original channel data for network training. Although CNN has strong feature extraction capability, it cannot complete classification tasks well due to the complexity of EEG features. Both BCNN and PDNN use 2-D EEG maps as training sets, and perform classification tasks well. It shows that the 2-D EEG map based on FFT in this paper can well transform the spatial features of EEG data. In the comparison of LSTM, BCNN, and PDNN, PDNN obtains the highest classification accuracy. The initial input of LSTM is the channel EEG signal, which is weak in extracting the EEG spatial features. In the process of converting the original data into the 2-D EEG map, EEG data will inevitably lose some time series features. Therefore, BCNN cannot achieve the best classification performance. PDNN first uses CNN to extract the spatial features of EEG data and LSTM to extract the temporal features of EEG data, and then uses feature mosaicing to fuse the temporal and spatial features to achieve better classification results.

In order to further illustrate the effectiveness of the proposed algorithm, comparison experiments have been conducted for PNDD and other mainstream EEG classification algorithms, including support vector machine (SVM) [10], stacked automatic encoder

(SAE) [15], deep stacking network (DSN) [12] and hidden-layer visible deep stacking network (HVDSN) [12]. The experimental results are shown in Table 3. The kernel function of SVM is RBF. SAE adopts a 5-layer structure, and the number of nodes in each layer is 200. For DSN and HVDSN, the stacking module is set at 5 layers, and the hidden layer increases by 50 in each stack. All the accuracy rates in Table 3 are from multiple training, and the highest accuracy rate is taken.

Table 3. The recognition accuracy (%) of different algorithms

Model	Test accuracy (%)
PDNN	**82.33**
HVDSN	74.33
DSN	70.83
SAE	62.50
RBF SVM	62.50

From Table 3, we can see that the recognition ability of the proposed PDNN is better than the other four mainstream MI EEG classification algorithms. It is verified that the proposed PDNN based on spatiotemporal features of EEG has superior classification performance. As the classic machine learning algorithm, SVM is commonly used for small sample data analysis. However, due to the high complexity of EEG features, the ability of SVM to extract EEG features is insufficient. SAE based on unsupervised training, is easy to fall into a local optimum due to the small sample size. DSN and HVDSN avoid the phenomenon of gradient disappearance caused by too many layers. But their initial input is the EEG signals of all channels, which is weak in extracting EEG spatial features.

4 Conclusion

Based on the deep learning theory and the spatiotemporal features of EEG, a parallel deep network is proposed to realize feature classification of EEG signals. This algorithm not only uses the CNN to extract the spatial features of the 2-D EEG maps generated by applying FFT on EEG time series data, but also extracts the EEG time series features based on LSTM. The proposed method is applied to the EEG data collected by Emotiv and evaluate its classification performance through multiple experiments. Experiments show that PDNN has superior performance on recognition tasks and good generalization ability. The PDNN provides a new idea for the recognition of motor imagery EEG signals. In the future, we will conduct more in-depth research from multiple aspects: The spatial and temporal features of EEG Under high noise background will be studied to cope with complex EEG signal analysis; Further expand the classification category and experimental samples to improve the application value of the algorithm.

References

1. Bajaj, V., Rai, K., Kumar, A., et al.: Rhythm based features for classification of focal and nonfocal EEG signals. IET Signal Process. **11**(6), 743–748 (2017)
2. Arunkumar, N., Ramkumar, K., Venkatraman, V., et al.: Classification of focal and non focal EEG using entropies. Pattern Recogn. Lett. **94**, 112–117 (2017)
3. Wu, D., King, J.T., Chuang, C.H., et al.: Spatial filtering for EEG-based regression problems in brain-computer interface (BCI). IEEE Trans. Fuzzy Syst. **26**(2), 771–781 (2017)
4. Hassan, A.R., Bhuiyan, M.I.H.: A decision support system for automatic sleep staging from EEG signals using tunable Q-factor wavelet transform and spectral features. J. Neurosci. Meth. **271**, 107–118 (2016)
5. Subasi, A.: Application of adaptive neuro-fuzzy inference system for epileptic seizure detection using wavelet feature extraction. Comput. Biol. Med. **37**(2), 227–244 (2019)
6. Shiu, K., Alok, S.: A new parameter tuning approach for enhanced motor imagery EEG signal classification. Med. Biol. Eng. Comput. **56**(10), 1861–1874 (2018)
7. Behrooz, N., Reza, B., Mansoor, Z.J.: An efficient hybrid linear and kernel CSP approach for EEG feature extraction. Neurocomputing **73**(1–3), 432–437 (2009)
8. Sharma, M., Deb, D., Acharya, U.R.: A novel three-band orthogonal wavelet filter bank method for an automated identification of alcoholic EEG signals. Appl. Intell. **48**(5), 1368–1378 (2018)
9. Purnamasari, P.D., Ratna, A.A.P., Kusumoputro, B.: EEG based emotion recognition system induced by video music using a wavelet feature vectors and an artificial neural networks. Adv. Sci. Lett. **23**(5), 4314–4319 (2017)
10. Sumit, S.: High performance EEG signal classification using classifiability and the Twin SVM. Appl. Soft Comput. **30**(3), 305–318 (2015)
11. Langkvist, M., Karlsson, L., Loutfi, A.: A review of unsupervised feature learning and deep learning for time series modeling. Pattern Recogn. Lett. **42**, 11–24 (2014)
12. Tang, X., Zhang, N., Zhou, J., et al.: Hidden-layer visible deep stacking network optimized by PSO for motor imagery EEG recognition. Neurocomputing **234**, 1–10 (2016)
13. Spampinato, C., Palazzo, S., Kavasidis, I., Giordano, D.: Deep learning human mind for automated visual classification. In: 2017 IEEE Conference on Computer Vision and Pattern Recognition, pp. 4503–4511. The Institute of Electrical and Electronics Engineers, Hawaii (2017)
14. Bashivan, P., Bidelman, G.M., Yeasin, M.: Spectrotemporal dynamics of the EEG during working memory encoding and maintenance predicts individual behavioral capacity. Eur. J. Neurosci. **40**(12), 3774–3784 (2014)
15. Li, J., Struzik, Z., Zhang, L., et al.: Feature learning from incomplete EEG with denoising autoencoder. Neurocomputing **165**, 23–31 (2015)

Using Simple Genetic Algorithm for a Hand Contour Classification: An Experimental Study

Jaroslav Moravec(✉)

Faculty of Electrical Engineering and Informatics, University of Pardubice, náměstí Čs. legií 565, 530 02 Pardubice, Czech Republic
jaroslav.moravec@student.upce.cz,
j.moravec.email@seznam.cz

Abstract. The area of biometric systems has passed through considerable advancement in the past two decades. Supporting of security provision plays a key role in many branches. There are large amount of the biometrical markers which can be utilized in the person identification process. One of the possible ways is a method which uses a hand shape contour classification. The presented paper solves the problem of hand contours classification with use of a Simple Genetic Algorithm (SGA). The foundations of the SGA were established in 1950's, but an improvement process of the SGA continues. The hand contour for the classification purposes is obtained from a color image from a biometric scanner. The biometric scanner has fixed pegs to hold the hand, or the hand can be freely placed on the scanning area. A core of the proposed estimator is an Iterative Closes Point algorithm which enables matching of the two point-clouds and expressing their dissimilarity regarding the elected metrics. In the experimental section, a large number of experiments were performed with a different setting of the SGA working parameters. Beside the capability to correctly align/match the hand contours, selected standard benchmark tests were performed with a corresponding number of dimensions. The presented estimator solves the thee-dimensional optimization task. Based on experimental results, it was proven that in the case of identical contours the proposed method, which utilizes the SGA optimizer, provides very accurate results.

Keywords: Simple genetic algorithm · Iterative closest point algorithm · Hand contour classification · Biometrics · Person identification

1 Introduction and Related Works

The area of Evolutionary Algorithms (EA) has its foundations in the Darwinian theory of natural selection. The origins of the EAs falls in the 40s and 50s of the last century. The Mote Carlo algorithm [32] falls into this interesting area too. The aim of proposals of such algorithms was to simulate the natural selection conditions under time-acceptable circumstances and to reach an optimal solution as fast as possible. The vigorous development of the EAs came in 80s and 90s thanks to wide expansion of computers which were affordably priced. A complex survey of the EA can be found in e.g. [50]. Thanks to gradual development, the EA area was divided into three main sub-areas:

R. Silhavy (Ed.): CSOC 2020, AISC 1225, pp. 93–109, 2020.
https://doi.org/10.1007/978-3-030-51971-1_8

1. Genetic Algorithms (GA/SGA) which were described by J. Holland [12–16, 18]. Interesting branch is as well the Genetic Programming (GP) which was described by John Koza [27, 28].
2. Evolutionary Programing (EP), described by L. Fogel [7–9].
3. Evolutionary Strategy (ES), described by the authors I. Rechenberg, H.P. Schwefel and P. Bienert [2, 39, 40].

All three main streams utilize the foundations which were defined in 50s. However, there are significant differences in the way how the individuals of the population are defined and how the population is processed. The GA usually use a binary coded chromosome. The ES and EP uses real numbers. Based on the Holland's SGA, many interesting metaheuristic iterative optimizers were proposed during the 90s e.g. [10, 11] which similarly to ES, use real numbers to code the individuals in the population. As the example can be mentioned: Simulated Annealing (SA) [26], Differential Evolution (DE) [44], Ant Colony Optimization (ACO) [6], Particle Swarm Optimization (PSO) [35], Tabu Search (TS), Memetic Algorithms (MA) [21]. For many real problems it is not necessary to find an exact solution of the solved task. An approximate solution can be sufficient enough.

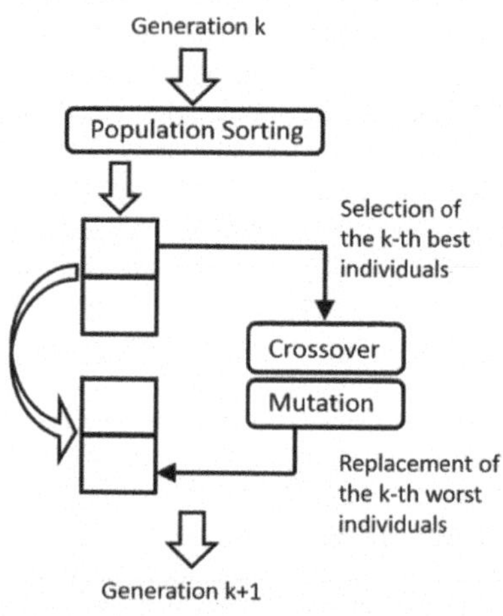

Fig. 1. Chart of simple genetic algorithm.

2 Simple Genetic Algorithm (SGA)

The Genetic Algorithms (GA) falls into wider groups of algorithms which are named Evolutionary Algorithms. The first official reports of the EA falls in 1954 and the author was Nils Aall Barricelli. His research stayed unnoticed for a long time. Alex Fraser published a series of his work in 1957. The theme of his work was "Artifical selectio" (artificial selection). Since then these very special type of algorithms became the foreground of interest of many researchers. The big popularizer of the GA was John

Holland which published his work "Adaptation in Natural and Artificial Systems" in 1975. This book was accessible for everyone and everyone can buy it still. Holland in his book, described so called "Holland's schema theorem" which is a theoretical consideration of why the GAs are capable of finding the correct solution.

The GA is a stochastic algorithm and serves to solve the optimization tasks. Many operations which are executed with the set of input data were adopted from the area of nature genetics. The GA consists of several elementary building blocks: 'Selection', 'Crossover', 'Mutation', 'Replacement Strategy' and 'Postprocessing' – see Fig. 1. All these operations are usually serially ordered. Every operation represents a certain set of handling of the individuals in the population. The population of individuals is a de-facto group of possible solutions of the given task. All individuals are, in the first step of the GA, randomly deployed across the whole space of possible solutions. The number of individuals is, in comparison to the number of possible solutions of the task, insignificant. The selection operation can select the best individuals or the average individuals or mixture of similar operations. A fixed set of operators does not exist. Many new operators arise every year. The GA can solve a relatively wide spectrum of problems and, because the GA is a relatively young scientific discipline, every innovation brings a certain surprise. There are a large number of modifications of the original SGA algorithm e.g. parallel GA, mesh-network GA, GA with micro-populations etc. Every group of the GA is designed to solve a concrete sort of problem with concrete time-demands. The GA's are an assembled group of very simple operations over a set of individuals of the population. Every individual usually represents one concrete solution of the given task and can be represented by any means – not only as the binary coded chain of numbers 1 and 0. The original GA in the 70's used only binary coded genes of the chromosome. In later times many different modifications arose which achieved significantly better results.

3 Description of Proposed Method

The proposed method utilizes the Simple Genetic Algorithm as the optimizer of an n-dimensional optimization task. A core of the estimator is an iterative method which enables matching/aligning of the two point-clouds.

3.1 The RGB Image of the Hand, the Hand Contour Obtaining

The RGB image $\boldsymbol{I_{RGB}}$ from the camera (right hand, back of the hand, the thumb leads to down, the fingers lead the left, all images in visible spectrum only) – is shown in Fig. 2. The image has size $I_w \times I_h$ and is first transformed with use of a transformation $\boldsymbol{T_{Gray}}$ converted from the RGB representation to the gray image $\boldsymbol{I_{Gray}}$ according to the rule (Non-linear Luma):

$$\boldsymbol{I_{RGB}} = \begin{bmatrix} P_{0,0}^{RGB} & \cdots & P_{x,0}^{RGB} \\ \cdots & \cdots & \cdots \\ P_{0,y}^{RGB} & \cdots & P_{x,y}^{RGB} \end{bmatrix} \overset{\boldsymbol{T_{Gray}}}{\longrightarrow} \boldsymbol{I_{Gray}} = \begin{bmatrix} P_{0,0}^{Gray} & \cdots & P_{x,0}^{Gray} \\ \cdots & \cdots & \cdots \\ P_{0,y}^{HSB} & \cdots & P_{x,y}^{Gray} \end{bmatrix} \quad (1)$$

$$T_{Gray} : P_{x,y}^{B\&W} = 0.299 \times R + 0.587 \times G + 0.114 \times B$$

$$R, G, B \in \langle 0, 255 \rangle$$

$$x, y \in \mathbb{N}^0; x \in \langle 0, I_w - 1 \rangle, y \in \langle 0, I_h - 1 \rangle, I_w, I_h \in \mathbb{N}^0$$

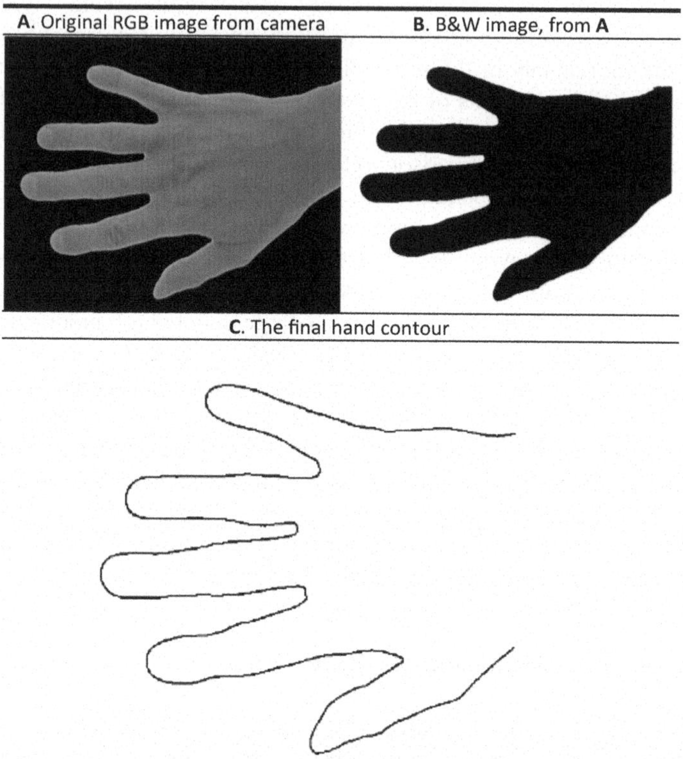

A – Original RGB image of the hand. **B** – B&W representation of the image which was obtained from the image in the point **A.** with use of the thresholding of the gray-scale image. **C.** – the hand contour ***C***, obtained with use of calculation according to the Alg. 1 and the image in **B.**

Fig. 2. Individual images RGB and B&W and the hand contour

where the $P_{x,y}^{RGB}$ means a pixel at coordinates (x, y) and R, G, B are the values of individual components of the pixel $P_{x,y}^{RGB}$. The origin of the coordinate system is placed in the top left. The horizontal axis is $+X$ (from left to the right), the vertical axis is $+Y$ (from top to bottom). The obtained image is named $\boldsymbol{I_{Gray}}$ and in the next step is, with use of the transformation $\boldsymbol{I_{Gray}} \overset{T_{threshold}}{\longrightarrow} \boldsymbol{I_{B\&W}}$, converted to the black and white representation:

$$I_{Gray} \overset{T_{threshold}}{\longrightarrow} I_{B\&W},\ I_{B\&W} = \begin{bmatrix} P_{0,0}^{B\&W} & \cdots & P_{x,0}^{B\&W} \\ \cdots & \cdots & \cdots \\ P_{0,y}^{B\&W} & \cdots & P_{x,y}^{B\&W} \end{bmatrix} \tag{2}$$

$$T_{threshold} : f(x) = \begin{cases} black, & if\ Gray < 127 \\ white, & if\ Gray \geq 127 \end{cases}$$

where the element *Gray* means the gray color level in the grayscale image I_{Gray} and for the *Gray* holds that $Gray \in \langle 0, 255 \rangle$. In the next step the hand contour C is computed with use the image $I_{B\&W}$ and the Algorithm 1. Algorithm 1 is a simple gradient algorithm which operates in the whole area of the image and uses the so-called look-up table in 8 surrounding neighbors of every pixel $P_{x,y}^{RGB}$. The "look-up-table" is marked as L_{TBL}. The contour C is defined as:

$$C, C \subset I_{B\&W}, \forall P_{x,y}^{B\&W}, P_{x,y}^{B\&W} \in C \wedge P_{x,y}^{B\&W} = 1 \tag{3}$$

$$C = \left(P_{x,y}^{B\&W}\right), j \in \mathbb{N}^+,\ P_{x,y}^{B\&W} \in I_{B\&W}$$

The result of the Algorithm 1 is the hand contour which can be trimmed to the expected length. A condition enabling successful execution of the Algorithm 1 is: the $I_{B\&W}$ has to be free of noise. If the image $I_{B\&W}$ is loaded with noise (large or small artefacts) it is necessary to filtrate it out and after that use the Algorithm 1. See also [23–25]. Algorithm 1 is also suitable to treat the images from the biometric scanner which uses the pegs to fixate the hand.

3.2 Use of the SGA, the Chromosome Assemblage, Objective Function, Evolutionary Process

The SGA is used as the optimizer of the three-dimensional objective function. Its task is to find as optimal solution in alignment of two hand contours which will be hereafter marked as $\mathcal{M}$ and S. At the time of calculation with use of the SGA, the chromosome is used according to the Fig. 3. The chromosome length is 33 bits. First 11 bits define the Δx value and second 11 bits define the Δy value and the third 11 bits defines the angle $\Delta\alpha$. The angle is in radians and the values Δx, Δy are in pixels. The image size is 640 × 480 pixels.

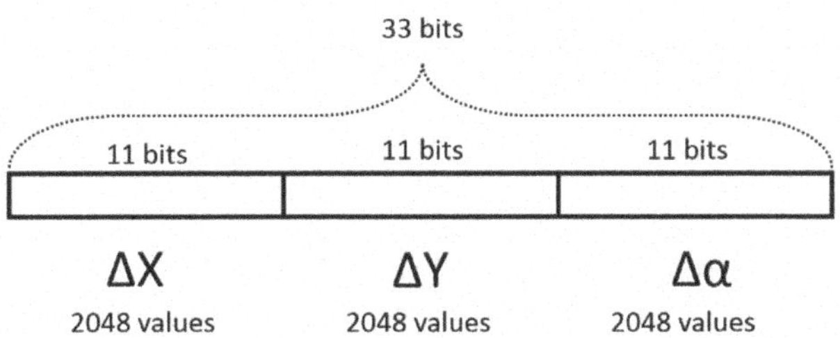

Fig. 3. A chromosome representation

In calculation the contour $\mathcal{M}$ is first placed on the drawing so that the center of mass of the contour $\mathcal{M}$ is placed in the center of the drawing. The axis of the contour $\mathcal{M}$ is parallel with axis of the coordinate system. The contour $\boldsymbol{S}$ is placed on the drawing so that the center of mass of the contour $\boldsymbol{S}$ is placed at a distance 1/6 of the width or height of the drawing far from the center of mass of the contour $\mathcal{M}$. The distance is elected randomly. The value of the angle $\Delta\alpha$ is elected randomly too in the range $\pm 30°$. The angle between the axis of the contour $\mathcal{M}$ and the axis of the contour $\boldsymbol{S}$ is marked as $\Delta\alpha$. The angle is defined in radians.

Algorithm 1.
The Algorithm which enables to calculate the hand contour $\boldsymbol{C}$ in the image $\boldsymbol{I_{B\&W}}$

```
Input: I_B&W
Output: C
C = (0);
for x=0; x < I_w; x=x+1
   for y=0; y < I_h; y=y+1
      if P^B&W_x,y = 1
         C_nt = 0; E_x = 0
         while (E_x = 0 ∧ C_nt < 8)
            x_t = x + L_TBL[C_nt].Δx; y_t = y + L_TBL[C_nt].Δy;
            if P^B&W_xt,yt ∈ I_B&W ∧ P^B&W_xt,yt = 0
               E_x = 1; exitwhile;
            endif
            C_nt = C_nt + 1;
         endwhile
         if E_x = 1 then
            C = C ∪ P^B&W_x,y
         endif
      endif
   endfor
endfor
```

$\boldsymbol{C}$- hand contour or set of points $P^{B\&W}_{x,y}$. The values $\Delta x, \Delta y$ defines the change in the axes X, Y for look-up table L_{TBL} for individual points or direction of the next seeking step C_{nt}:
0: $L_{TBL}[0].\Delta x = +1; L_{TBL}[0].\Delta y = +0;$ 1: $L_{TBL}[1].\Delta x = +1; L_{TBL}[1].\Delta y = +1;$
2: $L_{TBL}[2].\Delta x = +0; L_{TBL}[2].\Delta y = +1;$ 3: $L_{TBL}[3].\Delta x = -1; L_{TBL}[3].\Delta y = +1;$
4: $L_{TBL}[4].\Delta x = -1; L_{TBL}[4].\Delta y = +0;$ 5: $L_{TBL}[5].\Delta x = -1; L_{TBL}[5].\Delta y = -1;$
6: $L_{TBL}[6].\Delta x = +0; L_{TBL}[6].\Delta y = -1;$ 7: $L_{TBL}[7].\Delta x = +1; L_{TBL}[7].\Delta y = -1;$

Let's mark the evolutionary process with use of the selected evolutionary estimator as $\Im_{\mathrm{EA}}$ and now it is possible to write that:

$$\mathcal{E} = \arg\mathrm{opt}_{\mathcal{H}\in\mathbb{R},S}\Im_{\mathrm{EA}}(\mathcal{M}, S, \mathcal{H}). \tag{4}$$

The aim of the evolutionary process (task of the optimizer) is to find such value $\mathcal{E}$, for which holds that the $\mathcal{E} \cong 0$. The metrics of the *fitness* function was elected based on purely practical experiments and is given by distance of the point clouds $\mathcal{M}$ and $\boldsymbol{S}$ in the sense of the Taxicab geometry [29] or Manhattan distance as follows:

$$fitness = \begin{cases} \sum_{j=1}^{j=n} \left| P_j^S - P_{min}^{\mathcal{M}} \right|^6 & if\ \mathcal{L} = 1 \\ \sum_{j=1}^{j=n} \left| P_j^S - P_{min}^{\mathcal{M}} \right| & if\ \mathcal{L} = 0 \end{cases} \tag{5}$$

$$d\left(P_j^S, P_{min}^{\mathcal{M}}\right) = min\ d\left(P_j^S, P_i^{\mathcal{M}}\right)|_{i=1}^{i=n}$$

$$fitness \in \mathbb{N}^0, fitness_P \in \mathbb{N}^0, j \in \mathbb{N}^+$$

$$\mathcal{L} = \begin{cases} 1, & if\ x_{0,1} \notin \boldsymbol{\Omega} \\ 1 & if\ x_7 \notin \left\langle -0.40^{RAD};\ +0.40^{RAD} \right\rangle \\ 0, & otherwise \end{cases}$$

where the P_j^S is a point of the contour $\boldsymbol{S}$ and the $P_i^{\mathcal{M}}$ is a point of the contour $\mathcal{M}$. The point $P_{min}^{\mathcal{M}}$ is the point which is nearest to the point P_j^S in the sense of used metrics and holds that the $d\left(P_j^S, P_i^{\mathcal{M}}\right)$ is minimal. The limiting conditions given by the function $\mathcal{L}$, were elected based on practical experiments. The whole contour $\boldsymbol{S}$ moves in the space/plane XY in the space of possible solutions $\boldsymbol{\Omega}$ (coordinates in the plane XY), and the center of the $\boldsymbol{\Omega}$ is the point P_Ω. It is possible to write that:

$$P_\Omega(x_\Omega = 0.5I_w, y_\Omega = 0.5I_h) \tag{6}$$

and the space $\boldsymbol{\Omega}$ is expressed as follows:

$$\boldsymbol{\Omega} = [\langle x_\Omega - 0.15I_w, x_\Omega + 0.15I_w\rangle, \langle y_\Omega - 0.15I_h, y_\Omega + 0.15I_h\rangle]. \tag{7}$$

i.e.$[\langle 320 - 0.15 \times 640\rangle, \langle 320 + 0.15 \times 640\rangle, \langle 240 - 0.15 \times 480\rangle, \langle 240 + 0.15 \times 480\rangle]$

for the drawing of size 640×480 pixels. The whole processed task is three-dimensional and the objective function is nonlinear, non-continuous and non-separable.

3.3 Computational Core of the Estimator, The ICP Algorithm

As the core of the classifier the ICP algorithm – the Iterative Closest Point Algorithm [1] was elected. This algorithm was first unveiled in 1992 and is deduced from previously published work [19]. The task of the ICP is to unambiguously match/align two point-clouds in the 3 dimensional space with use of operation rotation and translation so as the Euclidean distance according to the (2) was as small as possible. The optimal alignment is given by the value $\mathcal{E} = 0$ and this value can be usually reached if two identical point clouds are matched. The process of optimization can be expressed as:

$$\varepsilon = \arg\ min_{\boldsymbol{R},\boldsymbol{t}} \sum\nolimits_{i=1}^{n} \left\| (\mathbf{R}P_i + \boldsymbol{t}) - P_j \right\|_2^2 \tag{8}$$

where $\mathbf{R}, \mathbf{R} \in \mathbb{R}$ is a rotational matrix 3×3, $\boldsymbol{t}$ is a translational vector. In [1] the process of alignment of the two point-cloud in 3D space is presented. The P_i and P_j represents the points of the both compared clouds. The number of points in the clouds can be different. The original ICP algorithm [1] consists of the three steps:

1) First, for every point P_i a closest point from the set P_j is found with use of the actual values $\mathbf{R}$ and $\boldsymbol{t}$.
2) Update step of the values $\mathbf{R}$ and $\boldsymbol{t}$ so as this rule was valid:

$$(\mathbf{R}, t) = min_{\mathbf{R},t} \sum_{i=1}^{n} \|(RP_i + t) - P_j\|_2^2 \tag{9}$$

3) If the value $\mathcal{E}$ is smaller than previously defined minimum the algorithm ends. The optimal values $\mathbf{R}, t$ were found. In other case go to the point 1

In the first step of the algorithm both clouds have to be matched to smallest possible distance. This step is usually solved in such way that the center of masses of the clouds are identified. A big advantage of the ICP algorithm is that the proof of capability of convergence of the ICP algorithm to only one local optimum was already published many times see e.g. [31, 36].

4 Experimental Results

Section of experimental results is divided into two parts:

A. tests of the SGA algorithm with use of the six standard benchmark functions which will be marked as F1, F2, F3, F4, F5, F6 see e.g. [4, 33, 37, 38] and selected number of dimensions will be 2, 3 and 6. **B.** Experimental results of efficiency of the SGA in hand contour matching process.

4.1 Tests of the SGA with Use of Six Different Benchmark Functions

One of the first benchmark functions are proposed in the publication [4] the function commonly marked as F1–F5. With regard to the publication [45], there is no guarantee that the standard benchmark functions will be capable of showing the usability of the tested optimizer for solving of our problem. The 'no free lunch' theorem says that if the two different optimization algorithms will be tested with use of all accessible benchmark functions, the performance will be on average identical. The benchmark functions are not capable of catching all theoretical or practical problems.

For basic experimental testing of the SGA, six standard benchmark functions were elected which are usually called as DeJong's functions F1, F2, F3, F4, F5 and the Rastrigin's function F6. Corresponding equations and 2-dimensional visualizations are recorded in Table 1. For experimental purposes the functions with the number of dimensions 2, 3 and 6 were elected, because the task of matching of hand contours is, indeed, three dimensional in 2D space. All results are recorded in Tables 2, 3, 4 and selected working parameters are recorded in Table 5.

In all cases the expected value of the fitness function is equal to zero. The SGA attained to the zero value for all tested functions. In Table 2, 3, 4 are also recorded the values of max., avg., and std. dev. of all individuals in the last generation. The working parameters (see Table 5) were elected so as the range was adequate to the optimization task – hand contour classification. If the number of dimensions is higher, the number of individuals have to be higher as well. The final results which are recorded in Table 2, 3, 4 are the best reached from a certain number of iterations – repeating of whole process. The biggest number of iterations was elected for the $D_{im} = 2$ problem; e.g. 1000. The number of necessary evaluations of the objective function is then computed as the number of individuals multiplied to the number of generations multiplied to the number of iterations

Table 1. Graphical representation of used benchmark functions, see [51, 52]

F1 – De Jong's function 1,(Sphere model)	**F2 – Axis parallel hyper-ellipsoid function**
$f(x_i)=\sum_{i=1}^{n} x_i^2$ $-500 \le x_i \le 500, i = 1..n,$ $f(x^*) = 0, x_i = 0$	$f(x_i)=\sum_{i=1}^{n} ix_i^2$ $-100 \le x_i \le 100, i = 1..n,$ $f(x^*) = 0, x_i = 0$
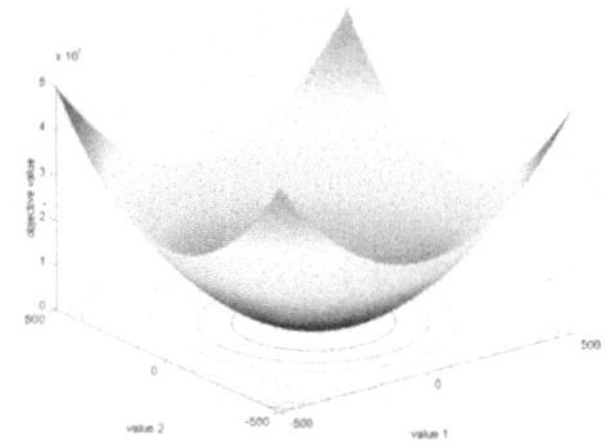	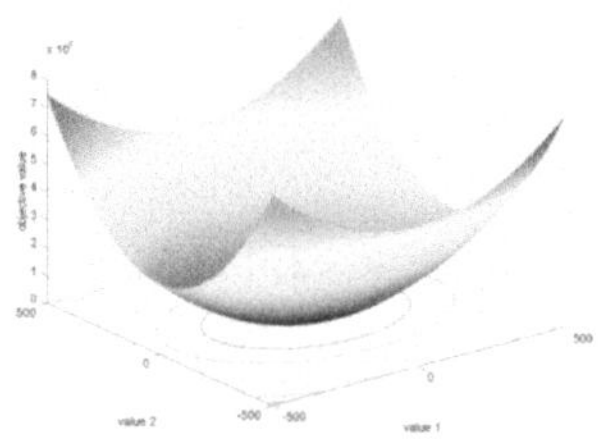
F3 – Rotated hyper-ellipsoid function (Schw.1.2)	**F4 – Moved axis parallel hyper-ellipsoid function**
$f(x_i)=\sum_{i=1}^{n}\left(\sum_{j=1}^{i} x_j\right)^2$ $-65 \le x_i \le 65, i = 1..n,$ $f(x^*) = 0, x_i = 0$	$f(x_i)=\sum_{i=1}^{n}\left(i(x_i-5i)\right)^2$ $-50 \le x_i \le 50, i = 1,2..n,$ $f(x^*) = 0, x_i = 5i$
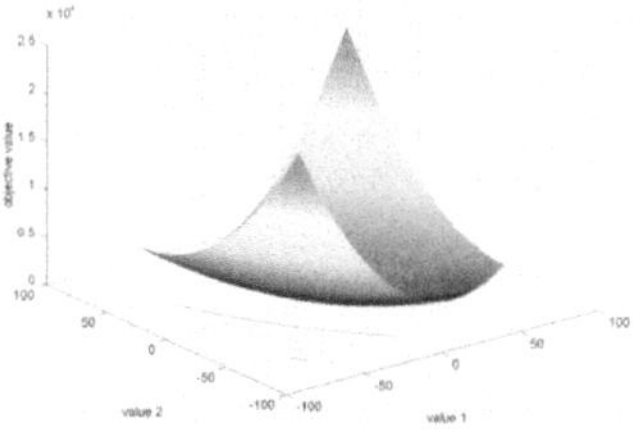	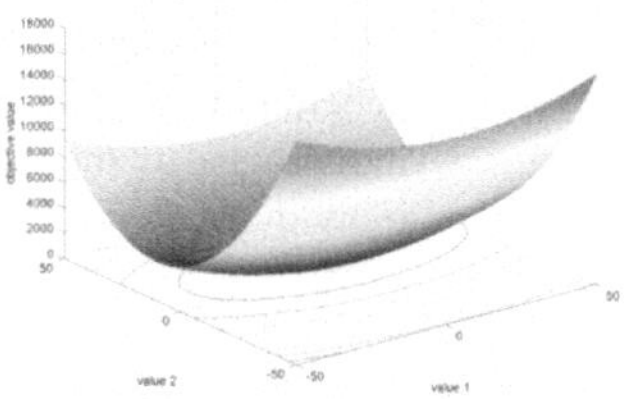
F5 – Rosenbrock's valley, (De Jong's function 2)	**F6 – Rastrigin's function 6**
$f(x_i)=\sum_{i=1}^{n-1}\left(100(x_{i+1}-x_i^2)^2+(1-x_i)^2\right)$ $-2.048 \le x_i \le 2.048, i = 1..n,$ $f(x^*) = 0, x_i = 1$	$f(x_i)=10n+\sum_{i=1}^{n}\left(x_i^2-10\cos(2\pi x_i)\right)$ $-5.12 \le x_i \le 5.12, i = 1..n,$ $f(x^*) = 0, x_i = 0$
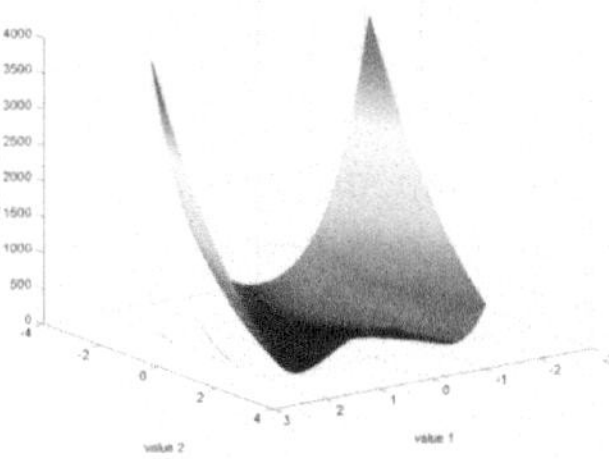	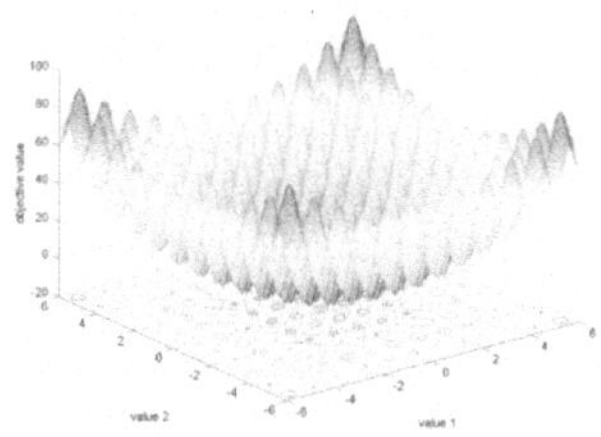

Table 2. Dim = 2; Results of the tasks F1–F6

Fn.	Corr.	Best	Max	Avg	Dev
F1	0	0	0.01000	0.00001	0.000316
F2	0	0	0.00010	6.0E−07	7.73E−06
F3	0	0	71.65781	2.57034	5.563141
F4	0	0	0.00010	6.6E−06	2.48E−05
F5	0	0	0.32708	0.060738	0.034173
F6	0	0	4.99665	0.533181	0.902755

Fn.-tested function, Corr.-correct result, best-the attained result with use of the SGA, max.-maxi. Value of an individual in the population in the last generation, avg-average value of the fitness of all individuals in the last generation, dev-std. dev of all individuals in the last generations.

Table 3. Dim = 3; Results of the tasks F1–F6

Fn.	Corr.	Best	Max	Avg	Dev
F1	0	0	0	0	0
F2	0	0	0	0	0
F3	0	0	1,00E−06	3,00E−08	1,71E−07
F4	0	0	0,0004	8E−06	5,63E−05
F5	0	0	0,010218	0,00097221	0,002086
F6	0	0	0	0	0

Fn.-tested function, Corr.-correct result, best-the attained result with use of the SGA, max.-maxi. Value of an individual in the population in the last generation, avg-average value of the fitness of all individuals in the last generation, dev-std. dev of all individuals in the last generations.

Table 4. Dim = 6; Results of the tasks F1–F6

Fn.	Corr.	Best	Max	Avg	Dev
F1	0	0	0	0	0
F2	0	0	0	0	0
F3	0	0	0	0	0
F4	0	0	0.002600	0.000510	0.000838
F5	0	0	0.112909	0.0155538	0.033652
F6	0	0	0	0	0

Fn.-tested function, Corr.-correct result, best-the attained result with use of the SGA, max.-maxi. Value of an individual in the population in the last generation, avg-average value of the fitness of all individuals in the last generation, dev-std. dev of all individuals in the last generations.

Table 5. Benchmark functions F1–F6, working parameters.

Function		T1	T2	T3	T4	T5	T6
Resolution SGA		1e−1	1e−2	1e−3	1e−2	1e−3	1e−2
Domain		+500 −500	+100 −100	+65 −65	+50 −50	+2.048 −2.048	+5.12 −5.12
Npop Gen. Iter.	D = 2 D = 3 D = 6	**A** **B** **C**					
Use	D = 2 D = 3 D = 6	* * *	* * *	* * *	* * *	* * *	* * *
Key: A, B, C variants of working parameters: **Variant: SGA: Npop/Gen./Iter** A: SGA: 100/1000/1000 B: SGA: 150/2000/100 C: SGA: 300/10000/20				'Domain' – searching area of the EA, 'Gen.' – the number of generations the test, 'Iter.' – the number of iterations, '*' – tested function if used for the given number of dimensions, Values 1.0e-1 represents 1×10^{-1} etc.			
F1: [3–5, 22, 33, 34, 46, 48] F2: [4, 33, 34, 48] F3: [4, 33, 34, 48] F4: [4, 33, 34]				F5: [3, 4, 22, 33, 34, 41–43, 46, 48] F6: [3, 20, 22, 33, 34, 37, 38, 41, 46–48, 49]			

– e.g. $100 \times 1000 \times 1000 = 1 \times 10^8$ evaluations – see Table 5. For $D_{im} = 2$ problems, only 20 iterations were elected.

It is well visible that the SGA was capable of optimizing all selected benchmark functions. All tested benchmark functions are, indeed, simpler but belong to the standard set of benchmark functions.

4.2 Experiments of Efficiency of the SGA in Hand Contour Classification

The SGA has a large number of working parameters. The wrong setting of these parameters usually leads to the optimizer malfunction. Moreover, some types of operators can not be used at the same time. The SGA has to use an optimal set of working parameters to be capable of providing optimal results. To ascertain the optimal values N_{POP}, N_{GEN}, the standard operators roulette wheel and elitist mutation were selected. Computational accuracy corresponds to the number of genes of the chromosome - see Fig. 3. Elected accuracy of the fitness function was 1×10^{-3}. The resolution is identical for both axes X and Y and also for the angle α. The size of the drawing is 640×480 pixels. The minimal resolution in axis X is 0.64 pixels and in axis Y is 0.48 pixels. The minimal resolution in heading is 0.00628 radians and it is 0.36 degree. Such values do not represent breathtaking accuracy but are good enough for correct N_{POP}, N_{GEN} values estimation. The result is recorded in Table 6. To reach

Table 6. Effect of the N_{POP}, N_{GEN} to, accuracy 1×10^{-3}

$N_{GEN} \rightarrow$ N_{POP}	5	10	15	20	30	40	50	75	100
4	1660.73	953.54	276.8	157.5	196.65	303.64	71.76	0.93	27.02
5	1735.85	3118.4	304.61	301.03	231.13	211.56	0.91	108.11	4.65
8	8.52	369.07	2.79	26.09	9.31	1.86	0.92	0	0
10	154.24	461.98	87.6	0	0	267.34	0.93	0	0
12	891.12	0	162.16	13.04	0	0.93	0	0	0
15	11.18	103.45	327.13	1.86	13.97	0	0	0	0
20	225.54	256.16	151.44	24.23	0	0	0	**0**	**0**
25	315.51	1601.53	11.18	11.18	0	0	0	**0**	**0**
30	100.65	151.27	0	0	0	0	0	**0**	**0**
50	70.83	150.98	0	0	0	0	0	**0**	**0**

Individual values are in pixels. Selection-roulette wheel, crossover-two points, mutation-is elitist and is applied to 2 random individuals in selection + a one bit changed randomly for two individuals in whole population except the best.

the expected fitness value zero it is necessary to set: $N_{POP} = 20, N_{GEN} = 75$, ideally $N_{POP} = 50, N_{GEN} = 100$.

The mutational operator has also significant effect to successful convergence. The elitist mutation de-facto represents the combination of SGA and Simulated Annealing [30] in area of the best individual. The secondary mutational operator then enables the random change of the two randomly selected individuals. By these steps, the original idea of the SGA [12, 13, 16, 17] is preserved. If the operator of the elitist mutation is omitted, worse results are reached. If the two-point crossover operator is replaced by a one-point crossover operator the result is worse. This is similar for the uniform crossover operator. The tournament selection operator provides identical results as the roulette wheel. The proposed estimator with the used working parameters setting, which is mentioned above, provide great results in the hand contour classification task.

There is still only one big problem, the resolution is equal to 0.36°. With regard to the max. length of a human hand of an adult which is approx. 25 cm - from wrist to tip of the middle finger, the inaccuracy in contour matching is equal to 0.15 pixels. An average length of the contour is equal to 1800 pixels for the image of size 640×480 pixels. This gives the final error of $1800 \times 0.15 = 270$ pixels. The value of 270 pixels is only an estimation and corresponds to the maximal error for a selected resolution 1×10^{-3} or 33 bits of the chromosome for the 3-dimensional problem. An average deviation of two contours can be e.g. 2500–6000 pixels. The deviation of the two non-identical contours of an identical person can be 500–2500 pixels, if the image capturing process was not conducted under ideal conditions e.g. in the biometrical chamber. The solution of this problem is, indeed, to increase the accuracy in the position estimation in the X and Y axis and in heading $\Delta\alpha$ to at least 1×10^{-5}. The length of the chromosome will be 51 bits and for every dimension, 17 bits is reserved. The angle resolution is equal to 0.0036°. For such setting it is not possible to use the previously estimated working parameters N_{POP}, N_{GEN}, because for $N_{POP} = 20, N_{GEN} = 75$ the SGA is not

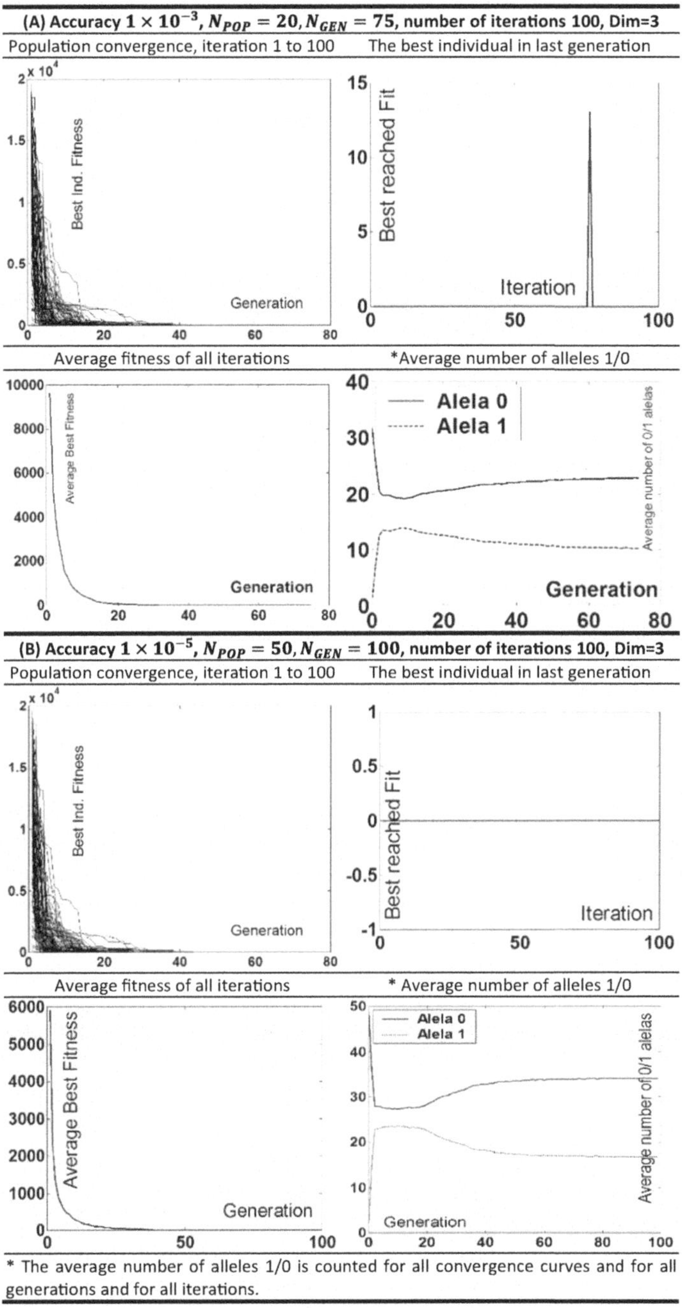

Fig. 4. Convergence curves for selected accuracy 1×10^{-3} and 1×10^{-5} and $D_{im} = 3$.

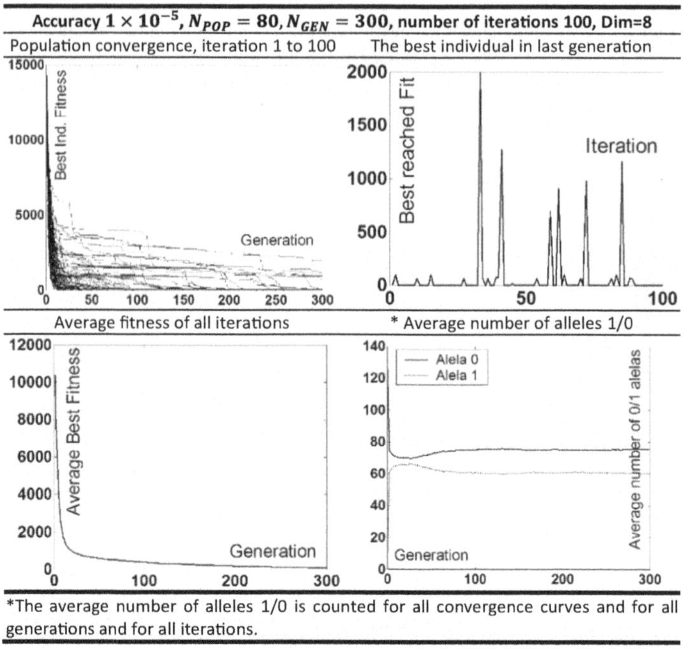

Fig. 5 Convergence curves for elected accuracy 1×10^{-5} a $D_{im} = 8$.

capable of estimating correct results and the value $\mathcal{E}$ is equal to $\mathcal{E} \cong 104.71$ pixels. Suitable selection of the N_{POP}, N_{GEN} will be now $N_{POP} = 50, N_{GEN} = 100$. For these values the estimator is again stable but usually suffers from random malfunction in the convergence process. Unfortunately, such random malfunctions are a common matter for SGA. Increasing the number of individuals and generations to reach better results means the time demands are adequately increased as well. Under such conditions, the estimator need not be suitable to the real conditions.

At the end of experiments the convergence curves and other important statistical markers will be shown. Selected accuracy is 1×10^{-3} and 1×10^{-5} for corresponding minimal values of $N_{POP} = 20, N_{GEN} = 75$ and $N_{POP} = 50, N_{GEN} = 100$. The total number of iterations is 100. The results are recorded in Fig. 4 and 5. It is well visible that the SGA has a big tendency to random malfunction. A chart of the best individual in the last generation shows that for one iteration out of 100, the malfunction was observed and instead of expected value $\mathcal{E} \cong 0$, the estimator calculated a worse result $\mathcal{E} \cong 14$. In this case it would be necessary to repeat whole the calculation. The percentage share of malfunction is for the SGA relatively high, and strongly depends on the type of solved task. In our case, the low-dimensional task is solved and the number of malfunctions is small. In the case that the SGA is solving a higher dimensional problem e.g. 8 or 9 dimensional, where every finger of the hand contour is moveable (i.e. the finger rotates around the corresponding knuckle), the situation become significantly worse. The result of such arrangement is recorded on Fig. 5. The number of malfunctions for every 100 iterations is really large. It is almost equal to 25% - see Fig. 5; for the best individual in last generation. Such an estimator is, indeed, absolutely unusable.

5 Conclusion

In the presented paper the evolutionary estimator which is capable to match two identical hand contours was proposed with use of the Iterative Closest Point algorithm and the Simple Genetic Algorithm. Based on extensive experimental results, it was ascertained that the proposed methodology is well usable for the images with small (640×480 pixels) and even large image resolution with a chromosome accuracy 1×10^{-3}. The disadvantage of the chromosome resolution 1×10^{-3} is a certain level of uncertainty in classification. This is especially where the contour heading is concerned. If higher chromosome accuracy is elected equal to 1×10^{-5}, the chromosome length is extended to 51 bits. The number of individuals and the number of generations have to be increased as well. The computational demands (time demands) will be larger too, but there is a guarantee the angle resolution will be equal to 0.0036°. Such accuracy is good enough. The SGA evolutionary estimator shows very captious behavior if an unsuitable set of working parameters is elected. This is the huge disadvantage of the SGA. The SGA also does not belong to the group of simplest optimizers from the code length and algorithmic complexity point of view.

Acknowledgment. The publication was supported by the funds of University of Pardubice, Czech Republic – Student grant competition project (SGS_2020_001).

Author would like to express cordial thanks to Mr. Paul Hooper for his careful English text correction, patience and stamina.

References

1. Besl, P.J., McKay, H.D.: A method for registration of 3-D shapes. IEEE Trans. Pattern Anal. Mach. Intell. **14**(2), 239–256 (1992)
2. Beyer, H.G., Schwefel, H.P.: Evolution strategies a comprehensive introduction. J. Nat. Comput. **1**(1), 3–52 (2002)
3. Botev, Z., Kroese, D., Liu, J., Nariai, S., Taimre, T.: The Cross-Entropy Toolbox. The University of Queensland, Australia (2004). http://www.maths.uq.edu.au/CEToolBox/. http://iew3.technion.ac.il/CE/
4. DeJong, K.: An analysis of the behavior of a class of genetic adaptive systems. Ph.D. thesis, University of Michigan (1975)
5. http://www.cs.cmu.edu/afs/cs/project/jair/pub/volume24/ortizboyer05a-html/node6.html
6. Dorigo, M.: Optimization, learning and natural algorithms. Ph.D. thesis, Politecnico di Milano, Italy (1992)
7. Fogel, L.J.: Artificial Intelligence Through Simulated Evolution, New York, USA (1966)
8. Fogel, L.J.: Intelligence Through Simulated Evolution: Forty Years of Evolutionary Programming. Wiley, Hoboken (1999)

9. Fogel, D.B., Chellapilla, K.: Revisiting evolutionary programming. Appl. Sci. Comput. Intell. **3390**, 2–11 (1998)
10. Glover, F.: Future paths for integer programming and links to artificial intelligence. Comput. Oper. Res. **13**, 533–549 (1986)
11. Glover, F., Kochenberger, G.A.: Handbook of Metaheuristics. Springer, Heidelberg (2003)
12. Goldberg, D.E.: Simple genetic algorithms and the minimal deceptive problem. In: Genetic Algorithms and Simulated Annealing, pp. 74–88. Pitman, London (1987)
13. Goldberg, D.E.: Genetic Algorithms in Search, Optimization, and Machine Learning. Addison-Wesley, Boston (1989)
14. Goldberg, D.E., Deb, K., Horn, J.: Massive multimodality, deception, and genetic algorithms. Technical report no. 92005, University of Illinois (1992)
15. Goldberg, D.E., Richardson, J.: Genetic algorithms with sharing for multimodal function optimization. In: 2nd ICGA Conference, pp. 41–49 (1987)
16. Holland, J.H.: Outline for a logical theory of adaptive systems. J. ACM **9**(3), 297–314 (1962)
17. Holland, J.H.: Adaptation in Natural and Artificial Systems. University of Michigan Press, Ann Arbor (1975)
18. Holland, J.H.: Adaptation in Natural and Artificial Systems. MIT Press, Cambridge (1992)
19. Horn, B.K.P.: Closest form solution of absolute orientation using unit quater nions. J. Opt. Soc. Am. **4**(4), 629–642 (1987)
20. Herrera, F., Lozano, M., Verdegay, J.L.: Tackling real-coded genetic algorithms: operators and tools for behavioural analysis. J. Artif. Intell. Rev. **12**, 265–319 (1998)
21. Chen, X.S., Ong, Y.S., Lim, M.H., Tan, K.C.: A multi-facet survey on memetic computation. IEEE Trans. Evol. Comput. **15**(5), 591–607 (2011)
22. Idoumghar, L., Melkemi, M., Schott, R.: A novel hybrid evolutionary algorithm for multi-modal fiction optimization and engineering applications. In: 13th IASTED International Conference on Artificial Intelligence and Soft Computing, pp. 87–93 (2009)
23. Jetenský, P., Marek, J., Rak, J.: Fingers segmentation and its approximation. In: 25th International Conference Radioelektronika, Radioelektronika 2015, pp. 431–434. IEEE (Institute of Electrical and Electronics Engineers), New York (2015). ISBN 978-1-4799-8117-5
24. Jetenský, P.: Human hand image analysis extracting finger coordinates using circular scanning. In: Proceedings of 25th International Conference Radioelektronika, Radioelektronika 2015, pp. 427–430. IEEE (Institute of Electrical and Electronics Engineers), New York (2015). ISBN 978-1-4799-8117-5
25. Jetenský, P.: Human hand image analysis extracting finger coordinates and axial vectors: finger axis detection using blob extraction and line fitting. In: 24th International Conference Radioelektronika, pp. 1–4. IEEE (Institute of Electrical and Electronics Engineers), New York (2014). ISBN 978-1-4799-3715-8
26. Kirkpatrick, S., Gelatt, C.D., Vecchi, M.P.: Optimization by simulated annealing. Science **220**(4598), 671–680 (1983)
27. Koza, J.: Genetic Programming: On the Programming of the Computers by Means of Natural Selection, 5th edn. MIT Press, Cambridge; London (1996)
28. Koza, J.R.: Genetic Programming – 2. MIT Press, Cambridge (1994)
29. Krause, E.F.: Taxicab Geometry: An Adventure in Non-Euclidean Geometry. Dover Publications, Mineola (1987)
30. Laarhoven, P.J.M.V., Aarts, E.H.L.: Simulated Annealing: Theory and applications. Reidel Publishing Company, Dordrech (1987)

31. Maier-Hein, L., Franz, A.M., Santos, T.R., Schmidt, M., Fangerau, M., Meinzer, H.P., Fitzpatrick, J.M.: Convergent iterative closest-point algorithm to accomodate anisotropic and inhomogenous localization error. IEEE Trans. Pattern Anal. Mach. Intell. **34**(8), 1520–1532 (2012)
32. Metropolis, N., Ulam, S.: The Monte Carlo method. J. Am. Stat. Aassoc. **44**, 335–341 (1949)
33. Molga, M., Smutnicki, C.: Test functions for optimization needs (2005). www.zsd.ict.pwr.wroc.pl/files/docs/functions.pdf
34. Pohlheim, H.: GEATbx.com, genetic and evolutionary algorithm toolbox for Matlab (2006). http://www.geatbx.com
35. Poli, R.: An analysis of publications on particle swarm optimisation applications. Department of Computer Science University of Essex, Technical report CSM-469 (2007). ISSN 1744-8050
36. Pottmann, H., Huang, Q.X., Yang, Y.L., Hu, S.M.: Geometry and convergence analysis of algorithms for registration of 3D shapes. Int. J. Comput. Vis. **67**(3), 277–296 (2006)
37. Rastrigin, L.A.: Statistical Search Methods. Nauka, Moscow (1968)
38. Rastrigin, L.A.: Extremal control systems. In: Theoretical Foundations of Engineering Cybernetics Series, Nauka, Russian (1974)
39. Rechenberg, I.: Evolutionsstrategies: Optimierung technischer Systeme nach Prinzipien der biologischen Evolution, Frommann-Holzboog, Stutgart, Germany (1973)
40. Rechenberg, I.: Evolutionsstrategie '94, Frommann-Holzboog, Stuttgar (1994)
41. Richards, M., Ventura, D.: Choosing a starting configuration for particle swarm optimization. In: IEEE International Conference on Neural Networks, vol. 3, pp. 2309–2312 (2004)
42. Rosenbrock, H.H.: An automatic method for finding the greatest or least value of a function. Comput. J. **3**, 175–184 (1960). http://web.ift.uib.no/~antonych/RGA4.html#dixon. http://www.applied-mathematics.net/optimization/rosenbrock.html
43. Schwefel, H.P.: Evolution and Optimum Seeking. Wiley, New York (1995)
44. Storn, R., Price, K.: Differential evolution—a simple and efficient heuristic for global optimization over continuous spaces. J. Glob. Optim. **11**, 341–359 (1997)
45. Wolpert, D.H., Macready, W.G.: No free-lunch theorems for search. Technical report 95-02-010, Santa Fe Institute (1995)
46. Zelinka, I., Onwubolu, G., Babu, V.: New Optimization Techniques in Engineering. Springer-Verlag, Department of Informatics and Artificial Intelligence, Faculty of Applied Informatics, Tomas Bata Univerzity in Zlin Czech Republic (2004)
47. Web1: Laboratory of Information Processing, Department of Information Technology, Lappeenranta University of Technology (LUT), P.O. Box 20, FI-53851 Lappeenranta, Finland. http://www.it.lut.fi/ip/evo/functions/functions.html
48. Web2. http://www.pg.gda.pl/~mkwies/dyd/geadocu/index.html
49. Web3. http://www.geatbx.com/docu/algindex-03.html. Accessed Mar 2020
50. Web4. http://www.ise.ufl.edu/pardalos/. (autor Dr. Panos M. Pardalos). Accessed Mar 2020
51. Web5. http://handwork.4fan.cz/. Accessed Mar 2020
52. Web6. http://robomap.4fan.cz/. Accessed Mar 2020

Bio-inspired Collaborative and Content Filtering Method for Online Recommendation Assistant Systems

Sergey Rodzin, Olga Rodzina, and Lada Rodzina(✉)

Southern Federal University, Nekrasovsky lane 44, 347928 Taganrog, Russia
srodzin@sfedu.ru, lada.rodzina@gmail.com

Abstract. The authors present a hybrid model of a recommender system. The system includes the characteristics of collaborative and content filtering. Also, the article describes a population filtering algorithm and the architecture of a recommendation system based on it. The results of experimental studies on an array of benchmarks and an estimation of filtering efficiency based on a hybrid model and a population algorithm are presented. The results are compared with the traditional method of collaborative filtering using the Pearson correlation coefficient.

Keywords: Recommendation system · Collaborative and content filtering · Population algorithm

1 Introduction

The amount of information, goods in online stores, books, films, videos, games, as well as the number of other users on social networks is so large that it is unrealistic for the user to consider all available offers. Due to this, we can meet more and more often with systems that analyze the user profile and try to predict what will be most interesting for them at a given time. For example, under each item in the online store there you can see such an option "similar products", or "maybe you want to purchase ", or "usually, buy with this product". Referral systems are widely used on e-commerce websites to offer customers items (goods, films, services, etc.). They help users to choose from a large number of offers. One of the easiest ways is to recommend items that are the most popular. However, such an approach does not take into account the fact that users may have different tastes. Recommender systems should solve the problems of targeted promotion of goods and services taking into account specific user preferences based on the logging of data on customer actions. This data can have a significant amount, changes rapidly, and updates over time. The task of adapting to a specific user is complicated by the inherent uncertainty in the choice of a specific Internet resource and the particularities of the Internet.

Traditionally, recommendation systems are divided into the following types [1]:

R. Silhavy (Ed.): CSOC 2020, AISC 1225, pp. 110–119, 2020.
https://doi.org/10.1007/978-3-030-51971-1_9

- collaborative filtering based on ratings from other users. This type is characterized by high accuracy. However, without knowing anything about the user's preferences, these recommendation systems have a high entry threshold;
- content filtering based on data collected about each item or service. This type of system can make recommendations to new users, even without the ratings of other users, but the accuracy of their forecasts is not high, and the development time increases;
- expert filtering based on knowledge about the subject area, and not on information about the subject. This type of system is distinguished by the high complexity of developing and collecting data;
- hybrid systems, the basis of which is the algorithmic composition of individual filtering strategies (weighted, a combination of features from various sources, cascading, meta-learning, etc.).

Collaborative filtering (imhonet.ru, *last.fm*) is based on the principle of searching for similar users and ascertaining their preferences or based on searching for goods similar to a given one and ascertaining their ratings previously made by the user. One of the promising methods in the field of collaborative filtering is the SVD-method based on singular matrix decompositions.

Content filtering (*Prismatic, Surfing bird*) is based on knowledge of any properties of objects, regardless of their evaluation by other users. At the same time, the user does not need to "train" the system for his/her preferences.

Expert filtering (large online stores) is based on manually created associative rules, created and edited by qualified experts.

The quality of recommendation systems is judged by accuracy and completeness [2]. Accuracy is the percentage of items or services that are preferred by the user among all suggested recommendations. Completeness is the proportion of items or services recommended by the user, among all preferred for him.

We have made a hypothesis that a combination of collaborative and content filtering results potentially improves recommendations accuracy. A hybrid approach is useful if collaborative filtering begins when data sparseness is significant. In this case, the hybrid approach allows you to first weigh the results according to content filtering and then shift these weights towards collaborative filtering as data about a particular client [3].

2 Materials and Methods

2.1 Problem Statement

Formally, the task of finding the recommended object can be represented as follows:

$$\forall u \in U, s'_u = \arg\max_{s \in S} h(u, s),$$

where U is a set of users, S is a set of objects that can be recommended for a user, h is a function that determines how much some object s is satisfied for some user u. Thus, it is necessary to select such an object $s' \in S$, at which satisfaction value for each user $u \in U$ is max.

The main issues of recommendation systems in solving the defined task are the following:

- sparse data. The recommendation system operates with a large amount of data, while users do not give ratings to most goods or services. As a result, the "product-user" matrix is very large and sparse, which makes it difficult to calculate recommendations and strengthens the problem of a "cold" start;
- "cold" start. New products and services, new users present a big problem for recommendation systems due to the lack of data on users or products recently appearing in the system;
- scalability. With the increase in the number of users in the system, the scalability problem appears. The collaborative filtering algorithm with complexity as $O(M * N)$, where M is a number of clients, N is a range of goods and services, where M = 107 and N = 106 are already too complicated for calculations. This requires greater scalability of the filtering algorithms since the recommender system must instantly respond to online requests from all customers, regardless of their purchase history and ratings;
- synonymy. Similar or identical products have different names, and the recommender system is not able to detect these hidden connections and assigns these products to different objects;
- fraud. Unscrupulous suppliers are trying to fraudulently raise the rating of their goods or services and lower the rating of their competitors;
- diversity. Algorithms based on sales and ratings create difficult conditions for the promotion of new and little-known products since they are replaced by popular goods or services that have long been on the market;
- "black sheep". These are users whose opinions do not always coincide with the opinions of most others. Due to their unique taste, it is impossible to recommend something related for them.

Almost all modern Internet resources, including search engines, online stores, forums, and social networks focused on working with a large number of users, collect and log information about their activity. As well, they process it to personalize their content for each specific client. Information and data on user activity are characterized by a large volume and heterogeneity. The main goals of processing user activity data are clustering users and resources, building behavioral and socio-demographic profiles of users, segmenting the customer base, providing recommendations, the most interesting information and marketing offers.

2.2 Hybrid Request Model

In recommendation systems [4, 5] we can see hybrid models based on collaborative and content filtering. These models can be classified as follows:

1. Building a unified model using the characteristics of collaborative and content filtering models.

2. Implementation of some characteristics of the content model in systems based on collaborative filtering.
3. Implementation of some characteristics of the collaborative filtering model in systems based on the content model.
4. Implementing collaborative and content filtering models separately and using combinations of recommendations received.

The basic principle of building a hybrid recommendation model is to use the filtering characteristics of content and collaborative filtering in one recommendation system [6].

The introduction of some characteristics of the content model into collaborative filtering systems implies that such hybrid recommender systems are not only built on a collaborative component but also include some content filtering data in the user profile. These data serve as the basis for calculating the similarity of user preferences instead of the overall evaluated objects [7, 8]. Not only objects with good reviews from other users are recommended for the user, but also those objects that the user may like based on his personal preferences.

The introduction of some characteristics of the collaborative filtering model into systems based on the content model implies a decrease in the dimension of profiles using content filtering.

Separate implementation of recommendation systems involves either a combination of ratings from two systems using a linear combination of ratings or the use of a system that should be better suited in a particular case.

The proposed hybrid filtration model is based on the application of the population algorithm as a machine learning method to solve the problem [9, 10]. A variety of input data in the hybrid model is supported by a population of custom "characteristics" encoded in a population algorithm.

Using a population of encoded solutions and special operators, the algorithm searches for a solution with the highest value of the objective function. The weighted values of the objects remain relevant, "noise" is eliminated. A method is proposed for solving the problems of sparse data and a cold start by encoding the "preference" additional component in solutions.

2.3 Population Algorithm for Collaborative Filtering

A population algorithm explores the search space, synthesizes the solutions that are points of this space, requests an assessment of their quality or "fitness", which is then used to make "natural selection". In this way, they learn which areas of the search space contain the best solutions. Imagine the general scheme of a population filtering algorithm.

Let us suppose that in a finite state space S we have some function $f(x)$, $x \in S$. We have to find $max\ \{f(x);\ x \in S\}$. As well, suppose that x^* is a state with the maximum of the function $f_{max} = f(x^*)$.

The population algorithm for solving the problem includes the following steps.

Step 1. *Initialization* of a population (randomly or heuristically) $\xi_0 = (x_1, \ldots, x_{2n})$ from $2n$ individuals (n is an integer). Assign $k = 0$. For each population ξ_k defines

$$\xi_k = max\{f(x_i) : x_i \in \xi_k\}.$$

Step 2. *Generation* of a population $\xi_{k+1/2}$ using special operators.
Step 3. *Selection* and *reproduction* 2*n* individuals from population $\xi_{k+1/2}$ and ξ_k and getting a new population ξ_{k+S}.
Step 4. If $f(\xi_{k+S}) = f_{max}$, then *stop*, otherwise $\xi_{k+1} = \xi_{k+S}$, $k = k+1$ and move to step 2.

Suppose x^* is an optimum point. Define $d(x, x^*)$ as distance between points x and x^*. If we have a set of optimums S^*, then $d(x, S^*) = min\{d(x, x^*): x^* \in S^*)\}$ is a distance between point x and the set S^*. Such distance let us define as $d(x)$. Then $d(x^*) = 0$, $d(x) > 0$ for each $x \notin S^*$.

Considering that populations $X = min\{x_1,\ldots, x_{2n}\}$, suppose that

$$d(X) = min\{d(x) : x \in X\}.$$

This formula measures the distance between a population and the optimal solution.

Sequence $\{d(\xi_k);\ k = 0,\ 1,\ 2,\ \ldots\}$, generated by the population algorithm is a random sequence that is modeled by a homogeneous Markov chain.

Then the drift of a random sequence at time k is defined as

$$\Delta(d(\xi_k)) = d(\xi_{k+1}) - d(\xi_k).$$

The algorithm shutdown time is estimated as

$$\tau = min\{k : d(\xi_k) = 0\}.$$

The task is to study the relationship between time τ and the dimension of the problem n. At what drift values $\Delta(d(\xi_k))$ we can evaluate the expected value $E[\tau]$? The key issue here is the evaluation of the ratio d and Δ.

The population algorithm can solve the problem in a polynomial average time under the following drift conditions:

- if there is a polynomial $h_0(n) > 0$ (n is dimension of the problem) such as $d(X) \leq h_0(n)$ for any defined population X, in other words, the distance from any population to the optimal solution is a polynomial function of the dimension of the problem;
- in any moment $k \geq 0$, if $d(\xi_k) > 0$, then the polynomial $h_1(n) > 0$ is exist, such as

$$E[d(\xi_k) - d(\xi_{k+1}) | d(\xi_k) > 0] \geq 1/h_1(n),$$

in other words, random sequence drift $\{d(\xi_k);\ k = 0,\ 1,\ 2,\ \ldots\}$ is always positive towards the optimal solution and bounded by the inverse polynomial.

In other words, the estimate of the drift value is converted into an estimate of the running time of the algorithm, and the local property (drift in one step) is converted to a global property (the running time of the algorithm until the optimum is found)! It's easier to evaluate the drift - using the drift analysis, the conditions are defined, the fulfillment of which guarantees the solution of the filtering problem in polynomial time.

Note that this result was obtained under the assumption that the number of generations (which is equivalent to the number of calculations of the objective function) is the most important factor in estimating the computation time of the algorithm. This is probably true for most applications of population-based algorithms because the estimation of the objective function there is the most time-consuming part of the algorithm, in contrast to the complexity of executing the operators of population-based algorithms, the complexity of which is estimated as $O(n) - O(n\ ln(n))$ [9].

How are solutions have encoded? Let us consider the structure of the coding solutions using the example of a recommendation system for watching movies. It has the form presented in Table 1.

Table 1. Coding solutions for a movie recommendation system

Rate	Year	Gender	Profession	9 parameters describing user profile			18 genres			9 additional movies' characteristics		
				genre preferences	…	language	thriller	…	western	director	…	movie language
w_1	w_2	w_3	w_4	w_5	…	w_{13}	w_{14}	…	w_{31}	w_{32}	…	w_{40}

2.4 Structure of Coding Solutions on the Example of a Recommendation System for Watching Films

The structure of coding solutions on the example of a recommendation system for watching films is presented in Table 1.

The structure includes 40 elements: w_1 – rate, w_2 – launch year, w_3 – user's gender, w_4 – profession, w_5 – w_{13} – nine genes that describe a user profile (from genre preferences to film language), w_{14} – w_{31} – 18 movie genres, w_{32} – w_{40} – 9 additional features. The more "weight" w_i, the more important this parameter is for evaluating user preferences.

This structure of decision coding in a population algorithm using the example of a recommendation system for watching films covers the entire range of input data.

2.5 Population Algorithm Architecture

For the experiments, a prototype recommender system was built based on the population algorithm. The system implements the collection, analysis and use of ratings of some users to predict interest in films of other users. The prototype architecture of the recommender system is shown in Fig. 1.

The recommendation system process is divided into three stages.

At the first stage, the user is registered, demographic data is collected, as well as information on the rating of films that he has already watched or bought.

In the second stage, the user profile is personalized. For those films that the user has already rated, a search is made from the database (DB) of the corresponding genres and characters of films that will be taken into account as preferences, along with ratings of other users. When recommending a film, the preferred types of elements are film genres, preferred directors, actresses and actors, producers, screenwriters, and language. This is especially effective when user ratings are not yet formed.

In the third stage, a population algorithm is used to search for the appropriate recommendation.

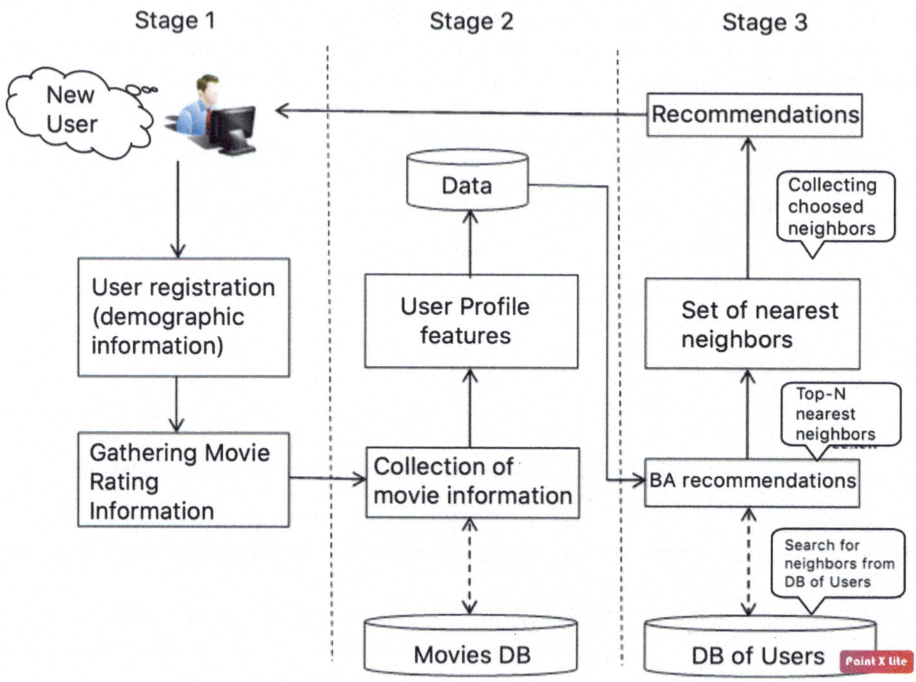

Fig. 1. Recommended system prototype architecture.

3 Results

For the experiments, a data set *MovieLens* 1*M* [11] was selected. This set contains data on 1,000,000 ratings 6,040 users 3952 films. As contextual descriptions of objects dataset were taken the dataset *TagGenome*. Estimation of filtration efficiency based on a hybrid model and population algorithm was carried out by comparison with the traditional method of collaborative filtering using the Pearson correlation coefficient [12, 13]:

$$corr(user1, user2) = \frac{\sum_{i=0}^{n}\left(user1_i - \overline{user1}\right)\left(user2_i - \overline{user2}\right)}{\sqrt{\sum_{i=0}^{n}\left(user1_i - \overline{user1}\right)^2}\sqrt{\sum_{i=0}^{n}\left(user2_i - \overline{user2}\right)^2}}$$

where *user1*, *user2* are users and their rates; *n* is a number of object, $\overline{user1}$ and $\overline{user2}$ – is average user rating. Correlation coefficient *corr* takes values from –1 (absolute mismatch) to 1 (absolute match).

The experiments were repeated using various sets of weight parameters to show how they influence the outcome of the recommendations. In a sample of 10 users, the average value of the weight parameters was about 0.2649. Moreover, the weights describing the user profile ($w_5 - w_{13}$) turned out to be more significant than many characteristics of the film.

To compare the population algorithm with the collaborative filtering algorithm using the Pearson correlation coefficient, a set of 5 decision coding structures with a different number of parameters was configured for 100 random users (8, 16, 24, 32, and 40). In Table 2 the composition of the parameters is given (√) for the Pearson algorithm and the population algorithm with a different number of parameters (BA(8) – BA(40)).

Table 2. The composition of the filtering parameters for the Pearson algorithm and the bio-inspired algorithm

	Parameters				
	User rating	Year, gender, profession	9 users profile's options	18 movie genres	9 additional movie features
Pearson algorithm	√	√		√	√
BA (8)	√		√		
BA (16)	√	√	√		
BA (24)	√			√	√
BA (32)	√	√		√	√
BA (40)	√	√	√	√	√

The prediction accuracy of the population algorithm for BA(8) – BA(40) turned out to be higher than the Pearson algorithm. In particular, the average prediction accuracy by the Pearson algorithm is about 70.1%, while the average accuracy of the population algorithm with different compositions of parameters ranged from 80% to 82%.

A comparison of the processing time of the population algorithm with a different composition of parameters showed that BA(8) exceeds BA(16), BA(24), BA(32) и BA(40). Its average operating time is about 20 s. The more parameters the algorithm includes, the longer the processing time.

4 Conclusion

Recommendation systems are a definite alternative to search algorithms because they allow you to detect objects that cannot be found last. Such systems try to predict which objects (movies, music, books, news, websites) will be of interest to the user, based on the content information about his/her profile and also using the preferences of other users to predict unknown preferences for this user.

The article proposes an approach based on a hybrid model that combines the results of collaborative and content filtering, which allows increasing the accuracy of recommendations to solve the problem of the sparseness of data and "cold" start. The model is based on the use of a population algorithm as a heuristic method for finding a solution. The drift analysis of the population algorithm showed that it solves the problem in polynomial time.

The experiments show that the developed population algorithm is superior in accuracy to the competing collaborative filtering algorithm based on the Pearson correlation coefficient.

The presented hybrid filtration model is static. In reality, the perception and popularity of certain products change over time, as do the tastes of users [14]. The recommendation system should take into account this temporal dynamic. The population approach, in our opinion, allows taking into account the temporal drift, which can improve the accuracy of recommendations.

Acknowledgments. The reported study was funded by RFBR according to the research project № 18-29-22019 and project № 19-07-00570.

References

1. Su, X., Khoshgoftaar, T.M.: A survey of collaborative filtering techniques a survey of collaborative filtering techniques. In: Advances in Artificial Intelligence Archive, pp. 1–19. Hindawi Publishing Corporation (2009). https://doi.org/10.1155/2009/421425
2. Shani, G., Gunawardana, A.: Evaluating recommendation systems. In: Recommender Systems Handbook, pp. 257–297. Springer, Heidelberg (2011). https://doi.org/10.1007/978-0-387-85820-3_8
3. Burke, R.: Hybrid Web recommender system. In: The Adaptive Web, pp. 377–408 (2007). https://doi.org/10.1007/978-3-540-72079-9_12
4. Lops, P., Gemmis, M., Semeraro, G.: Content-based recommender systems: state of the art and trends. In: Recommender Systems Handbook, pp. 73–105 (2011). https://doi.org/10.1007/978-0-387-85820-3_3
5. Shani, G., Gunawardana, A.: Evaluating recommendation systems. In: Recommender Systems Handbook, pp. 257–297 (2011). https://doi.org/10.1007/978-0-387-85820-3_8
6. Rodzin, S., Rodzina, O.: New computational models for big data and optimization. In: Proceedings of the 9th IEEE International Conference Application of Information and Communication Technologies (AICT), pp. 3–7 (2015). https://doi.org/10.1109/icaict.2015.7338504
7. Bova, V., Kravchenko, Y., Rodzin, S., Kuliev, E.: Hybrid method for prediction of users' information behavior in the internet based on bioinspired search. In: Journal of Physics: Conference Series, ITBI 2019, vol. 1333, p. 032008, pp. 1–7. IOP Publishing (2019). https://doi.org/10.1088/1742-6596/1333/3/032008
8. Kravchenko, Y., Kursitys, I., Bova, V.: The development of genetic algorithm for semantic similarity estimation in terms of knowledge management problems. In: Advances in Intelligent Systems and Computing, vol. 573, pp. 84–93 (2017). https://doi.org/10.1007/978-3-319-57261-1_9

9. El-Khatib, S., Rodzin, S.I., Skobtcov, Y.A.: Investigation of optimal heuristical parameters for mixed Aco-k-means segmentation algorithm for MRI images. In: Proceedings of the Conference on Information Technologies in Science, Management, Social Sphere and Medicine (ITSMSSM), pp. 216–221 (2016). https://doi.org/10.2991/itsmssm-16.2016.72
10. Rodzin, S., Rodzina, O.: Metaheuristics memes and biogeography for trans computational combinatorial optimization problems. In: Proceedings of the 6th International Conference - Cloud System and Big Data Engineering, pp. 1–5 (2016). https://doi.org/10.1109/confluence.2016.7508037
11. Harper, F.M., Konstan, J.A.: The movieLens datasets: history and context. ACM Trans. Interact. Intell. Syst. (TiiS) **5**(4), 19 (2016)
12. Koren, Y., Bell, R., Volinsky, C.: Matrix factorization techniques for recommender systems. IEEE Comput. **42**(8), 42–49 (2009). https://doi.org/10.1109/MC.2009.263
13. Hernando, A., Bobadilla, J., Ortega, F.: A non-negative matrix factorization for collaborative filtering recommender systems based on a Bayesian probabilistic model. In: Knowledge-Based Systems, vol. 97, pp. 188–202 (2016). https://doi.org/10.1007/978-3-030-26773-5_14
14. Boucher-Ryan, P., Bridge, D.: Collaborative recommending using formal concept analysis. Knowl.-Based Syst. **19**(5), 309–315 (2006). https://doi.org/10.1007/978-1-84628-226-3_16

Computer-Based Support for Searching Rational Strategies for Investors in Case of Insufficient Information on the Condition of the Counterparty

V. A. Lakhno[1], V. G. Malikov[1], D. Y. Kasatkin[1(✉)], A. I. Blozva[1], V. G. Saiko[2], and V. N. Domrachev[2]

[1] Department of Computer Systems and Networks, The National University of Life and Environmental Sciences of Ukraine, Kiev 03041, Ukraine
{valss21,dm_kasat}@ukr.net
[2] Department of Applied Information Systems, Taras Shevchenko National University of Kyiv, Kiev 01033, Ukraine

Abstract. The paper considers the model of the procedure for mutual investment into information technology and systems, in the case when the financial resources of the counterparty of the investment belong to a fuzzy set. The model is intended for the decision support system that is being developed for investors, in cases that require a choosing of rational strategies. As an apparatus for solving the problem, we use the tools of the theory of multi-step games with several terminal surfaces. Such an approach, in our opinion, allows us to find solutions to the problem under consideration. A feature of the approach proposed in the paper, and accordingly the model, was the assumption that the one side or a player-investor, does not have complete information about the actions of the opposite side. Such incomplete information may include, for example, the lack of data on the financial strategies of the investment counterparty, on the status of its financial resources, etc. It is shown that the proposed model can become the basis for the corresponding algorithmic and software implementation during the development of an intelligent decision support system for choosing rational investor strategies. During computational experiments, we have evaluated the possibility of applying the model to such complex investment objects as information technology and systems. Investing in the latter, which include, for example, 5G technology; the Internet of things; SMARTCITY artificial intelligence; cybersecurity intelligent systems and others are associated with high risk and a high degree of uncertainty.

Keywords: Decision support · Investing · Game theory · Fuzzy sets · Rational strategy

1 Introduction

At present, the solution of the problem of evaluating high-risk operations associated with investing in new promising information technologies and systems (hereinafter ITS) is impossible without the use of computer-based support in the decision-making

R. Silhavy (Ed.): CSOC 2020, AISC 1225, pp. 120–130, 2020.
https://doi.org/10.1007/978-3-030-51971-1_10

process. This applies, for example, to tasks related to the selection of optimal or rational investment strategies in such promising areas as: 5G development; Internet of Things (IoT); SMARTCITY artificial intelligence; cybersecurity intelligent systems, etc. However, the owners and managers of many corporations and firms [1, 2] are not yet fully aware of the need to solve the existing permanent problem of mutual investment in the newest ITSs. And, as a result, in the context of the growing need for the introduction of advanced ITS in various areas of human activity, the lack of a clear understanding by investors of the mechanisms for choosing rational investment strategies and the insufficient information about the status and actions of the counterparty, lead to situations in which there is a high degree of risk associated with possible disruptions in the implementation of promising high-tech ITSs in everyday use.

The main problems [3, 4] that companies and organizations are facing in assessing and choosing rational investment strategies for advanced ITSs are: the lack of a methodology for determining the exact values for assessing the risks associated with investing in the latest ITS; the complexity of the proposed models and methods, in particular when it comes to their algorithmization and subsequent implementation, (for example, in decision support systems (DSS) for choosing mutual investment strategies in ITS); the lack of a methodology and evaluation of counterparty strategies for mutual investment, (for example, when its financial resources are unlimited or its information is inaccurate, or, for example, described in the categories of fuzzy sets, etc.).

It should be noted that the concept of fuzzy sets provides a convenient starting point for constructing conceptual foundations, which are in-line in many respects to the fundamentals that are used in the case of ordinary sets. But at the same time they are more general than the latter, and potentially have a wider scope. In particular, this applies to areas such as: classification of images; data processing; game theory, etc. It is essential that such fundamentals provide a natural way to solve problems in which the source of inaccuracy is rather the absence of clearly defined criteria of belonging to the class, than the presence of random variables. Therefore, this article attempts to solve problems of a conflicting nature and in which the incompleteness of information is not stochastic, but has the nature of fuzzy information specified by fuzzy sets [5, 6].

All this together makes it relevant to continue research towards the development of new methods and models for choosing rational investment strategies in the face of fuzzy information, in particular for situations where additional requests for the development of advanced ITS arise, which causes a change in the level of risks for investors, thus requires them to review their own mutual investment strategies.

2 The Purpose of the Article

The purpose of the article is to develop a model for decision support systems for choosing rational strategies for mutual investment in the latest information technologies and systems used in various fields of human activity.

3 Literature Review

Currently, the growth of papers in the segment of the development of new methods and models to support decision-making at the choice of investment strategies in ITS is actively continuing. As shown in [1, 3] - making decisions on investing in ITS is a constant task. However, a drawback of many works, and in particular [5–7], is the lack of real recommendations on the development of mutual invest strategies in ITS.

Studies on the implementation of technologies of expert (ES) and decision support systems (DSS) in the processes of choosing strategies for investing in ITS have become a new direction [8–11]. However, one can note a drawback of these works is the lack of unambiguous modeling results when choosing a rational investor strategy. Most of the considered models [12–14] do not allow finding effective recommendations and strategies for mutual investment of the parties.

In [12, 13], the relevance of solving the problem of integration into the selection process of investment strategies of modern DSS and ES is analyzed. The authors have associated this with large volumes of initial data and iterative approaches when choosing investment strategies.

In [14, 15], methodologies for creating various modules for DSS in the field of investment were described in sufficient detail. One of the drawbacks of these works is the lack of models for choosing investment strategies in situations where one of the parties does not have complete information about the financial resources of the investment counterparty. It was shown in work [16] that this is an important characteristic in such an interaction.

The models proposed in [17, 18] also do not allow assessing the risk of loss of financial resources by investors. In [16, 18], models based on game theory were proposed for evaluating the performance of investment activities of the parties of the investment process. The authors, however, did not take into account many factors, for example, the change in the financial components of the counterparty investor. This drawback in previous studies of various authors can be eliminated by applying the methods of the theory of differential and multi-step quality games with several terminal surfaces [19]. This will increase the efficiency of forecast calculations by investors in assessing the risks of possible financial losses.

Thus, as the analysis of the performed studies has shown, the problem of the further development of models for DSS in the scope of mutual investment of ITS remains relevant. In particular, one should consider situations with incomplete information on the financial condition of the second investor in the process of finding sets of preference and optimal investment strategies.

4 Models and Methods

The problem considered in the article relates to the direction of research previously presented in works [16, 19]. In these publications, the authors used the apparatus of the theory of differential and multi-step games. In accordance with the approach used in these works, two sides are considered: player #1 - investor #1 (INV_1), player #2 - investor #2 (INV_2). Players use financial resources to achieve their goals.

We believe that for a given period of time $t = \{0, 1, \ldots, T\}$ (T is a natural number), players 1 and 2 are allocated, respectively, $h_1(0)$ and $h_2^{\xi}(0)$ financial resources for certain investment projects in the field of ITS. For example, it can be Smart City IT, information security, telecommunications, etc. Players interact. Interaction is described as a bilinear multi-step game. Moreover, the game is considered with alternate moves with incomplete information players have about each other. That is, unlike games with full information, the first player (a group of players whom we conditionally combine according to interests in a particular investment project) does not exactly know the initial state of the second player (second group). Player #1 knows that his condition belongs to the fuzzy set $\{X, m(.)\}$, where X subset $R+$ $m(.)$ – is the membership function of the state of $h_2^{\xi}(0)$ to the set of X, $m(h_2^{\xi}(0)) \in [0, 1]$ for $h_2^{\xi}(0) \in X$. Let us note, that in the context of the problem being solved, to fuzzy set $\{X, m(.)\}$ the following data can be attributed: financial resources of players, growth rates of resources, elasticity coefficients of investments of players, priority for the development of any particular information technology and system, which, for example, are determined by experts, and any other characteristics that describe the issue in question. The moves in such a game take place alternately. At even times, the first player makes a move, at odd times - the second.

Let $t = 2n, h_1(t), h_1(t+1)$ – be the states of the first player at time instants $t, t+1$. Also $h_2^{\xi}(t), h_2^{\xi}(t+1)$ – the state of the second player at time instants $t, t+1$. Then at the time moment $t+1, t+2$, the states of the players are determined from the relations:

$$h_1(t+1) = g_1(t) \cdot h_1(t) - u(t) \cdot g_1(t) \cdot h_1(t); \tag{1}$$

$$h_2^{\xi}(t+1) = h_2^{\xi}(t) - s_1 \cdot u(t) \cdot g_1(t) \cdot h_1(t); \tag{2}$$

$$h_2^{\xi}(t+2) = g_2(t+1) \cdot h_2^{\xi}(t+1) - v(t+1) \cdot g_2(t+1) \cdot h_2^{\xi}(t+1); \tag{3}$$

$$h_1(t+2) = h_1(t+1) - s_2(t+1) \cdot v(t+1) \cdot g_2(t+1) \cdot h_2^{\xi}(t+1); \tag{4}$$

Here $u(t), v(t) \in [0, 1], s_1 > 0, s_2 > 0$.

At the moment of time $t \in \{0, 2, \ldots 2 \cdot n\}$ the first player multiplies the value $h_1(t)$ by a coefficient (rate of change, growth rate) $g_1(t) \cdot h_1(t)$, which is allocated for investment in ITS at time t. And then the states of the players at a time $t+1$ are determined from relations (1) and (2). Due to the elasticity of mutual investments, the second investor will additionally allocate an amount $s_1 \cdot u(t) \cdot g_1(t) \cdot h_1(t)$ for investing in ITS. The elasticity of the investments of the second investor in relation to the investments of the first investor is characterized by a coefficient s_1.

If it turns out that the following condition is fulfilled: $m\{h_2^{\xi}(t+1) : (h_2^{\xi}(t+1) < 0))\geq p_0, (0 \leq p_0 \leq 1)$, then we will say that INV_1 invested in the ITS of its interest in such a way, that INV_2 has exhausted its investment resources with certainty p_0. Let us assume that on the part of the first player, the procedure for investing in ITS is completed, otherwise INV_1 continues the investment process.

Then INV_2 makes its move. It acts just like INV_1. And then the states of the players are determined from relations (3) and (4). If the following condition is met: $m\{h_1(t+2):(h_1(t+2)<0))<p_1,(0\leq p_1\leq 1)$, then we will say that INV_2, invested in ITS in such a way that INV_1 has exhausted its investment resources with certainty $1-p_1$. Let us assume that on the part of the second player, the procedure for investing in ITS is completed.

The first player seeks to find many of their initial states that satisfies the following requirement.

Requirement: if the game starts with such initial states, the first player can by choosing his control actions, $u(0),\ldots u(t)(t=2\cdot n)$ ensure that INV_2 has exhausted his investment resources with certainty p_0. Herewith, the strategy chosen by player 1 contributes to preventing INV_1 from exhausting its investment resources with certainty $1-p_1$.

The set of such states will be called the preference set of the first player.

Let us define a function $F(.):X\rightarrow R_+,F(x)=\{\sup m(y), for\, y\leq x\}$. Denote by Φ – the set of such functions, by $2\cdot n$ – the nearest to a T natural even number, $T^*=\{0,2,\ldots,2\cdot n\}$ – the set of natural even numbers.

Definition. A pure strategy $u(.,.,.)$ of the first player is a function $u(.,.,.):T^*\times R_+\times\Phi\rightarrow[0,1]$, such that $u(t,h_1,F)\in[0,1],(F\in\Phi)$.

So the strategy of the first player is a function that allows the first player to determine, based on the information available, the amount of financial resources that player No. 1 allocates for ITS investment.

Player number 2 chooses his strategy $v(.)$ based on any information.

The goal of the first player is to find a multitude of preferences. Also, the goal of INV_1 is to find strategies, applying which he will receive the fulfillment of conditions that allow him to complete the investment procedure. Such strategies of the first player will be optimal strategies. The formulated game model corresponds, according to the classification of decision theory to the task of making a decision under conditions of uncertainty. Note that such a model is a bilinear multi-step quality game with several terminal surfaces with alternate moves. Finding the preference sets of the first player and his optimal strategies depends on many parameters.

To describe the preference sets of the first player, two values must be entered:

$$\varphi(0)=\inf\{\varphi'\},\psi(0)=\inf\{\psi'\},F(\varphi')\geq p_0,F(\psi')\geq p_1. \tag{5}$$

The preference sets of the first player and his optimal strategies are found for $T=1,3,\ldots$ Let us introduce the notation for the preference sets: $Q_1^T(p_0,p_1)$ – the preferences set of the first player of which he in T turns successfully completes the investment procedure.

When $T=1$ we have the $Q_1^1(p_0)=\{h_1(0):s_1\cdot g_1\cdot h_1(0)\geq\varphi(0)\}$.

Optimal strategy

$$u_*(1,h_1,\varphi)=\{1, for\, s_1\cdot g_1\cdot h_1\geq\varphi;0, otherwise\} \tag{6}$$

Let us write down the sets of optimality and optimal strategies for various cases of the correlation of game parameters.

Case 1. $p_0 = p_1$.

1.1. $g_1 > g_2$.

Let $k_0 \in N$ (the set of natural numbers): $s_1 \cdot g_1 \cdot s_2 \leq (g_1/g_2)^{k_0}, s_1 \cdot g_1 \cdot s_2 > (g_1/g_2)^{k_0+1}$

Then

$$Q_1^T(p_0, p_0) = \{h_1(0) : \varphi(0) \leq s_1 \cdot g_1 \cdot (g_1/g_2)^k \cdot h_1(0);\\ \varphi(0) \succ s_1 \cdot g_1(g_1/g_2)^{k-1} h_1(0)\},$$

Where $T = 2 \cdot k + 1 \leq 2 \cdot k_0 + 3$.

Multitude

$$Q_1^{2k_0+5}(p_0, p_0) = \{h_1(0) : \varphi(0) \succ s_1 \cdot g_1 \cdot (g_1/g_2)^{k_0+1} h_1(0);\\ \varphi(0) \leq (g_1/(s_2 \cdot g_2)) \cdot h_1(0)\}, Q_1^T(p_0, p_0) = \emptyset.$$

For $T = 2 \cdot k + 1 \geq 2 \cdot k_0 + 7$.

Optimal Strategy

$$u_*(n, h_1, \varphi) = \{1, for\, s_1 \cdot g_1 \cdot h_1 \geq \varphi; 0, otherwise\} \tag{7}$$

A ray $\{h_1(0) : h_1(0) \in R_+, \varphi(0) = (g_1/(s_2 \cdot g_2)) \cdot h_1(0)\}$-is a barrier, in accordance with [20]. This means that from the states $\{h_1(0) : \varphi(0) \succ (g_1/(s_2 \cdot g_2)) \cdot h_1(0)\}$, the first player cannot achieve the goal with certainty $p \geq p_0$. This ray can be called a balance ray for the mutual investment procedure.

1.2. $g_1 \leq g_2$.

1.2.1. $s_1 \cdot g_1 \cdot s_2 \leq 1$.

In this case, we receive $Q_1^T(p_0, p_0) = \emptyset$.

For $T = 2 \cdot k + 1 \geq 3$.

1.2.2. $s_1 \cdot g_1 \cdot s_2 \succ 1$.

1.2.2.1. $s_1 \cdot g_2 \cdot s_2 \succ 1$.

In this case, receive $Q_1^T(p_0, p_0) = \emptyset$ for $T = 2 \cdot k + 1 \geq 3$.

1.2.2.2. $s_1 \cdot g_2 \cdot s_2 \leq 1$.

In this case, there will be

$$Q_1^3(p_0, p_0) = \{h_1(0) : \varphi(0) \leq (g_1/(s_2 \cdot g_2)) \cdot h_1(0), \varphi(0) \succ s_1 \cdot g_1 \cdot h_1(0)\}.$$

Optimal strategy:

$$u_*(n, h_1, \varphi) = \{1, for\, s_1 \cdot g_1 \cdot h_1 \geq \varphi; 0, otherwise\} \tag{8}$$

$Q_1^T(p_0, p_0) = \emptyset$ for $T = 2 \cdot k + 1 \geq 5$.

Case 2. $p_0 > p_1$. In this case $Q_1^T(p_0, p_1) = Q_1^T(p_0, p_0)$.

Case 3. $p_0 < p_1$. In this case, we receive the following result:

$Q_1^T(p_0, p_1) = Q_1^T(p_0, p_0) \cap \{h_1(0) : \varphi(0) \leq (g_1/(s_2 \cdot g_2)) \cdot h_1(0)\}$. Model proposed in the paper was tested in the MathCad environment, as well as in Qt.

5 Models and Methods

The main goals of computational experiments based on the proposed model are: 1) to verify the model's performance in comparison with models of the other authors who are engaged in research in the field of computer modeling of decision support processes for investing in ITS, associated with a high degree of uncertainty and risk of loss financial resources; 2) finding the sets of strategies of players 1 and 2 and assessing the reliability of events that are associated with the loss by players of their financial resources when investing in ITS. In accordance with the plan and methodology for conducting computational experiments, any values in the range from 0 to 100 conventional units (that is, investing players can operate with hundreds of thousands or millions in specific funds) of the financial resources of the players were considered. The interaction time of the players is not limited.

Figure 1 illustrates the results of computational experiments. In Fig. 1 such designations are accepted:

a) round markers (3 – Fig. 1) will show rays of balance;
b) preference zones will be under the rays of balance. Moreover, the zone of preference of the first investor is below the ray of balance, and the zone of preference of the second investor is above the ray of balance;
c) a dotted blue line (2 – Fig. 1) with triangular markers without shading indicates the path of the first investor. The dotted line with solid filled green triangular markers (1 – Fig. 1) indicates the trajectory of the second investor.

The restrictions that are imposed on the financial resources of investors are indicated by solid lines with square markers. Square markers without shading correspond to the restrictions of the first investor, respectively, for the second investor - with a solid fill. We should also note that with all the ratios of the parameters of the game, its solution is given. It allows you to find the optimal behavior of the first investor in case of incomplete information about the state of the financial resource of the second investor. It is assumed that they know only a fuzzy set of their financial resources.

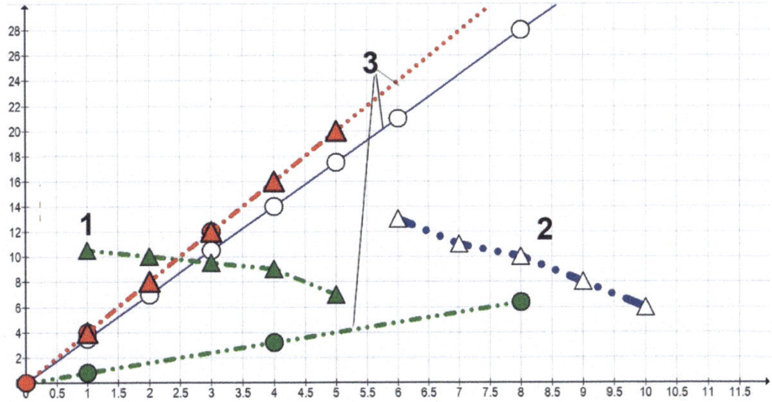

Fig. 1. Investors paths consistent with their rational investment strategies: 1 – The trajectory of the first investor; 2 – The trajectory of the second investor; 3 – Rays of balance of the investors.

6 Discussion of Simulation

The Results. Experimental Results for the First Investor

The ray of balance and the trajectory of the first investor are shown in blue. The figure illustrates an experiment in a positive orthant on a plane $(h_1(0), \varphi(0))$.

The set of rays emanating from the point (0,0) is considered. They are defined by the ratio: $\varphi = (3.5 - 1/n) \cdot h_1$. These rays define the preference sets of the first investor in n steps with confidence p_0. It is assumed that $p_0 = p_1$.

The set $Q_1^T(p_0, p_0)$ is defined as follows:

$$\{(h_1(0), \varphi(0)) : (h_1(0), \varphi(0)) \in R_+^2, (3.5 - 1/(n-1) \cdot h_1(0) \leq \varphi(0) \prec (3.5 - 1/n) \cdot h_1(0)\}.$$

If $n = 1$, then we have:

$$Q_1^1(p_0) = \{(h_1(0), \varphi(0)) : (h_1(0), \varphi(0)) \in R_+^2, 0 \leq \varphi(0) \prec (2.5) \cdot h_1(0)\}.$$

A ray of balance in uncertainty, defined by fuzzy sets, is given by: $\varphi(0) = (3.5) \cdot h_1(0)$.

The Results of the Experiment for the Second Investor

The motion paths and the ray of balance are shown by lines with solid and round markers. The symmetrical problem for the second player was considered. As a result of this, their sets of preferences and their trajectory of movement were given. As in the first experiment, rays emanating from the point (0,0) were considered in the positive orthant. They are given by the following ratio: $h_2(0) = (0.8 + 1/n) \cdot \varphi$. Using it, preference sets of the second player are set in n steps. In this experiment, the set is $Q_2^n(p_0, p_0)$ described as follows:

$$\{(\varphi(0), h_2(0)) : (\varphi(0), h_2(0)) \in R^2_+, (0.8 + 1/(n-1)) \cdot \varphi(0) \leq h_2(0) \\ \prec (0/8 + 1/n) \cdot \varphi(0)\}.$$

If $n = 1$, than we have:

$$Q^1_2(p_0) = \{(\varphi(0), h_2(0)) : (\varphi(0), h_2(0)) \in R^2_+, 0 \leq h_2(0) \prec (1.8) \cdot \varphi(0)\}.$$

The ray of balance in uncertainty, defined by a fuzzy set, is set as: $h_2(0) = (0.8) \cdot \varphi(0)$.

Experimental Results for Equilibrium Financial Strategies of Investors. The calculation shows the "movement" along the ray of balance: $h_2(0) = (3.5) \cdot \varphi(0)$.

Here we consider the initial task for the first player.

The results of the experiments show the ability of the model to provide effective decision support in the field of investment in information technology and systems (ITS). The studies conducted in this work continue to develop the studies cited in our publications [16, 19]. In these works, the theoretical and methodological foundations of designing DSS using bilinear multi-step quality game with several terminal surfaces were presented. The approach proposed in the work made it possible to eliminate the shortcomings of the early versions of the model, since this paper takes into account incomplete information on the financial condition of the second player, namely, incomplete information within the category of fuzzy sets, which distinguishes our study from the works of other authors [1–9, 13, 17, 18].

The disadvantage of the model is the fact that the data obtained during the experiment and the predictive assessment when choosing investor investment strategies did not always coincide with the actual data. The maximum deviation of the results of the simulation experiment from practical data was 9–14%.

7 Conclusions

Thus, in the course of research, the outcome of which are presented in the article, the following results were obtained:

- an updated model of mutual investment in various objects, and, first of all, in high-risk projects for the development of information technologies and systems (ITS), was proposed. The difference between the model and the similar ones is that for the first time it takes into account the lack of complete information about the financial condition of the second investor or group of investors, which can conditionally be attributed to the second group. In addition, for the first time in the proposed model, in comparison with the well-known approaches in game theory, the dynamic programming method was used to solve the problem with incomplete information defined by fuzzy sets. The latter gave a new stimulus for the search for approaches to the effective solution of problems in which the content is of the nature of fuzzy sets;

– computational experiments were performed. During the experiments, it was shown that the proposed model is able to provide effective decision support in the field of mutual investment, in particular, for such areas as investing in information technology and systems (ITS). The validity of the model was confirmed; the maximum deviation of the results of a computational experiment from practical data was 9–14%.

References

1. Irani, Z., Love, P.E.: The propagation of technology management taxonomies for evaluating investments in information systems. J. Manag. Inf. Syst. 161–177 (2000)
2. Parente, S.T., Van Horn, R.L.: Valuing hospital investment in information technology: does governance make a difference? Health Care Finan. Rev. **28**(2), 31 (2006)
3. Nguyen, T.H., Newby, M., Macaulay, M.J.: Information technology adoption in small business: confirmation of a proposed framework. J. Small Bus. Manag. **53**(1), 207–227 (2015)
4. Sabherwal, R., Jeyaraj, A.: Information technology impacts on firm performance: an extension of Kohli and Devaraj (2003). MIS Q. **39**(4), 809–836 (2015)
5. Kilic, M., Kaya, İ.: Investment project evaluation by a decision making methodology based on type-2 fuzzy sets. Appl. Soft Comput. **27**, 399–410 (2015)
6. Zeng, S., Xiao, Y.: TOPSIS method for intuitionistic fuzzy multiple-criteria decision making and its application to investment selection. Kybernetes **45**(2), 282–296 (2016)
7. Wang, R., Wang, J., Gao, H., Wei, G.: Methods for MADM with picture fuzzy muirhead mean operators and their application for evaluating the financial investment risk. Symmetry **11**(1), 6 (2019)
8. Sandström, J., Kyläheiko, K., Collan, M.: Managing uncertainty in long life cycle investments: unifying investment planning, management, and post-audit with a fuzzy DSS. Int. J. Bus. Innov. Res. **11**(1), 133–145 (2016)
9. Goztepe, K.: Designing fuzzy rule based expert system for cyber security. Int. J. Inf. Secur. Sci. **1**(1), 13–19 (2012)
10. Fielder, A., Panaousis, E., Malacaria, P., et al.: Decision support approaches for cyber security investment. Decis. Support Syst. **86**, 13–23 (2016)
11. Lakhno, V.A.: Development of a support system for managing the cyber security. Radio Electron. Comput. Sci. Control (2), 109–116 (2017)
12. Chen, L., Pan, W.: An intelligent decision support system for improving information integrity in procuring infrastructures in Hong Kong. In: Proceedings of the 21st International Symposium on Advancement of Construction Management and Real Estate, pp. 213–221. Springer, Singapore (2018)
13. Ribas, J.R., da Silva Rocha, M.: A decision support system for prioritizing investments in an energy efficiency program in favelas in the city of Rio de Janeiro. J. Multi-Criteria Decis. Anal. **22**(1–2), 89–99 (2015)
14. Dai, J., Wang, D., Yang, X., Wei, X.: Design and implementation of a group decision support system for university innovation projects evaluation. In: 2016 11th International Conference on Computer Science & Education (ICCSE), pp. 148–151. IEEE, August 2016
15. Ribeiro, M.I., Ferreira, F.A., Jalali, M.S., Meidutė-Kavaliauskienė, I.: A fuzzy knowledge-based framework for risk assessment of residential real estate investments. Technol. Econ. Dev. Econ. **23**(1), 140–156 (2017)

16. Lakhno, V., Malyukov, V., Gerasymchuk, N., et al.: Development of the decision making support system to control a procedure of financial investment. Eastern-Eur. J. Enterprise Technol. **6**(3), 24–41 (2017)
17. Shea, V.J., Dow, K.E., Chong, A.Y.L., Ngai, E.W.: An examination of the long-term business value of investments in information technology. Inf. Syst. Front. **21**(1), 213–227 (2019)
18. Manshaei, M.H., Zhu, Q., Alpcan, T., et al.: Game theory meets network security and privacy. ACM Comput. Surv. **45**(3), 1–39 (2013)
19. Akhmetov, B.S., Akhmetov, B.B., et al.: Adaptive model of mutual financial investment procedure control in cybersecurity systems of situational transport centers. News Natl. Acad. Sci. Repub. Kaz. Ser. Geol. Tech. Sci. **3**(435), 159–172 (2019)
20. Isaacs, R.: Differential games: a mathematical theory with applications to warfare and pursuit, control and optimization. Courier Corporation (1999)

Development of an Educational Device Based on a Legacy Blood Centrifuge

Mohamed Abdelkader Aboamer(✉)

Department of Medical Equipment Technology, College of Applied Medical Sciences, Majmaah University, Al Majmaah 11952, Saudi Arabia
m.aboamer@mu.edu.sa

Abstract. Lifetime of medical equipment in hospitals ranges between 1 to 16 years, depending on the usage and maintenance of the medical device. Reusing the obsolete medical devices, converting them to low-cost educational systems and cutting down the charge of new educational medical equipment is the aim of this study. The proposed system is retrofitted with new mechanical and electrical modifications which include system identification and Proportional-Integral (PI) controller based on MATLAB, incremental rotary encoder, zero crossing phase controlling dimming circuit and Arduino microcontroller. System identification was applied to represent the new modified system by z-transform. Acting stresses, displacement, and factor of safety tests were applied through motion study. The proposed new mechanical part integrated with the original system yielded a maximum stress of 11.09 MN/m2 which below the value of the yield strength of the plain carbon steel SAE 1030 which is 220 MN/m2, The maximum displacement was 0.2 mm and the minimum factor of safety was 3.7 leading to the fact that the newly designed shaft can support not only 11.09 MN/m2 but also 41.033 MN/m2 when rotating at 1000 RPM. The proposed system was designed, implemented and tested as the whole system was modelled by 97.7% best-fit using z-transform. Also, a clinical validation has achieved by means of four different blood parameters: glucose, cholesterol, creatinine, and turbidity.

Keywords: Obsolete equipment · Zero crossing · System identification · Z-transform · PI controller · T-test

1 Introduction

At first glance, the reusing process for medical and non-medical devices have a promising success point of view, and there are previous investigations discussing these issues, especially in industry. The reuse of the industrial robots shows major benefits, where the integration between machine engineering, electronics, and computerized management became popular. The core benefits of re-engineering are covering the maintenance of fitting main mechanical components and upgrading robot hardware and software in an optimal and highly effective approach to high-tech mechatronics such as industrial robots [1]. Reintroducing procedure is also needed in radiological medical devices where the speedy industrial growth created new modifications in medical imaging and approaches. The same growth speed caused in quicker technical and

The original version of this chapter was revised: The affiliation of author Mohamed Abdelkader Aboamer has been corrected. The correction to this chapter is available at https://doi.org/10.1007/978-3-030-51971-1_53

R. Silhavy (Ed.): CSOC 2020, AISC 1225, pp. 131–152, 2020.
https://doi.org/10.1007/978-3-030-51971-1_11

functional uselessness of the similar medical imaging instruments. Therefore, generating a demand for an overhauling or renewal. Devices or equipment older than 10 years are no longer state-of-the-art equipment and the changing or replacement is an essential issue [2]. Therefore, provisioning a legacy equipment and retrofitting processes for old medical equipment to low-cost educational equipment in different campuses can decrease the cost of buying new lab facilities.

The centrifuge is the workhorse of any medical diagnostics facility. From extraction of plasma from whole blood (for performing immunoassays or determining hematocrit value) to concentration of pathogens and parasites in biological fluids such as blood, urine, and stool (for microscopy), centrifugation is the first key-step for most diagnostic assays [3]. However, many investigations focused on various types of educational equipment such as: spectrophotometer [4], low-cost motion control [5] and New Educational Equipment for Networking Study by Physical Visualizations and Physical Direct Manipulations [6] but low-cost educational centrifuge system still rare and a retirement and conversion processes for old medical equipment to remote low-cost educational also still rare. Moreover, a rare number of patents were focused on the medical laboratories' educational devices but the number of them was talking about training devices for nursing such as United States patent number 5314339 it seemed like a dummy which, consists of all parts of the human body. Also, United States patent number 2016/0030952 A1 showed a centrifuge device but not for training where the user can control it as either manual or electrical.

With respect to different perspectives, Proportional-Integral-Derivative (PID) controllers is one of the best recognized controllers in the field of industrial control. In addition to its unpretentious structure and durable rendering in a widespread variety of operating circumstances. The creation and implementation of the controller needs a certain indispensable requirement of three parameters: gain of proportional, time constant of integral and time constant of derivative. Still now, tremendous exertion was devoted to improving approaches to decrease the optimizing time of choosing the parameters of the controller [7]. The context of tuning is the expression for choosing the appropriate PID parameters, Ziegler-Nichols formula is the classical tuning approach although, it is not at all times a good choice where, it possibly will wholly fail to tune the processes [8].

The detection of Zero-crossing approach is the practically popular technique for determining the frequency of a periodic signal. Literally the zero-crossing indicator notices the conversion of a signal waveform from positive side to negative side, in an ideal world it supplies a pulse with a narrow band that synchronizes accurately with the zero-voltage state [9–12]. There are numerous types to determine the zero-crossing such like: Simple Optical Isolated Semiconductor Devices [13] and Comparator Circuits with Fixed Hysteresis [14].

A numerous technique for determining the stresses levels in Ductile and Brittle materials such like: Von Mises Stress, Max Shear Stress, Mohr-Coulomb Stress, and Max Normal Stress [15] however, Von Mises Stress method for Ductile materials shows an accurate prediction for the Factor of Safety (FOS) conversely with the other methods and Law of cosines is a common approach for calculating displacement [16].

Black-box modelling is one of system identification methods and is valuable when the essential attention is in discovering the most suitable curve which can represent the data in any case of a mathematical equation. There are linear and nonlinear model constructions for introducing dynamic systems. The proposed method is in general a trial-and-error approach starting with linear to more composite constructions. There are different types of model such like: Transfer Function (TF) which represents by means of numbers of poles and zeros, linear ARX model which is a polynomial model and State-space model [17].

The aim of the work is to design and implement a new mechanical design to convert the obsolete medical equipment to a low cost simple educational Model based on MATLAB and decreasing the cost of buying a new educational medical equipment, especially in developing countries. The blood centrifuge uses an AC Motor up to 1000 RPM, Microcontroller ATmega328P, adjustable time up to 99 min and adjustable speed. System identification and Proportional-Integral (PI) by using MATLAB to open the door to new opportunities by allowing users to represent the system with minimum error. Also, a designed speed control circuit to control the speed of the motor depending on the pulse width modulation of the microcontroller and a new mechanical design is built up to attach the encoder with the rotor of the centrifuge for logging the actual speed. **Sample Heading (Third Level).** Only two levels of headings should be numbered. Lower level headings remain unnumbered; they are formatted as run-in headings.

2 Materials and Methods

The proposed system as shown in Fig. 1 is divided into eight phases: First is: Inspections and system preparations to find what the faults inside the device is and prepare it for the modification process. Second is: Mechanical modifications which include: selecting an appropriate material for the new mechanical part, mechanical drawing subsection where SOLIDWORKS program. Third is: Electrical modifications which include: the interfacing process between MATLAB program and the old device, speed sensor, dimming circuit, system identification and Proportional-Integral (PI) controller. Fourth is: Mechanical and Electrical analysis which include: Stress analysis using Von Mises stress technique, Displacement by means of law of cosines, and Factor of Safety tests were applied through a motion study by means of 1000 Revolution per Minute (RPM). Also time response for dimming circuit has been tested. Fifth is final prototype and sixth is the clinical validation where 11 blood samples have been collected to test four different blood parameters: glucose, cholesterol, creatinine, and turbidity by means of t-test statistical method.

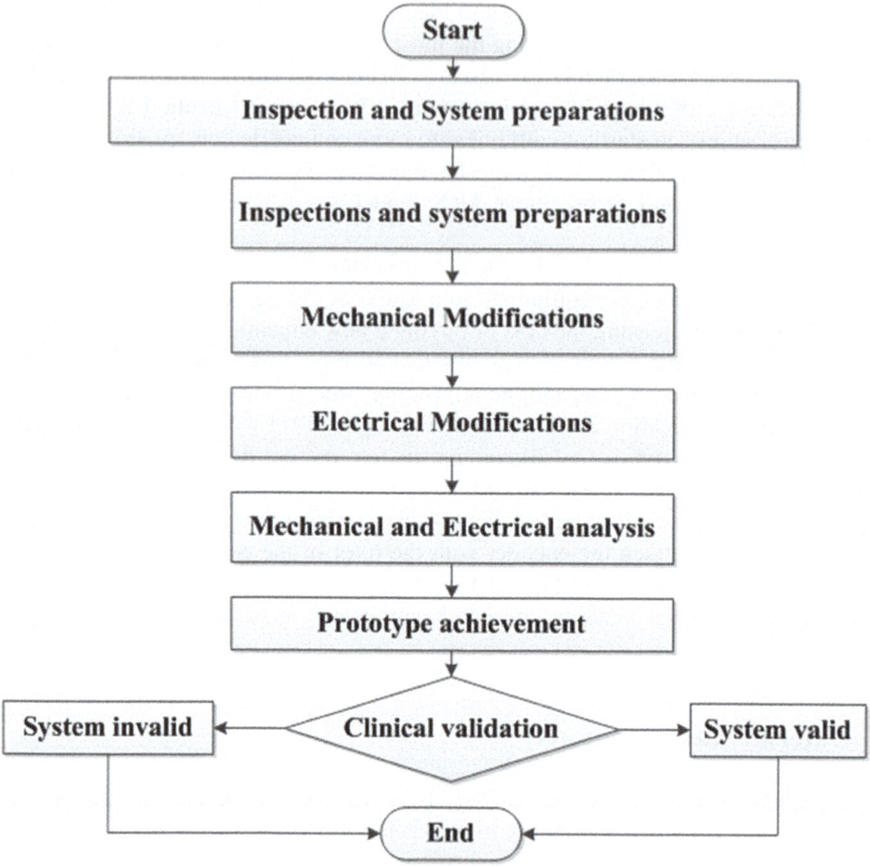

Fig. 1. Block diagram of the proposed approach.

2.1 Inspections and System Preparations

In this stage the old system has lied under maintenance procedures to prepare it for modifications. The model which has chosen for the blood centrifuge is "K CENTRIFUGE PLC SERIES" as shown in Fig. 2 where, speed and time have controlled by means of an analogy nob which means there is no suitable approach of monitoring the speed and the system has seemed like an open loop system.

Fig. 2. The obsolete device after repair, and clean process.

2.2 Mechanical Modifications

Long shaft has designed to attach three parts where: motor, the tray of samples and rotary encoder which worked as a speed sensor. The proposed mechanical part has designed by SOLIDWORKS Program. Plain carbon steel SAE 1030 is the material which has chosen for the proposed mechanical design. As the current investigation has focused on the low-cost Mechanical part, therefore, carbon steel is the most recommended material where, it has suitable mechanical properties and reasonable cost. The chosen material has Yield strength equal to 220 MN/m2, Tensile strength equal 399 MN/m2 and Poisson's ratio equal to 0.28 [18]. The model which has chosen for the blood centrifuge is "K CENTRIFUGE PLC SERIES" where, speed and time have controlled by means of an analogy nob which means there is no suitable approach of monitoring the speed and the system has seemed like an open loop system. The small shaft which has used to attach between the motor and the blood samples tray as shown in Fig. 2 showed that blood centrifuge after preparation process. Where the shaft length approximately 3 cm which attached the samples tray with the motor.

2.3 Electrical Modifications

The block diagram of the interfacing process has illustrated as shown in Fig. 3 which contains five stages. The first is the blood centrifuge device or an AC motor 1.1 A and 220 VAC which, modified by the new mechanical design to attach with a speed sensor "Omron Rotary Encoder 1000 Pulse per Revolution", second is the dimming circuit or an AC phase control circuit "load power 5A, working voltage 240VAC and auto detect 50 or 60 Hz" which enables the controller to control the speed of the centrifuge machine, third is the Arduino UNO microcontroller which, control the AC phase control circuit by using MATLAB included a PI controller and can read the signal from the speed sensor as a closed loop system, fourth part is the computer program using MATLAB which, control the Arduino and read the signals from it and the final section is the system identification Process to describe the system as a mathematical equation in Z domain.

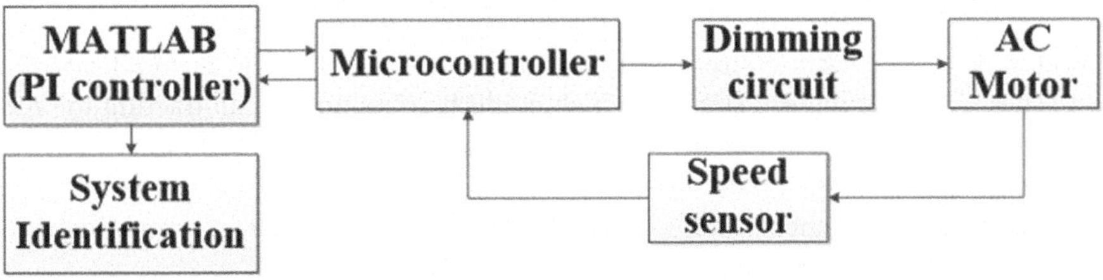

Fig. 3. The block diagram of a reengineering process.

2.4 Mechanical and Electrical Analysis

Stress, Displacement and Factor of Safety

Von Mises Stress was the method which has been chosen to test the different stresses levels during the new mechanical design rotate where the Von Mises Stress method for Ductile materials such as steel and it shows an accurate prediction for the Factor of Safety (FOS) conversely the other methods such as Max Shear Stress (Tresca) where it is for ductile but considered as lower predicted safety factors, Mohr-Coulomb Stress, and Max Normal Stress which were for Brittle [19]. The Von Mises Stress has been applied by fixed speed 1000 RPM has been applied for 1 min.

Scientifically the von-Mises yield criterion is illustrated as Eq. (1):

$$J_2 = K^2, \tag{1}$$

Where, k is the yield stress but in pure shear. Moreover, the shear yield stress in pure shear is equal to the tensile yield stress (δ_y) divided by ($\sqrt{3}$) therefore, the yield stress in pure shear cloud be described as Eq. (2).

$$K = \frac{\delta_y}{\sqrt{3}}, \tag{2}$$

Assume von Mises stress (δ_v) equal to the yield strength δ_y therefore, a new obtained equations cloud be illustrated in Eqs. (3 and 4).

$$\delta_v = \delta_y = \sqrt{3J_2}, \tag{3}$$

$$\delta_v^2 = 3J_2 = 3K^2, \tag{4}$$

Replacing J_2 with components of Cauchy stress tensor Eq. (5) will be obtained.

$$\delta_v^2 = \frac{1}{2}\left[(\delta_{11} - \delta_{22})^2 + (\delta_{22} - \delta_{33})^2 + \left(\delta_{33} - \delta_{11}\right)^2 + 6\left(\delta_{33}^2 + \delta_{31}^2 + \delta_{12}^2\right)\right] = \frac{3}{2} s_{ij} s_{ij}, \tag{5}$$

Where, s is a deviatoric stress or in the other meaning s is the unequal principal-stress. There are 3 deviatoric stresses which could be obtained by subtracting the mean stress components (δ^-) from each principal stress for example: $\delta_x - \delta^-$, $\delta_y - \delta^-$ *and* $\delta_z - \delta^-$.

Furthermore, the Cauchy stress tensor (δ) for infinite element in three dimensions could be described as Eq. (6).

$$\delta = \begin{bmatrix} \delta_{11} & \delta_{12} & \delta_{13} \\ \delta_{21} & \delta_{22} & \delta_{23} \\ \delta_{31} & \delta_{32} & \delta_{33} \end{bmatrix} \equiv \begin{bmatrix} \delta_{xx} & \delta_{xy} & \delta_{xz} \\ \delta_{yx} & \delta_{yy} & \delta_{yz} \\ \delta_{zx} & \delta_{zy} & \delta_{zz} \end{bmatrix} \equiv \begin{bmatrix} \delta_x & \tau_{xy} & \tau_{xz} \\ \tau_{yx} & \delta_y & \tau_{yz} \\ \tau_{zx} & \tau_{zy} & \delta_z \end{bmatrix} \tag{6}$$

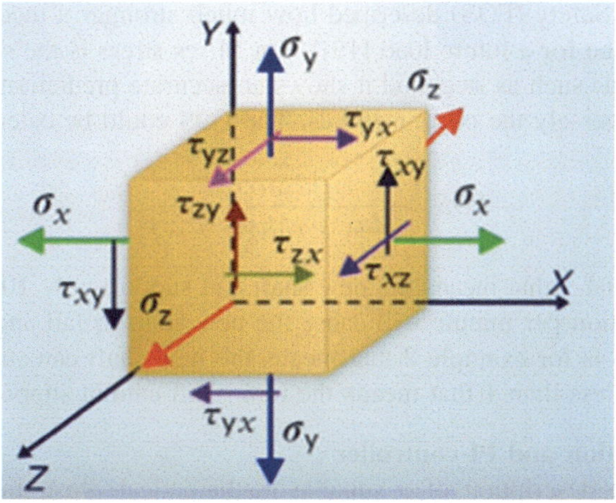

Fig. 4. Infinite element in three dimensions.

Where, $\delta_x, \delta_y \, and \, \delta_z$ are the 3 normal stresses components and $\tau_{xy}, \tau_{xz}, \tau_{yx}, \tau_{yz}, \tau_{zx} \, and \, \tau_{zy}$ are the 6 share stresses which could be illustrated as shown in Fig. 4.

The Von Mises Stress equation can be calculated from the Cauchy stress tensor matrix as:

$$\sigma_v = \sqrt{\frac{1}{2}[(\sigma_x - \sigma_y)^2 + (\sigma_y - \sigma_z)^2 + \left(\sigma_z - \sigma_x\right)^2\Big] + 3\left(\sigma_{xy}^2 + \sigma_{yz}^2 + \sigma_{zx}^2\right)} \quad (7)$$

Where σ_v is the Von Mises stress.

Displacement cloud be calculated by means of the law of cosines [19] as shown in Fig. 5. Where C is the displacement between start point and end by means of the length of b, length of a and angle c by the equation as:

$$c^2 = a^2 + b^2 - 2ab\cos c, \quad (8)$$

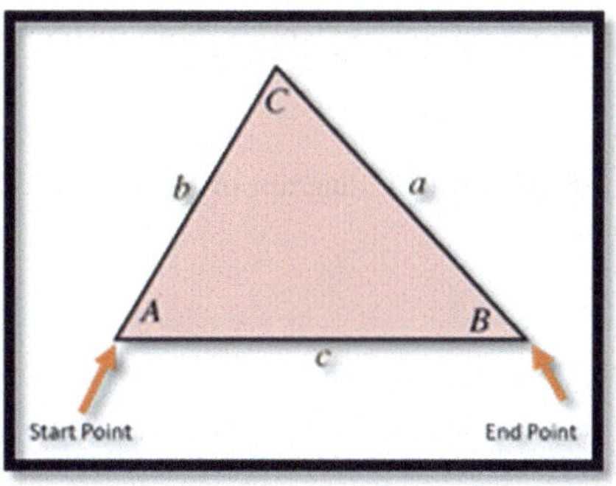

Fig. 5. Law of cosines to calculate the displacement between start point and end.

The Factor of Safety (FOS) described how much stronger a mechanical design is that it required to be for a future load [19]. Von Mises stress is the selected technique for ductile materials such as steel and it shows an accurate prediction for the Factor of Safety (FOS) conversely the other methods. The FOS could be calculated as:

$$FOS = \frac{Yield\,Strength}{Max\,von\,Mises\,stress}, \tag{9}$$

If the FOS equal 1 that means the new shaft can support only 1000 RPM and any additional Revolution per minute will cause the new shaft to fail and defamation will appear but if FOS is for example 2 that means the new shaft can support up to 2000 RPM and if FOS less than 1 that means the new shaft cannot support 1000 RPM.

System identification and PI controller

Academically, the description of a suitable mathematical illustration of a dynamic system by means of a transfer functions is termed as a system identification. Originally, System identification one of the topics of engineering control, but is also observed in other fields of engineering [20–24]. System identification will be a suitable chose if the input and output data are available as observed quantities. However, a large number of methods such like: transfer function, difference equations, differential equations and state space models could be used in system identification [17] but Z-Transform believed as a significant title role in the investigation of linear discrete system identification [17].

Black-box modelling is in general a trial-and-error approach starting with linear to more composite constructions. There are different types of model such like: Transfer Function (TF) which represents by means of numbers of poles and zeros, linear ARX model which is a polynomial model and State-space model.

The Single Input Single Output (SISO) system identification could be described as shown in Fig. 6 by the following equation:

$$Adjusted\,speed = V + Error \tag{10}$$

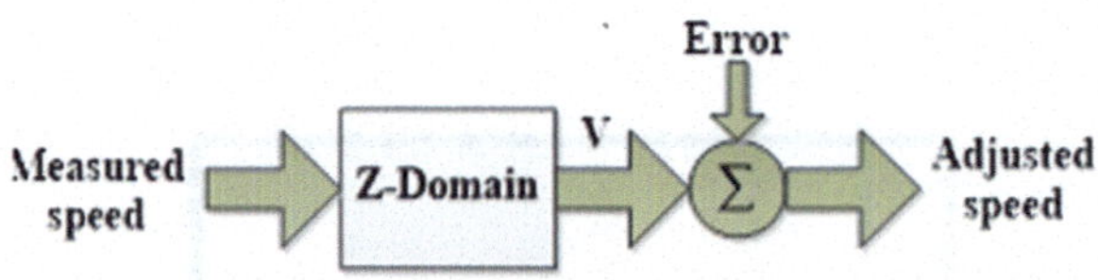

Fig. 6. Block diagram of the linear.

Where: V is the predicted measured speed.

Z-Transform domain has been chosen as a single input single output (SISO) system identification method as shown in Fig. 6 to describe the system as a mathematical equation [25]. The Z-Transform equation [25] has been illustrated by the following equation:

$$f(z) = \sum_{n=0}^{\infty} \frac{f(n)}{Z^n} \tag{11}$$

Where: $f(z)$ is the Z-Transform with respect to the variable Z, f(n) is a time series with respect to the variable n with sampling rate equal 50 ms.

In the current investigation the PI controller has applied to keep and regulate the Centrifuge speed. The introduced controller has illustrated as shown in Fig. 7. E(t) signifies the difference between the wanted speed R(t) and the measured speed B(t). G(t) shows the controller outcome, which is the summing of proportional and integral processes. The AC voltage is controlled by means of the dimmer circuit and is signified by C(t). The second Ziegler-Nichols tuning approach has used to calculate the PI controller parameters.

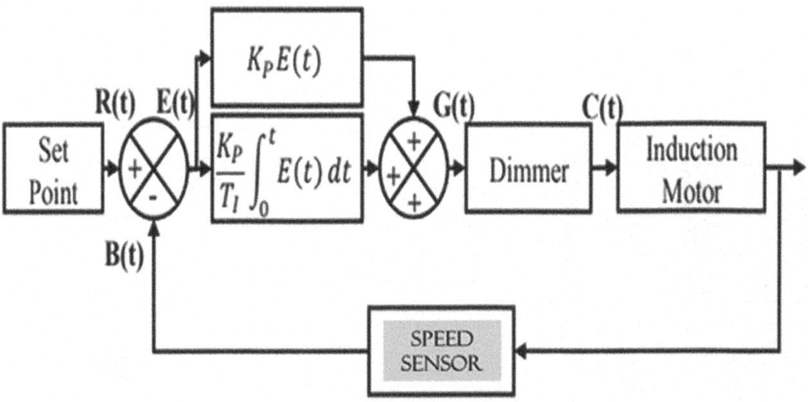

Fig. 7. The block diagram of a PI controller.

Interfacing between MATLAB and the device

The dimming schematic diagram that was utilized in the velocity controlling structure is explained in Figs. 8 and 9. The proposed circuit was provided from a previous published investigation [26]. This schematic diagram was created to regulate an AC supply to control the speed of the proposed centrifuge with single-phase induction motor (220V, 350 W, 50 Hz). The circuit is divided into two parts the first section is the zero-crossing detection circuit as shown in Fig. 10 which entails of an Operational amplifier LM358 with 2 mV input offset voltage work as a comparator to produce a train of pulses from zero to saturation volte, full wave rectification to fetch the positive cycles of the source supply and the optocoupler phototransistor output 4N35 isolate between the AC section and the brain part of the system.

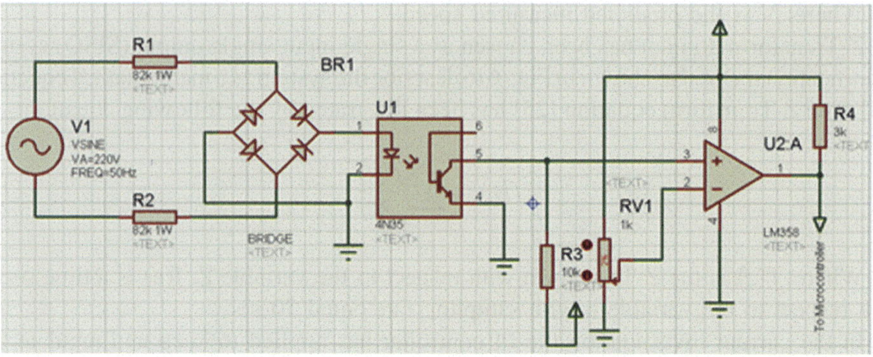

Fig. 8. The zero-crossing detection circuit.

The second is the speed control part as shown in Fig. 9. The phase control circuit include: OptoDiac MOC3021 which control the gate of the Triac Q6004L3 which is the main components of the second section. The Triac gate trigger voltage equals 1.3 V, gate trigger current 10 mA and maximum peak voltage that is allowed across it equals 400 V. PWM which produced from the microcontroller can control the amount of the power of the sin wave signal based on the zero-crossing trigger from the zero-crossing circuit.

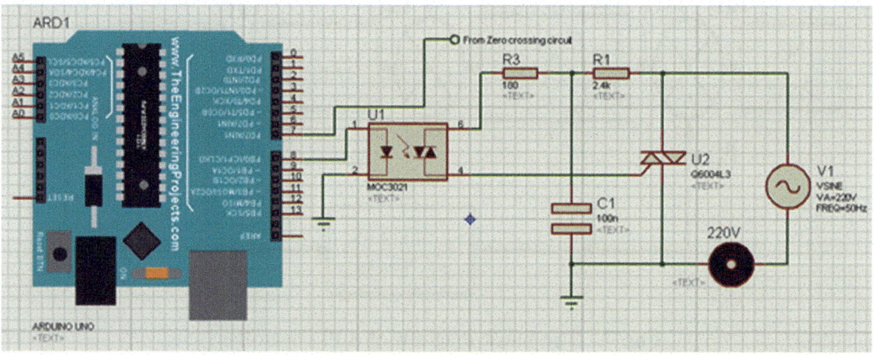

Fig. 9. The speed control circuit.

Omron incremental Rotary Encoder (1000 Pulse per Revolution) as shown in Fig. 10 has been used as a feedback speed application for a single induction motor by using two tracks: A and B. This is Omron quadrature incremental rotary encoder E6B2-CW6C with 1000 P/R (pulse per revolution). The encoder outputs gray code which you can interpret using a microcontroller or Arduino and find out which direction the shaft is turning and by how much. This allows you to add feedback to motor control systems. Encoders of this kind are often used in CNC machines, balancing robots and dead reckoning navigation. This encoder is of improved reliability with reverse connection and load short-circuits protection. Resolution: 1000 Pulse/Rotation, Encoding Method

Incremental, Input Voltage: 5 - 24VDC, Maximum Rotating Speed: 6000 rpm, Allowable Radial Load: 30 N, Allowable Axial Load: 20 N, Cable Length: 50 cm, Shaft Diameter: 6 mm [12]. The basic work principle of the proposed incremental rotary encoder is an optical mechanism by using discs which contains transparent sections than are equally spaced a light emitting diode serves a sender and its light is detected by a photo detector therefore the encoder can generate a sequence of pulses. The output is measured in pulses per revolution. To detect the motion sense the encoder needs two tracks as shown in Fig. 10.

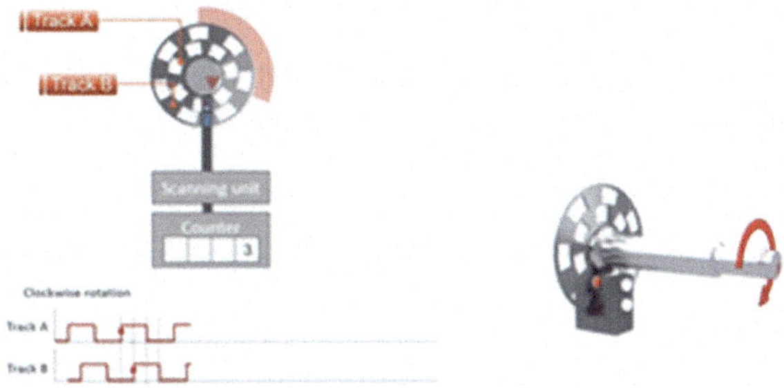

Fig. 10. The internal structure of the incremental rotary encoder.

Graphical User Interface (GUI)

The Graphical User interface (GUI) which, control the system has been achieved by using MATLAB program, where it can control the speed and the operation time of the motor and can read the motor speed and saved the speed records as an EXCEL file which can be used for system identification as shown in Fig. 11.

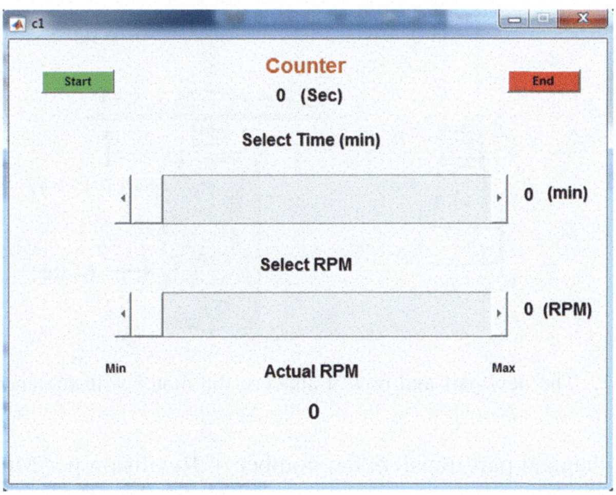

Fig. 11. MATLAB Graphical User interface.

Graphical User Interface (GUI)

The data has collected from 11 samples from Blood bags for volunteers from medical services of Majmaah University, where age in between 20 to 30 years old and all male type. Paired t test has chosen to discover the significant difference between the four blood parameters (Glucose, Cholesterol, creatinine and turbidity) has been tested and the results compared with a commercial blood centrifuge "K CENTRIFUGE PLC SERIES". The paired t test could be illustrated by the following equation:

$$t = \frac{m}{s/\sqrt{n}} \tag{12}$$

Where: m and s were the mean and the standard deviation of the difference between the two paired samples, respectively. n is the sample size [27].

3 Results

3.1 Mechanical Design

The new mechanical part has designed and implemented with length equal 100.8 cm which has used to attach the speed sensor to translate the rotation motion from the motor to the incremental encoder to calculate the speed of motor as shown in Fig. 12.

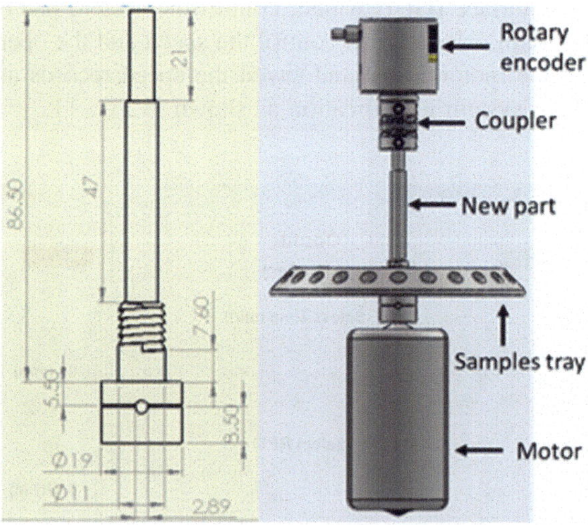

Fig. 12. The new part and how it attaches the motor with the encoder.

The new mechanical part transfers the number of Revulsion per Minutes (RPM) to the rotary encoder which works as a speed sensor where, the speed sensor has sent the RPM to the controller called Arduino which, has used to link the electrical circuit to MATLAB program via Universal Serial Bus (USB) port.

Stress, displacement or deformation and Factor of Safety analysis have been applied as a motion study by using SOLIDWORKS program.

The final view for the new mechanical part which has attached between the motor and the rotary encoder could be illustrated in Fig. 13. Also, the original cover of the blood centrifuge has cut into two halves to explain the internal structure of the blood centrifuge.

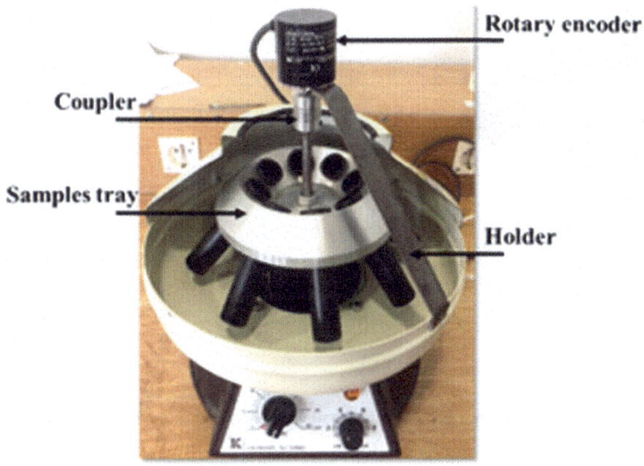

Fig. 13. The attachments between the motor and the encoder by means of the new mechanical design.

Stress, displacement or deformation and Factor of Safety analysis have been applied as a motion study by using SOLIDWORKS program.

After applying 1000 RPM and the motion study analysis the new designed shaft revealed a good support for the desired RPM where the stress analysis by means of Von Mises Stress showed that all stresses were below the yield strength that means the newly designed shaft worked in the elastic region and no deformation has happened as shown in Fig. 14 which showed that all color chart of the stress analysis is below the yield strength. As shown in Fig. 16 the max stress is 11.09 MN/m^2 which below the value of the yield strength which equal to 220 MN/m^2.

The maximum displacement is 0.2 mm as shown in Fig. 15 which revealed that there was no deformation where the color chart showed that the minimum displacement is 0.0003787 mm and maximum displacement is 0.2538 mm.

The minimum Factor of Safety is 3.7 as shown in Fig. 16 where color chart showed that the minimum FOS is 3.7 which means that the newly designed shaft can support 3.7 x 11.09 which equal 41.033 MN/m^2 during using 1000 RPM.

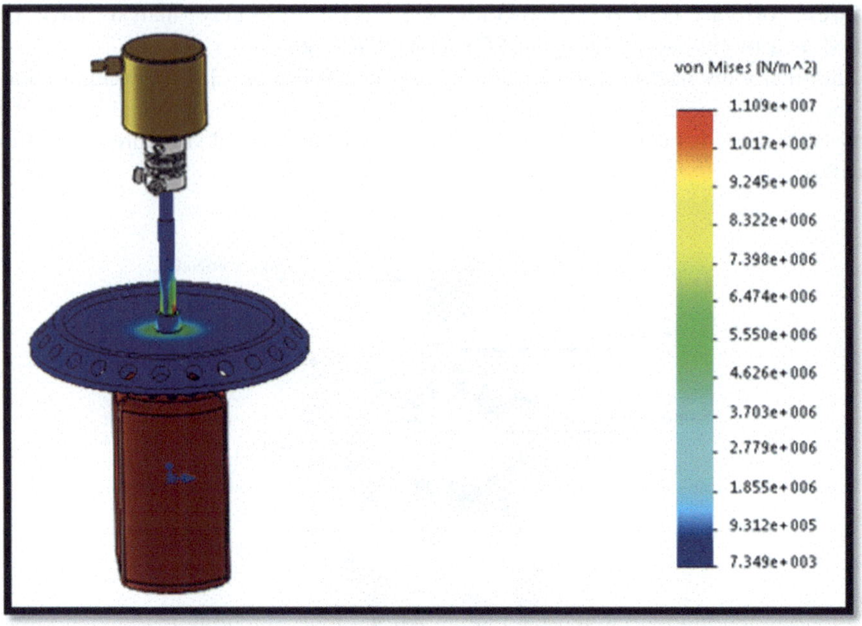

Fig. 14. Von Mises Stress after applying 1000 RPM.

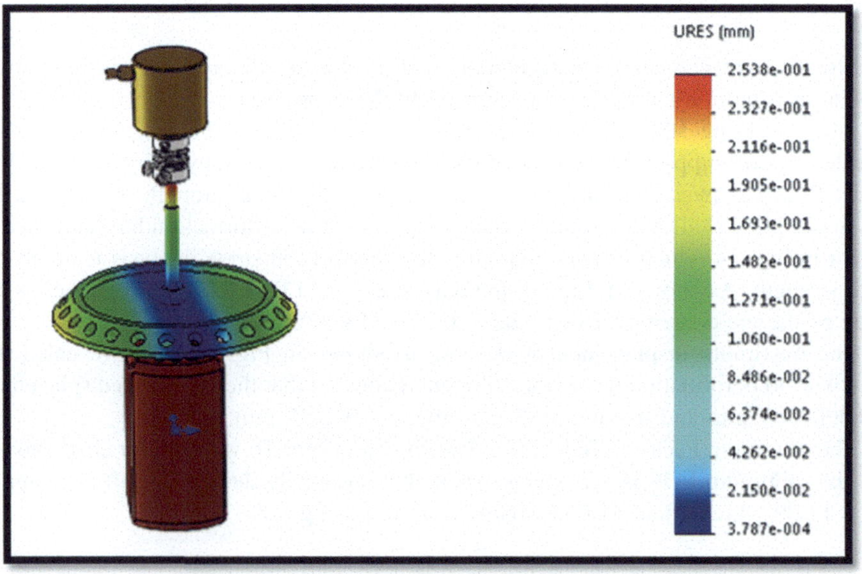

Fig. 15. Displacement after applying 1000 RPM.

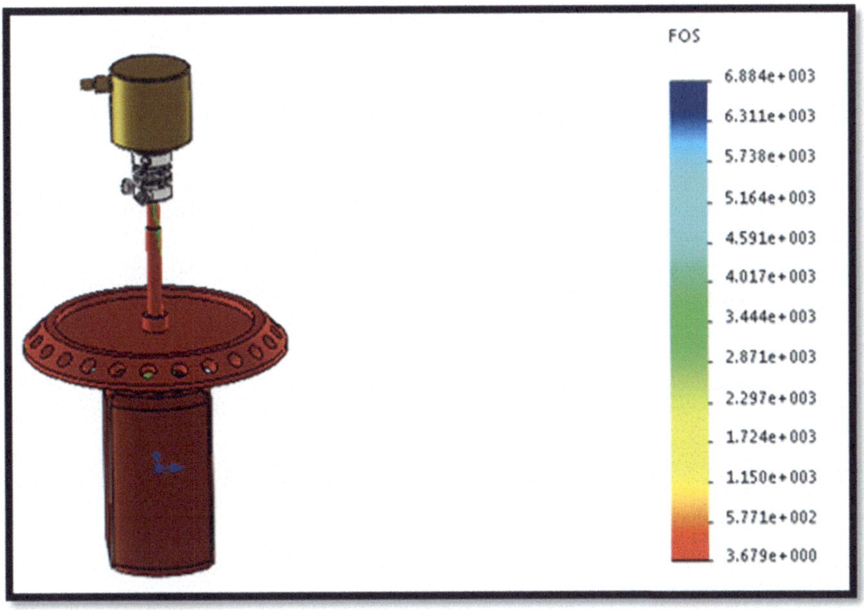

Fig. 16. The factor of Safety after applying 1000 RPM.

3.2 Interfacing and System Identification

After the process of permutations and combinations between numbers of poles and zeros for discovering the best Z model, the process of permutations and combinations have applied on 5 zeros and 5 poles where, Table 1 has illustrated.

Table 1. The best fit according to the number of zeros and poles.

S.	Number of zeros	Number of poles	Percentage of fitting (%)
1	1	1	97.26
2	1	2	**97.77**
3	1	3	97.45
4	1	4	92.5
5	1	5	92.1
6	2	1	97.3
7	2	2	**97.77**

(continued)

Table 1. (*continued*)

S.	Number of zeros	Number of poles	Percentage of fitting (%)
8	2	3	97.45
9	2	4	96.07
10	2	5	90.7
11	3	1	97.3
12	3	2	97.45
13	3	3	97.45
14	3	4	97.44
15	3	5	96.1
16	4	1	97.4
17	4	2	97.45
18	4	3	96.4
19	4	4	96.39
20	4	5	95.36
21	5	1	97.43
22	5	2	97.45
23	5	3	36.71
24	5	4	93.95
25	5	5	93.03

Based on the previous analysis as shown in Table 1, the two highest percentage of fitting equal 97.77% the first come zeros equal 1 and poles equal 2 and the second come from zeros equal 2 and poles equal 2. MSE has calculated to differentiate between the two highest percentage of fittings, MSE equals 9.5 are given by the two therefore, any one of them can be used to represent the system with the highest fitting. Moreover 2 poles and 2 zeros have chosen for creating the best model. The transfer function of 2 poles and 2 zero can be described by the following equation:

$$F(z) = \frac{0.001562z^{-1} - 0.001493z^{-2}}{1 - 1.981z^{-1} + 0.9807z^{-2}} \tag{13}$$

The actual recorded data has illustrated in Fig. 17 and the step response of the best model has illustrated in Fig. 18.

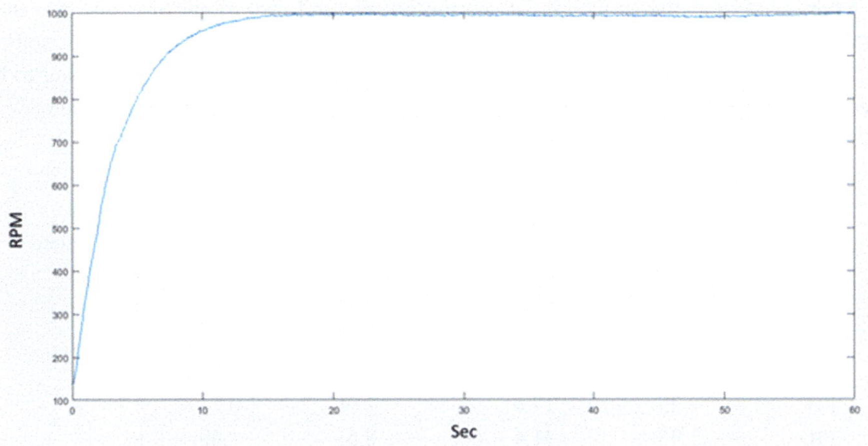

Fig. 17. The actual recorded data.

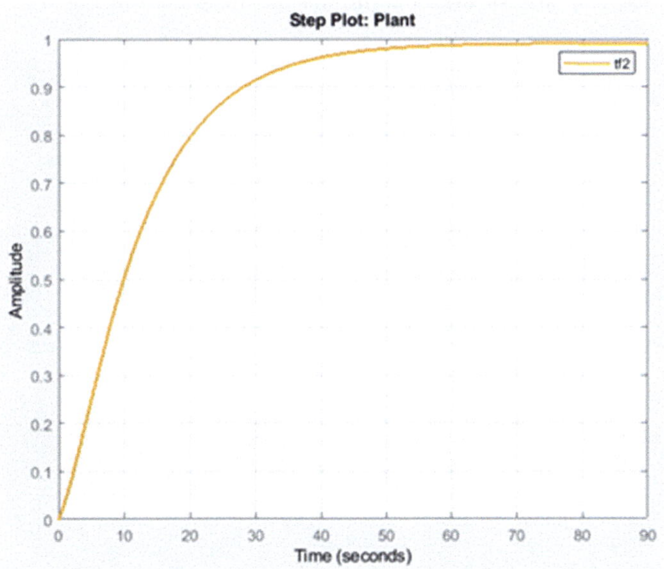

Fig. 18. The step response of the best model.

After applying the concept of permutations and combinations on the three parameters of PID controller, some features have collected such as: rise time, settling time, overshoot percentage, peak value and the steady state amplitude according to the step response. These features which have mentioned before illustrated in Table 2.

Table 2. Different features of best controller.

S.	Controller	Rise time (sec)	Settling time (sec)	Overshoot (%)	Peak	Steady state amplitude
1	P	7.225	17.7	2.74	0.71	0.7
2	I	7.05	256	70.9	1.71	1
3	**PI**	**8.95**	**29.3**	**6.8**	**1.07**	**1**
4	PD	7.25	17.6	2.68	0.713	0.7
5	PID	8.95	31.4	8.54	1.09	1

According to the numerical calculation in Table 2 the best controller is PI controller where; the tuned response of the PI controller has illustrated in Fig. 19.

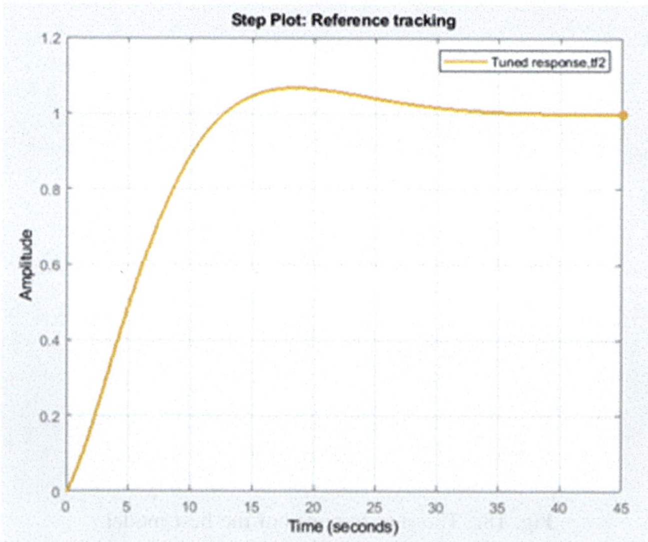

Fig. 19. The tuned response of PI controller.

The time response for the zero-crossing detection circuit which has produced a good detection for the zero-crossing state as shown in Fig. 20(a). The zero-crossing detection circuit provides the PWM circuit to control the amount of power as shown in Fig. 20(b) which used to power on the AC motor of the centrifuge.

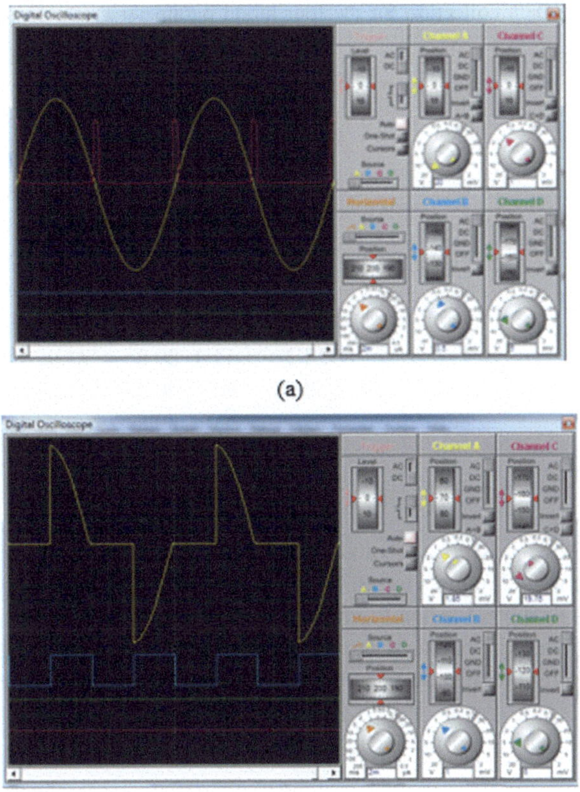

(a)

(b)

Fig. 20. (a) zero crossing circuit diagram. (b) PWM circuit to control the amount of signal power.

3.3 Clinical Validation

The equations are an exception to the prescribed specifications of this template. You will need to determine whether or not your equation should be typed using either the Times New Roman or the Symbol font (please no other font). To create multileveled equations, it may be necessary to treat the equation as a graphic and insert it into the text after your paper is styled.

A clinical validation has applied to test the proposed new system where four blood parameters (Glucose, Cholesterol, creatinine and turbidity) has been tested and the results compared with a commercial blood centrifuge. Table 3 showed the results for each blood parameter where control means commercial blood centrifuge and control means the new system.

Table 3. Data collection for the four blood parameters from 11 samples.

Number of samples	Glucose		Cholesterol		Creatinine		Turbidity	
	Control	Test	Control	Test	Control	Test	Control	Test
1	18.8	19.1	63.3	64.1	29.9	29.9	33	40
2	19.7	19.7	31.9	30.1	76.1	76	23.8	23.8
3	17.5	17.5	62.8	63.1	12.3	12.4	18	17
4	19.1	19	64.9	65.6	31.9	31.7	45	39
5	17.9	18	63.8	62.9	40.3	40.5	63	65
6	17.3	17.3	56.5	58.2	27.8	27.7	34	35
7	21.9	22.4	63.2	63.9	14	14.5	20	20
8	22.6	21.8	54.2	53.8	39.8	37.9	48	45
9	19.5	19.6	59.6	60	40.7	41	66	69
10	21.3	20.8	63.2	63.1	39.8	39.2	49	46
11	7.3	8	58.2	58.3	29.6	29.5	34	33

The paired t test showed a p-value equals to 0.8323 when the result of Glucose parameter has compared, equals to 0.6398 when the result of Cholesterol parameter has compared, equals to 0.3921 when the result of Creatinine parameter has compared and equals to 0.9321 when the result of Turbidity parameter has compared between the new system and the commercial blood centrifuge. The validation test has achieved in 4500 RPM for 5 min.

4 Conclusion

Reusing the obsolete medical equipment, converting it to a low-cost educational one and cutting down the charge of a new educational medical equipment is the aim of this study. The proposed system has been achieved and could be an effective tool for decreasing the cost amount of buying an educational device, where the obsolete devices could be converted to education by very low-cost. By this proposed model users will be able to apply system identification and use different estimating methods by using system identification and PI controllers' toolboxes inside MATLAB. The blood centrifuge uses an AC Motor up to 1000 RPM, Microcontroller ATmega328P, adjustable time up to 99 min and adjustable speed using a dimming circuit which include a zero-crossing detection system. System identification and Proportional-Integral (PI) by using MATLAB to open the door to new opportunities by allowing users to represent the system with minimum error. Also, a designed speed control circuit to control the speed of the motor depending on the pulse width modulation of the microcontroller and a new mechanical design is built up to attach the encoder with the rotor of the centrifuge for logging the actual speed. The P-values showed that there is no significant difference at the 5% significance level between the new system and commercial blood centrifuge for the four blood parameters (Glucose, Cholesterol, creatinine and turbidity) where all p-values were less than 0.05. That means there is no significant difference between the two systems based on the four parameters (glucose, cholesterol, creatinine and turbidity).

Acknowledgment. The authors are thankful to the (Department of Biomedical Equipment Technology, College of Applied Medical Sciences, Majmaaah University, 11952, Saudi Arabia) and the Deanship of Scientific Research for co-operative provision, guidance, and deliberations that enhanced the Project number: 38/148.

Supplementary Materials

The Data used in this study has been collected by the attached rotary encoder which worked as a speed sensor. The data used in the study is available in the additional file which attached as a file name: IO.mat but future researchers have to ask authors before using these data.

Author Contributions

The contribution of the first author is in the previous studies, final editing, designing, implementation procedures, and preparation of the draft of the manuscript.

Funding

This work wasn't supported.

References

1. Karastoyanov, D., Karastanev, S.: Reuse of Industrial Robots. IFAC (International Federation of Automatic Control) Hosting by Elsevier Ltd. 2405-8963
2. European Society of Radiology (ESR): Renewal of radiological equipment. Insights Imaging **5**, 543–546 (2014)
3. Bhamla, M.S., Benson, B., Chai, C., Katsikis, G., Johri, A., Prakash, M.: Hand-powered ultralow-cost paper centrifuge. Nat. Biomed. Eng. **1** (2017)
4. Andreev, A.I., Nikitenko, V.A., Pautkina, A.V.: Modern spectrometer equipment in educational laboratory of department of physics. In: 13th International Scientific-Technical Conference on Actual Problems of Electronics Instrument Engineering (APEIE), vol. 03, p. 1 (2016). https://doi.org/10.1109/apeie.2016.7807091
5. Montesinos-Miracle, D., Galceran-Arellano, S., Gomis-Bellmunt, O., Sudria-Andreu, A.: A new low-cost motion control educational equipment. In: European Conference on Power Electronics and Applications, pp. 1–6 (2007). https://doi.org/10.1109/epe.2007.4417308
6. Yoshihara, K., Watanabe, K., Iguchi, N.: New educational equipments for networking study by physical visualizations and physical direct manipurations. In: IEEE 30th International Conference on Advanced Information Networking and Applications (AINA), pp. 227–230 (2016). https://doi.org/10.1109/aina.2016.91
7. Zhao, Z.-y., Tomizuka, M., Isaka, S.: Fuzzy gain of PID controller. IEEE Members, 1392 p
8. He, S.Z., Tan, S.H., Xu, F.L., Wang, P.Z., Department of Electrical Engineering: NUS10 kent ridge crescent Singapore 0511. PID selftuning control using a fuzzy adaptive mechanism, p. 708
9. Weidenburg, R., Dawson, F.P., Bonert, R.: New synchronization method for thyristor power converters to weak AC-systems. IEEE Trans. Ind. Electron. **40**(5), 505–511 (1993)
10. Vainio, O., Ovaska, S.J.: Noise reduction in zero crossing dection by predictive digital filtering. IEEE Trans. Ind. Electron. **42**(1), pp. 58–62 (1995)
11. Vainio, O., Ovaska, S.J.: Digital filtering for robust zero crossing detectors. IEEE Trans. Instrument. Measur., 426–430 (1996)

12. Vainio, O., Ovaska, S.J.: Adaptive lowpass filters for zero- crossing detectors. In: Proceedings of the 28th Annual International Conference of IEEE, Sevilla, Spain, November 5–8, 2002
13. Hess, H.L., Wall, R.W., et al.: A microcontroller-based pulsed width modulated voltage source inverter. In: North American Power Symposium, Bozeman, Montana, October 2, 1995
14. LM193, LM293, LM393 Dual Differential Comparator data sheet, Texas Instruments
15. Nudehi, S., Steffen, J.: Analysis of Machine Elements Using SOLIDWORKS Simulation 2016, chap. 2 (2016). ISBN-10: 1630570044
16. Hazewinkel, M. (ed.) Laplace Transform. Encyclopedia of Mathematics. Springer Science + Business Media B.V./Kluwer Academic Publishers (2001) [1994]. ISBN 978-1-55608-010-4
17. Mohamed, I.I., Aboamer, M.A., Azar, A.T., Wahba, K., Schumann, A., Bär, K.J.: Nonlinear single-input single-output model-based estimation of cardiac output for normal and depressed cases. Neural Comput. Appl. **31**(7), 2955–2978 (2017). https://doi.org/10.1007/s00521-017-3245-8
18. SubsTech: Knowledge source on materials Engineering, Carbon steel SAE 1030. http://www.substech.com/dokuwiki/doku.php?id=carbon_steel_sae_1030. Accessed June 2012
19. Nudehi, S., Steffen, J.: Analysis of Machine Elements Using SOLIDWORKS Simulation 2016, 1st edn. Stephen Schroff, USA (2016)
20. Azar, A.T., Vaidyanathan, S.: Handbook of Research on Advanced Intelligent Control Engineering and Automation. Advances in Computational Intelligence and Robotics (ACIR). IGI Global, USA (2015). ISBN: 9781466672482
21. Azar, A.T., Vaidyanathan, S.: Computational Intelligence Applications in Modeling and Control. SCI, vol 575. Springer, Berlin (2015). ISBN: 978-3-319- 11016-5
22. Azar, A.T., Vaidyanathan, S.: Chaos Modeling and Control Systems Design, Studies in Computational Intelligence, vol. 581. Springer, Berlin (2015). ISBN 978-3-319-13131-3
23. Azar, A.T., Zhu, Q.: Advances and Applications in Sliding Mode Control Systems. SCI, vol. 576. Springer, Berlin (2015). ISBN: 978-3-319-11172-8
24. Zhu, Q., Azar, A.T.: Complex System Modelling and Control Through Intelligent Soft Computations. STUDFUZZ, vol. 319. Springer, Berlin (2015). ISBN: 978-3-319-12882-5
25. Love, J.: Process Automation Handbook, 1st edn. Springer, London (2007)
26. Sanjayaa, Y., Fauzia, A., Edikresnhaa, D., Munira, M.M., Khairurrijala: Single phase induction motor speed regulation using a PID controller for rotary forcespinning apparatus. Procedia Eng. **170**, 404–409 (2017)
27. Levine, D.M., Stephan, D.F., Szabat, K.A.: Statistics for Managers Using Microsoft Excel, 8th edn. (2016). ISBN-10: 0134173058, ISBN-13: 978-0134173054

Does Fertilizer Influence Shape and Asymmetry in Wheat Leaf?

S. G. Baranov[1,2(✉)], I. Y. Vinokurov[2], I. M. Schukin[2], V. I. Schukina[2], I. V. Malcev[2], I. E. Zykov[3], A. A. Ananieff[4], and L. V. Fedorova[5]

[1] Vladimir State University, Vladimir 600000, Russia
bar.serg58@gmail.com
[2] The Upper Volga Federal Agricultural Research Centre, Vladimir Region 601261, Russia
[3] State Humanitarian Technological University, Orekhovo-Zuevo 142611, Russia
[4] Timiryazev Agricultural Academy, Moscow 127550, Russia
[5] Sechenov First Moscow State Medical University, Moscow 119991, Russia

Abstract. The paper considers the bilaterally symmetrical genetic traits and their variability under the influence of environmental factor. The method of geometric morphometrics was applied. The tasks of the work included the comparing the shape and asymmetry of winter wheat leaf blade under cultivating with different doses of mineral fertilizer. From each individual from 20 plants 4–6 laminas were collected. Four levels of the widely used mineral fertilizer NPK were used. The statistical significance of centroid size in all groups of plants was at the same level ($p = 0.0001–0.001$). A clear fluctuating asymmetry was not detected. A statistically significant directional asymmetry was obtained (MS "side", $p = 0.0001$), and therefore a mixture of both types of asymmetry was found in all samples. An increase in the dose of fertilizer did not affect the stability of development and caused a higher variability in the symmetric, longitudinal component of the variation in the shape than in the asymmetric component of variability.

Keywords: Procrustes ANOVA · Fluctuating asymmetry · Geometric morphometrics · Shape · Developmental stability

1 Introduction

Fluctuating asymmetry (FA) is a random non-directional deviation from the strict symmetry of bilateral structures and serves indicating of ontogenetic noise, as an adaptation and co-adaptation of a population to a change in the environment. Fluctuating asymmetry is a popular indicator of the developmental stability in populations, both animals and plants [1–3].

Most of the work relates to the determination of FA of woody or shrubby plant forms as indicators of environmental impact over a long period of population ontogenesis. Studies on the developmental stability of cultivated plants, including annual or biennial forms, have

R. Silhavy (Ed.): CSOC 2020, AISC 1225, pp. 153–160, 2020.
https://doi.org/10.1007/978-3-030-51971-1_12

been conducted less. For organisms with high homozygosity of the genotype, a high deviation in the variance of the FA value was reported [4–8].

In cereal crops, a similar dependence is expected, but not studied in detail. Thus, there is a methodological gap in testing the magnitude of FA and other types of bilateral asymmetry in cultivated plants, including family Poaceae. Recent studies show that the most effective method for determining FA is the geometric morphometrics [3, 6, 9, 10].

Its advantage lies in an integrated approach to the object of study, for example, to a sheet plates sample, in which the shape is averaged and a high degree of freedom in Procrustes analysis of variance due to a significant set of landmarks with coordinates XY, which are usually applied along the contour of the sheet plate. Technically, the method consists of: a) labeling, b) mirroring the labels on the left and right half, c) finding a consensus shape and removing the effect of the size of the plate, and: d) determining the difference in the coordinates of the landmarks labeled on the edges of both sides of the plate [1–3, 6, 9, 11].

The geometric morphometrics was used in the study of some environmental factors, such as anthropogenic impact, climat, and geographical isolation [11–14]. The flag (upper) leaf plate of wheat has a linear structure, reaches a length up to 25 cm. It's an important source of photosynthesis assimilates. We have undertaken the comparison of the shape and asymmetry of winter wheat leaf blades depending on the dose of fertilizer application. The tasks of the work included: a) evaluating of the FA of sheet plates; b) a comparison of the features of the shape using different doses of fertilizer.

2 Methods

Wheat cultivation was carried out on a gray forest medium loamy soil on the plateau of the flat relief of the Suzdal district of the Vladimir region (Russia). Four doses of the widely used mineral fertilizer nitroammophos were used: 1) intensive mineral ($N_{90}P_{90}K_{90}$), 2) high-intensity mineral ($N_{120}P_{120}K_{120}$), 3) intensive organic-mineral (60t organic fertilizer + $N_{90}P_{90}K_{90}$), 4) high-intensity organic-mineral (80t organic fertilizer + $N_{120}P_{120}K_{120}$. From each site with an area of 35 m^2 in the third decade of June 2019 (the stage of wax ripeness), 20–25 specimens of wheat plants (*Triticum aestivum* L., "Poem" cultivar) were randomly selected. The collection was carried out twice in one week interval. The upper (flag) leaf blade of wheat is the main source of assimilants, supplying about 80% of the products of photosynthesis to the generative part of the plant.

From each individual, 4–6 sheet plates were collected and the smallest ones were rejected, as according to accepted standards, fluctuating asymmetry is tested in completely formed plates. The number of plants in 4 samples was 19, 22, 23, and 18. Thus, the number of plate samples from each site was ca from 80 to 100.

First, two true landmarks (LM) were labeled at the base and apex of the plate. Those LMs were lying on the axis of symmetry. An important place in FA testing is the choice of labels on the left and right margins of the leaf blade. These landmarks are not generally accepted, and are called landmarks of the second type. They are arranged regularly and are called sliding marks or semilandmark. The left and right margin of plates represented an uneven arc, on which points were divided successively, dividing it into equal segments.

We used a set of TPS softs (Rohlf, 2019). For regular placement of labels, the automatic splitting was used. First, a curve was formed (option "pencil", TPSDig2). Then 48 labels were set ("resample" option, right mouse button) and completely 50 LMs formed (2 landmarks of the 1st true type other 48 were semilandmarks). Semi LMs had the coordinates, possessed homologous properties, therefore had biological meaning and their coordinates could be compared (Fig. 1):

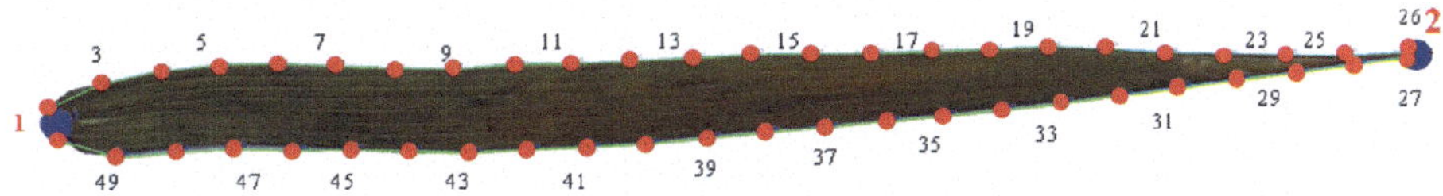

Fig. 1. The arrangement of 50 landmarks on the leaf blade. LMs №1 and №2 are not paired

The labeling was unified: the LMs № 3 and № 50 on the left and № 26 and № 27 on the right were placed as close as possible to the true ones (№ 1 and № 2). Each file was saved in TPS format, a common file was formed (option "append file", TPSUtil) and a resulting file was created with the coordinates of 50 LMs (option "append TPS curve to landmarks"). The resulting file was used in the MorphoJ program package (ver1.07a) to create a data set. Thus, 4 data sets were created, one to each soil fertilizer regime.

The some LM coordinates were removed, qualified as outliers ("find outlines" option). This made it possible to avoid the heterogeneity of the shape that interferes the assessment of FA. A consensus shape (Procrustes fit) was built. The TPSRelw provides visualization of the preferred direction of vectors showing LMs shift (Fig. 2):

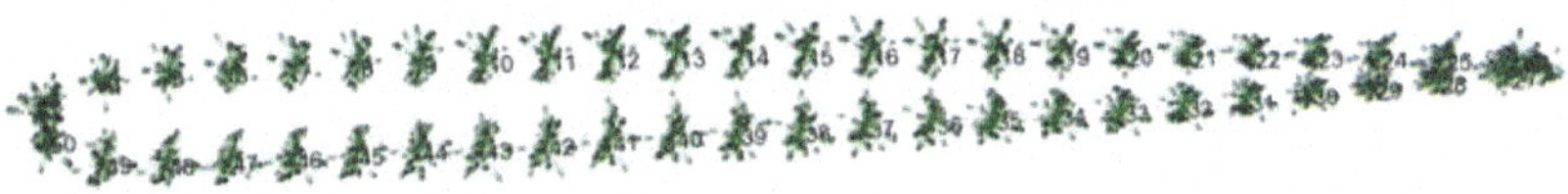

Fig. 2. Superimposed shape and vectors of LMs dispersion

The middle part of the leaf visually varied less than the distal parts. Allometric features of the form were studied using the size of the centroid. By centroid we meant the averaged shape in morphospace with a value equal to the square root of the sum of the squares of the landmarks coordinates along the X- and Y-axes. The centroid size is used as a common size unit in geometric morphometrics [1, 2, 8]. Four leaf samples were compared in individual plant variability, i.e. the category (classifier) "leaf" served as the conditional unit of study.

Statistical significance was tested in Generalized Procrustes Analysis. The interaction of the mean squares "side × leaf" factors indicated a deviation in fluctuating asymmetry in the plate, and accordingly, a deviation in development stability, the "side" factor indicated the presence of directional asymmetry. The supplementary research tools were: a) canonical covariance analysis of the shape symmetry and asymmetry matrices; b) thin spline plates with a total energy of deformation; c) analysis of the procrustes distances between morpho spaces in the first principal components.

The standard error of imaging was determined by repeated photographing. The error of labeling was calculated by re-applying landmarks on each image. All statistical analyzes were carried out using a significance level of 95%; in the analysis of the principal components), permuting reproduction of samples with a number of rounds n = 10,000 was used.

3 Results

A preliminary study included verification of common metric features. The width of the sheet plate varied equally in the samples: the standard error was 0.5 cm ± 0.01. The length of the sheet plate also did not differ in the samples ($F = 0.85$; $p > 0.05$) and had the following values (cm): 13.69 ± 0.31 (№ 1); 13.76 ± 0.46 (№ 2); 14.01 ± 0.36 (№ 3); 14.61 ± 0.48 (№ 4) with a noticeable, although not significant increase of value towards a high dose of organic-mineral fertilizer.

After building a consensus form, an analysis difference in each sample was carried out. A statistically significant difference was in the shape of leaf blades in all 4 samples, i.e. in all cases, the centroid value indicated a high heterogeneity in the shape of the leaf blades. The statistical significance in all groups was on the level $p < 0.0001$–0.001 (Table 1).

Table 1. Consensus shape difference

Dose of fertilizer, №	Source of variation	SS	MS	df	F
1	Centroid size	8116837	427201.9	19	2.9**
	Error	41855858	147379.8	284	
2	Centroid size	7758521	352660	22	1.79*
	Error	63208987	196912.7	321	
3	Centroid size	19990093	869134.5	23	4.2**
	Error	62857874	206769.3	304	
4	Centroid size	2897.822	160.9901	18	7.51**
	Error	5683.067	21.44553	265	

Note: *SS* – sum of square; *df* – degree of freedom; *MS* – mean square; *F* – criterion of Goodall; * – probability level $p < 0.001$;** – probability level $p < 0.0001$

Next, we determined the individual differences in plants ("leaf") and two types of asymmetry ("side" and "leaf × side", Table 2).

Table 2. Procrustes ANOVA results. The source of variation "leaf" indicates the individual variability of the shape of the lamina, "side" – the presence of directional asymmetry. The product "leaf × side" – indicates the presence of non-directional asymmetry, including FA

Dose of fertilizer, №	Source of variation	SS	MS	df	F
1	Leaf (1)	0.07	0.000	912	0.04 ^{ns}
	side (2)	1.18	0.025	48	11.93**
	leaf × side (1 × 2)	1.88	0.002	912	86.16**
	error (3)	0.65	0.000	27264	
2	(1)	0.04	0.0000	1056	1.4**
	(2)	0.00	0.0001	48	3.04**
	(1 × 2)	0.03	0.0000	1056	5.5**
	(3)	0.15	0.0000	30816	
3	(1)	0.03	0.0000	1104	0.69 ^{ns}
	(2)	0.01	0.0001	48	3.75**
	(1 × 2)	0.04	0.0000	1104	7.77**
	(3)	0.13	0.0000	29184	
4	(1)	0.03	0.0000	864	1.28*
	(2)	0.00	0.0001	48	3.61**
	(1 × 2)	0.02	0.0000	864	5.3**
	(3)	0.13	0.0000	25440	

Note: *SS* – sum of square; *df* – degree of freedom; *MS* – mean square; *F* – criterion of Goodall; * – probability level $p < 0.001$; ** – probability level $p < 0.0001$; *ns* – probability level $p > 0.05$(not significant)

A "pure" fluctuating asymmetry was not detected. In all samples a statistically significant directional asymmetry was obtained (MS "side", $p < 0.05$), and therefore a mixture of both types of asymmetry. Dose fertilizers №1 and №3 did not affect the difference in the size of the centroid. This indicated a uniform dispersion of landmarks and the absence of variation in sheet shape. The reduction of the experimental unit to lower levels: "photo" and "image" led to an increase in the degrees of freedom in the analysis and an increase in statistical significance, in all sources of variation $p < 0.0001$. The error in the magnitude of the Procrustes distances consisted of the error of photographing (11.5%) and the error of labeling (88.5%). The total error was 25% of the sum of the squares of the interaction "leaf × side".

To determine the difference, a covariance analysis was performed for the pull of leaves to attempt divide into 4 samples. The first and second canonical variates occupied a large part of the shape variation, which consisted of two components: symmetric and asymmetric. The first component explains the variation in the coordinates of the landmarks along the margins of the sheet plate, while the asymmetric component means the variation in pair landmarks (Fig. 3).

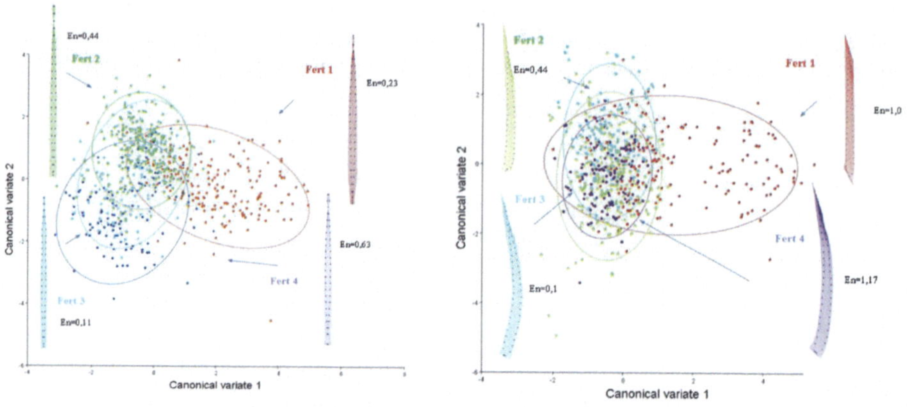

A B

Fig. 3. The canonical variate analysis of symmetric (A) and asymmetric (B) matrices. Fert1-4 – doses of fertilizer. Plot is showing 90% confidence ellipses of population means

The thin-plate spline method was used. Splines created showed the final distribution of vectors in the procrustes space in the first, most informative principle component, explaining 60–80% of the total dispersion of landmarks. Transformational grids demonstrated the conformation of a sheet plate with an affine or isometric rigid shift of landmarks uniform for all of them and in all directions [3, 9]. The energy of structural distortion of the splines was highest in the 4th group of leaves, both in the asymmetry (E = 1.17) and in the symmetry component (E = 0.63). In the Fig. 3B the splines of the asymmetry matrix in doses No. 3–4 have a high curvature of the right side with a characteristic sickle shape, could be explained by the heterogeneous presence of auxin in the leaf primordium.

In each sample, 48 principal components were generated (by the number of paired labels). The first component included from 49.8 to 66.6% of the longitudinal (symmetric) variability of the sheet plate and from 52.1 to 91.7% in the asymmetric (bilateral, mirror) variability, i.e. asymmetry varied more than symmetry. A comparison was made of the procrustes distances and the quadratic distances of the Mahalanobis, which determine the difference between the centers of the sets in the first principal component (Table 3).

Table 3. The difference in the shape of the leaf blade depending on 4 doses of fertilizer. Procrustes and Mahalanobis distances (in brackets) in the first principal component of the singular decomposition of the covariance matrix

Dose, №	Matrix of symmetry			Matrix of asymmetry		
	1	2	3	1	2	3
2	0.016^* (2.753^*)			0.058^* (1.85^*)		
3	0.015^* (2.80^*)	0.007^* (1.34^*)		0.057^* (1.89^*)	0.002^{ns} (1.09^*)	
4	0.019^* (3.179^*)	0.013^* (2.086^*)	0.01^* (1.618^*)	0.057^* (1.91^*)	0.002^{ns} (0.9^*)	0.002^{ns} (1.13^*)

Note: * – $p < 0.0001$; ns – $p > 0.05$

As follows from the table, with an increase in the dose of fertilizer, the symmetric component of the variation showed a stronger variability (everywhere $p < 0.0001$), thus, the variability of the coordinates of the landmarks along the length of the plate was higher than the bilaterally symmetric variability.

4 Discussions

An alternative to the described method is the LAMINA soft usage [15, 16], the advantage of which is in scanning speed and determining the area, shape and size of sheet plates. The authors believe that labeling and processing in the morpho space significantly increases the number of degrees of freedom in the analysis of variance, and, consequently, its power, despite the complexity of labeling, which takes part of the working time.

An increase in the dose of fertilizer affected the symmetric component of the variability of the leaf blade. Procrustes analysis showed that the variability was not allometric because the size of the sheet, which increased with the dose of fertilizer, did not affect the centroid value, as the main characteristic of the shape. The shape of the plate ranged from a statistically insignificant difference (fertilizer № 1 and № 3) to a statistically significant difference (fertilizer № 2 and № 4). At the same time, a high dispersion of variation in morpho-geometric characteristics should be recognized as natural among woody plants in various species and at the borders of their areal [17].

Bilateral asymmetry, defined in the Procrustes analysis, included both types of asymmetry – directional and fluctuating. Directional asymmetry (predominant landmarks shift to one side) revealed in all samples.

Serious arguments in favor of the variability of asymmetry depending on the dose of fertilizer were the following: a) difference in the shape of the asymmetry component (the first principal component, the distance of Mahalanobis); b) an increase in the energy of deformation (bending energy) in the thin splines of covariance matrices. Deviations in the stability of development were not obtained, but in the asymmetry there was a tendency increasing bilateral asymmetry in a mixture of 2 types of asymmetry.

A decrease in the number of landmarks from 50 to 25 did not lead to serious changes in the statistical results. We consider the following to be the final conclusions: a) wheat leaf plates had two types of bilateral asymmetry, an increase in the dose of fertilizer did not affect the stability of wheat development; b) an increase in the dose of fertilizer caused a higher variability in the symmetric, longitudinal variability of the sheet plate shape than in the asymmetric component of variability.

References

1. Auffray, J.C., Debat, V., Alibert, P.: Shape asymmetry and developmental stability. In: Chaplain, M.A.J., Singh, G.D., McLachlan, J.C. (eds.) On Growth and Form: Spatio-Temporal Pattern Formation in Biology, pp. 309–324. Wiley, Chichester (1999)
2. Graham, J.H., Raz, S., Hel-Or, H., Nevo, E.: Fluctuating asymmetry: methods, theory, and applications. Symmetry **2**, 466–540 (2010). https://doi.org/10.3390/sym2020466

3. Baranov, S., Vinokurov, I., Fedorova, L.: Environmental factors affecting the expression of bilateral-symmetrical traits in plants. In: Gene Expression and Phenotypic Traits, 13 December 2019. IntechOpen (2019). https://doi.org/10.5772/intechopen.89460
4. Baranov, S.G., Bibik, T.S., Vinokurov, I.Yu.: Wheat testing developmental stability measurement test in agrocenosis of Vladimir high plain. Adv. Curr. Nat. Sci. **12**, 272–276 (2018). https://doi.org/10.17513/use.37007
5. Gallaher, T.J., Adams, D.C., Attigala, L., Burke, S.V., Craine, J.M., Duvall, M.R., Klahs, P. C., Sherratt, E., Wysocki, W.P., Clark, L.G.: Leaf shape and size tracks habitat transitions across forest-grassland boundaries in the grass family (Poaceae). Evolution **73**(5), 927–946 (2019). https://doi.org/10.1111/evo.13722
6. Pavlinov, I.I., Mikeshina, N.G.: Principles and methods of geometric morphometrics. Zh. Obshch. Biol. **6**(63), 473–493 (2002)
7. Shi, P., Zheng, X., Ratkowsky, D.A., Li, Y., Wang, P., Cheng, L.: A simple method for measuring the bilateral symmetry of leaves. Symmetry **10**(4), 118 (2018). https://doi.org/10.3390/sym10040118
8. Ustyuzhanina, O.A., Sokolova, L.A., Golofteeva, A.S., Burlutsky, V.A.: Vliyanie raznyh mineral'nyh fonov na urozhajnost' i koehfficient fluktuiruyushchej asimmetrii dlya ozimoj i yarovoj pshenic. Problemy regional'noj ehkologii **3**, 99–102 (2017)
9. Klingenberg, C.P.: MorphoJ: an integrated software package for geometric morphometrics. Mol. Ecol. Resour. **11**(2), 353–357 (2011). https://doi.org/10.1111/j.1755-0998.2010.02924.x
10. Krieger, J.D.: Controlling for curvature in the quantification of leaf form. In: Morphometrics for nonmorphometricians 2010, pp. 27–71. Springer, Heidelberg (2010)
11. Vujić, V., Rubinjoni, L., Selaković, S., Cvetković, D.: Small-scale variations in leaf shape under anthropogenic disturbance in dioecious forest forb *Mercurialis perennis*: a geometric morphometric examination. Arch. Biol. Sci. **68**(4), 705–713 (2016). https://doi.org/10.2298/ABS151111011V
12. Andalo, C., Bazin, A., Shykoff, J.A.: Is there a genetic basis for fluctuating asymmetry and does it predict fitness in the plant *Lotus corniculatus* grown in different environmental conditions? Int. J. Plant Sci. **161**(2), 213–220 (2000)
13. Migicovsky, Z., Li, M., Chitwood, D.H., Myles S.: Morphometrics reveals complex and heritable apple leaf shapes. Front. Plant Sci. **4**(8) (2018). https://doi.org/10.3389/fpls.2017.02185
14. Vieira, M., Mayo, S.J., de Andrade, I.M.: Geometric morphometrics of leaves of *Anacardium microcarpum* Ducke and A. occidentale L.(Anacardiaceae) from the coastal region of Piauí, Brazil. Braz. J. Bot. **37**(3), 315–327 (2014). https://doi.org/10.1007/s40415-014-0072-3
15. Dornbusch, T., Andrieu, B.: Lamina2Shape–an image processing tool for an explicit description of lamina shape tested on winter wheat (*Triticum aestivum* L.). Comput. Electron. Agric. **70**(1), 217–224 (2010). https://doi.org/10.1016/j.compag.2009.10.009
16. Graham, J.H., Whitesell, M.J., II, M.F., Hel-Or, H., Nevo, E., Raz, S.: Fluctuating asymmetry of plant leaves: batch processing with LAMINA and continuous symmetry measures. Symmetry **7**(1), 255–268 (2015). https://doi.org/10.3390/sym7010255
17. Klimov, A.V., Proshkin, B.V.: Phenetic analysis of *Populus nigra, P. laurifolia* and *P.* × *jrtyschensis* in natural hybridization zone. Vavilov J. Genet. Breed. **22**(4), 468–475 (2018). https://doi.org/10.18699/vj18.384

Mobile Teleworking – Its Effects on Work/Life Balance, a Case Study from Austria

Michal Beno(✉)

VSM/CITY University of Seattle, Panonska Cesta 17, 85104 Bratislava, Slovakia
michal.beno@vsm-student.sk

Abstract. The nature of work, including how, where and when it is done, is changing rapidly these days. Employees are constantly engaging in modern ICT that permits them to work anywhere, at any time and in any way. Remote working, mobile working, flexible schedules and compressed workweeks are some of the many work adjustments or combinations of adjustments in use in different companies worldwide. ICT naturally reduces the need for physical mobility in a globalised world by means of telepresence and videoconferencing in the face of globalisation, which increases the physical mobility needs (consumerism, offshoring, global sourcing and others). Generally, it was found that work/life balance policies are a remarkable tool for the improvement of performance, high productivity, morale and commitment. The aim of this paper is to analyse the initial issue of mobile teleworking as a problem for work/life balance in an international sports company. The method used is Skype interviews with 12 mobile teleworkers. There is gender equality (six males and six females). The main objectives are to understand if issues in the work/life balance are obstacles to mobile working, in addition whether this kind of work is becoming more common and is expanding, and lastly if mobile telework intensifies the delimitation of work and life. Our results show that mobile teleworking involves readiness to travel and is equated with a high level of autonomy in work and self-discipline. Further, work/life balance is endangered by high mobility, which can be decreased when customers are educated. Moreover, mobile life is not a long-term option, and mobile life is a suitable tool to fit in with family arrangements. Generally, focus on work content is more important than the workplace.

Keywords: Mobile teleworking · Work/life balance · Austria

1 Introduction

There is no doubt that we are living in a world of transformation in the way people work. The need to be mobile is becoming more popular and important. It is not only managers and employees who are on the move these days, but also jobholders in technical and administrative occupations; skilled workers and craftsmen have to travel to subsidiaries or directly to customers.

The traditional work paradigm consists of various factors, such as employment on a contract basis, lifetime employment, standardised working hours, full-time employment (35–40 h/week), state social security, and centralised workplaces (offices, factories,

R. Silhavy (Ed.): CSOC 2020, AISC 1225, pp. 161–171, 2020.
https://doi.org/10.1007/978-3-030-51971-1_13

etc.) [1]. Being present at work physically and being present for face-to-face meetings are the main features of the workplace in the 20th century.

The future of work is changing dramatically. The traditional work paradigm is being replaced by the (mobile) work paradigm of the 21st century, which is characterised by engaging with digital technology, flexibility, mobility, teleworkers and mobile teleworkers, and freelancers who work remotely. The workplace used to be about being in the office, having separate work spaces and being able to meet other workers at just about any time. Nowadays, the boundaries between work, life and free time are disappearing rapidly. This trend started with the increased popularity of flexible working time models and was significantly enhanced by information and communications technology (ICT), such as Wi-Fi, smartphones and tablets. Today we can be reached anywhere at any time by phone or e-mail.

The desire by many people for a work/life balance through modern work forms such as telework, home office, mobile working and flexible schedules is becoming more pressing. According to Luk & Brown [2], continued globalisation will drive the growth of mobile workers among executives, consultants, sales and field professionals and other mobile professionals in all areas.

We interpret telework as a tool providing greater flexibility in the way paid work is done anywhere and at any time. Telework, also known as work from home or mobile working, has attracted renewed attention due to the deployment of ICTs that make it possible to work wherever possible and practical [3]. We explain mobile working as a programme of flexibility including not only telework at home, but also work done at different places (at a client's premises, at a satellite office, in a telecentre or telecottage, etc.) serving different employers, companies or customers, depending on the place and time.

As reported by the Flexible Working Study [4], home office and mobile working is mostly a phenomenon of individual cases in Austrian business society. Only 20% of Austrian employers provide employees with a mobile teleworking option. As stated by Vyslozil [5], every second employee in Austria commutes (52.6%). Having to travel a long way to work upsets the work/life balance (WLB). Furthermore, the physical presence in the office is considered important. We believe that the reduction of commuters by means of flexibility improves the balance between work and life as well as the environment. There are immediate high-impact benefits, such as the reduction of fine dust and less air pollution. Because of the combination of spatially mobile work and modern office concepts, work comes to the people. This generates income and career opportunities and increases well-being.

The intention of this paper is to provide a concise case study of some recent theoretical and empirical research on the basis of data obtained from mobile teleworkers in an Austrian international sports company and the effects of telework on WLB. The aim of the paper is to analyse the initial issue of mobile teleworking as a problem for WLB. The data from the sports company are obtained through Skype interviews with 12 mobile workers; there is gender equality (six males and six females). The evidence gives an insight into questions about whether mobile telework is the solution for WLB, whether new obstacles arise, whether this kind of work is becoming more common and is expanding, and lastly if mobile telework intensifies the delimitation of work and life.

In the following section, we briefly outline the methodology used in our research. The third section gives a short overview of the concept of teleworking and mobile teleworking. The fourth section provides an account of the effect of WLB on telework. In the fifth section, a case study is presented to analyse the results of the Skype interviews. The sixth section is a discussion, and the paper closes with our conclusions.

2 Methodology

We carried out a two-step research project to study teleworking, mobile teleworking and WLB in Austria. The first step was a literature review of the extent and nature of telework, mobile telework and WLB. The second step was Skype interviews for quantitative research among mobile teleworkers in an international sports company in Austria. The data used in this study were collected from 7 February to 25 February 2019. The sample size of the interviewees who participated in the mobile teleworker evaluation study and whose interviews were included in this research was six males and six females to ensure gender equality. It was also necessary that they must all be parents.

The mobile teleworkers' job area included sales. In accordance with a grounded theoretical approach, we used the semi-structured interview format to collect focused data. All mobile teleworkers had completed at least their secondary school education. For additional selected demographics, see Table 1.

Table 1. Description of study informants.

Private sector	Gender	Age in years	Job tenure	Parental status
Mobile teleworkers	6 male 6 female	24–38 (millennials)	1–4 years	1–2 children

The advanced state of ICT offers modern methods of data collection and analysis, for example telephone surveys, computerised data analysis, web-based surveys, e-mail interviews or telepresence interviews (Viber, Skype, Hang-out).

In our study, we decided to use Skype interviews because of the easy setup, avoiding having to meet the workers in person, and enabling us to interview the right person. Interviewing is the most widely used form of data collection in qualitative research [6]. Lo Iacono et al. [7] emphasise that "Skype opens up new possibilities by allowing us to contact participants worldwide in a time efficient and financially affordable manner." Additionally, Skype offers the possibility of audio or video calls, instant messaging, videoconferencing/group discussions and file transfer [8]. As Markham [9] indicates, with the use of the Internet for research "a researcher's reach is potentially global, data collection is economical, and transcribing is no more difficult than cutting and pasting". In 2016, the total number of Skype users was 4.9 million [10]. In Skype interviews, ethical issues are treated the same as in face-to-face and online interviews. Janghorban et al. conclude that this tool of interviewing "offers an alternative or supplemental choice to researchers who want to change from conventional face-to-face interviews" [11].

3 From Teleworking to Mobile Working

Basically, telework describes working at a distance. The idea of working remotely arose in the 1970s [12] during the oil crisis [13]. Based on the interview results in our paper, telework refers to the home office model – working at home. By contrast, mobile telework is a combined model of telework, work at the customer's premises, work on the move (e.g. in a hotel, in the train, working during the trip, etc.) and business trips (meetings, conferences, congresses, events, etc.). Generally, teleworking does not fundamentally mean working from home; different arrangements are possible.

Telework, also known as telecommuting, can be defined simply as when employees work at some place other than the traditional workplace [14]. Traditionally, telework is one of a number of options organisations can choose from to adapt to changing circumstances (market conditions, balancing working life, personal preferences or obligations). A major characteristic of telework is freedom from some of the time and space restrictions that have been associated with work in the past, but are now already being integrated in many kinds of work practice [1]. Typical (regular) and atypical (non-regular) employment were affected by diversification of the labour market and the demands of the 21st century. The typical employment model is characterised by employment in terms of open-ended employment contracts, fixed working hours, the provision of social security and a specific job with definite remuneration [1, 15]. Atypical employment, on the other hand, is based on a fixed-term labour contract; part-time rather full-time work; work that falls outside labour relations and is covered by civil law; new ideas such as working at home, outwork and teleworking; and the distribution of working hours that is adapted to the needs of the employer [15]. We include here cases where the boundary between employment and self-employment is less clear. As stated in the introduction, today's modern economy is built on a mobile paradigm. Valenduc & Vendramin distinguish six major categories of telework: 1) home-based telework; 2) telework in satellite offices; 3) telework in telecentres or telecottages; 4) distance working companies; 5) mobile telework and 6) mixed telework [16, 17].

Mobile telework is one of the growing forms of telework, as stated by Vendramin [17]. She further explains it as being tasks performed at a distance and anywhere – at the client's premises, at home, at a subsidiary, at a colleague's home or even in the train or in a hotel room. Mobile teleworkers are distinguished from traditional fieldworkers by their use of online connections (e-mail). According to EcaTT (electronic commerce and telework trends) norms, mobile workers are defined as those who work at least 10 h per week away from home and their main place of work, e.g. on business trips, in the field, travelling or at the customer's premises [1]. The mobile worker can therefore be thought of as being virtually connected while travelling. This kind of work is characterised by the flexibility of the workplace and the working hours. These travelling workers regularly link up with the company by means of ICT when their work requires business trips, a great deal of travel and the frequent interchange information with the office.

On the basis of the data in Acas [18], we summarise the potential benefits and challenges in Table 2.

Table 2. Mobile working – potential benefits and challenges.

Potential benefits	• can provide savings on office expenses and other overheads • can improve productivity as there is no need to travel to employer's workplace • can result in lower absenteeism and turnover rates
Potential challenges	• can be expensive and some people may feel socially isolated • managers may find it difficult to communicate with and manage remote workers • career development may suffer if away from office often • the recording of working time can be problematic

As stated by Global Workplace Analytics research, Fortune 1000 companies around the globe are entirely revamping their space because employees are already mobile. Studies repeatedly show the employee is not at his/her desk 50%–60% of the time [19]. Based on the recent Strategy Analytics Global Mobile Workforce Forecast [2], the global mobile workforce is set to increase to 1.87 billion people in 2022, accounting for 42.5% of the global workforce. According to the Sixth European Working Conditions Survey [20], most workers (62% of men and 78% of women) have a single main place of work almost all the time. Nearly a third of workers (30%) divide their working time across multiple locations. Despite the popular image of mobile workers as young knowledge workers typing away on their laptops in a park or a café, it is more common in the construction (57%), transport (49%) and agriculture (50%) sectors to have more than one regular place of work.

Among the 28 EU member states, 60.4% of employees in Austria work on computers. Of these, 35% work on computers at the employer's premises, while 25.4% of them work on computers away from the employer's premises [20].

4 The Effect of Work/Life Balance on Telework

Work/life balance (WLB) means a modern and intelligent interconnection between work and private life against the background of changing working and living conditions and taking into account private, social, cultural and health needs. A very central aspect is the balance of family and work life. Integrated WLB concepts include different working time models, adapted work organisation, workplace flexibility models and other supportive and health-protection services for the workplace. We think that WLB is primarily to be understood as an economic issue in which a favourable outcome is guaranteed for everyone involved – individuals, the organisation and society – with economic and social benefits.

The term WLB has become widely used in different fields of debate in recent years, and it is dealt with in various ways. In Fagan et al. [21], the authors refer to WLB as "work and personal life since work is part of life, and therefore to see it in terms of a work/life interface is misleading; and personal life captures the range of commitments and duties which an individual may have, and which can vary across the life course, while still allowing family to be a large part of personal life for most people".

There are clearly two groups of WLB effects of telework. The first (positive) one seems prevalent, and this involves the way that teleworking provides the means to achieve a better WLB [22–26]. This is possible because it helps the individual achieve the preferred balance between work and private life. He/she has the freedom to determine when and where to work, that is temporal and spatial flexibility [27, 28]. We are of the view that the positive features are the flexibility of work/life balance, the time saved by teleworkers because of not having to commute daily and the freedom of being able to perform work and home-related tasks at the same time.

The second (negative) effect is conflict between teleworkers and management of both work and home environments [22, 29, 30]. Kurland and Bailey [3] state that conflict in these environments will lead to blurred work and family boundaries. Furthermore, teleworking individuals have greater difficulty in separating work and family activities [31–33]. Additionally, some theorists indicate the possibility of overload, stress and depression [34, 35].

5 Mobile Teleworking and WLB Results – Case Study

Mobile work is multifaceted. Generally, case studies provide a deeper insight into the descriptions of problems and their solutions, as well as the relevant arrangements and how to overcome obstacles. The workers interviewed for our research are employed by a company that is part of a leading international sports corporation with worldwide subsidiaries. The type of work done is covered by a company agreement that allows work to be performed outside the office. In order to improve competitiveness, and in particular to react quickly to customer requests, all business fields are dependent on ICT. These include devices/systems that enable location-independent information processing. Mobile telework in this company therefore means, firstly, a combination of telework, work at the customer's location and work on the move.

The employers, employees, team members and customers communicate with each other via telephone, smartphone or online via the internal global corporate network and e-mail. For direct exchange, teleworkers use real-time tools for chatting, such as Skype, Hang-out and WhatsApp. Internal meetings with face-to-face contact are mostly the exception, not the rule. Meetings take place instead via so-called conference calls. Meetings with customers are done virtually or face-to-face, depending on the customer's preference.

According to the mobile teleworkers, the open-plan office is not very popular. One of the employees expressed it as follow: "We have both possibilities, we have a large office and a shared version. This means that you have a fixed workplace and no fixed workplace, but only a roller container where you have to empty the desk completely and lock everything away for the next day so another colleague can use it in the meantime". The free choice of a place to work can lead to conflict. A mobile teleworker described it as "not being visible".

In terms of the organisation's agreement, the employee has the right to apply for permission to do mobile telework or to stop doing it, but the final decision is up to the supervisor. The arrangement of working time is not free of conflict. The mobile teleworker allocates time to suit himself/herself, nevertheless the worker is often expected to be available at all times. Conflicts of this nature arise between employees or between employees and supervisors, although nobody is obliged to remain in contact outside working hours.

Working at a customer's premises does not have the benefit of a free working station. The employee works at the customer's place, uses the Internet connection and finishes the tasks under whatever conditions are available. The employee is also at the customer's beck and call at all times. In this case, travel and other expenses are usually incurred according to the tax guidelines. There is no additional remuneration for this. The travel time is recorded and is later given as comp time or is paid out.

5.1 Work-Life Balance from the Employee's Perspective

As shown by our results, all the interviewees agree that WLB is endangered by high mobility, such as travelling to customers and seeing them at their premises. One correspondent reveals that "at home, I do not have this option any more. Most of my work time I spend on the way to the customers." Another one said that "the majority of my work I spent in a hotel, at the airport… You have to be careful that you do not consider that as home." One female respondent summarised a few things the teleworker can do to maintain his/her WLB in the midst of business travels. "Being on your way somewhere or being abroad seems as if you are contracted for eight to five, which becomes online 24/7. If possible, take your family with you adding a few days to your trip."

It is important to mention that as stated by a male respondent: "To educate customers is very important in mobile teleworking." Another male mobile teleworker adds: "Make appointments more rational, so that you are not so often in the office or at the customer's premises, but at home or on the way, where you can often work without interruption and concentrate more." Another one says that "each worker is responsible for the balance between work/life and family". Older respondents emphasise that "our expectations are not to practice this kind of work for a long period… after that one has to rethink the future perspectives." One female Skype respondent says that mobile telework can be seen as a good tool for arranging family affairs: "I live in the countryside and it is hard to combine work and family life due to many circumstances. This kind of work brought me more flexibility." But all the mobile teleworkers agree that one must focus on work content: "It does not matter where you work; it is more about what, when and how you do it." Mutual cooperation with colleagues is important: "We work in virtual teams rather than in real teams," a younger mobile teleworker stressed.

Generally, we discovered that, on the one hand, mobile teleworkers complain of steadily increasing burdens, including the importance of social, health and psychological issues, but others, such as mobility, travel and flexibility are described as a source of autonomy, responsibility and social recognition. We show, on the other hand, that this kind of work is not only an issue for employees, but also for employers, who face problems in connection with the highly specialised knowledge required, the importance of experience, long-term commitment by employees, and the individual circumstances for each case (that is, there are no general solutions).

6 Discussion

The teleworker's workday is more flexible compared with the situation in the past; there are some benefits and some risks. The workday is much more fluid and more outcome-focused rather than time-focused. The tasks and devices operate independently of the location. In our opinion, ICT has been implementing tools that make people more efficient, but of course the workday is no shorter, though modern technologies make it more flexible. Based on the interview data, workers enjoy splitting the day up, which was not possible when work was narrowly defined by location and schedule. In the 20th century, work was just work. There was no separation of activity, outcomes, social aspects, etc., which are characterised by finishing work and going home. Nowadays, all kinds of arrangements for finishing tasks at home, working alone or in collaboration with others at any time and in any place are possible.

Can mobile telework work? Should employees go back to the way we were, with more time spent commuting and in the office, or do something else? We are of the opinion that this kind of work does work, as stated in our paper. Furthermore, we see this issue as a constantly evolving aspect of work life in this century, impacting on the public and private sector and organisations. Each organisation has to revisit this topic annually by asking what is working and what is not working and updating policies and procedures.

7 Conclusions

The image often evoked of the employee working in a traditional office, compared with one who works at home, in a hotel or somewhere on the road, is not a prototype, but rather a special case of mobile working. This kind of work is diverse in Austria. The potential of mobile forms of work has not yet been exhausted. According to the data of the Deloitte survey, the flexitime with core time model is well established in Austria and is used by at least half the employees in 61% of Austrian companies [36].

Mobile working is widespread, because the desire for a satisfactory balance between work and free time is becoming more important for many, and work is less dependent on a single location. The global mobile workforce is set to increase from 1.52 billion in 2017, which accounted for 39.3% of the global workforce, to 1.88 billion in 2023, which will be 43.3% of the global workforce [37].

But in Austria, a large proportion of employees still prefer to work exclusively at the employer's premises. This could change as digitisation progresses. According to the latest data in January 2019, the global digital population represents 4.4 billion active Internet users and nearly 4,0 billion mobile Internet users [38]. An important factor influencing flexibility is ICT, with the large proportion of work done by computers, laptops and smartphones.

In summary, mobile teleworking has both positive and negative aspects. While work and family roles are very important, managing work and family demands is a matter of great concern for many employees. Based on our results, WLB is not achieved purely on an individual level. Mobile telework is often associated with high levels of autonomy and self-determined work. We emphasise that WLB is at risk because of the high mobility (the large amount of travelling), but this can be solved by good customer education. This kind of work is not seen as a long-term prospect, but probably just more of a transitional stage to future work. The tool is useful for balancing family arrangements and work. Mobile working releases you from being tied down regarding where you work; the employee knows what to do and when to do it, but where to do it is his/her own choice.

In the end, however, the employee has to decide individually to what extent to make use of the potential of flexibility. Additionally, we realised that mobile teleworking is not just a problem for employees, but for employers too. We have come to the realisation that the best jobholder leaves the company when he/she is fully competent. Since such a worker is highly sought after in the labour market, he/she is often lured by others. This type of headhunting of skilled employees is a lucrative source of human capital for a company's rivals. A company that suspects its employees are being approached by headhunters should draw up a policy to deal with the problem.

Teleworking is an undeniable reality in today's world, and associated with it is the striving for work/life balance by all role players. Investigating the effects of mobile telework on WLB is essential because, as we have shown in our paper, mobile telework impacts on the individual's WLB in a number of ways.

References

1. ECaTT Final Report (2000). https://web.fhnw.ch/personenseiten/najib.harabi/publications/books/benchmarking-progress-of-telework-and-electronic-commerce-in-europe. Accessed 21 Feb 2019
2. Luk, G., Brown, A.: The Global Mobile Workforce is Set to Increase to 1.87 Billion People in 2022, Accounting for 42.5% of the Global Workforce. https://www.strategyanalytics.com/strategy-analytics/news/strategy-analytics-press-releases/strategy-analytics-press-release/2016/11/09/the-global-mobile-workforce-is-set-to-increase-to-1.87-billion-people-in-2022-accounting-for-42.5-of-the-global-workforce#.WeSldBOCxE5m. Accessed 21 Feb 2019
3. Kurland, N.B., Bailey, D.E.: Telework: the advantage and challenge of working here, there, anywhere, and anytime. Organ. Dyn. Autumn **28**(2), 53–68 (1999)
4. Flexible working Studie 2017, Deloitte (2017). https://www2.deloitte.com/content/dam/Deloitte/at/Documents/human-capital/deloitte-oesterreich-studie-flexible-working-2017.pdf. Accessed 21 Feb 2019
5. Vyslozil, A.: Leere Kilometer: We weit pendelt, ist unglücklich. https://kurier.at/wirtschaft/karriere/leere-kilometer-pendeln-macht-ungluecklich/400036705. Accessed 21 Feb 2019
6. Edwards, R., Holland, J.: What is Qualitative Interviewing? Bloomsbury Academic, London (2013)
7. Lo Iacono, V., Symonds, P., Brown, D.H.K.: Skype as a tool for qualitative research interviews. Sociol. Res. **21**(2), 12 (2016). http://www.socresonline.org.uk/21/2/12.html. Accessed 22 Feb 2019
8. Weller, S.: The Potentials and pitfalls of using Skype for qualitative (longitudinal) interviews. http://eprints.ncrm.ac.uk/3757/1/Susie%20Weller.pdf. Accessed 22 Feb 2019
9. Markham, A.N.: The methods, politics, and ethics of representation in online ethnography. In: Denzin, N., Lincoln, Y. (eds.) Collecting and Interpreting Qualitative Materials, 3rd edn. SAGE Publications, Los Angeles (2008)

10. 26 Amazing Skype Statistics and Facts, November 2017. https://expandedramblings.com/index.php/skype-statistics/. Accessed 22 Feb 2019
11. Janghorban, R., Roudsari, R.L., Taghipour, A.: Skype interviewing: the new generation of online synchronous interview in qualitative research (2014). https://pdfs.semanticscholar.org/902c/f289abd362f7304a2971c9065ee90813d7d6.pdf. Accessed 22 Feb 2019
12. Nilles, J.M., Carlson, R.F., Gay, P., Hanneman, G.J.: The Telecommunications-Transportation Tradeoff. Options for Tomorrow. Wiley, New York (1976)
13. Scholefield, G., Peel, S.: Managers' attitudes to teleworking. NZ J. Employ. Relat. **34**(3), 1–13 (2009)
14. Beno, M.: Transformation of human labour from stone age to information age. In: Younas, M., Awan, I., Ghinea, G., Catalan Cid, M. (eds.) Mobile Web and Intelligent Information Systems. MobiWIS 2018. Lecture Notes in Computer Science, vol. 10995, pp. 205–216. Springer, Cham (2018)
15. Liptak, K.: Is atypical typical? – atypical employment in Central Eastern European countries. http://www.emecon.eu/fileadmin/articles/1_2011/emecon%201_2011%20Liptak.pdf. Accessed 22 Feb 2019
16. Valenduc, G., Vendramin, P.: Le travail à distance dans la société de l'information. http://www.ftu-namur.org/fichiers/Emerit17.pdf. Accessed 22 Feb 2019
17. Vendramin, P.: Telework in the scenarios for the future of work. http://www.ftu-namur.org/fichiers/TW98-pvgv.pdf. Accessed 22 Feb 2019
18. Flexible working and work-life balance. http://m.acas.org.uk/media/pdf/3/1/Flexible_working_and_work_life_balance_Nov.pdf. Accessed 25 Feb 2019
19. GlobalWorkplaceAnalytics.com. https://globalworkplaceanalytics.com/telecommuting-statistics. Accessed 22 Feb 2019
20. Eurofound, Sixth European Working Conditions Survey (2016). https://www.eurofound.europa.eu/sites/default/files/ef_publication/field_ef_document/ef1634en.pdf. Accessed 24 Feb 2019
21. Fagan, C., Lyonette, C., Smith, M., Saldana-Tejeda, A.: The influence of working time arrangements on work-life integration or 'balance': a review of the international evidence. https://www.ilo.org/wcmsp5/groups/public/@ed_protect/@protrav/@travail/documents/publication/wcms_187306.pdf. Accessed 25 Feb 2019
22. Ng, F.Ch.: Academics telecommuting in open and distance education universities: issues, challenges, and opportunities. Int. Rev. Res. Open Distance Learn. **7**(2), 16 p. (2006)
23. Grint, K.: The Sociology of Work. Polity Press, Cambridge (2005)
24. Pettinger, L.: Friends, relations and colleagues: the blurred boundaries of the workplace. In: Parry, J., Taylor, R., Pettinger, L., Glucksman, M. (eds.) A New Sociology of Work. Blackwell Publishing, London (2005)
25. Jeffrey, H.E., Hawkins, A.J., Ferris, M., Weitzman, M.: Finding an extra day a week: the positive influence of perceived job flexibility on work and family life balance. Fam. Relat. **50**(1), 49–58 (2001)
26. Beno, M.: Managing telework from an Austrian manager's perspective. In: Rocha, Á., Adeli, H., Reis, L.P., Costanzo, S. (eds.) Trends and Advances in Information Systems and Technologies. WorldCIST 2018. Advances in Intelligent Systems and Computing, vol. 745, pp. 16–29. Springer, Cham (2018)
27. Schots, M., Taskin, L.: Flexible working times: towards a new employment relationship? https://dial.uclouvain.be/pr/boreal/object/boreal%3A17513/datastream/PDF_01/view. Accessed 25 Feb 2019
28. Peters, P., Den Dulk, L., Tanja Van der Lippe, T.: The effects of time-spatial flexibility and new working conditions on employees' work-life balance: the Dutch case, community. Work Fam. **12**(3), 279–297 (2009)

29. Tremblay, D.G.: Balancing work and family with telework? Organizational issues and challenges for women and managers. Women Manag. **17**, 155–170 (2003/2004)
30. The Guardian, Work-life balance: flexible working can make you ill, experts say. https://www.theguardian.com/money/2016/jan/02/work-life-balance-flexible-working-can-make-you-ill-experts-say. Accessed 25 Feb 2019
31. Hill, E.J., Miller, B.C., Weiner, S.P.: Influences of the virtual office on aspects of work and work/life balance. Pers. Psychol. **51**(3), 667–683 (1998)
32. Lewis, S., Cooper, C.L.: The work-family research agenda in changing contexts. J. Occup. Health Psychol. **4**(4), 382–393 (1999)
33. Tietze, S., Musson, G.: Recasting the home-work relationship: a case of mutual adjustment? Organ. Stud. **26**(9), 1331–1352 (2005)
34. Duxbury, L.E., Higgins, Ch.A., Thomas, D.R.: Work and family environments and the adoption of computer-supported supplemental work-at-home. J. Vocat. Behav. **49**(1), 1–23 (1996)
35. Tavares, A.I.: Telework and health effects review, and a research framework proposal. MPRA (2015). https://mpra.ub.uni-muenchen.de/71648/1/MPRA_paper_71648.pdf. Accessed 25 Feb 2019
36. Deloitte. https://www2.deloitte.com/content/dam/Deloitte/at/Documents/human-capital/deloitte-oesterreich-studie-flexible-working-2017.pdf. Accessed 25 Feb 2019
37. Luk, G.: https://www.strategyanalytics.com/access-services/enterprise/mobile-workforce/market-data/report-detail/global-mobile-workforce-forecast-update-2017-2023. Accessed 25 Feb 2019
38. Global digital population as of January 2019 (in millions). https://www.statista.com/statistics/617136/digital-population-worldwide/. Accessed 25 Feb 2019

A Binary Bat Algorithm Applied to Knapsack Problem

Lorena Jorquera[1(✉)], Gabriel Villavicencio[1], Leonardo Causa[2], Luis Lopez[1], and Andrés Fernández[1]

[1] Escuela de Ingeniería en Construcción, Pontificia Universidad Católica de Valparaíso, Valparaíso, Chile
{lorena.jorquera,gabriel.villavicencio, luis.lopez,andres.fernandez}@pucv.cl

[2] Escuela de Tecnologías de la Información, Facultad de Ingeniería y Negocios, Universidad de las Américas, Santiago, Chile
lcausamo@edu.udla.cl

Abstract. Combinatorial problems with NP-hard complexity appear frequently in operational research. Making robust algorithms that solve these combinatorial problems is of interest in operational research. In this article, a binarization mechanism is proposed so that continuous metaheuristics can solve combinatorial problems. The binarization mechanism uses the concept of percentile. This percentile mechanism is applied to the bat algorithm. The NP-hard knapsack problem (MKP) was used to verify our algorithm. Additionally, the binary percentile algorithm was compared with other algorithms that have recently has solved the MKP, observing that the percentile algorithm produces competitive results.

Keywords: Combinatorial optimization · Multidimensional knapsack problem · Metaheuristics

1 Introduction

Combinatorial problems frequently appear in the industry. For example we found combinatorial problems in Bio-informatics [1], Operations Research [2–5], allocation problems [6–8], vehicle routing problems [9], robust optimization, and problems of scheduling [10–12], among others.

In recent years we have witnessed the generation of a large number of nature-inspired algorithms that are capable of solving optimization problems efficiently. Examples of these algorithms are a black hole, gravitational search, and multi-verse algorithm, among others. Due to the nature of the phenomenon, we find that many of these algorithms work in continuous search spaces and must be adapted to solve the MKP since the latter is a combinatorial problem. The adaptation process must ensure that the exploitation and exploration mechanisms are preserved so that the efficiency of the algorithm is maintained.

R. Silhavy (Ed.): CSOC 2020, AISC 1225, pp. 172–182, 2020.
https://doi.org/10.1007/978-3-030-51971-1_14

When performing a state of the art, it is found that there are several binarization techniques that allow adapting algorithms that work in continuous spaces based on binary spaces. Among the main binarization methods used are the functions of transfer, angular modulation, and quantum focus; The reader can consult [13,14] for more details. In this article, we developed a binarization method based on the percentile technique to perform the binarization process.

Bat metaheuristic algorithm was selected to verify the effectiveness of the percentile technique in the binarization process. Bat algorithm was proposed by [] and has been successfully applied to various engineering and operations research problems. The binary version of the bat algorithm will be denoted by BPBA and this version will be applied to MKP. Experiments were developed that account for the percentile operator contribution in the binarization process. For this, a random operator was built that serves as a baseline to measure the contribution. Additionally, to improve the results, a local search operator was designed. In the comparison, two KMTR algorithms proposed by [15] and BAAA developed in [16] were selected. The first uses the k-means technique to execute the binarization transformation and the second uses transfer functions. The results show that the percentile algorithm contributes substantially to the binarization process and that binarization generates competitive results.

2 KnapSack Problem

The allocation of resources which is modeled by MKP is a frequent problem at the industrial level. The objective of MKP is to identify a subset of objects, which produce the greatest benefit by satisfying a set of restrictions specific to each problem. MKP corresponds to one of the most studied problems that fall in the NP-hard category. When we look at the literature, we see that MKP has been solved using different techniques. For example, in [15] an unsupervised learning technique was applied. A quantum binarization was designed in [17] to address MKP. In [18] a fruit fly algorithm was adapted, and in [19] a differential algorithm was used. Formally MKP is defined as:

$$\text{maximize} \sum_{j=1}^{n} p_j x_j \tag{1}$$

$$\text{subjected to} \sum_{j=1}^{n} c_{ij} x_j \leq b_i, \; i \in \{1, ..., m\}. \tag{2}$$

With $x_j \in \{0, 1\}$, $j \in \{1, ..., n\}$. p_j represents the profit of the item j. c_{ij} corresponds a cost related to dimension i and element j. In each dimension, the constraints i are noted by b_i. The solution can be conceptualized using a binary representation, in this way a 0 means absence of the element inside the knapsack.

3 Binary Percentile Bat Algorithm

The areas of data mining and statistics are cross-cutting disciplines and we can find applications of them in different situations, such as transports, agriculture, cities, and mathematics [9,15,20–24]. The percentile technique applied to the binarization of metaheuristics that work in continuous spaces is explored in this section. In Fig. 1, a general diagram of the solution is shown. As a first stage, the solutions must be initialized, then it is determined if the criterion of the maximum number of iterations has been met. In the case that it has not been fulfilled, the bat algorithm is executed and using the solutions obtained, the percentile algorithm is applied to perform its binarization and subsequently a repair algorithm for the case that there is any restriction that is not complied with. Once the binarization has been carried out through the percentile technique, the solutions obtained are compared with the best. In the case of having a superior, a local search operator is executed. Otherwise, the previous steps are repeated.

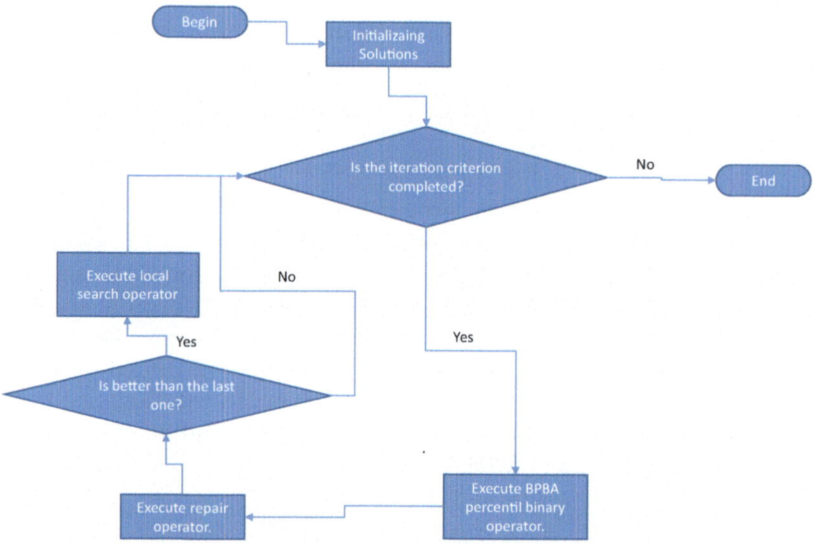

Fig. 1. Flowchart of the binary percentile bat algorithm.

3.1 Initialization and Element Weighting

Since the bat algorithm is a swarm algorithm, to begin the search for the best solution, it is necessary to initialize the list of solutions. The mechanism used in the generation of each solution is as follows: first, a random element is chosen. Then, it should be verified if it is feasible to incorporate new elements, therefore, it is necessary to evaluate the restrictions of the problem. In the selection of the new element, if it is feasible, a list of possible candidates is generated that is consistent with the restrictions. Each item is assigned a previously calculated weight and the one with the best weight is selected. This procedure is repeated until no additional element can be incorporated.

In the calculation of the weight of an element, several heuristics have been developed. For example, in [25] a pseudo-utility was proposed. The equation that allows the calculation is written in Eq. 3. In this equation, the variable w_j corresponds to the surrogate multiplier whose value is between 0 and 1. This multiplier can be interpreted intuitively as a shadow price of the constraint j.

$$\delta_i = \frac{p_i}{\sum_{j=1}^{m} w_j c_{ij}} \tag{3}$$

A more intuitive measure focused on the average resource occupancy was proposed by [26]. It is shown in Eq. 4.

$$\delta_i = \frac{\sum_{j=1}^{m} \frac{c_{ij}}{m b_j}}{p_i} \tag{4}$$

In this article, a variation based on the average occupation will be used. This variation was proposed in [15]. In this proposal, the existing elements in the backpacks are also considered to calculate the average occupancy. Therefore, in each iteration, according to the elements selected for each solution, the measurement must be calculated again. The mathematical expression that defines this measure is shown in Eq. 5.

$$\delta_i = \frac{\sum_{j=1}^{m} \frac{c_{ij}}{m(b_j - \sum_{i \in S} c_{ij})}}{p_i} \tag{5}$$

3.2 Percentile Operator

Considering that the bat algorithm works in a continuous space and also it is iterative. Then, it results in the speed and position of each solution being updated in $\mathbb{R}^n$. In Eq. 6 the above is written in general. In this equation x_{t+1} corresponds to the position of the particle x at the time $t+1$. To calculate the position, the *Delta* function, which is specific to each algorithm, must be considered additionally.

$$x_{t+1} = x_t + \Delta_{t+1}(x(t)) \tag{6}$$

In the case of the bat algorithm, binarization is done through a binary percentile algorithm. This algorithm considers that, given a solution x, we calculate the magnitude of its displacement $\Delta^i(x)$ in the i-th component and then group all the displacements in order to obtain the values for the percentiles {20, 40, 60, 80, 100}. In each percentile, a transition probability is associated that corresponds to the values shown in Eq. 7. Using these transition probabilities together with Eq. 8 the binarization of the solutions is performed. The algorithm is detailed in Algorithm 1.

$$P_{tr}(x^i) = \begin{cases} 0.1, & \text{if } x^i \in \text{group } \{0,1\} \\ 0.5, & \text{if } x^i \in \text{group } \{2,3,4\} \end{cases} \tag{7}$$

$$x^i(t+1) := \begin{cases} \hat{x}^i(t), & \text{if } rand < P_{tg}(x^i) \\ x^i(t), & \text{otherwise} \end{cases} \tag{8}$$

Algorithm 1. Percentile operator

```
Function pbinary(vLst, pLst)
Input vList, pList
Output pGroupValue
pValue = getPValue(vLst, pLst)
for each value in vList do
    pGroupValue = getPGroupValue(pValue,vLst)
end for
return pGroupValue
```

3.3 Repair Operator

Because the percentile operator and the local search operator can generate solutions that do not satisfy any of the constraints, it is necessary to apply a repair operator. The measure described in Eq. 5 was used to select the elements that perform the repair. In the case that repair is required, the element with the maximum measure is chosen and removed from the solution. This process is iterated until a solution that satisfies all constraints is obtained. The pseudocode is shown in Algorithm 2.

Algorithm 2. Repair Algorithm

```
Function Rep(Sl_in)
Input Input solution Sl_in
Output The Rep solution Sl_out
S ← S_in
while needRepair(S) == True do
    s_max ← MaxWeight(S)
    S ← removeElement(S, s_max)
end while
state ← False
while state == False do
    s_min ← MinWeight(S)
    if s_min == Null then
        state ← True
    else
        S ← addElement(S, s_min)
    end if
end while
S_out ← S
return S_out
```

4 Results

4.1 Understanding BPBA Contribution

In this section, the experiments are developed to determine the contribution of the percentile binary operator in the binarization process. To determine this contribution, problems cb.5.250 of the OR library were selected. As comparison artifacts, violin plots, tables, and statistical comparisons were used. Particularly, the non-parametric Wilcoxon test was used to perform analyses. In the plots, the X-axis identifies the analyzed instances and the Y-axis the %-Gap defined in Eq. 9. The Wilcoxon test aims to verify whether the difference between the results obtained by BPBA is significant compared to the other algorithms. The exploration of the parameters is detailed in Table 1.

$$\% - Gap = 100\frac{BestKnown - SolutionValue}{BestKnown} \tag{9}$$

Table 1. Setting of parameters for binary percentile bat search algorithm.

Parameters	Description	Value	Range
N	Solutions	30	[20, 25, 30]
G	Percentiles	5	[4, 5, 6]
Iteration number	Maximum number of iterations	1000	[1000]

Percentile Binary Operator Contribution. To understand the contribution of the percentile operator to the binarization process a random operator was constructed to be considered as a baseline. This random operator was configured to perform transitions with a fixed probability of 0.5 without considering in which percentile the speed variable is located. Two configurations were analyzed: the first configuration includes the local search operator and the second configuration does not consider this operator. The above with the objective of decoupling the contribution of the local search operator and the percentile. BPBA corresponds to our standard algorithm. *rd.ls* is the random variant that includes the local search operator. *BPBA.wls* corresponds to the version with a binary percentile operator without a local search operator. Finally, *rd.wls* describes the random algorithm without a local search operator.

The first comparison corresponds to comparing the best values between BPBA and *rd.ls*. These results are shown in Table 2. BPBA exceeds *rd.ls*. Despite being higher, the values are close. Our second comparison corresponds to considering the averages. In this case, BPBA exceeds *rd.ls* in almost all problems bluntly. The comparison of distributions through the violin plots shown in Fig. 2, indicates that the dispersion of the distributions *rd.ls* is greater than the dispersions of BPBA. In particular, this is notorious in problems 1, 2, 4, 5, 6 and 9. Therefore, the percentile and local search operators as a whole contribute to

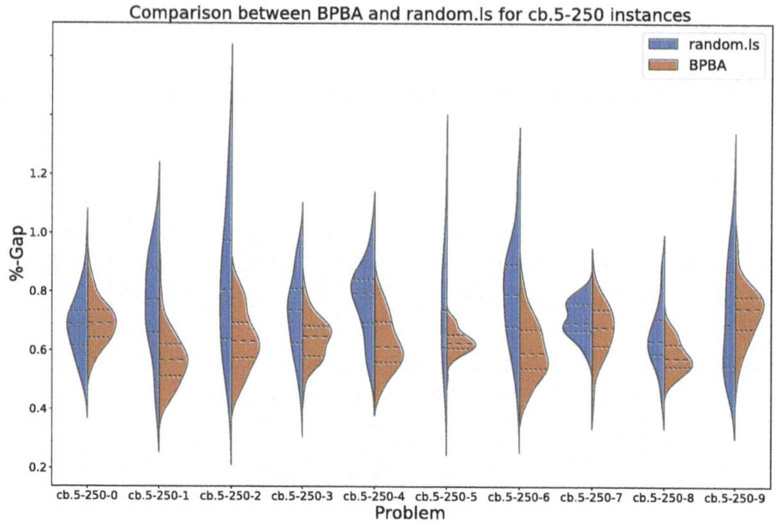

Fig. 2. Evaluation of percentile binary operator with local search operator

improving and making the results more robust. Finally, BPBA distributions are closer to zero than the *rd.ls* distributions, indicating that BPBA has consistently better results than *random.ls*. When we evaluate the behavior of the algorithms through the Wilcoxon test, this indicates that the difference is significant.

The following analysis aims to separate the contribution of the local search operator from the binary percentile operator, in order to be able to size the contribution of the percentile operator to the binarization process. For this, the comparison between the *wls* and *rd.wls* algorithms is made.

Table 2. Evaluation of percentile binary operator

Set	Best Known	Best *rd.ls*	Best BPBA	Best *rd.wls*	Best *wls*	Avg *rd.ls*	Avg BPBA	Avg *rd.wls*	Avg *wls*
cb.5.250-0	59312	59211	59225	59158	59175	59124	59131	59038.3	59101.7
cb.5.250-1	61472	61435	61472	61409	61409	61276.1	61335.9	61284.9	61344.1
cb.5.250-2	62130	62036	62074	61969	61969	61862.4	61969.9	61771.4	61900.7
cb.5.250-3	59463	59367	59446	59365	59349	59205.5	59260.3	59113.4	59214.5
cb.5.250-4	58951	58914	58951	58883	58930	58671.3	58809.7	58664.8	58747.5
cb.5.250-5	60077	60015	60056	59990	60015	59874.9	59911.1	59823.6	59924.1
cb.5.250-6	60414	60355	60355	60348	60349	60196.1	60283.7	60210.6	60266.2
cb.5.250-7	61472	61436	61472	61407	61407	61253.6	61286.8	61209.3	61341
cb.5.250-8	61885	61829	61885	61790	61782	61686.3	61738.1	61615	61683
cb.5.250-9	58959	58832	58866	58822	58787	58762.7	58671.7	58631.5	58735.5
Average	60413.5	60343.0	60380.2	60314.1	60317.2	60191.3	60239.8	60136.3	60225.8
p-value							2.67 e–05		1.45 e–04

The results are shown in Table 2: when analyzing the best results, we observe that *wpe* gets better results than *05.wpe* in all problems except 1, 6 and 7. However, the results are again very close. When we analyze the average, *wpe* exceeds *05.wpe* in all problems. The Wilcoxon test indicates that the difference is significant. This indicates that *wpe* is consistently better than *05.wpe*. Note that in this case we are not using the local search operator. The violin plots are shown in Fig. 3. This plots show that the dispersion of the solutions for the case of *05.wpe* is much greater than in the case of *wpe*. This indicates that the percentile binary operator applied to the bat algorithm plays an important role in the accuracy of the results.

4.2 BPBA Comparisons

This section aims to evaluate the performance of BPBA with the BAAA algorithm. The BAAA algorithm uses transfer functions as a general binarization mechanism. In particular, BAAA used the $\tanh = \frac{e^{\tau|x|}-1}{e^{\tau|x|}+1}$ to make the transfer. The τ parameter of the *tanh* function was set to a value of 1.5. In addition, BAAA also uses a local search procedure to improve solutions. As a maximum number of iterations, BAAA used 35,000 very different from the case of BPBA. In our BPBA algorithm, the configurations are the same used in the previous experiments. In addition to the comparison with BAAA, a comparison was also made with the KMTR-BH and KMTR-Cuckoo binarizations. KMTR uses the K-means learning technique without supervision to perform the binarization process.

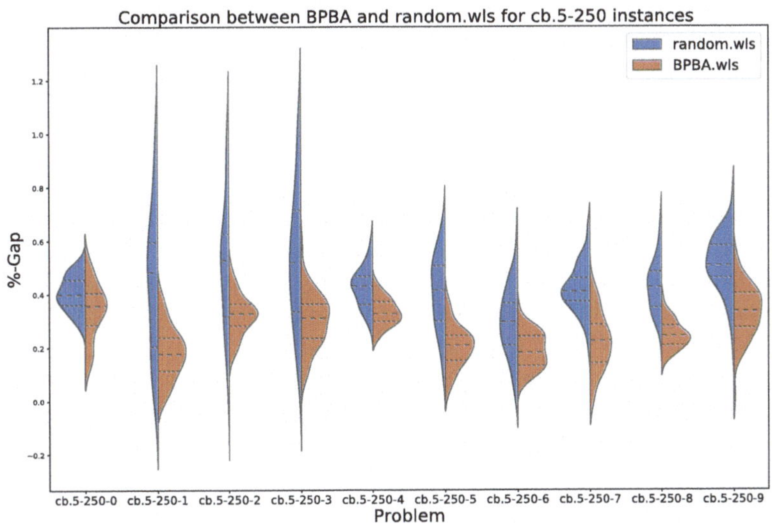

Fig. 3. Evaluation of percentile binary operator without local search operator

Table 3. OR-Library benchmarks MKP cb.5.500

Instance	Best Known	BAAA Best	Avg	KMTR-BH Best	Avg	KMTR-Cuckoo Best	Avg	BPBA Best	Avg	Std
0	120148	120066	120013.7	120096	120029.9	120082	120036.8	120082	120003.4	62.4
1	117879	117702	117560.5	117730	117617.5	117656	117570.6	117656	117569.3	80.9
2	121131	120951	120782.9	121039	120937.9	120923	120855.1	120923	120757.2	75.4
3	120804	120572	120340.6	120683	120522.8	120683	120455.7	120683	120458	76
4	122319	122231	122101.8	122280	122165.2	122212	122136.4	122212	122011.1	64.3
5	122024	121957	121741.8	121982	121868.7	121946	121824.6	121982	121684.3	56.4
6	119127	119070	118913.4	119068	118950	118956	118895.5	118956	118793.7	55.2
7	120568	120472	120331.2	120463	120336.6	120392	120320.4	120487	120189.3	65.8
8	121586	121052	120683.6	121377	121161.9	121201	121126.3	121295	121120.4	64.2
9	120717	120499	120296.3	120524	120362.9	120467	120335.5	120467	120201.7	68
10	218428	218185	217984.7	218296	218163.7	218291	218208.9	218291	218088.1	88.4
11	221202	220852	220527.5	220951	220813.9	220969	220862.3	220951	220688.2	64.1
12	217542	217258	217056.7	217349	217254.3	217356	217293	217356	217187.8	54.6
13	223560	223510	223450.9	223518	223455.2	223516	223455.6	223516	223441.8	81
14	218966	218811	218634.3	218848	218771.5	218884	218794	218848	218799.3	72.5
15	220530	220429	220375.9	220441	220342.2	220433	220352.7	220410	220358.8	81.2
16	219989	219785	219619.3	219858	219717.9	219943	219732.8	219858	219750.2	76.6
17	218215	218032	217813.2	218010	217890.1	218094	217928.7	218010	217939.5	70.1
18	216976	216940	216862	216866	216798.8	216873	216829.8	216866	216790.2	86.3
19	219719	219602	219435.1	219631	219520	219693	219558.9	219631	219522.2	80.8
20	295828	295652	295505	295717	295628.4	295688	295608.8	295688	295622.3	71.1
21	308086	307783	307577.5	307924	307860.6	308065	307914.8	307924	307797.6	62.9
22	299796	299727	299664.1	299796	299717.8	299684	299660.9	299796	299613	83.7
23	306480	306469	306385	306480	306445.2	306415	306397.3	306415	306284.9	61.5
24	300342	300240	300136.7	300245	300202.5	300207	300184.4	300207	300051.9	76.4
25	302571	302492	302376	302481	302442.3	302474	302435.6	302481	302435	64.8
26	301339	301272	301158	301284	301238.3	301284	301239.7	301272	301196.7	63
27	306454	306290	306138.4	306325	306264.2	306331	306276.4	306331	306096.4	65.4
28	302828	302769	302690.1	302749	302721.4	302781	302716.9	302771	302631.8	47
29	299910	299757	299702.3	299774	299722.7	299828	299766	299757	299709.9	54.7
Average	214168.8	214014.2	213862	214059.5	213964.1	214044.2	213959.1	214037.4	213893.1	69.2

The results of the experiment are detailed in Table 3. For comparison, the problem set cb.5.500 of the OR library was used. The results for BPBA were obtained from 30 executions for each problem. To distinguish the best results, these were marked in black. For the best value indicator, BAAA was greater in 4, KMTR-BH in 11, KMTR-Cuckoo in 8 and BPBA in 8. We must bear in mind that the sum is greater than 30 because in some cases there was a tie between some of the algorithms In the average indicator, BAAA was higher in 2 cases, KMTR-BH in 12, KMTR-Cuckoo in 7 and BPBA in 5. We must also take into account that the standard deviation In most of the problems it was quite low, indicating that BPBA has good accuracy.

5 Conclusions

In this article, the percentile technique was explored as a method to perform the binarization of the bat algorithm. We must emphasize that the percentile technique is a general binarization technique and therefore can be applied in the binarization of any continuous swarm intelligence algorithm. MKP was used to evaluate the performance of the binary algorithm. The contribution of the percentile binary operator was analyzed, and it was concluded that this operator contributes to the accuracy and quality of the solutions obtained. Additionally, a comparison was made with the newly developed BAAA and KMTR algorithms and the results show that BPBA has a good performance.

As a future line of research, it is interesting to use the bat algorithm in the three binarization mechanisms: Percentile, k-means, and transfer functions. In addition, it is interesting to binarize other metaheuristics with the percentile technique along with solving other NP-hard problems. Another line of longer-term research corresponds to exploring adaptive techniques to dynamically change the parameterization of metaheuristics, in this sense it is interesting to explore reinforced learning techniques.

References

1. Barman, S., Kwon, Y.-K.: A novel mutual information-based boolean network inference method from time-series gene expression data. PLoS ONE **12**(2), e0171097 (2017)
2. Crawford, B., Soto, R., Astorga, G., García, J.: Constructive metaheuristics for the set covering problem. In: International Conference on Bioinspired Methods and Their Applications, pp. 88–99. Springer, Cham (2018)
3. García, J., Crawford, B., Soto, R., Astorga, G.: A percentile transition ranking algorithm applied to binarization of continuous swarm intelligence metaheuristics. In: International Conference on Soft Computing and Data Mining, pp. 3–13. Springer, Cham (2018)
4. Crawford, B., Soto, R., Monfroy, E., Astorga, G., García, J., Cortes, E.: A meta-optimization approach for covering problems in facility location. In: Workshop on Engineering Applications, pp. 565–578. Springer, Cham (2017)
5. García, J., Crawford, B., Soto, R., Astorga, G.: A clustering algorithm applied to the binarization of swarm intelligence continuous metaheuristics. Swarm Evol. Comput. **44**, 646–664 (2019). http://www.sciencedirect.com/science/article/pii/S221065021730528X
6. Garcia, J., Crawford, B., Soto, R., Astorga, G.: A percentile transition ranking algorithm applied to knapsack problem. In: Proceedings of the Computational Methods in Systems and Software, pp. 126–138. Springer, Cham (2017)
7. Astorga, G., Crawford, B., Soto, R., Monfroy, E., García, J., Cortes, E.: A meta-optimization approach to solve the set covering problem. Ingeniería **23**(3) (2018)
8. García, J., Lalla-Ruiz, E., Voß, S., Droguett, E.L.: Enhancing a machine learning binarization framework by perturbation operators: analysis on the multidimensional knapsack problem. Int. J. Mach. Learn. Cybern. 1–20 (2020)
9. García, J., Peña, A.: Robust optimization: concepts and applications. In: Nature-Inspired Methods for Stochastic, Robust and Dynamic Optimization, p. 7 (2018)

10. García, J., Crawford, B., Soto, R., García, P.: A multi dynamic binary black hole algorithm applied to set covering problem. In: International Conference on Harmony Search Algorithm, pp. 42–51. Springer, Singapore (2017)
11. García, J., Altimiras, F., Peña, A., Astorga, G., Peredo, O.: A binary cuckoo search big data algorithm applied to large-scale crew scheduling problems. Complexity **2018** (2018)
12. Crawford, B., Soto, R., Monfroy, E., Astorga, G., García, J., Cortes, E.: A meta-optimization approach to solve the set covering problem. Ingeniería **23**(3), 274–288 (2018)
13. Crawford, B., Soto, R., Astorga, G., García, J., Castro, C., Paredes, F.: Putting continuous metaheuristics to work in binary search spaces. Complexity **2017** (2017)
14. García, J., Moraga, P., Valenzuela, M., Crawford, B., Soto, R., Pinto, H., Peña, A., Altimiras, F., Astorga, G.: A Db-scan binarization algorithm applied to matrix covering problems. Comput. Intell. Neurosci. **2019** (2019)
15. García, J., Crawford, B., Soto, R., Castro, C., Paredes, F.: A k-means binarization framework applied to multidimensional knapsack problem. Appl. Intell. **48**(2), 357–380 (2018)
16. Zhang, X., Wu, C., Li, J., Wang, X., Yang, Z., Lee, J.-M., Jung, K.-H.: Binary artificial algae algorithm for multidimensional knapsack problems. Appl. Soft Comput. **43**, 583–595 (2016)
17. Haddar, B., Khemakhem, M., Hanafi, S., Wilbaut, C.: A hybrid quantum particle swarm optimization for the multidimensional knapsack problem. Eng. Appl. Artif. Intell. **55**, 1–13 (2016)
18. Meng, T., Pan, Q.-K.: An improved fruit fly optimization algorithm for solving the multidimensional knapsack problem. Appl. Soft Comput. **50**, 79–93 (2017)
19. Liu, J., Wu, C., Cao, J., Wang, X., Teo, K.L.: A binary differential search algorithm for the 0–1 multidimensional knapsack problem. Appl. Math. Model. **40**(23–24), 9788–9805 (2016)
20. García, J., Pope, C., Altimiras, F.: A distributed k-means segmentation algorithm applied to lobesia botrana recognition. Complexity **2017** (2017)
21. Graells-Garrido, E., García, J.: Visual exploration of urban dynamics using mobile data. In: International Conference on Ubiquitous Computing and Ambient Intelligence, pp. 480–491. Springer, Cham (2015)
22. Garcia, J., Măntoiu, M.: Localization results for zero order pseudodifferential operators. J. Pseudo-Differ. Oper. Appl. **5**(2), 255–276 (2014)
23. Graells-Garrido, E., Peredo, O., García, J.: Sensing urban patterns with antenna mappings: the case of santiago, chile. Sensors **16**(7), 1098 (2016)
24. Peredo, O.F., García, J.A., Stuven, R., Ortiz, J.M.: Urban dynamic estimation using mobile phone logs and locally varying anisotropy. In: Geostatistics Valencia 2016, pp. 949–964. Springer, Cham (2017)
25. Pirkul, H.: A heuristic solution procedure for the multiconstraint zero? One knapsack problem. Naval Res. Logist. **34**(2), 161–172 (1987)
26. Kong, X., Gao, L., Ouyang, H., Li, S.: Solving large-scale multidimensional knapsack problems with a new binary harmony search algorithm. Comput. Oper. Res. **63**, 7–22 (2015)

Comparative Analysis of DoS and DDoS Attacks in Internet of Things Environment

Abdulrahman Aminu Ghali(✉), Rohiza Ahmad, and Hitham Seddiq Alhassan Alhussian

Computer and Information Sciences Department, Universiti Teknologi PETRONAS, 32610 Seri Iskandar, Perak Darul Ridzuan, Malaysia
aminuabdulrahman81@yahoo.com, {rohiza_ahmad, seddig.alhussian}@utp.edu.my

Abstract. With the high rate of accepting Internet of Things (IoT) from individual nowadays. The IoT devices comprises television sets (TV), voice controller, smart lock, smart light, speakers, nest smoke alarm, air conditioning, ring doorbell and smart plug etc. Thus, the impact of IoT provides many industries with new ways of handling business operations such as procurement, manufacturing, distribution of goods and service, etc. With the aid of intelligent sensors, actuators, etc., the industries have revolutionized to an era known as Industry 4.0 (IR4.0), where business operations can be done and potentially optimized, by controlling interconnecting "things" via data communication channels. Similar to other electronic communication platforms in IoT ensuring security is paramount important especially in detecting the menace of the security challenges in the IoT environment. This paper aims to perform a comparative analysis between Denial of service (DoS) and Distributed denial of service (DDoS) attacks in the IoT environment. In essence, the comparative analysis reveals that the DDoS attack is the most significant attack to address in the IoT environment with about 68% Gbps requests compared to the DoS attack with 58% Gbps request.

Keywords: DoS · DDoS · IoT environment · Security threat

1 Introduction

The advent of IoT has provided many product-based industries such as transportation, sports, agriculture and health care system with new ways of handling operations such as procurement, manufacturing and distribution of goods and services to save individual time [1–3]. In IoT, not only devices or "Things" can send and receive data from one another when connected to the internet, however, they can also remotely control the operations of others. For instance, doctors can now monitor their patient's health remotely with the aid of sensors attached to the patient's body. The sensor reads several data from the body and send to the doctor's mobile phone who can diagnose the ailment of the patient in case of an emergency [4, 5]. This enabled due to the networking of the IoT devices that is linked with each other to perform a specific task.

R. Silhavy (Ed.): CSOC 2020, AISC 1225, pp. 183–194, 2020.
https://doi.org/10.1007/978-3-030-51971-1_15

Despite, the tremendous achievement brought by the IoT that ease individuals live, it suffers several challenges such as Denial of Service (DoS) and Distributed Denial of Service (DDoS) [6–9].

In view of the challenges of these attacks in the IoT environment, there is need for research to study the trend of these challenge. The paper aims to perform a comparative analysis of the DoS and DDoS attacks to identify area for further researcher. Although the aforementioned attacks are common in other areas, but the impact is bigger in the IoT environment as communicated data are not just received, but they can act as triggers for subsequent actions.

Furthermore, denying a legit device from communicating may lead to the catastrophic result in the IoT environment. For instance, when an alarm is set to wake up a person at home for medication, the connected devices are not just being connected but rather continues ringing until the person wakes up and take medicine, if the sender of the message or any device being denied to send such important message it may lead to a catastrophic situation to the patient's health [10, 11]. Therefore, this study investigates the most significant attack between the two forms of attacks with a view to propose solution that can improve the IoT environment. Figure 1 further described the simple IoT environment. From the figure, the cloud is connected with the various IoT devices with the aim of performing specific task. Once the devices connected, it collects, sends and receive data without the human intervention.

The paper is organized as follows: Sect. 2 discusses the DoS attacks. Section 3 describes the DDoS attacks. Section 4 provides a comparative analysis. Section 5 covers the conclusion of the paper.

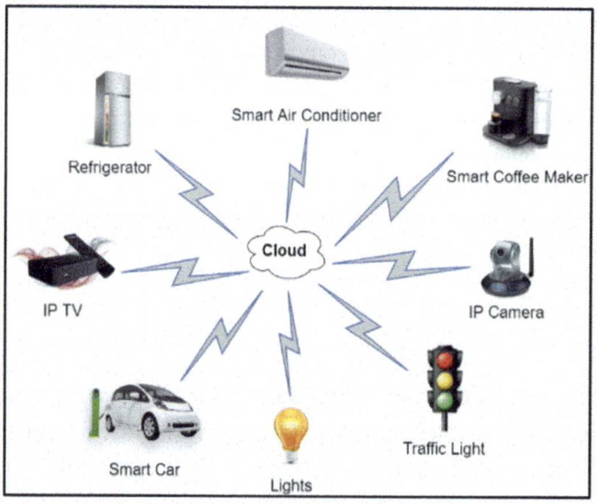

Fig. 1. IoT environment

2 Denial of Service Attack

The DoS attack remains an issue in the IoT environment due to the fact that it stands very difficult to neutralize by the IoT users. Thus, such an attack needs a quick solution for IoT users. There are various types of DoS attacks in the IoT environment. These are the denial of sleep attack, path-based DoS attack, jamming, wormhole, vampire attack, carousel attack, and stretch attack etc. [12]. The main aim of instigating the DoS attacks in the IoT environment is to make the system not available by making the network and the server unreachable. This attack lead to the untrustworthiness of the IoT data and reduces system availability [13]. The effect of these vulnerabilities is that it can allow attacks from a remote place using tools and command by the attacker which generates more damages to the IoT environment. For instance, in 2013 over 120 Gbps of network traffic occurred from unknown attackers instigating Spamhaus DoS attack to the UK industries, such an attack is considered as the largest attack in 2013 [14]. In addition, a DoS attack can be seen as a major vulnerability in the IoT environment that needs urgent attention. Hence, it is very important to prevent DoS attacks in the IoT environment in order to safeguard the entire devices and the IoT data. The examples given below indicate the damages of DoS attacks not only in the IoT environment but also in the related environment. The intruder commits various damages through DoS attack on the IoT devices with the intention of denying or modifying data from the IoT environment. For instance, in 2013 the Chinese internet went down because of a DoS attack in which China wasn't capable of defending itself from the attacks [14]. Similarly, in 2014, unnamed internet service provider experiences an attack called network time protocol (NTP) the attack reaches 400 Gbps network traffic which seen as the largest attack in history so far caused by DoS attacks. Also, UK-based phone carrier Carphone warehouse fall as a victim of DoS attack in which intruders took millions of customer's data in 2015. In the same year, threats of a DoS attack occurred in Microsoft's, Xbox Live service was also attacked by the intruders in which the successfully weak the security of Microsoft's services. Likewise, in 2016 HSBC customers were also victims of DoS attack by losing access to their online banking account in the United Kingdom [14]. Furthermore, in 2017 and 2018, the intruders launched various attacks on the IoT devices with the intention of denying users to access their data [15]. Similarly, it is being reported that the menace of DoS attacks worldwide exceeds 12,000 attacks in every three weeks according 2001 reports [16, 17].

2.1 How DoS Attacks Works

As earlier mentioned, the DoS attack denies a legitimate user from accessing data in the IoT devices. In such attack, the intruder usually sends several requests to target the IoT server, in order to make the server overloaded with heavy traffic. These will make the devices no longer communicating with each other. In this attack, a single request is sent to change the address of the devices and overloads the IoT server not to function [18, 19]. Besides, the attacker takes the advantages of using the botnet for the malicious attack environment. While the botnet is a zombies that compromised the devices by running malicious software under command and control of the attacker [20]. From the figure. The intruder sends a single request to the IoT server with the aim of shutting down the server or unavailable to the user. Figure 2 shows how the connected devices were not able to receive or sends data to the user due to heavy request that were sends. Figure 2 will illustrate how DoS attacks works.

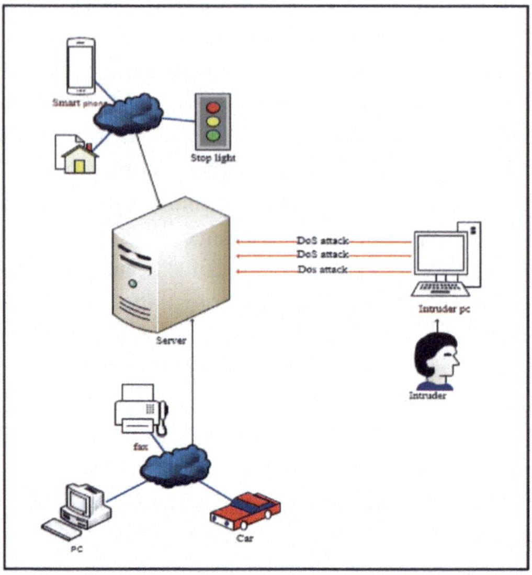

Fig. 2. How DoS attacks works

2.2 Sinkhole Attack

The inspiration of DoS attacks by the intruders leads to the untrustworthiness of the data, where intruder attracts the IoT devices with false routing update. The sinkhole attack presents a very serious threat to the IoT environment. The consequences of this attack, make the routing protocols to be vulnerable to destroy the connectivity of the network [21, 22]. The severe consequences of such attack leads to the privacy violation and data modification. For example, the sinkhole attack works when the intruder attacks the devices by sending a request to the nodes, the attacker nodes respond to the request by sending the routed packets via the attacker nodes.

2.3 Triggered Attacks by DoS

DoS attackers have launched several attacks in the IoT environment with the aim deny users access. The trigger of the attack can take off the network as mentioned in the previous sections. The stimulated attacks can be described as follows.

Application Layer Attacks. The application layer is concerned with security issues such as confidentiality, integrity, authentication and availability. In this layer, the attacks target the server. Therefore, is very hard to detect or circumvent such attack in this layer, while at the same time the attackers use various machines to deliver attacks in this layer [23, 24]. Among the protocol in this layer include Hypertext Transfer

Protocol (HTTP), Constrained Application Protocol (COAP) and Message Queuing Telemetry Transport (MQTT) attacks to mention a few [25].

Fragmentation Attack. Fragmentation attacks are the attacks usually occurs in the network layer to alter the packets. The attacker sends a deceived packet to the network to overwhelms with false packets. The network infected due to the heavy packet in the header than intended [26].

Volumetric Attack. In this kind of attack, the network bandwidth is crushed by the attack. The attacks are sent based on Internet Control Message Protocol (ICMP) as a form of a request to put down the network capabilities. Such attacks aim to stop and reduce the speed services rendered by the IoT network [27].

TCP-State Exhaustion Attack. TCP state exhaustion attacks run down the webserver connections. The risk of the attack limited the connection of the devices to less number of selected device by the attacker's choice [28].

SYN Flood Attack. Synchronization (SYN) flood is an attack sends to target the server. The severe of the attack hog the server by sending fake traffic. Besides, the network will overload with a heavy request, and prevents the user from connecting to the IoT network [29].

Buffer Over Flow Attack. This attack happens in the other environment while in the IoT environment also occurs. The attacks overload the network by malfunctioning the data to unveil the private information. Once the network is malfunctioned series of data will leak [30].

Teardrop Attack. The attack is sends to breakdown the network by sending complicated IP data packets. In the process of recompiling the fragmented packets into original packets the devices will be denied user by accessing the network. The resultant of these attacks obscure the devices by not communicating with each other [31].

3 Distributed Denial of Service Attack

The DDoS attack brings a lot of setback in the IoT environment. As such the attacks gained the utmost attention from IoT users. In 2016, the attacks continue having impacts on the IoT devices which recorded as the highest attacks in the year mentioned above. In the same year of Mirai the attack destroyed thousands of connected devices in the history of IoT [32–34]. The DDoS attacks can be described as an attack that send in several requests to compromise the IoT system. The request is targeting the server to disrupt the normal traffic flow of the data. Figure 3 displays the DDoS attacks in the IoT environment.

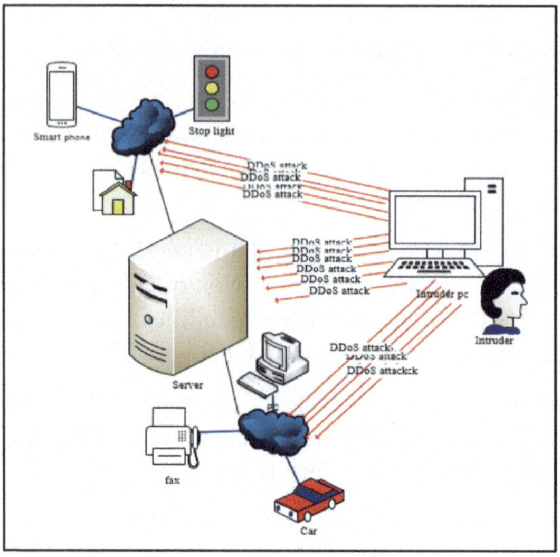

Fig. 3. How DDoS attacks sends

3.1 Triggered Attacks by the DDoS

The next section will provide various attacks that instigated by DDoS attackers in the IoT environment.

Slowloris Attack. This type of attack is launched to crush down the webserver. The menace of this attack is based on the HTTP header requests to hog down the network. The concerns of the attack make the server not connecting to the user application in the phone. Thus, the attack does not require any bandwidth to be active [35].

UDP Flood Attack. User datagram protocol flood is an attack with the aim of denying users for accessing the IoT networks. The attempt is to send a huge number of packets to the remote hosts. This will enable the host not to communicate to the other ports, therefore no application will send or receive packets [36].

Ping Flood Attack. This type of attack is similar to the UDP floods attack; the difference is, ping flood attack uses Internet control message protocol (ICMP) echo request to launch the attack in the network. The nature of the attack consumes the network bandwidth in which the server can only reply with the attackers ICMP echo packets. The fears of the attack contribute to the network being weak [37].

HTTP Flood Attack. Hypertext transfer protocol attack is one of the attacks uses HTTP GET or POST request to denied webserver from communicating in the application layer. Thus, the attack completely terminates the network connection between the IoT devices [38].

3.2 Possibility of DDoS Attacks in the IoT Environment

There are number of DDoS attacks possibility in the IoT environment. Among these includes:

Network Security is Extremely Interdependent. In this type of attack, the attackers aim is to target the network in other to crushed down the network performance. The attackers use various way to launched the attacks due to the internet interdependent. Thus, the attacks are vulnerable to DoS and DDoS but depends on the security mechanism provided by the user to prevent the system.

Lack of Software Update. Lack software updates contribute in several attacks. The attacks are always possible due to the patches of the software that were not updated. Thus, updating the software removes the perilous areas that may cause destruction to the software. Whereas, updating the software enhances various features and removes outdated software that may cause damages to the IoT software.

Devices Security is Interdependent. The attack is similar to the network security interdependent attack. In this attack, the attacker targets the IoT devices to acquire sensitive information from the targeted devices. The attack succeeds in many cases because it relies on the IoT systems, when a single device victimizes then others fall as victims.

3.3 Classification of DoS/DDoS Attacks in the IoT Environment

The following section will provide the classification of DoS and DDoS attacks in the IoT environment. Whereas, the classification is based on the motivation by the aforesaid attack.

Financial Reasons. The attacks from the financial aspect are described as a worrisome in the IoT environment. Usually, these attacks are based on financial benefits that the attackers gain. Such attacks require experience and technical skills to perform the attacks.

Revenge Motive. The attack is usually based on revenge purposes, various attackers of this sort have weaker technical skills and are eager to launch an attack if they have been allowed to do so because of the perceived injustice.

Intellectual Motive. In this type of attack, the attackers are inspired to target the IoT system either for the experimentation reasons or learning purposes. Thus, the attacks are motivated by the young attackers who want to show their skills based on the attacking motives on the IoT environment.

Experience Motive. The attackers are considered as the most dangerous attackers in the IoT environment. Most of these attackers have high technical skills with excellent experience in targeting a system. Considering their skills, the attackers have the motive to attack the IoT environment for personal gain.

Competitive Motive. In this category of attack, the attackers are motivated based on competition reasons. The motivation of the attack is seen as the riskiest attack in the IoT environment due to the highest inspirations by other attackers. Once the attacker gets the motive by other attackers, the next is to attack he IoT systems.

4 Comparative Analysis

This study has employed the use of Wireshark version 3.0.4 with Windows 7 Intel(R) Xeon(R) Processor 3.40 GHz, 8 GB RAM to achieved the analysis. Thus, the dataset use for the analysis is free and available in public domain for easy access and use. The dataset use for the analysis is based on Canadian institute for cybersecurity (CICDDoS 2019 and CICDoS 2017). Based on the results of the analysis most significant attacks in the IoT environment are shown on Fig. 4. Whereas, the overall result will indicate the focal contributions of the study based on DoS and DDoS attacks in the IoT environment. Figure 4 will display the timeline of the attacks in the IoT environment.

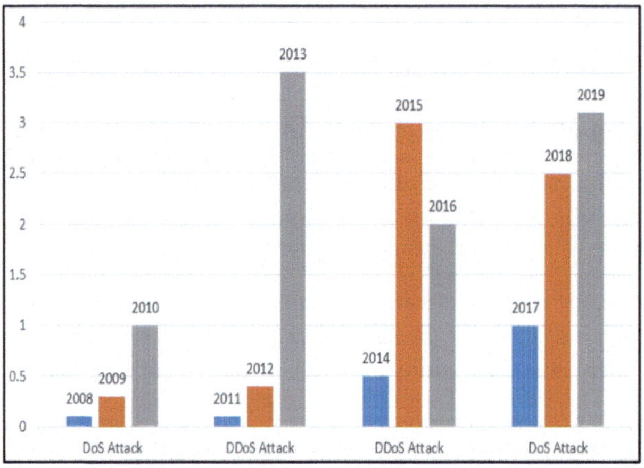

Fig. 4. The timeline of DoS/DDoS attacks

As shown in Fig. 4, the statistical analysis of the DoS and DDoS timeline indicates that each of the year the attacks fluctuate. This specifies that there are no permanent attacks on specific timeline and urgent attention need to be considered such as confidentiality, integrity and authentication in addressing the aforesaid problems. Figure 5 illustrate the comparison between the DoS and DDoS attacks. The figure indicating the most significant attack that will be addressed in the IoT environment. Whereas, there are many studies review the IoT security issues with the aim of providing solutions such as methods and framework for preventing DoS and DDoS attacks in the IoT environment. As such, none of the work have provided the most significant attacks between the two aforesaid attacks.

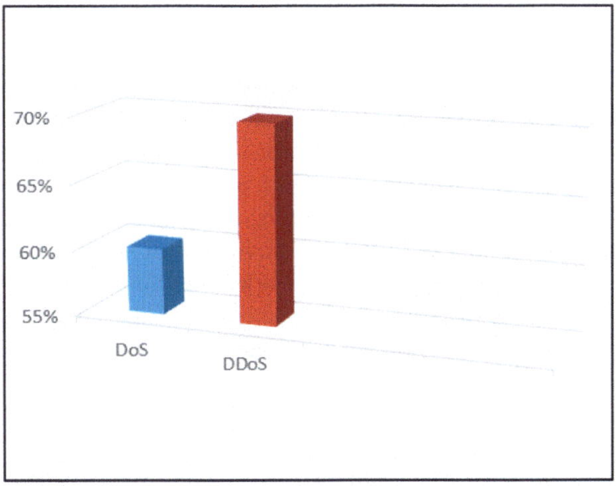

Fig. 5. Comparison between DoS/DDoS attacks

Furthermore, the study analysis indicates that DDoS attacks is the most significant attacks that need urgent attention to address in the IoT environment with the highest attention. Thus, the analysis indicates that the DDoS attack has 68% Gbps request compared to the DoS attack with about 58% Gbps request. The threshold based on the analysis indicates that the higher the percentage of the request the more the impact is made, and the lower the percentage of the request the less impact of the attacks. This analysis expected to provides a novel way for researchers to develop a new method, model and framework for addressing the identified attacks in the IoT environment. Figure 6 illustrate the threshold summary.

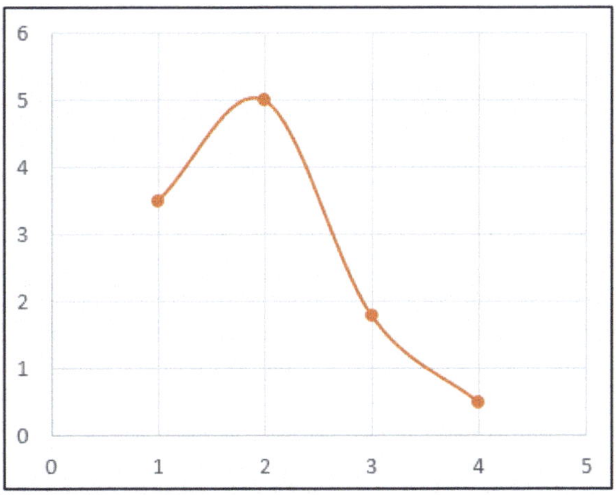

Fig. 6. Threshold summary

5 Conclusion and Future Work

The study performs a comparative analysis between the DoS and DDoS attacks in the IoT environment. The effectiveness of both attacks were discussed. The main contribution of the study is to identify the most significant attacks in the IoT environment that will aid the researchers in providing a robust model in preventing aforementioned attacks in an IoT environment. Based on the results presented the DDoS attack indicates the most significant attacks in the IoT environment with the highest request in terms of attracting the server or the IoT network. Therefore, there is urgent need for enhancing the IoT models in preventing the menace of the attacks in the IoT environment. This will provide the efficiency of the model in covering the unexplored characteristics of the attacks. Hence, the study is exploring a new way to provide a robust hybrid approach to eliminate the recurrence of the DDoS attacks in the IoT environment. This will increase the effectiveness of the model in providing high quality model that will prevent the attack. Additionally, the study aims to develop an algorithm based on the proposed approach that can enhance the efficiency of the battery life at the perception layer even when the DDoS attack occurs.

Acknowledgement. The research was fully supported by the Centre of Graduate Studies (CGS). The authors fully acknowledge the Universiti Teknologi PETRONAS (UTP) for financial support, which has made this research possible.

References

1. Pirmagomedov, R., Koucheryavy, Y.: IoT Technologies for Augmented Human: a Survey. Internet of Things, p. 100120 (2019)
2. Boubiche, D.E., Pathan, A.S.K., Lloret, J., Zhou, H., Hong, S., Amin, S.O., Feki, M.A.: Advanced industrial wireless sensor networks and intelligent IoT. IEEE Commun. Mag. **56** (2), 14–15 (2018)
3. Serpanos, D., Wolf, M.: Industrial internet of things. In: Internet-of-Things (IoT) Systems, pp. 37–54. Springer, Cham (2018)
4. Debauche, O., Mahmoudi, S., Manneback, P., Assila, A.: Fog IoT for health: a new architecture for patients and elderly monitoring. Procedia Comput. Sci. **160**, 289–297 (2019)
5. Aghili, S.F., Mala, H., Shojafar, M., Peris-Lopez, P.: LACO: lightweight three-factor authentication, access control and ownership transfer scheme for e-health systems in IoT. Future Gener. Comput. Syst. **96**, 410–424 (2019)
6. Peraković, D., Periša, M., Cvitić, I.: Analysis of the IoT impact on volume of DDoS attacks. In: XXXIII Simpozijum o novim tehnologijama u poštanskom i telekomunikacionom saobraćaju–PosTel, 2015, pp. 295–304 (2015)
7. Sicari, S., Rizzardi, A., Miorandi, D., Coen-Porisini, A.: REATO: REActing TO denial of service attacks in the internet of things. Comput. Netw. **137**, 37–48 (2018)
8. Liu, G., Quan, W., Cheng, N., Zhang, H., Yu, S.: Efficient DDoS attacks mitigation for stateful forwarding in internet of things. J. Netw. Comput. Appl. **130**, 1–13 (2019)
9. Roohi, A., Adeel, M., Shah, M.A.: DDoS in IoT: a roadmap towards security & countermeasures. In: 2019 25th International Conference on Automation and Computing (ICAC), pp. 1–6. IEEE, September 2019

10. Kajwadkar, S., Jain, V.K.: A novel algorithm for DoS and DDoS attack detection in internet of things. In: 2018 Conference on Information and Communication Technology (CICT), pp. 1–4. IEEE, October 2018
11. Boussada, R., Hamdane, B., Elhdhili, M.E., Saidane, L.A.: Privacy-preserving aware data transmission for IoT-based e-health. Comput. Netw. **162**, 106866 (2019)
12. Patil, A., Gaikwad, R.: Comparative analysis of the prevention techniques of denial of service attacks in wireless sensor network. Procedia Comput. Sci. **48**, 387–393 (2015)
13. Carl, G., Kesidis, G., Brooks, R.R., Rai, S.: Denial-of-service attack-detection techniques. IEEE Internet Comput. **10**(1), 82–89 (2006)
14. Lachance, L., et al.: The Impact of Denial of Service Attacks in the IoT (2016). https://www.globalsign.com/en/blog/denial-of-service-in-the-iot/
15. Khan, M.A., Salah, K.: IoT security: review, blockchain solutions, and open challenges. Future Gener. Comput. Syst. **82**, 395–411 (2018)
16. Moore, D., Voelker, G., Savage, S.: Inferring internet denial-of-service activity. In: 10th USENIX Security Symposium, Washington DC (2001)
17. Park, H., Li, P., Gao, D., Lee, H., Deng, R.H.: Distinguishing between FE and DDoS using randomness check. In: International Conference on Information Security, pp. 131–145. Springer, Heidelberg, September 2008
18. Baig, Z.A., Sanguanpong, S., Firdous, S.N., Nguyen, T.G., So-In, C.: Averaged dependence estimators for DoS attack detection in IoT networks. Future Gener. Comput. Syst. **102**, 198–209 (2020)
19. Jazi, H.H., Gonzalez, H., Stakhanova, N., Ghorbani, A.A.: Detecting HTTP-based application layer DoS attacks on web servers in the presence of sampling. Comput. Netw. **121**, 25–36 (2017)
20. Bertino, E., Islam, N.: Botnets and internet of things security. Computer **2**, 76–79 (2017)
21. Salehi, S.A., Razzaque, M.A., Naraei, P., Farrokhtala, A.: Detection of sinkhole attack in wireless sensor networks. In: 2013 IEEE International Conference on Space Science and Communication (IconSpace), pp. 361–365. IEEE, July 2013
22. Bugeja, J., Jacobsson, A., Davidsson, P.: On privacy and security challenges in smart connected homes. In: 2016 European Intelligence and Security Informatics Conference (EISIC), pp. 172–175. IEEE, August 2016
23. Mahmoud, R., Yousuf, T., Aloul, F., Zualkernan, I.: Internet of things (IoT) security: current status, challenges and prospective measures. In: 2015 10th International Conference for Internet Technology and Secured Transactions (ICITST), pp. 336–341. IEEE, December 2015
24. HaddadPajouh, H., Dehghantanha, A., Parizi, R.M., Aledhari, M., Karimipour, H.: A survey on internet of things security: requirements, challenges, and solutions. Internet of Things, p. 100129 (2019)
25. Da Cruz, M.A., Rodrigues, J.J., Lorenz, P., Solic, P., Al-Muhtadi, J., Albuquerque, V.H.C.: A proposal for bridging application layer protocols to HTTP on IoT solutions. Future Gener. Comput. Syst. **97**, 145–152 (2019)
26. Pongle, P., Chavan, G.: A survey: attacks on RPL and 6LoWPAN in IoT. In: 2015 International Conference on Pervasive Computing (ICPC), pp. 1–6. IEEE, January 2015
27. Ojeda Adan, M.F.: Designing an Internet of Things Attack Simulator (2019)
28. Keary, T.: Dos vs DDoS Attacks. Blog (2018). https://www.comparitech.com/net-admin/dos-vs-ddos-attacks-differences-prevention/?fbclid=IwAR3QQtEzPjiK8aHdn1TmXgJGcKIvEwXXWKNZixmukQ0ztUvi0sJkgeZJLB8#Broad_Types_of_DOS_and_DDOS_Attacks
29. Chen, Q., Chen, H., Cai, Y., Zhang, Y., Huang, X.: Denial of service attack on IoT system. In: 2018 9th International Conference on Information Technology in Medicine and Education (ITME), pp. 755–758. IEEE, October 2018

30. Mullen, G., Meany, L.: Assessment of buffer overflow based attacks on an IoT operating system. In: 2019 Global IoT Summit (GIoTS), pp. 1–6. IEEE, June 2019
31. Bao, C., Guan, X., Sheng, Q., Zheng, K., Huang, X.: A tool for denial of service attack testing in IoT. Presented at the 1st Conference on Emerging Topics in Interactive Systems (2016)
32. Kolias, C., Kambourakis, G., Stavrou, A., Voas, J.: DDoS in the IoT: Mirai and other botnets. Computer **50**(7), 80–84 (2017)
33. Hallman, R., Bryan, J., Palavicini, G., Divita, J., Romero-Mariona, J.: IoDDoS—the internet of distributed denial of service attacks. In: IoTBDS, pp. 47–58 (2017)
34. Tripathi, N., Mehtre, B.M.: DoS and DDoS attacks: impact, analysis and countermeasures. In: National Conference on Advances in Computing, Networking and Security, Nanded, India (2013). https://docs.google.com/viewer
35. Hirakawa, T., Ogura, K., Bista, B.B., Takata, T.: A defense method against distributed slow HTTP DoS attack. In: 2016 19th International Conference on Network-Based Information Systems (NBiS), pp. 152–158. IEEE, September 2016
36. Huraj, L., Simon, M., Horák, T.: IoT measuring of UDP-based distributed reflective DoS attack. In: 2018 IEEE 16th International Symposium on Intelligent Systems and Informatics (SISY), pp. 000209–000214. IEEE, September 2018
37. Pajila, P.B., Julie, E.G.: Detection of DDoS attack using SDN in IoT: a survey. In: Intelligent Communication Technologies and Virtual Mobile Networks, pp. 438–452. Springer, Cham (2019)
38. Kambourakis, G., Kolias, C., Stavrou, A.: The mirai botnet and the IoT zombie armies. In: MILCOM 2017–2017 IEEE Military Communications Conference (MILCOM), pp. 267–272. IEEE (2017)

Horse Optimization Algorithm: A Novel Bio-Inspired Algorithm for Solving Global Optimization Problems

Dorin Moldovan(✉)

Faculty of Automation and Computer Science, Technical University of Cluj-Napoca, Cluj-Napoca, Romania
dorin.moldovan@cs.utcluj.ro

Abstract. We introduce a novel bio-inspired algorithm named Horse Optimization Algorithm (HOA) which has as primary source of inspiration the hierarchical organization of the horse herds. The article presents the main principles of the proposed algorithm, the pseudo-code version of the algorithm and a modified version named Discrete Binary Horse Optimization Algorithm (DBHOA). HOA is evaluated and validated using six objective functions and is compared with Chicken Swarm Optimization (CSO) and Cat Swarm Optimization Algorithm (CSOA). Finally, the article presents the application of DBHOA in features selection for data generated in a smart grid for the classification of the stability of the system and it considers as experimental support the Electrical Grid Stability Simulated Dataset.

Keywords: Horse Optimization Algorithm · Swarm Intelligence · Smart grid · Machine learning · Features selection

1 Introduction

Bio-inspired computing refers to a class of optimization algorithms which apply the intelligence of the nature and one of their main applications is the solving of the complex engineering optimization problems. Even though the bio-inspired algorithms are very popular due to the large diversity of the sources of inspiration such as the echolocation behavior of the bats [1], the foraging behavior of the ravens [2], the climbing of the wild goats [3], the hunting mechanism of the owls in the dark [4] and the food search and the mating behavior of the butterflies [5], the research literature does not contain a bio-inspired algorithm which returns the best results in the case of all optimization problems. Some bio-inspired algorithms are better in the exploration of the search space while others are better in the exploitation of the search space. The primary objective of this research article is to introduce an algorithm which approaches both the exploration and the exploitation of the search space, is simple to implement and

R. Silhavy (Ed.): CSOC 2020, AISC 1225, pp. 195–209, 2020.
https://doi.org/10.1007/978-3-030-51971-1_16

is comparable to other bio-inspired algorithms in terms of performance. Moreover, this article presents the adaptation of the newly introduced algorithm in features selection for high dimensional datasets.

The main contributions of this research article are:

(1) The development of an algorithm inspired from the hierarchical organization of the horse herds called HOA;
(2) The adaptation of HOA for optimization problems in which the search space is characterized by arrays of zeros and ones;
(3) The application of HOA in a representative engineering problem, namely the classification of the smart grid stability;

The article is organized as follows: Sect. 2 presents the research background, Sect. 3 presents the main research methods including the general biology of the horse herds and the pseudo-code of HOA, Sect. 4 describes the evaluation and the validation of HOA, Sect. 5 presents the application of HOA in engineering problems with a particular use case in the smart grid stability classification and Sect. 6 presents the main conclusions and the future research work.

2 Research Background

The first subsection presents features selection approaches from literature based on bio-inspired algorithms, in particular those bio-inspired algorithms similar to HOA, namely CSO and CSOA. The second subsection considers approaches from literature applied in the classification of the smart grid stability.

2.1 Features Selection Based on Bio-Inspired Algorithms

In [6] is proposed a system for features selection that is based on CSO and which searches the optimal combination of features maximizing the performance of the classification while minimizing the number of the selected features. Similar to the approach presented in this research article that method was benchmarked on data from UCI repository. However, that approach considers the K-Nearest Neighbor (K-NN) [7] classifier in the definition of the fitness function while in this article the Random Forest (RF) [8] classifier is considered because even if K-NN works very well with multi class datasets, the dataset considered in this article has only two classes and RF is extremely flexible, characterized by high accuracy and deals very well with over fitting.

The authors of [9] consider an approach in which the features are selected using an improved version of the CSOA. For the improved CSOA version (ICSO), in that article are presented two methods, one in which the current solution is directed to the experience of the reference group before moving towards the solution which is optimal and one in which the mechanism applied in the changing of the positions of the cats is modified considering the value of the parameter SRD (seeking range of the selected dimension). That approach was tested and evaluated on datasets from UCI machine learning repository, but the classifier applied in that approach is Support Vector Machines (SVM) [10]. Some disadvantages of SVM compared to RF are the long training time, the feature scaling requirement and the requirement for extensive memory.

2.2 Smart Grid Stability Classification

Even though the analysis of the stability of the smart grids was approached in literature before from different perspectives such as the proposal of a novel delay-adaptive control strategy which enhances the transient stability [11] of the system and the development of a quantitative framework applied in the assessment of the voltage stability in the case of smart power networks [12], there are relatively few studies which consider the classification of the smart grid stability. However, the prediction of the smart grid stability was approached in [13] where the authors propose a new real-time model order reduction technique for predicting the stability of the smart grid. That method is capable of predicting the limit of the stability, the transient stability and unstable machines. That method was tested on three test systems and the results show that it is practical for large-scale power systems. A drawback of that method is that it should be adapted in power networks that are characterized by high penetration of Renewable Energy Sources (RES) and therefore there are still major issues which require further investigation.

3 Research Methods

3.1 General Biology of Horses

In this subsection are presented the main characteristics of the horses which are considered in the development of HOA:

(1) Horses are social animals organized in herds that have one dominant stallion or mare and a number of mares, colts, fillies, yearlings and foals.
(2) When a rival stallion defeats the dominant stallion of a herd, it takes control over all the other members of the herd.
(3) Horses have different types of gaits such as walk, trot, canter and gallop.
(4) The best defense method of the horses is running.
(5) Horses have a very good long term memory.
(6) The hierarchy of the herd sets up the order in which the resources such as food, drink and shelter are accessed.
(7) Some horses do not fit in the herd because they are too low in rankings.

3.2 HOA

Next parts detail the main principles of HOA. The algorithm does not discriminate between males and females and thus the dominant stallions or mares of each horse herd are generically called stallions and the horses that do not belong to any horse herd are also called stallions.

Horse Herd Hierarchy. A horse herd has a dominant stallion or mare and the hierarchical order of the horses in a herd specifies the priority access to resources. The hierarchy of the horses in a herd is computed in the initial phase of the algorithm considering the fitness values of the horses from that herd and it is updated every M iterations where M is a positive integer value. Let $herd = \{h_1, ..., h_k\}$ be a herd of k horses and $P : herd \rightarrow \{1, ..., k\}$ a function such that:

(1) If $Fitness(h_i) < Fitness(h_j)$ where $i \neq j$ and $i, j \in \{1, ..., k\}$ then:

$$P(h_i) > P(h_j) \tag{1}$$

(2) If $Fitness(h_i) = Fitness(h_j)$ where $i \neq j$ and $i, j \in \{1, ..., k\}$ then:

$$[P(h_i) - P(h_j)] \times (i - j) > 0 \tag{2}$$

The rank of each horse h_i such that $i \in \{1, ..., k\}$ is described by the formula:

$$h_i.rank = \frac{P(h_i)}{k} \tag{3}$$

Each herd has a center that is equal to the weighted average of the positions of the horses from the herd such that the weights are represented by the ranks of the horses. Using that approach, the center of the herd is closer to the dominant stallion of the herd than to the horses that have lower ranks. The center of the herd is calculated using the formula:

$$herd_{center} = \frac{\sum_{i=1}^{k} (x_i \times h_i.rank)}{\sum_{i=1}^{k} h_i.rank} \tag{4}$$

Stallion Dynamics. The stallion is attracted by the center of the herd because the herd permits access to many resources such as shelter, drink and food.

The distance between the position of the stallion and the center of the horse herd H is computed using the Euclidean distance and has the following formula:

$$d(stallion, herd) = \sqrt{\sum_{j=1}^{D} (stallion_j - herd_{center,j})^2} \tag{5}$$

where D is the number of dimensions of the search space, $stallion_j$ is the value of the position of the stallion for the j-th dimension and $herd_{center,j}$ is the value of the position of the center of the herd for the j-th dimension.

Horse Velocity Update. If the horse belongs to the set *herd* of horses then it updates its velocity using the formula:

$$v_{i,j}^{t+1} = v_{i,j}^{t} + h_i.rank \times h_i.gait \times \left(herd_{center,j}^{t} - x_{i,j}^{t}\right) \tag{6}$$

If the horse is a single stallion that does not belong to any horse herd then it updates its velocity using the formula:

$$v_{i,j}^{t+1} = v_{i,j}^{t} + r \times h_i.gait \times \left(nherd_{center,j}^{t} - x_{i,j}^{t}\right) \tag{7}$$

where r is a random number from $[0, 1]$.

In both cases $h_i.gait$ is a random number from the interval $[1, 2]$. The horses that belong to a herd update their velocities according to the rank in the herd, the gait and the closeness to the center of the herd. In the case of the single stallions the equivalent of the rank is a random number from the interval $[0, 1]$ and the considered herd is the nearest herd $nherd$.

Horse Position Update. The formula that is used for updating the position of a horse is:

$$x_{i,j}^{t+1} = x_{i,j}^{t} + v_{i,j}^{t+1} \tag{8}$$

where t is the current iteration, $t+1$ is the new iteration, $i \in \{1, ..., N\}$ is the index of the horse for which the position is updated and $j \in \{1, ..., D\}$ specifies the index of the dimension that is updated.

Horse Memory Update. The memory of a horse is a matrix that has a number of rows equal to the value of the Horse Memory Pool (HMP) of the horse and D columns.

$$M_i^{t+1} = \begin{bmatrix} m_{1,i,1}^{t+1} & \cdots & m_{1,i,D}^{t+1} \\ \cdots & \cdots & \cdots \\ m_{HMP,i,1}^{t+1} & \cdots & m_{HMP,i,D}^{t+1} \end{bmatrix}$$

The formula that is applied in order to update the cells of the memory matrix is:

$$m_{k,i,j}^{t+1} = x_{i,j}^{t+1} \times \mathcal{N}(0, sd) \tag{9}$$

such that $k \in \{1, ..., HMP\}$ and $\mathcal{N}(0, 1)$ is a normal distribution with mean 0 and standard deviation sd.

HOA Pseudo-code. The pseudo-code of HOA is presented in Algorithm 1.

Algorithm 1: Horse Optimization Algorithm

```
Data: N_horses, N_iterations, M, DSP, SSP, HMP, HDR, D
Result: h_gbest
initialize H = {h_1, ..., h_N_horses} in the D - dimensional search space;
for h ∈ H do
    compute Fitness(h) and update h_gbest;
end
iteration = 0;
while iteration < N_iterations do
    if iteration % M == 0 then
        top N_horses × DSP horses are dominant stallions;
        for each dominant stallion initialize a herd in T;
        next best N_horses × SSP horses form the set S;
        distribute the remaining horses in herds from T randomly;
        bottom N_horses × HDR horses are positioned randomly;
        compute the fitness value of each bottom horse;
    end
    evaluate the rank of each horse from H − S;
    determine the center of each herd from T;
    for h ∈ H do
        h.gait = Random(1, 2);
        if h ∈ S then
            nherd = SelectNearestHerd(s, T);
            UpdateHorseVelocity(h, nherd);
        else
            UpdateHorseVelocity(h, h.herd);
        end
        UpdateHorsePosition(h);
        UpdateHorseMemory(h, HMP);
        compute Fitness(h) and update h_gbest;
    end
    evaluate the rank of each horse from H − S;
    determine the center of each herd from T;
    for s ∈ S do
        nherd = SelectNearestHerd(s, T);
        if Fitness(s) < Fitness(nherd.stallion) then
            swap s and nherd.stallion;
        end
    end
    iteration = iteration + 1;
end
return h_gbest;
```

The **input** of HOA is N_{horses} - the total number of horses, $N_{iterations}$ - the number of iterations, M - a parameter used when the hierarchy of the horses is updated, DSP - the dominant stallions percent, SSP - the single stallions percent, HMP - the horse memory pool, HDR - the horses distribution rate and D - the number of dimensions of the search space. The positions and the velocities of the horses can be restricted to take values from the intervals $[P_{min}, P_{max}]$ and $[v_{min}, v_{max}]$, respectively, such that P_{min} is the minimum possible value of the position, P_{max} is the maximum possible value of the position, v_{min} is the minimum possible value of the velocity and v_{max} is the maximum possible value of the velocity. The **output** of HOA is h_{gbest} - the position of the best horse after a number of iterations equal to $N_{iterations}$.

The algorithm starts with the random initialization of the positions of the horses in the D - dimensional search space in **line 1**. Then for each horse from the herd $H = \{h_1, ..., h_{N_{horses}}\}$ the fitness value is computed using the objective function $Fitness$ and the value of h_{gbest} is updated to the position of the best horse **lines 2–4**. The value of the current iteration is initialized to 0 in **line 5**. For a number of iterations equal to $N_{iterations}$ the steps from **lines 7–37** are repeated. If the value of the current iteration is divisible by M **line 7** then the horses are distributed as follows: the top $N_{horses} \times DSP$ horses according to the value of the fitness function are labeled as dominant stallions **line 8**, a new herd is initialized for each dominant stallion in the set T such that $|T| = N_{horses} \times DSP$ **line 9**, the next best $N_{horses} \times SSP$ horses according to the value of the fitness function are distributed in the set S **line 10**, the remaining horses are distributed in herds from T randomly **line 11**, the bottom $N_{horses} \times HDR$ are positioned randomly in the D - dimensional search space keeping their current velocities **line 12** and for each bottom horse the new fitness value is computed in **line 13**.

The rank of each horse from $H - S$ is evaluated in **line 15** using the Eqs. (1), (2) and (3), and then the center of each herd is computed in **line 16** using the Eq. (4).

For each horse h from H **line 17** the instructions from **lines 18–27** are performed. The gait of the horse h is selected randomly from the interval $[1, 2]$ in **line 18**. If the horse is a single stallion **line 19** then the nearest herd $nherd$ is selected from the set of all herds T in **line 20** such that the value computed using the Eq. (5) is minimal and the velocity is updated using the Eq. (6). If the horse belongs to a herd then its velocity is updated using the Eq. (7). The position of the horse h is updated in **line 25** using the Eq. (8), the memory of the horse h is updated in **line 26** using the Eq. (9) and in **line 27** the new fitness value of the horse h is computed using the function $Fitness$ and the value of h_{gbest} is updated.

The rank of each horse is reevaluated in **line 29** due to the changes in the fitness values of the horses and the center of each herd from T is computed again in **line 30**. For each single stallion s from S **line 31** the nearest herd $nherd$ is selected in **line 32** and if the fitness of s is better than the fitness of the stallion of the nearest herd $nherd.stallion$ **line 33** then the current positions and the roles of s and of the stallion of that herd are swapped in **line 34** as follows: the stallion s becomes the new dominant stallion of the herd $nherd$ and the stallion $nherd.stallion$ becomes single and therefore an element of the set S.

The current iteration of the algorithm is incremented with 1 in **line 37** and the final value of h_{gbest} is returned in **line 39**.

4 HOA Evaluation and Validation

In this subsection HOA is evaluated and validated using six objective functions presented in Table 1. The number of dimensions in each case is 10 and the optimal value of each objective function is 0.

Table 1. Benchmark objective functions for HOA evaluation

Objective function	Range
$OF_1(x) = \sum_{i=1}^{N} x_i^2$	$[-100, 100]$
$OF_2(x) = max(\lvert x_i \rvert, 1 \leq i \leq N)$	$[-100, 100]$
$OF_3(x) = \sum_{i=1}^{N-1} \left[100 \times \left(x_{i+1} - x_i^2\right)^2 + (x_i - 1)^2\right]$	$[-30, 30]$
$OF_4(x) = \sum_{i=1}^{N} \lvert x_i \rvert + \prod_{i=1}^{N} \lvert x_i \rvert$	$[-10, 10]$
$OF_5(x) = \frac{1}{4000} \times \sum_{i=1}^{N} x_i^2 - \prod_{i=1}^{N} cos\left(\frac{x_i}{\sqrt{i}}\right) + 1$	$[-600, 600]$
$OF_6(x) = \sum_{i=1}^{N} \left(\sum_{j=1}^{i} x_j\right)^2$	$[-100, 100]$

The results obtained when HOA is evaluated using the six benchmark objective functions are compared to the results obtained using the following bio-inspired algorithms which present similarities with HOA: Chicken Swarm Optimization (CSO) [14] and Cat Swarm Optimization Algorithm (CSOA) [15]. For each bio-inspired algorithm the following configurable parameters are considered:

(1) HOA: N_{horses} - the number of horses, $N_{iterations}$ - the number of iterations, M - the horses hierarchy update frequency, DSP - the dominant stallions percent, SSP - the single stallions percent, HMP - the horse memory pool, HDR - the horses distribution rate, v_{min} - the minimum velocity, v_{max} - the maximum velocity and sd - the standard deviation applied in the equations used for the updating of the memories of the horses;
(2) CSO: $N_{chickens}$ - the number of chickens, $N_{iterations}$ - the number of iterations, RP - the roosters percent, HP - the hens percent, CP - the chicks percent, G - the chickens swarm update frequency, FL_{min} - the minimum flight length and FL_{max} - the maximum flight length;
(3) CSOA: N_{cats} - the number of cats, $N_{iterations}$ - the number of iterations, MR - the mixture ratio, SMP - the seeking memory pool, SRD - the seeking range of the selected dimension, CDC - the counts of dimensions to change, SPC - the self position considering, c_1 - a constant used when the positions of the cats are updated, v_{min} - the minimum velocity and v_{max} - the maximum velocity;

In Table 2 are summarized the values of the configurable parameters used in experiments.

Table 2. Summary of the configurable parameters used in experiments

Algorithm	Configurable parameters
HOA	$N_{iterations} = 2500$, $N_{horses} = 50$
	$M = 10$, $DSP = 10$, $SSP = 10$, $HMP = 10$, $HDR = 10$
	$v_{min} = -0.1$, $v_{max} = 0.1$, $sd = 1$
CSO	$N_{iterations} = 2500$, $N_{chickens} = 50$
	$RP = 20$, $HP = 20$, $CP = 60$, $G = 10$
	$FL_{min} = 0.5$, $FL_{max} = 0.9$
CSOA	$N_{iterations} = 2500$, $N_{cats} = 50$
	$MR = 0.2$, $SMP = 10$, $SRD = 0.2$, $CDC = 8$, $SPC = 0$
	$v_{min} = -0.0001$, $v_{max} = 0.0001$, $c_1 = 0.5$

Table 3 presents a summary of the final results obtained in the experiments by each bio-inspired algorithm for each objective function from the set of objective functions $\{OF_1, OF_2, OF_3, OF_4, OF_5, OF_6\}$.

Table 3. Summary of the results obtained by each bio-inspired algorithm for each objective function

Objective function	HOA	CSO	CSOA
OF_1	1.16793E−005	0	2.04818E−005
OF_2	0.00140535	5.60519E−045	0.00124068
OF_3	8.65968	8.98673	8.73878
OF_4	0.00355079	5.46506E−44	0.000528397
OF_5	0.70933	0	3.33861E−010
OF_6	1.98137E−005	0	1.30817E−005

In Fig. 1 is presented in more details the evolution of the objective function for each experiment in each iteration. As can be seen in the figure CSO finds the optimal solution fast while the algorithms CSOA and HOA perform better in the exploration of the search space.

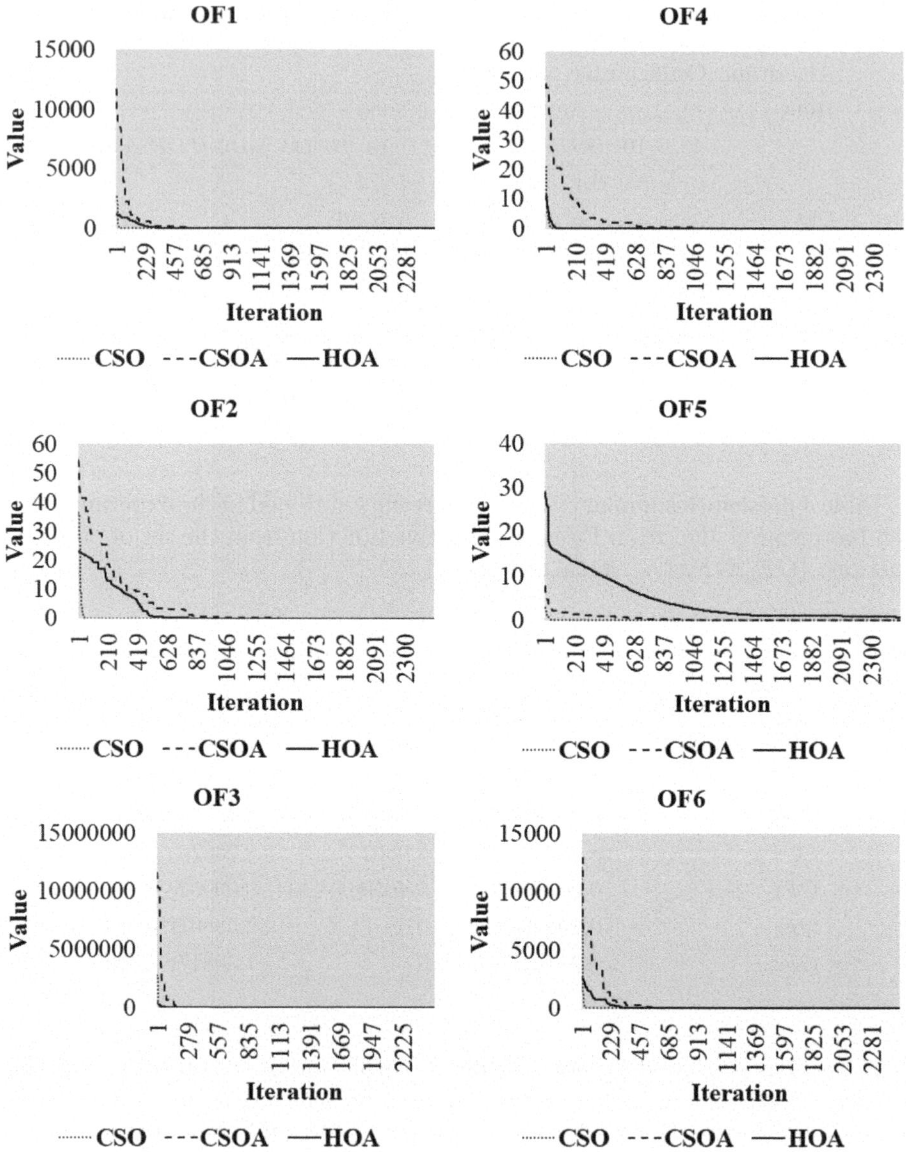

Fig. 1. Evolution of the objective function in each experiment

5 HOA Engineering Application - A Case Study on Smart Grid Stability Classification

5.1 HOA Approach for High Dimensional Data Features Selection

HOA must be adapted to discrete optimization problems prior to its application in features selection for high dimensional data. The search space of the optimization problem approached in this article as illustrative research use case is represented by vectors of size N where N is the number of dimensions of the data samples and the values of the elements of those vectors are from the set $\{0, 1\}$. The value 0 means that a feature is not selected and the value 1 means that a feature is selected. Consequently a discrete binary version of HOA is considered and the following modification is introduced in order to convert the position x which is represented by continuous values to the binary version z represented by discrete values:

$$z = \begin{cases} 1 \text{ if } r < \frac{1}{1+e^{-x}} \\ 0 \text{ otherwise} \end{cases} \tag{10}$$

where r is a random number from the interval $[0, 1]$.
The fitness function is defined by the formula:

$$\begin{aligned} Fitness(z) &= \frac{1}{3} \times (1 - RF_{accuracy}(z_{selected\ features})) \\ &+ \frac{1}{3} \times \frac{|z_{selected\ features}|}{|z_{all\ features}|} + \frac{1}{3} \times \frac{RF_{avg\ max\ depth}(z_{selected\ features})}{RF_{avg\ max\ depth}(z_{all\ features})} \end{aligned} \tag{11}$$

where:

(1) $RF_{accuracy}(z_{selected\ features})$ is the accuracy of the RF model trained using the features selected according to the position of z;
(2) $|z_{selected\ features}|$ is the number of selected features indicated by the position of z;
(3) $|z_{all\ features}|$ represents the maximum possible value of features that can be selected according to the position of z;
(4) $RF_{avg\ max\ depth}(z_{selected\ features})$ is the average maximum depth of the trees of the RF model trained using the features indicated by $z_{selected\ features}$;
(5) $RF_{avg\ max\ depth}(z_{all\ features})$ is the average maximum depth of the trees of the RF model trained using the features indicated by $z_{all\ features}$;

The objective of the fitness function is to maximize the accuracy of the RF classification model, to minimize the number of the selected features and to minimize the average maximum depth of the trees of the RF model. The ideal value that can be returned by the fitness function is 0 and the worst value that can be returned by the fitness function is 1.

5.2 HOA Based Methodology for Smart Grid Stability Classification

The machine learning methodology applied in the classification of the stability of the smart grid is presented in Fig. 2.

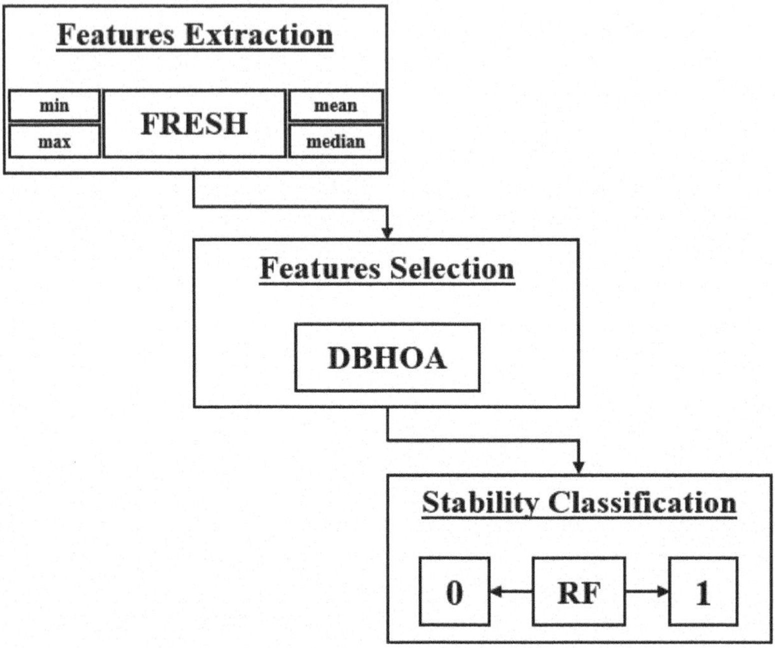

Fig. 2. Machine learning methodology for smart grid stability classification

The methodology consists of three main steps: (1) the extraction of the features using the Feature Extraction on basis of Scalable Hypothesis tests (FRESH) algorithm [16], (2) the selection of the features using the discrete binary version of HOA, namely Discrete Binary Horse Optimization Algorithm (DBHOA) and (3) the classification of the stability of the smart grid using an approach based on RF.

5.3 HOA Based Smart Grid Stability Classification Results

Experimental Dataset Characteristics. The main characteristics of the Electrical Grid Stability Simulated Dataset [17] which is used in experiments are summarized in Table 4. More details about that dataset can be found in [18].

Features Extraction Results. The number of features after the features extraction step is 208. Some of the most representative features which are extracted using the FRESH algorithm are the minimum, the maximum, the mean and the median.

Table 4. Electrical grid stability simulated dataset characteristics summary

Characteristic	Value
Labels number	2
Samples number	10000
Features (f_1, f_2, f_3, f_4)	**Reaction time** (τ) of the electricity producer and of the first, the second and the third electricity consumers
Features (f_5, f_6, f_7, f_8)	**Nominal power** (P) produced by the electricity producer and consumed by the first, the second and the third electricity consumers
Features (f_9, f_{10}, f_{11}, f_{12})	**Gamma coefficient** (γ) of the electricity producer and of the first, the second and the third electricity consumers

Features Selection Results. The configurable parameters used in the features selection experiments are the ones presented in Table 2 with the additional modifications: $N_{iterations} = 50$, $N_{horses} = 20$, $N_{chickens} = 20$ and $N_{cats} = 20$. The positions are represented by arrays of size $D = 208$ with values from the interval $[-100, 100]$ and the discretization of the positions in arrays of zeros and ones is performed each time the fitness function is computed.

In Table 5 are summarized the features selection results. In addition to DBHOA, the article presents the results when the features are selected using Discrete Binary Chicken Swarm Optimization (DBCSO) and Discrete Binary Cat Swarm Optimization Algorithm (DBCSOA).

Table 5. Summary of the features selection results

Features selection approach	Best fitness value	Random forest accuracy	Number of selected features	Average maximum depth
DBHOA	0.300	90.3%	85	19.0
DBCSO	0.297	90.4%	81	19.6
DBCSOA	0.328	90.5%	101	19.4

Stability Classification Results. In Table 6 are presented the classification results when the methodology applied in the smart grid stability classification selects the features using the approaches based on DBHOA, DBCSO and DBCSOA, respectively.

Table 6. Summary of the classification results

Features selection approach	Actual stable & Predicted stable	Actual unstable & Predicted stable	Actual stable & Predicted unstable	Actual unstable & Predicted unstable
DBHOA	80	632	108	1180
DBCSO	76	629	111	1184
DBCSOA	76	631	109	1184

6 Conclusions and Future Research Work

This article introduced HOA, a novel algorithm inspired from the hierarchical organization of the horse herds. HOA was evaluated and validated using six objective functions and was compared to CSO and CSOA. The results show that the performance of HOA is comparable to the performance of those two algorithms. A modified version of HOA, namely DBHOA, was applied in features selection for high dimensional datasets with a case study on smart grid stability classification. As future research work the following directions are proposed: (1) the improvement of the current version of HOA, (2) the development of novel modified versions of HOA and (3) the application of HOA in other research areas.

References

1. Yang, X.-S.: A new metaheuristic bat-inspired algorithm. In: Gonzalez, J.R., Pelta, D.A., Cruz, C., Terrazas, G., Krasnogor, N. (eds) Nature Inspired Cooperative Strategies for Optimization (NICSO 2010). Studies in Computational Intelligence, vol. 284, pp. 65–74. Springer, Heidelberg (2010). https://doi.org/10.1007/978-3-642-12538-6_6
2. Brabazon, A., Cui, W., O'Neill, M.: The Raven roosting optimisation algorithm. Soft Comput. **20**(2), 525–545 (2016). https://doi.org/10.1007/s00500-014-1520-5
3. Shefaei, A., Mohammadi-Ivatloo, B.: Wild goats algorithm: an evolutionary algorithm to solve the real-world optimization problems. IEEE Trans. Ind. Inf. **14**(7), 2951–2961 (2018). https://doi.org/10.1109/TII.2017.2779239
4. Jain, M., Maurya, S., Rani, A., Singh, V.: OWL search algorithm: a novel nature-inspired heuristic paradigm for global optimization. J. Intell. Fuzzy Syst. **34**(3), 1573–1582 (2018). https://doi.org/10.3233/JIFS-169452
5. Arora, S., Singh, S.: Butterfly optimization algorithm: a novel approach for global optimization. Soft Comput. **23**, 715–734 (2019). https://doi.org/10.1007/s00500-018-3102-4
6. Hafez, A.I., Zawbaa, H.M., Emary, E., Mahmoud, H.A., Hassanien, A.E.: An innovative approach for feature selection based on chicken swarm optimization. In: Proceedings of the 7th International Conference on Soft Computing and Pattern Recognition (SoCPaR), Fukuoka, Japan, pp. 19–24 (2015). https://doi.org/10.1109/SOCPAR.2015.7492775
7. Pestov, V.: Is the k-NN classifier in high dimensions affected by the curse of dimensionality? Comput. Math. Appl. **65**(10), 1427–1437 (2013). https://doi.org/10.1016/j.camwa.2012.09.011
8. Chen, W., Li, Y., Xue, W., Shahabi, H., Li, S., Hong, H., Wang, X., Bian, H., Zhang, S., Pradhan, B., Ahmad, B.B.: Modeling flood susceptibility using data-driven approaches of naive Bayes tree, alternating decision tree, and random forest methods. Sci. Total Environ. **701**, 134979 (2020). https://doi.org/10.1016/j.scitotenv.2019.134979
9. Lin, K.-C., Zhang, K.-Y., Huang, Y.-H., Hung, J.C., Yen, N.: Feature selection based on an improved cat swarm optimization algorithm for big data classification. J. Supercomput. **72**, 3210–3221 (2016). https://doi.org/10.1007/s11227-016-1631-0

10. Lin, K.-C., Hsieh, Y.-H.: Classification of medical datasets using SVMs with hybrid evolutionary algorithms based on endocrine-based particle swarm optimization and artificial bee colony algorithms. J. Med. Syst. **39**, 119 (2015). https://doi.org/10.1007/s10916-015-0306-3
11. Wang, Z., Wang, J.: A delay-adaptive control scheme for enhancing smart grid stability and resilience. Int. J. Electr. Power Energy Syst. **110**, 477–486 (2019). https://doi.org/10.1016/j.ijepes.2019.03.030
12. Aldeen, M., Saha, S., Alpcan, T.: Voltage stability margins and risk assessment in smart power grids. IFAC Proc. Vol. **47**(3), 8188–8195 (2014). https://doi.org/10.3182/20140824-6-ZA-1003.02102
13. Shamisa, A., Majidi, B., Patra, J.C.: Sliding-window-based real-time model order reduction for stability prediction in smart grid. IEEE Trans. Power Syst. **34**(1), 326–337 (2019). https://doi.org/10.1109/TPWRS.2018.2868850
14. Meng, X., Liu, Y., Gao, X., Zhang, H.: A new bio-inspired algorithm: chicken swarm optimization. In: Tan, Y., Shi, Y., Coello, C.A.C. (eds) Advances in Swarm Intelligence. ICSI 2004. Lecture Notes in Computer Science, vol. 8794, pp. 86–94. Springer, Cham (2014). https://doi.org/10.1007/978-3-319-11857-4_10
15. Chu, S.-C., Tsai, P.-W., Pan, J.-S.: Cat Swarm Optimization. In: Yang, Q., Webb, G. (eds) PRICAI 2006: Trends in Artificial Intelligence. PRICAI 2006. Lecture Notes in Computer Science, vol. 4099, pp. 854–858. Springer, Heidelberg (2006). https://doi.org/10.1007/978-3-540-36668-3_94
16. Christ, M., Braun, N., Neuffer, J., Kempa-Liehr, A.W.: Time series feature extraction on basis of scalable hypothesis tests (tsfresh - a Python package). Neurocomputing **307**, 72–77 (2018). https://doi.org/10.1016/j.neucom.2018.03.067
17. Dua, D., Graff, C.: UCI Machine Learning Repository. University of California, School of Information and Computer Science, Irvine, CA (2019). https://archive.ics.uci.edu/ml/citation_policy.html
18. Arzamasov, V., Bohm, K., Jochem, P.: Towards Concise Models of Grid Stability. In: Proceedings of the 2018 IEEE International Conference on Communications, Control, and Computing Technologies for Smart Grids (SmartGridComm), Aalborg, Denmark, pp. 1–6 (2018). https://doi.org/10.1109/SmartGridComm.2018.8587498

Mathematical Model of the Influence of Transnationalization on the Russian Agricultural Machinery Market

Eugeny V. Lutsenko[1], Ksenia A. Semenenko[1,2(✉)], Irina V. Snimschikova[1], Valery I. Loiko[1], and Marina P. Semenenko[2]

[1] Kuban State Agrarian University Named After I.T. Trubilin, Krasnodar, Russia
ksenia.semenenko@gmail.com
[2] Krasnodar Research Center for Animal Husbandry and Veterinary Medicine, Krasnodar, Russia

Abstract. The article presents a mathematical model of the influence of transnationalization on the Russian agricultural machinery market that provides multi-parameter typing and system identification based on empirical data with the use of automated system-cognitive analysis and its own software tool – an intelligent system called "Eidos". The ASC-analysis makes it possible to justify the dual effects of transnationalization on the formation and development of the Russian agricultural machinery market. An analysis of the results of a numerical experiment shows that the "Eidos" intelligent system can be used to solve the problems of forecasting, making decisions, and researching a simulated subject area by studying its model in other subject areas. #CSOC1120

Keywords: Transnationalization · Automated system-cognitive analysis · "Eidos" intellectual system · Agricultural machinery market

1 Introduction

Russia has its own sufficiently developed production potential, which ensures the production of agricultural equipment in the volumes and nomenclature demanded by society. However, the Russian agricultural machinery market is not closed, and it is significantly affected by multinational corporations specializing in the development, production and supply of the high-quality agricultural machinery.

In some cases, agricultural machinery samples that produced in Russia have the similar functional and cost characteristic with foreign analogues. Thus, in the Russian agricultural machinery market there is a direct competition between Russian and foreign models of equipment [1]. In addition, Russia itself is a supplier of agricultural machinery to other countries. In these conditions:

- the buyer and consumer of this equipment faces a difficult situation in which they need to make decisions associated with the presence of many alternatives, the degree of preference of which they need to be compared by a large number of criteria;

R. Silhavy (Ed.): CSOC 2020, AISC 1225, pp. 210–222, 2020.
https://doi.org/10.1007/978-3-030-51971-1_17

- Russian producers of agricultural machinery should clearly understand their specific advantages over foreign competitors, both in general and specifically for various types and even specific models (brands) of agricultural machinery [2].

However, to solve these problems, it is necessary:

- on the one hand, to develop a mathematical model based on empirical data that reflects the strength and direction of the influence of various factors on the Russian agricultural machinery market,
- on the other hand, it is not entirely clear how to do this, namely with the help of which mathematical method and software tools to implement it.

To create such a model, a mathematical method is needed that provides multi-parameter typing and system identification based on empirical data. As a method, it is proposed to use automated system-cognitive analysis [3–5], which is briefly described in the next section.

2 Materials and Methods

2.1 Justification of the Expediency of the Method Application

As a method of the research, we used automated system-cognitive analysis (ASC-analysis), which is a new innovative method of artificial intelligence: it also has its own software tool – an intelligent system called "Eidos" (open source software) [3–5].

The "Eidos-X++" system differs from the other artificial intelligence systems in the following parameters:

- it was developed in a universal setting, independent of the subject area; therefore, it is universal and can be applied in many subject areas (http://lc.kubagro.ru/aidos/index.htm);
- it is in full open free access (http://lc.kubagro.ru/aidos/_Aidos-X.htm), and with the relevant source texts (http://lc.kubagro.ru/__AIDOS-X.txt);
- it is one of the first domestic systems of artificial intelligence of the personal level, i.e. it does not need special training from the user in the field of technologies of artificial intelligence (there is an act of introduction of "Eidos" system in 1987: http://lc.kubagro.ru/aidos/aidos02/PR-4.htm);
- it provides stable identification in a comparable form of strength and direction of cause-effect relationships in incomplete noisy interdependent (nonlinear) data of a very large dimension of numerical and non-numerical nature, measured in different types of scales (nominal, ordinal and numerical) and in the different units of measurement (i.e. does not impose strict requirements to the data that cannot be performed);
- it contains a large number of local (supplied with the installation) and cloud educational and scientific applications (currently 31 and 188, respectively; http://lc.kubagro.ru/aidos/Presentation_Aidos-online.pdf);
- it provides multilingual interface support in 44 languages; language databases are included in the installation and can be replenished automatically;

- it supports on-line environment of knowledge accumulation and it is widely used all over the world (http://aidos.byethost5.com/map5.php);
- it is the most time-consuming computationally system where the operations of the synthesis models and implements recognition made by using graphic processing unit (GPU) that some tasks can only support up to the solution of these tasks several thousand times that really provides intelligent processing of big data, big information and big knowledge;
- it provides transformation of the initial empirical data into information, and its knowledge and solution using this knowledge of classification problems, decision support and research of the subject area by studying its system-cognitive model, generating a very large number of tabular and graphical output forms (development of cognitive graphics), many of which have no analogues in other systems (examples of forms can be found in: http://lc.kubagro.ru/aidos/aidos18_LLS/aidos18_LLS.pdf);
- it well imitates the human style of thinking: gives the results of the analysis, understandable to experts on the basis of their experience, intuition and professional competence.

The essence of the ASC-analysis method consists in increase of the degree of model formalization and in transformation of the data into information, and information - into the knowledge and solutions based on the data of identification problems (recognition, classification and forecasting), decision support and research of the simulated domain [3].

2.2 Formalization of the Subject Area

The mathematical model of ASC-analysis is based on the system theory of information [3] and provides comparable processing of large volumes of fragmented and noisy interdependent data presented in various types of scales (nominal, ordinal and numerical) and various units of measurement [3–6]. It should be noted that ASC-analysis has its own software tool - the "Eidos" intellectual system, which implements this mathematical model and the corresponding numerical calculation technique (data structures and processing algorithms).

Based on the source data [7], the "Eidos" system generates classification and descriptive scales and gradations (Tables 1 and 2). The source data is encoded using classification and descriptive scales and gradations, as a result of which a training sample is formed (Table 3).

Table 1. Classification scales and gradations (fragment)

Code	Name
1	PRODUCED IN RUSSIA TRACTORS, UNITS-1/3-{6493.0, 10365.0}
2	PRODUCED IN RUSSIA TRACTORS, UNITS-2/3-{10365.0, 14237.0}
3	PRODUCED IN RUSSIA TRACTORS, UNITS-3/3-{14237.0, 18109.0}
4	PRODUCED IN RUSSIA COMBINE HARVESTERS, UNITS-1/3-{4273.0, 5273.0}
5	PRODUCED IN RUSSIA COMBINE HARVESTERS, UNITS-2/3-{5273.0, 6273.0}
6	PRODUCED IN RUSSIA COMBINE HARVESTERS, UNITS-3/3-{6273.0, 7273.0}
7	PRODUCED IN RUSSIA FORAGE HARVESTERS, UNITS-1/3-{268.0, 587.0}
8	PRODUCED IN RUSSIA FORAGE HARVESTERS, UNITS-2/3-{587.0, 906.0}
9	PRODUCED IN RUSSIA FORAGE HARVESTERS, UNITS-3/3-{906.0, 1225.0}
10	PRODUCED IN RUSSIA PLOUGHS, UNITS-1/3-{1506.0, 6711.0}

Table 2. Descriptive scales and gradations (fragment)

Code	Name
1	IMPORT GERMANY, MLN. US$-1/3-{319.0000000, 493.0000000}
2	IMPORT GERMANY, MLN. US$-2/3-{493.0000000, 667.0000000}
3	IMPORT GERMANY, MLN. US$-3/3-{667.0000000, 841.0000000}
4	IMPORT NETHERLANDS, MLN. US$-1/3-{61.0000000, 114.0000000}
5	IMPORT NETHERLANDS, MLN. US$-2/3-{114.0000000, 167.0000000}
6	IMPORT NETHERLANDS, MLN. US$-3/3-{167.0000000, 220.0000000}
7	IMPORT BELARUS, MLN. US$-1/3-{110.0000000, 175.6666667}
8	IMPORT BELARUS, MLN. US$-2/3-{175.6666667, 241.3333333}
9	IMPORT BELARUS, MLN. US$-3/3-{241.3333333, 307.0000000}
10	IMPORT USA, MLN. US$-1/3-{65.0000000, 115.3333333}

Training sample is a normalized source database.

Table 3. Training sample (fragment)

Object	2010	2011	2012	2013	2014	2015	2016	2017
N1	1	3	2	1	1	1	1	1
N2	4	6	5	5	5	4	6	6
N3	7	9	9	8	7	8	8	8
N4	10	10	10	10	10	10	10	12
N5	13	13	13	13	13	14	15	15
N6	16	17	16	16	16	17	18	18
N7	19	19	19	20	20	20	20	21
N8	24	24	23	22	22	22	22	22
N9	1	3	3	2	2	1	1	1
N10	4	6	6	4	4	4	5	6

2.3 Calculation of Statistical Models

The matrix of absolute frequencies (Table 4) and the matrix of conditional and unconditional percentage distributions (Table 5) are calculated directly on the basis of normalized empirical data (Table 3).

Table 4. Matrix of absolute frequencies

		Classes					Sum
		1	...	j	...	W	
Factor values	1	N_{11}		N_{1j}		N_{1W}	
	...						
	i	N_{i1}		N_{ij}		N_{iW}	$N_{i\sum} = \sum_{j=1}^{W} N_{ij}$
	...						
	M	N_{M1}		N_{Mj}		N_{MW}	
Total number of features by class				$N_{\sum j} = \sum_{i=1}^{M} N_{ij}$			$N_{\sum\sum} = \sum_{i=1}^{W}\sum_{j=1}^{M} N_{ij}$
Total number of training sample objects by class				$N_{\sum j}$			$N_{\sum\sum} = \sum_{j=1}^{W} N_{\sum j}$

Table 5. Matrix of conditional and unconditional percentage distributions

		Classes					Unconditional probability of a sign
		1	...	j	...	W	
Factor values	1	P_{11}		P_{1j}		P_{1W}	
	...						
	i	P_{i1}		$P_{ij} = \frac{N_{ij}}{N_{\sum j}}$		P_{iW}	$P_{i\sum} = \frac{N_{i\sum}}{N_{\sum\sum}}$
	...						
	M	P_{M1}		P_{Mj}		P_{MW}	
Unconditional class probability				$P_{\sum j}$			

It should be noted that the ASC-analysis and its software tool, the "Eidos" intellectual system, use two methods for calculating matrices of conditional and unconditional percentage distributions:

1. as $N_{\sum j}$ the total number of characteristics by class is used;
2. as $N_{\sum j}$ the total number of objects in the training sample by class is used.

2.4 Particular Criteria of Knowledge and the Calculation of System-Cognitive Models

Then, based on Table 5, using particular criteria, the matrices of system-cognitive models are calculated (Table 6).

Table 6. System-cognitive model matrix

		Classes					The significance of the factor
		1	…	j	…	W	
Factor values	1	I_{11}		I_{1j}		I_{1W}	$\sigma_{1\sum} = \sqrt[2]{\frac{1}{W-1}\sum_{j=1}^{W}(I_{1j}-\overline{I_1})^2}$
	…						
	i	I_{i1}		I_{ij}		I_{iW}	$\sigma_{1\sum} = \sqrt[2]{\frac{1}{W-1}\sum_{j=1}^{W}(I_{ij}-\overline{I_i})^2}$
	…						
	M	I_{M1}		I_{Mj}		I_{MW}	$\sigma_{M\sum} = \sqrt[2]{\frac{1}{W-1}\sum_{j=1}^{W}(I_{Mj}-\overline{I_M})^2}$
Grade of class reduction		$\sigma_{\sum 1}$		$\sigma_{\sum j}$		$\sigma_{\sum W}$	$H = \sqrt[2]{\frac{1}{(W \cdot M-1)}\sum_{j=1}^{W}\sum_{i=1}^{M}(I_{ij}-\bar{I})^2}$

The essence of these methods is that the amount of information is calculated in the value of the factor that the object of modeling will pass under its action into a certain state corresponding to the class. This allows to comparatively and correctly process heterogeneous information about observations of the modeling object, presented in various types of measuring scales and various units of measurement [6].

Based on the system-cognitive models, the problems of identification (classification, recognition, diagnostics, forecasting), decision support (inverse forecasting problem) [8], as well as the task of studying a simulated subject area by studying its system-cognitive model [9, 10] are solved.

To solve these problems, two additive integral criteria are used in the ASC-analysis and the "Eidos" system.

2.5 Integral Criteria and Solution of the Problems of System Identification and Making Management Decisions

The task of making management decisions is the inverse problem of forecasting. When forecasting based on the values of factors affecting the control object, it is determined what state it will go under their influence, but when making decisions, on the contrary, by the desired (target) state of the control object, a system of values of factors determining the transition of the object to this target state is allocated.

Currently, the “Eidos” system uses two additive integral criteria: the sum of knowledge and the resonance of knowledge.

The first integral criterion “Sum of knowledge” is the total amount of knowledge contained in the system of factors values of various nature that characterize the control object itself, control factors and the environment, about the transition of the object to future target or undesirable states.

The integral criterion is an additive function of particular knowledge criteria:

$$I_j = \left(\overrightarrow{I_{ij}}, \overrightarrow{L_i}\right).$$

In the expression, parentheses denote the scalar product. In coordinate form, this expression has such form:

$$I_j = \sum_{i=1}^{M} I_{ij} L_i,$$

Where:

M – is the number of gradations of descriptive scales (features);
$\overrightarrow{I_{ij}} = \{I_{ij}\}$ – is the state vector of the j-th class;
$\overrightarrow{L_i} = \{L_i\}$ – is the state vector of the recognizable object, including all kinds of factors characterizing the object itself, control actions and the environment (array-locator), i.e.:

$$\overrightarrow{L_i} = \begin{cases} 1 & if\ i - th\,factor\,acts; \\ n,\ where\,n > 0, & if\ i - th\,factor\,acts\,with\,the\,truth\,n; \\ 0, & if\ i - th\,factor\,does\,not\,act \end{cases}$$

In the current version of the “Eidos-X++” system, the coordinates of the state vector of the recognized object were taken to be either 0 if there is no sign, or n if it is present in the object with intensity n, i.e. represented n times.

The second integral criterion “Semantic resonance of knowledge” is the normalized total amount of knowledge contained in the system of factors of various nature that characterize the control object itself, control factors and the environment, about the transition of the object to future target or undesirable states.

The integral criterion is an additive function of particular criteria of knowledge and has such form:

$$I_j = \frac{1}{\sigma_j \sigma_i M} \sum_{i=1}^{M} (I_{ij} - \bar{I_j})(L_i - \bar{L})$$

Where:
M – is the number of gradations of descriptive scales (features);

$\overline{I_j}$ – is the average informativeness of the class vector;
$\bar{L}$ – is the average of the vector object;
σ_j – is the standard deviation of particular class vector knowledge criteria;
σ_1 – is the standard deviation of the recognized object vector.
$\overrightarrow{I_{ij}} = \{I_{ij}\}$ – is the state vector of the j-th class;
$\overrightarrow{L_i} = \{L_i\}$ – is the state vector of the recognizable object, including all kinds of factors characterizing the object itself, control actions and the environment (array-locator), i.e.:

$$\overrightarrow{L_i} = \begin{cases} 1 & if\, i - th\, factor\, acts; \\ n,\ where\, n > 0, & if\, i - th\, factor\, acts\, with\, the\, truth\, n; \\ 0, & if\, i - th\, factor\, does\, not\, act \end{cases}$$

The "Eidos" system provides the construction of intelligent information-measuring systems in various subject areas [6, 8–12] due to the large number of software interfaces that provide automated input of various types of data into the system: text, tabular and graphic.

By means of multiparametric typing, a system-cognitive model is created in the system, using which, if the model is sufficiently reliable, problems of system identification, prediction, classification, decision support and research of the modeled object can be solved by studying its system-cognitive model [6, 8–12].

3 Results and Discussion

3.1 Reliability Assessment of Created System-Cognitive Models

The most reliable of the created system-cognitive models was the INF3 model (Table 7).

Table 7. Analytical form of particular criteria of knowledge

Name of knowledge model and particular criterion	Expression for the private criterion	
	Through relative frequencies	Through absolute frequencies
INF3, particular criterion: Chi-square: differences between actual and theoretically expected absolute frequencies	–	$I_{ij} = N_{ij} - \frac{N_i N_j}{N}$

Where:

I_{ij} – particular criterion of knowledge: the amount of knowledge in the fact of observing the i-th value of the past parameter that the object will go into the state corresponding to the j-th value of the future parameter;
N_{ij} – the number of meetings of the j-th value of the future parameter with the i-th value of the past parameter;
N_i – the number of meetings of the i-th value of the last parameter over the entire sample;
N_j – the number of meetings of the j-th value of the future parameter over the entire sample;
N – the number of meetings of the j-th value of the future parameter at the i-th value of the passed parameter over the entire sample.
According to the L1 criterion [9], its reliability was 0.976 with a maximum of 1.000, which is a very good result (Fig. 1).

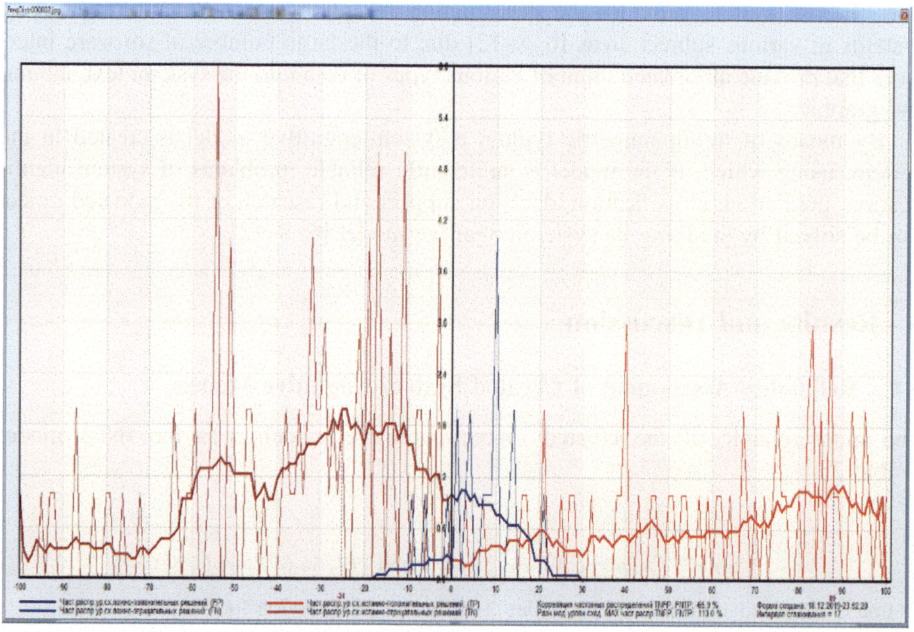

Fig. 1. Output forms for the reliability of statistical and system-cognitive models at ASC-analysis of the Russian agricultural machinery market

3.2 Simplified Solution to the Decision Support Task (SWOT-Analysis)

When making decisions, the strength and directions of the influence of factors on the states of the modeling object to one or another class are determined. In fact, this is a solution to the problem of SWOT-analysis [8]. The “Eidos” system in 4.4.8 mode supports the solution of this problem. In this case, the determination system of a given class is revealed, i.e. a system of values of factors determining the transition of an object of modeling and control to a state corresponding to this class (Fig. 2).

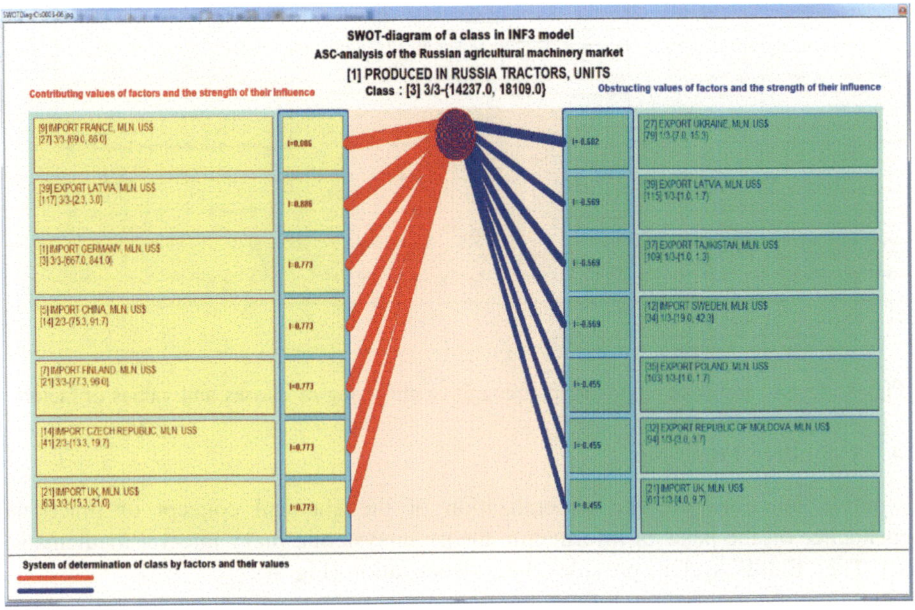

Fig. 2. SWOT-diagram reflecting a class determination system

Information can be presented not only in the above-mentioned forms, but also in many other tabular and graphical forms, which are not given in this work only because of the limited size. In particular, information that is much more complete can be displayed in these forms (including, in general, everything that has in the model).

In conclusion, it should be noted that the "Eidos" system provides automated SWOT-analysis without the involvement of experts.

3.3 Cluster-Constructive Analysis of Classes and Values of Factors

The "Eidos" system provides agglomerative cognitive clustering of classes (Fig. 3) and factor values [9]. The clusters combine the states of the modeling object that are similar in the system of the factors that determine them (compatible). Alternative (incompatible) states are in different clusters.

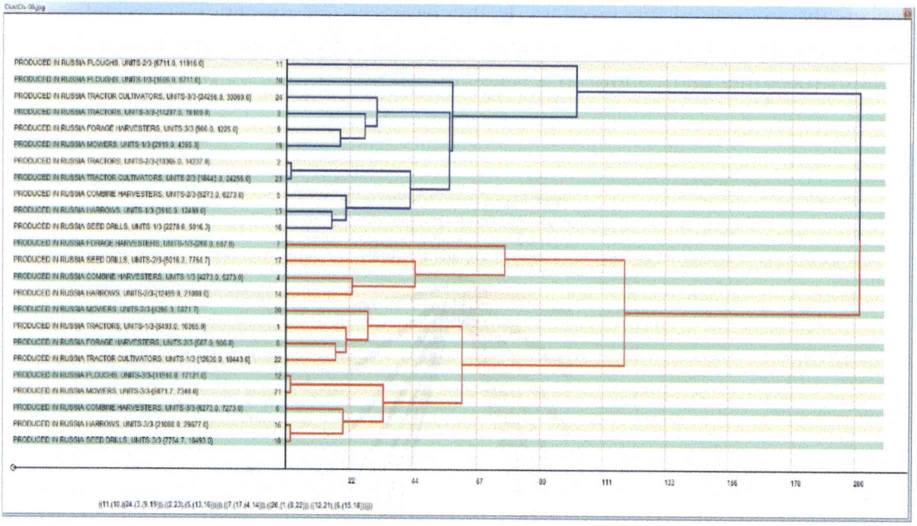

Fig. 3. Output forms of agglomerative cognitive clustering of classes and values of factors

3.4 Cognitive Functions

Cognitive functions are the generalization of the classical concept of functional dependence on the basis of information theory in systemic fuzzy interval mathematics [10]. The "Eidos" system provides their visualization (Fig. 4).

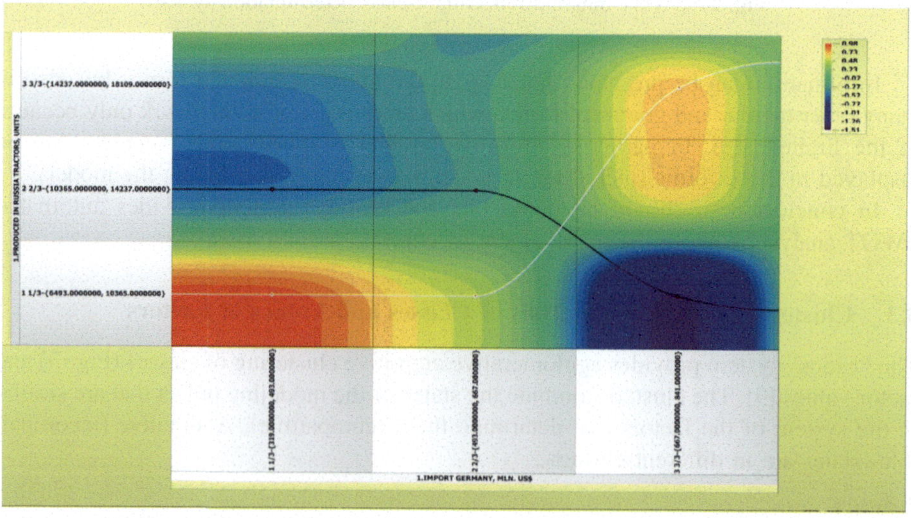

Fig. 4. An example of a cognitive function that reflects the strength and direction of influence of gradations of a descriptive scale (i.e., factor values) on the transition of a modeling object to states corresponding to gradations of a classification scale (classes) when analyzing the agricultural machinery market

3.5 Raising the Status of Research Results

This paper briefly describes how system-cognitive models are developed and applied in the ASC-analysis, reflecting how much information is contained in different values of factors about the transition of the modeling object to various future states. In the system-cognitive analysis, the hypothesis is formulated that this amount of information and its sign reflect, respectively, the strength and direction of action of causal-effect patterns that actually exist in the modeled subject area. The paper substantiates that system-cognitive models have the status of substantial phenomenological models [8]. For further increase of the status to the level of empirical laws, it is necessary to expand the empirical field and to create appropriate models. If after this, we reveal the mechanisms and causal actions of these laws and give their meaningful interpretation, then we can expand the scope of empirical laws to the entire subject area in which the same causal and mechanisms operate, and thus formulate scientific laws [11].

4 Conclusion

An analysis of the results of the numerical experiment shows that the solution of the problems posed and implemented in the "Eidos" system is quite effective, which makes it possible to reasonably state that the goal of the work has been achieved.

Transnationalization processes have a significant impact on the agricultural machinery market, which today is rather uneven due to the presence of large transnational corporations on it, controlling most of it and enhancing the development of mutual competition by increasing production, introducing various technological innovations, as well as further mergers and acquisitions.

The ASC-analysis makes it possible to justify the dual effects of transnationalization on the formation and development of the Russian agricultural machinery market, due to structural deformation of the market and openness to import; as well as to identify the problems of competition between domestic producers and major global players in terms of quality, price and technology, which allowed to identify the main drivers for further development.

The methods of the ASC-analysis and its software tool, the "Eidos" intelligent system, can also be used to solve the problems of forecasting, making decisions, and researching a simulated subject area by studying its model in other subject areas [3–10].

Thus, mathematical models (particular and integral criteria), methods of numerical calculations (data structures and algorithms for their processing), screen forms of process control, a program interface for entering data into the "Eidos" system and increasing the degree of formalization of initial data from verbalization to normalized databases (API), screen forms of text and graphic output forms based on the results of solving forecasting problems, decision making and research, software implementation of mathematical models, methods of numerical calculation the products, interface, and cognitive graphics in the "Eidos" system are an adequate means to solve the problem posed in the article.

References

1. Semenenko, K.A., Becker, T.K.: The level of dealer's satisfaction with the producers of agricultural machinery: German experience. Sci. J. KubSAU **114**(10), 1–14 (2015)
2. Snimschikova, I., Semenenko, K.: Transnational corporations at the agricultural machinery market in integration process context. In: International Scientific Conference "The Global Economic Crisis and the Future of European Integration", Niš, Serbia, pp. 363–369 (2013)
3. Lutsenko, E.V.: Automated system-cognitive analysis in the management of active objects (system theory of information and its application in the study of economic, socio-psychological, technological and organizational-technical systems). Monograph (scientific publication). KubSAU, Krasnodar (2002)
4. Lutsenko, E.V.: Open scalable interactive intelligent on-line environment "Eidos" ("Eidos-online"). Certificate of RosPatent of the Russian Federation for a computer program. Application No. 2017618053 from 08 July 2017, State registration number 2017661153, registered 10 April 2017
5. Lutsenko, E.V.: Intelligent scalable open interactive online environment for teaching and researching on the basis of ASC-analysis and "Eidos" system. Sci. J. KubSAU **06**(130), 1–55 (2017)
6. Lutsenko, E.V.: Metrization of measuring scales of different types and joint comparable quantitative processing of heterogeneous factors in system-cognitive analysis and the Eidos system. Sci. J. KubSAU **08**(092), 859–883 (2013)
7. Uncomtrade analytics. https://comtrade.un.org. Accessed 10 Dec 2019
8. Lutsenko, E.V.: Quantitative automated SWOT and PEST analysis using ASC-analysis and "Eidos X++" intelligent systems. Sci. J. KubSAU **07**(101), 1367–1409 (2014)
9. Lutsenko, E.V., Korzhakov, V.E.: Method of cognitive clusterization or clusterization on the basis of knowledge (Clusterization in system-cognitive analysis and intellectual system "Eidos"). Sci. J. KubSAU **07**(071), 528–576 (2011)
10. Lutsenko, E.V., Orlov, A.I.: Cognitive functions as a generalization of the classical concept of functional dependence on the basis of information theory in ASC-analysis and system fuzzy interval mathematics. Sci. J. KubSAU **01**(095), 122–183 (2014)
11. Lutsenko, E.V.: Problems and prospects of the theory and the methodology of scientific cognition and the automated system-cognitive analysis as an automated method of scientific knowledge, providing meaningful phenomenological modeling. Sci. J. KubSAU **03**(127), 1–60 (2017)
12. Lutsenko, E.V.: Invariant to volumes of data, a fuzzy multiclass generalization of F-measure of plausibility in Van Rijsbergen models in ASC-analysis and in the "Eidos" system. Sci. J. KubSAU **02**(126), 1–32 (2017)

A Percentil Bat Algorithm an Application to the Set Covering Problem

Lorena Jorquera[1(✉)], Pamela Valenzuela[1], Francisco Altimiras[2], Paola Moraga[1], and Gabriel Villavicencio[1]

[1] Escuela de Ingeniería en Construcción, Pontificia Universidad Católica de Valparaíso, Valparaíso, Chile
{lorena.jorquera,pamela.valenzuela,paola.moraga, gabriel.villavicencio}@pucv.cl

[2] Facultad de Ingeniería y Negocios, Universidad de le Américas, Santiago, Chile
francisco.altimiras@gmail.com

Abstract. The study and understanding of algorithms that solve combinatorial problems based on swarm intelligence continuous metaheuristics, is an area of interest at the level of basic and applied science. This is due to the fact that many of the problems addressed at industrial level are of a combinatorial type and a subset no less than these are of the NP-hard type. In this article, a mechanism of binarization of continuous metaheuristics that uses the concept of the percentile is proposed. This percentile concept is applied to the An Lion optimization algorithm, solving the set covering problem (SCP). Experiments were designed to demonstrate the importance of the percentile concept in the binarization process. Subsequently, the efficiency of the algorithm is verified through reference instances. The results indicate that the binary percentile bat Algorithm (BPBA) obtains adequate results when evaluated with a combinatorial problem such as the SCP.

1 Introduction

In recent years, there is a research line of algorithms inspired by natural phenomena focused on solving optimization problems. As examples of these algorithms we have Cuckoo Search [1], Black Hole [2], Bat Algorithm [3] and Sine Cosine Algorithm [4] among others. Due to the type of phenomena in which these algorithms are inspired, a large part of these algorithms operate naturally in continuous spaces. On the other hand, the combinatorial problems have a great relevance at the scientific and industrial level. We find them in different areas, such as Civil Engineering [5], Bio-Informatics [6], Operational Research [7–10], Big Data [11], resource allocation [12–14], programming problems [15,16], routing problems [17–20] among others. If we want to apply continuous algorithms to combinatorial problems, these algorithms must adapt. In the process of adaptation, the mechanisms of exploration and exploitation of the algorithm can be altered, having consequences in the efficiency of the algorithm.

R. Silhavy (Ed.): CSOC 2020, AISC 1225, pp. 223–233, 2020.
https://doi.org/10.1007/978-3-030-51971-1_18

Several binarization techniques have been developed to address this situation. In a literature search, the main binarization methods used correspond to transfer functions, angle modulation and quantum approach. In this article we present a binarization method that uses the percentile concept to group the solutions and then perform the binarization process. To verify the efficiency of our method, we used the BPBA. This algorithm was proposed in [3].

To check our binary percentile bat optimisation algorithm (BPBA), we use the well-known set covering problem. Experiments were developed using a random operator to validate the contribution of the percentile technique in the binarization process of the sine-cosine algorithm. In addition, to verify our results, the JPSO algorithm developed by [21] and MDBBH developed in [15] were chosen. The results show that BPBA algorithm obtain competitive results.

2 Set Covering Problem

The set covering problem corresponds to a classical problem in combinatorics, and complexity theory. The problem aims to find subsets of a set. Given a set and its elements, we want to find subsets that completely cover the set at a minimum cost. The SCP is one of the oldest and most studied optimization problems. It is well-known to be NP-hard [22].

SCP is an active problem because medium and large instances often become intractable and cannot be solved any more using exact algorithms. Additionally, SCP due to its large number of instances, is used to verify the behavior of proposed new algorithms. In recent years, SCP has been approached by various continuous Swarm intelligence metaheuristics and using different methods to binarize metaheuristics. In [23], they used the teaching-learning-based optimisation with a specific binarization scheme to solve medium and large size instances. An SCP in fault diagnosis application was developed by [24]. In this article, the Gravitational Search algorithm was used and transfer functions were applied to perform the binarization. A rail scheduling problem application using SCP was developed in [15]. In [21] a Jumping PSO was used to solve SCP and in [12] CS metaheuristics were binarized using a percentile algorithm.

SCP has many practical applications in engineer, e.g., vehicle routing, facility location, railway, and airline crew scheduling [7,25–27] problems.

The SCP can be formally defined as follows. Let $A = (a_{ij})$, be a $n \times m$ zero-one matrix, where a column j cover a row i if $a_{ij} = 1$, besides a column j is associated with a non-negative real cost c_j. Let $I = \{1, ..., n\}$ and $J = \{1, ...m\}$, be the row and column set of A, respectively. The SCP consists in searching a minimum cost subset $S \subset J$ for which every row $i \in I$ is covered by at least one column $j \in J$, i.e.:

$$\text{Minimize } f(x) = \sum_{j=1}^{m} c_j x_j \tag{1}$$

$$\text{Subject to } \sum_{j=1}^{m} a_{ij} x_j \geq 1, \forall i \in I, \text{ and } x_j \in \{0, 1\}, \ \forall j \in J \tag{2}$$

where $x_j = 1$ if $j \in S$, $x_j = 0$ otherwise.

3 Binary Percentil Bat Algorithm

The application of statistics and machine learning techniques, are used in many areas such as transports, smart cities, agriculture and computational intelligence, examples of applications are found in [28–32]. In this article, we explore the percentile concept applied to binarization of continuous metaheuristics. As an algorithm to binarize, bat algorithm is used and as problem to solve, SCP. The Proposed BPBA has two modules. The first module corresponds to the initialization of the feasible solutions, and is detailed in the Sect. 3.1. Once the initialization of the particles is performed, it is consulted if the maximum if iterations is satisfied. Subsequently if the criterion is not satisfied, The BPBA is then run in conjunction with the percentile operator, this is detailed in Sect. 3.2). Once the transitions of the different solutions are made, we compare the resulting solutions with the best solution previously obtained. In the event that a superior solution is found, this replaces the previous one. The general algorithm scheme is detailed in Fig. 1.

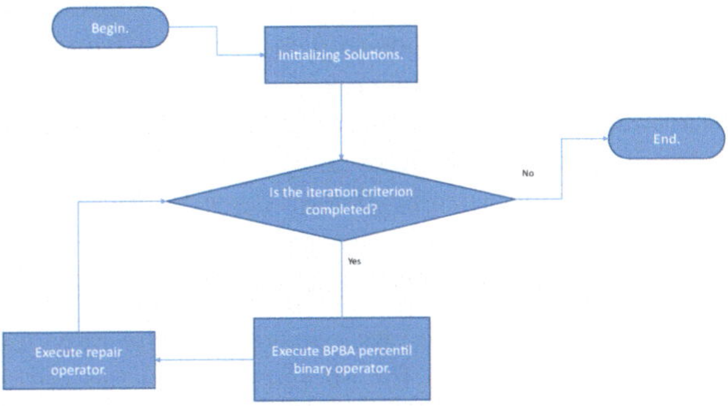

Fig. 1. Flowchart of the BPBA algorithm.

3.1 Initialization

For initialization of a new solution, a column is randomly chosen. It is then queried whether the current solution covers all rows. In the case of the solution does not meet the coverage condition, the heuristic operator is called (Sect. 3.3). This operator aims to select a new column. This heuristic operation is iterated until all rows are covered. Once the coverage condition is met, the solution is optimized. The optimization consists of eliminating columns where all their rows are covered by more than one column. The detail of the procedure is shown in Algorithm 1.

Algorithm 1. Initialization Operator

Function Initialization()
Input
Output Initialized solution S_{out}
$S \leftarrow$ SRandomColumn()
while All row are not covered **do**
  S.append(Heuristic(S))
end while
$S \leftarrow$ dRepeatedItem(S)
$S_{out} \leftarrow S$
return S_{out}

3.2 Percentile Binary Operator

Since our bat algorithm is continuous swarm intelligence metaheuristic, it works in an iterative way by updating the position and velocity of the particles in each iteration. As bat is continuous, the update is done in $\mathbb{R}^n$. In Eq. 3, the position update is written in a general way. The $x(t+1)$ variable represents the x position of the particle at time $t + 1$. This position is obtained from the position x at time t plus a Δ function calculated at time $t + 1$. The function Δ is proper to each metaheuristic and produces values in $\mathbb{R}^n$. For example in Cuckoo Search $\Delta(x) = \alpha \oplus Levy(\lambda)(x)$, in Black Hole $\Delta(x) = \text{rand} \times (x_{bh}(t) - x(t))$ and in the Firefly, Bat and PSO algorithms Δ can be written in simplified form as $\Delta(x) = v(x)$.

$$x(t+1) = x(t) + \Delta_{t+1}(x(t)) \tag{3}$$

In the percentile binary operator, we considering the movements generated by the algorithm in each dimension for all particles. $\Delta^i(x)$ corresponds to the magnitude of the displacement $\Delta(x)$ in the i-th position for the particle x. Subsequently these displacement are grouped using $\Delta^i(x)$, the magnitude of the displacement. This grouping is done using the percentile list. In our case the percentile list used the values {20, 40, 60, 80, 100}.

The percentile operator has as entry the parameters percentile list (pList) and the list of values (vList). Given an iteration, the list of values corresponds to the magnitude $\Delta^i(x)$ of the displacements of the particles in each dimension. As a first step the operator uses the vList and obtains the values of the percentiles given in the pList. Later, each value in the vList is assigned the group of the smallest percentile to which the value belongs. Finally, the list of the percentile to which each value belongs is returned (pGroupValue). The algorithm is shown in Algorithm 2.

A transition probability through the function P_{tr} is assigned to each element of the vList. This assignment is done using the percentile group assigned to each value. For the case of this study, we particularly use the Step function given in Eq. 4.

$$P_{tr}(x^i) = \begin{cases} 0.1, & \text{if}\, x^i \in \text{group}\, \{0, 1\} \\ 0.5, & \text{if}\, x^i \in \text{group}\, \{2, 3, 4\} \end{cases} \tag{4}$$

Afterwards the transition of each particle is performed. In the case of bat search the Eq. 5 is used to perform the transition, where $\hat{x}^i$ is the complement of x^i. Finally, each solution is repaired using the heuristic operator.

$$x^i(t+1) := \begin{cases} \hat{x}^i(t), & \text{if } rand < P_{tg}(x^i) \\ x^i(t), & \text{otherwise} \end{cases} \tag{5}$$

Algorithm 2. percentile binary operator

```
1: Function percentileBinary(vList, pList)
2: Input vList, pList
3: Output pGroupValue
4: percentileValue = getPValue(vList, pList)
5: for each value in vList do
6:     pGroupValue = getPGroupValue(pValue,vList)
7: end for
8: return pGroupValue
```

3.3 Heuristic Operator

The goal of the Heuristic operator is to select a new column for cases where a solution needs to be built or repaired. As input variables, the heuristic operator considers the incomplete solution S_{in} which must be completed. The operator obtains the columns that belong to S_in, then obtains the rows R that are not covered by the solution to S_in. With the set of rows not covered and using Eq. 6 we obtain in line 4 the best 5 rows to be covered. With this list of rows ($lRows$) on line 5 we obtain the list of the best columns according to the heuristic indicated in Eq. 7. Finally randomly in line 6 we obtain the column to incorporate.

$$WeightRow(i) = \frac{1}{L_i}. \tag{6}$$

where L_i is the sum of all ones in row i

$$WeightColumn(j) = \frac{c_j}{|R \cap M_j|}. \tag{7}$$

where M_j is the set of rows covered by Col j

Algorithm 3. Heuristic operator

1: **Function** Heuristic(S_{in})
2: **Input** Input solution S_{in}
3: **Output** The new column *colOut*
4: $lRows \leftarrow$ getBestRows(S_{in}, N=10)
5: $listcolumsnOut \leftarrow$ getBestCols($LRows$, M=5)
6: $colOut \leftarrow$ getCol($lcolsOut$)
7: **return** *colOut*

4 Results

In this section we detail the behavior of BPBA when it is applied to SCP. The contribution of the binary percentile operator was studied when solving the different SCP instances. Additionally, BPBA was compared with other algorithms that have recently resolved SCP. To solve the different SCP instances, a PC with Windows 10, core i5 processor and 16GB in RAM was used. The program was coded in Python 3.5. For the statistical analysis, the non-parametric Wilcoxon signed-rank test was used in addition to violin charts. The analysis of violin charts is performed by comparing the dispersion, median and the interquartile range of the distributions.

4.1 Insight of BPBA Algorithm

In this module we developed experiments that allowed us to study the contribution of the binary percentile operator with respect to the quality of the solutions obtained and the iterations used when solving SCP instances. To perform this evaluation, we used the instances of Balas and Carrera. In the comparison, a random operator was designed which replaces our binary percentile operator. This random operator instead of executing transitions assigning transition probabilities per group, uses a fixed value of 0.5. To compare the distributions of the results of the different experiments we use violin Chart. The horizontal axis X corresponds to the problems, while Y axis uses the measure % - Gap defined in Eq. 8

$$\% - Gap = 100\frac{BestKnown - SolutionValue}{BestKnown} \tag{8}$$

Furthermore, a non-parametric test, Wilcoxon signed-rank test is carried out to determine if the results of BPBA with respect to the random algorithm have significant difference or not. The parameter settings and browser ranges are shown in Table 1.

BPBA corresponds to our standard algorithm. *random.0.5* is the random variant. The results are shown in Table 2 and Fig. 2. When we compared the Best Values between BPBA and *rnd.0.5* which are shown in Table 2. In the Average comparison, BPBA outperforms *rnd.0.5* in all problems. The comparison of distributions is shown in Fig. 2. We see the dispersion of the *rnd.0.5*

Table 1. Setting of parameters for Binary percentile bat Algorithm.

Parameters	Description	Value	Range
N	Number of solutions	25	[20, 25, 30]
G	Number of percentiles	5	[4, 5, 6]
Iteration number	Maximum iterations	800	[700, 800, 900]

distributions are bigger than the dispersions of BPBA. Therefore, the percentile binary operator, contribute to the precision of the results. Finally, the BPBA distributions are closer to zero than *rnd.0.5* distributions, indicating that BPBA has consistently better results than *random.0.5*. When we evaluate the behaviour of the algorithms through the Wilcoxon test, this test indicates that there is a significant difference between the two algorithms.

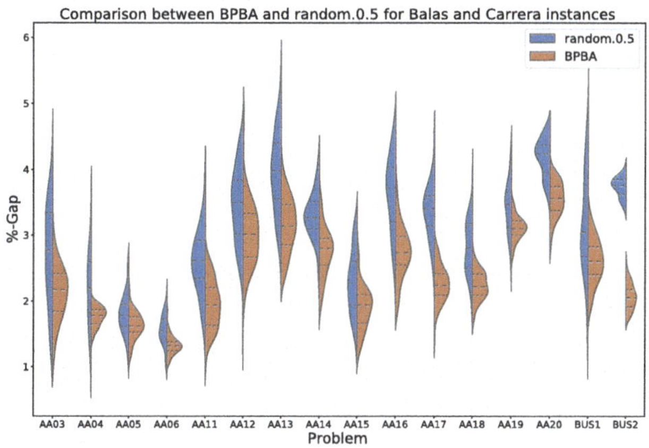

Fig. 2. Evaluation of percentile binary operator.

4.2 BPBA Compared with JPSO and MDBBH

In this section, we detail the comparisons made with the objective of evaluating the performance of our BPBA algorithm. For the evaluation, the larger problems of the OR-library were chosen. To develop the comparison, two algorithms were selected. The first one corresponds to a discretization Particle Swarm Optimization (PSO) technique called Jumping PSO (JPSO), [21]. The second one, is a binarization of the Black Hole technique called Multi Dynamic Binary Black Hole (MDBBH) Algorithm, [15]. JPSO uses a discrete PSO based on the frog jump. JPSO works without the concept of velocity, replacing this one by a component of random jump which allows to perform the movement in the discrete search space. On the other hand, MDBBH uses a binarization mechanism specific to the Black Hole algorithm. This is based on the concept of closeness to the Black

Table 2. Balas and Carrera instances.

Instance	Row	Col	Density	Best known	BPBA (avg)	Std	(Secs)	*rnd.0.5* (avg)	Std
AA03	106	8661	4.05%	33155	33579.5	127.1	209	33693.5	241.1
AA04	106	8002	4.05%	34573	34867.6	58.5	200	34931	231.2
AA05	105	7435	4.05%	31623	31885.8	65.3	203	31889.2	103.2
AA06	105	6951	4.11%	37464	37658.2	59.1	201	37651.3	97.1
AA11	271	4413	2.53%	35384	35764.6	134.7	190	35901.2	197.1
AA12	272	4208	2.52%	30809	31451.3	153.3	210	31631.3	210.2
AA13	265	4025	2.60%	33211	33995.5	136.5	190	34201.4	231.4
AA14	266	3868	2.50%	33219	33869	126	145	33967.3	121.1
AA15	267	3701	2.58%	34409	34821.2	112.9	142	34846.8	161.3
AA16	265	3558	2.63%	32752	33328.3	139.3	188	33601.3	183.9
AA17	264	3425	2.61%	31612	32002.9	89.2	164	32386.2	149.3
AA18	271	3314	2.55%	36782	37290.5	78.7	134	37417.6	148.1
AA19	263	3202	2.63%	32317	33092.5	77.8	147	33109.1	136.1
AA20	269	3095	2.58%	34912	35848.2	93.7	142	35963.1	97.8
BUS1	454	2241	1.88%	27947	28482.8	86.5	190	28756.1	231.6
BUS2	681	9524	0.51%	67760	68540.5	84.5	175	68678.9	101.2
Average				35495.56	36029.9	101.44	176.8	36164.08	165.11
P-value								1.52e−03	

Hole (BH). BH corresponds to the solution that has obtained the best value. When a solution is close to BH, the transitions are less likely to be performed. In the case of being away from the BH, the probability of transition is greater.

For instances E and F which are of medium size, the results of the MDBBH and BPBA algorithms are similar when comparing their Best Value. Only in instance E.2 BPBA was superior to MDBBH. When analyzing the average, JPSO obtained better results, however, the values of BPBA were close, being MDBBH the one that obtained the worse performance. For the case of problems G and H, BPBA was superior in problems G.1, G.2 and G.3 and MDBBH in problem H.3 when comparing their Best Value. In the case of averages, JPSO was superior in all cases. However, BPBA obtained results quite close, leaving MDBBH behind. In this point we must emphasize that the percentile technique used in BPBA allows binarizing any continuous swarm intelligence algorithm unlike JPSO that is specific for PSO (Table 3).

Table 3. OR-Library benchmarks.

Instance	Row	Col	Density	Best Known	JPSO Avg	MDBBH Best	Avg	BPBA Best	Avg	Time (s)
E.1	500	5000	10%	29	29	29	29	29	30,7	12,9
E.2	500	5000	10%	30	30	31	31,6	30	33,6	18
E.3	500	5000	10%	27	27	27	27,4	27	28,9	21,3
E.4	500	5000	10%	28	28	28	29,1	28	30,2	20,1
E.5	500	5000	10%	28	28	28	28	28	30	20,4
F.1	500	5000	20%	14	14	14	14,1	14	14,6	23,3
F.2	500	5000	20%	15	15	15	15,3	15	17,7	25,3
F.3	500	5000	20%	14	14	14	14,8	14	15,6	24,2
F.4	500	5000	20%	14	14	14	14,9	14	15,2	22,9
F.5	500	5000	20%	13	13	14	14,1	14	16,5	27,7
G.1	1000	10000	2%	176	176	177	178,5	177	179,5	204,5
G.2	1000	10000	2%	154	155	157	160,6	156	160,7	219,9
G.3	1000	10000	2%	166	167,2	168	170,4	168	171	237,9
G.4	1000	10000	2%	168	168,2	169	170,9	170	173,2	236,7
G.5	1000	10000	2%	168	168	168	169,8	168	171,8	264,7
H.1	1000	10000	5%	63	64	64	64,9	65	67	200,8
H.2	1000	10000	5%	63	63	64	64	65	65,9	191,1
H.3	1000	10000	5%	59	59,2	59	60	60	61,1	192
H.4	1000	10000	5%	58	58,3	59	60,4	59	62,8	184,2
H.5	1000	10000	5%	55	55	55	56,4	56	57,4	178,5
Average				67.1	67.30	67.7	68.71	67.85	70.17	116.32

5 Conclusions

In this work, the BPBA algorithm based on the percentile concept was proposed to perform binarization of bat metaheuristics. We must emphasize that this percentile concept can be applied in the binarization of any continuous metaheuristics of the swarm intelligence type. To evaluate the performance of Our percentile concept the set covering problem was used together with the bat algorithm. Experiments were designed to evaluate the contribution of the binary operator percentile of the algorithm. It was found that the operator contributes significantly to improve the accuracy and quality of the solutions, all these visualized through violin graphics. Finally, compared to the best metaheuristic algorithm that has solved SCP, our algorithm had a lower yield of 1.74 %, which is not a big difference considering that JPSO uses a particular adaptation mechanism for PSO and the percentile concept It can be easily adapted to any technique. As future works, we believe that the autonomous search for efficient configuration is a little-explored area and it could help to improve the performance of the binarizations both in the quality of the solutions and in the times of convergence. Additionally, it is interesting to see how this percentile technique works with other metaheuristics and other NP-hard problems. Finally, we believe that using machine learning we can generate indicators or obtain correlations that allow us to understand how the percentile algorithm translates the exploration and exploitation properties from the continuous space to the binary space.

References

1. Yang, X.-S., Deb, S.: Cuckoo search via lévy flights. In: World Congress on Nature & Biologically Inspired Computing, 2009. NaBIC 2009, pp. 210–214. IEEE (2009)
2. Hatamlou, A.: Black hole: a new heuristic optimization approach for data clustering. Inf. Sci. **222**, 175–184 (2013)
3. Yang, X.-S.: A new metaheuristic bat-inspired algorithm. In: Gonzáez, J.R., Pelta, D.A., Cruz, C., Terrazas, G., Krasnogor, N. (eds.) Nature inspired cooperative strategies for optimization (NICSO 2010), pp. 65–74. Springer, Berlin (2010)
4. Mirjalili, S.: Sca: a sine cosine algorithm for solving optimization problems. Knowl. Based Syst. **96**, 120–133 (2016)
5. Khatibinia, M., Yazdani, H.: Accelerated multi-gravitational search algorithm for size optimization of truss structures. Swarm Evol. Comput. **38**, 109–119 (2017)
6. Barman, S., Kwon, Y.-K.: A novel mutual information-based boolean network inference method from time-series gene expression data. PLoS ONE **12**(2), e0171097 (2017)
7. Crawford, B., Soto, R., Monfroy, E., Astorga, G., García, J., Cortes, E.: A meta-optimization approach for covering problems in facility location. In: Workshop on Engineering Applications, pp. 565–578. Springer, Cham (2017)
8. Crawford, B., Soto, R., Astorga, G., García, J.: Constructive metaheuristics for the set covering problem. In: International Conference on Bioinspired Methods and their Applications, pp. 88–99. Springer, Cham (2018)
9. García, J., Peña, A.: Robust optimization: concepts and applications. In: Lorente, J.D.S., Osaba, E. (eds.) Nature-Inspired Methods for Stochastic, Robust and Dynamic Optimization. IntechOpen, London (2018)
10. García, J., Moraga, P., Valenzuela, M., Crawford, B., Soto, R., Pinto, H., Peña, A., Altimiras, F., Astorga, G.: A db-scan binarization algorithm applied to matrix covering problems. Comput. Intell. Neurosci. **2019**, 16 (2019)
11. García, J., Altimiras, F., Peña, A., Astorga, G., Peredo, O.: A binary cuckoo search big data algorithm applied to large-scale crew scheduling problems. Complexity **2018**, 15 (2018)
12. García, J., Crawford, B., Soto, R., Astorga, G.: A percentile transition ranking algorithm applied to knapsack problem In: Proceedings of the Computational Methods in Systems and Software, pp. 126–138, Springer, Cham (2017)
13. Astorga, G., Crawford, B., Soto, R., Monfroy, E., García, J., Cortes, E.: A meta-optimization approach to solve the set covering problem. Ingeniería, **23**(3), 274–288 (2018)
14. García, J., Lalla-Ruiz, E., Voß, S., Droguett, E.L.: Enhancing a machine learning binarization framework by perturbation operators: analysis on the multidimensional knapsack problem, Int. J. Mach. Learn. Cybern. 1–20 (2020)
15. García, J., Crawford, B., Soto, R., García, P.: A multi dynamic binary black hole algorithm applied to set covering problem. In: International Conference on Harmony Search Algorithm, pp. 42–51. Springer, Cham (2017)

16. García, J., Crawford, B., Soto, R., Astorga, G.: A percentile transition ranking algorithm applied to binarization of continuous swarm intelligence metaheuristics. In: International Conference on Soft Computing and Data Mining, pp. 3–13. Springer, Cham (2018)
17. Franceschetti, A., Demir, E., Honhon, D., Van Woensel, T., Laporte, G., Stobbe, M.: A metaheuristic for the time-dependent pollution-routing problem. Eur. J. Oper. Res. **259**(3), 972–991 (2017)
18. Crawford, B., Soto, R., Astorga, G., García, J., Castro, C., Paredes, F.: Putting continuous metaheuristics to work in binary search spaces. Complexity **2017**, 19 (2017)
19. García, J., Crawford, B., Soto, R., Astorga, G.: A clustering algorithm applied to the binarization of swarm intelligence continuous metaheuristics. Swarm Evol. Comput. **44**, 646–664 (2019)
20. Garcia, J., Măntoiu, M.: Localization results for zero order pseudodifferential operators. J. Pseudo Differ. Oper. Appl. **5**(2), 255–276 (2014)
21. Balaji, S., Revathi, N.: A new approach for solving set covering problem using jumping particle swarm optimization method. Nat. Comput. **15**(3), 503–517 (2016)
22. Gary, M.R., Johnson, D.S.: Computers and Intractability: A Guide to the Theory of NP-Completeness. WH Freeman and Company, New York (1979)
23. Lu, Y., Vasko, F.J.: An or practitioner's solution approach for the set covering problem. Int. J. Appl. Metaheuristic Comput. (IJAMC) **6**(4), 1–13 (2015)
24. Li, Y., Cai, Z.: Gravity-based heuristic for set covering problems and its application in fault diagnosis. J. Syst. Eng. Electron. **23**(3), 391–398 (2012)
25. Kasirzadeh, A., Saddoune, M., Soumis, F.: Airline crew scheduling: models, algorithms, and data sets. EURO J. Transp. Logist. **6**(2), 111–137 (2017)
26. Horváth, M., Kis, T.: Computing strong lower and upper bounds for the integrated multiple-depot vehicle and crew scheduling problem with branch-and-price. CEJOR **27**, 1–29 (2017)
27. Stojković, M.: The operational flight and multi-crew scheduling problem. Yugoslav J. Oper. Res. **151**, 25–48 (2016)
28. García, J., Crawford, B., Soto, R., Carlos, C., Paredes, F.: A k-means binarization framework applied to multidimensional knapsack problem. Appl. Intell. **48**(2), 1–24 (2017)
29. García, J., Pope, C., Altimiras, F.: A distributed k-means segmentation algorithm applied to lobesia botrana recognition. Complexity **2017**, 14 (2017)
30. Graells-Garrido, E., García, J.: Visual exploration of urban dynamics using mobile data. In: International Conference on Ubiquitous Computing and Ambient Intelligence, pp. 480–491, Springer, Cham (2015)
31. Graells-Garrido, E., Peredo, O., García, J.: Sensing urban patterns with antenna mappings: the case of santiago, chile. Sensors **16**(7), 1098 (2016)
32. Peredo, O.F., García, J.A., Stuven, R., Ortiz, J.M.: Urban dynamic estimation using mobile phone logs and locally varying anisotropy. In: Gómez-Hernández, J., Rodrigo-Ilarri, J., Rodrigo-Clavero, M., Cassiraga, E., Vargas-Guzmán, J., (eds.) Geostatistics Valencia 2016, pp. 949–964, Springer, Cham (2017)

A K-means Grasshopper Algorithm Applied to the Knapsack Problem

Hernan Pinto[1], Alvaro Peña[1], Leonardo Causa[2], Matías Valenzuela[1], and Gabriel Villavicencio[1](✉)

[1] Escuela de ingeniería en construcción, Pontificia Universidad Católica de Valparaíso, Valparaíso, Chile
{hernan.pinto,alvaro.pena,matias.valenzuela, gabriel.villavicencio}@pucv.cl

[2] Escuela de Tecnologías de la Información, Facultad de Ingeniería y Negocios, Universidad de las Américas, Santiago, Chile
lcausamo@edu.udla.cl

Abstract. In engineering and science, there are many combinatorial optimization problems. A lot of these problems are NP-hard and can hardly be addressed by full techniques. Therefore, designing binary algorithms based on swarm intelligence continuous metaheuristics is an area of interest in operational research. In this paper we use a general binarization mechanism based on the k-means technique. We apply the k-means technique to grasshopper algorithm to solve multidimensional knapsack problem (MKP). Experiments are designed to demonstrate the utility of the k-means technique in binarization. Additionally we verify the efficiency of our algorithm through benchmark instances, showing that binary k-means grasshopper algorithm (BKGOA) obtains adequate results when it is evaluated against another state of the art algorithm.

Keywords: Combinatorial optimization · KnapSack · Metaheuristics · K-means

1 Introduction

The combinatorial problems have great relevance at industrial level, we find them in various areas such as Civil Engineering, Bio Informatics [1], Operational Research [2,3], resource allocation [4–6], scheduling problems [7–9], routing problems [10–13] among others. On the other hand, in recent years, algorithms inspired by nature phenomena to solve optimization problems have been generated. As examples of these algorithms we have Cuckoo Search, Black Hole, Bat Algorithm, and grasshopper Algorithms [14] among others. Many of these algorithms work naturally in continuous spaces and therefore must be adapted to solve combinatorial problems. In the process of adaptation, the mechanisms of exploration and exploitation of the algorithm can be altered, having consequences in the efficiency of the algorithm.

R. Silhavy (Ed.): CSOC 2020, AISC 1225, pp. 234–244, 2020.
https://doi.org/10.1007/978-3-030-51971-1_19

Several binarization techniques have been developed to address this situation. In a literature search, the main binarization methods used correspond to transfer functions, angle modulation and quantum approach, for more detail refer to [15]. In this article we present a new binarization method that uses the k-means technique to group the solutions and then perform the binarization process. To verify the efficiency of our method, we used the grasshopper (GOA) algorithm.

Grasshoppers are insects and under certain conditions can be considered a pest, the main characteristic of these insects is related to their movement. When it is in a larval state, its movement is slow in contrast with the large and abrupt movements of the adults.

The mathematical model to simulate the movement of grasshoppers are presented in Eq. 1

$$X_i^d = c\Big(\sum_{j=1, j\neq i}^{N} c\frac{ub_d - lb_d}{2} s(|x_j^d - x_i^d|)\frac{x_j - x_i}{d_{ij}}\Big) + \hat{T}_d \tag{1}$$

The first c reduces the movements of grasshoppers around the target. In other words, this parameter balances exploration and exploitation of the entire swarm around the target. The second c decreases the attraction zone, comfort zone, and repulsion zone between grasshoppers. The c coefficient is calculated as: $cmax - l\frac{cmax-cmin}{L}$, where L= 500 is de maximum number of iteration. The cmax, and cmin coefficients are 1 and 0.00001 respectively. The component $s(r) = fe^{\frac{-r}{l} - e^{-r}}$ indicates if a grasshopper should be repelled from (explore) or attracted to (exploitation) the target. For this case $f = 0.5$ and $l = 1.5$. Finally $\hat{T}_d$ represents the best solution at this iterations.

To check our binary Grasshopper algorithm (BKGOA), we use the well-known knapsack problem. Experiments were developed using a random operator to validate the contribution of the k-means technique in the binarization process of the grasshopper algorithm. In addition, local search operator is used to strengthen the results. Moreover, the binary artificial algae (BAAA) algorithm was used to compare our results. BAAA was developed in [16] and uses transfer functions to perform the binarization process. The results show that the k-means technique obtains results superior to those obtained by the random operator and that our BKGOA algorithm shows competitive results against the BAAA algorithm.

2 KnapSack Problem

The multidimensional knapsack problem (MKP), models resource allocation situations. The goal is to find the subset of objects that produce the greatest benefit by satisfying a set of constraints. MKP is one of the most studied NP-hard class combinatorial problems, however it remains as a challenge due to the difficulties encountered in solving medium and large size instances. A search for the last few years in the literature, we find that MKP has been solved for example in

[17] using a quantum binarization technique, in [18] applied the k-means clustering technique to perform binarization, in [19] used a differential algorithm with transfer functions, and in [20] a modification of the PSO equations was used. MKP can be set as:

$$\text{maximize} \sum_{j=1}^{n} p_j x_j \tag{2}$$

$$\text{subjected to} \sum_{j=1}^{n} c_{ij} x_j \leq b_i \ , \ i \in \{1, ..., m\}. \tag{3}$$

With $x_j \in \{0,1\}$, $j \in \{1, ..., n\}$. p_j corresponds to the profit of element j. c_{ij} represents a cost associated with dimension i and element j. The constraints in each dimension i are represented by b_i. The solution can be modelled using a binary representation, in this representation a 0 means the element is not included in the knapsack.

3 Binary Grasshopper Algorithm

The application of statistics and data mining, can be found in many areas such as transports, smart cities, agriculture and computational intelligence [18,21–24]. In this section, we explore the application of the k-means technique to the binarization of continuous metaheuristics.

The binary grasshopper algorithm (BKGOA) uses 4 operators to achieve its goal. An initialization operator, A local search operator, the binary operator using the k-means technique, and a repair operator for the case of solutions that do not meet any of constraints. The general scheme of the algorithm is shown in Fig. 1. In the first step, BKGOA initializes the solutions, this is detailed in Sect. 3.1. Once we have all the solutions initialized, we check the compliance of the stopping criterion of the algorithm associated to the maximum number of iterations. The grasshopper algorithm is then run in conjunction with the k-means operator, this is described in Sect. 3.2. Subsequently once the binarization was performed, the results are compared with the best solution obtained so far. In the case of finding a superior solution, it replaces the previous solution and is submitted to a local search operator. In the following subsection we will explain in detail the initialization method, the binary operator and the repair operator.

3.1 Initialization and Element Weighting

As BKGOA is a swarm algorithm, to begin the exploration and exploitation of the search space, the list of solutions must be initialized. For the generation of each solution, an element is randomly chosen first. The next step is to check if other elements can be incorporated, for this we must evaluate the constraints of our problem. To select the new element, we generate a list of possible elements that comply with the constraints. Each element is calculated a weight and the

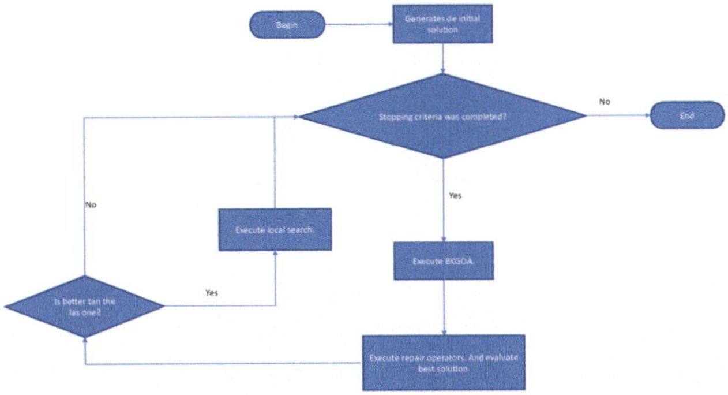

Fig. 1. Flowchart of the binary grasshopper algorithm.

best weight element is selected. This procedure is repeated until no additional element can be incorporated.

To calculate the weight of each element, several methods have been developed. In [25] a pseudo-utility in the surrogate duality approach was proposed. The way to calculate it is shown in Eq. 4. In this equation the variable w_j corresponds to the surrogate multiplier whose value is between 0 and 1. This multiplier can be interpreted as a shadow prices of the j-th constraint.

$$\delta_i = \frac{p_i}{\sum_{j=1}^{m} w_j c_{ij}} \tag{4}$$

A more intuitive measure focused on the average resource occupancy was proposed by [26]. It is shown in Eq. 5.

$$\delta_i = \frac{\sum_{j=1}^{m} \frac{c_{ij}}{m b_j}}{p_i} \tag{5}$$

In this article we used a variation of this last measure focused on the average occupation and proposed in [18]. This variation considers the elements that exist in knapsacks to calculate the average occupancy. In each iteration depending on the selected items in the solution the measure is calculated again. The expression of this new measure is shown in Eq. 6.

$$\delta_i = \frac{\sum_{j=1}^{m} \frac{c_{ij}}{m(b_j - \sum_{i \in S} c_{ij})}}{p_i} \tag{6}$$

3.2 K-means Transition Ranking Operator

Consider our metaheuristic is a continuous and population based. Due to its iterative nature, it needs to update the position of particles at each iteration. When the metaheuristic is continuous, this update is performed in $\mathbb{R}^n$ space. In Eq. 7, the update position is presented in a general manner. The x_{t+1} variable represents the x position of the particle at time t+1. This position is obtained from the position x at time t plus a Δ function calculated at time t+1. The function Δ is proper to each metaheuristic and produces values in $\mathbb{R}^n$. For example in Cuckoo Search $\Delta(x) = \alpha \oplus Levy(\lambda)(x)$, in Black Hole $\Delta(x) = \text{rand} \times (x_{bh}(t) - x(t))$ and in the Firefly, Bat and PSO algorithms Δ can be written in simplified form as $\Delta(x) = v(x)$.

$$x_{t+1} = x_t + \Delta_{t+1}(x(t)) \tag{7}$$

In the k-means transition ranking operator, a model for transitions in a discrete space is proposed. This model is based on considering the movements generated by the metaheuristic in each dimension for all particles. $\Delta^i(x)$ corresponds to the magnitude of the displacement $\Delta(x)$ in the i-th position. Subsequently these displacement are grouped using the magnitude of the displacement $\Delta^i(x)$. This grouping is done using the k-means technique where k represents the number of clusters used. Finally, a generic P_{tr} function given by the Eq. 8 is proposed to assign a transition probability.

$$P_{tr} : \mathbb{Z}/k\mathbb{Z} \rightarrow [0, 1] \tag{8}$$

A transition probability through the function P_{tr} is assigned to each group. Naturally, this P_{tr} function is modelled as a cumulative probability function. For the case of this study, we particularly use the linear function given in Eq. 9. In this equation, $\mathrm{N}(x^i)$ indicates the location of the group to which $\Delta^i(x)$ belongs. The α coefficient, corresponds to the transition probability and β to the transition separation coefficient. Both parameters are estimated in each metaheuristic. For our particular case, $\mathrm{N}(x^i) = 0$ corresponds to elements belonging to the group that has the lowest Δ^i values. $\mathrm{N}(x^i) = 7$ corresponds to the group of elements that have the greatest Δ^i values.

$$P_{tr}(x^i) = P_{tr}(N(x^i)) = \alpha + \beta N(x^i)\alpha \tag{9}$$

The k-means transition ranking operator starts calculating Δ^i for each of the particles. This step is specific in each metaheuristic. Subsequently the particles are grouped using k-means clustering technique and the magnitude of Δ^i as distance. With the group assigned to each particle we obtain the probability of transition using Eq. 9. Afterwards the transition of each particle is performed. In the case of grasshopper algorithm the rule 10 is used to perform the transition, where $\hat{x}^i$ is the complement of x^i. Finally, each solution is repaired using the repair operator.

$$x^i(t+1) := \begin{cases} \hat{x}^i(t), & \text{if } rand < P_{tg}(x^i) \\ x^i(t), & \text{otherwise} \end{cases} \tag{10}$$

3.3 Repair Operator

Local search and binary k-means technique can generate infeasible solutions. There are different mechanisms to address these infeasible solutions. In this article, the repair of the solutions was considered. To perform the repair, the measure described in the Eq. 6 was used. If the solution requires repair, the element with the maximum measure is chosen and it is removed from the solution, this process is iterated until a valid solution is obtained. The solution is then improved by exploring the possibility of incorporating new elements. This stage of improvement is iterated until there are no elements that can be incorporated without violating the constraints.

4 Results

4.1 Insight of BKGOA Algorithm

This section aims to find out the contribution of the binary k-means operator to the performance of the algorithm. To carry out the comparison problems cb.5.250 from the OR-library were selected. Violin charts and the Wilcoxon non-parametric signed-rank test were used to perform the statistical analyzes. In the charts, the X axis identifies the studied instances and the Y axis the % -Gap described in the Eq. 11. The Wilcoxon test is run to determine if the results obtained by BKGOA have significant difference with respect to other algorithms. The parameter settings and browser ranges are shown in Table 1.

$$\% - Gap = 100 \frac{BestKnown - SolutionValue}{BestKnown} \tag{11}$$

Table 1. Setting of parameters for Binary grasshopper Algorithm.

Parameters	Description	Value	Range
N	Number of solutions	30	[20, 25, 30]
G	Number of clusters	5	[4,5,6]
Iteration number	Maximum iterations	1000	[1000]
Cmax	Maximum for c coefficient	1	1
Cmin	Minimum for c coefficient	0.00001	0.00001
f	f parameter for s function	0.5	0.5
l	l parameter for s function	1.5	1.5

Evaluation of K-means Binary Operator. A random operator was designed to evaluate the contribution of the k-means binary operator. This random operator has the peculiarity of performing the transitions with a fixed probability of 0.5 without considering in which cluster the variable is located. Two configurations were studied: The first one where the local search operator is included and the second where the local operator is not considered. BKGOA corresponds to our standard algorithm. *random.ls* is the random variant that includes the local search operator. *BKGOA.wls* corresponds to the version with k-means binary operator without local search operator. Finally *random.wls* describes the random algorithm without local search operator.

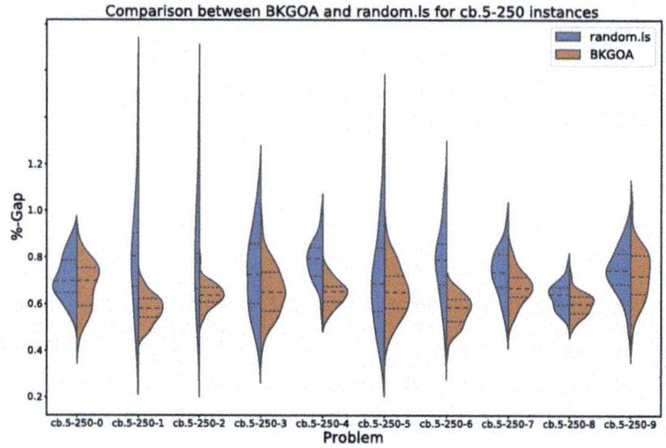

Fig. 2. Evaluation of k-means binary operator with Local Search operator

When we compared the Best Values between BKGOA and *random.ls* which are shown in Table 2. BKGOA outperforms to *random.ls*. However the Best Values between both algorithms are very close. In the Average comparison, BKGOA outperforms *random.ls* almost in all problems. The comparison of distributions is shown in Fig. 2. We see the dispersion of the *random.ls* distributions are bigger than the dispersions of BKGOA. In particular this can be appreciated in the problems 1, 2, 3, 5, 6 and 7. Therefore, the k-means binary operator together with local search operators, contribute to the precision of the results. Finally, the BKGOA distributions are closer to zero than *random.ls* distributions, indicating that BKGOA has consistently better results than *random.ls*. When we evaluate the behaviour of the algorithms through the Wilcoxon test, this indicates that there is a significant difference between the two algorithms.

Table 2. Evaluation of k-means binary operator

Set	Best Known	Best *random.ls*	Best BKGOA	Best *random.wls*	Best *wls*	Avg *random.ls*	Avg BKGOA	Avg *random.wls*	Avg *wls*
cb.5.250-0	59312	59211	59225	59158	59175	59132.1	59152.4	59071.8	59132.1
cb.5.250-1	61472	61435	61435	61409	61409	61324.6	61387.1	61288.3	61370.7
cb.5.250-2	62130	62036	62074	61969	61969	61894.4	61978.5	61801.6	61924.6
cb.5.250-3	59463	59367	59446	59365	59349	59257.8	59322.1	59136.1	59269.3
cb.5.250-4	58951	58914	58951	58883	58930	58725.6	58824.6	58693.6	58758.5
cb.5.250-5	60077	60015	60056	59990	60015	59904.6	59952.3	59837.8	59945.9
cb.5.250-6	60414	60355	60355	60348	60349	60208.2	60332.1	60230.6	60311.1
cb.5.250-7	61472	61436	61436	61407	61407	61290.8	61331.9	61233.9	61342.3
cb.5.250-8	61885	61829	61885	61790	61782	61737.1	61787.7	61644.9	61738.1
cb.5.250-9	58959	58832	58866	58822	58787	58769.1	58782.8	58653.7	58770.4
Average	60413.5	60343	60372.9	60314.1	60317.2	60224.4	60285.15	60159.1	60256.3
p-value							1.14 e−05		2.43 e−06

4.2 BKGOA Comparisons

In this section we evaluate the performance of our BKGOA with the algorithm BAAA developed in [16]. BAAA uses transfer functions as a general mechanism of binarization. In particular BAAA used the tanh $= \frac{e^{\tau|x|}-1}{e^{\tau|x|}+1}$ function to perform the transference. The parameter τ of the tanh function was set to a value 1.5. Additionally, a elite local search procedure was used by BAAA to improve solutions. As maximum number of iterations BAAA used 35000. The computer configuration used to run the BAAA algorithm was: PC Intel Core(TM) 2 dual CPU Q9300@2.5 GHz, 4 GB RAM and 64-bit Windows 7 operating system. In our BKGOA algorithm, the configurations are the same used in the previous experiments. These are described in the Table 1.

The results are shown in Table 3. The comparison was performed for the set cb.5.500 of the OR-library. The results for BKGOA were obtained from 30 executions for each problem. In black, the best results are marked for both indicators the Best Value and the Average. For the best value indicator BAAA was higher in 4, and BKGOA in 26. In the Average indicator BAAA was higher in 2 instances, and BKGOA in 28. We should also note that the standard deviation in most problems was quite low, indicating that BKGOA has good accuracy.

Table 3. OR-Library benchmarks MKP cb.5.500

Instance	Best Known	BAAA Best	Avg	BKGOA Best	Avg	Time(s)	std
0	120148	120066	120013.7	120096	119468.9	454	42.6
1	117879	117702	117560.5	117656	116962.2	503	41.4
2	121131	120951	120782.9	120923	119612.8	525	45
3	120804	120572	120340.6	120683	120280.7	509	51.7
4	122319	122231	122101.8	122212	121193.3	516	48.7
5	122024	121957	121741.8	121982	120871.4	530	58.5
6	119127	119070	118913.4	118956	117664.6	512	44.4
7	120568	120472	120331.2	120463	119045.8	500	57.5
8	121586	121052	120683.6	121295	119899.5	546	52.1
9	120717	120499	120296.3	120467	120083.1	507	51.4
10	218428	218185	217984.7	218296	216644.1	483	48.1
11	221202	220852	220527.5	220951	219189.5	495	56.4
12	217542	217258	217056.7	217356	216865.9	472	48.3
13	223560	223510	223450.9	223516	221477.7	496	41.9
14	218966	218811	218634.3	218848	216924.2	541	49.9
15	220530	220429	220375.9	220410	218579.3	510	44.9
16	219989	219785	219619.3	219858	218218.6	500	62.3
17	218215	218032	217813.2	218010	215843.4	478	59.8
18	216976	216940	216862	216866	215690.9	431	38
19	219719	219602	219435.1	219631	219328.6	496	34
20	295828	295652	295505	295688	294356.8	401	36.6
21	308086	307783	307577.5	307924	306964.5	443	33.2
22	299796	299727	299664.1	299796	298487.7	404	41.9
23	306480	306469	306385	306480	302992.2	403	37.9
24	300342	300240	300136.7	300245	298723.6	410	37.3
25	302571	302492	302376	302481	301451.7	398	30
26	301339	301272	301158	301284	299240	403	27.1
27	306454	306290	306138.4	306331	303189.5	389	29.6
28	302828	302769	302690.1	302771	301330.7	368	26.4
29	299910	299757	299702.3	299828	298514.7	379	28.6
Average	214168.8	214014.2	213862.0	214040.8	213975.2	464.4	43.3

5 Conclusions

In this article, the k-means technique is used to perform the binarization of the GOA algorithm. It should be noted that the k-means technique can be applied in the binarization of any continuous swarm-intelligence algorithm. We used the knapsack problem to evaluate our binary algorithm. The contribution of the k-means binary operator was studied, observing that this operator contributes to the precision and quality of the solutions obtained. Additionally, we develop the comparison with the BAAA algorithm.

As a future line of research, it is interesting to compare the performance of the k-means operator with the transfer functions used in BAAA, all applied in the binarization of the GOA algorithm. Also, it is interesting to use the k-means operator to binarize other algorithms in addition to solving other NP-hard problems. From the theoretical point of view, it would also be interesting to understand how the exploration and exploitation of space are altered when we introduce the k-means operator. Another interesting research line is to explore adaptive techniques to automate the selection of parameters.

References

1. Barman, S., Kwon, Y.-K.: A novel mutual information-based boolean network inference method from time-series gene expression data. PLoS ONE **12**(2), e0171097 (2017)
2. Crawford, B., Soto, R., Monfroy, E., Astorga, G., García, J., Cortes, E.: A meta-optimization approach for covering problems in facility location. In: Workshop on Engineering Applications, pp. 565–578. Springer, Cham (2017)
3. García, J., Crawford, B., Soto, R., Astorga, G.: A percentile transition ranking algorithm applied to binarization of continuous swarm intelligence metaheuristics. In: International Conference on Soft Computing and Data Mining, pp. 3–13. Springer, Cham (2018)
4. Garcia, J., Crawford, B., Soto, R., Astorga, G.: A percentile transition ranking algorithm applied to knapsack problem. In: Proceedings of the Computational Methods in Systems and Software, pp. 126–138. Springer, Cham (2017)
5. García, J., Altimiras, F., Peña, A., Astorga, G., Peredo, O.: A binary cuckoo search big data algorithm applied to large-scale crew scheduling problems. Complexity **2018**, 15 (2018)
6. García, J., Lalla-Ruiz, E., Voß, S., Droguett, E.L.: Enhancing a machine learning binarization framework by perturbation operators: analysis on the multidimensional knapsack problem, Int. J. Mach. Learn. Cybern. 1–20 (2020)
7. García, J., Crawford, B., Soto, R., García, P.: A multi dynamic binary black hole algorithm applied to set covering problem. In: International Conference on Harmony Search Algorithm, pp. 42–51. Springer, Cham (2017)
8. García, J., Crawford, B., Soto, R., Astorga, G.: A clustering algorithm applied to the binarization of swarm intelligence continuous metaheuristics. Swarm Evol. Comput. **44**, 646–664 (2019)
9. Crawford, B., Soto, R., Astorga, G., García, J.: Constructive metaheuristics for the set covering problem. In: International Conference on Bioinspired Methods and Their Applications, pp. 88–99. Springer, Singapore (2018)

10. García, J., Peña, A.: Robust optimization: concepts and applications. In: Lorente, J.D.S., Osaba, E. (eds.) Nature-Inspired Methods for Stochastic, Robust and Dynamic Optimization, p. 7. IntechOpen, London (2018)
11. Astorga, G., Crawford, B., Soto, R., Monfroy, E., García, J., Cortes, E.: A meta-optimization approach to solve the set covering problem. Ingeniería, **23**(3), 274–288 (2018)
12. García, J., Moraga, P., Valenzuela, M., Crawford, B., Soto, R., Pinto, H., Peña, A., Altimiras, F., Astorga, G.: A db-scan binarization algorithm applied to matrix covering problems. Comput. Intell. Neurosci. **2019**, 16 (2019)
13. Garcia, J., Măntoiu, M.: Localization results for zero order pseudodifferential operators. J. Pseudo Differ. Oper. Appl. **5**(2), 255–276 (2014)
14. Saremi, S., Mirjalili, S., Lewis, A.: Grasshopper optimisation algorithm: theory and application. Adv. Eng. Softw. **105**, 30–47 (2017)
15. Crawford, B., Soto, R., Astorga, G., García, J., Castro, C., Paredes, F.: Putting continuous metaheuristics to work in binary search spaces. Complexity **2017**, 19 (2017)
16. Zhang, X., Wu, C., Li, J., Wang, X., Yang, Z., Lee, J.-M., Jung, K.-H.: Binary artificial algae algorithm for multidimensional knapsack problems. Appl. Soft Comput. **43**, 583–595 (2016)
17. Haddar, B., Khemakhem, M., Hanafi, S., Wilbaut, C.: A hybrid quantum particle swarm optimization for the multidimensional knapsack problem. Eng. Appl. Artif. Intell. **55**, 1–13 (2016)
18. García, J., Crawford, B., Soto, R., Carlos, C., Paredes, F.: A k-means binarization framework applied to multidimensional knapsack problem. Appl. Intell. **48**(2), 1–24 (2017)
19. Liu, J., Wu, C., Cao, J., Wang, X., Teo, K.L.: A binary differential search algorithm for the 0–1 multidimensional knapsack problem. Appl. Math. Model. **40**, 9788–9805 (2016)
20. Bansal, J.C., Deep, K.: A modified binary particle swarm optimization for knapsack problems. Appl. Math. Comput. **218**(22), 11042–11061 (2012)
21. García, J., Pope, C., Altimiras, F.: A distributed k-means segmentation algorithm applied to lobesia botrana recognition. Complexity **2017**, 14 (2017)
22. Graells-Garrido, E., García, J.: Visual exploration of urban dynamics using mobile data. In: International Conference on Ubiquitous Computing and Ambient Intelligence, pp. 480–491, Springer, Cham (2015)
23. Peredo, O.F., García, J.A., Stuven, R., Ortiz, J.M.: Urban dynamic estimation using mobile phone logs and locally varying anisotropy. In: Gómez-Hernández, J., Rodrigo-Ilarri, J., Rodrigo-Clavero, M., Cassiraga, E., Vargas-Guzmán, J. (eds.) Geostatistics Valencia 2016, pp. 949–964. Springer, Cham (2017)
24. Graells-Garrido, E., Peredo, O., García, J.: Sensing urban patterns with antenna mappings: the case of santiago, chile. Sensors **16**(7), 1098 (2016)
25. Pirkul, H.: A heuristic solution procedure for the multiconstraint zero? One knapsack problem. Naval Res. Logist. **34**(2), 161–172 (1987)
26. Kong, X., Gao, L., Ouyang, H., Li, S.: Solving large-scale multidimensional knapsack problems with a new binary harmony search algorithm. Comput. Oper. Res. **63**, 7–22 (2015)

Hierarchical Approach Towards High Fidelity Image Generation

Arindam Chaudhuri[1(✉)] and Soumya K. Ghosh[2]

[1] Samsung R&D Institute, Noida 201304, Delhi, India
arindam_chau@yahoo.co.in

[2] Department of Computer Science Engineering, Indian Institute of Technology Kharagpur, Kharagpur 701302, India
skg@iitkgp.ac.in

Abstract. The high-fidelity image generation has been a subject of active research in the recent past. It provides benchmark towards image decoder's performance. The autoregressive image models generate small images successfully but scalability has been a problem. The challenges include vast encoding of previous context and learning distribution which maintains global semantic coherence and exactness. These issues have been addressed through subscale pixel network (SPN) and multidimensional upscaling. To improve the accuracy further, in this work a hierarchical version of image generation model is presented. It disentangles background, object shape and appearance to hierarchically generate images of fine-grained object categories. To achieve this information theory associates a factor to latent code and condition relationships between codes to induce hierarchy. The hierarchical model's learned features are used to cluster real images. The experimental results on ImageNet and CelebAHQ datasets for different image sizes highlight hierarchical model's superiority against the benchmarks. The images are generated with better fidelity with respect to large scale samples.

Keywords: Image generation · High fidelity · Semantic coherence · Image scalability · Hierarchical model · Image benchmarks · Real images

1 Introduction

The success of generative model is highlighted in terms of two aspects viz targets with high fidelity and generalization on held-out data. Autoregressive (AR) models trained through traditional maximum likelihood estimation (MLE) have produced appreciable scores over wide domain spectrum [1–5]. These scores represent model's generalization ability. With respect to sample fidelity, outputs generated through AR models have reached state-of-the-art fidelity in above domains. However, in unconditional large-scale image generation domain, AR samples still need to address long-range structure and semantic coherence issues. An important difficulty impeding high-fidelity image generation is multi-faceted relationship between model's MLE scores and its sample

R. Silhavy (Ed.): CSOC 2020, AISC 1225, pp. 245–256, 2020.
https://doi.org/10.1007/978-3-030-51971-1_20

fidelity. MLE is well-defined measure as held-out scores' improvements produce visual samples' fidelity enhancements. On the contrary, unlike adversarial methods [6] MLE forces model towards entire empirical distribution. This guarantees model's ability to generalize at allotting capacity costs towards distribution parts. Another issue is large images' high dimensionality which requires architectural dependencies and expression capacity of surrounding contexts. It requires requirements translation to high memory and computation. The learning is performed over full distribution with samples achieving high fidelity. The model is guided first to focus on visually more salient bits and then less salient bits. There are two visually salient distribution subsets (a) subset determined through smaller size sub-images sub-sampled at all positions from original image and (b) subset determined through few most significant bits/RGB channel in image. The multi-dimensional upscaling maps from one distribution subset to other one by upscaling images in size or depth. Thus three networks are trained: (a) a decoder on small size, low depth image slices subsampled at every *n* pixels from original image (b) a size-upscaling decoder that generates large size, low depth image conditioned on small size, low depth image and (c) a depth-upscaling decoder that generates large size, high depth image conditioned on large size, low depth image. The Fig. 1 shows this process. The decoders in (b) and (c) training is taken care of through subscale pixel network (SPN) architecture [7].

Fig. 1. A representation of multidimensional upscaling [7]

In this paper based on the success achieved from high fidelity image generation in [7], a hierarchical version of the fine-grained image generation model is presented. The model hierarchy is chosen to achieve the best possible results. The model's fine-grained object categories hierarchically disentangle background, object's shape and its appearance [8]. The major motivations being complexity reduction in annotation process and latent data structure in unsupervised learning. The fine-grained image datasets are blessed with inherent hierarchical organization. With AR models there has been a challenge to generate large images with high fidelity. Another issue has been the encoding of previous context and learning the distribution which maintains both global semantic coherence and detail exactness. SPN addresses initial issue by compactly capturing image-wide spatial dependencies with lesser memory and computation requirements. The multi-dimensional upscaling grows image both in terms of size and depth via intermediate stages utilising distinct SPNs. The state-of-the-art experimental results in multiple settings on ImageNet and CelebAHQ datasets for different image sizes highlight the superiority of the proposed method with respect to the benchmarks set up in [8]. The images are generated with very high fidelity towards large scale samples on the experimental datasets. This paper is organized as follows. In Sect. 2 computational method of hierarchical version of high-fidelity image generation is highlighted. This is followed by experiments and results in Sect. 3. Finally, in Sect. 4 conclusions are given.

2 Computational Method

In this section framework of proposed hierarchical version of high-fidelity image generation model is presented. The schematic representation of system is given in Fig. 2.

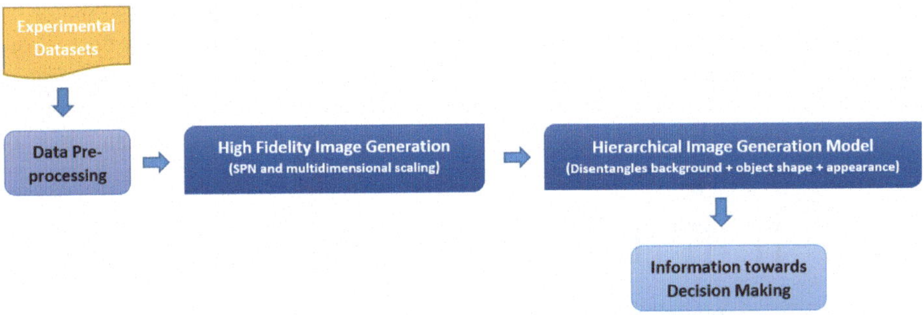

Fig. 2. Image generation framework through hierarchical model

2.1 Problem Definition

The research problem entails in generating images with high fidelity. In order to achieve this, we propose hierarchical deep learning-based image generator to generate scalable high-fidelity images. The generation task on subscale pixel networks and multi-dimensional scaling aspects provide information towards several decision-making activities.

2.2 Datasets

All the experiments are performed on ImageNet [9] and CelebAHQ [10] datasets. ImageNet dataset is organized according to the WordNet hierarchy where each node of the hierarchy is depicted by hundreds and thousands of images. Presently there are over an average of five hundred images/node. CelebAHQ is a large-scale face attributes dataset with more than 200K celebrity images, each with 40 attribute annotations. The images in this dataset cover large pose variations and background clutter. The dataset has large diversities, large quantities and rich annotations.

2.3 Basic Image Generation Ordering Model

In this section we discuss aspects related to basic image generation ordering [7]. We highlight conventional generation ordering with subscale ordering, size and depth upscaling. A standard AR model such as PixelCNN [3] generates $HI \times WI$ color image spanning from top-left and bottom-right positions. It fully generates three 8-bit channels/pixel in given position:

$$Prob(x) = \prod_{ht=1}^{HI} \prod_{wt=1}^{WI} \prod_{ct=1}^{\{R,G,B\}} Prob(x_{ht,wt,ct} | \mathrm{x}_{pgi}) \tag{1}$$

Here $\mathbf{x}_{pgi}$ represents all previously generated ordered intensity values and ht, wt, ct are row, column, colour channel indices. Each conditional distribution $Prob(x_{h,w,c} | \mathbf{x}_{pgi})$ is parametrized through deep neural network [3].

An alternative ordering segregates a large image towards sequence of equally sized slices with various core aspects. It is easy to compactly encode long-range dependencies across pixels in large images. It induces spatial structure over original image by aligning subsampled slices. This also has an implicit size upscaling side effect. From neural architecture perspective, it is possible for same decoder within SPN to be applied towards all slices as they have identical structure. The smaller slices allow self-attention [1] in SPN without local contexts [4]. The ordering is two-dimensional analogue of one-dimensional subscale ordering [5]. The subscale ordering is defined as follows:

$$Prob(x) = \prod_{i=1}^{SI} \prod_{j=1}^{SI} \prod_{ht=1}^{HI/SI} \prod_{wt=1}^{WI/SI} \prod_{ct}^{\{R,G,B\}} Prob(x_{i+SI*ht,j+SI*wt,ct} | \mathrm{x}_{pgi}) \tag{2}$$

Here $\mathbf{x}_{pgi}$ represents all previously generated ordered intensity values. With scaling factor SI each slice $HI/SI \times WI/SI$ is obtained by selecting pixel per SI pixels in both height and width. There are SI^2 interleaved slices in original image, specified by row and column meta-position offset (i, j). The subscale ordering captures size upscaling implicitly. The size upscaling is performed explicitly by training single slice decoder on sub-images. It generates first slice of subscale ordering from single slice decoder itself [3]. The rest of the image is generated through subscale ordering by actual network. The single-slice model is trained on first image slices or on slices at all positions in all images with respect to shared spatial structure. As a result of this, same SPN that captures subscale ordering acts simultaneously as full-blown image model and size upscaling model is initialized with outputs of single-slice decoder. The multi-dimensional upscaling applies in terms of image height, width and channel depth. This is done in stages such that network first generates s_1 most significant image bits through subscale ordering. Then second network generates next s_2 most significant image bits conditioned on all s_1 image bits and so on. The weights among networks at different stages of depth upscaling are not shared. In depth upscaling lower significance bits are only generated when more significant bits at all positions have been generated in previous stage. The multidimensional upscaling allows model focus on visually salient image bits not affected by less salient and less predictable image bits.

2.4 The Conventional Image Generation Model

The conventional image generation model [7, 8] addresses challenges involved in training conventional decoders on large images. The previous models such as PixelRNN, PixelCNN [3] and Image Transformer [4] construct a generated context representation for each dimension/pixel. The existing AR approaches requires an appreciable computation and memory. The quadratic memory self-attention requirements limit severely for larger images. On mitigating memory and computational encoding requirements, dependencies among large number of variables often comes at global context expense. The modeling choices such as cropping images within decoders [5] and performing self-attention over local neighborhoods [4] neglect global dependencies. The model parallelism with the joint use of very large number of accelerators does not overcome challenges in global structure learning. To address these challenges SPN is devised which embodies subscale ordering. SPN architecture has an embedding part for slices at preceding meta-positions. It allows decoder for current slice being generated. The embedding part is a convolutional neural network with residual blocks. It takes input preceding slices concatenated along depth dimension. The slices are ordered along channel dimension when concatenated. The empty padding slices preserves relative meta-positions of each preceding slice with respect to target slice. The slice above any target in 2D meta-grid is aligned in same position along input depth axis. This achieves equivariance in embedding architecture. The padding slices also ensures that input slice tensor depth remains same for all target slices. The embedding part also receives meta-position of target slice as input with embedding tiled spatially across slice tensor. The pixel intensity values are embedded as one-hot indices. The context embedding network passes its input through self-attention layers and then residual blocks finally emits slice-sized feature map. The decoder takes position preserving encoded slice tensor as input. Here each position in target slice is given as input the encoded pixels' representation at that same position in preceding slices. In addition, it processes target slice in raster-scan order. The decoder is a hybrid architecture combining masked convolution and self-attention. An initial one-dimensional self-attention network is used to gather entire available slice context. The slice is reshaped into one-dimensional tensor before it is given as input to masked one-dimensional self-attention layers. The masking is performed over previous pixels only in self-attention layers. Then layers' output is reshaped back into 2D tensor concatenated depth-wise with slice output embedding network and is taken as conditioning input to GatedPixelCNN. PixelCNN network models target slice with full masking over pixels and channel dimensions. The memory requirements are significantly lower due to smaller spatial slice size and their compact concatenation along channel dimension of input tensor. With this entire previously generated context is captured at each decoder position. The log-likelihood derived above decomposes as sum over slices. An unbiased estimator of log-loss is obtained by uniformly sampling target slice choice. The maximum likelihood learning through stochastic gradient descent on this monte carlo estimate with all gradients are computed by backpropagation. SPN serves as size-upscaling network when first slice of input tensor is initialized with an externally generated sub image. It is ensured that smaller sub images used for initialization and images used in training of SPN decoder are identical to each other. The upscaled image in depth is itself divided into slices and they are concatenated along channel dimension into slice tensor for conditioning image. The latter is then added as fixed additional input towards SPN embedding. The conditioning model is normal SPN trained on low bit depth data.

2.5 Hierarchical Image Generation Model Architecture

The hierarchical version of high-fidelity image generation model [8] is proposed here. The computational benefits serve the major motivation. A generative model which hierarchically generates images with fine-grained details is useful towards fine-grained real image grouping. The model generates a fine-grained image through hierarchically generating and stitching together background image, parent image capturing one object variation factor and child image capturing another factor. The variation object factors are disentangled without any supervision information theory. High mutual information is enforced between parent latent code and parent image and child latent code conditioned on parent code and child image. By imposing constraints on relationship between parent and child latent codes, the parent and child codes are induced to capture object's shape and color/texture details respectively. The hierarchical model automatically generates masks at both parent and child stages through which latent codes are conditioned to focus on relevant object factors and stitch together generated images across stages. The features learned through this unsupervised hierarchical image generation process can be used to cluster real images into their fine-grained classes. The hierarchical model is different from non-hierarchical version in terms of accuracy based on similarities and running time when data volume grows. The model architecture is shown in Fig. 3. The rest of this section highlights the working of hierarchical architecture. Let us consider $Y = \{y_1, \ldots ..., y_N\}$ as dataset which contains unlabeled images with fine-grained object categories. The unsupervised generative model is learned such that high-quality images are produced matching true data distribution. The learning process involves disentanglement towards relevant factors of image variations in Y. The object's background, shape, appearance and pose are considered as variation factors. The hierarchical model successfully associates each latent code to particular fine-grained category aspects where learned features are used towards grouping real images in Y. The process consists of background, parent and child interacting stages. The background stage generates realistic background image BI. The parent stage generates object's shape stitches it onto BI to produce parent image PI. The child stage fills in object's outline with appropriate color and texture to produce final child image CI. Thus, the objective function becomes:

$$Obj_{fn} = \alpha Obj\_fn_{bi} + \beta Obj\_fn_{pi} + \vartheta Obj\_fn_{ci} \quad (3)$$

Here Obj_fn_{bi}, Obj_fn_{pi} and Obj_fn_{ci} denote objectives towards background, parent, and child stage respectively. The weights are denoted by α, β and ϑ. An end-to-end training is performed for all stages. The latent codes provide conditioning for different stages. The hierarchical model has input as a continuous noise vector v, categorical background bi, categorical parent pi and categorical child ci codes. $[v \sim N(0,1)$; $bi \sim category\left(N_{bi}, \frac{1}{N_{bi}}\right)$; $pi \sim category\left(N_{pi}, \frac{1}{N_{pi}}\right)$; $bi \sim category\left(N_{ci}, \frac{1}{N_{ci}}\right)]$. Now parent and background codes relationship with child code is presented. The parent and child code assume an implicit hierarchy presence in Y. The fine-grained categories are grouped and differentiated considering identical shape and appearance. The hierarchy is identified when parent code categories is less than child code categories and multiple child codes share same parent code. These conditions help push parent and child codes to capture shape and appearance. The background and child code consider some correlation between an object and background in which it is positioned. To avoid conflicting object-background pairs, during training background code is set as child code. During testing this condition is removed. The background stage synthesizes background image which is a canvas for parent and child stages to stitch foreground aspects. The aim is to disentangle background as separate variation factor. The background stage shares common feature pipeline with parent and child stages. This stage consists of generator and discriminator pair. Generator is conditioned on latent background code which controls different background classes and on latent code controls intra-class background details. To generate background, it is assumed to access an object bounding box detector that detect instances of super category. The detector is used to locate non-object background patches in each real image. Then generator and detector are then trained using adversarial and classification loss objectives. The auxiliary classification loss makes background generation task more explicit and is computed on patch level. The patches inside and outside detect object in real images. They constitute training set for foreground and background respectively. It is used to train binary classifier with cross-entropy loss. Then classifier trains the generator. This loss updates generator so that classifier assigns high probability to generated background patches.

The real data distribution is modeled through two-level foreground generation process via parent and child stages. The parent stage can be viewed as modeling higher-level information about an object and child stage conditioned on parent stage models lower-level information. Capturing multi-level information gives overall image generation process simple interpretation and for fine-grained generation it is easy for the model to generate appearance details conditioned on object's shape, without having to worry about the background and other variations. The generator here consists of a series of convolutional layers and residual blocks and performs mapping to feature representation. This stage generates foreground entity and stitches it to existing background. The two generators transform feature space into parent foreground and mask respectively to obtain the parent image. The discriminator induces parent code to represent hierarchical concept. With no supervision from image labels, information theory helps to discover this concept in a completely unsupervised manner. The mutual information is maximized while approximating discriminator posterior. The parent stage's result is an image composed of background and object's outline. The task that remains is filling outline with appropriate texture/color to generate final fine-grained object image. The color/texture information about object is encoded with child code conditioned on parent code. On concatenating with feature representation, resulting feature chunk is fed into generator with series of convolutional and residual blocks. The

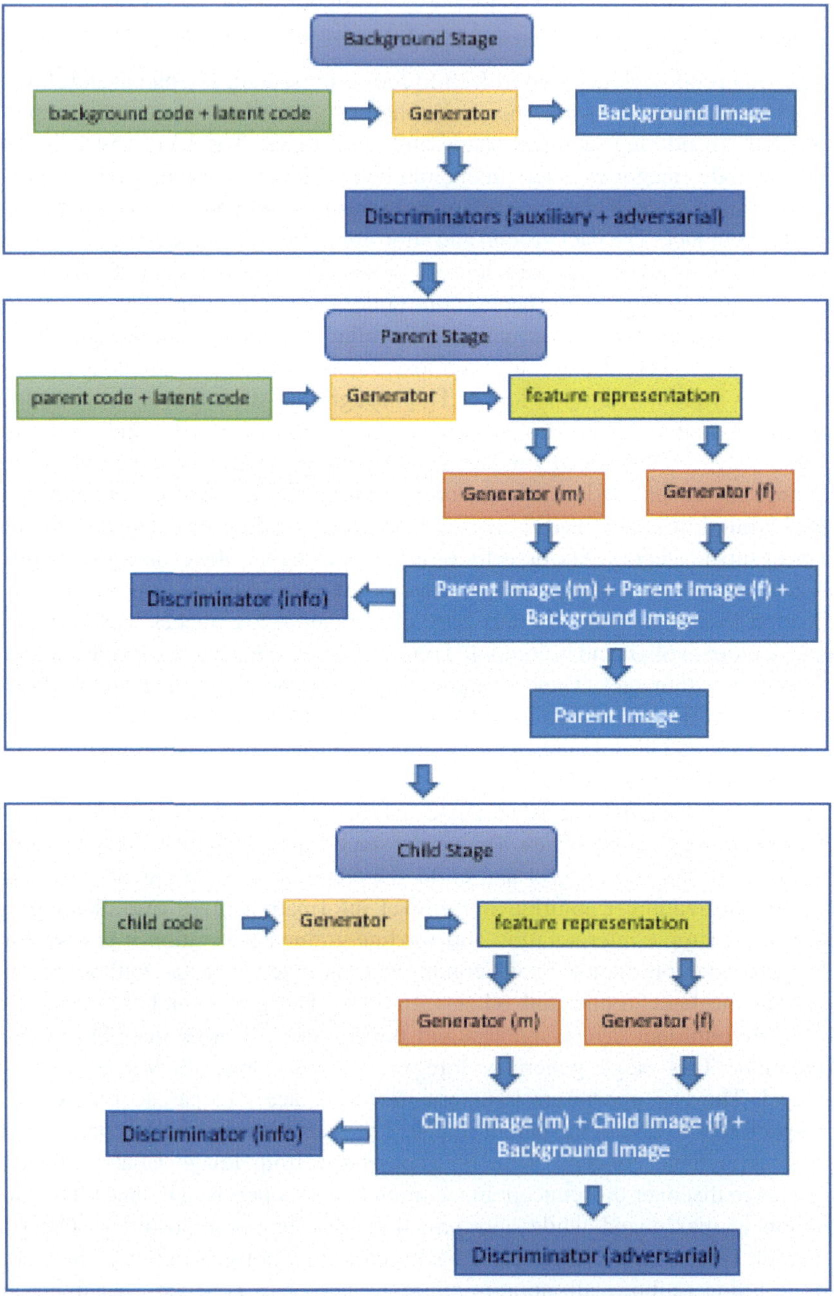

Fig. 3. The hierarchical architecture of high-fidelity image generation

two generators map resulting features into child foreground and child mask respectively. The stitching process obtains the complete child image.

3 Experiments and Results

In this section experimental results are presented. The investigations highlight hierarchical model's capability in terms of high-fidelity samples at high resolution and produces unconditional CelebAHQ better quality samples than [8, 11] and improved MLE scores. The results are extended towards high resolution ImageNet images at log-likelihoods. On these resolutions the unconditional samples are attributed by unprecedented global coherence. Here large networks are trained both in terms of hidden units and network depth. The context embedding network has 7 convolutional layers and 8-10 self-attention layers per layer. The masked decoder has PixelCNN with 20 layers. The1D transformer in decoder has 8–10 layers. The other dataset specific hyperparameter details are available in [8]. The hybrid decoder's performance is taken as benchmark. It compares well at state-of-the-art models 32×32 down sampled ImageNet as shown in Table 1. SPN affects in low resolution setting values. This happens as resulting image slices become very small and image coarse grained. On 64×64 down sampled ImageNet state-of-the-art log likelihood of 3.52 bits/dim. At this resolution SPN has a score of 3.53 bits/dim. The hierarchical image generation model has log likelihood of 3.96 bits/dim. The negative Log-likelihood scores for 5-bit datasets are shown in Table 2. For ImageNet at 128×128, standard dataset from [12] is used. SPN has log-likelihood over this model as 3.08 bits/dim. The hierarchical image generation model has better log-likelihood of 3.05 bit/dim as shown in Table 3. The Fig. 4 shows 128×128 8-bit ImageNet samples for depth upscaling and complete multidimensional upscaling. These settings do not affect negative log likelihood but samples with depth upscaling show significant semantic coherence that is not present in samples without upscaling. Also, multidimensional upscaling increases the overall success rate of samples. Additional intermediate ImageNet samples are available in [8]. For CelebAHQ at 256×256 high fidelity samples of celebrity faces are produced. It compares well with other model samples' [8]. In Table 2 significant

Table 1. Negative log-likelihood scores for down sampled ImageNet in bits/dim [8]

Networks used	ImageNet 32×32	ImageNet 64×64
Gated PixelCNN	3.83	3.57
Parallel multiscale	3.95	3.70
PixelSNAIL	3.80	–
Image transformer	3.77	–
Glow	4.09	3.81
Decoder baseline	3.79	3.52
SPN	–	3.53
SPN + depth upscaling	–	3.53
Hierarchical image generation model	4.12	3.96

Table 2. Negative Log-likelihood scores for 5-bit datasets in bits/dim [8]

Networks used	ImageNet 64 × 64 (5 bit)	CelebAHQ 256 × 256 (5bit)
Glow	1.76	1.03
SPN	1.41	0.61
Hierarchical image generation model	1.35	0.47

Table 3. Negative log-likelihood scores for high-resolution ImageNet in bits/dim [8]

Networks used	ImageNet 128 × 128	ImageNet 256 × 256
Parallel multiscale	3.55	–
SPN	3.08	2.97
SPN + depth upscaling	3.08	3.01
Hierarchical image generation model	3.05	2.96

(a)

(b)

Fig. 4. Hierarchical Image Generation model (RGB ImageNet): (a) 8-bit 128 × 128 samples with depth upscaling (b) 8-bit 128 × 128 samples with full-blown multidimensional upscaling

Fig. 5. Hierarchical Image Generation model: 8-bit 256 × 256 RGB CelebAHQ samples with depth upscaling

improvement in MLE scores are highlighted. The Fig. 5 shows samples for 8-bit CelebAHQ-256. Some significant results for both 256 × 256 and 512 × 512 images are available in [8].

4 Conclusion

In this work, learning the distribution of complex natural images with high sample fidelity through hierarchical generative models is addressed. The model disentangles background, object shape and appearance towards hierarchical generation of fine-grained object categories. The information theory associates a factor to latent code and condition relationships between codes to induce hierarchy. The hierarchical model's learned features are used to cluster real images. The proposed model attains both state-of-the-art MLE scores on large-scale images from complex domains and generates high fidelity full 8-bit samples from resulting distributions without altering the sampling process. The generated samples show an appreciable semantic coherence and exactness at large scale size of full 8-bit 128 × 128, 256 × 256 and 512 × 512 images.

References

1. Vaswani, A., et al: Attention is all you need. arXiv 1706.03762 (2017)
2. Wu, Y., et al: Google's neural machine translation system: bridging the gap between human and machine translation. arXiv 1609.08144 (2016)
3. Van den Oord, A., et al: Wavenet: a generative model for raw audio. arXiv 1609.03499 (2016)
4. Parmar, N., et al: Image transformer. arXiv 1802.05751 (2018)
5. Kalchbrenner, N.: Video pixel networks. arXiv 1610.00527 (2016)
6. Arora, S., et al: Do GANs actually learn the distribution? An empirical study. arXiv 1706.08224 (2017)
7. Menick, J., et al: Generating high fidelity images with subscale pixel networks and multidimensional upscaling. In: 7th International Conference on Learning Representations (2019)
8. Chaudhuri, A.: Some experiments on generating high fidelity images. Technical report, Samsung R&D Institute Delhi, India (2019)
9. ImageNet. http://www.image-net.org/
10. CelebAHQ. http://mmlab.ie.cuhk.edu.hk/projects/CelebA.html
11. Kingma, D.P., et al: Generative flow with invertible 1 × 1 convolutions. arXiv 1807.03039 (2018)
12. Kolesnikov, A., et al: Deep probabilistic modeling of natural images using a pyramid decomposition. arXiv 1612.08185 (2016)

Evaluation of a Novel Intelligent Firewall Simulator for Dynamic Cyber Attack Lab

Irfan Syamsuddin(✉), Rini Nur, Meylanie Olivya, Irmawati, and Zawiah Saharuna

Center for Applied ICT Research (CAIR),
Department of Computer and Networking Engineering,
Politeknik Negeri Ujung Pandang, Makassar, Indonesia
{irfans,rini,meylanie,irmawati,zawiah}@poliupg.ac.id

Abstract. Firewall administration is an important topic in network security courses. However, most teaching aids of the topic commonly deal with static firewall analyses that does not adequately covers current dynamic and variety of cyber security attacks. This study presents our approach to introduce an intelligent firewall simulator to fill the gap. The simulator is based on iNetwork software and it has three different dynamic security attacks scenarios or modules. Each module is equipped with handbook or manual to aid student performs the simulation procedures. Evaluation of intelligent firewall simulator usage is performed by expert and student. Expert evaluation is focused on the content validity of the modules and its accompanying handbook, whilst student evaluation measures the perception of students after using the simulator. Finally, both evaluations show positive results that the content and manual of intelligent firewall modules are valid and also effective to improve student skills and understanding.

Keywords: Intelligent firewall · Simulation · Teaching module · Evaluation

1 Introduction

Vocational higher education such as polytechnics applies specific model of higher education which provides greater portion of hands on practices in comparison to theoretical concepts in the fields of science and technology. Hence it is commonly found that polytechnic students spend more time doing practical assignments than theoretical lectures in the classroom. As the result, vocational students have stronger problem solving abilities which equip them well to face most problems when they get job which is very much demanded by the industry. However, the challenges faced by higher education in developing countries are the ability of the institution to provide appropriate teaching facilities to students due to amongst other funding limitation. This impacts on the capability of instructor to provide adequate learning aids and supports to students.

Similar situation might be found at the Department of Computer and Networking Engineering (NCE) Department at State Polytechnic of Ujung Pandang (PNUP) where students deal with practicum-based assignments within 4 years of study period. One of

R. Silhavy (Ed.): CSOC 2020, AISC 1225, pp. 257–267, 2020.
https://doi.org/10.1007/978-3-030-51971-1_21

the big challenges in the process of teaching hands on practice course such as Computer Network Security (CNS) is the need for hardware to let the students practicing various practical exercises in a special laboratory. The provision of special laboratory with particular hardware, such as firewalls, routers and switches, requires very large cost, which is generally difficult to satisfy by higher education institutions due to limited budget allocation [1]. Moreover, for advanced topics such as CNS course, the designated laboratory must be strictly isolated from existing computer networks in campus while still have access to the Internet. This logical separation is urgent in order to block any possibilities of misuse by students during practical courses that could affect or even damage computer network infrastructure, partly or even entirely [2].

Considering the issues, several simulators software were introduced to enable students to perform practical assignments in CNS course without extra hardware installation in the laboratory. There are several simulator software that have been applied so far, such as Cisco Packet Tracer, GNS3, NS2 and iNetwork as reported in a working in progress paper [3]. We have developed a novel Intelligent Firewall Simulator (IFS) to fill the gap of unavailability of firewall simulator to intelligently deal with dynamics and various types of cyber attacks. The main purpose of current study is to evaluate the effectiveness of our approach. The evaluation is performed by considering expert and student views towards three main research questions as follows:

- How valid is the content of intelligent firewall simulator (IFS) modules with respect to the curriculum?
- How valid is the IFS manual for guiding the user?
- Does the use of IFS improve the understanding and skills of students?

The rest of this paper is structured as follows. Section 2 describes literature review, while the methodology and tools used to conduct the research are presented in Sect. 3. Then, results and discussion are given in Sect. 4. Finally, the study is summarized in the last section.

2 Literature Review

Using simulation tool in learning computer network security is considered beneficial from many aspects, such as design, implementation and evaluation. From design perspective, network simulations allow student to establish simple to complex network topology without extra cost and physical space. From implementation point of view, student could apply any scenarios easily and they are fully under controlled. Then, in terms of evaluation, it is easier to perform evaluation on simulator environments rather than in real physical ones. As a results, simulation approaches have been gaining popularity in recent decades [2, 3].

The use of computer network simulator has been regarded as an appropriate answer to deal with limited physical laboratory facilities. Although, to a certain extent, computer network simulator could not fully present ideal look and feel of the actual network behaviour. In case of a firewall, basic concept of how firewall is configured or how it works in ruling the network flow as well as data flow analysis behind the work of the firewall can be learned by using simulator software [4]. Pastor et al. [4] argue that implementation of computer network software in which several firewalls combined has brought many benefits in enhancing the learning process, and significantly reduces cost of laboratory development in case of open source software as exemplified in recent studies [3, 5].

Another approach to enable firewall simulation was described by Yan [6] using cloud computing technology. However, the approach requires the existence of a physical network to the cloud that not all institutions could provide. Hajdarevic, et al. [7] discuss the use of GNS3 simulator in assessing firewall rules through a dedicated training of ethical hacker. They assert that GNS3 offers a flexible way to learn network security particularly firewall mostly in Linux environment [7]. Web based firewall was also proposed by Trabelsi and Mustafa [8]. This is a unique approach in which user needs only a browser to run the application, create topology, connecting all devices, define firewall rules and analyze the simulation results. Similar approach is presented by Losilla [9] with assessment tool in wireless network.

In addition, Opnet based simulation for firewall was reported by Ameen and Nourildean [10]. The authors apply a cloud topology to investigate firewall impact on the network over Opnet simulator. Similarly, Al-Hilfi et al. [11] also employed Opnet to assess firewall functions in terms of wireless LAN security. Both papers argue that Opnet is a powerful tool in tackling firewall simulations. Likewise, Cisco's packet tracer is considered as the most widely used simulator in learning firewall and network security topics [12]. It has been used in teaching computer security at undergraduate level by Airi and Anderson [13]. Also, a study by Dengping et al. [14] shows that the simulator is appropriate simulator for computer network management course. More recently, Zhang et al. [15] report their positive results in using simulator for teaching information security course. However, the software is vendor oriented for Cisco products.

In [16], a novel network simulator called iNetwork was introduced for the first time. It is a very lightweight application that does not need heavy hardware requirements. This Java-based simulator could be run on various versions of Windows machines without the installation process just by copying the given folder. Besides, the GUI design is very interactive and easy to use even by beginners. The most important feature of iNetwork is its' built in tutorial covering wide range computer network topics including firewall [16, 17].

Moreover, iNetwork offers flexibility and portability in firewall simulation by providing two options to perform firewall simulation and testing. The first option is by using the existing iNetwork Simulator utilities such as ping command through windows command line. Secondly, the test can be done via packets GUI for managing ICMP, TCP or UDP packets data sending. In addition, the simulation results are viewed using the existing Activity Log interface [17].

Among the aforementioned simulation tools, most of them cover firewall simulation in static way of firewall administration. In other words, they not adequately cover current dynamic and variety of cyber security attacks. This is why we need to introduce an intelligent firewall simulator to fill the gap. Based on our experiences, iNetwork simulator has the capacity to apply intelligent firewall simulation. Therefore, we extend the simulator by developing a novel simulator called Intelligent Firewall Simulator (IFS) that offers capabilities to fill the gap found in current literature. The IFS consist of three scenarios called modules to illustrate different cyber-attacks and how an intelligent firewall deals with them. Each module is accompanied by tutorial to guide student performing hands on lab activities.

3 Tools and Method

Main tool used in the study is iNetwork simulator which has various networking simulation topics from peer to peer networking to firewall and wireless simulations [16, 17]. Its rich features with low computing requirements makes it fit to our curriculum and laboratory needs. Unfortunately, iNetwork only provides simple and static firewall simulation which could not support the dynamic one. Therefore, Intelligent Firewall Simulator (IFS) is developed as an extension of iNetwork simulator to fill the gap of dynamic firewall simulation.

The main feature of IFS is its ability to simulate advanced dynamic firewall which is adaptive to different attacks scenarios. Such ability is known as intelligent firewall that can be simulated o deal with recent dynamic cyber attacks should be well understood by students during hands on practical lab session.

Figure 1 shows the main interface of Intelligent Firewall Simulator (IFS) running over iNetwork simulator. IFS's core function is to apply log analysis from previous blocked packets and sources in order to understand whether there were current access in an attack or not. This feature is not available in standard iNetwork simulations. It works by mining log data (not saved by iNetwork) and use them as input which firewall can reference when making decision. Therefore IFS showcases novel simulation processes on the dynamic prevention of various forms of attacks. We structure the IFS simulation with three scenarios represented as beginner, intermediate and advanced modules with increasing filtering rule complexity, respectively. In order to help students practicing the lab independently each module has manual or lab guide which is clear enough to assist them doing the simulation and understand how intelligent firewall works effectively.

The beginner module has 3 rules that might be applied dynamically by firewall administration which expected to be finished within 30 min. The next one called intermediate module has 6 rules applied to the intelligent firewall simulator. The last module is a bit more complex with 10 rules which requires approximately 90 min for students to practice and evaluate it (see Table 1).

The following figures show the network structures of three modules of IFS. For beginner and intermediate modules, the rules are similar that it should be done individually. Once students finishing all tasks as mentioned in lab manual, they should submit written reports describing their lab activity results found in both modules (Fig. 2).

While in the advanced module (Fig. 4), students work in groups instead individual. At the end of lab activity, every group make presentation on their simulation process and also defend their findings and opinions during the question and answer session (Fig. 3).

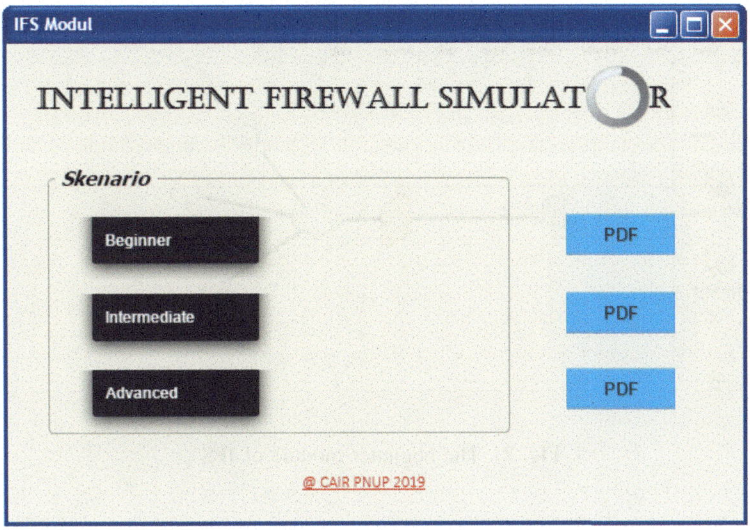

Fig. 1. Interface of Intelligent Firewall Simulator

Table 1. Firewall simulation modules.

Heading level	Number of rules	Network components	Expected time to finish
Beginner	3	PC (4), Switches (1)	30 min
Intermediate	6	PC (3), Switches (2), Router (2), Web Server (1)	60 min
Advanced	19	PC (3), WirelessPC (2), Access Point (1), Switches (3), Router (3), Web Server (1), DNS Server (2), Mail Server (2)	90 min 10 point, bold

In order to deal with three research questions aforementioned in previous section, three surveys were constructed based on previous related studies. Participants of the survey are classified into expert and student groups.

The first survey was developed in accordance with studies by Muharram, et al. [18] and Syamsuddin [19]. It was a content validity survey which was aimed at evaluating the validity of IFS modules which consists of five questions such as connection with the topics, educational aspects, tool usability, practicability of module and learning module development value. This survey was administered to the expert group.

The second survey, which was also given to the expert group, was structured according to study in [19]. Its aim was to evaluate validity of accompanying tutorial or module from four aspects namely, content, language, layout and learning.

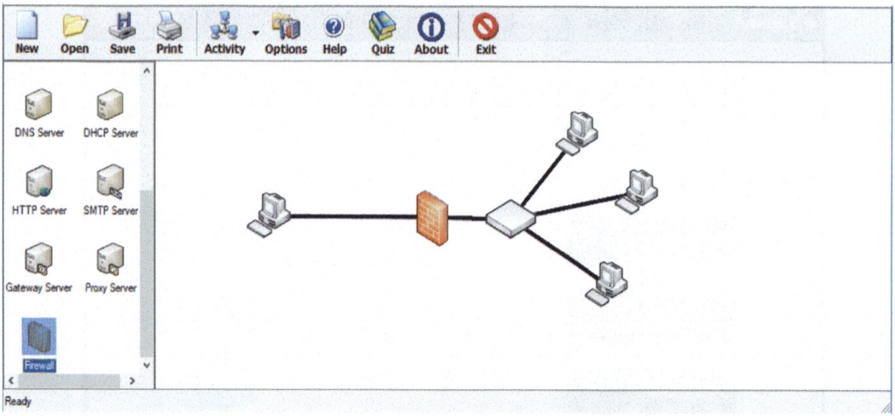

Fig. 2. The beginner module of IFS

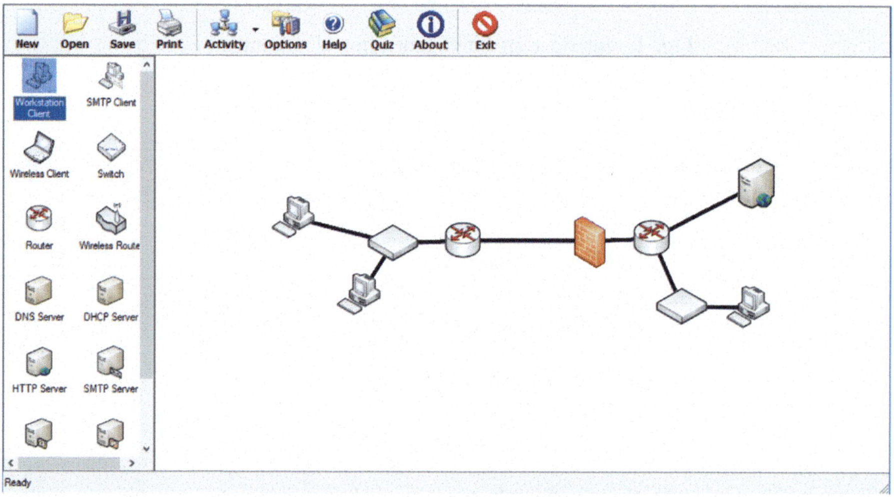

Fig. 3. The intermediate module of IFS.

The third survey was given to students. The survey was based on Muharram, et al. [18] intended to obtain users' perception after using IFS modules. Through the last survey, effectiveness of the modules in enhancing student knowledge and skills in related course might be measured.

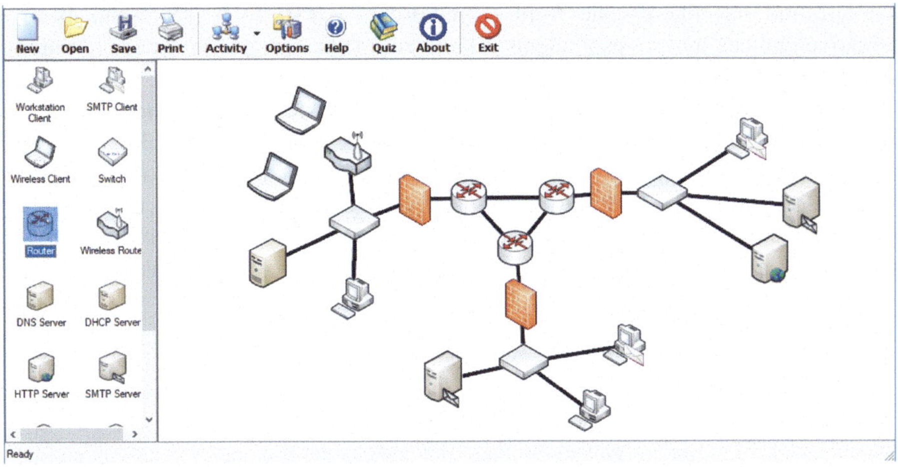

Fig. 4. The advanced module of IFS

4 Results and Discussion

Five experts involved in the first survey on content validity which consists of five questions. The question uses likert scales, ranging from 1 (represents the worst) to 5 (represents the best). The process is as follows. After having all responses from experts then the process is continued by calculating the mean score of each criterion. Finally, it is finalized by calculating validity score (VS) which is actually the mean value of all five criteria.

The mean score of the five criteria are 3.7, 3.96, 3.76, 4.08 and 3.54 respectively. The highest score goes to practicability of the module, while the lowest score goes to learning development value. In short, these yield validity score of 3.8 for the content validity. Based on this value, it is clearly indicated that the experts agree to acknowledge that the content of iNetwork based firewall simulation modules are acceptable and valid (Table 2 and Fig. 5).

Table 2. First survey on content validity.

No.	Validity criteria	E1	E2	E3	E4	E5	M
1	Connection with the topics	3.9	3.7	3.5	3.8	3.6	3.70
2	Educational aspects	3.8	4.5	4.5	3.4	3.6	3.96
3	Tool usability	3.6	3.8	4	3.8	3.6	3.76
4	Practicability of module	4	3.8	4.6	3.8	4.2	4.08
5	Learning module development value	3.2	3.4	3.9	3.8	3.4	3.54

Validity Score (VS) 3.8

The second evaluation is to measure validity of the manual or handbook accompanying each module. The same experts are involved in this survey which covers four aspects, namely content, language, layout and learning. Similar to previous survey, each question has likert scale from 1 (represents worst) to 5 (represents best) as alternative choices by the respondents. The result of the second survey is presented in the following table (Fig. 6).

Likewise, after all experts select their preferences on each question, then we calculate the mean score of each question. It is revealed that the mean scores are 3.92, 3.78, 4.02 and 3.88 respectively. The highest score goes to layout while the lowest one is language. These yield validity score of 3.9 for the handbook. The findings show that all experts agree on the validity of the handbook accompanying the three modules of IFS since they provide clear and easy to understand practical guideline for student (Table 3).

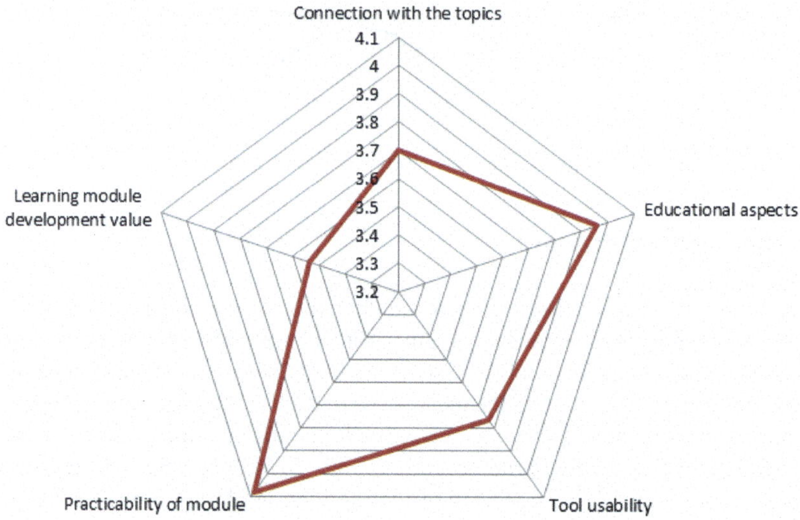

Fig. 5. Result of content validity.

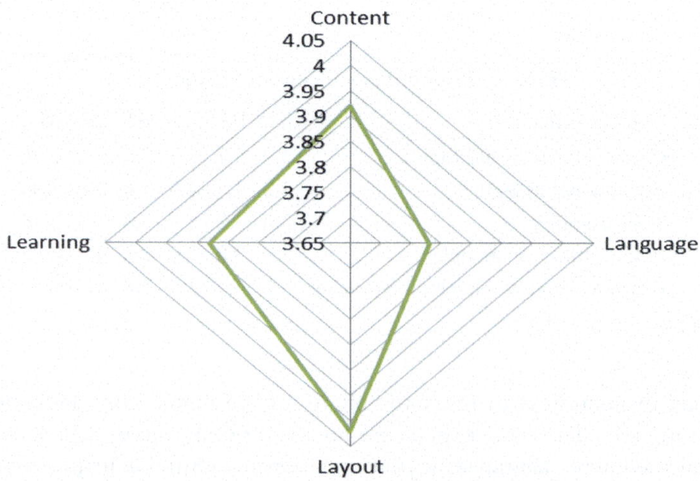

Fig. 6. Result of handbook validity.

Table 3. Second survey on handbook validity.

No.	Validity criteria	E1	E2	E3	E4	E5	M
1	Content	3.6	3.9	4.4	4.1	3.6	3.92
2	Language	3.4	4	4.1	4.1	3.3	3.78
3	Layout	3.8	4	4.4	4.2	3.7	4.02
4	Learning	3.8	3.8	4.3	4	3.5	3.88

Validity Score (VS) 3.9

The last evaluation was carried out with the purpose of obtaining students' perceptions in using IFS. The survey also uses a typical likert scale with the same options of Strongly Disagree (SD), Disagree (D), Neutral (N), Agree (A), and Strongly Agree (SA) and assign 1, 2, 3, 4, 5 values respectively. All students participated in this voluntary survey and the results are shown in the following figure (Fig. 7).

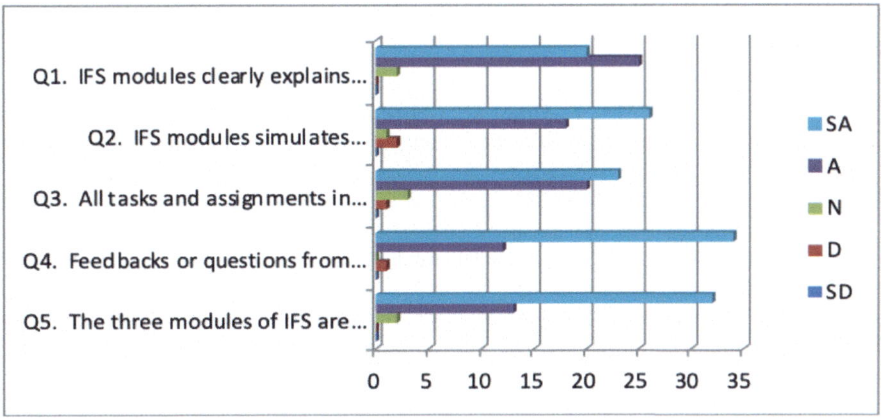

Fig. 7. Result of the last survey.

In the first question which addresses clarity of the module to explain the principles firewall concept in network security, 25 students agree and 20 of them strongly agree, while only 2 students are neutral. In the second question concerning their understanding through intelligent firewall operations, 26 students strongly agree and 18 others agree while two students do not agree and one of them neutral. In terms of assignments of the third question, 23 students strongly agree and 20 of them agree, while a student disagree and three of them neutral. For the fourth question regarding students' feedback and questions, 34 students strongly agree and 12 others agree while only one disagrees. Then in terms of the final question, 32 students strongly agree and 13 of them agree while two students neutral.

5 Conclusion

A novel Intelligent Firewall Simulator (IFS), simulation processes on the dynamic prevention of various forms of attacks described in this paper. IFS consists of beginner, intermediate and advanced modules to facilitate effective students learning by following the given step-by-step lab manuals. The implementation of IFS is evaluated from expert and student point of views.

In short, experts agree that both content and handbook are valid and acceptable as seen by total validity score of 3.9 for content and total validity score of 3.8 out of 5. Similarly, the students in the last survey conclude that the use IFS in learning is perceived useful to enhance their understanding.

It is expected for future research to adopt gamification approaches in order to enrich IFS learning experience and to improve student engagements. Another promising future research direction is to apply problem based learning with real network security scenarios.

References

1. Van Leeuwen, B., Eldridge, J., Urias, V.: Cyber analysis emulation platform for wireless communication network protocols. In: International Carnahan Conference on Security Technology (ICCST), pp. 1–6 (2017)
2. Gaffer, S.M., Alghazzawi, D.M.: Using virtual security lab in teaching cryptography. Int. J. Modern Educ. Comput. Sci. **1**, 26–32 (2012)
3. Syamsuddin, I.: VILARITY: a virtual lab for teaching computer security. TEM J. Tech. Educ. Manage. **8**(3), 125–130 (2019)
4. Pastor, V., Díaz,G., Castro, M.: State-of-the-art simulation systems for information security education, training and awareness. In: Education Engineering (EDUCON), pp. 1907–1916 (2010)
5. Montero, Á. M., Manzano, D. R.: Design and deployment of hands-on network lab experiments for computer science engineers. Int. J. Eng. Educ. **33**(2), 142–149 (2017)
6. Yan, C.: Build a laboratory cloud for computer network education. In: Proceedings of the 6th International Conference on Computer Science & Education (ICCSE 2011), pp. 1013–1018 (2011)
7. Hajdarevic, K., Kozic, A., Avdagic, I., Masetic, Z., Dogru, N.: Training network managers in ethical hacking techniques to manage resource starvation attacks using GNS3 simulator. In: XXVI International Conference on Information, Communication and Automation Technologies (ICAT), pp. 1–6 (2017)
8. Trabelsi, Z., Mustafa, U.: A web-based firewall simulator tool for information security education. In: Proceedings of the Sixteenth Australasian Computing Education Conference, vol. 148, pp. 83–90 (2014)
9. Losilla, F.: A web-based design and assessment tool for educational wireless networking projects. Comput. Appl. Eng. Educ. **25**(6), 992–1000 (2017)
10. Ameen, S.Y., Nourildean, S.W.: Firewall and VPN investigation on cloud computing performance. Int. J. Comput. Sci. Eng. Surv. **5**(2), 15–21 (2014)
11. Al-Hilfi, H.M.T., Salih, B.A., Marghescu, I.: Design of secured WLAN by using packet filtering firewall. In: International Conference on Wireless Communications, Signal Processing and Networking (WiSPNET), pp. 1857–1862 (2017)

12. Javid, S.R.: Role of packet tracer in learning computer networks. Int. J. Adv. Res. Comput. Commun. Eng. **3**(5), 6508–6511 (2014)
13. Airi, P., Anderson, P.K.: Cisco packet tracer as a teaching and learning tool for computer networks. DWU Res. J. **26**, 67–72 (2017)
14. Dengping, T., Yanqin, Z., Zhe, Y., Guoping, C., Guangdi, X.: Design of firewall experiment based on computer network management simulation platform. Exp. Tech. Manage. **4**, 39–47 (2015)
15. Zhang, T., Yin, M., Yang, Y.: An experiment teaching mode for an Introduction to Information Security course. World Trans. Eng. Tech. Educ. **12**(4), 725–728 (2014)
16. Sandrasegaran, K., Trieu, M.: iNetwork: an interactive learning tool for communication networks. In: Tools for Teaching Computer Networking and Hardware Concepts, pp. 39–61 (2006)
17. Ye, M., Sandrasegaran, K.: Teaching about firewall concepts using the iNetwork Simulator. In: 7th International Conference on Information Technology Based Higher Education and Training, ITHET 2006, pp. 889–892 (2006)
18. Muharram, Adnan, Sudding: The development of an enzyme catalase kit for engineering students at technical vocational schools. Global J. Eng. Educ. **19**(2), 168–172 (2017)
19. Syamsuddin, I.: Evaluation of NgeXTEA - a cryptography learning module. Global J. Eng. Educ. **20**(3), 196–200 (2018)

Schoolteacher Preference of Cyber-Safety Awareness Delivery Methods: A South African Study

Kagisho Mabitle(✉) and Elmarie Kritzinger

School of Computing, University of South Africa, Roodepoort, South Africa
mabitk@unisa.ac.za

Abstract. School learners, which are mostly young people, enjoy using the Internet - yet they remain naïve about the associated risks of using the Internet. Learners are particularly susceptible to cyber-risks and hazards such as anxiety, cyber-bullying, and human trafficking, to name but a few. Therefore, raising cyber-safety awareness in schools is absolutely essential. Cyber-safety awareness is about the safe and responsible use of information and communication technology (ICT). From the viewpoint of facilitating and enhancing cyber-safety awareness in diverse South African schools, this study ranks various types of delivery methods which are preferred by schoolteachers, in order to make learners more aware of cyber-safety. Based on a survey of 109 schoolteachers across 18 schools in 2 districts, the research reveals that conventional delivery methods (i.e. events and posters) were the most preferred delivery methods for raising cyber-safety awareness in schools. Social media (an online delivery method), and the lecture (an instructor-led delivery method), were also highly preferred by schoolteachers. In general, online delivery methods (i.e. email, mLearning, and the blog) were the least preferred delivery methods for raising cyber-safety awareness. The research duly suggests that a combination of select conventional, online and instructor-led delivery methods (i.e. events, posters, social media and the lecture) should be used for raising cyber-safety awareness in South African schools.

Keywords: Internet · Cyber-safety · Awareness · Schools · Learners · Teachers · Resource inequality · Preference · Delivery methods · Factors · Strategies

1 Introduction

Information and Communication Technology (ICT) has become a part of everyday modern life (Herrero et al. 2018; Kritzinger 2016). ICT influences and transforms economic, political, social and cultural structures (Romero-Ruiz et al. 2017). ICT is used in a variety of settings; these include health (Mogoba et al. 2019; Haux et al. 2002), transportation (Givoni and Thomopoulos 2015), business (Hynes 2013), energy (Morán et al. 2016), and of course education (Asongu and Odhiambo 2019; Islam et al. 2019; Nath 2019; Wu et al. 2019; Semerci and Aydin 2018; McFarlane and Sakellariou 2002). Given the catholic application of ICT, and the rampant usage of the Internet

R. Silhavy (Ed.): CSOC 2020, AISC 1225, pp. 268–283, 2020.
https://doi.org/10.1007/978-3-030-51971-1_22

(Çikrıkci 2016), it appears that a strong need then arises to educate Internet users and potential users (and school learners especially) about cyber-safety measures. In order to effectively raise awareness about safe Internet usage, it is important to rank the various types of delivery methods which schoolteachers might prefer to use, so that these delivery methods may be noted, created, and made readily available for usage. This is particularly important in the diverse South African school environment, which is characterised by unequal access to teaching and learning resources. Ranking the factors and strategies which inform the schoolteachers' choice of delivery method/s is equally important, as this insight will reveal the constraints and/or opportunities of raising cyber-safety awareness in heterogeneous South African schools, and consequently inform future endeavours.

2 School Learners and the Internet

Cyber-risks and threats have now entered the classrooms, schoolyards and homes of children and young people (Goodstein 2007; Myers et al. 2011). A significant proportion of school learners in South Africa are now able to access the Internet using their mobile phones. In a survey with 503 respondents, which assessed the current situation of cyber-safety awareness among school learners in South Africa, it was found that 98% of the learners had access to the Internet through their mobile phones (Kritzinger 2017). With such easy and convenient access to the Internet for school learners, and consequential exposure to cyber-risks, it then becomes crucial that school learners are made aware of cyber-safety measures. Kritzinger (2017) concludes, justifiably, that awareness of cyber-safety could ensure that learners have the necessary attitude towards cyber-safety, as well as the knowledge and skills to deal with real-life cyber-risks.

Computing technology and the Internet have become entangled in young people's lives (Guan and Subrahmanyam 2009). Youngsters and school learners are going online and are using the Internet at ever younger and tender ages. In Europe for example, 93% of 9-16-year-old users go online at least weekly, and 60% go online every single day. 59% of this age group also has a social media/networking profile (Livingstone et al. 2011). Consistently, a 2008 World Internet Project survey of 13 countries revealed that among youth aged 12-14-years-old, 88% in the United States used the Internet. The percentage of Internet users in this age group was 100% in the United Kingdom, 98% in Israel, 95% in Canada, and over 70% in Singapore (Reuters 2008). The average age of initial Internet use is seven in Denmark and Sweden, and eight in several Northern European countries (Livingstone et al. 2011).

Younger children are said to lack the skills and the confidence in areas of Internet use that are especially important for cyber-safety, such as the blocking of unwanted contacts. Therefore, there needs to be a new policy focus on promoting cyber-safety awareness amongst younger Internet users, and particularly in schools, since school is the second most common location of Internet usage (63%) after home use (87%) (Livingstone et al. 2011). This does not mean simply promoting cyber-safety awareness at the secondary school level exclusively, where the traditional cyber-safety awareness focus has been. But rather, attention should be focused on the primary school level, with younger users of the Internet, by providing them with age-appropriate cyber-safety training and guidance. Online cyber-safety resources aimed at younger children, for instance, must not assume reading proficiency. Teacher training also needs to equip teachers, particularly at the primary school level, with the knowledge and skills to support younger children (Livingstone et al. 2011), in enlightening them about cyber-risks.

Cyber-safety awareness and educational programmes are perceived as a defence mechanism for protecting information and other assets (Kruger et al. 2011). Given the importance of raising cyber-safety awareness, schools and teachers in particular are required to participate in teaching school learners about cyber-safety (Kouttis 2016). However, teachers and parents are themselves ill-prepared to educate learners about cyber-safety (De Lange and Von Solms 2012; Govender and Skea 2015), and learners' awareness of security matters continues to be poor (McQuade 2007). Considering that school learners are attractive candidates for cyber-safety attacks (Rezgui and Marks 2008) because of their gullibility, learners need to be educated on how to use ICT and associated devices in a safe and accountable manner (Kritzinger 2017), in order to mitigate cyber-risks.

2.1 School Learners and Cyber-Risks

School learners are susceptible to cyber-risks such as anxiety (Przybylski et al. 2013); cyber-bullying (Li 2007); exposure to explicit content (Strassberg et al. 2014); human trafficking (Sykiotou 2007); identity theft (Lai et al. 2012); insider threat (Kim et al. 2017); Internet addiction (Kandell 1998); malware (Feizollah et al. 2015); phishing (Arachchilage and Love 2013); and spam (Chakraborty et al. 2016). Ubiquitous Internet access can exacerbate exposure to these cyber-risks. Estimates from India (Telecom Regulatory Authority of India 2007) imply that the mobile Internet connection is more popular than traditional PC Internet connection in that country. Recent studies have shown that the rate of mobile communication penetration in Africa is on the rise (Haftu 2019), and that most school learners have access to a mobile phone (Porter et al. 2015; De Lange and Von Solms 2012). It is highly likely that such a phone has access to an Internet connection. Other studies have also highlighted the importance of ensuring cyber-safety in the school environment in countries with Internet access (Hanewald 2008; Perren et al. 2012) such as South Africa, using specific delivery methods.

3 Delivery Methods for Raising Cyber-Safety Awareness

Abawajy (2014) and Wilson and Hash (2003) consider numerous techniques or delivery methods for raising cyber-safety or ICT security awareness. These delivery methods, which were included in our survey questionnaire, are; *Award programmes, Events, Flyers or leaflets, Mouse pads, Newsletters, Posters, Stationery, Crossword puzzles, Video games, Lecture, Formal presentations, Animation, Blog, eLearning, Email, mLearning, Screensavers, Social media, Simulations* and *Interactive videos*. The research then grouped these 20 delivery methods into 6 categories, namely;

Conventional, Game-based, Instructor-led, Online, Simulation-led and *Video-based.* A description of these delivery methods is depicted in Appendix A.

3.1 Factors Influencing the Preference of Delivery Methods

Strauss and Frost (1999) and Walker (1965) consider numerous factors which influence the selection of instructional media or training/delivery methods. These factors, also included in our survey questionnaire, are; *Availability of hardware, Availability of software, Availability of network connectivity, Development costs, Teaching and learning costs, Maintenance costs, Time to produce, Length of lesson, Number of students to be taught, Student engagement, Student feedback, Student motivation, Student participation, Number of instructors to develop, Number of instructors to instruct, Ease of administration, Ease of evaluating students, Ease of presentation, Flexibility of medium* and *Facilities needed for presentation.* The research then similarly grouped these 20 media selection criteria or factors into 5 categories, namely; *Technology, Cost, Time, People* and *Operation.* A description of these factors is depicted in Appendix B.

3.2 Strategies Used for the Selection of Delivery Methods

Newman (2015) studied the strategies that instruction librarians use when selecting delivery methods or media to enhance active learning. Three main themes emerged from that study, namely: *professional development, learner's needs* and *media characteristics.* These strategies were also included in our survey questionnaire. The characteristics of the media and learner needs were noted as the top media selection strategies in that study. Under the professional development theme, attending workshops, peer reviews, read the literature and networking, were identified as strategies for selecting media. Under the learner needs theme, learning outcomes, learner analysis and the consideration of the audience, were identified as strategies for selecting media. Under the characteristics of media theme, cost, technology support, time, ease, accessibility, comfort level and training, were identified as strategies for selecting media (Newman 2015). The themes and corresponding strategies are depicted in Appendix C.

4 Research Methodology

An exploratory quantitative approach (using descriptive statistics and frequencies) was employed in this study, to rank the delivery methods teachers would prefer to use, for raising cyber-safety awareness in their schools. The research study also ranked the factors and strategies which influence the schoolteachers' preference of cyber-safety awareness delivery methods. The study was confined to the Sedibeng West and East school districts of the Gauteng Province of South Africa. A survey was used as the research strategy, with the questionnaire as the data gathering method. A total of 180 surveys were distributed (20 via email and 160 in person), with a net result of 109 completed, usable responses (a 61% positive response rate; none from the email distribution). 72% of the responses were from public, government schools; whilst 28% were from private, independent schools. 77% of the teachers were female, with the balance of 23% being male. Schoolteachers were essentially asked to respond to the following statements, and they could choose (X) as many responses as they wish. These responses were subsequently ranked:

A. I would use (X) the following delivery methods for raising cyber-safety awareness at my school.
B. The following factors (X) would influence my delivery method choices when it comes to raising cyber-safety awareness at my school.
C. I would use (X) the following strategies to guide my delivery method choices in raising cyber-safety awareness at my school.

5 Results

The ranking of the responses corresponding to Appendices A, B and C follows.

5.1 Delivery Methods for Raising Cyber-Safety Awareness

Below is a ranking of the delivery methods which teachers would prefer to use.

Figure 1 depicts that events (63%) and posters (62%), which are conventional delivery methods, were the top two most preferred delivery methods for raising cyber-safety awareness in the surveyed schools. Social media (51%), which is a form of online delivery method, ranked third most preferred. The lecture (50%), which is a form of instructor-led delivery method, ranked fourth most preferred. These top 4 delivery methods are relatively quick to produce, and are generally easily accessible. Online delivery methods such as email (12%), mLearning (12%) and the blog (10%), ranked the least preferred delivery methods - so did the mouse pad (13%), a conventional delivery method. These results could be attributable to the fact that Internet connection and computer hardware are largely inaccessible in public, government schools - which formed the majority (72%) of the survey responses.

5.2 Factors Influencing the Preference of Delivery Methods

Below is a ranking of the factors which influence the schoolteachers' preference of cyber-safety awareness delivery methods.

Figure 2 demonstrates that the time to produce (57%) mostly influences the schoolteachers' preference of a cyber-safety awareness delivery method. This is understandable, as teachers are pressed with time, and would prefer a delivery method which is quick to put together. The availability of technology is the second most influential set of factors. In particular, the availability of hardware (53%), the availability of software (53%), and the availability of network connectivity (51%), are important considerations for schoolteachers when exploring delivery methods to use

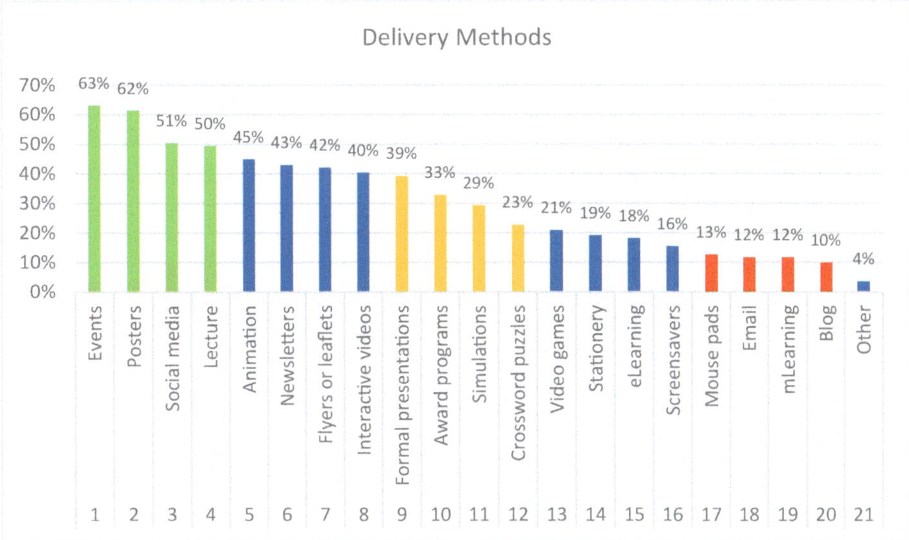

Fig. 1. Delivery methods for raising cyber-safety awareness

for raising cyber-safety awareness. The ease of administration (22%), ease of presentation (19%), flexibility of medium (16%) and the ease of evaluating students (15%) influences the schoolteachers' preference of a cyber-safety awareness delivery method the least.

5.3 Strategies Used for the Selection of Delivery Methods

Below is a ranking of the strategies which influence the schoolteachers' preference or selection of cyber-safety awareness delivery methods.

Figure 3 portrays that the consideration of the learner's needs (69%) and age (68%) are the top two strategies influencing the schoolteachers' preference or selection of cyber-safety awareness delivery methods. Considering the learner's needs refers to understanding how learners prefer to learn, as this has a direct impact on the type of delivery method introduced or preferred. For example, some learners are visual learners, and so using delivery methods with many images and pictures would be ideal in communicating a cyber-safety message to such learners. Studying the literature (42%) and the consideration of the characteristics and capability of the delivery methods (35%) ranked the bottom strategies influencing the schoolteachers' preference or selection of cyber-safety awareness delivery methods.

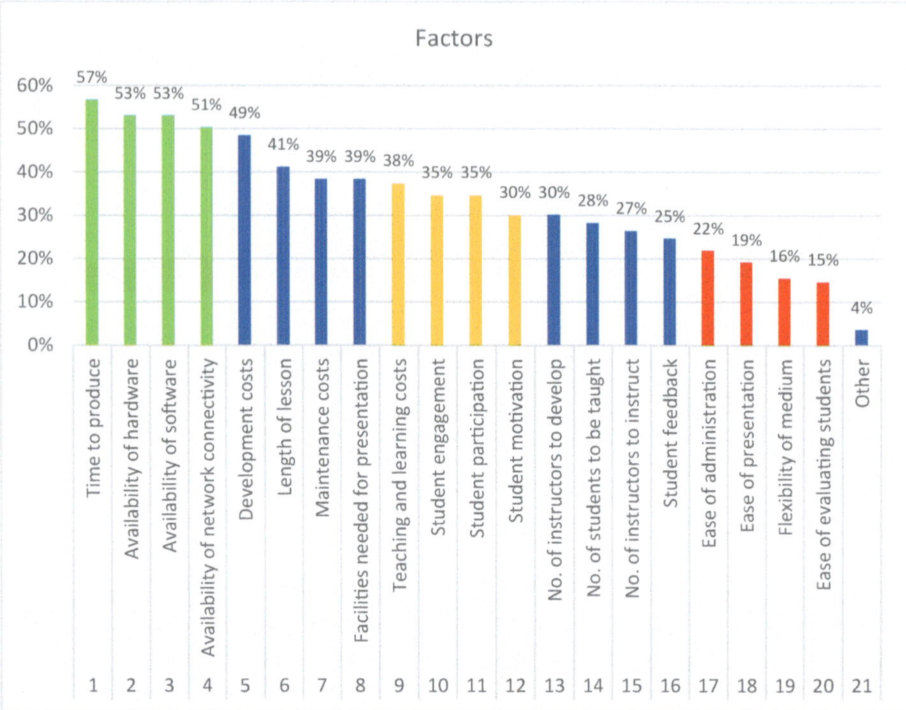

Fig. 2. Factors influencing the preference of delivery methods

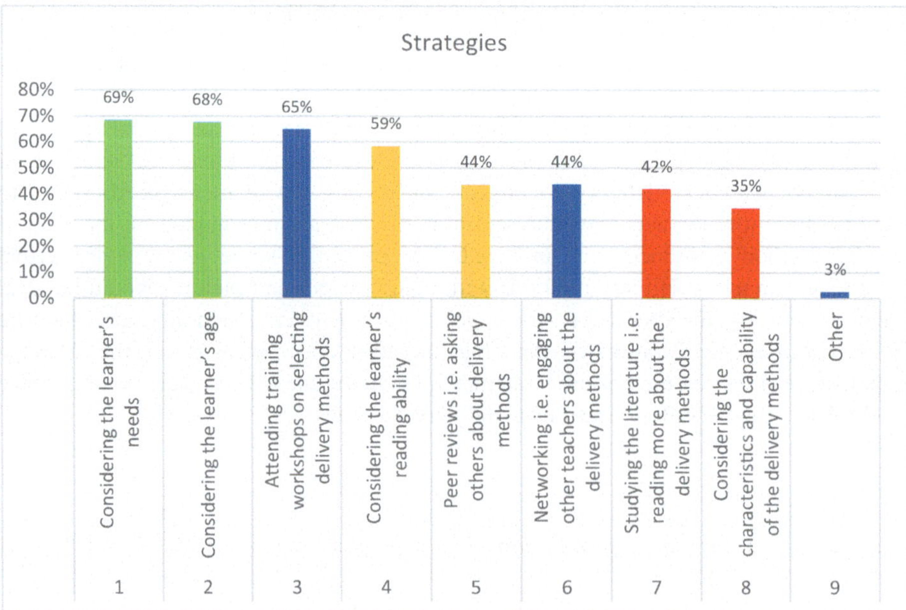

Fig. 3. Strategies used for the selection of delivery methods

6 Critical Discussion

Conventional delivery methods were five (5) of the top ten (10) most preferred delivery methods for raising cyber-safety awareness. Therefore, because of their accessibility; events, posters, newsletters, flyers or leaflets, and award programmes, should be used by schoolteachers for raising cyber-safety awareness. As the most (1st) preferred delivery method, events such as school assemblies in the mornings, and "civvies" or casual days on Fridays, must be used at least monthly for the promotion of cyber-safety awareness. Posters can be put up and flyers handed out at the cyber-safety awareness events. Online delivery methods were two (2) of the top ten (10) most preferred delivery methods. Therefore, because of their popularity, appeal and accessibility, social media and animation should be used by schoolteachers for raising cyber-safety awareness. Since there are significantly more mobile Internet connections than traditional PC Internet connections in South Africa (Joubert 2008), plus a majority of South African school learners are now able to access the Internet through their mobile phones (Kritzinger 2017), social media must be used at least monthly for the promotion of cyber-safety awareness in schools. As the third (3rd) most preferred delivery method, social media in the form of multimedia WhatsApp messages must be used. WhatsApp is specifically recommended as it is a popular instant messaging service, which is widely used by school learners, their parents, and teachers. WhatsApp will ensure that school learners can receive the communicated cyber-safety awareness lesson or message instantaneously.

Instructor-led delivery methods were also two (2) of the top ten (10) most preferred delivery methods. Therefore, because of their accessibility, interesting and interactive lectures and formal presentations during lunch breaks should be used by schoolteachers for raising cyber-safety awareness. In addition to making cyber-safety awareness more interesting and current, Wilson and Hash (2003) suggest repeating an awareness message and using a variety of delivery methods to present that particular message. This is said to greatly increase the users' retention of the cyber-safety awareness lesson or issue. As such, a combination of select conventional, online and instructor-led delivery methods should be used to repeat a cyber-safety awareness message. The South African Department of Education (DoE), and the Gauteng Department of Education (GDE) should alternatively create these cyber-safety awareness delivery methods on behalf of schoolteachers, for dissemination and use in Gauteng schools, especially public schools.

Game-based delivery methods are desirable learning methods and should be used for raising cyber-safety awareness, as they can make learning more enjoyable (Franklin et al. 2003; Weisskirch 2006; Crossman and Crossman 1983) and fun. Crossword puzzles, ranked twelfth (12th), and video games, ranked thirteenth (13th) as preferred delivery methods, are ideal for enhancing the learners' engagement, participation, and motivation in cyber-safety awareness programmes. Crossword puzzles are not only fun, but they are effective learning methods. They have a positive ripple effect on the learner. For example, by identifying correct answers in a puzzle, the learner's confidence increases (Weisskirch 2006). This increased confidence can have a positive effect on performance (Cassidy 2004), and the satisfaction resulting from performance can reinforce learning (Childers

1996). In an educational settings, the efficacy of video games is demonstrated by reduced training time and instructor load, and enhanced knowledge acquisition and retention (Brownfield and Vik 1983; Ricci 1994). Furthermore, educational video games are said to be particularly effective when designed to address a specific problem (Griffiths 2002), such as cyber-safety awareness in schools.

Media characteristics ranked the bottom, eighth (8th) strategy which guides the schoolteachers' choice or preference of cyber-safety awareness delivery methods. However, schoolteachers must consider this strategy alongside the learner's needs, learner's age, and learner's reading ability strategies, which ranked first (1st), second (2nd), and fourth (4th) respectively. This is because media characteristics such as visual displays (i.e. 3-Dimensionality); printed words; sound (i.e. verbal or nonverbal); motion (i.e. still, limited movement, or full movement visuals); colour; and learner feedback, are correlated with the learner's needs, age and reading ability. For instance, a very colourful animation which has 3D motion picture, sound, few or no printed words, and captures the learner's feedback, would be ideal for raising cyber-safety awareness to a group of Grade R or Grade 1 school learners. Similarly, the use of WhatsApp messages in cyber-safety awareness programmes would be ideal for school learners in Grade 4 onwards. These learners are likely to have their own smartphones, and are likely to be active users of this popular instant messaging service. However, an accessible event such as a school drama on cyber-safety awareness, could be less discriminant, and more inclusive, across the learner's needs, ages and reading abilities. Furthermore, such an event could introduce learners to the performing arts.

7 Conclusion

Information and Communication Technology (ICT) has become a part of daily life. However, ICT and associated advances present their own set of benefits and risks. School learners are especially susceptible to cyber-risks, and they need to be made aware of such risks, using a combination of select conventional, online and instructor-led delivery methods, as preferred by the surveyed South African schoolteachers. It is easy to see why an event such as a school assembly for example, would be an ideal delivery method for raising cyber-safety awareness. The delivery method (i.e. event/assembly) is most convenient, as it is quick to produce, and requires little or no technology. The same applies to posters, social media posts and lectures. The availability of technology is an influential set of factors, and its lack impedes teachers from using modern, contemporary delivery methods. The call is thus for the South African government to make such technology (i.e. computing hardware, software and network connectivity) available in public schools, for the partial purpose of raising cyber-safety awareness in South African schools.

Learner's needs and age are the top two strategies influencing the schoolteachers' preference or selection of cyber-safety awareness delivery methods. With regards to the learner's needs, a delivery method using spoken words may be superior over another, because poor readers can more easily understand the medium, especially if the spoken words are in the learner's own language. With regards to the learner's age, Dale (1969) suggests that it is more efficient to use abstract media with older learners, and concrete media and experiences with younger learners. Similarly, many educators believe that media or delivery methods may be differentially effective for different types of learners, and therefore, delivery methods that are best suited for various learner types should be explored and identified; this being an exciting opportunity for future research.

Appendix A: Delivery Methods

See Tables 1 and 2.

Table 1. Delivery methods for raising cyber-safety awareness

1. **Award programs** - giving learners awards/certificates for demonstrating cyber-safety awareness
2. **Events** - using school assemblies, fairs, walks etc. to raise cyber-safety awareness
3. **Flyers or leaflets** - communicating cyber-safety awareness on printed sheets of paper
4. **Mouse pads** - communicating cyber-safety, pads can be placed inside computer labs at schools
5. **Newsletters** - periodic cyber-safety communication available in electronic or print format
6. **Posters** - made visible in areas where there is significant learner traffic i.e. tuck shop/noticeboards
7. **Stationery** - pens, pencils, writing pads, memory sticks etc. conveying cyber-safety messages
8. **Lecture** - a person teaching cyber-safety awareness in a classroom
9. **Formal presentations** - expert-based approach directed at a large audience
10. **Animation** - simulated motion picture delivering a cyber-safety message
11. **Blog** - raising cyber-safety awareness using diary-style text entries called posts online
12. **eLearning** - teaching cyber-safety awareness utilising electronic technologies such as a website
13. **Email** - communicating cyber-safety using messages distributed through electronic technology
14. **mLearning** - communicating cyber-safety using laptops, tablets, phablets and smartphones
15. **Screensavers** - animation or images replacing a static computer display after a period of inactivity
16. **Social media** - use of Facebook, WhatsApp, YouTube etc. to raise cyber-safety awareness
17. **Interactive videos** - responsive cyber-safety videos which support user interaction and input
18. **Crossword puzzles** - communicating cyber-safety awareness using classical crossword puzzles
19. **Video games** - communicating cyber-safety awareness using engaging video games
20. **Simulations** - using technology to imitate a real-life situation, in delivering a cyber-safety message

Table 2. Delivery methods classification

Cyber-safety awareness delivery methods																			
Conventional							Game-based		Instructor-led		Online							Simulation-led	Video-based
Award programmes	Events	Flyers or leaflets	Mouse pads	Newsletters	Posters	Stationery	Crossword puzzles	Video games	Lecture	Formal presentations	Animation	Blog	eLearning	Email	mLearning	Screensavers	Social media	Simulations	Interactive videos

Appendix B: Factors

See Tables 3 and 4.

Table 3. Factors influencing the preference of delivery methods

1. **Availability of hardware** - physical components of a computer i.e. monitor, keyboard, and mouse
2. **Availability of software** - programs used to operate computers and related devices i.e. MS Office
3. **Availability of network connectivity** - infrastructure for connecting computers i.e. LAN/Internet
4. **Development costs** - costs incurred in the production or manufacturing of the delivery method
5. **Teaching and learning costs** - costs incurred in teaching the learner using the delivery method
6. **Maintenance costs** - costs incurred in keeping the delivery method in good condition/working order
7. **Time to produce** - the time it takes to produce the delivery method with cyber-safety content
8. **Length of lesson** - the duration or the amount of time available for raising cyber-safety awareness
9. **No. of students to be taught** - the number of students to be taught using the delivery method
10. **Student engagement** - the ability of the delivery method to ignite learning passion in a learner
11. **Student feedback** - the ability of the learner to provide feedback using the delivery method
12. **Student motivation** - the ability of the delivery method to create a desire for the learner to learn
13. **Student participation** - extent to which the learners' involvement in the learning process is supported
14. **No. of instructors to develop** - the number of teachers required to develop the delivery method
15. **No. of instructors to instruct** - the number of teachers required to teach using the delivery method
16. **Ease of administration** - the level of effort or difficulty in managing the delivery method
17. **Ease of evaluating students** - the level of effort or difficulty in assessing the learners
18. **Ease of presentation** - the level of effort or difficulty in presenting the lesson using the delivery method
19. **Flexibility of medium** - the extent to which the delivery method can be modified or changed as required
20. **Facilities needed for presentation** - the requirement of facilities such as classrooms, labs etc

Table 4. Influential themes and factors

Themes	Factors
Technology	Availability of hardware
	Availability of software
	Availability of network connectivity
Cost	Development costs
	Teaching and learning costs
	Maintenance costs
Time	Time to produce
	Length of lesson
People	No. of students to be taught
	Student engagement
	Student feedback
	Student motivation
	Student participation
	No. of instructors to develop
	No. of instructors to instruct
Operation	Ease of administration
	Ease of evaluating students
	Ease of presentation
	Flexibility of medium
	Facilities needed for presentation

Appendix C: Strategies

See Table 5.

Table 5. Strategies used for the selection of delivery methods

Themes	Strategies
Learner's needs	Considering the learner's needs
	Considering the learner's age
	Considering the learner's reading ability
Professional development	Attending training workshops on selecting delivery methods
	Peer reviews i.e. asking others about delivery methods
	Networking i.e. engaging other teachers about the delivery methods
	Studying the literature i.e. reading more about the delivery methods
Media characteristics	Considering the characteristics and capability of the delivery methods

References

Abawajy, J.: User preference of cyber security awareness delivery methods. Behav. Inf. Technol. **33**(3), 237–248 (2014)

Arachchilage, N.A.G., Love, S.: A Game design framework for avoiding phishing attacks. Comput. Hum. Behav. **29**(3), 706–714 (2013)

Asongu, S.A., Odhiambo, N.M.: Enhancing ICT for quality education in sub-Saharan Africa. Educ. Inf. Technol. **24**, 1–17 (2019)

Brownfield, S., Vik, G.: Teaching basic skills with computer games. Train. Dev. J. **37**(12), 53–56 (1983)

Cassidy, S.: Learning styles: an overview of theories, models, and measures. Educ. Psychol. **24** (4), 419–444 (2004)

Chakraborty, M., Pal, S., Pramanik, R., Chowdary, C.R.: Recent developments in social spam detection and combating techniques: a survey. Inf. Process. Manag. **52**(6), 1053–1073 (2016)

Childers, C.D.: Using crossword puzzles as an aid to studying sociological concepts. Teach. Sociol. **24**(2), 231–235 (1996)

Çikrıkci, Ö.: The effect of Internet use on well-being: meta-analysis. Comput. Hum. Behav. **65**, 560–566 (2016)

Crossman, E.K., Crossman, S.M.: The crossword puzzle as a teaching tool. Teach. Psychol. **10** (2), 98–99 (1983)

Dale, E.A.: Audiovisual Methods in Teaching, 3rd edn. Holt, Rinehart & Winston, New York (1969)

De Lange, M., Von Solms, R.: An e-Safety educational framework in South Africa. In: Proceedings of the Southern Africa Telecommunication Networks and Applications Conference (SATNAC), Western Cape, South Africa (2012)

Feizollah, A., Anuar, N.B., Salleh, R., Wahid, A., Wahab, A.: A review on feature selection in mobile malware detection. Digit. Invest. **3**, 22–37 (2015)

Franklin, S., Peat, M., Lewis, A.: Non-traditional interventions to stimulate discussion: the use of games and puzzles. J. Biol. Educ. **37**(2), 79–84 (2003)

Givoni, M., Thomopoulos, N.: ICT for Transport: Opportunities and Threats. Edward Elgar Publishing, Cheltenham (2015)

Goodstein, A.: Totally Wired: What Teens and Tweens are Really Doing Online. St Martin's Press, New York (2007)

Govender, I., Skea, B.: Teachers' understanding of E-Safety: an exploratory case in KZN South Africa. Electron. J. Inf. Syst. Dev. Ctries. **70**(1), 1–17 (2015)

Griffiths, M.: The educational benefits of video games. Educ. Health **20**(3), 47–51 (2002)

Guan, S.S.A., Subrahmanyam, K.: Youth internet use: risks and opportunities. Curr. Opinion Psychiatry **22**(4), 351–356 (2009)

Haftu, G.G.: Information communications technology and economic growth in Sub-Saharan Africa: a panel data approach. Telecommun. Policy **43**(1), 88–99 (2019)

Hanewald, R.: Confronting the pedagogical challenge of cyber-safety. Aust. J. Teach. Educ. **33** (3), 1–16 (2008)

Haux, R., Ammenwerth, E., Herzog, W., Knaup, P.: Health care in the information society. A prognosis for the year 2013. Int. J. Med. Inf. **66**(1), 3–21 (2002)

Herrero, S.T., Nicholls, L., Strengers, Y.: Smart home technologies in everyday life: do they address key energy challenges in households? Curr. Opin. Environ. Sustain. **31**, 65–70 (2018)

Hynes, B.: At the heart or on the periphery: the role of information communications technology (ICT) in small firms. Int. J. Manag. Pract. **6**(1), 24–39 (2013)

Islam, A.A., Mok, M.M.C., Gu, X., Spector, J., Hai-Leng, C.: ICT in higher education: an exploration of practices in Malaysian universities. IEEE Access **7**, 16892–16908 (2019)

Joubert, R.: The mobile web: an untapped opportunity for publishers (2008). http://www.bizcommunity.com/Article/196/15/30508.html. Accessed 7 May 2018

Kandell, J.J.: Internet addiction on campus: the vulnerability of college students. Cyberpsychol. Behav. **1**(1), 11–17 (1998)

Kim, K.N., Yim, M.S., Schneider, E.: A study of insider threat in nuclear security analysis using game theoretic modelling. Ann. Nucl. Energy **108**, 301–309 (2017)

Kouttis, S.: Improving security knowledge, skills and safety. Comput. Fraud Secur. **4**, 12–14 (2016)

Kritzinger, E.: Short-term initiatives for enhancing cyber-safety within South African schools. S. Afr. Comput. J. **28**(1), 1–17 (2016)

Kritzinger, E.: Cultivating a cyber-safety culture among school learners in South Africa. Afr. Educ. Rev. **14**, 1–20 (2017)

Kruger, H.A., Flowerday, S., Drevin, L., Steyn, T.: An assessment of the role of cultural factors in information security awareness. In: Information Security South Africa (ISSA), pp. 1–7. IEEE (2011)

Lai, F., Li, D., Hsieh, C.T.: Fighting identity theft: the coping perspective. Decis. Support Syst. **52**(2), 353–363 (2012)

Li, Q.: New bottle but old wine: a research of cyberbullying in schools. Comput. Hum. Behav. **23**, 1777–1791 (2007)

Livingstone, S., Haddon, L., Görzig, A., Ólafsson, K.: Risks and Safety on the Internet: The Perspective of European Children. EU Kids Online Full Findings. LSE, London (2011)

McFarlane, A., Sakellariou, S.: The role of ICT in science education. Camb. J. Educ. **32**(2), 219–232 (2002)

McQuade, S.C.: We must educate young people about cybercrime before they start college. Chron. High. Educ. **53**(18), B29–B31 (2007)

Mogoba, P., Phillips, T.K., Myer, L., Ndlovu, L., Were, M.C., Clouse, K.: Smartphone usage and preferences among postpartum HIV-positive women in South Africa. AIDS Care **31**(6), 723–729 (2019)

Morán, A.J., Profaizer, P., Zapater, M.H., Valdavida, M.A., Bribián, I.Z.: Information and communications technologies (ICTs) for energy efficiency in buildings: review and analysis of results from EU pilot projects. Energy Build. **127**, 128–137 (2016)

Myers, J., McCaw, D., Hemphill, L.: Responding to Cyber Bullying: An Action Tool for School Leaders. Sage Publications, Thousand Oaks (2011)

Nath, S.: ICT integration in Fiji schools: a case of in-service teachers. Educ. Inf. Technol. **24**(2), 963–972 (2019)

Newman, E.: Media selection and active learning: design practices of designers-by-assignment within the academic library. Doctoral dissertation, Capella University (2015)

Perren, S., Corcoran, L., Cowie, H., Dehue, F., Mc Guckin, C., Sevcikova, A., Tsatsou, P., Völlink, T.: Tackling cyberbullying: review of empirical evidence regarding successful responses by students, parents, and schools. Int. J. Confl. Violence **6**(2), 283–292 (2012)

Porter, G., Hampshire, K., Abane, A., Munthali, A., Robson, E., Bango, A., de Lannoy, A., Gunguluza, N., Tanle, A., Owusu, S., Milner, J.: Intergenerational relations and the power of the cell phone: perspectives on young people's phone usage in sub-Saharan Africa. Geoforum **64**, 37–46 (2015)

Przybylski, A.K., Murayama, K., DeHaan, C.R., Gladwell, V.: Motivational, emotional, and behavioral correlates of fear of missing out. Comput. Hum. Behav. **29**(4), 1841–1848 (2013)

Reuters: American youth trail peers in Internet usage (2008). http://www.pcmag.com/article2/0,2817,2335412,00.asp. Accessed 15 Oct 2019

Rezgui, Y., Marks, A.: Information security awareness in higher education: an exploratory study. Comput. Secur. **27**(7), 241–253 (2008)

Ricci, K.E.: The use of computer-based videogames in knowledge acquisition and retention. J. Interact. Instr. Dev. **7**(1), 17–22 (1994)

Romero-Ruiz, K., Echeverri-Sánchez, L., Peña-Plata, J., Vásquez-Giraldo, S., Aguilera-Cardona, M., Herazo-Avendaño, C., Valencia-Arias, A., Bran-Piedrahita, L.: Information and communication technologies impact on family relationship. Procedia-Soc. Behav. Sci. **237**, 30–37 (2017)

Semerci, A., Aydin, M.K.: Examining high school teachers' attitudes towards ICT use in education. Int. J. Progress. Educ. **14**(2), 93–105 (2018)

Strassberg, D.S., Rullo, J.E., Mackaronis, J.E.: The sending and receiving of sexually explicit cell phone photos ("Sexting") while in high school: one college's students' retrospective reports. Comput. Hum. Behav. **41**, 177–183 (2014)

Strauss, J., Frost, R.D.: Selecting instructional technology media for the marketing classroom. Mark. Educ. Rev. **9**(1), 11–20 (1999)

Sykiotou, A.P.: Trafficking in human beings: internet recruitment. Misuse of the Internet for the recruitment of victims of trafficking in human beings, DG of Human Rights and Legal Affairs, Council of Europe (2007)

Telecom Regulatory Authority of India: The Indian telecom services performance indicators April - June 2007, TRAI, New Delhi, India (2007)

Walker, R.W.: An evaluation of training methods and their characteristics. Hum. Factors **7**(4), 347–354 (1965)

Weisskirch, R.S.: An analysis of instructor-created crossword puzzles for student review. Coll. Teach. **54**(1), 198–201 (2006)

Wilson, M., Hash, J.: Building an information technology security awareness and training program. NIST Spec. Publ. **800**(50), 1–39 (2003)

Wu, D., Li, C.-C., Zhou, W.-T., Tsai, C.-C., Lu, C.: Relationship between ICT supporting conditions and ICT application in Chinese urban and rural basic education. Asia Pac. Educ. Rev. **20**(1), 147–157 (2019)

An Ontology Model for Interoperability and Multi-organization Data Exchange

Andrei Tara[1], Alex Butean[1,2(✉)], Constantin Zamfirescu[2], and Robert Learney[3]

[1] Effective Decisions, 16 Iosif Veleceanu, 550057 Sibiu, Romania
research@andreitara.com, alex@butean.com
[2] Lucian Blaga University of Sibiu, 10 Victoriei, 550024 Sibiu, Romania
zbc@acm.org
[3] Digital Catapult, 101 Euston Road, London NW1 2RA, UK
robert.learney@digicatapult.org.uk

Abstract. Progress in the uptake and use of technologies such as Artificial Intelligence and the Internet of Things seems to be stalled by the colossal fragmentation of information and data standards. This complexity is compounded by issues of inter-organisational differences, hindering effective collaboration. There is a growing demand for cross-organizational integrations in regulated but decentralized environments. This paper introduces an ontology architecture where information is sliced into independent semantic layers, each focusing on a specific aspect of the data. By dissolving traditional monolithic data structures into layered, light semantic components, the necessity to maintain contextual metadata is diminished. The model proposes an ontology structure based on decentralized technologies to ensure an open environment and to avoid the flaws of centralized systems.

Keywords: Decentralized · Ontology · Model

1 Introduction

Language has been the most influential progress factor of all time because it allows individual human brains to connect, like neurons, to form an extended thinking system. In modern society, data is the language of progress [8]. There is a growing demand to allow separate organizations to exchange data between themselves. In order to suit a variety of individual company policies, software agents have to cooperate and organize into cross-organizational systems. The first component that fosters such collaboration is the data ontology. Existing industry models [16] promote heavy structures with multiple dependencies that are hard to manipulate in open and distributed ecosystems. Advancements introduced by decentralized technologies like DLT [7], DHT [25] and IPFS [4] have opened up new possibilities to share ontology schemas in multi-organization environments. The current paper presents a practical design that goes beyond horizontal fragmentation by introducing the possibility to divide the semantic information vertically into independent layers.

R. Silhavy (Ed.): CSOC 2020, AISC 1225, pp. 284–296, 2020.
https://doi.org/10.1007/978-3-030-51971-1_23

2 Related Work

2.1 Ontology-Based Data Access Frameworks

The most frequent method to establish common-sense rules regarding data ontologies is to offer access to standardised frameworks. Decentralised approaches in this area [2, 14] are proposing federated solutions that work with vast amounts of data formalised by ontologies. One of the most representative frameworks in this field is the OBDAIR framework [26]. It offers an efficient four-tier architecture capable of integrating data from disparate and heterogeneous data sources. Although it provides an original way to incorporate data from disparate data sources, OBDAIR requires constant adjustments from domain specialists. This reduces the possibility to distribute the processed information in an accessible, standardised format.

2.2 Ontologies and Distributed Reasoning Models

The adoption and improvement of standards are supported by high-level abstract syntax languages such as RDF [30], OWL [24], and RuleML [28]. These languages were designed for modelling and reasoning with ontologies in the context of the Semantic Web [18]. An improvement in this area is represented by enhanced formalized models for distributed reasoning [19]. These models constrain data exchange between agents to rigid properties of the ontologies for a deterministic output. By using techniques of order-sorted logic with rigidity and sort predicates between multiple ontologies, new insights can be obtained. The mathematics behind the distributed reasoning model is robust, and the overall concept simplifies a lot the practical usage of ontologies. Still, it overlooks the natural evolution of the same ontology while opening the possibility to transfer information in a backwards-compatible way.

2.3 Ontologies on Top of Distributed Hash Tables (DHT)

Some of the most common big data challenges [1] include the need to ensure the privacy of the data and the need to store the data structure in a distributed manner. DHTs can be used to store data that needs to be protected and shared at the same time. The deployment of ontologies over DHTs [21] is a significant improvement that serves the purpose of privacy and increases fault tolerance. In this direction, many pioneering approaches [6, 11, 15, 20, 27] have already proved the efficiency of the technique. The existing implementations show increased query throughput and better data management. Currently, the data structure used by DHTs gets distributed vertically: an ontology is considered a unified asset that is stored over a single storage location or replicated over multiple storage locations. The next evolution is to distribute the ontology over multiple storage locations using horizontal partitioning (sharding) [3, 29] by separating internal ontology dependencies over multiple storage locations.

3 Proposed Model

3.1 Overall Architecture

The current paper proposes an ontology-based data exchange protocol that allows consistent communication between distributed software agents in a multi-organization environment. The protocol is composed of interoperable ontology models [5] that represent the input and output of the agents. Although each organization has its own semantics, context and understanding of the data, this protocol will act as a translator/abstractor that will foster internal and external collaborations. The abstraction of the communication layer [10] opens up new possibilities to create high-level solutions by combining existing agents and technologies. In previous approaches that adopt standard ontology schemas [17], users are empowered to create elaborate scenarios where the underlying components can be connected dynamically. All these approaches keep the information concentrated into monolithic and rigid structures. Our new model instead proposes a multi-layered architecture to increase flexibility and decentralize the storage structures. Each layer is composed of machine-readable semantic data structures that provide context on a particular dimension of the ontology. The most important aspect of this approach is that data structures can be used, deployed and updated in a decentralized manner.

From a high-level perspective, the architecture is composed of the following layers:

- Structural layer - the formal specification of ontology in its purest form composed of concepts, properties and relations;
- Connection layer - contains information about the location of concepts from external ontologies and mappings between multiple versions of the same ontology;
- Encoding layer - specify the used encoding format like UTF-8, ISO or any other chosen format;
- Defaults layer - used to define the fallback values for specific properties;
- Validation layer - used to add formal validation rules on schema properties;
- Restriction layer - contains a set of contextual restrictions between schema properties;
- Naming layer - tags schema properties classes and relations in human-readable format;
- Instruction layer - includes guidance information on how the user should provide the input data;
- Versioning layer - contains community proposals about schema structure changes and future evolution;
- Template layer - used for contextual fragmentation of the schema.

The definition of an ontology is acknowledged as being viable only if the model provides all the data details required by the machine to operate. Figure 1. displays the high-level architecture of the model. The core layers are designed to store semantic information that is machine-targeted. The optional layers provide human-readable information that is not always mandatory in the communication between agents but it is imperative in cases involving human-computer interactions.

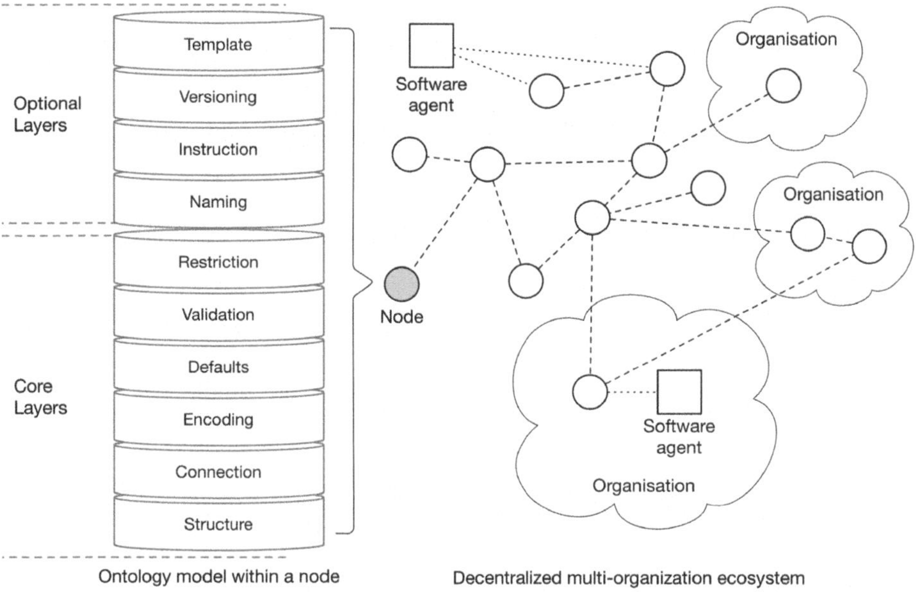

Fig. 1. Overall architecture

3.2 Structural Layer

The structural layer contains the architectural definition of ontology in its purest form as depicted in Fig. 2. From a semantic perspective, the structural layer is divided into 2 spaces: concept [12] and classification [9]. The concept space provides knowledge about concepts, properties and relations between them. All concepts are grouped and organized by a hierarchical system defined in the classification space. The hierarchy contains abstract notions on the root that becomes progressively more refined and specific down towards the leaves. Each concept contains a set of fundamental properties referencing basic data types such as numbers or strings, or advanced properties referencing other concepts through the instance of relation. Properties withholding private information are marked as *protected* and their value is encrypted at runtime to match the requirements of data privacy [23] in frameworks like GDPR [13]. Generic concepts expressing abstract structures are refined into specific ones by the use of inheritance relationships, ensuring flexibility, extensibility and modularity.

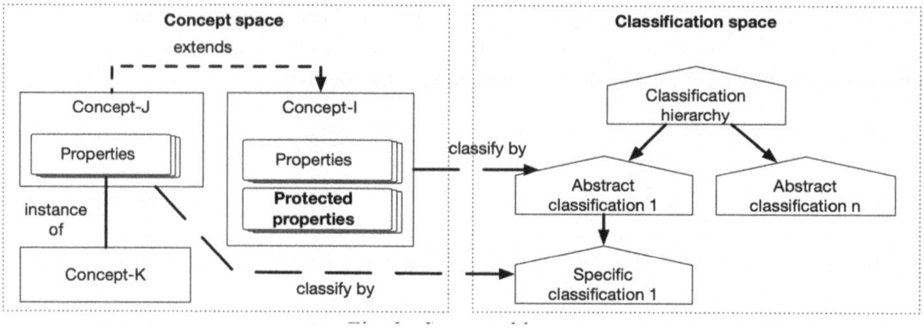

Fig. 2. Structural layer

3.3 Connection Layer

The connection layer ensures the connectivity of the current ontology from a structural and a temporal perspective as depicted in Fig. 3. The structural connection describes the link with a concept from an external ontology. It contains a link to the location of the ontology and the connectivity details with the referenced concept.

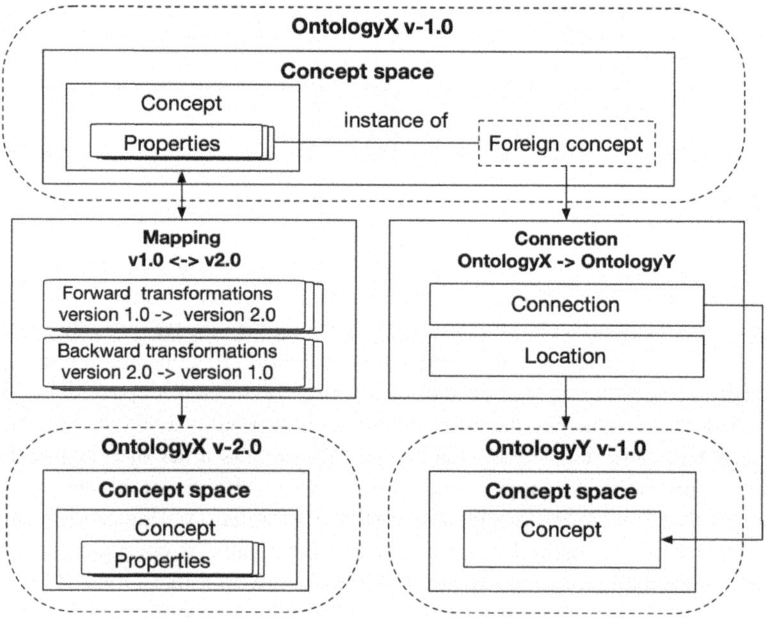

Fig. 3. Connection layer

Efficient ontologies need to be plastic, subject to constant change and improvement but at the same time need to be sufficiently stable for consistent communication. The current architecture ensures the plasticity and stability of the structure by using versioning and property mapping mechanisms.

Further, the formal model of the transformations and the specific characteristics are defined from a mathematical perspective. Considering the following notations:

- property P with definition domain D on version X
- property P' with definition domain D' on version Y
- from a temporal perspective $X < Y$
- f is the forward transformation and f' is the backwards transformation

mapping M can be defined as:

$$M(P,P') = \{ <f,f'> | f : D \rightarrow D', f' : D' \rightarrow D\}$$

Depending on the evolution of the domain between versions X and Y, the following cases emerge:

1. $f(P) = P' \; and f'(P') = P \Rightarrow$ no loss of information between versions X and Y then $M(P,P')$ is called a stable mapping noted as $Ms(P,P')$;
2. $f(P) = P' \; and f'(P') \neq P \Rightarrow$ there is a loss of information on the backward transformation between versions X and Y then $M(P,P')$ is called a forward-stable mapping noted as $Mf(P,P')$;
3. $f(P) \neq P' \; and f'(P') = P \Rightarrow$ there is a loss of information on the forward transformation between versions X and Y then $M(P,P')$ is called a backwards-stable mapping noted as $Mb(P,P')$;

Consider $T = \{M(P,P')\}$ as the set of all mappings between versions X and Y, the information between versions may be transported bidirectionally with ease when

$$T = \{M(P,P') | M(P,P') \simeq Ms(P,P')\}$$

Even though in practice, implementing a full set of stable mappings is a difficult task, it provides an excellent ideal goal to aim for.

Of course, the ideal case presented above is unlikely to be found in practical cases, which is why the case of partially stable mappings is closer to reality:

$T = \{M(P,P') | M(P,P') \in Ms(P,P') \cap Mb(P,P')\}$ partially backwards stable
$T = \{M(P,P') | M(P,P') \in Ms(P,P') \cap Mf(P,P')\}$ partially forward stable

Information might be lost in the partial stability cases, which is why it is necessary to consider a contextual and gradual approach, tailored according to the specific needs of a domain, system or consortium.

3.4 Encoding Layer

As displayed in Fig. 4, the encoding layer provides information about the encoding format like UTF-8, ISO, etc. for the properties contained by concepts. At the machine level, information about encoding is crucial; without it, useful machine-to-machine communication is not possible. If there is no specification for the encoding of a property, it is inherited from the default encoding of the concept or the default encoding of the ontology.

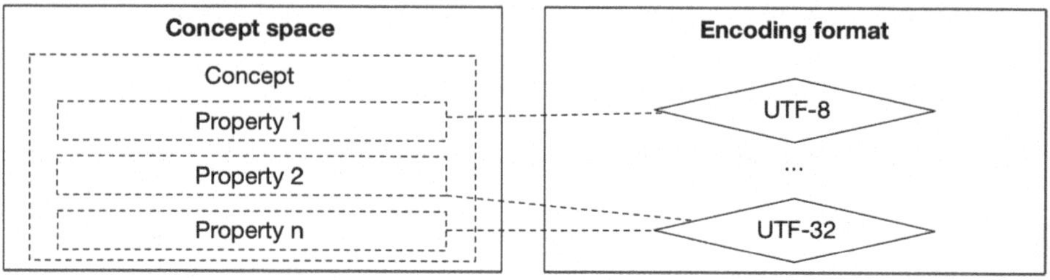

Fig. 4. Encoding layer

3.5 Defaults Layer

In other architectures [22], the ontology contains only the static definition of the concepts held, without focusing on the runtime details. In practice, dealing with heavy models providing explicit values for all properties represents an overhead for the end-user, especially when only a subset is relevant in a particular context. The current architecture allows encoding of fallback values and to limit the domain of properties by explicitly specifying the permitted set of values. Both fallback values and the value-set are contained in a contextual profile dependent on a language/locale as depicted in Fig. 5. Additionally, it is possible to create a new profile that inherits definitions from an existing one.

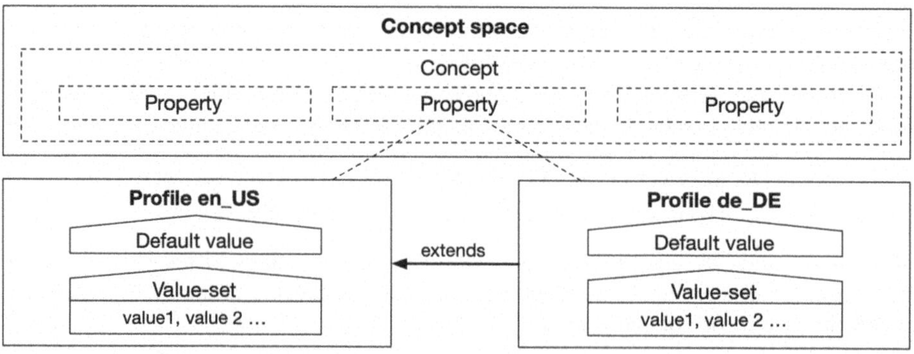

Fig. 5. Defaults layer

3.6 Validation Layer

Cross-organizational communication requires particular attention over data validation, integrity and privacy [23]. In current enterprise architectures [16], validation is decoupled from the structural definition of the data. The validation process is assured by external business logic coded on active computing components. This approach is suboptimal since data validation can only be performed after ingestion, and needs to be implemented by all the organizations using explicit programs. The validation layer (Fig. 6) provides a set of predicates used in the validation process of the properties. Predicates rely on descriptive definitions encoded at the ontology level, allowing an ad-hoc validation of data without the need to ingest it into an organization's processing system. To make things even easier to instantiate, there are predefined predicates associated with all primitive concepts.

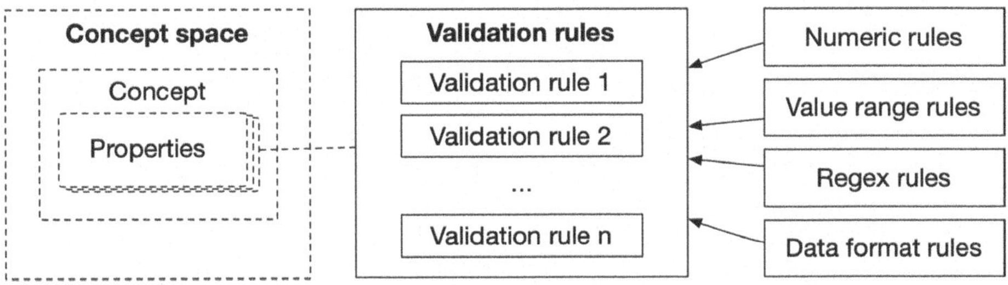

Fig. 6. Validation layer

3.7 Restriction Layer

The restriction layer works as a primitive yet powerful tool, used to add contextual restrictions inside a concept. It allows dynamic modification of a property's values using simple if/then/else logic. As shown in Fig. 7, accumulating a set of such simple rules can create an efficient way to enforce a particular behaviour at the ontology level.

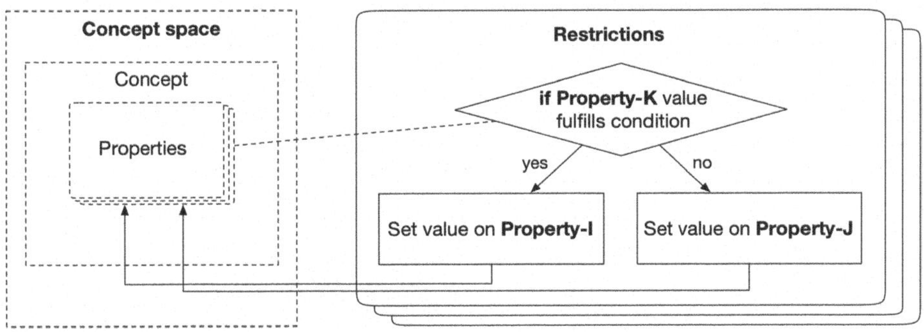

Fig. 7. Restriction layer

3.8 Naming Layer

The naming layer tags schema, properties, classes, and relations in a human-readable format. Organized in contextual profiles, these are dependent on a language/locale as displayed in Fig. 8. Being an optional layer, defining it is not mandatory unless the specified concept is used in direct interactions with the end-user.

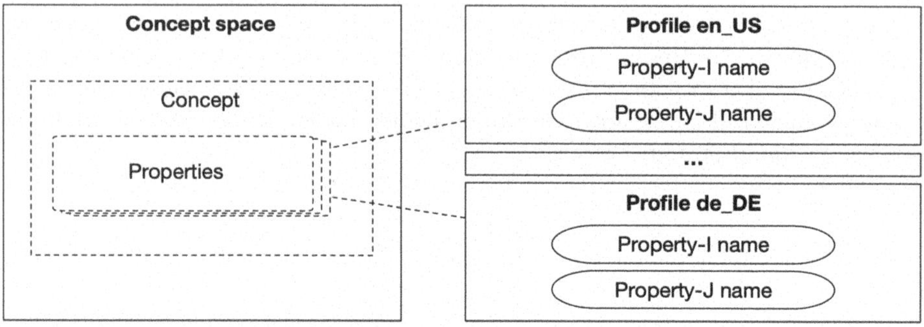

Fig. 8. Naming layer

3.9 Instruction Layer

In most cases, machine to machine communication doesn't need any human interaction and the data exchange is only relevant from a high-level perspective. However, the final beneficiary of any software system is a human being and even though most of the properties are for machines, some of the properties are defining a user input and they need human-friendly instructions to prevent input errors. The current architecture inserts the instructions about the data directly at the ontology level. As depicted in Fig. 9 the instruction layer includes guidance information on how the user should provide values for each property.

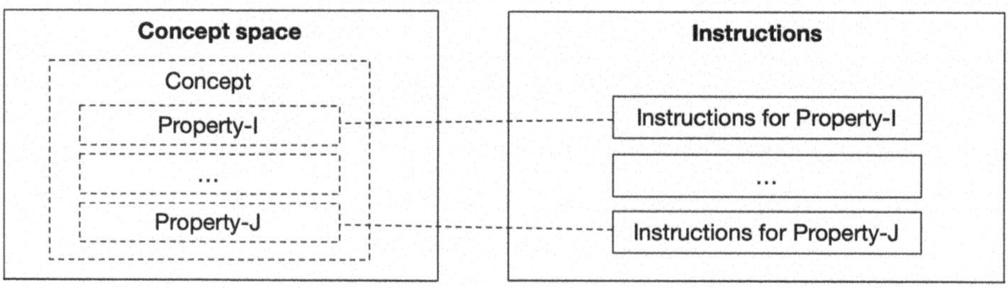

Fig. 9. Instruction layer

3.10 Versioning Layer

The versioning layer (Fig. 10) provides a formal framework to handle the evolution of the ontology model in a multi-organization context. It offers a collection of tools to facilitate the collaborative changes that need to be performed towards a new version. Each contributor proposes a modification in the concept or classification space by submitting a change request. Beside structural change requests, the layer also contains information about forward and backward mappings of the modified structure as needed. With the support of a voting system, relevant change requests and associated mappings are selected, processed and used as the input for the new version.

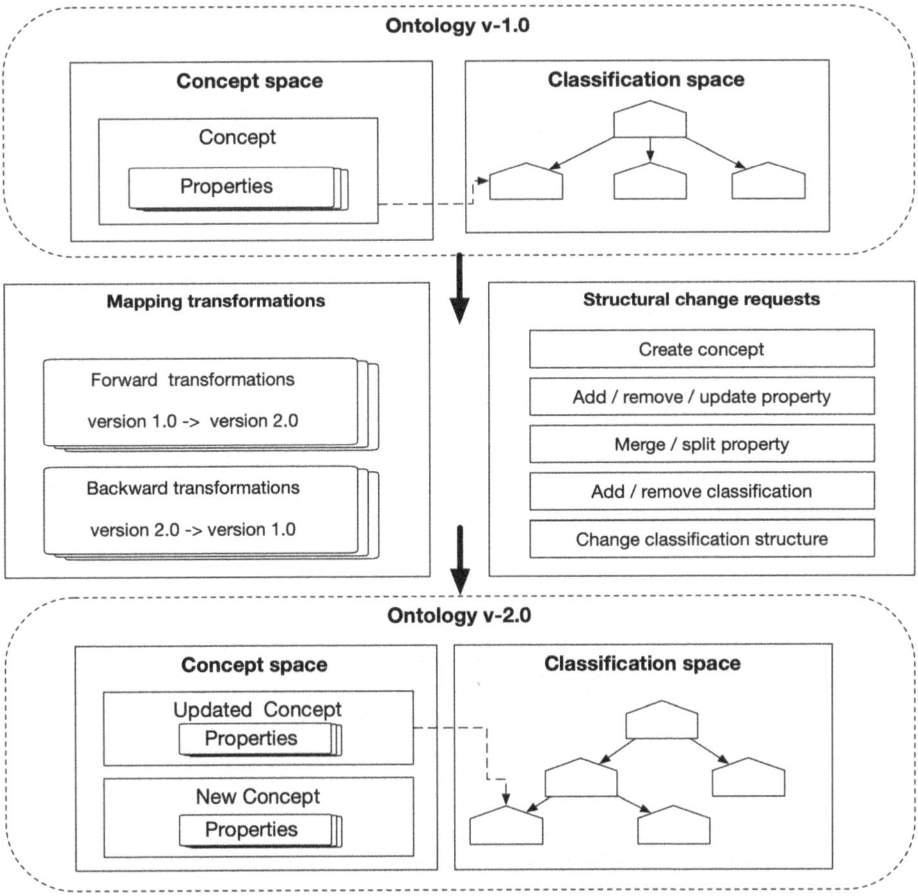

Fig. 10. Versioning layer

3.11 Template Layer

The structure layer describes the concepts using the maximal set of properties that covers all possible business cases. From a practical perspective, not all properties are required in every circumstance. The template layer (Fig. 11) implements a mechanism that groups properties into subsets. This way, each specific scenario can use only certain subsets, thus reducing the data size and processing time.

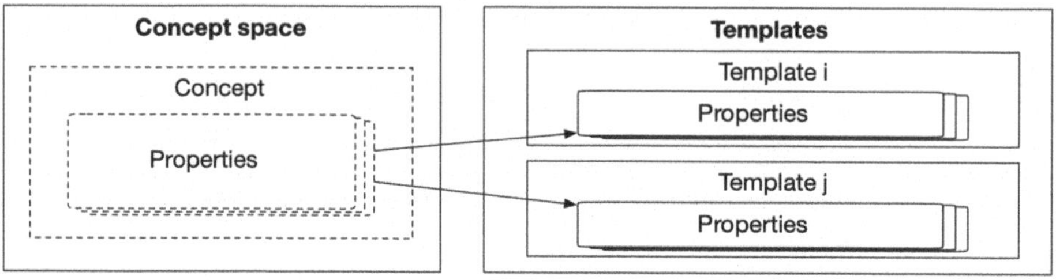

Fig. 11. Template layer

4 Conclusion

Collaboration beyond company borders remains an open issue mostly because the existing tools confine data into centralized silos, and existing data structures are insufficiently flexible. Existing ontology models were created to work with client-server architectures. Using the same models to facilitate serverless architectures has proven to be suboptimal because cross-organizational data exchange demands open standards and collaborative governance.

The current paper presents a new ontology model with the following unique characteristics:

- a decentralized storage distribution is now possible due to multiple, independent vertical layers that replace the old monolithic structure
- native support for encryption and data privacy
- interoperability between versions of the same ontology
- organizations can cooperate to build new data bridges

Having a technical model that was designed to be deployed on software agents inside a decentralized environment represents an important step forward. However, from a multi-dimensional perspective (scalability, compatibility, performance), the model has to be tested in a practical ecosystem during the next iteration of this approach.

Acknowledgement. This project has received funding from the European Union's Horizon 2020 research and innovation programme under grant agreement No. 870148 - "DIY4U".

References

1. Acharjya, D.P., Kauser, A.P.: A survey on big data analytics: challenges, open research issues and tools. Int. J. Adv. Comput. Sci. Appl. (IJACSA) **7**(2), 511–518 (2016)
2. Akerkar, R. (ed.): Scalable End-User Access to Big Data. Chapman and Hall/CRC Press, Boca Raton (2019). ISBN 9780367379117
3. Bagui, S., Nguyen, L.T.: Database sharding: to provide fault tolerance and scalability of big data on the cloud. Int. J. Cloud Appl. Comput. (IJCAC) **5**, 36–52 (2015)
4. Benet, J.: IPFS - Content Addressed, Versioned, P2P File System. https://arxiv.org/pdf/1407.3561.pdf
5. Baqa, H., Bauer, M., Bilbao, S., Corchero, A., et al.: Towards semantic interoperability standards based on ontologies. Semantic Interoperability White Paper, 26 p (2019)
6. Cai, M., Frank, M.R., Yan, B., MacGregor, R.M.: A subscribable peer-to-peer RDF repository for distributed metadata management. J. Web Semant. **2**(2), 109–130 (2004)
7. Casino, F., Dasaklis, T.K., Patsakis, C.: A systematic literature review of blockchain-based applications: current status, classification and open issues. Telematics Inf. **36**, 55–81 (2019). ISSN 0736-5853
8. Diyotta: Modern Data Integration Whitepaper (2016). https://www.diyotta.com/resources. Accessed 01 Jan 2020
9. Dumais, S., Chen, H.: Hierarchical classification of web content. In: The 23rd Annual International ACM SIGIR Conference on Research and Development in Information Retrieval, pp. 256–263 (2000)
10. Elghamrawy, S., El-Desouky, A.: Distributed multi-agent communication system based on dynamic ontology mapping. Int. J. Commun. Netw. Distrib. Syst. **10**(1), 1–24 (2013)
11. Fang, Q., Zhao, Y., Yang, G., Zheng, W.: Scalable distributed ontology reasoning using DHT-based partitioning. In: The Semantic Web (ASWC 2008). Lecture Notes in Computer Science, vol. 5367, pp. 91–105. Springer, Bangkok (2008)
12. Fudholi, D.H., Rahayu, W., Pardede, E., Hendrik: A data-driven approach toward building dynamic ontology. In: Mustofa, K., Neuhold, E.J., Tjoa, A.M., Weippl, E., You, I. (eds.) Information and Communication Technology. ICT-EurAsia 2013. Lecture Notes in Computer Science, vol. 7804. Springer, Heidelberg (2013)
13. General Data Protection Regulation (GDPR). https://gdpr-info.eu/. Accessed 01 Jan 2020
14. Haase, P., Horrocks, I., Hovland, D., Hubauer, T., et al.: Optique system: towards ontology and mapping management in OBDA solutions. In: Third International Workshop on Debugging Ontologies and Ontology Mappings, CEUR, Anissaras/Hersonissou - Greece, pp. 21–32 (2013)
15. Heine, F.: Scalable P2P based RDF querying. In: International Conference on Scalable Information Systems, Infoscale (2006)
16. Hepp, M., Leukel, J., Schmitz, V.: A quantitative analysis of eCl@ss, UNSPSC, eOTD, and RNTD content, coverage, and maintenance. In: International Conference on e-Business Engineering (ICEBE 2005), Beijing - China, pp. 572–581. IEEE (2005)
17. ISO 13584-32:2010 Industrial automation systems and integration - Parts library - Part 32: Implementation resources: OntoML: Product ontology markup language. https://www.iso.org/standard/50639.html. Accessed 01 Jan 2020
18. Jensen, J.: A systematic literature review of the use of semantic web technologies in formal education. Br. J. Edu. Technol. **50**, 505–517 (2019)
19. Kaneiwa, K., Mizoguchi, R.: Distributed reasoning with ontologies and rules in order-sorted logic programming. Web Semant. Sci. Serv. Agents World Wide Web **7**(3), 252–270 (2009)

20. Kaoudi, Z., Koubarakis, M., Kyzirakos, K., Magiridou, M., Miliaraki, I., Papadakis-Pesaresi, A.: Publishing, discovering and updating semantic grid resources using DHTs. In: CoreGRID Workshop on Grid Programming Model, Grid and P2P Systems Architecture (2006)
21. Kaoudi, Z., Kyzirakos, K., Koubarakis, M.: SPARQL query optimization on top of DHTs. In: Patel-Schneider, P.F. et al. (eds.) The Semantic Web – ISWC 2010. ISWC 2010. Lecture Notes in Computer Science, vol. 6496. Springer, Heidelberg (2010)
22. Khouri, S., Boukhari, I., Bellatreche, L., Sardet, E., Jean, S., Baron, M.: Ontology-based structured web data warehouses for sustainable interoperability: requirement modeling, design methodology and tool. Comput. Ind. **63**(8), 799–812 (2012)
23. Pandit, H.J., O'Sullivan, D., Lewis, D.: An ontology design pattern for describing personal data in privacy policies. In: The 9th Workshop on Ontology Design and Patterns, vol. 2195, pp. 29–39. CEUR, Monterey (2018)
24. Patel-Schneider, P.F., Hayes, P., Horrocks, I.: OWL Web Ontology Language Semantics and Abstract Syntax. W3C Recommendation. http://www.w3.org/tr/2004/rec-owl-semantics-20040210. Accessed 01 Jan 2020
25. Rosen, A.: Towards a Framework for DHT Distributed Computing, Georgia State University, Department of Computer Science. https://scholarworks.gsu.edu/cs_diss/107/. Accessed 01 Jan 2020
26. Santipantakis, G., Kotis, K., Vouros, G.A.: OBDAIR: ontology-based distributed framework for accessing, integrating and reasoning with data in disparate data sources. Expert Syst. Appl. **90**, 464–483 (2017). ISSN 0957-4174
27. Stoica, I., Morris, R., Liben-Nowell, D., Karger, D.R., Kaashoek, F.M., Dabek, F., Balakrishnan, H.: Chord: a scalable peer-to-peer lookup protocol for internet applications. IEEE/ACM Trans. Netw. (TON) **11**(1), 17–32 (2003)
28. The Rule Markup Initiative. http://www.ruleml.org. Accessed 01 Jan 2020
29. Wang, G., Jerry, Z., Shi, Nixon, M., Han, S.: SoK: sharding on blockchain. In: ACM Conference on Advances in Financial Technologies (AFT 2019), pp. 41–61. ACM, New York (2019)
30. World Wide Web Consortium - W3C: Resource Description Framework (RDF). https://www.w3.org/RDF. Accessed 01 Jan 2020

A Novel Approach for Intrusion Detection Based on Deep Belief Network

Cao Tien Thanh[✉]

Ho Chi Minh City University of Foreign Languages and Information Technology, Ho Chi Minh City, Vietnam
thanh.ct@huflit.edu.vn

Abstract. In recent years, cyber-attacks have many new forms with larger scale and much more complexity. This requires many network protection solutions and amplify the need for robust cybersecurity practices. One of the effective method to prevent network attacks is to use to Intrusion Detection Systems (IDSs), which can detect attacks never seen before. Many researchers have tried to produce anomaly - based IDSs, but they have not been able to detect malicious network traffic enough to ensure implementation in real networks. Clearly, it is still a challenge for the security community produce IDS suitable for implementation in real world. In this paper, we propose a new approach using a Deep Belief Network called Workflow-based Collaborative Learning with a combination of supervised and unsupervised machine learning methods for network attacks detection. Our proposed approach will be tested with network security datasets and compared with previously existing methods. The experimental evaluation shows that the valid of our approach.

Keywords: Cyber-attack · Deep Belief Network · Intrusion Detection System

1 Introduction

Along with the strong development of the network globally, security and information security issues has become a big challenge for companies, organizations, businesses and countries. The study of network intrusion forms and network intrusion detection techniques is one of the areas that security experts, researchers are very interested in recently. Currently there are many forms of network intrusion and network attack which can be divided into four main categories, namely, Probe (scan and probe), DoS (denial of service attack), User to Root (U2R - illegal access to local superuser), and Remote to Local (R2L - unauthorized remote access). The researchers in [1] presented a taxonomy on distributed cyber intrusion detection techniques. The solutions are divided into four categories, namely, statistical, algorithmic, mathematical and heuristical.

Statistical solutions include techniques such as statistical characteristics of data sample, use extrapolation or interpolation of data on some of the most suitable properties, estimating errors in the observed data, spectrum analysis data model [1].

Algorithmic Solutions: Techniques using processes (step by step) to calculate, process data, and automatically deduce [1].

R. Silhavy (Ed.): CSOC 2020, AISC 1225, pp. 297–311, 2020.
https://doi.org/10.1007/978-3-030-51971-1_24

Mathematical Solutions: Techniques using mathematical models, state mechanisms, geometric and algebraic techniques to achieve intrusion detection purposes [1].

Heursitical Solutions: Techniques using the knowledge of experts to analyze and evaluate data. In addition to using visual techniques, experience filtering, and combining various techniques [1].

In recent times, there are many research groups in the world have applied artificial intelligence models and intelligent computing to intrusion detection systems to find abnormal signs, from which warn to the system as well as the network administrator to have timely preventive methods [2, 4]. In this paper, we will propose a method using Deep Belief Networks to process network data to detect the signatures of network attacks. With the ability to efficiently analyze the larger of multidimensional data, we will be implementing and testing with the NSL-KDD and UNSW-NB15 datasets, the proposed method get better results than previously machine learning methods.

The next section of the paper will have the following structure: Part II will briefly introduce some network intrusion and attack techniques and recent studies in detection. In Part III, Deep Belief Networks which is a deep learning model and the method for applying network data processing to detect network attack will be introduced. Part IV presents and analyzes the experiment results obtained. The final section will be some conclusions and reviews.

2 Overview of Network Intrusion and Methods for Detection

In this section we will introduce common network intrusion and attack techniques, presenting the principles of operation together with the strengths and weaknesses of each technique. Finally, we will provide a summary of some of the intrusion techniques that have been investigated and used recently.

2.1 Intrusion and Attack Techniques

Probe Attack

Probing is a type of attack in which the intruder scans network devices to determine weakness in topology design or some opened ports and then use them in the future for illegal access to personal information.

Open Scan: Is the simples scanning technique [1]. This technique utilizes the TCP protocol and the SYN flag to detect TCP ports. This simple technique is easily detected by a firewall.

Half-Open Scan: Commonly dubbed as the TCP SYN scan, is a common method for port identification that allows the scanner to gather information about open ports without completing the TCP handshake process.

Stealth Scans: The aforementioned cyber scanning techniques only use the typical SYN flag to investigate open ports. Hence, they are easily detected and logged by IDSs [1].

Sweep Scan: Which do not aim to identify active ports but rather identify active hosts. They are characterized as performing sweeps, since their purpose is to identify the status of as many hosts as possible instead of focusing on an individual host.

Miscellaneous Scans: Cyber scanning insights by shedding light on scans that deal with various protocols. These include the FTP bounce, UDP, IP protocol and RPC scans.

There are many examples that represent probing over a network, such as nmap, portsweep, ipsweep, satan [5].

Denial of Service (DoS)

Denial of service (DoS) is one of the most frequent cases representing attacks on network resources and making network services unavailable for their users. There are many types of DoS attacks, and every type has it is own behavior on consuming network resources to achieve the intruder's aim, which is to render the network unavailable for its users.

UDP Flood, A UDP flood, by definition, is any DDoS attack that floods a target with User Datagram Protocol (UDP) packets. The goal of the attack is to flood random ports on a remote host. This causes the host to repeatedly check for the application listening at that port, and (when no application is found) reply with an ICMP 'Destination Unreachable' packet. This process saps host resources, which can ultimately lead to inaccessibility [6].

ICMP (Ping) Flood: Similar in principle to the UDP flood attack, an ICMP flood overwhelms the target resource with ICMP Echo Request (ping) packets, generally sending packets as fast as possible without waiting for replies. This type of attack can consume both outgoing and incoming bandwidth, since the victim's servers will often attempt to respond with ICMP Echo Reply packets, resulting a significant overall system slowdown [6].

SYN Flood, A SYN flood DDoS attack exploits a known weakness in the TCP connection sequence (the "three-way handshake"), wherein a SYN request to initiate a TCP connection with a host must be answered by a SYN-ACK response from that host, and then confirmed by an ACK response from the requester. In a SYN flood scenario, the requester sends multiple SYN requests, but either does not respond to the host's SYN-ACK response, or sends the SYN requests from a spoofed IP address. Either way, the host system continues to wait for acknowledgement for each of the requests, binding resources until no new connections can be made, and ultimately resulting in denial of service [6].

Ping of Death (PoD), A ping of death ("POD") attack involves the attacker sending multiple malformed or malicious pings to a computer. The maximum packet length of an IP packet (including header) is 65,535 bytes. However, the Data Link Layer usually poses limits to the maximum frame size – for example 1500 bytes over an Ethernet network. In this case, a large IP packet is split across multiple IP packets (known as fragments), and the recipient host reassembles the IP fragments into the complete packet. In a Ping of Death scenario, following malicious manipulation of fragment content, the recipient ends up with an IP packet which is larger than 65,535 bytes when reassembled. This can overflow memory buffers allocated for the packet, causing denial of service for legitimate packets [6].

There are some of examples DoS attack such as Back, Land, Neptune, Ping of Death (PoD), Smurf, Teardrop [5].

Remote to User (R2L)
Remote to user (R2L) is one type of computer network attacks, in which an intruder sends set of packets to another computer or server over a network where he/she does not have permission to access as a local user.

There are some of examples Remote to Local (R2L) attack such as Ftp_write, Guess_passwd, Imap, Multihop, Phf, Spy, Warezmaster [5].

User to Root Attacks (U2R)
User to root attacks is a type of attack where the intruder tries to access the network resources as a normal user, and after several attempts, the intruder becomes as a full access user

There are some of examples User to root (U2R) attack such as buffer_overflow, loadmodule, perl, rootkit [5].

2.2 Network Intrusion and Attack Detection Method

In this section we will present some of the network attack detection techniques that have been proposed recently by some research groups. Authors in [3] used a feature extraction technique and a Support Vector Machine (SVM) to classify attacks. The author proposes a feature extraction method for the KDD99 dataset using the Consistency Sybset Evaluation algorithm and the Best First method. The test results show that using SVM with finite datasets is more accurate with the original dataset. The authors achieved a fairly high error of 99.9185% [3]. However, with the asymmetric data, other research groups often use the measure associated with the attack data.

Authors in [2] present an IDS based on the Support Vector Machine (SVM). It uses evolutionary algorithms to optimize SVM parameters to improve accuracy during intrusion detection. The author of [8] has provided an AOCD (An Adaptive Outlier Based Scope-Based Scalarized Scan Detection Approach) solution for early detection of high-precision network scanning when tested with KDD99 datasets. In this work, they presented a solution for converting network traffic data into a format that filters and classifiers can operate. It is worth noting that the authors selected random samples in the database, thereby identifying the set of features for clustering detection used in early detection of network scanning techniques. The team tested on the real data collected from the Tezpur University Intrusion Detection System (TUIDS), which confirmed that the solution is capable of detecting cyber-attacks highly accurately [8].

3 Instruction Detection Method with Improved Deep Belief Network (DBN) Algorithm

3.1 The Overview and Related Research

DBN is a classic model in deep learning. By learning a deep nonlinear network structure, the approximation of complex functions is realized. The purpose is to improve the accuracy of classification or prediction. Commonly used models in deep learning are mainly convolutional neural networks and automatic encoders, etc. [7]. At the same time, DBN is also a deep neural network classifier composed of multiple Restricted Boltzmann Machines (RBMs) and a Back Propagation (BP) Neural Network [2, 7]. It is widely used in speech recognition, image recognition, network intrusion detection and other fields. Differentiating the DBN from other neural network models is the two-phase training algorithm: the unrestricted training mode based on the Restricted Boltzmann Machine (RBM), and the refining phase is the process of supervised learning. RBM is a neural network with a hidden layer and a layer represented in Fig. 1 [9].

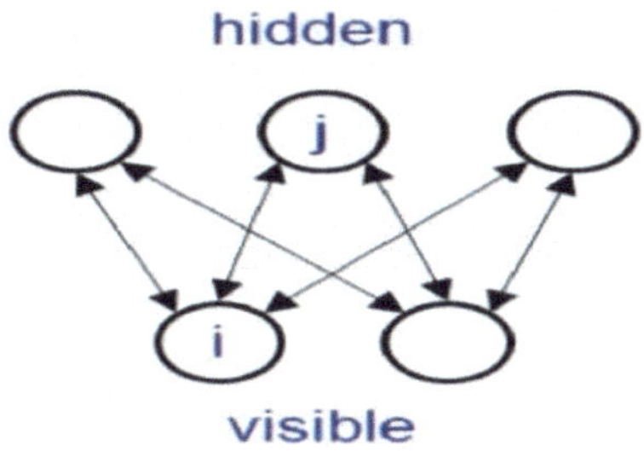

Fig. 1. Restricted Boltzmann Machine (RBM) [9]

The RBM, inherited from the Hopfield network model, uses the energy function to evaluate the configuration of the network, including the linkage between neurons, the neuronal freedom coefficient, the state of the neurons (initial value out of neurons) [10]. However, RBMs differ from Hopfield without the links between neurons in the same class, only having no direction between the hidden neurons and the neurons of the class. The energy function of the network in terms of parameters θ determined to be represented by Eq. 1.

$$E(v, h|\theta) = -\sum\nolimits_{i=0}^{V} \sum\nolimits_{i=0}^{H} w_{ij} v_i h_j - \sum\nolimits_{i=0}^{V} b_i v_i \sum\nolimits_{i=0}^{H} a_j h_j \tag{1}$$

where, θ = **(w, b, a)** and demonstrating the relationship between the neurons in the visible layer i and hidden layer j, (b_i, a_j is the coefficient of freedom of these two neurons); V and H are the total number of neurons in the visible layer and the hidden layer. Through the energy function, the state probability distribution of the network is represented by the Eq. 2.

$$p(v, h) = \frac{e^{-E(v,h)}}{\sum_{v,h} e^{-E(v,h)}} \tag{2}$$

In case the network configuration is known θ, the probability of the visible layer v is:

$$p(V|\theta) = \frac{\sum_h e^{-E(v,h)}}{\sum_u \sum_h e^{-E(u,h)}} \tag{3}$$

Considering the case that the RBM is binary, the values *v, h* only receive 0 or 1. Then, the probability that i-neuron of visible layer with activation state in the case that states of hidden layers are known *h* is determined by the Eq. 4 [9, 10].

$$p(v_i = 1|h, \theta) = \frac{e^{-E(v_i=1,h)}}{\sum_{v_i} \sum_h e^{-E(v_i=1,h)}} \tag{4}$$

v_i only receive 0 or 1 so from formula E(v, h) we have:

$$p(v_i = 1|h, \theta) = \frac{e^{\left(\sum_{j=1}^{H} w_{ij}h_j + b_i + \sum_{j=1}^{H} a_j h_j\right)}}{e^{\left(\sum_{j=1}^{H} w_{ij}h_j + b_i + \sum_{j=1}^{H} a_j h_j\right)} + e^{\left(\sum_{j=1}^{H} a_j h_j\right)}} = \frac{1}{1 + e^{-\left(\sum_{j=1}^{H} w_{ij}h_j + b_i\right)}} \tag{5}$$

Or:

$$p(v_i = 1|\mathbf{h}, \theta) = \sigma\left(b_i + \sum_{j=1}^{H} w_{ij}h_j\right) \tag{6}$$

Here:

$$\sigma(x) = (1 + e^{-x})^{-1}$$

According to the authors [9, 10], the network will be trained by updating w_{ij} with the formula (7)

$$p(h_j = 1|\mathbf{V}, \theta) = \sigma\left(a_j + \sum_{i=1}^{V} w_{ij}h_j\right) \tag{7}$$

The classifying approach is designed to consist of three modules: user data preprocessing module, DBN model training modules and collaboration module. The user data pre-processing module convert the original user descriptive information to vector, and divide the user data set into two parts: DBN model training data and test data; DBN model training module trains DBN model with training data set which is get in user data pre-processing module, includes two processes: model pre-training process and model fine tuning process; Collaboration module defines the feedback process of the convergence and accuracy of the first two modules in the form of workflow. Schematic overview of the approach is shown in Fig. 2.

3.2 Manipulation and Development

Based on the DBN algorithm described in Sect. 3.1. We propose an new approach to solve the intrusion detection problem called workflow-based Collaborative Leaning.

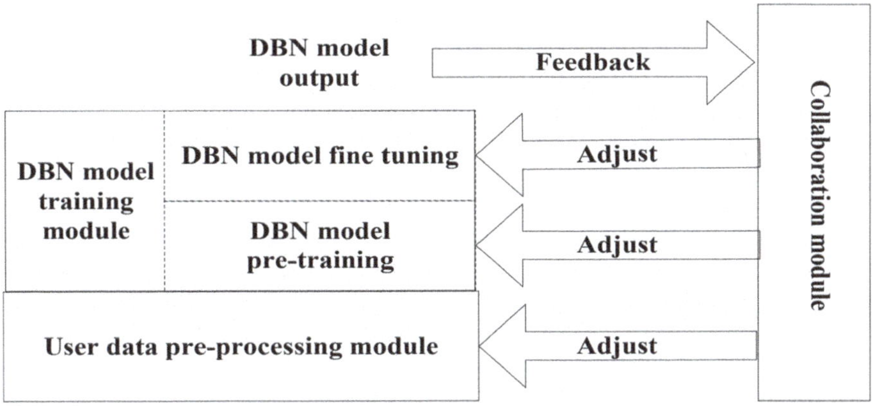

Fig. 2. Schematic approach overview

This solution makes use of the provided network security data sets: NSL-KDD, UNSW-NB15, in addition to decreased pre-training time of a deep belief network (DBN) by using an adaptive step size technique to enhance the convergence of the contrastive divergence (CD) algorithm, thereby reducing the number of epochs to train the restricted Boltzmann machine (RBM) that supports the DBN infrastructure.

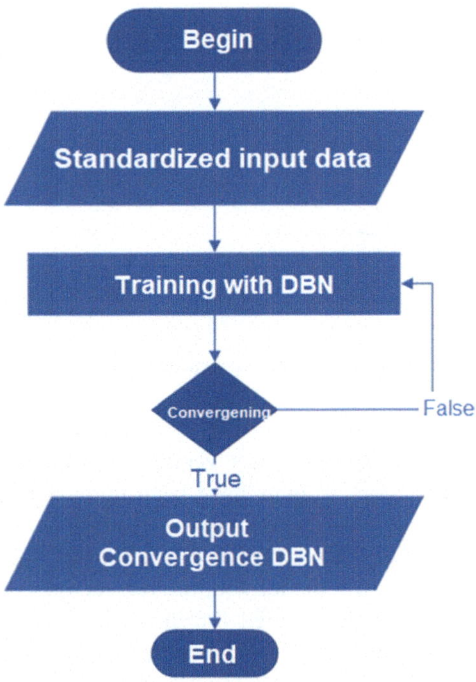

Fig. 3. Improved DBN algorithm diagram

Algorithm diagram of DBN network training process in intrusion detection problem is shown in Fig. 3. The input of the DBN network is the network security dataset that has been vectorized according to the features that have been extracted in the database.

In the model training process, different settings of parameters in each module may affect the output of subsequent module so that it will affect the accuracy of the final DBN model. For example, if the proportion of the training data set is too low, it will prevent the intrusion detection feature extraction, resulting in a low accuracy; if the maximum number of epochs is too low, it will make the network immature. In the collaborative module, we define the feedback process of the convergence and accuracy of the first two modules in the form of workflow. According to the convergence and accuracy of DBN model, we determine whether to make a reverse adjustment to the parameters of the user data pre-processing module and DBN training module or not, finally, improving the performance of the DBN model.

Workflow is a class of business processes that can be completely or partially performed automatically. Documents, information, and tasks can be passed and executed between different actors based on a series of procedural rules. The workflow we defined is shown as Fig. 4.

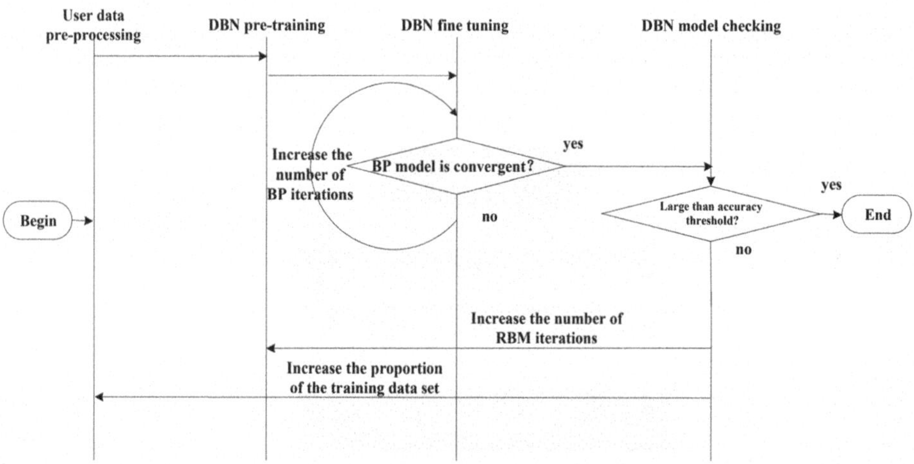

Fig. 4. Collaborative learning workflow

In this workflow, the data pre-processing module uses the original user data to generate a set of normalized user descriptive vector. After the completion of the DBN training process and the network weights have been initialized, the flow enters DBN fine tuning stage. If the BP network is not convergent in fine tuning stage, increase the number of epochs of BP until the convergence is achieved, then you get the trained DBN model. After checking DBN model with the test data set, and if the accuracy of DBN model does not achieve the desired threshold, increase the proportion of the training data set in classified data set in the user data pre-processing stage and increase the number of epochs of RBM in DBN pre-training stage. Retrain DBN model until the DBN model has a desired accuracy.

4 Experiment and Results

4.1 Description of Using Datasets

4.1.1 Dataset NSL-KDD

NSL-KDD is a widely used dataset of current IDSs [13]. The NSL-KDD is an improved dataset from the KDD99 dataset, which eliminates duplicate data, redundant data, training data and more experimental data. It has 41 features as in KDD99, collected and labeled from various test attacks. There are four types of attacks in 8 main forms that are labeled corresponding to attack techniques or tools that use for detection, namely, probe attack includes: *portsweep*, *satan;* DoS attack includes: *Neptune, Smurf*; R2L includes: *Imap, Warezmaster*; and U2R includes: *buffer_overflow, loadmodule* with the numbers are described in the Table 1.

Table 1. Network attacks data in NSL-KDD

Type of attacks	Training	Testing
Portsweep	2931	157
Satan	3633	735
Neptune	41214	4567
Smurf	2646	665
Imap	658	307
Warezmaster	20	944
buffer_overflow	30	20
loadmodule	9	2
Total	51141	7397

4.1.2 Dataset UNSW-NB15

The KDD98, KDDCUP99, NSL-KDD datasets have been used for a long time, so many of the attack techniques used to generate data are outdated, without any updating existing attack techniques. The UNSW-NB15 dataset [11] has overcome this and is being used extensively in intrusion detection studies. This dataset was released in 2015 at the Cyber Range Lab in New South Wales. The attack data is generated automatically from the IXIA Perfect Storm system with many new attack techniques. Packets are captured, preprocessed and extracted into 49 features. There are a total of 175,341 instances in the training dataset and 82,332 instances in the test dataset. In the training dataset, there are 10491 instances, while in the test dataset there are 3496 instances [12].

4.2 Experimentation and Evaluation of Results

The training program for network intrusion detection was built in Python using the TensorFlow library. The experiment was implemented on a 2.6 GHz Intel Core i7 PC with 16 GB of RAM and a NVIDIA GeForce GT 750 M GPU. The final configuration of the DBN was selected through experimental processes, and it has 2 hidden layers, each with 256 nodes (Fig. 5). The input vector of the network is dimensioned equal to the numbers of features of the datasets described above.

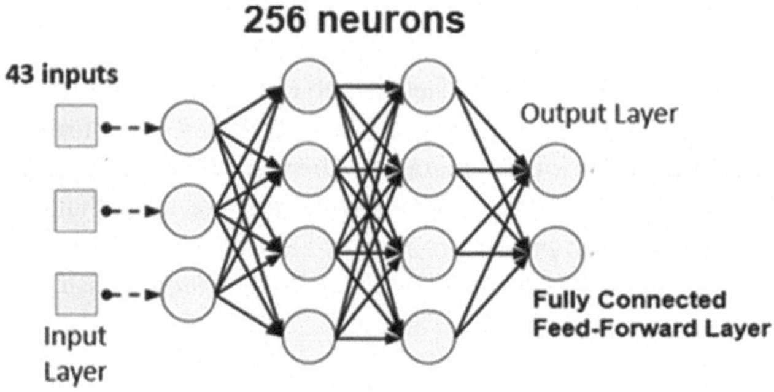

Fig. 5. DBN network

Table 2. Confusion Matrix - NSL-KDD

		DBN-prediction	
		Attacks	*Normal*
Real data	Attacks	7210	187
	Normal	162	9548

Unsupervised training is done by using greedy algorithm of author [10], and the refined training phase using the back propagation algorithm. For the NSL-KDD datasets in addition to the described port scan scanners we used unlabeled data "normal" for training (67343 instances) and test (9710 instances). With NSL-KDD, we tested two classifiers: the first is the binary classifier to determine whether the network attack detection is active. During training and testing, we labelled all attacks data as "Attacks". The second classifier is trained to classify four types with 8 main forms of network attacks in NSL-KDD. For the UNSW-NB15 dataset, because the scanning types are labelled together, we only apply a binary classification to determine whether the data is an attack or not. We also used the "normal" data to train and test the model. The results of the classifiers are often evaluated through the Confusion Matrix and some of the measurements are based on Confusion Matrix. Table 3 describes how to calculate the confusion matrix for a binary classifier.

Table 3. Confusion matrix of binary classifiers

		Predicted class	
		Class 1	*Class 0*
Real data	Class 1	f_{11}	f_{10}
	Class 0	f_{01}	f_{00}

The following measures will be used to evaluate the performance of intrusion detection models [10, 14].

$$Accuracy = \frac{f_{11} + f_{00}}{f_{11} + f_{10} + f_{01} + f_{00}} \tag{8}$$

$$Error\_rate = \frac{f_{10} + f_{01}}{f_{11} + f_{10} + f_{01} + f_{00}} \tag{9}$$

$$TPR = \frac{f_{11}}{f_{11} + f_{10}} \tag{10}$$

$$FAR = \frac{f_{01}}{f_{01} + f_{00}} \tag{11}$$

Table 2 is a Confusion Matrix of network layer detection detectors using DBN for the NSL-KDD data. Table 4 compares the True Positive Rate (TPR) and False Accept Rate (FAR) of the proposed method with the results of several other methods presented in the articles [3, 8] and [12]. The results showed that the DBN detection rate was higher than other methods, while the false alarm rate was low. Table 5 compares Accuracy (acc), Precision (π) and Recall (ρ) (Fig. 6) of DBN and other machine learning algorithms [14]. The results show that the DBN archives a more accurate classification of algorithms that have been applied previously. For the UNSW-NB15 datasets, as there are no experimental intrusion attack with this dataset, we just compare the results of the DBN with two other powerful algorithms for data classification: SVM and Random Forest. Table 6 shows the results, which indicates that DBN classification results are better than compared algorithms.

TPR: is used to measure the percentage of actual positives which are correctly identified.

FAR: is stated as the ratio of the number of false acceptances divided by the number of identification attempts

Precision: is defined as the percentage of True Positive samples among those classified as Positive (TP + FP).

Recall: is defined as the ratio of True Positive to the true number of samples (TP + FN).

Table 4. Comparison result of TPR and FAR in NSL-KDD

Method	TPR	FAR
SVM	0.93134	0.00012
AOCD	0.97889	0.00613
ANN-MLP	0.82567	0.02910
DBN	0.99241	0.02714

Table 5. Comparison results of 8 types of network attack detection

Method	(%)	Type of attack							
		Portsweep	Satan	*Neptune*	*Smurf*	*Imap*	*Warezmaster*	*Buffer_overflow*	Loadmodule
Nave Bayes	acc	93.49							
	π	78.86	90.31	92.17	89.12	93.53	93.91	96.67	92.33
	ρ	92.11	93.54	98.09	93.78	95.86	91.73	91.88	96.11
k-NN	acc	91.71							
	π	97.99	90.31	92.17	90.03	91.41	90.16	93.75	90.03
	ρ	91.61	90.68	92.30	91.78	95.86	91.73	91.88	89.39
SVM	acc	97.12							
	π	98.13	90.31	92.17	89.12	91.49	100	96.67	95.46
	ρ	97.12	98.16	96.13	93.78	95.86	91.73	91.88	92.88
DT	acc	96.45							
	π	89.79	94.02	98.05	100	98.15	94.17	97.44	96.01
	ρ	96.01	97.77	99.98	93.78	95.86	100	93.66	95.82
Random forest	acc	89.86							
	π	75.86	90.31	92.17	95.48	93.57	83.91	86.67	100
	ρ	88.45	93.49	98.09	90.78	85.86	91.73	91.88	97.55
MLP	acc	98.79							
	π	88.86	90.31	100	99.12	97.09	93.98	97.67	98.55
	ρ	98.78	93.86	98.11	93.78	95.55	93.39	94.06	95.37
DBN	acc	99.16							
	π	98.68	94.13	96.56	99.43	99.84	100	99.44	99.17
	ρ	100	99.53	98.07	98.52	99.86	98.35	95.19	97.46

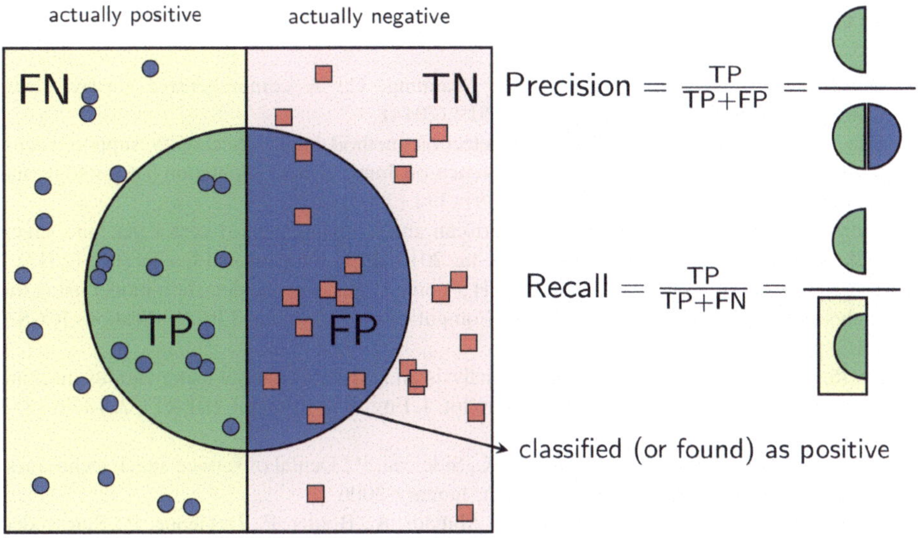

Fig. 6. Precision (π) and Recall (ρ)

Table 6. Comparison of results of TPR and FAR in UNSW-NB15

Algorithms	TPR	FAR
SVM	0.9874	3.32
Random forest	0.9913	3.45
DBN	0.9964	2.74

5 Conclusion

With the development of data as well as computing power, machine learning and deep learning have been used in many areas such as bioinformatics [15], finance [16], healthcare [17] and business [18–20] to help professionals save time and effort. In this paper, we present a new approach for intrusion detection based on Deep Belief Network and named is Workflow-based Collaborative Learning with a combination of supervised and unsupervised machine learning model for detecting network attacks. According to the convergence and accuracy of DBN model, we determine whether to make a reverse adjustment to the parameters of the user data pre-processing module and DBN training module or not, finally, improving the performance of the DBN model. Experiments with the NSL-KDD dataset and the UNSW-NB15 datasets showed that the DBN algorithm had high detection rates for network attack detection, while ensuring a lower false alarm rate than the study results before. In the future, we will be experimenting with more and more data networks to test the algorithm's ability to work in real-time intrusion detection systems.

References

1. Bou-Harb, E., Debbabi, M., Assi, C., Scanning, C.: A comprehensive survey. IEEE Commun. Surv. Tutorials **16**(3), 1496–1519 (2014)
2. Gao, M., Tian, J., Xia, M.: Intrusion detection method based on classify support vector machine. In: Second International Conference on Intelligent Computation Technology and Automation, ICICTA 2009, vol. 2, pp. 391–394 (2009)
3. Vidhya, M.: Efficient classification of portscan attacks using support vector machine, green high performance computing (ICGHPC). In: 2013 IEEE International Conference (2013)
4. Lee, S.Y, Kim, Y.S., Lee, B.H., Kang, S.H., Youn, C.H.: A probe detection model using the analysis of the fuzzy cognitive maps. In: Computational Science and Its Applications ICCSA 2005, pp. 287–291 (2005)
5. Dr. Revathi, S., Malathi, A.: A detailed analysis on NSLKDD dataset using various machine learning techniques for intrusion detection. Int. J. Eng. Res. Technol. (IJERT), **2**, 1848–1853 (2013)
6. Elleithy, K.M., Blagovic, D., Cheng, W.K., Sideleau, P.: Denial of service attack techniques: analysis, implementation and comparison, January 2006
7. Fatahi, M., Shahsavari, M., Ahmadi, M., Ahmadi, A., Boulet, P., Devienne, P.: Rate-coded DBN: an online strategy for spike-based deep belief networks. Biological. Inspired Cogn. Archit. **24**, 59–69 (2018)
8. Bhuyan, M.H., Bhattacharyya, D.K., Kalita, J.K.: AOCD: an adaptive outlier based coordinated scan detection approach. IJ Netw. Secur. **14**(6), 339–351 (2012)
9. Hinton, G.E.: Learning multiple layers of representation, Trends Cogn. Sci. **11**, 428–434 (2007). https://doi.org/10.1016/j.tics.2007.09.004
10. Hinton, G.E., Osindero, S., Teh, Y.W.: A fast learning algorithm for deep belief nets. Neural Comput. **18**, 1527–1554 (2016)
11. Moustafa, N., Slay, J.: UNSW-NB15: a comprehensive data set for network intrusion detection systems (UNSW-NB15 network data set). In: Military Communications and Information Systems Conference (MilCIS). IEEE (2015)
12. Wei, P., Li, Y., Zhang, Z., Hu, T., Li, Z., Liu, D.: An optimization method for intrusion detection classification model based on deep belief network, IEEE Access, **7**, 87593–87605 (2019). https://doi.org/10.1109/access.2019.2925828
13. Dr. Revathi, S., Malathi. A.: A detailed analysis on NSLKDD dataset using various machine learning techniques for intrusion detection. Int. J. Eng. Res. Technol. (IJERT), **2**, 1848–1853 (2013)
14. Ambedkar, C., Kishore Babu, V.: Detection of probe attacks using machine learning techniques. Int. J. Res. Stud. Comput. Sci. Eng. (IJRSCSE) **2**(3), 25–29 (2015)
15. Verma A., Mehta, S.: A comparative study of ensemble learning methods for classification in bioinformatics. In: 2017 7th International Conference on Cloud Computing, Data Science & Engineering - Confluence, pp. 155–158 (2017). https://doi.org/10.1109/confluence.2017.7943141
16. Xie, G., Zhao, Y., Jiang, M., Zhang, N.: A novel ensemble learning approach for corporate financial distress forecasting in fashion and textiles supply chains. Math. Probl. Eng. **2013**, 1–9 (2013). https://doi.org/10.1155/2013/493931
17. Li, Y., Bai, C., Reddy, C.K.: A distributed ensemble approach for mining healthcare data under privacy constraints. Inf. Sci. (Ny) **330**, 245–259 (2016). https://doi.org/10.1016/j.ins.2015.10.011
18. Tran, T.K., Phan, T.T.: A hybrid approach for building a Vietnamese sentiment dictionary. J. Intell. Fuzzy Syst. **35**(1), 967–978 (2018). https://doi.org/10.3233/JIFS-172053

19. Tran, T.K., Phan, T.T.: Deep learning application to ensemble learning—the simple, but effective, approach to sentiment classifying. Appl. Sci. **9**(13), 2760 (2019). https://doi.org/10.3390/app9132760
20. Tran, T.K., Phan, T.T.: Mining opinion targets and opinion words from online reviews. Int. J. Inf. Technol. **9**(3), 239–249 (2017)

A K-Means Grasshopper Optimisation Algorithm Applied to the Set Covering Problem

Gabriel Villavicencio[1(✉)], Matias Valenzuela[1], Francisco Altimiras[2], Paola Moraga[1], and Hernan Pinto[1]

[1] Escuela de ingeniería en construcción, Pontificia Universidad Católica de Valparaíso, Valparaíso, Chile
{gabriel.villavicencio,matias.valenzuela,paola.moraga, hernan.pinto}@puv.cl

[2] Facultad de Ingeniería y Negocios, Universidad de las Américas, Santiago, Chile
fjaltimiras@gmail.com

Abstract. Many of the problems addressed at the industrial level are of a combinatorial type and a sub-assembly not less than these are of the NP-hard type. The design of algorithms that solve combinatorial problems based on the continuous metaheuristic of swarm intelligence is an area of interest at an industrial level. In this article, we explore a general binarization mechanism of continuous metaheuristics based on the k-means technique. In particular, we apply the k-means technique to the Grasshopper optimization algorithm in order to solve the set covering problem (SCP). The experiments are designed with the aim of demonstrating the usefulness of the k-means technique in binarization. Additionally, we verify the effectiveness of our algorithm through reference instances. The results indicate the K-means binary grasshopper optimization algorithm (KBGOA) obtains adequate results when evaluated with a combinatorial problem such as the SCP.

1 Introduction

The combinatorial problems have great relevance at industrial level, we find them in various areas such as Civil Engineering [1], Bio Informatics [2], Operational Research[3,4], resource allocation [5–7], scheduling problems[8–10], routing problems [11–14] among others. On the other hand, in recent years, algorithms inspired by nature phenomena to solve optimization problems have been generated. As examples of these algorithms we have Cuckoo Search [15], Black Hole [16], Bat Algorithm [17], and Grasshopper Optimisation Algorithm [18] among others. Many of these algorithms work naturally in continuous spaces and therefore must be adapted to solve combinatorial problems. In the process of adaptation, the mechanisms of exploration and exploitation of the algorithm can be altered, having consequences in the efficiency of the algorithm.

Several binarization techniques have been developed to address this situation. In a literature search, the main binarization methods used correspond to transfer

R. Silhavy (Ed.): CSOC 2020, AISC 1225, pp. 312–323, 2020.
https://doi.org/10.1007/978-3-030-51971-1_25

functions, angle modulation and quantum approach. In this article we present a binarization method that uses the k-means technique to group the solutions and then perform the binarization process. To verify the efficiency of our method, we used the KBGOA. This algorithm was proposed in [18] and was applied to test functions in addition to structural engineering problems.

Grasshoppers are insects and under certain conditions can be considered a pest, the main characteristic of these insects is related to their movement. When it is in a larval state, its movement is slow in contrast with the large and abrupt movements of the adults.

When we look at different areas of engineering and science, we see that combinatorial problems appear frequently. We find them in various areas such as Civil Engineering [1], Bio Informatics [2], Operational Research[3], resource allocation [5], scheduling problems[8], routing problems [19,20] among others. On the other hand, in recent years, algorithms inspired by natural phenomena have been able to solve NP-hard combinatorial problems of significant size. As an example of algorithms inspired by nature, we have Cuckoo Search [15], Black Hole [16], Bat Algorithm [17], and Grasshopper Optimisation Algorithm [18], among others. Many of these algorithms work naturally in continuous spaces and, therefore, must be adapted to solve combinatorial problems. In the process of adaptation, the mechanisms of exploration and exploitation of the algorithm can be altered, having consequences in the efficiency of the algorithm. Several binarization techniques have been developed to address this situation. In a bibliographic search, the main binarization methods used are: transfer functions, angle modulation, and quantum approach. In this article, we develop a binarization method that uses the k-means technique to cluster the solutions and then perform the binarization process. To verify the effectiveness of our method, we use KBGOA. This algorithm was proposed in [11] and applied to test functions in addition to structural engineering problems. Grasshoppers are insects and under certain conditions can be considered a pest, the main characteristic of these insects is related to their movement. When it is in a larval state, its movement is slow in contrast to the large and abrupt movements of adults.

The mathematical model to simulate the movement of grasshoppers are presented in Eq. 1

$$X_i^d = c\Big(\sum_{j=1, j\neq i}^{N} c\frac{ub_d - lb_d}{2} s(|x_j^d - x_i^d|)\frac{x_j - x_i}{d_{ij}}\Big) + \hat{T}_d \tag{1}$$

The first c reduces the movements of grasshoppers around the target. In other words, this parameter balances exploration and exploitation of the entire swarm around the target. The second c decreases the attraction zone, comfort zone, and repulsion zone between grasshoppers. The c coefficient is calculated as: $cmax - l\frac{cmax-cmin}{L}$, where L= 500 is de maximum number of iteration. The cmax, and cmin coefficients are 1 and 0.00001 respectively. The component $s(r) = fe^{\frac{-r}{l}} - e^{-r}$ indicates if a grasshopper should be repelled from (explore) or attracted to (exploitation) the target. For this case $f = 0.5$ and $l = 1.5$. Finally $\hat{T}_d$ represents the best solution at this iterations.

To check our k-means binary grasshopper algorithm (KBGOA), we use the well-known set covering problem. Experiments were developed using a random operator to validate the contribution of the k-means technique in the binarization process of the Grasshopper algorithm. In addition, to verify our results, the JPSO algorithm developed by [21] and MDBBH developed in [8] were chosen. The results show that KBGOA algorithm obtain competitive results.

2 Set Covering Problem

The set covering problem corresponds to a classical problem in combinatorics, and complexity theory. The problem aims to find subsets of a set. Given a set and its elements, we want to find subsets that completely cover the set at a minimum cost. The SCP is one of the oldest and most studied optimization problems. It is well-known to be NP-hard [22].

SCP is an active problem because medium and large instances often become intractable and cannot be solved any more using exact algorithms. Additionally, SCP due to its large number of instances, is used to verify the behavior of proposed new algorithms. In recent years, SCP has been approached by various continuous Swarm intelligence metaheuristics and using different methods to binarize metaheuristics. In [23], they used the teaching-learning-based optimisation with a specific binarization scheme to solve medium and large size instances. An SCP in fault diagnosis application was developed by [24]. In this article, the Gravitational Search algorithm was used and transfer functions were applied to perform the binarization. A rail scheduling problem application using SCP was developed in [8]. In [21] a Jumping PSO was used to solve SCP and in [5] CS metaheuristics were binarized using a k-means algorithm.

SCP has many practical applications in engineer, e.g., vehicle routing, facility location, railway, and airline crew scheduling [3,25–27] problems.

The SCP can be formally defined as follows. Let $A = (a_{ij})$, be a $n \times m$ zero-one matrix, where a column j cover a row i if $a_{ij} = 1$, besides a column j is associated with a non-negative real cost c_j. Let $I = \{1, ..., n\}$ and $J = \{1, ...m\}$, be the row and column set of A, respectively. The SCP consists in searching a minimum cost subset $S \subset J$ for which every row $i \in I$ is covered by at least one column $j \in J$, i.e,:

$$\text{Minimize } f(x) = \sum_{j=1}^{m} c_j x_j \tag{2}$$

$$\text{Subject to } \sum_{j=1}^{m} a_{ij} x_j \geq 1, \forall i \in I, \text{ and } x_j \in \{0,1\}, \ \forall j \in J \tag{3}$$

where $x_j = 1$ if $j \in S$, $x_j = 0$ otherwise.

3 Binary GrassHoppers Algorithm

The application of statistics and machine learning techniques, are used in many areas such as transports, smart cities, agriculture and computational intelligence, examples of applications are found in [28–32]. In this article, we explore the k-means technique applied to binarization of continuous metaheuristics. As an algorithm to binarize, Grasshopper is used and as problem to solve, SCP. The Proposed KBGOA has two modules. The first module corresponds to the initialization of the feasible solutions, and is detailed in the Sect. 3.1. Once the initialization of the particles is performed, it is consulted if the maximum if iterations is satisfied. Subsequently if the criterion is not satisfied, The KBGOA is then run in conjunction with the k-means technique, this is detailed in Sect. 3.2). Once the transitions of the different solutions are made, we compare the resulting solutions with the best solution previously obtained. In the event that a superior solution is found, this replaces the previous one. The general algorithm scheme is detailed in Fig. 1.

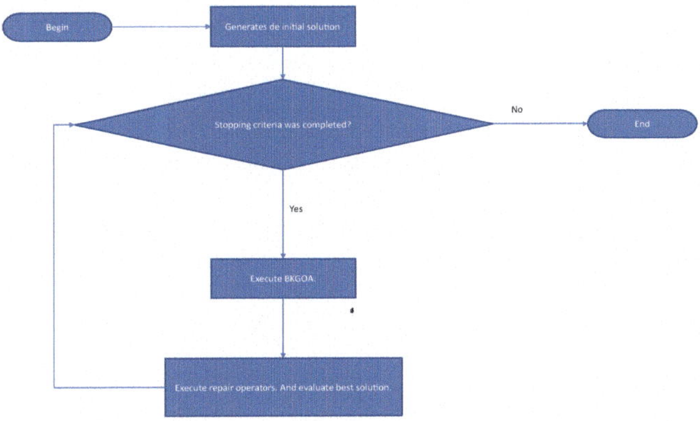

Fig. 1. Flowchart of the KBGOA algorithm.

3.1 Initialization

For initialization of a new solution, a column is randomly chosen. It is then queried whether the current solution covers all rows. In the case of the solution does not meet the coverage condition, the heuristic operator is called (Sect. 3.3). This operator aims to select a new column. This heuristic operation is iterated until all rows are covered. Once the coverage condition is met, the solution is optimized. The optimization consists of eliminating columns where all their rows are covered by more than one column. The detail of the procedure is shown in Algorithm 1.

Algorithm 1. Initialization Operator

Function Initialization()
Input
Output Initialized solution S_{out}
$S \leftarrow$ SRandomColumn()
while All row are not covered **do**
    S.append(Heuristic(S))
end while
$S \leftarrow$ dRepeatedItem(S)
$S_{out} \leftarrow S$
return S_{out}

3.2 K-Means Transition Ranking Operator

Consider our metaheuristic is a continuous and population based. Due to its iterative nature, it needs to update the position of particles at each iteration. When the metaheuristic is continuous, this update is performed in $\mathbb{R}^n$ space. In Eq. 4, the update position is presented in a general manner. The x_{t+1} variable represents the x position of the particle at time $t+1$. This position is obtained from the position x at time t plus a Δ function calculated at time $t+1$. The function Δ is proper to each metaheuristic and produces values in $\mathbb{R}^n$. For example in Cuckoo Search $\Delta(x) = \alpha \oplus Levy(\lambda)(x)$, in Black Hole $\Delta(x) =$ rand $\times$ $(x_{bh}(t) - x(t))$ and in the Firefly, Bat and PSO algorithms Δ can be written in simplified form as $\Delta(x) = v(x)$.

$$x_{t+1} = x_t + \Delta_{t+1}(x(t)) \tag{4}$$

In the k-means transition ranking operator, a model for transitions in a discrete space is proposed. This model is based on considering the movements generated by the metaheuristic in each dimension for all particles. $\Delta^i(x)$ corresponds to the magnitude of the displacement $\Delta(x)$ in the i-th position. Subsequently these displacement are grouped using the magnitude of the displacement $\Delta^i(x)$. This grouping is done using the k-means technique where k represents the number of clusters used. Finally, a generic P_{tr} function given by the Eq. 5 is proposed to assign a transition probability.

$$P_{tr} : \mathbb{Z}/k\mathbb{Z} \rightarrow [0,1] \tag{5}$$

A transition probability through the function P_{tr} is assigned to each group. Naturally, this P_{tr} function is modelled as a cumulative probability function. For the case of this study, we particularly use the linear function given in Eq. 6. In this equation, $\mathrm{N}(x^i)$ indicates the location of the group to which $\Delta^i(x)$ belongs. The α coefficient, corresponds to the transition probability and β to the transition separation coefficient. Both parameters are estimated in each metaheuristic. For our particular case, $\mathrm{N}(x^i) = 0$ corresponds to elements belonging to the group that has the lowest Δ^i values. N(x^i) = 7 corresponds to the group of elements that have the greatest Δ^i values.

$$P_{tr}(x^i) = P_{tr}(N(x^i)) = \alpha + \beta N(x^i)\alpha \tag{6}$$

The k-means transition ranking operator starts calculating Δ^i for each of the particles. This step is specific in each metaheuristic. Subsequently the particles are grouped using k-means clustering technique and the magnitude of Δ^i as distance. With the group assigned to each particle we obtain the probability of transition using Eq. 6. Afterwards the transition of each particle is performed. In the case of grasshopper algorithm the rule 7 is used to perform the transition, where $\hat{x}^i$ is the complement of x^i. Finally, each solution is repaired using the repair operator.

$$x^i(t+1) := \begin{cases} \hat{x}^i(t), \text{ if } rand < P_{tg}(x^i) \\ x^i(t), \quad \text{otherwise} \end{cases} \tag{7}$$

3.3 Heuristic Operator

The goal of the Heuristic operator is to select a new column for cases where a solution needs to be built or repaired. As input variables, the heuristic operator considers the incomplete solution S_{in} which must be completed. The operator obtains the columns that belong to S_in, then obtains the rows R that are not covered by the solution to S_in. With the set of rows not covered and using Eq. 8 we obtain in line 4 the best 5 rows to be covered. With this list of rows ($lRows$) on line 5 we obtain the list of the best columns according to the heuristic indicated in Eq. 9. Finally randomly in line 6 we obtain the column to incorporate.

$$WeightRow(i) = \frac{1}{L_i}. \tag{8}$$

Where L_i is the sum of all ones in row i.

$$WeightColumn(j) = \frac{c_j}{|R \cap M_j|}. \tag{9}$$

Where M_j is the set of rows covered by Col j.

Algorithm 2. Heuristic operator

1: **Function** Heuristic(S_{in})
2: **Input** Input solution S_{in}
3: **Output** The new column *colOut*
4: $lRows \leftarrow$ getBestRows(S_{in}, N=10)
5: $listcolumsnOut \leftarrow$ getBestCols($LRows$, M=5)
6: $colOut \leftarrow$ getCol($lcolsOut$)
7: **return** *colOut*

4 Results

In this section we detail the behavior of KBGOA when it is applied to SCP. The contribution of the binary k-means operator was studied when solving the different SCP instances. Additionally, KBGOA was compared with other algorithms that have recently resolved SCP. To solve the different SCP instances, a PC with Windows 10, core i5 processor and 16 GB in RAM was used. The program was coded in Python 3.5. For the statistical analysis, the non-parametric Wilcoxon signed-rank test was used in addition to violin charts. The analysis of violin charts is performed by comparing the dispersion, median and the interquartile range of the distributions.

4.1 Insight of KBGOA Algorithm

In this module we developed experiments that allowed us to study the contribution of the binary k-means operator with respect to the quality of the solutions obtained and the iterations used when solving SCP instances. To perform this evaluation, we used the instances of Balas and Carrera. In the comparison, a random operator was designed which replaces our binary k-means operator. This random operator instead of executing transitions assigning transition probabilities per group, uses a fixed value of 0.5. To compare the distributions of the results of the different experiments we use violin Chart. The horizontal axis X corresponds to the problems, while Y axis uses the measure % - Gap defined in Eq. 10

$$\% - Gap = 100\frac{BestKnown - SolutionValue}{BestKnown} \tag{10}$$

Furthermore, a non-parametric test, Wilcoxon signed-rank test is carried out to determine if the results of KBGOA with respect to the random algorithm have significant difference or not. The parameter settings and browser ranges are shown in Table 1.

Table 1. Setting of parameters for k-means Binary GrassHopper Search Algorithm.

Parameters	Description	Value	Range
N	Number of solutions	25	[20, 25, 30]
G	Number of clusters	5	[4, 5, 6]
Iteration Number	Maximum iterations	500	[400, 500, 600]
cmax	Maximum for c coefficient	1	1
cmin	Minimum for c coefficient	0.00001	0.00001
f	f parameter for s function	0.5	0.5
l	l parameter for s function	1.5	1.5

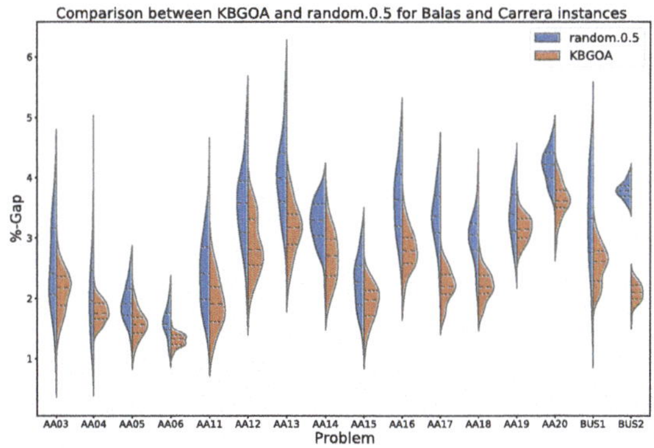

Fig. 2. Evaluation of k-means binary operator.

Table 2. Balas and Carrera instances.

Instance	Row	Col	Density	Best Known	KBGOA (avg)	Std	(Secs)	*rnd.0.5* (avg)	Std
AA03	106	8661	4.05%	33155	33356.2	125.1	208	33693.5	241.1
AA04	106	8002	4.05%	34573	34988	58.5	199	34931	231.2
AA05	105	7435	4.05%	31623	31721.9	65	201	31889.2	103.2
AA06	105	6951	4.11%	37464	37676.6	56.4	201	37651.3	97.1
AA11	271	4413	2.53%	35384	35800.6	132.2	188	35901.2	197.1
AA12	272	4208	2.52%	30809	31429.5	153.2	210	31631.3	210.2
AA13	265	4025	2.60%	33211	34030.4	136	191	34201.4	231.4
AA14	266	3868	2.50%	33219	33764.5	125.7	144	33967.3	121.1
AA15	267	3701	2.58%	34409	35012.6	112.9	142	34846.8	161.3
AA16	265	3558	2.63%	32752	33203.4	138.3	189	33601.3	183.9
AA17	264	3425	2.61%	31612	32000.9	89.4	163	32386.2	149.3
AA18	271	3314	2.55%	36782	37178.6	78.4	133	37417.6	148.1
AA19	263	3202	2.63%	32317	33200.5	78.3	146	33109.1	136.1
AA20	269	3095	2.58%	34912	35849.6	91.5	142	35963.1	97.8
BUS1	454	2241	1.88%	27947	28402.6	86.7	190	28756.1	231.6
BUS2	681	9524	0.51%	67760	68831.8	83.1	176	68678.9	101.2
Average				35495.56	36027.9	100.7	176.4	36164.1	165.1
p-value								1.1e7−04	

KBGOA corresponds to our standard algorithm. *random.0.5* is the random variant. The results are shown in Table 2 and Fig. 2.

When we compared the Best Values between KBGOA and *rnd.0.5* which are shown in Table 2. In the Average comparison, KBGOA outperforms *rnd.0.5* in all problems. The comparison of distributions is shown in Fig. 2. We see the dispersion of the *rnd.0.5* distributions are bigger than the dispersions of KBGOA. Therefore, the k-means binary operator, contribute to the precision of the results. Finally, the KBGOA distributions are closer to zero than *rnd.0.5* distributions, indicating that KBGOA has consistently better results than *random.0.5*.

Table 3. OR-Library benchmarks.

Instance	row	col	Density	Best Known	JPSO Avg	MDBBH Best	Avg	KBGOA Best	Avg	Time(s)
E.1	500	5000	10%	29	29	29	29	29	29.2	8.8
E.2	500	5000	10%	30	30	31	31.6	30	30.9	12.6
E.3	500	5000	10%	27	27	27	27.4	27	27.5	17.2
E.4	500	5000	10%	28	28	28	29.1	28	28.3	11.4
E.5	500	5000	10%	28	28	28	28	28	28.3	14.4
F.1	500	5000	20%	14	14	14	14.1	14	14	17.9
F.2	500	5000	20%	15	15	15	15.3	15	15.1	16.2
F.3	500	5000	20%	14	14	14	14.8	14	15.2	17.8
F.4	500	5000	20%	14	14	14	14.9	14	14.4	18.4
F.5	500	5000	20%	13	13	14	14.1	14	14.5	17.1
G.1	1000	10000	2%	176	176	177	178.5	176	178.1	173.9
G.2	1000	10000	2%	154	155	157	160.6	155	156	172.4
G.3	1000	10000	2%	166	167.2	168	170.4	167	170.1	178.4
G.4	1000	10000	2%	168	168.2	169	170.9	169	172.3	176.5
G.5	1000	10000	2%	168	168	168	169.8	168	170.4	168.3
H.1	1000	10000	5%	63	64	64	64.9	64	65.2	152.9
H.2	1000	10000	5%	63	63	64	64	64	64.7	156.1
H.3	1000	10000	5%	59	59.2	59	60	60	60.9	143.7
H.4	1000	10000	5%	58	58.3	59	60.4	59	59.7	165
H.5	1000	10000	5%	55	55	55	56.4	55	56.1	139
Average				67.1	67.30	67.7	68.71	67.5	68.54	88.9

When we evaluate the behaviour of the algorithms through the Wilcoxon test, this test indicates that there is a significant difference between the two algorithms (Table 3).

4.2 KBGOA Compared with JPSO and MDBBH

In this section, we detail the comparisons made with the objective of evaluating the performance of our KBGOA algorithm. For the evaluation, the larger problems of the OR-library were chosen. To develop the comparison, two algorithms were selected. The first one corresponds to a discretization Particle Swarm Optimization (PSO) technique called Jumping PSO (JPSO), [21]. The second one, is a binarization of the Black Hole technique called Multi Dynamic Binary Black Hole (MDBBH) Algorithm, [8]. JPSO uses a discrete PSO based on the frog jump. JPSO works without the concept of velocity, replacing this one by a component of random jump which allows to perform the movement in the discrete search space. On the other hand, MDBBH uses a binarization mechanism specific to the Black Hole algorithm. This is based on the concept of closeness to the Black Hole (BH). BH corresponds to the solution that has obtained the best value. When a solution is close to BH, the transitions are less likely to be performed. In the case of being away from the BH, the probability of transition is greater.

For instances E and F which are of medium size, the results of the MDBBH and KBGOA algorithms are similar when comparing their Best Value. Only in instance E.2 KBGOA was superior to MDBBH. When analyzing the average,

JPSO obtained better results, however, the values of KBGOA were close, being MDBBH the one that obtained the worse performance. For the case of problems G and H, KBGOA was superior in problems G.1, G.2 and G.3 and MDBBH in problem H.3 when comparing their Best Value. In the case of averages, JPSO was superior in all cases. However, KBGOA obtained results quite close, leaving MDBBH behind. In this point we must emphasize that the k-means technique used in KBGOA allows binarizing any continuous swarm intelligence algorithm unlike JPSO that is specific for PSO.

5 Conclusions

In this work, the KBGOA algorithm based on the k-means technique was proposed to perform binarization of GOA metaheuristics. We must emphasize that this k-means technique can be applied in the binarization of any continuous metaheuristics of the swarm intelligence type. To evaluate the performance of Our k-means technique the set covering problem was used together with the Grasshopper algorithm. Experiments were designed to evaluate the contribution of the binary operator k-means of the algorithm. It was found that the operator contributes significantly to improve the accuracy and quality of the solutions, all these visualized through violin graphics. Finally, compared to the best metaheuristic algorithm that has solved SCP, our algorithm had a lower yield of 1.26%, which is not a big difference considering that JPSO uses a particular adaptation mechanism for PSO and the k-means technique It can be easily adapted to any technique. As future works, we believe that the autonomous search for efficient configuration is a little-explored area and it could help to improve the performance of the binarizations both in the quality of the solutions and in the times of convergence. Additionally, it is interesting to see how this k-means technique works with other metaheuristics and other NP-hard problems. Finally, we believe that using machine learning we can generate indicators or obtain correlations that allow us to understand how the k-means algorithm translates the exploration and exploitation properties from the continuous space to the binary space.

References

1. Khatibinia, M., Yazdani, H.: Accelerated multi-gravitational search algorithm for size optimization of truss structures. Swarm Evol. Comput. **38**, 109–119 (2017)
2. Barman, S., Kwon, Y.-K.: A novel mutual information-based boolean network inference method from time-series gene expression data. PloS One **12**(2), e0171097 (2017)
3. Crawford, B., Soto, R., Monfroy, E., Astorga, G., García, J., Cortes, E.: A meta-optimization approach for covering problems in facility location. In: Workshop on Engineering Applications, pp. 565–578. Springer (2017)
4. García, J., Crawford, B., Soto, R., Astorga, G.: A percentile transition ranking algorithm applied to binarization of continuous swarm intelligence metaheuristics. In: International Conference on Soft Computing and Data Mining

5. Garcia, J., Crawford, B., Soto, R., Astorga, G.: A percentile transition ranking algorithm applied to knapsack problem. In: Proceedings of the Computational Methods in Systems and Software, pp. 126–138. Springer (2017)
6. García, J., Altimiras, F., Peña, A., Astorga, G., Peredo, O.: A binary cuckoo search big data algorithm applied to large-scale crew scheduling problems. Complexity **2018** (2018)
7. García, J., Lalla-Ruiz, E., Voß, S., Droguett, E.L.: Enhancing a machine learning binarization framework by perturbation operators: analysis on the multidimensional knapsack problem. Int. J. Mach. Learn. Cybern. 1–20 (2020)
8. García, J., Crawford, B., Soto, R., García, P.: A multi dynamic binary black hole algorithm applied to set covering problem. In: International Conference on Harmony Search Algorithm, pp. 42–51. Springer (2017)
9. García, J., Crawford, B., Soto, R., Astorga, G.: A clustering algorithm applied to the binarization of swarm intelligence continuous metaheuristics. Swarm Evol. Comput. **44**, 646–664 (2019)
10. Crawford, B., Soto, R., Astorga, G., García, J.: Constructive metaheuristics for the set covering problem. In: International Conference on Bioinspired Methods and Their Applications, pp. 88–99. Springer (2018)
11. García, J., Peña, A.: Robust optimization: concepts and applications. In: Nature-Inspired Methods for Stochastic, Robust and Dynamic Optimization, p. 7 (2018)
12. Astorga, G., Crawford, B., Soto, R., Monfroy, E., García, J., Cortes, E.: A meta-optimization approach to solve the set covering problem. Ingeniería **23**(3), 274–288 (2018)
13. García, J., Moraga, P., Valenzuela, M., Crawford, B., Soto, R., Pinto, H., Peña, A., Altimiras, F., Astorga, G.: A DB-scan binarization algorithm applied to matrix covering problems. Comput. Intell. Neurosci. **2019** (2019)
14. Garcia, J., Măntoiu, M.: Localization results for zero order pseudodifferential operators. J. Pseudo Diff. Oper. Appl. **5**(2), 255–276 (2014)
15. Yang, X.-S., Deb, S.: Cuckoo search via lévy flights. In: World Congress on Nature & Biologically Inspired Computing. NaBIC 2009, pp. 210–214. IEEE (2009)
16. Hatamlou, A.: Black hole: a new heuristic optimization approach for data clustering. Inf. Sci. **222**, 175–184 (2013)
17. Yang, X.-S.: A new metaheuristic bat-inspired algorithm. In: Nature inspired cooperative strategies for optimization (NICSO 2010), pp. 65–74 (2010)
18. Saremi, S., Mirjalili, S., Lewis, A.: Grasshopper optimisation algorithm: theory and application. Adv. Eng. Softw. **105**, 30–47 (2017)
19. Franceschetti, A., Demir, E., Honhon, D., Van Woensel, T., Laporte, G., Stobbe, M.: A metaheuristic for the time-dependent pollution-routing problem. Eur. J. Oper. Res. **259**(3), 972–991 (2017)
20. Crawford, B., Soto, R., Astorga, G., García, J., Castro, C., Paredes, F.: Putting continuous metaheuristics to work in binary search spaces. Complexity **2017** (2017)
21. Balaji, S., Revathi, N.: A new approach for solving set covering problem using jumping particle swarm optimization method. Nat. Comput. **15**(3), 503–517 (2016)
22. Gary, M.R., Johnson, D.S.: Computers and intractability. A Guide to the Theory of NP-Completeness. Freeman, San Francisco (1979)
23. Lu, Y., Vasko, F.J.: An or practitioner's solution approach for the set covering problem. Int. J. Appl. Metaheuristic Comput. (IJAMC) **6**(4), 1–13 (2015)
24. Li, Y., Cai, Z.: Gravity-based heuristic for set covering problems and its application in fault diagnosis. J. Syst. Eng. Electr. **23**(3), 391–398 (2012)
25. Kasirzadeh, A., Saddoune, M., Soumis, F.: Airline crew scheduling: models, algorithms, and data sets. Euro. J. Trans. Logist. **6**(2), 111–137 (2017)

26. Horváth, M.: Computing strong lower and upper bounds for the integrated multipledepot vehicle and crew scheduling problem with branch-and-price. CEJOR **27**, 39–67 (2017)
27. Stojković, M.: The operational flight and multi-crew scheduling problem. Yugoslav J. Oper. Res. **15**(1), 25–48 (2016)
28. García, J., Crawford, B., Soto, R., Carlos, C., Paredes, F.: A k-means binarization framework applied to multidimensional knapsack problem. Appl. Intell. **48**, 357–380 (2017)
29. García, J., Pope, C., Altimiras, F.: A distributed k-means segmentation algorithm applied to Lobesia botrana recognition. Complexity **2017** (2017)
30. Graells-Garrido, E., García, J.: Visual exploration of urban dynamics using mobile data. In: International Conference on Ubiquitous Computing and Ambient Intelligence, pp. 480–491. Springer (2015)
31. Graells-Garrido, E., Peredo, O., García, J.: Sensing urban patterns with antenna mappings: the case of Santiago, Chile. Sensors **16**(7), 1098 (2016)
32. Peredo, O.F., García, J.A., Stuven, R., Ortiz, J.M.: Urban dynamic estimation using mobile phone logs and locally varying anisotropy. In: Geostatistics Valencia 2016, pp. 949–964. Springer (2017)

Management of Behavior of a Swarm of Robots Applicable to the Tasks of Monitoring a Some Territory

Gennady E. Veselov, Boris K. Lebedev, and Oleg B. Lebedev(✉)

Southern Federal University, Rostov-on-Don, Russia
gev@sfedu.ru, lebedev.b.k@gmail.com,
lebedev.ob@mail.ru

Abstract. The work is devoted to the design of bio-inspired methods and algorithms for controlling the collective behavior of robots based on adaptive behavior models of multi-agent biological systems of the region: detection of harmful substances; search and rescue operations in areas of natural and technological disasters, as well as combat zones, patrolling the borders of a certain area. The fundamental idea of swarm control is "swarm intelligence". The control of a homogeneous swarm of robots is based on the principle of force relaxation. The search for a solution is carried out in an affine space whose elements are n-dimensional points (positions). Each robot r_i calculates the objective function $f(X_i(t))$ - the value of the desired substance at the point $X_i(t)$. The fundamental problem that was solved in this paper is related to the development of the structure of an affine position space that allows one to display and search for interpretations of solutions with integer parameter values. Coordinate values must be discrete and independent of each other. In contrast to the canonical particle swarm method, in order to reduce the weight of affine bonds, by moving the p_i particle to a new position in the affine solution space, a directed mutation operator has been developed, the essence of which is to change the integer values of the genes in the chromosome. New chromosome structures have been developed for representing solutions. The temporal complexity of the algorithm lies in the range $O(n^2)$–$O(n^3)$, where n is the number of robots.

Keywords: Swarm of robots · Swarm of particles · Intelligent agents · Search space · Behavior of a group of robots · Cyberorganism

1 Introduction

In recent years, a new area of robotics, swarm robotics, has been developing intensively. The general idea of this direction can be formulated as follows: some tasks can be better solved not by one large and complex robot, but by a large number of small and simple robots that can act in concert. A swarm of robots can be seen as a cyber organism. In other words, we can say that a developed swarm of robots is a collective artificial intelligence [1–3].

R. Silhavy (Ed.): CSOC 2020, AISC 1225, pp. 324–332, 2020.
https://doi.org/10.1007/978-3-030-51971-1_26

Swarm robotics will allow in the future to create swarms of robots that can collectively solve a large number of problems, physically or informationally uniting on the basis of the principle of self-organization. One of the key problems associated with the design of swarm robotic systems is the problem of developing mathematical models of these devices [4, 5]. Designing control algorithms for sophisticated modern robots requires the development of mathematical models of robots in general and their subsystems, solving optimization problems and optimal control and other complex computational problems.

Swarm management also solves the problem of collecting environmental data. Due to the low carrying capacity of unmanned aerial vehicles (UAVs), they do not have enough energy to install powerful sensors with an acceptable detection area. This leads to the fact that each device is able to receive information only from a small radius around it. However, due to the use of the swarm model, the information collected by each device is transferred to its neighbors during the exchange process, supplementing their data on air flows, obstacles and other important environmental parameters [6, 7].

A number of studies on the tasks of constructing algorithms for the behavior of a group of robots are based on the use of bio-inspired approaches or approaches based on the use of analogies of the behavior of living creatures, in particular ant colonies, bees and other flock animals [8]. In this direction, models of biological objects (a kind of bionic paradigm) are taken as a basis. There are many examples of flocking behavior of insects and animals [9, 10]. It is believed that obtaining adequate models of biological systems will make it possible to naturally implement their flocking (swarm) behavior.

Swarm management is a new approach that explores the possibilities of building a system from a set of autonomous intelligent agents (robots) to achieve collective goals that cannot be achieved by a separate robot or for which collective execution of the task is more effective. The fundamental idea of swarm control is "Swarm Intelligence" [11–20], widely observed in the natural world: schools of fish, colonies of bees, ants, and others. The term "swarm intelligence" was introduced in 1989 in the context of the study of the system of cellular "robots". Note that not all population algorithms can be used to control a swarm of robots, and in some cases their modification is required.

2 Statement of the Problem

The search space D (the studied area) is divided into cells d_i (sections) of equal sizes $D = \{d_i | i = 1,2,\ldots,n_d\}$. Cells are the same size. The size of the cell is determined by the ability of the agent to process it. The number of cells is significantly greater than the number of agents (robots).

Let there be a linear vector space (LVP) whose elements are n-dimensional points (positions). For any two points p and q of this space, we uniquely associate a unique ordered pair of these points, which in what follows will be called a geometric vector (vector). $p,q \in V(p,q)$ is a geometric vector (ordered pair). The set of all points of a linear vector space, supplemented by geometric vectors, is called point-vector or affine space. An affine space is n-dimensional if the corresponding linear vector space is also n-dimensional.

An affine-relaxation model (ARM) of a particle swarm is a vertex graph which corresponds to the positions of a particle swarm, and arcs correspond to affine bonds between positions (points) in an affine space. Affinity is a measure of the proximity of two agents (particles). At each iteration, each agent p_k goes into a new state (position) in the affine space, in which the weight of the affinity bond between agent p_i and the base (best) agent $p*$ decreases. The transition of agent p_i to a new position $X_i(t + 1)$ from $X_i(t)$ is carried out using the relaxation procedure.

In the first step, agents are distributed around the periphery of the search space. Figure 1 shows the search space. Cells processed by agents are shaded in black. The aim of the swarm of robots is to study some territory in order to detect a site with an extreme value of a given substance.

Suppose that a two-dimensional column vector of variable parameters X form the x, y coordinates of the center of the cell (point) on the surface of, for example, the Earth in the Cartesian coordinate system $0xy$, so that $X = (x,y)^T$, where T is the transpose symbol; the objective function $f\ (X)$ is the value of the sought substance at point X. Note that the possible coordinates of the location of robots in the search space coincide with the coordinates of the cells over which the robots are located, are integer. Above one cell is only one robot.

The control of a homogeneous swarm of robots is based on the principle of force relaxation.

Denote by $R = \{r_i | i = 1,2,\ldots,s\}$ the swarm of robots under consideration, where s is the number of robots in the swarm; $X_i(t) = x_i(t),\ y_i(t))$ – the current position of the robot r_i. Let the initial position of the swarm robots be determined by the set of vectors $X_i\ (0)$, $i \in [1{:}s]$. Each robot r_i calculates the objective function $f(X_i(t))$ – the value of the desired substance at the point $X_i(t)$. At each iteration, the following actions are performed. We introduce the notation $W\ (t)$ – the set of coordinates of cells (points) over which robots are located at time t. The point $X_i^*(t) \in W(t)$ with the best value of the function $f\ (X_i(t))$, which is considered as an attractor, is determined.

In accordance with the force relaxation algorithm, the position of the robot r_i at the next time $t + 1$ is determined by the formula:

$$X_i(t+1) = X_i(t) + V_i(t). \quad (1)$$

The vector $V_i(t)$ has the meaning of the movement that the robot r_i makes when moving from the current position $X_i(t)$ to the position $X_i(t + 1)$. Often this vector is called the speed of the robot r_i at time t.

In the simplest case, the robot moves to a neighboring cell towards the best cell selected as an attractor.

$$\begin{aligned} &V_i(t) = (\delta x_i(t), \delta y_i(t)), \\ &\delta x_i(t) = x_i^*(t) - x_i(t); \delta y_i(t)) = y_i^*(t) - y_i(t). \\ &x_i(t+1) = (x_i(t) + sgn(\delta x_i(t)); \ y_i(t+1) = y_i(t) + sgn(\delta y_i(t)). \end{aligned} \quad (2)$$

Example.

Let $x_i(t) = 3, y_i(t) = 8; x_i^*(t) = 6, y_i^*(t) = 4.$
$\delta x_i(t) = 6 - 3 = 3, \text{sgn}\,(\delta\, x_i(t)) = +1;\ \delta\, y_i(t)) = 4 - 8 = -4; \text{sgn}\,(\delta\, y_i(t)) = -1$
$x_i(t+1) = 3 + 1 = 4; y_i(t+1) = 8 - 1 = 7.$

Figure 1 shows the search space D of size 25 × 6. Cells with the original location of the robots are shaded in solid black. The cell with the best value of the objective function is marked with a sign ◎. After the first iteration, robots move into cells shaded in a dark color (Fig. 1).

Figure 2 shows the search space D with the initial location of robots after the first iteration and a new location after the second iteration. The attractor cell with the best value of the objective function is marked with a sign ◎. Figure 3 shows the search space D with the initial location of robots after the second iteration. The cell with the best value of the objective function is marked with a sign ◎.The best cell found in the search process is marked with a sign ◇ (Fig. 3).

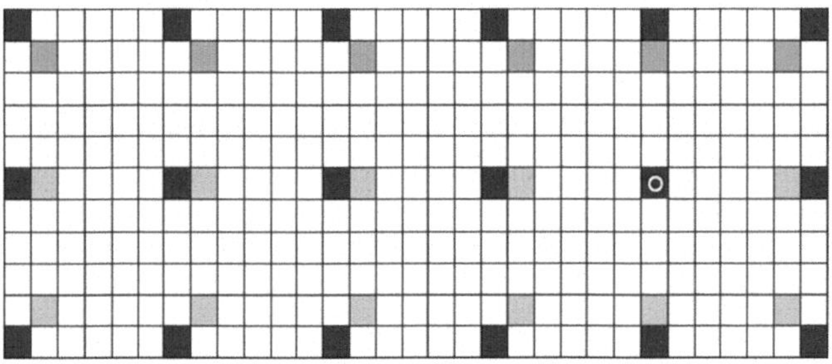

Fig. 1. Moving robots in the first iteration

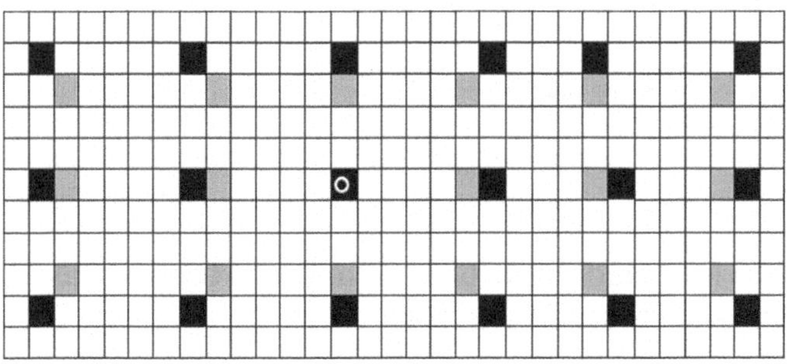

Fig. 2. Moving robots in the second iteration

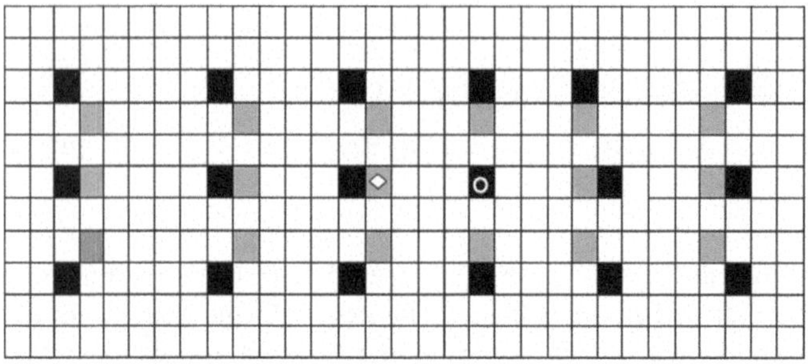

Fig. 3. Moving robots in the third iteration

By choosing an alternative $A(P, <a,b>, a \leq b)$ we mean a probabilistic event having, as possible outcomes, an integer c such that $a \leq c \leq b$, and the probability of occurrence c is defined as $P = (1/|a - b + 1|)$. In other words, with probability $P = (1/|a - b + 1|)$ the probability event $A(P, <a,b>, a \leq b) = c$.

$$\begin{aligned} &\text{Then, if } a_x = x_i^*(t),\ b_x = x_i(t), \text{ then } \delta x_i(t) = c_x = A_x(P_x, <a_x, b_x>, a_x \leq b_x); \\ &\text{if } a_y = y_i^*(t),\ b_y = y_i(t), \text{ then } \delta y_i(t) = c_y = A_y(P_y, <a_y, b_y>, a_y \leq b_y). \end{aligned} \tag{3}$$

If the number of outcomes $K = |a - b + 1|$ large, then it decreases by decreasing the parameter b. The value $K^* < K$ is set, hence the new value $b^* = (b + (K - K^*))$. Reducing the parameter K is done to limit the maximum value of the robot movement.

Example.

$$\text{Let } x_i(t) = 3,\ y_i(t) = 8; x_i^*(t) = 6,\ y_i^*(t) = 4.$$
$$a_x = 3,\ b_x = 6.\ P_x = (1/|a_x - b_x + 1|) = 1/4.\ a_y = 4,\ b_y = 8.$$
$$P_y = (1/|a_y - b_y + 1|) = 1/5.$$

Let the random outcomes of the events Ax and Ay be integers:

$$c_x = \delta x_i(t) = 4; \qquad c_y = \delta y_i(t)) = 5.$$

Then $x_i(t + 1) = 3 + 4 = 7$; $y_i(t + 1) = 8 + 5 = 13$.

3 The Behavior Algorithm of a Group of Robots Based on the Particle Swarm Method

In the third approach, the organization of the movement of the robot is carried out by analogy with the particle swarm method. Particle Swarm Optimization (PSO) is a method of stochastic optimization in some ways similar to evolutionary algorithms. This method does not model evolution, but flock behavior of animals [11–20].

The canonical paradigm of a swarm of particles involves the use of real parameter values in multidimensional, real, metric spaces. In contrast to the canonical particle swarm method, in our case the speed $V_i(t)$ and the coordinates of the point $X_i(t+1)$ cannot be represented as an analytical expression with real values of the variables. If a two-dimensional vector $V_i(t)$is used as a position in the Cartesian coordinate system $0xy$, then the number of parameters determining the position of the particle in the solution space (position) should be equal to two. The value of each coordinate is plotted on the corresponding axis of the decision space. In this case, some requirements arise for the chromosome structure and coordinate values. Coordinate values must be discrete and independent of each other. The directional mutation operator (DMO) acts as an analog of the speed $V_i(t)$ the essence of which is to change the integer values of the coordinates. Moving the particle p_i to a new position means a transition from $V_i(t)$ to a new one – $V_i(t+1)$ with new integer values of the coordinate genes obtained after applying the DMO.

The displacement vector $V_i(t)$, which the robot r_i makes when moving from the current position $X_i(t)$ to the position $X_i(t+1)$, has the form.

$$V_i(t) = V_i^I(t-1) + V_i^C(t-1) + V_i^S(t-1). \tag{4}$$

Let $X_i{**}(t)$ be the robot coordinate vector r_i, corresponding to its best value of the function $f_i(X)$ during the search time [0:t].

$V_i^I(t-1)$ – calculated using expressions (2). Note that $V_i^I(0) = 0$.
$V_i^S(t-1)$ – the amount of displacement from the point $V_i^S(t\text{-}1)$ in the direction of the point $X_i^*(t)$, calculated using expressions (3).
$V_i^C(t-1)$ – the amount of movement from the point $X_i(t)$ in the direction of the point $X_i{**}(t)$, calculated using expressions (3).
$V_i^I(t-1)$ – called the inertial component that performs the function of the robot's memory about its movement on the previous search cycle.
$V_i^S(t-1)$ – is called a social component that reflects the desire of the robot r_i to move towards the point $X_j{*}(t) \in W(t)$ of the robot r_j with the best value of the function $f(X_j(t))$.
$V_i^C(t-1)$ – called the cognitive component (by social analogy), because it formalizes the desire of the robot r_i to return from its current position $X_i(t)$ to position $X_i{**}(t)$, in which it had the best value in previous cycles functions $f(X_i(t))$ This component implements the robot's memory function about its best position during the search.

The algorithm for controlling a swarm of robots has the form.

1. $t = 1$.
2. Formation of the set $W(t)$ of points over which robots are located.
3. Calculation of quality functions $f(X_i(t))$ for each point $X_i(t) \in W(t)$.
4. For each robot r_i, the value $f(X_i(t))$ is entered into the array F_i.
5. The choice of the point $X_i{*}(t) \in W(t)$ with the best value of the quality function $f(X_i(t))$ among the swarm robots at time t.

6. For each robot r_i in the array F_i, a point $X_i{*}(t) \in F_i$ is found with the best value of the quality function f (Xi (t)) from the $t + 1$: it starts.
7. For each robot r_i, the parameters are calculated: $V_i^I(t-1)$; $V_i^C(t-1)$; $V_i^S(t-1)$.
8. For each robot r_i, its position at the next time $t + 1$: is determined:

$$X_i(t+1) = X_i(t) + V_i(t).$$

9. Robots are moving to new positions.
10. If $t < T$, then $t = t + 1$ and go to step 1, otherwise go to step 11.
11. The end of the algorithm.

4 Experimental Studies

To conduct experimental studies, a program for controlling the behavior of a group of robots was developed by emulation methods to detect a section with an extreme value of a given substance – ***GR***. C++ was used in Microsoft Visual Studio 2010 for Windows.

Based on the results of experimental studies, an average dependence of the degree of quality on the number of iterations was constructed (Fig. 4).

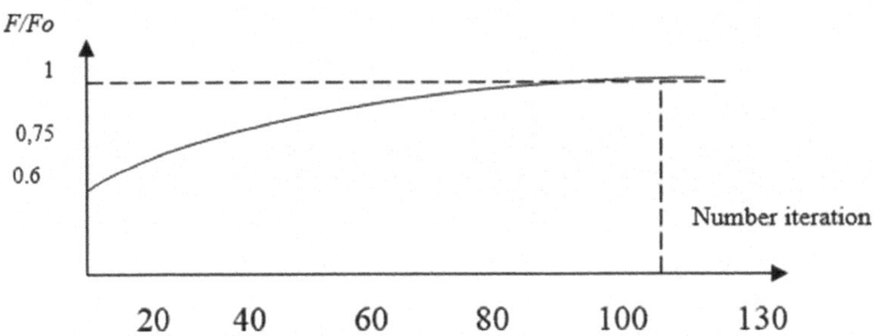

Fig. 4. Dependence of the degree of quality of the SR algorithm on the number of iterations

Each example was tested on 50 independent runs. When studying the convergence of algorithms for each experiment, the generation number was remembered, after which no improvement in the estimate was observed. In each series of 50 trials, the average value of the number of generations was determined, after which no improvement in the estimate was observed. For each test series, the best solution was determined. Based on the processing of experimental studies for algorithms, an average dependence of the quality of solutions on the number of iterations was constructed.

Studies have shown that the number of iterations at which the algorithm found the best solution lies in the range of 120–135. The algorithm converges on average to 130 iterations. On average, launching a program provides solutions that differ from the optimal solution by less than 2%. The time complexity of the algorithm for fixed values of M – the size of the swarm population and T – the number of iterations lies within $O(n)$. A general estimate of the time complexity lies within $O(n^2)$–$O(n^3)$.

5 Conclusion

The work is devoted to the development of algorithms for controlling the behavior of a group of robots. The work of a swarm of robots is to study a certain territory in order to detect a site with an extreme value of a given substance.

The control of a homogeneous swarm of robots is based on the principle of force relaxation. The search for a solution is carried out in an affine space whose elements are n-dimensional points (positions). The fundamental idea of swarm control is "swarm intelligence". The *PSO* method works with one static population whose members are gradually improving with the advent of information about the search space. This method is a type of directed mutation. *PSO* solutions mutate towards the best solutions found. Particles never die (because there is no selection).

The key problem that was solved in this paper is related to the development of the structure of an affine position space that allows us to display and search for interpretations of solutions with integer parameter values. In contrast to the canonical particle swarm method, in order to reduce the weight of affine bonds, by moving the particle p_i to a new position in the affine solution space, the directed mutation operator is developed, the essence of which is to change the integer values of the position coordinates. The experiments showed that the probability of obtaining a global optimum was 0.9. A general assessment of the time complexity for any approach to the organization of movement lies within the limits $O(n^2)$–$O(n^3)$.The swarm management method considered allows us to solve the problem of collecting environmental data. The solution to this problem involves the presence of control over the movement of agents in space. Robots must: maintain a distance between themselves and other objects, including other individuals of the swarm, as well as adhere to a given construction, and adjust the course according to the movement of neighboring devices and the swarm as a whole. Many of these problems are not addressed in this paper and are the goal of further research by the authors.

Acknowledgements. This research is supported by grants of the Russian Foundation for Basic Research of the Russian Federation, the project № 19-07-00645.

References

1. Karpenko, A.P.: Robotics and computer-aided design systems: Textbook. Publishing House MSTU them N.E. Bauman, Moscow, 71 p. (2014)
2. Syryamkina, V.I.: Teams of intelligent robots. Scopes of application. In: Syryamkina, V.I. (ed.) (Series: Intelligent Technical Systems) (sub-series: Cognitive Robotics). STT, Tomsk, 140 p. (2018)
3. Bondarchuk, A.S., Borovik, V.S., Gutsul, V.I.: Intelligent Robotic Systems: Textbook. Syryamkina, V.I. (ed.) (Series: Intelligent Technical Systems). Tomsk, 256 p. (2017)
4. Kalyaev, I.A., Gaiduk, A.R., Kapustyan, S.G. Models and Algorithms of Collective Control in Groups of Robots. Fizmatlit, Moscow, 280 p. (2009)
5. Karpov, V.E.: The collective behavior of robots. Desired and actual, modern mechatronics. In: Sat scientific Proceedings of the All-Russian Scientific School,132 p. (2011)

6. Ivanov, D.Ya.: Swarm intelligence methods for controlling groups of small unmanned aerial vehicles. In: Izvestiya SFU. Technical Science, Taganrog, vol. 3, no. 116, pp. 221–229 (2011)
7. Kalyaev, I.A., Gaiduk, A.R.: Flocking management principles in a group of objects. Mechatron. Autom. Manage. **12**, 27–38 (2004)
8. Karpenko, A.P.: Modern Search Engine Optimization Algorithms. In: Algorithms inspired by nature: a tutorial. Publishing House MSTU them N.E. Bauman, Moscow, 448 p. (2014)
9. Lebedev, B.K., Lebedev, O.B., Lebedeva, E.O.: Swarm algorithm for planning the operation of multiprocessor computing systems. Electron. Sci. J. Eng. Bull. Don **3** (2017)
10. Lebedev, B.K., Lebedev, O.B., Lebedeva, E.M.: Resource allocation based on hybrid swarm intelligence models. Sci. Tech. J. Inf. Tech. Mech. Opt. **17**(6), 1063–1073 (2017)
11. Clerc, M.: Particle Swarm Optimization. ISTE, London, UK, 187 p. (2006)
12. Kennedy, J., Eberhart, R.C.: Particle swarm optimization. In: Proceedings of IEEE International Conference on Neural Networks, pp. 1942–1948 (1995)
13. Cong, J., Romesis, M., Xie, M.: Optimality, scalability and stability study of partitioning and placement algorithms. In: Proceedings of the International Symposium on Physical Design, Monterey, CA, pp. 88–94 (2003)
14. Lebedev, B.K., Lebedev, O.B.: Hybrid bioinspired algorithm based on the integration of the branch and bound method and the ant colony method. Bull. Rostov State Transp. Univ. **2** (70), 77–88 (2018)
15. Vorobeva, E.Yu., Karpenko, A.P., Seliverstov, E.Yu.: Co-hybridization of particle swarm algorithms. Sci. Educ. Electr. J. **4** (2012)
16. Agasiev, T.A., Karpenko, A.P.: Modern technology of global optimization. Inf. Tech. **6**, 370–386 (2018)
17. Wang, X.: Hybrid nature-inspired computation method for optimization. Doctoral dissertation. Helsinki University of Technology, TKK Dissertations, Espoo, 161 p. (2009)
18. Blum, C., Roli, A.: Metaheuristics in combinatorial optimization: overview and conceptual comparison. ACM Comput. Surv. **35**, 268–308 (2003)
19. Lebedev, V.B.: Construction of the shortest connecting networks on the basis of swarm intelligence. Izvestiya SFU. Publishing house TTI SFedU, vol. 7, pp. 37–44 (2011)
20. Sha, D.Y., Hsu, C.-Y.: A hybrid particle swarm optimization for job shop scheduling problem. Comput. Ind. Eng. **51**, 791–808 (2006)
21. OR-Library is collection of test data for a variety of OR problem. http://mscmga.ms.ic.ac.uk

Smart Technologies for Smart Tourism Development

Tomáš Gajdošík(✉) and Andrea Orelová

Faculty of Economics, Matej Bel University,
Tajovského 10, 975 90 Banská Bystrica, Slovakia
{tomas.gajdosik, andrea.oreova}@umb.sk

Abstract. Development of technologies have also influenced tourism sector introducing the concept of Tourism 4.0 and thus emphasising the smart approach. Besides the fact that smart tourism technologies are recognised to have a positive impact on tourists' experience, effectiveness and competitiveness of businesses and destinations, their profound evaluation from tourists' and industry point of view is still lacking. The aim of the article is to analyse the level of expected impact and period of maximum impact of smart tourism technologies in Central Europe. The findings reveal that technologies providing information have reached the maturity, while the maximum impact is being soon awaited from technologies applied in data management, followed by technologies enhancing the experience. The true development of smart tourism in Central Europe can be expected within 3 to 4 years, as sensing technologies powered by Internet of Things will reach their plateau of productivity.

Keywords: Information technology · Smart tourism · Tourist experience · Tourism 4.0

1 Introduction

In recent years, development of technologies leads to replacement of work force with mechanisation and automation, which is being recognised as the 4th industrial revolution (Industry 4.0). Originally initiated in Germany, Industry 4.0, has attracted much attention as it is closely related with technologies such as the Internet of Things (IoT), Cyber Physical System (CPS), Enterprise Architecture (EA) Cloud-based Manufacturing, and Enterprise Integration (EI) [1]. Industry 4.0 is characterized by complexity, digitalisation, and by merging biological, physical and digital realms. From technological point of view, it is influenced mainly by artificial intelligence, robotics, Internet of Things, biotechnologies, nanotechnologies or 3D printing [2]. Its goal is to achieve higher level of productivity, automation and efficiency by digitalisation, customised production and human-machine interaction.

As tourism sector has always been at the forefront of technology [3, 4], similar pathways can be recognised, and the concept of "Tourism 4.0" is nowadays being introduced (Fig. 1). The advent of Industry 4.0 with new technologies have a profound impact on tourism, qualifying the current period as the "era of Tourism 4.0", where

R. Silhavy (Ed.): CSOC 2020, AISC 1225, pp. 333–343, 2020.
https://doi.org/10.1007/978-3-030-51971-1_27

technologies are modifying the behaviour of tourists, businesses and destinations, projecting them towards a smart perspective [5].

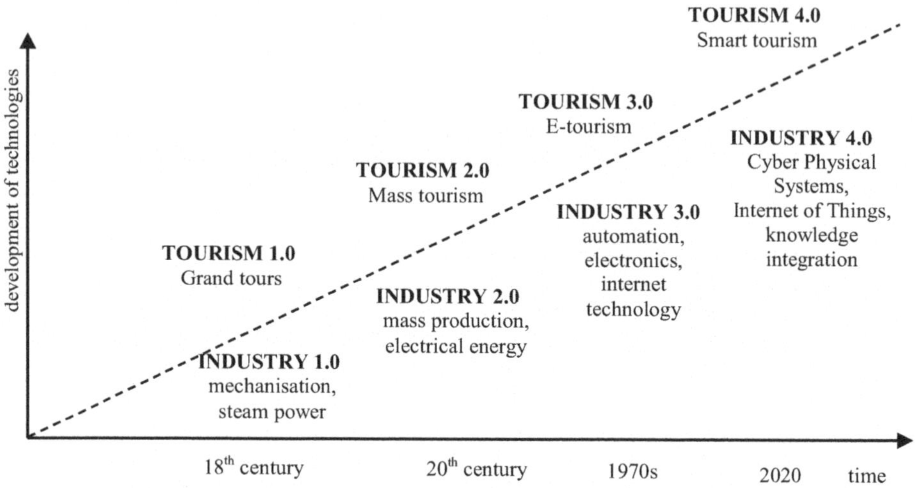

Fig. 1. Development of Industry 4.0 and Tourism 4.0

The smart perspective is based on the application of technologies that integrate hardware, software and network technologies to provide real-time awareness of the real world and advanced analytics that optimise business processes and performances [6]. Smart tourism development describes the current stage of tourism influenced by the evolution of technologies in which the physical and governance dimensions of tourism are entering the digital playing field and new levels of intelligence are achieved [7]. The concept of smartness in tourism is often discussed from two aspects: development of technologies and sustainability [8].

2 Smart Tourism Technologies

New digital technologies, or smart technologies, have introduced important innovations in various sectors and industries. The term "smart technologies" does not indicate that the technology itself is smart, rather individuals and industry can become smarter when using it, as it enables to quickly correspond, sensitively analyse and predict occurring situations [9]. It is a term describing a product, condition, or motion of a technology that entails a variety of functionalities that can be adapted to certain functions or is tailored to specific circumstances [10]. For example smart technologies for governments transform governance and organizational issues, and address economic, social, and environmental challenges [11]. Smart technologies for healthcare enables real-time monitoring of health without the need of visiting hospital [12] and smart technologies for smart cities contribute to higher energy efficiency, value co-creation and quality of residents' life [13]. Smart technologies are also applied in tourism sector as they are tightly knitted into the fabric of the travel experience and management of tourism product [14].

Smart tourism technologies (STTs) refer to general and specific applications that are able to enhance tourists' experiences, as well as generate added value and increase competitiveness [15]. They offer significant opportunities for management of businesses and destinations to enhance their competitiveness and offer personalised services for tourists to maximise their satisfaction. Smart tourism technologies enable to acquire, utilise and share tourism-related information while travelling [13]. They aggregate and harness data derived from physical infrastructure, social connections and other various sources and they are supported with the use of advanced technologies (e.g. machine learning, AI). Smart tourism technologies is an ambiguous "umbrella term" for many advanced technologies that take connectivity to a further step [16]. Moreover, other several terms can be used along with smart tourism technologies (Table 1).

Table 1. Example of smart tourism technologies

Expression	Example of technologies
Disruptive technologies [17]	Internet of Things, Internet of Everything, 5G mobile network, radio frequency identification, mobile devices, wearables smartphones and devices, 3D printing, apps along with APIs, cryptocurrency and Blockchain, sensor and beacon networks, pervasive computing, gamification, analytical capabilities supported by Artificial Intelligence
Smart tourism technologies [18]	Ubiquitous computing, Internet of Things, cloud computing, ubiquitous connectivity through Wi-Fi, near field communication, radio frequency identification, sensors, smartphones, mobile connected devices, beacons, virtual reality, augmented reality, mobile apps, integrated payment methods, smart cards, social network sites, etc.
Technological advances and tools [19]	Machine learning, artificial intelligence, industrial web, big data, Internet of Things, smart devices, robots, drones, beacons, virtual and augmented reality, near field communications, ubiquitous computing, etc.
Smart tourism technologies [20]	Destination apps, augmented and virtual reality, sensors, near field communication, QR codes, iBeacons, ubiquitous connectivity through Wi-Fi, latest generation websites, social networks, chatbots, etc.
Emerging technologies [21]	QR codes, ambient intelligence, Fi-Ware, wearables, 3D printing, robots, open data, deep learning, real-time databases, Wi-Fi, mobile connection payments, BLE, augmented reality, Internet of Things, virtual reality, recommendations systems, cryptocurrencies, etc.
New technology trends in digital tourism [22]	Wearables (watches, bracelets, glasses, lenses), augmented and virtual reality, 3D printing, artificial intelligence, e-agents, 3D holographic projection, accelerated mobile websites, cloud computing, big data analytics
Smart information and communication technologies [23]	Sensors, big data, open data, Internet of Things, radio Frequency identification, near field communication, recommender systems, context-aware systems, autonomous agents, augmented reality, mobile and wearable devices, etc.

In recent years, tourists have been making decision through information acquisition with the development of smart technologies and they use acquired information for taking on various experiences through entire tourism process [24]. Smart tourism technologies thus affect tourists in every part of their travel as they enhance tourist experience by giving all the related real-time information about travelling and services in the planning phase, enhance access to information to assist tourists in exploring the destination during the trip and prolong engagement to relieve experience by providing descent feedback after the trip [25].

In the hospitality industry, using smart technologies maximise the yield, enables to monitor competitors' performance, expand distribution channels, improve corporate image, setup best strategies [26] and improve management efficiency [27]. The impact of these technologies is therefore in four major areas: (1) strategic planning and revenue management, (2) operations, (3) marketing distribution and communication and (4) customer service and relationship management [26, 28]

By embedding smart technologies in destinations, destination marketers are striving to enrich tourist experience as well as to enhance destination competitiveness [18]. The competitive advantage of a destination is closely related to the quality of products offered, especially with smart technologies that give opportunities for tourists to participate in on-site co-creation of experience [29]. Thus the smart technologies can help destination to maintain and enhance its global position by influencing a plethora of travel-related activities and providing unique personalised experiences [30].

While the presented studies suggest the positive influence of smart tourism technologies on tourism development, the development of understanding about the degree and time of perceived impact of STTs on tourism development by tourists [31] hospitality industry [15] and destination management [18] is still lacking.

3 Material and Methods

The aim of the paper is to analyse the level of expected impact and period of maximum impact of smart tourism technologies in Central Europe.

The analysis is based on the primary data collected by a questionnaire survey among 499 tourists, 23 hotel managers and 6 destination managers in Slovakia. These responders were asked to give their opinion regarding the expected degree of impact of the selected technologies on tourism development (0 – no impact, 4 high impact) and anticipate the period of maximum impact in tourism (0 – today, 2 – short term /2–5 years/, 5 – medium term /5–10 years/, 10 – long term /10 years and more/.

In order to get a comprehensive understanding of smart technologies, the purposive sampling was used. Only those tourists that use technologies during the travel for experience enrichment, co-creation and sharing purposes were selected. Moreover, hotel managers were selected from those hotels that have a strategy to collect data on customers and external environment, and use information technology to meet the needs of guests and improve efficiency. Destination managers were selected from destinations using a destination management system, as this technological platform is a precondition for improving efficiency, facilitates co-creation of experiences and interconnects the destination stakeholders. Slovakia was chosen as a reference country, where the use of Internet services and information technologies are at the average of the European Union countries [32] and the country is recognised as a digital challenger [33].

4 Results

As the term "smart" has become a buzz word not only in the academic sphere, but also in the praxis, and the development of technologies is more and more powerful, it is important to identify those technologies that have the greatest impact on tourism development and to estimate when the highest impact will take place.

From the theoretical point of view, smart tourism technologies create value for tourists, enable tourism businesses and destinations to share knowledge, better understand tourists' needs, create personalised products, and thus to be more efficient and competitive. The results of the survey reveal that the maximum impact of the smart tourism technologies will occur in 2,33 years. Moreover, the highest expectations are from technologies with short-term period of maximum impact (Fig. 2).

Social media, public Wi-Fi, latest generation websites, as well as mobile apps are nowadays considered as technologies with the highest impact. These **technologies providing information** have already reached their maturity, with the ability to rejuvenate by enabling the use of location-based services, real-time conversation and AI-based applications. In particular, the emergence of user-generated content (UGC) by social media sites have enabled tourists and managers to communicate and interact with each other. Public Wi-Fi enables tourists to stay freely connected outside their usual environment and keep them up-to-date through Wi-Fi landing pages. Latest generation websites and mobile apps enhance the overall experience of tourists by providing all relevant personalised information and enabling co-creation.

Technologies related to data management and storage, such as destination management system, reputation management, cloud computing, Blockchain and open data are considered as technologies with medium expected impact within 2 or 3 years. DMSs facilitate the dissemination of data and knowledge among different stakeholders within a destination and thus improving coordination and integration of tourism product. Cloud computing enables data storage and access through the Internet and therefore exchange of information, which is crucial for smart tourism development. Moreover, Blockchain is one of the latest network-based technologies with major implications for smart tourism development mainly in five areas [34]: The use of cryptocurrencies by tourists and locals; smart contracts enabling businesses and destinations to automatically place orders, perform tasks and issue payments without the need of approvals and instructions; clearer, easier and faster processes in public services (e.g. blockchain passports); records and rights management (e.g. loyalty programmes) and tracking of products. Blockchain can bring also more security to

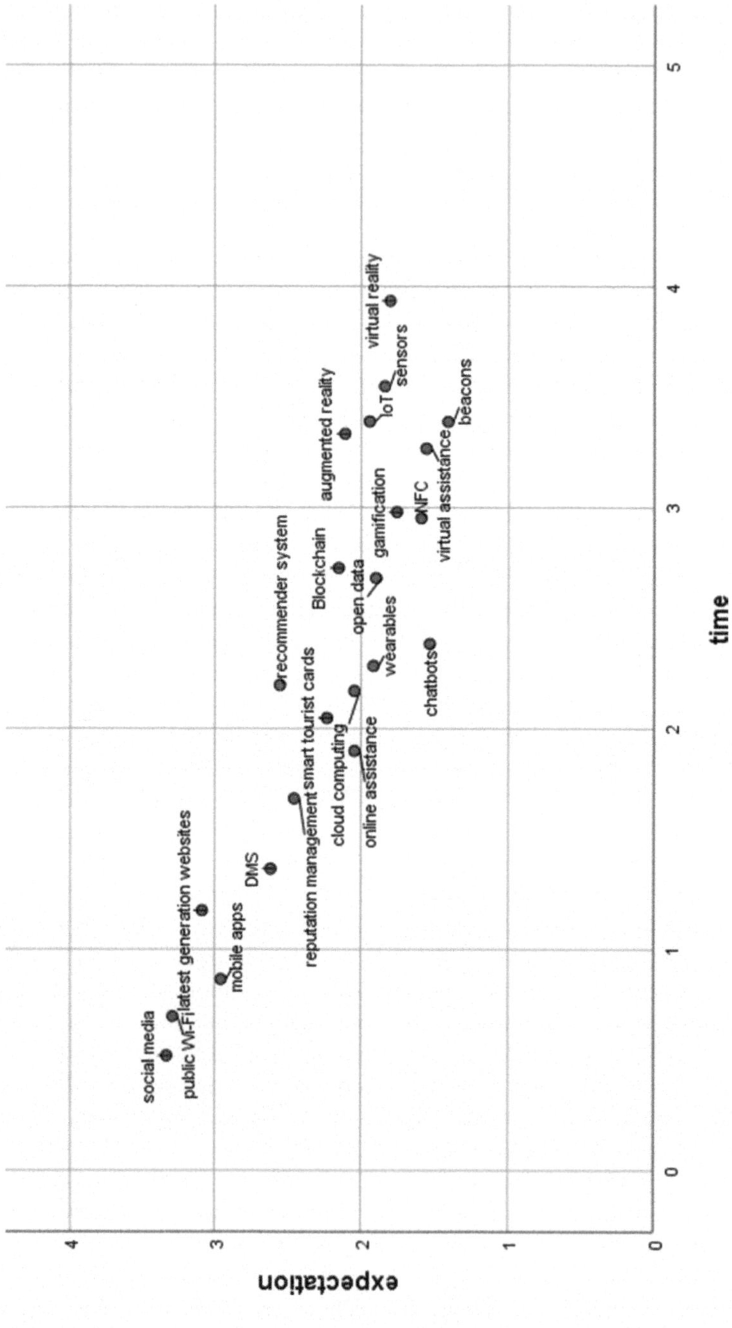

Fig. 2. Overall evaluation of smart tourism technologies

open data, which are used for planning and management. For smart tourism development, the real-time open data with standard APIs, instead of static (historic) open data, are useful, as they are used for developing mobile applications for tourists. These data management and storage technologies are prerequisites of big data analytics, that can analyse and interpret relevant data and to overcome the challenges of tourism development.

Technologies enhancing tourist experience have slightly lower expected impact and maturity within 2 to 4 years. Smart cards motivate tourists to consume services in destinations allowing authorisation access, ticketing or loyalty programmes. With the massive use of social media, sharing the experience in forms of status updates, comments, photos and videos is gaining more and more importance. Therefore there is a need for reputation management system. The tourist feedback can be analysed using text analytics, especially opinion mining and sentiment detection. Recommender systems allow to understand needs, budget, and travel preferences, and suggest the right offer at the right time for a right tourist. Moreover, with the help of AI, they provide automated, personalised and intelligent travel services. Online assistance (e.g. through WhatsApp, WeChat) enables co-creation, affects tourist experience and can be automated using chatbots. AI empowered chatbots can enhance tourist experience, starting from automated reminders before the trip, to suggesting nearby activities in a destination and sending feedback after the trip. Along with chatbots, virtual service agents care for tourist 24/7 and are based on machine learning algorithms, most of which fall under the supervised learning category. With data, skills and AI, the virtual assistant can synchronise with tourist's calendar, interact with smartphone, make recommendations or use any third-party applications. Wearables (e.g. smart watches, glasses, lenses) affect the interaction with the surroundings of a destination and enable augmented or virtual reality to enhance the experience. Augmented reality adds additional information to real environment and can be used to show objects and information that are not present now (e.g. reconstruction of historical sites). Virtual reality simulates 3D surroundings interfacing the real world with virtual environment helping tourists to try the intangible tourism product and assist them to make smarter decisions. The shift from products and services towards experiences while travelling can be also seen in terms of gamification. Through games and game elements, such as collecting points, achieving goals and different levels, tourists can find new horizons of experience in a tourism destination.

Sensing and network technologies will reach their plateau of productivity within 3 to 4 years. Near field communication (NFC) allows short-distance data exchange and can be used for contactless payments and authorisation. It can be embedded in a smartphone or smart card to facilitate the travel. Beacons, based on Bluetooth Low Energy, allow to send push notifications to tourists' smartphones and are used mainly indoors (e.g. hotels, museums) to assist guest with orientation and prepare tailor-made offers. Sensors embedded in a destination environment can monitor traffic, energy consumption, air quality or tourists' behaviour. The effect of these sensing technologies can be augmented by the use of Internet of Things, which allows collected data to be transferred to a data management system.

5 Conclusion

Smart technologies are changing the rules of the game of tourism development. Technology is a means to an end, not an end. Using smart technologies for tourism development leads to smarter decisions and is becoming more and more crucial, as tourism market is changing rapidly, because of more complex requirements of tourists and tougher competition among tourism businesses or destinations. If a business or a destination wants to be competitive and differentiate in terms of innovation and knowledge, it should embrace the smart approach. Moreover, if a tourist does not want to be confused in an unfamiliar environment and want to have a personalised experience, the use of smart technologies is inevitable.

Technologies providing information have nowadays reached the highest impact. These technologies are easy to use and are more cost-efficient to implement, however in order to stay up-to-date, they have to innovate and implement functions enabling personalisation, real-time interaction and co-creation. Data management technologies are becoming more important, allowing to store, interchange and analyse the large volumes of data. As tourist is becoming more powerful, soon (within 2 to 4 years) technologies will help him to be fully involved in co-creation of its own experiences. Therefore, the tourism industry has to prepare to be able to recognise tourist needs in order to co-crate the wonderful experience. Sensing technologies together with the Internet of Things will reach their plateau of productivity within 3 to 4 years and enabling the real smart tourism development in Central Europe.

In overall, the expected degree of impact of technologies is becoming lower with the predicted time to reach the maximum. Moreover, it should be stated that the opinions of tourists, hotel managers and destination managers differ in certain elements (Fig. 3). While tourists prefer mainly technologies providing information and technologies enhancing the experience, hotel managers rated better also data management technologies. In overall, destination managers have much higher expectation from technologies, as all technologies were rated with higher than medium impact. These findings reveal that the smart approach is being pushed mainly by destinations initiatives. The hospitality industry is lacking behind, as also tourists are more enthusiastic with the use of technologies. Therefore more attention should be paid to smart technologies applied in hospitality industry in the following years.

Smart tourism technologies are pushing forward the smart tourism concept, which is recognised as a further step in tourism development. By enhancing tourist experiences, securing effectiveness and competitiveness of businesses and destinations, smart tourism technologies can lead to overall sustainable competitiveness of tourism sector.

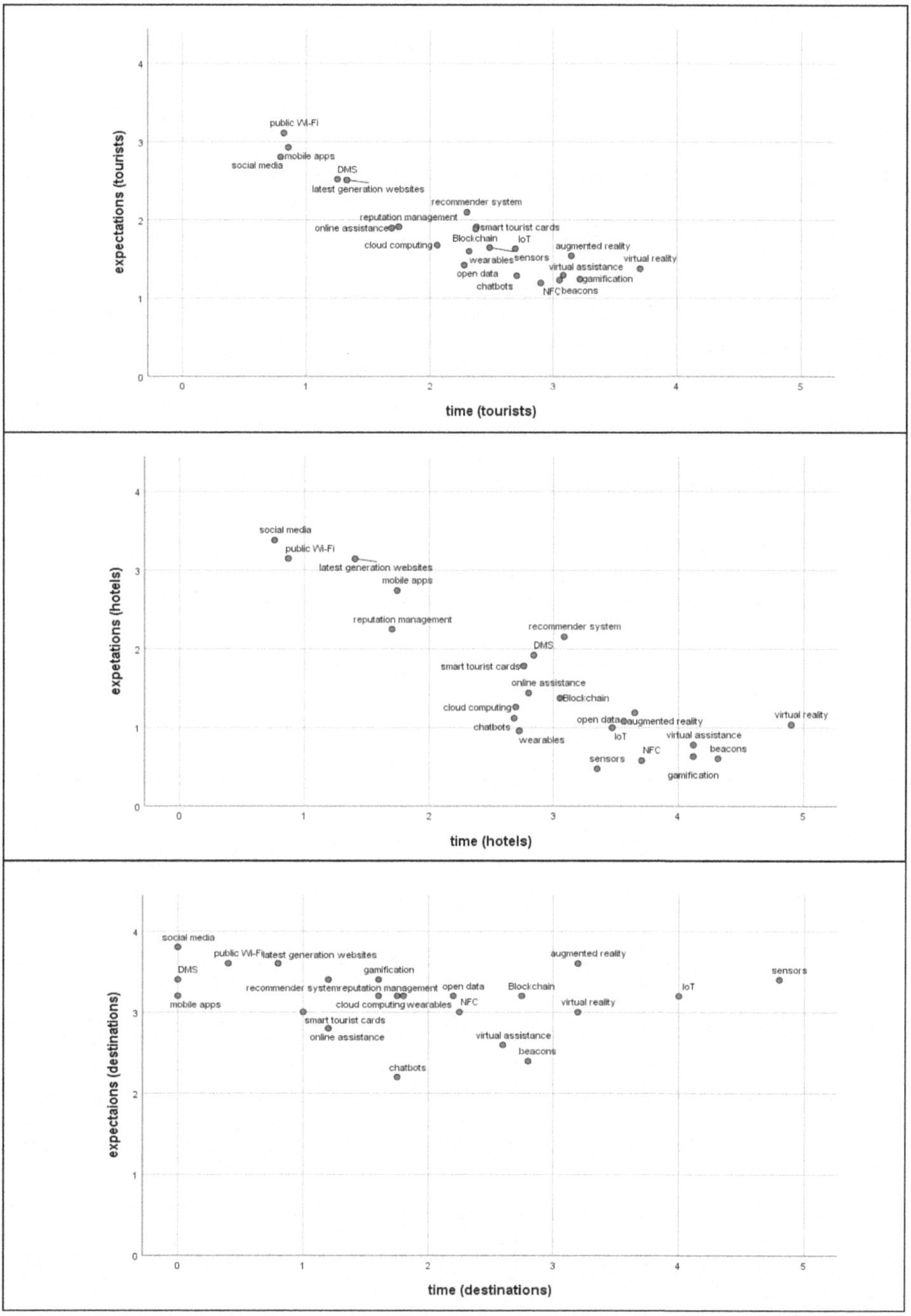

Fig. 3. Evaluation of smart tourism technologies by tourists, hotel managers and destination managers

Acknowledgments. The research was supported by the research project VEGA 1/0237/20 Tourism 4.0: Smart and Sustainable Tourism Development in the Competitive Environment.

References

1. Lu, Y.: Industry 4.0: A survey on technologies, applications and open research issues. J. Ind. Inf. Integr. **6**(1), 1–10 (2017). https://doi.org/10.1016/j.jii.2017.04.005
2. Grigelova, V., Abasova, J., Tanuska, P.: Proposal of data pre-processing for purpose of analysis in accordance with the concept industry 4.0. In: Silhavy, R. (ed.) Artificial Intelligence Methods in Intelligent Algorithms. AISC, vol. 985, pp. 324–331. Springer, Cham (2019). https://doi.org/10.1007/978-3-030-19810-7_32
3. Sheldon, P.J.: Tourism Information Technology. CAB International, New York (1997)
4. Buhalis, D.: eTourism. Information Technology for Strategic Tourism Management. Pearson Education Limited, Edinburgh (2003)
5. Pencarelli, T.: The digital revolution in the travel and tourism industry. Inf. Technol. Tourism (2019). https://doi.org/10.1007/s40558-019-00160-3
6. Washburn, D., et al.: Helping CIOs Understand "Smart City" Initiatives: Defining the Smart City, Its Drivers, and the Role of the CIO. Cambridge, MA (2010)
7. Gretzel, U.: Intelligent systems in tourism: a social science perspective. Ann. Tourism Res. Pergamon **38**(3), 757–779 (2011). https://doi.org/10.1016/j.annals.2011.04.014
8. Mehraliyev, F., Choi, Y., Köseoglu, M.A.: Progress on smart tourism research. J. Hosp. Tourism Tech. **10**(4), 522–538 (2019). https://doi.org/10.1108/jhtt-08-2018-0076
9. Lee, S.-H.: Smart industries based on smart technologies in convergence. J. Adv. Inf. Tech. Converg. **3**(1), 13–20 (2013)
10. Worden, K., Bullough, W.A., Haywood, J.: The smart approach. an introduction to smart technologies. In: Worden, K., Bullough, W.A., Haywood, J. (eds.) Smart Technologies, pp. 1–6. World Scientific Publishing, Danvers (2003)
11. López-Quiles, J.M., Rodríguez Bolívar, M.P.: Smart technologies for smart governments: a review of technological tools in smart cities. In: Rodríguez Bolívar, M.P. (ed.) Smart Technologies for Smart Governments, vol. 24, pp. 1–18. Springer, Cham (2018). https://doi.org/10.1007/978-3-319-58577-2_1
12. Chen, T., Chiu, M.-C.: Smart technologies for assisting the life quality of persons in a mobile environment: a review. J. Ambient Intell. Hum. Comput. **9**(2), 319–327 (2016). https://doi.org/10.1007/s12652-016-0396-x
13. Um, T., Chung, N.: Does smart tourism technology matter? Lessons from three smart tourism cities in South Korea. Asia Pac. J. Tourism Res. **0**(0), 1–19 (2019). https://doi.org/10.1080/10941665.2019.1595691
14. Xiang, Z., Tussyadiah, I., Buhalis, D.: Smart destinations: foundations, analytics, and applications. J. Destination Market. Manage. **4**(3), 143–144 (2015). https://doi.org/10.1016/j.jdmm.2015.07.001
15. Neuhofer, B., Buhalis, D., Ladkin, A.: Smart technologies for personalized experiences: a case study in the hospitality domain. Electr. Markets **25**(3), 243–254 (2015). https://doi.org/10.1007/s12525-015-0182-1
16. Femenia-Serra, F., Perles-Ribes, J.F., Ivars-Baidal, J.A.: Smart destinations and tech-savvy millennial tourists : hype versus reality. Tourism Rev. **2** (2018). https://doi.org/10.1108/tr-02-2018-0018

17. Buhalis, D.: Technology in tourism-from information communication technologies to eTourism and smart tourism towards ambient intelligence tourism: a perspective article. Tourism Rev. (2019). https://doi.org/10.1108/tr-06-2019-0258
18. Jeong, M. Shin, H.H.: 'Tourists' experiences with smart tourism technology at smart destinations and their behavior intentions. J. Travel Res. (2019). https://doi.org/10.1177/0047287519883034
19. Sigala, M.: New technologies in tourism: from multi-disciplinary to anti-disciplinary advances and trajectories. Tourism Manage. Pers. **25**, 151–155 (2018). https://doi.org/10.1016/j.tmp.2017.12.003. December 2017
20. Huang, C.D., et al.: Smart tourism technologies in travel planning: The role of exploration and exploitation. Inform. Manage. **54**(6), 757–770 (2017)
21. Ivars-Baidal, J.A., et al.: Smart destinations and the evolution of ICTs: a new scenario for destination management?. Curr. Issues Tourism, 1–20 (2017). https://doi.org/10.1080/13683500.2017.1388771
22. Gelter, H.: Digital tourism - An analysis of digital trends in tourism and customer digital mobile behaviour (2017)
23. Gretzel, U., Sigala, M., Xiang, Z., Koo, C.: Smart tourism: foundations and developments. Electr. Markets **25**(3), 179–188 (2015). https://doi.org/10.1007/s12525-015-0196-8
24. Koo, C., et al.: The role of IT in tourism. Asia Pac. J. Inf. Syst. **25**(1), 99–104 (2015). https://doi.org/10.14329/apjis.2015.25.1.099
25. Buhalis, D., Amaranggana, A.: Smart tourism destinations enhancing tourism experience through personalisation of services. In: Tussyadiah, I., Inversini, A. (eds.) Information and Communication Technologies in Tourism 2015. PAIT, pp. 377–389. Springer, Cham (2015). https://doi.org/10.1007/978-3-319-14343-9_28
26. Buhalis, D., Leung, R.: Smart hospitality—interconnectivity and interoperability towards an ecosystem. Int. J. Hosp. Manage. **71**, 41–50 (2018). https://doi.org/10.1016/j.ijhm.2017.11.011
27. Law, R., Sun, S., Chan, I.C.C.: Hotel technology: a perspective article. Tourism Rev. 5–8 (2019). https://doi.org/10.1108/tr-05-2019-0150
28. DiPietro, R.B., Wang (Raymond), Y.: Key issues for ICT applications: Impacts and implications for hospitality operations. Worldwide Hosp. Tourism Themes **2**(1), 49–67 (2010). https://doi.org/10.1108/17554211011012595
29. Neuhofer, B., Buhalis, D., Ladkin, A.: Conceptualising technology enhanced destination experiences. J. Destination Market. Manage. **1**(1–2), 36–46 (2012). https://doi.org/10.1016/j.jdmm.2012.08.001
30. Cimbaljević, M., Stankov, U., Pavluković, V.: Going beyond the traditional destination competitiveness–reflections on a smart destination in the current research. Curr. Issues Tourism **0**(0), 1–6 (2018). https://doi.org/10.1080/13683500.2018.1529149
31. Yoo, C.W., et al.: Improving travel decision support satisfaction with smart tourism technologies: a framework of tourist elaboration likelihood and self-efficacy. Technol. Forecast. Soc. Change **123**, 330–341 (2017). https://doi.org/10.1016/j.techfore.2016.10.071
32. European Commission: Digital economy and society index 2018. Country report Slovakia (2018). http://ec.europa.eu/information_society/newsroom/image/document/2018-20/sk-desi_2018-country-profile_eng_B4415E7E-9154-E26E-7B403212919F3F7C_52238.pdf
33. Novak, J., et al.: The rise of digital challengers. How digitization can become the next growth engine for Central and Eastern Europe (2018)
34. Nam, K., et al.: Blockchain technology for smart city and smart tourism: latest trends and challenges. Asia Pac. J. Tourism Res. **0**(0), 1–15 (2019). https://doi.org/10.1080/10941665.2019.1585376

Applying Computer Vision Methods for Fencing Constructions Monitoring

Alexey Smagin[1] and Konstantin Dubrovin[2(✉)]

[1] Mining Institute of Far East Branch of the Russian Academy of Sciences, Khabarovsk 680000, Russia
[2] Computing Center of Far East Branch of the Russian Academy of Sciences, Khabarovsk 680000, Russia
nobforward@gmail.com

Abstract. The issues of automated monitoring of the condition of mesh enclosing structures used on farms producing marine biological resources are considered. An algorithm is proposed for identifying breaks in mesh fencing underwater conditions and is implemented as a complex of programs in Python using the OpenCV computer vision library. The results of testing the algorithm are presented. It is shown that computer vision effectively copes with the classification of network cells in low-noisy and medium-noisy images. To work in more complex optical conditions, it is proposed to include a neural network module in the software package.

Keywords: Algorithm · Computer vision · Edges detection · Machine learning · Software

1 Introduction

The object of the study is specialized mesh fences used for production of marine bio resources at aquaculture enterprises. The purpose of this work is to develop and study image processing algorithms for the computer vision system of the Autonomous unmanned underwater vehicle (AUV), which provide a means for automating the process of monitoring the condition of the mesh fence, and thus exclude the presence of a man in the aggressive underwater environment.

Artificial mesh fences in the frames of the video stream received by the AUV cameras often contrast poorly with the surrounding background. In addition, these video frames contain various noises. They occur when there is insufficient light or optical distortions in the water environment, and, especially, when various representatives of underwater flora, fauna, silt and underwater debris appear in the frame.

One of the main tasks of image processing and analysis is segmentation, i.e. dividing the image into areas that meet certain uniformity criteria (brightness can be such a criterion). The main classical segmentation method is threshold image processing. It is one of the simplest and most important procedures for element-by-element transformations and almost always precedes the process of image analysis and recognition [1]. This operation involves matching the brightness value of each pixel in the image with the specified threshold value. Choosing a threshold value makes it

R. Silhavy (Ed.): CSOC 2020, AISC 1225, pp. 344–351, 2020.
https://doi.org/10.1007/978-3-030-51971-1_28

possible to select areas of a certain type in the image. The use of pre-threshold local filtering and binarization allows us to cut off extraneous noise and quickly determine the spatial characteristics of the grid cells on a "clean" image in order to determine the integrity of the mesh fence.

2 Image Filtering

Accurate color reproduction of images, its improvement, and restoration of color saturation in photo and video images play an important role in computer vision systems. Problems with color rendition arise at shooting in conditions of insufficient lighting or distortion. A number of artifacts, such as a lack or excess of light, make it difficult to process images with a view to determining the boundaries of objects of interest [2].

In the developed software package, the search for the boundaries of the mesh fence is performed with a two-stage image preprocessing algorithm. In the first stage, the compensation of artifacts related to lighting is performed using the Retinex algorithm [3, 4].

The Retinex algorithm is based on the principles of simulating the mechanism of human retinal receptors. At the first step of this algorithm, the source image I is divided into RGB channels C_i, i =1, 2, 3. Then, each RGB channel of the original image is divided into two components: the high-frequency R-component (reflectance image), containing the details of the image, and the low-frequency L-component (illumination image), containing the background of the image. The value of each pixel of the channel with coordinates (x,y) is represented as the product of the corresponding pixels of the L and R-components [5]:

$$C_i(x, y) = L_i(x, y) \times R_i(x, y). \tag{1}$$

A low-frequency Gauss filter is used to isolate the L-component:

$$L_i(x, y) = C_i(x, y) \otimes G(x, y), \tag{2}$$

where $G = \frac{1}{\sqrt{2\pi c}} e^{-(x^2+y^2)/2c^2}$ – the Gauss filter with kernel dimension c, (x, y) are the coordinates of the current pixel,$\otimes$ is a convolution operator.

At the fourth and last step, the natural logarithm of the selected L-component is subtracted from the natural logarithm of each color channel:

$$C_i^{'}(x, y) = lnC_i(x, y) - lnL_i(x, y), \tag{3}$$

where $C_i^{'}(x, y)$ is the i-th component of the output image.

After applying the Retinex algorithm, the main part of the channel pixel values belongs to the range [−1.1]. To render an image, the pixel values are normalized using the formula (4):

$$S_i(x, y) = C_i'(x, y) \times 255 + 127.5, \quad (4)$$

where S_i is the i-th component of the final image.

The MultiScale Retinex with Color Restoration modification used in the software package simulates the biological mechanisms of the eye, which adapts to complex conditions and copes well with the task of increasing the brightness and contrast of the image, showing the details of the image hidden due to low light and distortion.

In multiscale Retinex the output value of the color channel is a weighted sum of the output Retinex values with different dimensions of the Gauss filter core:

$$MSR_i(x, y) = \sum_{n=1}^{N} w_n s_{ni}(x, y),$$

where $s_{ni}(x,y)$ – the value of the pixel component (x,y) after applying the Retinex algorithm, w_n is the weight of the Retinex value. Color Restoration involves multiplying the values of all pixels of each color channel obtained in multiscale Retinex by the value defined as follows:

$$CR_i(x, y) = b \times ln(aMSR_i(x, y)) - \ln(\sum_{i=1}^{3} MSR_i(x, y))),$$

where $a = 46$, $b = 25$.

The Figs. 1 and 2 show the original image taken underwater [6] and the result of the algorithm at the first stage of preprocessing. There is visible improvement in contrast and brightness in the dark areas of the image, without quality loss in the light areas.

Fig. 1. Original image

Fig. 2. Result of multi-scale Retinex restoration

In the second stage, the image is grayed out and binarized using the Otsu threshold filter [7]. Let's look at how this filter works.

Let the image pixels have L levels of brightness, h_i is the number of image elements that have I brightness, and H is the total number of image pixels. Then $p_i = h_i/H$ – the frequency of occurrence of pixels having brightness i, where $i = 0,1,..,L\text{-}1$. Image pixel brightness values can be interpreted as random values, and their histogram can be interpreted as an estimate of the probability distribution density. The initial value $t = 1$ is set.

Image pixels are divided into two classes: class B, which includes pixels whose brightness is less than t, and class O, which includes pixels with the brightness of t and above. then the mathematical expectations and variances of each class are calculated:

$$\mu_B(t) = \sum_{i=0}^{t-1} ip(i), \mu_O(t) = \sum_{i=t}^{L-1} ip(i), \tag{5}$$

$$\sigma_B^2(t) = \sum_{i=0}^{t-1} (i - \mu_B(t))^2 p(i), \sigma_O^2(t) = \sum_{i=0}^{t-1} (i - \mu_O(t))^2 p(i). \tag{6}$$

The intra-class variance is defined as the weighted sum of the variances of two classes:

$$\sigma_\omega^2(t) = n_B(t)\sigma_B^2(t) + n_O(t)\sigma_O^2(t), \tag{7}$$

where $n_B(t) = \sum_{i=0}^{t-1} p(i)$ – the proportion of pixels assigned to class B, $n_O(t) = \sum_{i=t}^{L-1} p(i)$ – the proportion of pixels assigned to class O.

By iterating through all possible thresholds $t = 1,2,\ldots,L$, we can determine the optimal threshold for dividing the image into the "useful" and "background" part, which σ_{ω}^2 takes the lowest value. Calculating second-order statistics (variances) for each t value requires a large number of calculations; therefore, Otsu proposed a simpler method, which involves calculating another type of statistic– the interclass variance:

$$\sigma_M^2(t) = n_B(t)n_O(t)(\mu_O(t) - \mu_B(t))^2. \quad (8)$$

The calculation $\sigma_M^2(t)$ requires calculating only 1st-order statistics (mathematical expectations). The optimal t value corresponds to the maximum interclass variance.

Since variance is a measure of the spread of brightness levels around the average value, a large value indicates a large deviation from the average. In this regard, the operation of finding the maximum for determining the threshold means increasing the isolation of the two classes in the binarized image. At the optimal threshold value t, pixels that fall into class B (lying below the threshold) take a zero-brightness value (black color) and indicate the content of the image - the result of filtering by the Otsu method. This filter is chosen due to the fact that the threshold for it is calculated automatically for each image and the method is contained in the OpenCV computer vision library. Figure 3 shows the gray-scale image after the first filtering step, and Fig. 4 shows the result of binarization using the Otsu method.

Fig. 3. Grayscale image

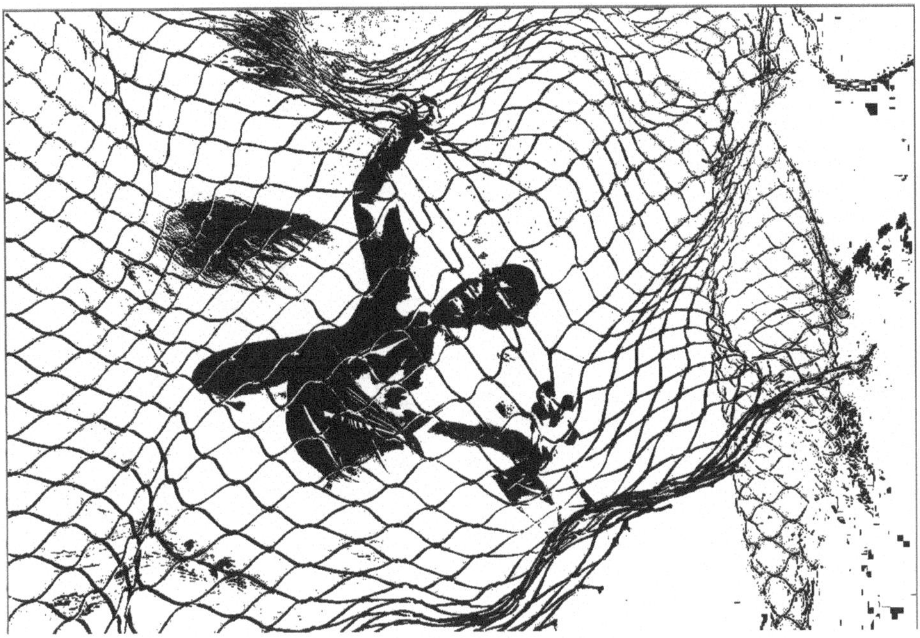

Fig. 4. Otsu filtered image

3 Detecting Tears and Determining Boundaries of a Mesh Fence

Checking the integrity of a mesh fence is performed after filtering is completed and its image is not noisy. At this stage, all contours in the image are searched for using the FindContours function of the OpenCV library. Then they are classified. If the contour is a closed polygon and its perimeter is in the range from 0.5 P to 2 P, where P is the average perimeter, this contour is classified as an unbroken cell and marked in green. An open loop or closed loop with a perimeter bigger than 2 P is marked in red and classified as a tear. A closed loop with a small perimeter (less than 0.5 P) is perceived by the system as seriously deformed (for such cells there is a high probability of tear in the future) and is marked in blue. Figures 5 and 6 present the result of the program's work, determining the integrity of the mesh fence.

Fig. 5. Filtered image

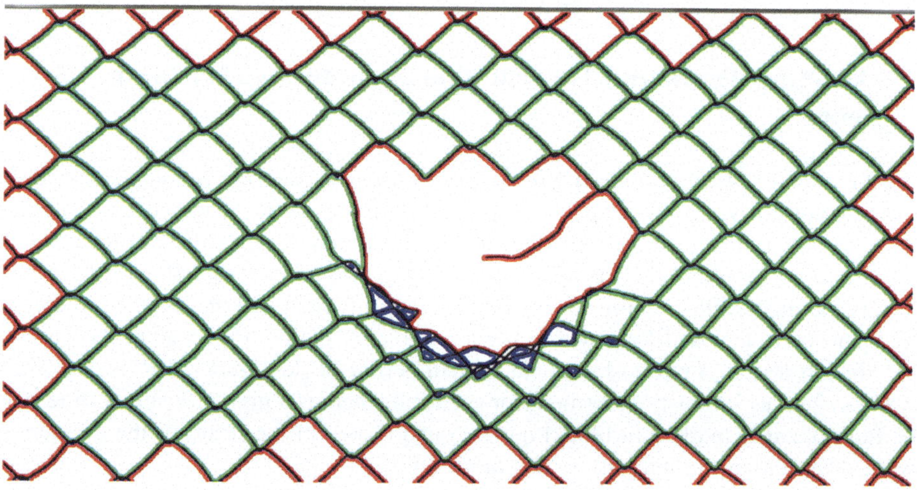

Fig. 6. Result of grid integrity detection

4 Conclusion

The analysis of the software package results have shown that it successfully copes with the classification of grid cells in low-noise images. However, the transformations performed at the preprocessing stage are not enough to completely exclude extraneous objects. For example, the silhouette of the person behind the mesh (Fig. 4) merges with it, which makes it possible to detect false tears and deformations of the fence. In this regard, the work on improving and expanding the functionality of the software package will continue. The final result of the study will be a software package with a neural network module for improved filtering of the external background and more effective detection of problem areas of a mesh fence.

Acknowledgments. The reported study was funded by RFBR, project number 19-37-90046\19 and it was supported in through computational resources provided by the Shared Services Center "Data Center of FEB RAS" (Khabarovsk).

References

1. Gonzalez, R., Woods, R.: Digital Image Processing, 3rd edn. Prentice Hall, Upper Saddle River (2008)
2. Pratt, W.: Digital Image Processing. CRC Press, Boca Ratton (1978)
3. Zhang, S., Wang, T., Dong, J.: Underwater image enhancement via extended multi-scale Retinex. Neurocomputing **245**, 1–9 (2017). https://doi.org/10.1016/j.neucom.2017.03.029
4. Loh, Y., Chan, C.: Getting to know low-light images with the exclusively dark dataset. Comput. Vis. Image Underst. **178**, 30–42 (2018). https://doi.org/10.1016/j.cviu.2018.10.010
5. Doo, H., Ick, H., Mi, H.: Color image enhancement using single-scale Retinex based on an improved image formation model. In: Proceedings of 16th European Signal Processing Conference (EUSIPCO 2008), Lausanne, Switzerland, pp. 1–5. IEEE, Piscataway (2008)
6. The Underwater Woman Homepage. http://www.theunderwaterwoman.com/science-storytelling. Accessed 12 Oct 2019
7. Otsu, N.: A Threshold selection method from gray-level histograms. IEEE Trans. Syst. Man Cybern. **9**(1), 62–66 (1979). https://doi.org/10.1109/tsmc.1979.4310076

The Structural Analysis of the World Gold Prices Dynamics

R. I. Dzerjinsky[1(✉)], E. N. Pronina[1], and M. R. Dzerzhinskaya[2]

[1] MIREA – Russian Technological University, 78, Vernadsky pros., Moscow 119454, Russia
9015111295@mail.ru, pvi173@rambler.ru

[2] Moscow State Technical University named after N.E. Bauman, 5, 2nd Baumanskaya St., Moscow 105005, Russia

Abstract. The article is devoted to analysis of the dynamics of world prices for gold. We consider the problem of detecting hidden periodicities and establish them on the basis of regularities in the structure of the analyzed time series. As a mathematical tool of solving the problem involved the machine structure functions A. N. Kolmogorov. In structural analysis have developed two approaches: the first one is the only study self-similar properties, the second one oscillates components with structural functions. In this paper an attempt was made to combine these two approaches. For example the dynamics of world prices for gold shows that the boundaries of the ranges of self-similar growth structure functions are interrelated with their local lows and been in aggregate they account for almost periods on the basis of which establishes the rhythmic structure of time series. Structural analysis of the dynamics of world prices for gold showed the presence of a secular cycle, within which is formed the structure of the three cycles of thirty years duration. Thirty years of rhythm manifested in the dynamics of prices and dynamics of world production of gold. Unstable phase of development falls on the period 2017–2041 years Dynamics of prices for gold in these years will be different fluctuations. #CSOC1120

Keywords: World gold market · Hidden periodicity · Structure functions · Scaling · Self-similarity

1 Introduction

Features of the world market of gold associated with a particular function that gold performs in the global economy, acting as both a commodity and global money.

As a commodity, gold is traditionally widely used in the jewelry industry. Due to its unique properties, it also became indispensable in the manufacture of various devices, high-tech industries, for example, in the manufacture of parts and components of jet engines, nuclear reactors, spacecraft. ZRotie elements are used as conductors and contacts high reliability in computers, phones, electronic devices, aviation equipment. The volume of gold consumption indicates the level of development of high technologies in the country.

R. Silhavy (Ed.): CSOC 2020, AISC 1225, pp. 352–365, 2020.
https://doi.org/10.1007/978-3-030-51971-1_29

In the modern world gold was de-monetized: after 1975 the national currency lost its gold backing, gold has ceased to perform monetary functions and directly used in international payments. In spite of this, it continues to be one of the most popular financial instruments becoming a particular investment product – the extraordinary world of money. Pon-former, the amount of gold reserves reflects the monetary and financial position of the country, gold forms an integral part of the financial reserves and insurance Fund of the state.

At the same time gold is a limited natural resource, its reserves is limited in the subsoil. From the history of the development of industries producing non-reproducible natural resources limited, well-known profile curve extraction. At the initial stage of development of its resource production is growing rapidly, then the growth slows down and the curve reaches a plateau, and then starts falling production phase, which is characterized by the deterioration of mountain-geological conditions of development and production costs are rising.

In recent years, reports by various expert institutions (Thomson Reuters GFMS, Silver Institute, WGC, CPM Group) expressed an opinion about the approaching reduction in the production of gold and silver [1]. The main reasons are: the almost complete absence of discoveries of new large fields; the lack of profitability of existing mines due to ore depletion, and the possible future long-term price reduction that will make many of the mines are unprofitable and will cause a further reduction in exploration.

Thus, gold, along with fuel and energy resources, may be another limiting factor in the development of the world economy and its financial system.

Prognostic assessment of crisis situations in the development of world market of gold, the timing of their onset, duration, and depth of the recession and is the subject of this study. It includes two parts:

The analysis of price dynamics in the global gold market, with the goal of identifying hidden periodicities and

Modeling long-term growth of world gold production, aimed at obtaining predictive estimates of the limits to growth of production and the timing of their achievement.

The work is performed on the basis of statistical data covering the period of over 500 years, from 1495 to the present. The unknown authors of similar studies which includes data on the dynamics of the global gold market over such a long period of time.

This article represents the first part of the study is devoted to the analysis of the dynamics of world prices for gold.

2 Statement of the Problem

An object of the research is the time series of world prices for gold, then we have the sequence of price levels for gold

$$X_1, X_2, \ldots, X_N,$$

related to equidistant time points, the duration of the intervals between them Δt can be one year, one month or one day depending on the level of review of data. It is assumed that the analyzed time series contain two components – trend and fluctuations about it.

The subject of this research relates to the identification of hidden periodicities on the basis of which regularities in the structure and rhythmic structure, manifested in the dynamics of world prices for gold.

Traditionally, the determination of hidden periodicities is based on the spectral analysis, despite the fact that the appointment of a harmonic Fourier analysis was originally associated with a completely different task, with the task of approximation of functions.

Apparatus for harmonic Fourier analysis allows to represent the function that satisfies the Dirichlet conditions in the form of trigonometric series in the General case is infinite. In the base period, the period of the lowest fundamental harmonic is taken of the length of the approximation interval, which is set to approximate the function, and the periods of all subsequent harmonics are divisors of the base period. However, the dissemination of the results outside the measuring interval leads to their distortion.

Incorrect assessment also arise when the shape of the periodic component of the real process significantly differ from the harmonic.

For this reason, preferable are methods for identifying periodicities aimed to determine the periods of the periodic component and free from a priori assumptions about the form of vibrations and other process parameters [2, 3]. One of such methods, based on the device's structural features and their modifications, is discussed later.

3 Structure Functions and Their Modifications

Fundamentals of structural analysis was laid in the works of academician A. N. Kolmogorov [4, 5]. In the early 1940-ies by A. N. Kolmogorov introduced in consideration of random processes with stationary increments. The concept of a process with stationary increments is a generalization of the notion of stationary process. It is obvious that any stationary random process is also a process with stationary increments. However, there are nonstationary processes with stationary increments. The simplest examples of such processes are linear functions

$$u(r) = c_1 r + c_0,$$

where c_1 and c_0 are constants.

Tool for describing stochastic processes with stationary increments serve structural functions. In the article, devoted to the study of turbulence [5], the structural function is introduced as the mathematical expectation of squared difference of instantaneous values of velocities measured at the neighboring points of a random velocity field:

$$S_u(\Delta r) = M[(u(r_1) - u(r_2))^2].$$

Slow change of speed there will be little impact on the values of the increments:

$$\Delta u = u(r_1) - u(r_2).$$

During the transition from speed $u(r)$ the difference $\Delta u(r)$ process components with large periods are suppressed, and the increment of velocity Δu be stationary, depending only on the distance between the points Δr where measured speed. Random function $u(r)$ is called a random function with stationary first increments.

For processes with stationary second increments random function $u(r)$ is the polynomial above the second degree, $u(r) = c_2r^2 + c_1r + c_0$ etc.

Structural feature is also important for the theory of random processes with stationary increments, like the autocorrelation function to the theory of stationary processes. In the case of an ergodic stationary process, structural and autocorrelation function are interrelated, and one of them clearly expressed in terms of another [6, 7]. The result for stationary ergodic random process can be used as a correlation function, and structural function. In General, the structure functions have a broader scope of applicability than correlation.

In the inertial range, which is much more dissipative and less global theory A. N. Kolmogorov predicts a power law dependence of the structure function of scale, which in order of magnitude not greater than the distance between two points of the considered fields. The result

$$S(\tau) = A \cdot \tau^{\gamma}$$

where A, γ are positive parameters. In this case, the graph of the structure function depending on the magnitude of the increment τ, constructed in the logarithmic scale, straighten up:

$$\ln(S) = \ln(A) + \gamma \ln(\tau)$$

Power laws have the interesting property that for them there is a group of transformations of similarity. The transformation of the time scale and scale transformation values u leave structural function $S(\tau)$ is unchanged:

$$\begin{aligned} S(\tau) = M[U \cdot u(t \cdot T + \tau \cdot T) - U \cdot u(t \cdot T)]^q = U^q \cdot M[u(t \cdot T + \tau \cdot T) - u(t \cdot T)]^q \\ = U^q \cdot S(\tau \cdot T) = A \cdot U^q \cdot T^{\gamma} \cdot \tau^{\gamma}. \end{aligned}$$

The scaling of the signal on the x-axis in T times, and the axis of ordinate to $U = T^{-\gamma/q}$ times does not change the structural function,

$$S(\tau) = A \cdot \tau^{\gamma}$$

Functions possessing such properties called *self-similar*. Self-similarity means *scale invariance*, *scaling* or *self-similarity*. In contrast to the structure function, the autocorrelation function of a stationary process cannot be self-similar [6].

The meaning of self-similarity is expressed in the constancy of the relative velocity changes of the structure function,

$$\frac{\Delta S}{S} : \frac{\Delta t}{t} = \gamma = const.$$

However, the self-similar growth can not be infinite. The interval τ, within which power-law changes of the structure function, determines the range of scale invariance.

Been close to the apparatus structure functions is D method of detecting hidden periodicities [3]. Basic working design D method is a function

$$D_X^n(\tau) = \frac{1}{\overline{T}} \int_0^{\overline{T}} |X^n(t+\tau) - X^n(t)| dt, \quad \overline{T} = T - \tau,$$

where [0,*T*] is the interval of the assignment process, *X*(*t*) – studied process, $X^n(t)$ – n − I derivative of the function *X*(*t*), τ – independent auxiliary variable dimensions *t* taking nonnegative real values. Functions are *D* image of a random process *X*(*t*). Here, the differentiation decreases, and in the case of a large *n* eliminates the effect of non-periodic (polynomial) component.

After identifying and filtering a periodic component of the higher the remainder of the process is again subject to *D*-transformation to identify the next adjustment periodicals.

A discrete interpretation of the structure function of order q for a time series X_1, X_2, …, X_N leads to the expression:

$$S(\tau) = \frac{1}{N-\tau} \sum_{t=1}^{N-\tau} |X_{t+\tau} - X_t|^q, \tag{1}$$

where X_t and $X_{t+\tau}$ the samples of the signal *X* at time *t* and (*t* + τ); *N* – number of samples [8].

The shift function is considered in [9], is essentially a structural function *S*(τ) of the first order.

In the behavior of the structure functions and their modifications are observed as the areas of self-growth, and oscillatory components. In this regard, the structural analysis has developed two approaches. The first approach is associated with the study only the self-similar properties [6–8], the second – with their oscillatory components [3, 9].

Since points that are multiples of almost periods of the investigated process design *S*(τ) and have the lows, they can be used to determine the parameters of hidden periodicities. The operation of taking the modulus gives the minima of the structural features characteristic the acute form, which is easily identificating and allows to locate the minimum with high accuracy [3].

It seems appropriate to combine these two approaches and to establish the linkages between local minima of the structure functions and the boundaries of the ranges of self-similarity. A similar attempt made in this paper on the example of the structural analysis of the dynamics of world prices for gold.

4 The Dynamics of World Prices for Gold and Its Features

Sources of statistical data, which were composed of time series of world prices for gold are two works: the monograph [10] and the official website of the Canadian company - gold dealer KITCO [11].

Annual dynamics of prices for gold are shown in Fig. 1. The graph resembles a step function: within a single step occur minor price fluctuations relative to its average level when the stairs are connected by short sections of rapid growth, which can be considered a transition from one price level to another.

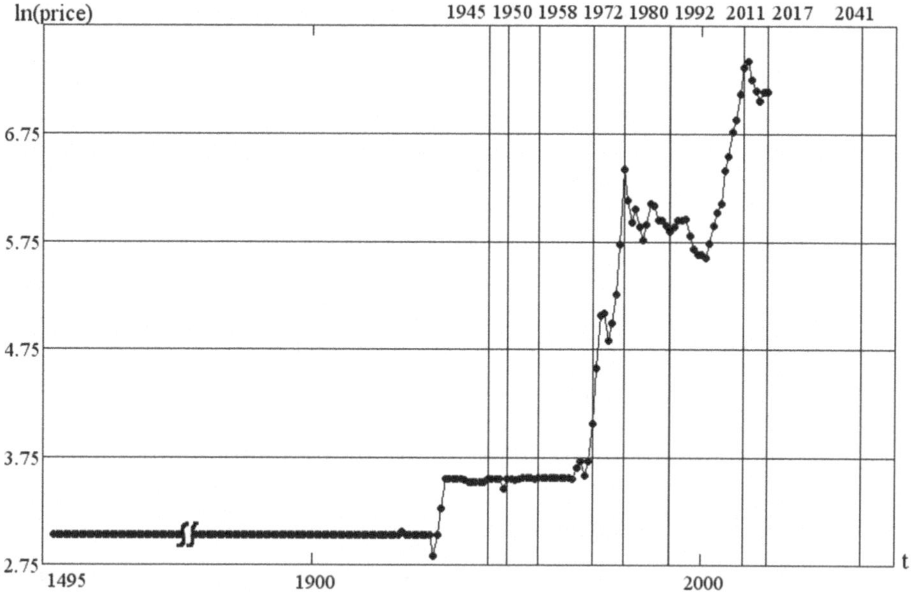

Fig. 1. Annual dynamics of prices for gold, USD/Troy ounce, 1500–2018 years. In the period 1500–1900 prices remained unchanged.

There are four "steps". The first was held more than four years (1493–1929 years), at 20 dollars per ounce. The second step was shorter – 1932–1968, at 35 dollars per ounce. Third step with average prices about $ 400 per ounce refers to a twenty-year period 1981–2003, and finally, the fourth step, since 2012 to present, with average prices of $ 1,300 per ounce.

Detailed data on changes in the price of gold has been available since 1968, so next will be considered two steps, two plots of price dynamics. The first refers to the period from 1968 to 1980, the second one - for the period from 2012 to 2018. For the first period we analyze daily data and monthly; for the second, in view of its short duration, only daily.

Changes in the price of gold during these periods of time were slow, so they have little impact on increment rates. As a result, in one step, the process itself and its increment can be considered stationary.

5 The Construction and Analysis of Structural Functions Time Series of World Prices for Gold

5.1 The Calculation of the Structure Functions

Calculation of structural functions time series of gold prices is provided by formula (1) in the software package Mathcad. The listing of the corresponding program is presented in Fig. 2.

```
S := | N ← last(X) + 1
     | for q ∈ 1.. 10
     |   for τ ∈ 1.. N − 1
     |                  N−1−τ
     |                    Σ   (|X_{t+τ} − X_t|)^q
     |                  t = 0
     |     K_{τ,q} ← ─────────────────────────────
     |                       N − τ
     | K
```

Fig. 2. The program of calculation of structure functions of order q.

5.2 Daily Data (The Period from 2012 to the Present)

A fragment of the dynamics of gold prices in the period from 2012 to the present are shown on picture 3. This period is characterized by slow change, the coefficient of variation is v = 0,143 (Fig. 3).

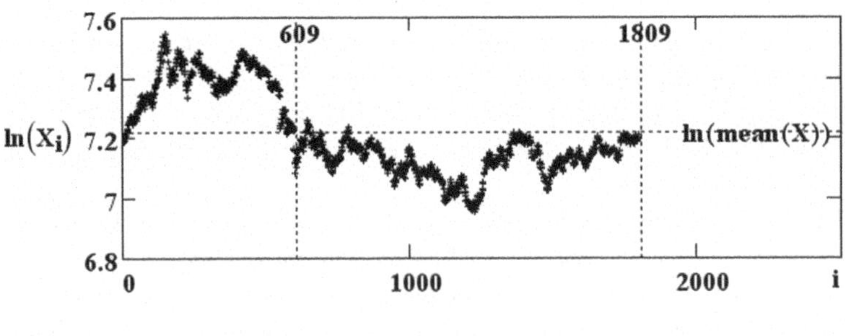

Fig. 3. The dynamics of daily prices gold, USD/Troy ounce, the working days in the period from 2012 to the present, a fragment of the graph Fig. 1.

Graphs of the structure functions of the price of gold in the logarithmic coordinate system shown in Fig. 4. The presence of linear plots in these graphs demonstrates the property of self-similarity.

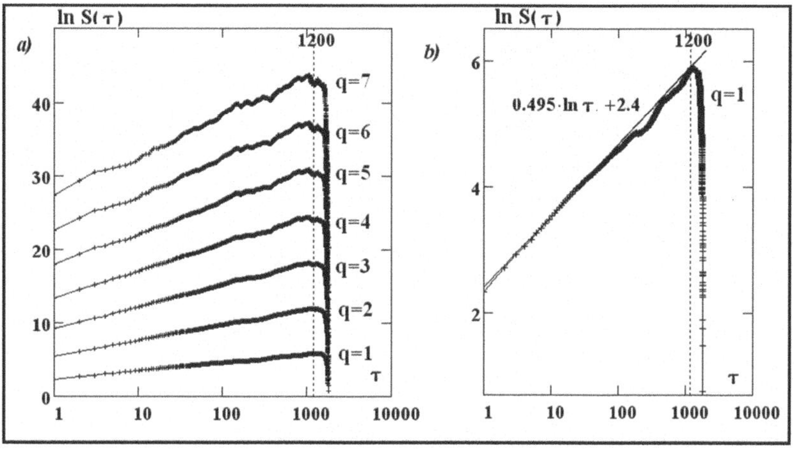

Fig. 4. The structural features built by the daily prices in the period from 2012 to the present.

With increasing order of structural functions, the self-similarity ranges within which the data are straightened, narrowed from about 1400 to 1100 days (Fig. 4, left). Let us take the upper bound of the range of the self-similarity an average rating of 1,200 business days. In terms of months, this value will be approximately 57 months, and years – 4.8 years. The slope of the structure function the first order close to 0.5 (Fig. 4, right). The same value of the tangent of the angle of inclination is, for example, for the model of classical Brownian motion. Is Hausdorff dimension of the trajectory of a Brownian particle is more than its topological dimension, which speaks about the fractal properties of a signal [12].

When plotting the dynamics of the price of gold in polar and cylindrical coordinate systems during a full rotation of 2π radians will take a range of the self-similarity of 1,200 business days. The corresponding graphs are shown in Fig. 5.

The profile graph in the polar coordinate system resembles a maple leaf, it is very uneven and has a palmate shape (Fig. 5, a). Accurately determine the number of "fingers" is difficult, the number 6 to 8, so obviously there is a period of smaller duration, which is 150–200 days.

In cylindrical coordinates every next turn of the spiral curve reproduces the qualitative features of the loop of the previous level. The observed regularity of the dynamics indicates the presence in the source data hidden periodicities and their phase coincidence component (Fig. 5, *b*).

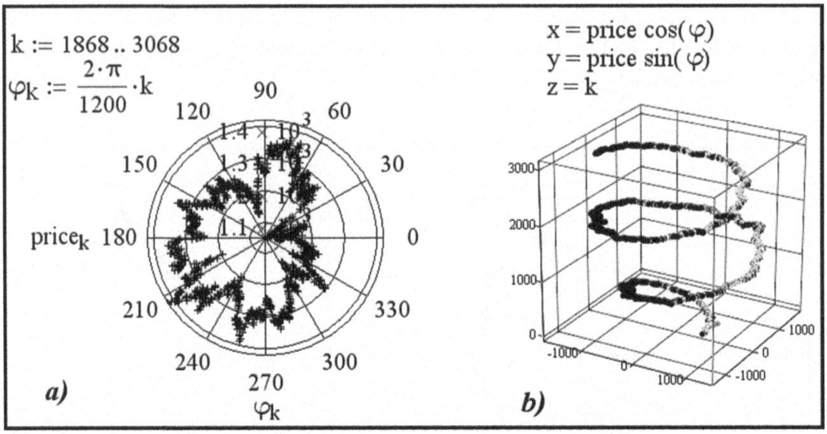

Fig. 5. The Dynamics of daily gold prices: a) polar coordinate system (2012–2018), b) cylindrical coordinates (2006–2018).

5.3 Monthly Data (The Period 1982 to 1997)

Carry out the same study for the monthly price of another time interval in the period 1982–1997, corresponding graph is shown in Fig. 6. The coefficient of variation in the interval from 160 338 for the month amounted to $v = 0.107$.

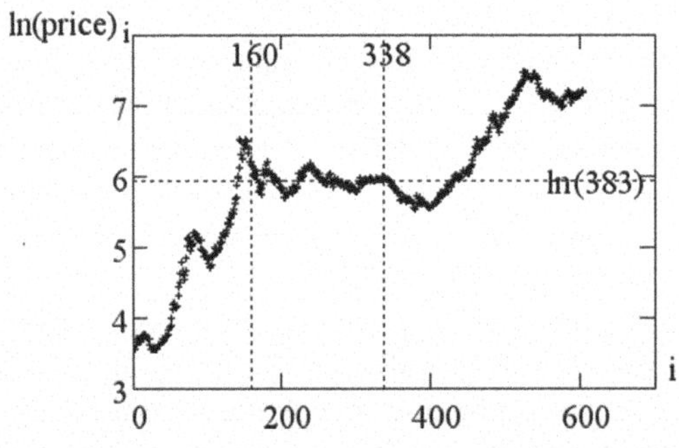

Fig. 6. Dynamics of monthly gold prices since 1968 to present; selected time period 1982–1997 years.

Graphs of the structure functions in logarithmic scale for the selected time interval are shown in Fig. 7. Here the structural functions include not only self-similar plot, but fluctuating. For the structural function of the first order upper bound of the range of scale invariance is blurred, it falls in $\tau_0 = 25 \div 36$ months (2–3 years). The increased order q reduces this interval $q = 7$ the upper bound is $\tau_0 \approx 27$ months. The slope of the dependence of the structure function the first order from the offset τ is again around 0.5.

Local minimum of the structure function of the first order occur when $\tau_1 = 86$ and $\tau_2 = 144$ months (7 and 12). It is known that for periodic functions, along with periods

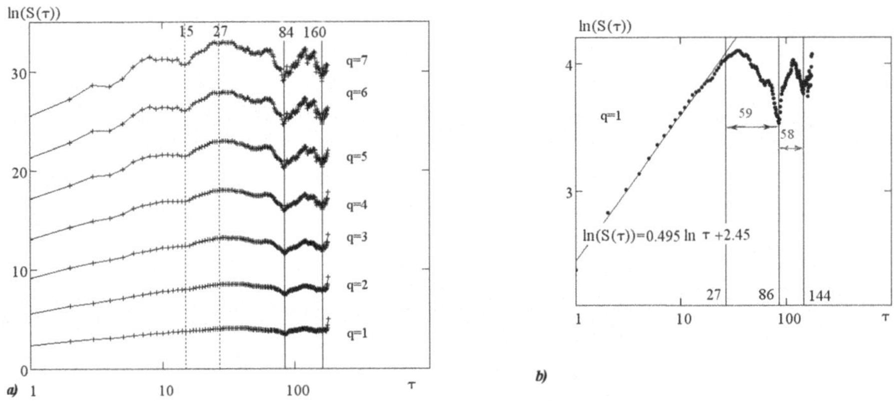

Fig. 7. Structure functions, constructed for the monthly price of gold 1982–1997 years.

of τ_1 and $\tau_2$the sum and difference of τ_1 and τ_2 there will also be periods. In our case, the difference $(\tau_2 - \tau_1)$ is equal to 58 months, whereas the duration of the interval between the upper limit of the range of invariance and the first local minimum of the structure function $(\tau_1 - \tau_0)$ is 59 months. A close value was obtained previously as the characteristic size of the range of scale invariance in the analysis of daily data (1200 days, 57 months). In other words, the characteristic size of the range of scale invariance of τ_0 structure functions, constructed on weekdays coincides with the durations of the intervals $\Delta\tau$ between the local minima of the structure functions for monthly data, and thus represents one of the almost periods of the oscillations.

5.4 Daily Data (The Period 1982 to 1997)

Analyses data of the same period as in the previous section, not only monthly, but daily. A fragment corresponding to the specified period, highlighted in Fig. 8. The coefficient of variation is $v = 0{,}165$.

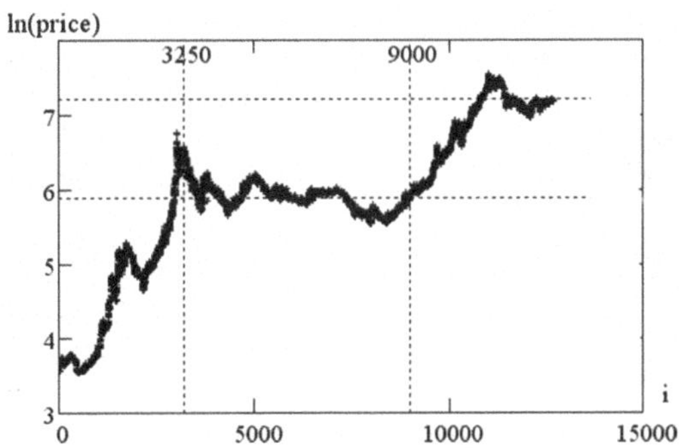

Fig. 8. The dynamics of daily prices of gold, 1968–2018 years.

For selection built structure functions of different orders (Fig. 9, *a*). The structural function of the first order, built on daily data for the period 1982–1997, he has a wide range of scale invariance of $\tau_0 = 900$ days (Fig. 9, *b*). The graphs of the structure functions of order higher than the first range of scale invariance is reduced due to the initial segment, which is cut off after 30 days. There are also local minima, $\tau_1 = 86$, $t_2 = 201$, $t_3 = 450$, $\tau_4 = 1800$ $\tau_5 = 3900$ days (Fig. 9, *a*). Note that the value of $\tau_2 \approx 200$ days obtained above as an estimate of the period almost when considering daily data in a polar coordinate system.

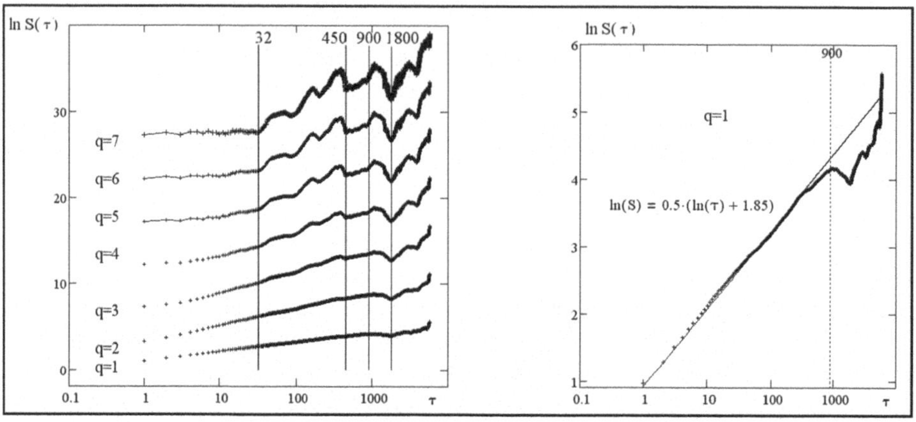

Fig. 9. The structural features built by daily data in 1982–1997 years.

Least common multiple of two characteristic sizes, 900 and 1200 working days is 3600 days. Schedule of price dynamics in a polar coordinate system built on the assumption that a complete revolution in 2π radian is 3600 days, is shown in Fig. 10.

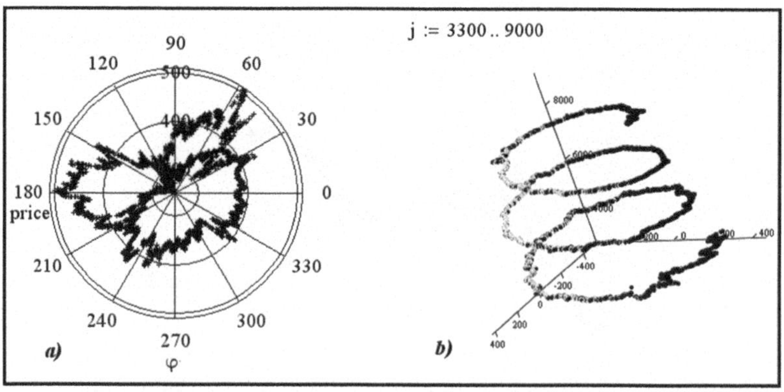

Fig. 10. Dynamics of daily gold prices a) the polar coordinate system; b) cylindrical coordinates, 1984–1997 years.

The profile in Fig. 10 forms a three-petal figure that indicates the presence period duration 3600/3 = 1,200 days, times the period of 900 days. On the chart of price dynamics in cylindrical coordinates (a full rotation is τ_4 = 1,800 days) clearly shows the coincidence of oscillations of different periods in phase (Fig. 10, *b*).

Thus, when determining the parameters of hidden periodicities of the process under study, along with the local minima of the structure functions, important are also the characteristic sizes of the plots are self-similar growth. The boundaries of the self-similarity of the structure function multiple local minima or coincide with the durations of the intervals between them.

6 The Rhythmic Structure of the Dynamics of Gold Prices

The width of the range of scale invariance of structural functions constructed for the analyzed piece of data can be taken as a major unit of structure, matching two rhythmic sequences. One of them is constructed according to the law of arithmetic progression is even and the other uneven, founded on the law of geometrical progression with the denominator equal to the number of Napier *e*. A model for coordination of rhythmic sequences of various types proposed in [9], it was called "cell development".

For the calculation of critical frontiers of cell structure it is important to determine the timeline position of the beginning of counting. Let us take the starting point in 1945, the year of the end of the Second world war, marks a time of drastic change tendencies of world development. Scale unit will be considered 1200 working days (4.8 years). Then the critical frontiers

$$0,\ 1,\ e,\ e^{e-1}e^2e^{2{,}293},\ e^e e^3$$

"development cells" will match 1945, 1950, 1958, 1972, 1980, 1992, 2011, 2017, 2041 the years (Fig. 11). In General, there is the age-old cycle, within which is formed the structure of the three cycles of the same duration. The first stroke in a model of time limited by the boundaries from 1 to e2the second beat from e^2 to third from to $e^3$where is the Fibonacci number. The duration of each of these intervals in real time is about one year.

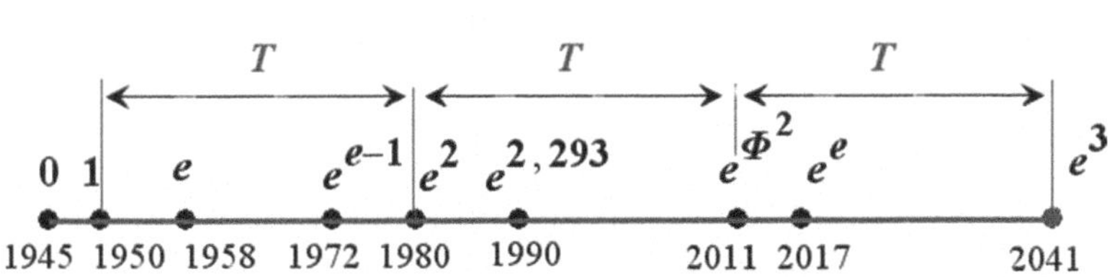

Fig. 11. Structure of the time series of price dynamics in the period 1945–2041 years, T $\approx$ 30.4 years.

In the resulting split four turn combined with real significance for the development of the world gold market over time: the origin is combined with that of 1945; import e^{e-1} – in 1972, synchronized irregular boundary e^2 – 1980; not synchronized uniform abroad – with 2011.

Historically gold almost five hundred years remained the universal equivalent of value, a benchmark to measure the value of other goods. In the early 1970-ies of the world financial system is such a model was not a rapid increase in the price of gold has destroyed a stable financial mechanism for the management of the global economy, for 10 years, prices rose more than 20 times.

In 1980 began the "correction", followed by a relatively stable twenty-year period of development. In the early 2000s, the situation repeats itself, the next rise in prices led in 2011 to the historic peak price of gold in 1895 dollars per ounce. Again, the correction (Fig. 1).

At the moment the process has entered a phase of restructuring, it falls in 2017–2041. In this period, to complete the old and new trends are formed, resulting in the development of the world financial system will be different unstable character, one manifestation of which will be sharp fluctuations in gold prices.

7 Conclusions

The presence in the dynamics of world prices for gold parts of the slow changes allowed be used to detect hidden in the dynamics of the periodicity of the device structure functions.

When determining the parameters of hidden periodicities, along with the local minima of the structural features are also important the characteristic size of the self-similar growth. The boundaries of the self-similarity multiple local minima or coincide with the durations of the intervals between them.

Structural analysis of time series world price of gold showed the presence of a secular cycle, within which is formed the structure of the three cycles of thirty years duration. Thirty years of rhythm manifested in the dynamics of prices and dynamics of world production of gold.

Further study suggests modeling long-term growth of world production of gold, it will focus on the prediction of the limits to growth in gold production and the timing of their achievement.

References

1. World Gold Council. www.gold.org. Accessed 21 Jan 2020
2. Serebrennikov, M.G., Pervozvansky, A.A.: Revealing of the latent periodicity. Science, Moscow, Russia (1965)
3. Veniaminov, S.S.: Revealing Latent Structure Laws in Processes and Signals. URSS, Moscow, Russia

4. Kolmogorov, A.N.: Curve in Hilbert's space, invariant in relation to one-parametrical group of movements. In: Proceedings of the Academy of Sciences USSR, vol. 26, no. 1, pp. 6–89 (1940)
5. Kolmogorov, A.N.: Local structure of turbulence in an incompressible liquid at Reynolds's very greater numbers. In: Proceedings of the Academy of Sciences USSR, vol. 30, no. 4, pp. 299–303 (1941)
6. Monin, A.S., Yglom, A.M.: Statistical a Hydromechanics. The Theory of Turbulence, vol. 2. Publishing office Hidrometeo, Saint-Petersburg (1996)
7. Rytov, S.M., Kravtsov, Y.A., Tatarsky, V.I.: Introduction in the Statistical Radiophysics, Part 2. Casual fields. Science, Moscow, Russia (1978)
8. Sychev, V.N.: Distant of correlation in catalogues of earthquakes and in measurements of seismic and seismoacoustic noise. Bull. KRSU **16**(1), 168–171 (2016)
9. Kuzmin, V.I., Gadzaov, A.F.: Method of Construction of Models on Empirical Data. The Manual. MGTU MIREA, Moscow, Russia (2012)
10. Laughlin, J.L.: The history of bimetallism in the United States. In: Appleton, D. (ed.), 4th edn., New York, US (1898). Online Library of Liberty. http://oll.libertyfund.org/titles/1167. Accessed 21 Jan 2020
11. Training dataset. www.kitco.com/charts. Accessed 21 Jan 2020
12. Vovk, I.V., Grinchenko, V.T., Matsypura, V.T., Snarsky, A.A.: Dozen of lectures on fractals. In: From Object of Admiration to the Tool of Knowledge: The Manual. Lenand, Moscow, Russia (2018)

Comparison of Key Points Clouds of Images Using Intuitionistic Fuzzy Sets

Stanislav Belyakov[1(✉)], Alexander Bozhenyuk[1], Kirill Morev[1], and Igor Rozenberg[2]

[1] Southern Federal University, 44, Nekrasovsky, 347922 Taganrog, Russia
beliacov@yandex.ru, avb002@yandex.ru, morev-ki@yandex.ru

[2] Public Corporation "Research and Development Institute of Railway Engineers", 27/1, Nizhegorodskaya Str., Moscow 109029, Russia
i.yarosh@vniias.ru

Abstract. The work is devoted to solving the problem of tracking objects with modern robotic systems. An algorithm for tracking objects based on the representation of images in the form of clouds of singular points using intuitionistic fuzzy sets is considered. The developed method takes into account the geometric structure of the tracked object. The truth of the found comparisons of the singular points of the images is evaluated by categories such as "not quite accurate", "almost exactly", "absolutely accurate". The results of the experiments, as well as estimates of the stability and speed of the algorithm, demonstrate the high quality of the task of tracking objects in a video sequence.

Keywords: Tracking · Accompaniment of objects · Special points of images · Intuitionistic fuzzy set

1 Introduction

Computer vision, as an area of scientific knowledge, is actively developing at present. The development of new methods of computer vision and their implementation is the main direction of development of modern intellectual complexes of technical vision. However, to perform complex tasks, it is necessary to analyze a huge amount of information in real time, which imposes significant restrictions on the ability to implement intellectual behavior. One of the methods for processing optical information that allows you to represent the image in a more compact form is the representation of this image in the form of a cloud of key points (KP). In this paper, we consider an approach to comparing two clouds of KP images based on intuitionistic fuzzy sets [1].

The task of tracking, that is, tracking an object in a video sequence, consists in finding a known template of the object in question in the frame of the video sequence and localizing its center [2]. There are many approaches to solving the problem of pattern matching. In this paper, we use the representation of images in the form of KP clouds.

In the general case, the task of comparing two KP clouds is reduced to finding the most exact correspondence between the descriptions of the neighborhoods (direction of

R. Silhavy (Ed.): CSOC 2020, AISC 1225, pp. 366–374, 2020.
https://doi.org/10.1007/978-3-030-51971-1_30

the brightness gradient, brightness histogram, etc.) of two points belonging to different images [3, 4]. An example of such a comparison is shown in Fig. 1.

Fig. 1. False matching example

It is supposed to solve the problem of searching for the image of an object in the image with large changes in the scale of the object, rotation angle or changes in projections. For this, it is necessary to determine the truth of the correspondence between two KP of different images. True is a correspondence that assigns a point from one image to the same point in another image, regardless of its spatial displacement, change in brightness or size. A false comparison is the opposite of the true.

In Fig. 1 shows a large number of false matches. Each yellow line connects two points of different images of one object and is a correspondence. There are lines connecting completely different points of the 3D scene, for example, the cat nose and background.

This paper proposes a new approach to evaluating the truth of correspondences. So, when studying some of the correspondences, estimates often arose like the following: "not quite right", "it would be a little to the left", "a little lower", etc. In an effort to recreate precisely human evaluations, it is proposed to use the theory of intuitionistic fuzzy sets.

2 Search Method

2.1 Intuitionistic Fuzzy Set

The main difference between an intuitionistic fuzzy set and a fuzzy set is the presence of a degree of uncertainty, in addition to the degree of belonging and non-belonging [1]. Thus, the intuitionistic fuzzy set can be represented in the form [5]:

$$A = \{ <x, \mu(x), v(x), \pi(x) > | x \in E \},$$

where x is an element of the set E, $\mu(x)$ is the membership degree of the element x to the intuitionistic fuzzy set A, $v(x)$ is the degree of non-membership, $\pi(x)$ is the degree of uncertainty.

Moreover:

$$\pi(x) = 1 - \mu(x) - v(x).$$

The main condition of the intuitionistic fuzzy set:

$$\mu(x) + v(x) \leq 1, \quad (1)$$

and if $\mu(x) + v(x) = 1$, then the intuitionistic fuzzy set becomes just a fuzzy set.

In Fig. 2 shows the interpretation of an intuitionistic fuzzy set in the form of an isosceles right triangle [1, 6]:

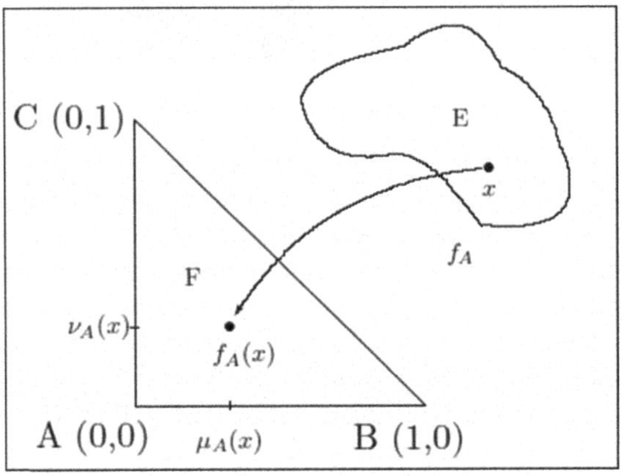

Fig. 2. Geometric interpretation of intuitionistic fuzzy set

2.2 Proposed Method

In this paper, we consider an algorithm for detecting the desired object in the search window (frame of a video sequence) that is resistant to changes in the rotation angle and changes in scale using the tools of intuitionistic fuzzy logic.

Traditionally, the task of tracking an object using KP images is divided into three stages.

First Stage. Key points of the image are found. The results of this step are KP positions and their sizes on the images of the object template and the search window.

Second Stage. A description of the surroundings of each of KP is carried out and their orientation (direction of brightness change) is calculated. The result of the stage is a description of the neighborhood of each point that is invariant to changes in scale and rotation.

Based on the analysis of existing methods, the BRISK algorithm [7] was chosen as the most suitable for work with significant changes in scale, rotation and projection changes in real time on noisy images. The BRISK method involves working with a binary descriptor, which greatly speeds up its work and the special deterministic structure of the template for testing gives it high resistance to changes in the rotation angle and scale of the observed scene.

The Third Stage. There is a comparison of KP of two images, the search for matches. In the new method, it is proposed to use the KP structure of the object image (distances and angles between all KP images) and search for KP in the same structure on another image. A similar approach greatly improves sustainability.

Further, using the method of exhaustive search, descriptions of the neighborhoods of all points are compared. We obtain a quantitative estimate of the difference in descriptors for all KP of two images. The resulting set of correspondence pairs is filtered by a relative threshold:

$$f(dif) = 1 - \frac{dif}{desc\_len},$$

$$\begin{cases} match = true, & if\ f(dif) \geq 0.8; \\ match = false, & if\ f(dif) < 0.8, \end{cases}$$

where *dif* is the size of the difference in descriptions; *desc_len* is the length of the descriptor; and *match* is the checked match. Thus, from the entire set of correspondences, pairs are selected that differ by no more than 20% in their descriptions.

The value of the difference in the descriptions of points is used in determining the non-membership degree of points to each other. The type of non-membership function is shown in Fig. 3, the maximum value of the difference between the descriptions is 512 (the length of the descriptor used). Non- membership function has the form:

$$\nu(x) = 1 - \frac{a\ XOR\ b}{512},$$

where a is the descriptor KP in the object template from the correspondence x; b is the descriptor KP in the search box from the same correspondence; x is the mapping being evaluated.

Further, for correspondences that have passed the threshold, the structure of the mapped points is calculated: a randomly taken match is assigned as a reference correspondence. Then, each point of the template and the search window is assigned a distance and orientation (angle) from the corresponding reference point.

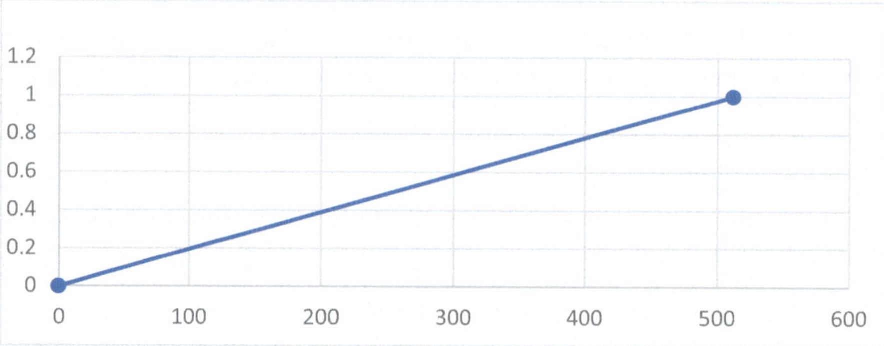

Fig. 3. Non-membership function ν(x)

The distance between points is defined as Euclidean:

$$D = \sqrt[2]{(X_i - X_j)^2 + (Y_i - Y_j)^2},$$

where D is the distance value, i and j are the indices of various points, and X and Y are the coordinates of KP centers in the pixel representation of the images.

The angle between points is defined as the angle between the vector connecting the two points (estimated and reference) and the positive direction of the Ox axis. The reference point is taken as the beginning of the vector.

In Fig. 4 shows the structure determination process. The yellow lines show the matches that have passed the threshold filtering. Red lines indicate the structure. The yellow circles show KP centers. The structure is determined from the reference point marked by a blue star by determining the distance and angles to all other points that have passed the threshold filtering.

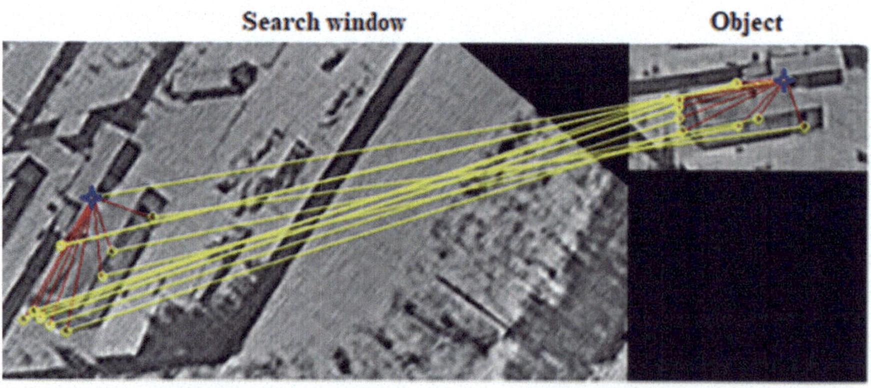

Fig. 4. Example of determining the structure of KP object and a search box

For both points of each filtered correspondence, there is distance and orientation. Based on the assumption that a true correspondence connects the same point of one object on two different scenes, it is fair to assume that when you move an object in the search window, change its scale and angle of rotation, all true matches form a cluster of points that are equally moving in space. The structure of the points will not be violated and, therefore, the distances and orientations of the points of true correspondences will change in direct proportion. In the ideal case, true correspondences form a cluster for which it will be true:

$$\begin{cases} \frac{M_1 dist_{obj}}{M_1 dist_{win}} = \frac{M_2 dist_{obj}}{M_2 dist_{win}} = \ldots = \frac{M_n dist_{obj}}{M_n dist_{win}}; \\ M_1 angle_{obj} - M_1 angle_{win} = \ldots = M_n angle_{obj} - M_n angle_{win}, \end{cases}$$

where $M_i dist_{obj}$ and $M_i dist_{win}$ are the distances KP of the object and the search box, respectively, from the *i*-th comparison, and $M_i angle_{obj}$ and $M_i angle_{obj}$ are the values of the KP orientations.

Thus, evaluating changes in orientation and distance for true matches, the change in the angle of rotation of the object and its scale is determined. The change in distance KP is defined as the ratio of its distance in the template to the distance in the search window. A rotation is the difference between the orientation of a point in a template and a search box. After finding the true correspondences and determining the parameters for changing the object, the localization of its center in the search window is not difficult. However, in real-life conditions, you have to work with noisy data. The noise is caused by the features of the system's detectors, errors in the sub-pixel localization of the KP center, and errors in comparing the KP neighborhoods. Based on the assumption of a normal noise distribution, correspondences will be considered true for which changes in scale and orientation will be within:

$$\begin{cases} \delta_{dist} \pm \sigma_{\text{dist}} \\ \delta_{angle} \pm \sigma_{\text{angle}} \end{cases},$$

where δ is the mean value, and σ is the standard deviation for the distribution of changes in distances (*dist*) and rotation angle (*angle*).

Thus, having received changes in the distance and the angle of rotation for all the tested correspondences, the average value of δ and σ for the distribution is considered. If the change in scale or orientation of the correspondence deviates from the average value by more than σ, the correspondence is considered false. A relative deviation less than σ is considered a degree of belonging. In this case, the membership function will look like:

$$\mu(\text{x}) = \max\left(1 - \frac{|dist - \delta_{dist}|}{\sigma_{\text{dist}}}, 1 - \frac{|angle - \delta_{angle}|}{\sigma_{\text{angle}}}\right),$$

where *dist* is the change in distance KP; *angle* - change of its rotation angle; δ_{scale} and δ_{rotate} - the average distribution values change the scale and angle of rotation, respectively; σ_{dist} and σ_{angle} are the corresponding standard deviations.

Thus, for each correspondence, the degree of membership function and the degree of non-membership function are determined.

However, according to property (1) of the intuitionistic fuzzy set, the values of the membership function and non-membership function in the total should not exceed one.

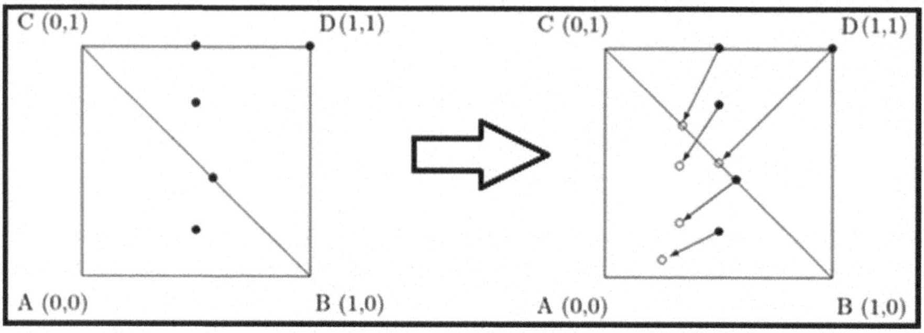

Fig. 5. Interpretation of an intuitive fuzzy set for the given functions μ (x) and ν (x) (left); Converted intuitive fuzzy set to the correct form (right)

And when using the above functions, cases of violation of the above condition are possible (see Fig. 5 on the left).

To solve this problem, an approach is used [8], which consists in switching from a single square (Fig. 5 on the left) to a classical triangular interpretation of an intuitionistic fuzzy set (Fig. 5 on the right).

The transition is carried out using continuous bijective transformation:

$$F(n,m) = \begin{cases} <0,0>, \ \mathit{if}\ n = m = 0; \\ <\frac{m^2}{m+n}, \frac{mn}{m+n}>, \ \mathit{if}\ n,m \in [0,1]\ \mathit{and}\ n \geq m\ \mathit{and}\ n+m \neq 0; \\ <\frac{mn}{m+n}, \frac{m^2}{m+n}>, \ \mathit{if}\ n,m \in [0,1]\ \mathit{and}\ n \leq m\ \mathit{and}\ n+m \neq 0. \end{cases}$$

Further, applying the above transformation to the obtained values of the membership degree and non-membership, estimates of the correspondence between pairs of points are calculated. Ultimately, iteratively assigning each correspondence as reference, pairs of mutually corresponding points are determined that retain the structure in the largest number of iterations. Thus, the true correspondence is determined, as well as the parameters for changing the scale, angle of rotation and displacement of the object.

Next, the obtained correspondence estimates are processed to determine their truth and falsity [9, 10]. For this, the following rules are introduced:

a) If $\mu(x) > \nu(x) + \pi(x)$, then the correspondence is true;
b) If $\nu(x) \geq \mu(x) + \pi(x)$, then the correspondence is false;
c) If $\mu(x) + \pi(x) > \nu(x)$, then an additional check is carried out.

When rule “c” is fulfilled, it is necessary to conduct an additional study of the neighborhood of the search box point from the pair being checked. Two different cases are considered here.

Case 1: $\mu(x) + \nu(x) < \pi(x)$.

This suggests that the structure of the singular points is weakly preserved, and the neighborhoods coincide strongly. This indicates either a false correspondence or a change in the shape of the object that violates the structure of points (for example, a projective transformation with a change in the parallelism of lines). To eliminate this uncertainty, we consider all points in the radius $R = 0.25m$, where m is the largest dimension of the object (using the scale obtained after evaluating the conservation of the structure of the singular points). If more than half of the key points located in the evaluated areas correspond to each other, then the match is considered true, and the object is considered changed in the search window.

Case 2: $\mu(x) + \nu(x) > \pi(x)$.

This case indicates the preservation of the structure of the object with a difference of neighborhoods of points from the estimated pair. A similar situation occurs when the object is blocked by obstacles (tree crowns, walls, landscape elements, etc.). In this case, to eliminate the uncertainty, a repeated search is made for the point of the object from the estimated pair. The radius of the search area is assigned $R = 0.1m$, where m is also the largest size of the object. If a singular point is found in the indicated neighborhood that corresponds to the point of the object by at least 80%, then the point of the object is replaced from the pair being evaluated by a new, just found point. Otherwise, the match is considered false.

3 Experiment Results

The developed object tracking method was implemented in the C++ programming language. In order to assess the stability and speed of the proposed method, experiments were conducted. Used platform: processor Intel Core i7-920 with a clock frequency of 2.66 GHz and a RAM capacity of 4 GB. The experimental video consisted of a clockwise rotating image in increments of 5 degrees per frame and increased in scale by 16% per frame. The size of the tracked object was 150 pixels wide and 100 high, the size of the search window was 768 × 768 pixels. During the experiments, it was found that the average speed of searching for an object throughout the frame was 0.2 s. However, when the search window is reduced to 300 × 300 pixels, operation is

Table 1. Characteristics of sustainability of the tracking method.

Parameter	Rating
Permissible changes in the angle of rotation of the object	0°–360°
Permissible object scale changes	Decrease and increase by more than 2 times
Permissible changes in the position of the object	Within the whole frame

achieved at a speed of 0.05 s, which corresponds to a processing speed of 25 frames per second. Table 1 shows the stability characteristics of the tracking method.

The resistance of the method to projective and perspective changes is low, however, it can be improved by timely updating of the object template when detecting the slightest distortions.

4 Conclusion

The presented algorithm for comparing KP clouds does not require high computing power, which allows for its operation in real time. The high-quality work of the method in noisy, low-quality images confirms the possibility of its use in modern systems of technical vision. The method is weakly resistant to perspective and projective distortions of the object in the search window. However, regular updating of the object template eliminates this weakness. The work of the method is based on assessing the truth of correspondences using intuitionistic fuzzy sets. The fuzzy conclusion used makes it possible to determine the truth and falsity of correspondences with high reliability. Thus, the use of intuitionistic fuzzy sets opens up great opportunities for the development of the intelligence of modern robotic systems.

Acknowledgments. The reported study was funded by RFBR according to the research projects #19-07-00074, #20-01-00197

References

1. Atanasov, K.: On Intuitionistic Fuzzy Sets Theory. Springer, Heidelberg (2012)
2. Tassov, K.L., Bekasov, D.E.: Overlapping solving in tracking tasks. Eng. Mag. Sci. Innov. **6** (18), 1–18 (2013)
3. Salahat, E., Qasaimeh, M.: Recent advances in feature extraction and description algorithms: hardware designs and algorithmic derivatives. In: Computer Vision: Concepts, Methodologies, Tools, and Applications, pp. 27–57. IGI Global (2018). https://doi.org/10.4018/978-1-5225-5204-8
4. Bay, H., Tuytelaars, T., Van Gool, L.: Surf: speeded up robust features. In: Computer Vision - ECCV, pp. 404–417. Springer, Heidelberg (2006)
5. Atanassov, K.: Intuitionistic fuzzy sets. Fuzzy Sets Syst. **20**(1), 87–96 (1986)
6. Atanassov, K.: More on intuitionistic fuzzy sets. Fuzzy Sets Syst. **33**(1), 37–45 (1989)
7. Leutenegger, S., Chli, M., Siegwart, R.: BRISK: binary robust invariant scalable keypoints. In: IEEE International Conference on Computer Vision (ICCV), pp. 2548–2555 (2011). https://doi.org/10.1109/fcvpr.2011.5995387
8. Atanassov, K.: Remark on a property of the intuitionistic fuzzy interpretation triangle. Not. Intuitionistic Fuzzy Sets **8**(1), 34–36 (2002)
9. Bian, J., Lin, W.Y., Matsushita, Y., Yeung, S.K., Nguyen, T.D., Cheng, M.M.: GMS: grid-based motion statistics for fast, ultra-robust feature correspondence. In: 2017 IEEE Conference on Computer Vision and Pattern Recognition (CVPR), pp. 2828–2837 (2017)
10. Yang, M., Wu, Y., Hua, G.: Context-aware visual tracking. IEEE Trans. Pattern Anal. Mach. Intell. **31**(7), 1195–1209 (2009)

Spatial Analysis Management Using Inconsistent Data Sources

Stanislav Belyakov[1(✉)], Alexander Bozhenyuk[1], Andrey Glushkov[1], and Igor Rozenberg[2]

[1] Southern Federal University, 44, Nekrasovsky, 347922 Taganrog, Russia
beliacov@yandex.ru, avb002@yandex.ru, andrey@glushkov.net

[2] Public Corporation "Research and Development Institute of Railway Engineers", 27/1, Nizhegorodskaya Str., Moscow 109029, Russia
i.yarosh@vniias.ru

Abstract. Cartographic spatial analysis plays an important role in decision making process formed by a user-analyst during a dialogue with geographic information system. The validity of the analysis result is determined by the content of the analysis workspace constructed by the analyst. Fragments from sources of inconsistent spatial data are included in the work area. The quality of such area mostly is far from satisfactory, but significant for solving the researched problem. As a result, display defects occur. It negatively affects the perception of the map and complicates the analysis process. In this paper, we consider the problem of controlling the analysis process for visual anomalies representation occurred due to the use of inconsistent data. A particular task mental image defects impact and situational awareness of analyst are analyzed. Semantic orientation analysis concept and the need for using analysis contexts associated with it are studied. An analytic model is proposed. The control problem is formulated as the process of choosing the closest meaningful context when an abnormal level of work area defective objects number occurs. The proposed method for solving the problem is based on looking for context sequences during an analytic session that meets the requirement of semantic proximity.

Keywords: Map data analysis management · Intelligent systems · Inconsistent data sources · Geoservice

1 Introduction

The need to obtain geodata regarding a particular problem situation in order to predict its expansion, research possible alternative solutions in various areas of business, production and planning. Being part of a complex system control loop, analysts solve hard-to-formalize problems using spatial data and data relationships. The search for a solution to be executed in real time. To find a way out of this situation, the analyst needs to study the terrain, evaluate the roots of the current situation, classify it and begin to search for a solution. To perform all the actions above, the expert analyst builds an analysis workspace using the geographic information system or geoservice.

R. Silhavy (Ed.): CSOC 2020, AISC 1225, pp. 375–384, 2020.
https://doi.org/10.1007/978-3-030-51971-1_31

The built area is a general map fragment stored by the geoservice. The analyst selects data which is useful for further analysis for the workspace. These steps are repeated until an acceptable solution to the problem is there.

Map data may contain errors, inaccuracies that spawn display defects of map fragments [1]. The presence of defects in the work area is not considered as abnormal, however, it carries the risk of decreasing the situational awareness of the analyst. This may be followed by making a wrong decision.

Shaping the situational awareness by the analyst is the result of understanding the analysis workspace map visual representations. Emotions play a significant role in this process. In particular, the desire to obtain important information forces one to turn to spatial data sources that have not undergone the necessary cartographic processing [2, 3]. A good example is data from social networks. Having potentially a high value, this kind of data provoke not only visual analysis errors, but also the loss of analysis semantic direction due to a raise of cognitive and emotional load. The example on Fig. 1 shows the case when an attempt to study GPS tracks leads to anomalies during the display of the work area.

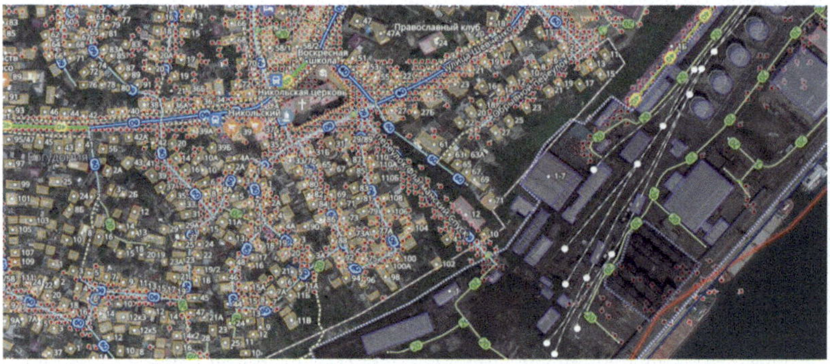

Fig. 1. Example of display anomaly

It may seem like an obvious solution applying a "rollback" to the previous analysis step in order to restore the quality of the workspace and avoid critical distortion of the analyzed area. Indeed, technically this is not so difficult. However, the impact on the mental representation of the progressing task in the analyst's mind remains irreversible. The information obtained cannot be ignored and this inevitably affects the further course of the interactive research. It turns out that the usefulness of the workspace has declined and a new flow of cartographic data should compensate for the loss of meaning. The analyst should be able to continue the analysis in the chosen sense direction, using a different combination of objects and relationships.

This paper analyzes the approach to detect anomalies in visual analysis and developing an appropriate strategy to continue the analysis.

2 Method Search

2.1 Overview of Known Approaches to Solve the Problem

Selecting useful cartographic data for analysis to solve the most important problem during geoservices operation [2]. The use of irrelevant and inaccurate information sources is inevitable and requires the application of appropriate measures to be safe from this obvious danger. Modern geo services are considered to be big data systems that accumulate spatial data from various sources. They usually include data monitoring tools, sensor networks, aerospace surveillance systems, even social networks messages. Information has a different quality, but carries a spatio-temporal reference and can be used for spatial analysis. The risk of using corrupt and inaccurate data is generally reduced by certification of sources and the use of regulated templates and data selection techniques. However, this approach does not give the expected positive effect for solving non-common problems. The creativity of the analyst and important, but insufficiently verified data are required. There are no such analysis anomaly detection tools in modern systems implemented.

The earth's surface maps representation reliability decreases over the time. This process is inevitable. Moreover, the normative period for updating cartographic data may not match to the speed of real world processes. For example, change in topography caused by a natural disaster can appear on maps only after a few years. To reduce the risk of such errors, operational mapping is used [3]. It is performed either automatically or manually by volunteer cartographers. Due to this, objects appear on the maps, the images of which are inaccurate and not consistent with the surrounding basic elements of the geographical basis. Similar objects generate visually visible defects in cartographic images. Despite the distortions, visualizing them on map is still useful for most cases, since it increases the level of situational awareness of the analyst. As the analysis of publications showed, the distortion effect on the subsequent use of cartographic information is not considered.

Image defects on maps are detected using software from the common set of geoservice functions [4, 5]. Defect detection algorithms are based on geometric relationships check. For example, overlapping lines on top of each other, duplication of the same objects, unacceptable intersections of shapes of certain layers, and others. However, geometric constraints are not the only factor determining the meaning of the image. The meaning of different geometric shapes combinations most of the time could not be formally followed from their shape ratio and size. In addition, defect detection functions are easy to use in map design mode, but are difficult to apply for interactive mode. Therefore, it is impossible to solve the problem with existing approaches without semantic control of the map image.

Analyzing modern cyber-physical systems security approaches [6], it should be noted that the threat of informational impact on the decision-making process belongs to the application layer. Already known methods of protection can reduce the risk to some level. For example, authentication of inconsistent data sources guarantees a certain quality of geodata. However, it is impossible to exclude non-authenticated sources from the analysis process, the presence of which critically changes the work area.

It turns out to be insufficient to establish a level of trust for each source, which is designed to display the level of risk of using information. An attempt to study and evaluate each data source that arises in the network is enormous and cannot be practically used.

There is a known protection approach against the considered information threat. It is based on the classification [7, 8]. The essence of the approach is to collect data on the precedents of impacts observed in reality and the subsequent application of one of the machine learning models to obtain a classifier of impacts. This approach gives a good result when there are informative signs of data inconsistency. In practice, this is difficult to do, since a significant role is played by visual analysis, where the mechanisms of the psychophysical perception of the user-analyst are involved.

2.2 Context-Based Analysis Management

By context during context-aware system research most often considered any information that allows you to identify the current situation and take adequate actions to solve the problem [10]. We assume that during interactive data interaction with a geoservice, the context description is defined by the boundaries of the spatial, temporal and semantic areas of analysis:

$$c = \{L_S, L_T, L_R\}$$

Spatial (L_S) and temporal (L_T) boundaries are continuous quantities that can be set either directly cartographically, or parameterized analytical relationships, or expert rules. Semantic (L_R) border is a set of instances and classes of objects and relationships, as well as patterns of communication with external data sources [1, 2].

The context interpreter program included into the geoservice executes requests for querying and visualizing objects according to the boundaries of the context.

The context concept is quite reasonably associated with the concept of the cartographic image meaning. The place, time and types of cartographic objects on the map integrally characterize the image and content that arises in the mind of the researcher. User of the geoservice working in the selected context is protected from redundant data that does not correspond to the semantic content of the analysis. At the same time, the less obvious is the preservation of meaningfulness in the sequence of changing contexts that is generated when solving informal problems. The search in this case consists not only on a visual terrain research, but also on a perception change of semantic angles. For example, the real estate cost forecast in a given city area is based not only on information about various types of buildings in the area and the existing infrastructure, but also, possibly, on other objects, which affect the political, investment, psychological and physical climate of the area. Like a stability of demographic flows, sustainability of the ecological development of the territory, etc. The analysis process in this case is filled with a new meaning, which cannot be obtained by simply summing up the meanings of standalone contexts. The experience of executing the analysis is valuable due to the fact that the discrepancy between the context and the meaning of the situation may arise in the analysis process with a fairly high probability. Its significance is the higher, the more uncertain the statement of the initial problem and the more noticeable the "knowledge gap" in assessing the situation [11]. The key issue of using contexts is identifying the one that most closely matches the meaning of the current situation and can be effectively used to continue the analysis process. There is a need to use the proximity metric, reflecting the researched objects observation experience [12]. Metrics are closely related to the conceptual model of the analysis process.

Let us consider the features of context-based analysis management. We assume that a geoservice operates many contexts

$$C = \{c_1, c_2, ..c_M\}$$

each can be initialized a random number of times during the session. For example, a corporate geoservice may include the "Engineering Networks" context, containing rules for the selection of different types of pipelines, wired and radio communications, and related equipment; "Logistics" context, could contain rules for selecting transport systems of an enterprise, warehouses and storage sites, packaging and loading of products. The set of contexts used in the session forms a sequence $\bar{c}_k$

$$\bar{c}_k = (\bar{c}_{1k}, \bar{c}_{2k}, \ldots, \bar{c}_{nk}), \bar{c}_{ik} \in C, \bar{C} = \bigcup_k \bar{c}_k.$$

Sets of context sequences $\bar{C}$ reflects the experience of analysis gained by geoservice. By $\bar{c}_w$ we assume the current sequence of contexts at the moment of analysis process. We will consider the analysis process successful if the analyst has reached the desired level of situational awareness, using the minimum possible number of transitions from one context to another one:

$$|\bar{c}_w| \rightarrow min \tag{1}$$

To achieve this goal, we should keep in mind that:

- the level of situational awareness cannot be measured directly. It can only be claimed that it does not decrease with the increasing sequence length $\bar{c}_w$;
- the selection of context that will be used in the analysis process in the next step is subjective. We can only state that elongation $\bar{c}_w$ happens not randomly, but rationally. The meaning and reason for adding a context is clear only to the analyst;
- Observation experience of the analysis process (many sequences $\bar{C}$) reliably presents possible options for the semantic direction of analysis.

The listed factors are leading to a control problem to be solved (1) as follows: at each step of the analysis, which is actually a context change. The best one is considered to be the one that best matches the accumulated experience. This means the new context $c_n \in C$ previously used in the shortest known sequence:

$$c_n \in \bar{c}_k, \bar{c}_k \in \bar{C},$$

$$c_w \subset \bar{c}_k \,\&\, |\bar{c}_k| = \min_{\bar{c}_j \in \bar{C} \& |\bar{c}_j| > |c_w|} |\bar{c}_j|, \tag{2}$$

Thus, in the process of analysis, the geoservice predicts the best context in the sense of (1). This context should be considered as a meaningful continuation of the analysis, since it is chosen on the basis of experience.

When anomalies in the analysis process occur due to the use of inconsistent data sources, the context $c_n \in C$, defined from (2) provides a logical continuation of the process. Maintaining the integrity of the analysis should ultimately lead to better decisions.

2.3 Prediction of the Context

The computational complexity of $c_n \in C$ is related to a lack of information about meaningful analysis sequences. The experience accumulated by the geoservice only partially covers the possible implementation of the process. It is difficult to develop a deterministic algorithm due to the inaccessibility of reliable data on the level of the analyst's current situational awareness. As the basic method to predict the context, we are going to use decision trees [9]. You can accumulate knowledge about the analysis with it. Let's consider the features of the application of this method.

Imagine the experience data from the set $\bar{C}$ in the form of subsets containing sequences of length $1, 2, 3, \ldots N$ elements:

$$\bar{C} = \bigcup_i \bar{C}^{(i)},$$

where $\bar{C}^{(i)}$ - is a set with the length equals to i.

Last element of any sequence $c_n \in C$ to be considered as a forecast value. Then for each $\bar{C}^{(i)}$ a decision tree can be built at the vertices of which elements of the current sequence $\bar{c}_w$ are compared with the subsets of contexts established when constructing the tree. The search result in the tree is the value $c_n \in C$.

Since the sequences have a different length, the geoservice experience seems to be a decision tree forest. Using the well-known forecasting mechanisms using the forest, one can obtain the predicted value $c_n \in C$.

As an example, consider a forest of two trees in Fig. 2. The numbers indicate the context numbers. The selection predicate takes a true value if the current set of contexts which include at least one element of the set indicated to the left of the selection vertex. Otherwise, the transition on the right branch should be performed. End vertices contain predicted context numbers. The trees are single-tier and have a high generalizing ability. For $\bar{c}_w = \{7, 1, 4\}$ the forecasted context will be $c_n = 5$ as the most common forecast value. For $\bar{c}_w = \{5\}$ the forecast is ambiguous; the following may be either context 2 or context 4.

We expect that the complementing $\bar{C}$ with the new sequence instances will improve the quality of forecast [9]. However, an additional opportunity to improve quality is to structure experience. Let us filter out the following two parts from each sequence $c_n \in C$:

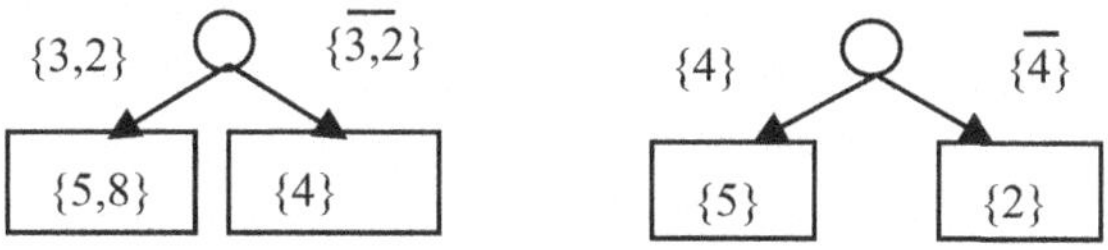

Fig. 2. «Forest» for context prediction example

$$c_n = c_n^l \cup c_n^o, \tag{3}$$

From the content point of view, the first part (c_n^l) represents a "long-term" component of analysis. It includes contexts used throughout the analysis session with the exception of a certain ending time interval τ. The component c_n^l content is an experimental data reflecting the increase in situational awareness of the user in the analysis process. They carry information about what is called "level, intercept" in the analysis of time series [13]. A component c_n^o reflects the "operational" stage of analysis. It includes the contexts used to complete the analysis. These contexts characterize the "trand, slope" [13] of the process, since they were the result of the analyst's previous thoughts. Criteria for selecting contexts in c_n^o is their falling into the time interval τ, which is around several minutes [14]. Importance of component selection c_n^o in that the contexts included in it are very likely lead to the end of the analysis session and indicate the achievement of the necessary situational awareness.

Representation (3) allows to improve forecasting by preliminary determination of the operational component c_n^o. Observed sequence of contexts $\bar{c}_w$ is used to identify the most promising areas of further analysis with the help of decision trees. As the contexts inside c_n^o are ordered, the analyst is invited to follow this sequence. If he does not, then a new trend is predicted taking into account the selected context.

2.4 The Control Implementation

The control problem (1) is solved in the process of interactive solution of the problem posed to the analyst. Geoservice acts as a recommendation system. At each stage of the analysis, it offers explicitly or implicitly the most promising context for further analysis. Analyst decides whether to follow or not to follow recommendations.

To make recommendations, data is used on sequences of contexts were used previously. This information is accumulated in geoservice log files filled with sequence

$$\{ < t_i, C(t_i), F(t_i) > \}, i = 0, 1, 2, \ldots, t_i \in [0, T]$$

where T is an analysis session duration, $C(t_i)$ - name of the context used at time t_i, $F(t_i)$ - geoservice function called through the client software interface at a time t_i. Using well-known algorithms [9], based on such data, decision trees can be constructed to predict the trend.

A recommendation to continue the analysis is formed at the time of fixating the display anomaly. The reason is communication with the source of inconsistent data.

Anomaly detection is based on the following principle: we assume that for each analyst the maximum number of objects N^* is subjectively set (in his profile), which is comfortably perceived by visual analysis of the cartographic image. Let

$$\alpha = n_d / N^* \tag{4}$$

be a proportion of objects in the analysis workspace containing a defect (n_d). α is an indicator for detecting anomalies: if it is greater than some threshold value α^*, then the number of objects with defects is anomalously high. Thus, the analysis management function must continuously monitor the number of objects with defects in the field of visual analysis.

3 Discussion

The final goal uncertainty for which the user-analyst uses geoservice is the main feature of analysis process management. So, it is difficult to construct a clear and definite model of interaction that allows us to evaluate and manage its usefulness [16, 17]. Extracting data from the logs of function calls and changing contexts in a session allow to solve this problem. It is could be achieved by accumulating experience from the process observation. Modern geo-services tools allow you to accumulate such data. Accordingly, the possibility of reducing uncertainty through the use of experience increases. Using sources of inconsistent data forces us to study the analysis process from meaning of user actions point. The formal definitions of the concepts "meaning" and "semantic direction" proposed in this paper made it possible to solve the analysis control problem in a new way. The practical utility of the proposed control method is determined by the accuracy of context prediction. For an experimental assessment of this value, a corporate geoservice was used, which included about 30 contexts. A random forest model was built for 200 implementations of analysis sessions using the scikit-learn library [15]. Only context identifiers were used as variables. On the test sample, the error was 18%. This value should be considered satisfactory, since a higher accuracy of prediction would rather indicate a retraining of the model. A promising way to increase accuracy may be to add variables that characterize the spatial and temporal boundaries of contexts. A distinctive feature of the considered analysis management method is the orientation to the decision-making process protection. Visual analysis of cartographic images greatly affects the situation image that arises in the analyst's mind. As a result, defects of displayed data reduce situational awareness. Therefore, the threat is identified not only by the data attributes, but also by the possible user reactions on them. As a result, specificity acquires a mechanism for restoring the analysis process. It is not a simple "rollback" to the previous point, but involves advancement in the most promising semantic direction.

4 Conclusion

The nature of visual presentation of the analysis area has a major impact on the decision quality. The need to obtain up-to-date information makes analyst users turn to sources of information which are inconsistent with the map, this data gives rise to cartographic display defects. At the same time has a high value for decision-making. On the one hand, this technique helps to find effective and innovative solutions. On the other hand, with an unacceptably high level of defects, there is a danger of losing the analysis semantic orientation. The result may be an inadequate decision. In this paper, we propose an analysis management method that reduces the likelihood of such a result. The further direction of the research is improvement of forecasting models that take into account the spatial and temporal boundaries of contexts.

Acknowledgments. The reported study was funded by RFBR according to the research projects #19-07-00074, #20-01-00197.

References

1. Longley, P.A., Goodchild, M., Maguire, D.J., Rhind, D.W.: Geographic Information Systems and Sciences, 3rd edn. Wiley, Hoboken (2011)
2. Shashi, S., Hui, S.: Encyclopedia of GIS. Springer, New York (2008)
3. Goodchild, M.F.: Citizens as sensors: the world of volunteered geography. GeoJournal **69** (4), 211–221 (2007)
4. https://www.autodesk.com/products/autocad/included-toolsets
5. https://www.esri.com/en-us/arcgis/about-arcgis/overview
6. Alguliyev, R., Imamverdiyev, Y., Sukhostat, L.: Cyber-physical systems and their security issues. Comput. Ind. **100**(9), 212–223 (2018)
7. Ntalampiras, S.: Automatic identification of integrity attacks in cyber-physical systems. Expert Syst. Appl. **58**(10), 164–173 (2016)
8. Lun, Y.Z., D'Innocenzo, A., Smarra, F., Malavolta, I., Di Benedetto, M.D.: State of the art of cyber-physical systems security: an automatic control perspective. J. Syst. Softw. **149**(3), 174–216 (2019)
9. Hastie, T., Tibshirani, R., Friedman, J.: The Elements of Statistical Learning. Springer, New York (2009)
10. Dey, A., Abowd, G.: Towards a better understanding of context and context-awareness. In: CHI 2000 Workshop on the What, Who, Where, When, and How of Context-Awareness, pp. 304–307 (2000)
11. Berndtsson, M., Mellin, J.: Active database knowledge model. In: Liu, L., Özsu, M.T. (eds.) Encyclopedia of Database Systems. Springer, New York (2018)
12. Kotkov, D., Wang, S., Veijalainen, J.: Survey of serendipity in recommender systems. Knowl. Based Syst. **111**(8), 180–192 (2016)
13. Kane, M.J., Price, N., Scotch, M., et al.: Comparison of ARIMA and Random Forest time series models for prediction of avian influenza H5N1 outbreaks. BMC Bioinform. **15**, 276 (2014)
14. Gibson, J.J.: A Theory of direct visual perception. In: Royce, J., Rozenboom, W. (eds.) The Psychology of Knowing. Gordon & Breach, New York (1972)
15. https://scikit-learn.org

16. Belyakov, S., Bozhenyuk, A., Rozenberg, I.: The intuitive cartographic representation in decision-making. In: World Scientific Proceeding Series on Computer Engineering and Information Science, vol. 10, pp. 13–18 (2016)
17. Belyakov, S., Belyakova, M., Savelyeva, M., Rozenberg, I.: The synthesis of reliable solutions of the logistics problems using geographic information systems. In: 10th International Conference on Application of Information and Communication Technologies (AICT), pp. 371–375. IEEE Press, New York (2016)

Correlation-Extreme Systems of Defect Search in Pipeline Networks

Sergey G. Frolov[1(✉)] and Anatoly M. Korikov[1,2]

[1] Tomsk State University of Control Systems and Radioelectronics, Tomsk, Russia
sgf2@tpu.ru, korikov@asu.tusur.ru
[2] Tomsk Branch of the Institute of Computing Technologies SB RAS, Tomsk, Russia

Abstract. Problems and problems arising during development of correlation-extreme systems (CES) for search of fluid leaks from pipelines under pressure are considered. Detection of defects in pipeline networks is a current scientific and practical task, as timely and accurate determination of pipeline defect coordinates and prompt elimination of this defect depends on the quality of the service provided to the consumer. The example of a heating network shows that the automated system based on the CES is able to monitor the defects of the heating network and notify the dispatcher of any abnormal situations to decide on further actions of the personnel serving the heating network. Technical means for implementation of such CES are proposed. #CSOC1120

Keywords: Correlation-extreme systems · Pipeline networks · Heat supply · Leakage localization · Correlation method

1 Introduction

Correlation-extreme systems (CES) are systems for processing information presented in the form of implementations of random processes (random fields) [1]. The CEC operation algorithm contains calculation of the cross-correlation function (CCF) (or autocorrelation function (ACF)) of random processes, which characterize the state of the control object (observation object, research object), and determination of coordinates of the position of the main extremum (global maximum) of the CCF or ACF [1], which in turn determine the state of the control object (observation, research). The theories and practice of CES are devoted to the works of many scientists, among which it is necessary to mention the works of A. A. Krasovsky [2, 3] and V. P. Tarasenko [1, 3]. In the cited works [1–3] there is an extensive bibliography on the topics of CES. It should be noted from this bibliography that the first All-Union Conference on CES was held in Tomsk from September 11 to September 14, 1979. A description of many applications of the CES is contained in the above-mentioned editions. Note the most important applications of the CES [1–3]:

- navigation, targeting and radar-location; These CES are referred to as Correlation Extreme Navigation Systems (CENS);

R. Silhavy (Ed.): CSOC 2020, AISC 1225, pp. 385–394, 2020.
https://doi.org/10.1007/978-3-030-51971-1_32

- management of technological processes; Such CES are referred to as Correlation Extreme Process Systems (CEPS);
- research of natural resources and environment; CES of this type are referred to as correlation-extreme geophysical systems (CEGS).

From the analysis of later literary sources, see for example [4], it follows that in the practice of application of the CES there is not only the emergence of new fields of their application, but also a significant development of classical fields of application of the CES. (It is sufficient to note the use of CEC in cruise missile navigation and guidance, robot control, etc.) Among the new promising applications of CES, we note the use of CES for detection and localization of defects in pipeline networks [5].

2 Correlation Method of Pipeline Defect Detection

Correlation method of defect detection in the pipeline section is a fairly common method to date [6]. This method is used to find defects in the pipeline section using industrial correlation leak detectors (CLD) [5, 6]. These instruments determine defects on the pipeline network with an error of up to 5 m. However, the use of these instruments in the automated defect detection system (for example, leakage of the heat carrier) on the pipeline is not possible due to the absence of their autonomous operation mode (CLD require the presence of the operator), as well as the high cost of CLD. Another bottleneck of this method is external interference, since a sound signal of up to 2000 Hz is used as the output signal in the CLD. The closely located infrastructure of the city (heavy roads, industrial zones with noisy production, construction works, etc.) can interfere with the signal removed by the CLD, as they all produce noise in the frequency band used by the CLD.

CLD are a type of CEC, more precisely a CEPS. In these systems, the correlation method is implemented as a cross-correlation analysis of two signals removed from a radiation source (in our case, a section of a pipeline with a defect), simultaneously at a known distance from each other. Applying a cross-correlation algorithm to the obtained data produces a curve for the correlation coefficient. By analyzing this curve, it can be understood whether there is a statistically significant relationship between the signals (in which case there are peaks on this curve corresponding to the time samples in which the signals are dependent on each other) or whether there is no such relationship.

In case of cross correlation of two signals, correlation coefficient graph has correlation coefficient dependence on time of displacement of one signal relative to the other [6]. Knowing the distance between the sensors from which the signal was removed, as well as the process parameters of the heat carrier in the pipeline affecting the speed of sound propagation in the liquid medium (such as pressure and temperature), it is possible to calculate the distance from the sensors in the section of the pipeline where the defect was detected. Figure 1 shows the cross-correlation of the two signals.

The equation for the velocity of sound propagation in the liquid medium is known [7], therefore, based on the known equations [7] and cross-correlation algorithms [6], it

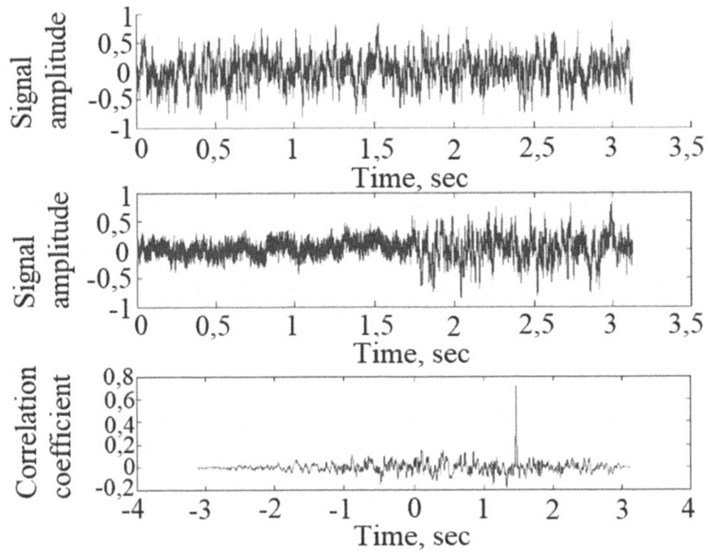

Fig. 1. The result of applying the cross correlation algorithm to two generated signals

is possible to create a CES for detecting a defect (leakage of the heat carrier) in the pipeline network.

3 Development of Automated Control and Accounting System for Heat Supply Based on CES

It has been noted above that the method based on the correlation analysis of two signals recorded from the pipeline section by sensors installed at a known distance between them is used in industrial CT. CT is expensive equipment and requires direct operator involvement with the site of the supposed defective site [6]. These circumstances make it impossible to use this equipment as a hardware part of the automated system for detecting and determining leakage coordinates in heat and water supply systems. Next, the use of CES in the automated control and accounting system (ACAS) of heat supply, shown in Fig. 2 below, is discussed.

The system structure includes hardware (intermediate and main accounting nodes, servers), software (data storage, processing, analysis and visualization subsystems) and dispatching control. Backup of ADCS servers ensures continuous operation of the system by hot switching of the system from the main server to the standby server. The developed automated system provides early detection of leaks, reduces heat losses and improves the quality of service to the population. The detection time depends on the methods of detecting leakage in the heat supply system.

In order to detect the site of pipeline damage in a timely manner and to quickly eliminate the fault in the heat supply system, various methods are used, both to detect the leakage of the heat carrier at an early stage of its appearance and to determine the

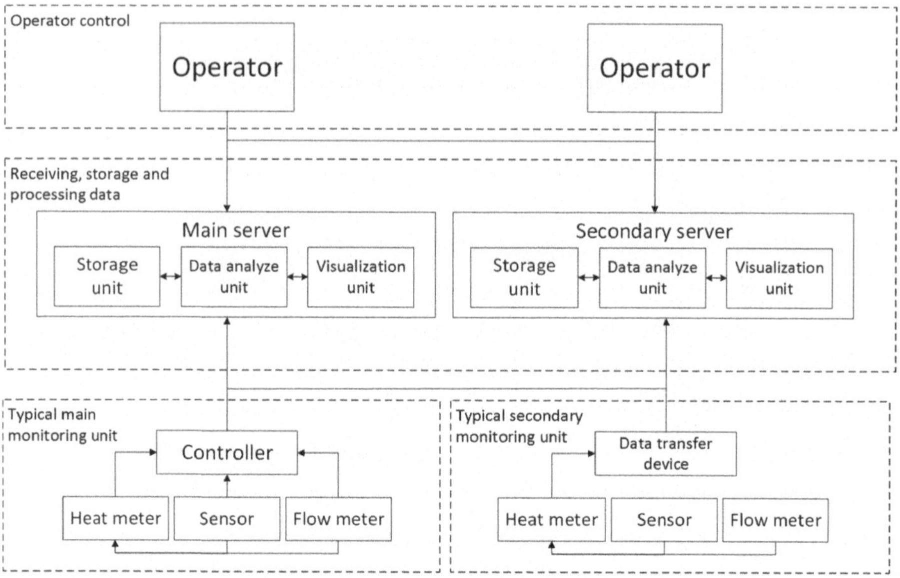

Fig. 2. Automated control and accounting system for heat supply

site of pipeline damage sufficiently accurately. Two groups of leak (pipeline damage) detection methods are known:

1) methods based on continuous monitoring of process parameters of the heat carrier in the heat supply system;
2) methods based on control of transient processes in the heat supply system.

We will briefly analyze the methods of leak detection and assess the possibility of their practical application in heat supply.

3.1 Methods Based on Monitoring of Heat Carrier Parameters

Methods of this group ensure timely detection of the defect at an early stage of its occurrence, but cannot determine with sufficient accuracy the specific location of leakage, as this requires tight coating of the heat supply system with sensors, controllers and network modems [8]. One such method is automated analysis of current parameters of the heat carrier. This method is used in the first leak detection line because it uses continuously monitored process parameters of the heat carrier (pressure, flow rate and temperature of the heat carrier before and after the consumer) as the data source. These indicators are constantly monitored by the manager, who in turn performs an analysis of the compliance of the received data with the data that is inherent in the current temperature and network configuration. The process data in the heat supply system are collected constantly under any conditions, so this method does not require special costs for automated data analysis to detect defects in the system. This method also allows to detect leakage in the heat supply area during the period of least activity of consumers. Such a period includes, for example, a time interval from 2 a.m. to 3 a.m. If during this period of time the water flow in the monitored area exceeds reasonable limits, there is a defect in the monitored area of the network. However, this method has a big disadvantage: dependence of flow rate and pressure measurement error on pipeline diameter. The larger the diameter of the monitored pipeline, the greater the variation in sensor readings can be interpreted as an error.

3.2 Methods Based on Control of Transient Processes in the Heat Supply System

Methods related to the second group provide not only detection, but also determination of coordinates of leakage place with sufficient accuracy. These methods require expensive equipment to obtain the necessary data, since in order to implement the method, it is necessary to perform some inherent manipulation on the observed network section (for example, to send a signal to the pipeline or to record a signal by a special device from the pipeline section and to process it). Such methods include a leak detection and localization method based on standing wave difference, as well as a correlation method. The correlation method of detecting heat carrier leakage is by far a fairly common method and is discussed by us above, so consider briefly the features of the method based on standing wave difference.

The leak detection and localization method based on the standing wave difference is based on the principles used in electrical and communication engineering to determine where the cable line breaks. Sinusoidal excitation of the cable by the generator at one end and simultaneous measurement of voltage and current is used. When a cable breaks, its resistance changes, that is, the relationship between voltage and current changes, incoming waves are reflected, and residual standing waves are created. The distance from the signal generation point to the cable break is determined from the analysis of the respective resonant frequencies.

In order to implement the method of difference of standing waves in the system of pipelines, steady-state oscillatory flow is induced by means of movement of the gate valve with low amplitude and frequency characteristic of maximum change of pressure in the area of disturbance for a certain frequency range is analyzed.

Each defect (feature) of the heat supply system generates secondary waves, which introduce changes in the pressure amplitude at the disturbance area [9]. By examining these changes and determining the frequency of the resonance oscillations, it is possible to obtain the distance from the signal generator to the site of the pipeline defect. This method is useful for finding a pipeline defect, but it has a significant disadvantage - the need for a signal source. In order to determine the coordinates of the leakage point by the method of standing wave difference in the pipeline, a sufficiently powerful signal is required, which can be generated by micro-movement of the shut-off valves of the network. However, there are significant sections of the network in the piping system that lack isolation valves. In addition to the mentioned disadvantage, there are difficulties of automation of this method, which are caused by the need to obtain special permits from state structures for remote management of technological facilities of critical infrastructure [10, 11]. The correlation method of detecting heat carrier leakage has no noted disadvantages, so it is a fairly common method to date. But there are features of application of this method in ACAS, which should be taken into account in implementation of ACAS subsystems.

When implementing the ACAS subsystems responsible for recording data directly from the pipeline, it is necessary to take into account that the pipeline defects produce noise in the range of up to 2000 Hz [6]. According to Kotelnikov's theorem [12], in order to process an analog signal with such a frequency range, a device having a sampling frequency of at least twice the maximum frequency of the input signal (in our case, not less than 4000 Hz) is necessary. As a result of the study, an apparatus for recording data from a potentially defective pipeline section was developed. Using this device, the ACAS in the heat supply system receives the necessary data, analyzing which, it is possible to determine the presence or absence of a defect on the investigated section of the heat supply system. When designing a device for recording data from a potentially defective pipeline section, questions arise as to the choice of technical means for implementing such a system, which are discussed below.

4 Select Device to Receive Signal from Potentially Defective Pipeline Section

Based on the price-quality ratio, the following two types of devices are the most suitable for the development of CES: Raspberry Pi family devices and Arduino family devices (Nano, Uno, Mega).

The Raspberry Pi family of devices are positioned on the market as single-board computers and have the following disadvantages:

- The size of the device, the Raspberry Pi is larger than Arduino, which negatively affects the compactness of the device;
- Cost;
- Modularity - in case of failure Raspberry Pi will have to change the whole device;
- Power consumption - a large number of redundant peripherals (network card, video outputs, USB ports) consume energy, which negatively affects the autonomy of the operation.

As the basis of the sensor capable of digitizing the signal in the required range and then transmitting it to the processing server, it is proposed to use Arduino Nano with an Atmel Atmega 328P microprocessor. This device is capable of processing input signal with frequency up to 8000 Hz. It also has the following advantages:

- Low cost - the cost is extremely low compared to competitors in the budget price range;
- Modularity - in case of failure of any module it can be quickly replaced;
- Scalability - Arduino platform has a huge number of modules for various purposes, which allows to develop the necessary functionality without spending a large amount of resources on the prototype;
- Compact - Arduino Nano is small;
- Ease of firmware development - Arduino has its own IDE for sketch development [13].

These advantages make it possible to use the device under development as a stand-alone sensor for the CES, capable of removing data on request from the system and transmitting it for processing. However, microprocessor devices of industrial manufacture, such as Adruino, have the following disadvantage: excessive power consumption due to the use of a power stabilizer and a USB controller (CH340 in the case of the analog and FTDI in the case of the original device). With these modules, Arduino Nano consumes 20 mA alone and 0.5 to 1 mA in the absence. This disadvantage can be compensated for by the larger capacity of the battery pack.

5 Description of the Prototype Signal Receiving Device for the CES

The prototype device includes the following components:

- Microprocessor board - the basis of the whole device, contains firmware, which is responsible for logic of device operation;
- Battery pack to provide battery life
- Flash module - data is stored in the memory, which will be subsequently sent to the server;
- Signal amplifier - amplifies the signal removed from the removing head;
- Removing head - reads the signal directly from the pipeline;
- Module for data transfer to the server.

A general diagram of the prototype signal acquisition device for the CES is shown in Fig. 3.

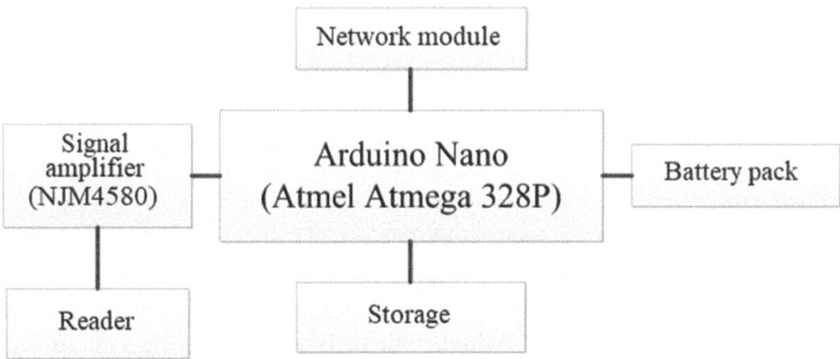

Fig. 3. Prototype of signal receiving device for CES

The principle of operation of the prototype signal receiving device for the CES is as follows:

After receiving the signal from the CES server, the device goes out of sleep and starts a recording cycle. The removal head, consisting of a piezoelectric element attached directly to the pipeline, records data that, passing through the amplifier, is read by the processor and stored in the flash memory. After recording some data, the device stops recording and, using the data transmission module, sends the received data to the CES server for processing and goes into sleep mode. Sleep mode requires a prototype to increase the battery life of the device. In sleep mode, energy is consumed only to check the presence of a packet from the server to start the operation cycle.

The prototypes operate independently of each other and have no communication with each other, but the CES server polls pairs of devices installed in one section of the pipeline at a known distance between them.

This prototype is part of the hardware and software complex of the CES for detecting and localizing leaks in the heat supply system, the use of which will allow to quickly determine the places of leaks (defects) on the pipeline section and eliminate them.

6 Hardware and Software Complex of CES for Detection and Localization of Leaks in the Heat Supply System

ACAS hardware and software system consists of several large modules:

1) primary monitoring module of network parameters;
2) module of communication with sensors;
3) data processing and analysis module;
4) network status display module

The primary monitoring module of network parameters is based on the already existing SCADA system [14], which continuously monitors the necessary technological parameters of the heat supply system. If they are not normal, the module sends a request for data from a potentially defective heat supply area using the sensor communication module.

The sensor communication module includes communication equipment, by means of which the process system sends a request for data from the sensor at the required moment of time and receives data, and a software component performing primary processing of the received signal (conversion into the necessary data format for subsequent processing).

The data processing and analysis module uses the obtained data to determine the presence of a fault in the area from which the signals were recorded. This module analyzes the results using the correlation method and, based on the analysis, outputs the necessary message to the manager (whether there is a defect in the system section or not).

The data processing and analysis module uses the obtained data to determine the presence of a fault in the area from which the signals were recorded. This module analyzes the results using the correlation method and, based on the analysis, outputs the necessary message to the manager (whether there is a defect in the system section or not).

Set of network of sensors installed on sections of heat supply system and technological system capable of data collection and processing as required forms software and hardware complex, which implements automated system of detection and localization of leaks in heat supply system.

7 Conclusion

As a result of the study, the structure of the system of automated dispatching control and accounting (ACAS) of heat supply is proposed, short analysis of methods of detection of leakage (pipeline damage) is carried out and preference is given to the correlation method of detection of leakage heat carrier. The correlation method provides sufficient accuracy in the localization of defects in pipeline networks and, in particular, leakage sites in heat supply systems, so can be applied in practice as a method for the localization of leakage sites. In order to automate the localization of leakage sites, the methods of the first and second groups noted above should be applied in conjunction with each other, as well as with technological systems already existing in the enterprise. Continuous monitoring of the heat carrier parameters is used as a method of detecting a defect in the heating network, and the correlation method after detecting a defective area is used as a method of localizing a particular leakage site. An apparatus for recording data from a potentially defective pipeline section has been developed for implementing the CES-based ACAS. With the help of this device the CES detects and localizes leaks in the heat supply system receives the necessary data, analyzing which, it is possible to determine the fact of presence or absence of a defect in the investigated heat supply area. The automated system based on the CES is able to monitor the defects of the heating network and notify the dispatcher of the possible occurrence of any emergency situations in order to make a decision on their elimination.

References

1. Strangul', O.N., Tarasenko, V.P.: Correlation-extremal systems for navigation and location of mobile objects. Autom. Remote Control **62**(7), 1204–1211 (2001)
2. Ishlinskii, A.Y., Kuznetsov, N.A., Fedosov, E.A., Chertok, B.E.: Dedicated to academician Aleksandr Arkad'evich Krasovskii on his 80th birthday. Autom. Remote Control **62**(7), 1027–1041 (2001)
3. Goritov, A.N., Korikov, A.M.: Optimality in robot design and control. Autom. Remote Control **62**(7), 1097–1103 (2001)
4. Korikov, A.: Artificial intelligence in robot control systems. In: IOP Conference Series: Materials Science and Engineering, vol. 363, no. 1, p. 012013 (2nd International Conference on Cognitive Robotics; Tomsk; Russian Federation; 22 November 2017) (2018)
5. Korrelyatsionnye techeiskateli [Correlation leak detectors]. https://www.z-tec.ru/product-category/categories/teleinspektsiya-techeiskateli-trassoiskateli/techeiskateli/korrelyatsionnye/. Accessed 19 Jan 2020 (in Russ.)
6. Korrelyatsiya signalov [Correlation of signals]. http://bourabai.ru/signals/ts08.htm. Accessed 19 Jan 2020 (in Russ.)

7. Skorost' zvuka v zhidkostyakh [The speed of sound in liquids], https://www.fxyz.ru/формулы_по_физике/акустика/распространение_звука/скорость_звука/скорость_звука_в_жидкостях/. Accessed 19 Jan 2020 (in Russ.)
8. Pravila tekhnicheskoy ekspluatatsii teplovykh energoustanovok [Technical operation rules of thermal power plants]. https://www.e-reading.club/bookrea-der.php/129367/Pravila_tehnicheskoii_ekspluatacii_teplovyh_energoustanovok.html. Accessed 19 Jan 2020 (in Russ.)
9. Covas, D., Ramos, H.: Standing wave difference method for leak detection in pipeline systems. J. Hydraulic Eng. ASCE 2005. http://www.civil.ist.utl.pt/~didia/Publications/RI_05%20(2005)%20Covas%20et%20al.%20(JHE_ASCE).pdf. Accessed 19 Jan 2020 (in Russ.)
10. O bezopasnosti kriticheskoy informatsionnoy infrastruktury Rossiyskoy Federatsii. Feder. zakon ot 26.07.2017 № 187-FZ [On security of the critical information infrastructure of the Russian Federation. Feder. Law of 26.07.2017 No. 187-FZ]. https://fstec.ru/tekhnicheskaya-zashchita-informatsii/obespechenie-bezopasnosti-kriticheskoj-informatsionnoj-infrastruktury/285-zakony/1610-federalnyj-zakon-ot-26-iyulya-2017-g-n-187-fz. Accessed 19 Jan 2020 (in Russ.)
11. O vnesenii izmeneniy v otdel'nye zakonodatel'nye akty Rossiyskoy Federatsii v svyazi s prinyatiem Federal'nogo zakona «O bezopasnosti kriticheskoy infrastruktury Rossiyskoy Federatsii» . Feder. zakon ot 26.07.2017 № 193-FZ [On Amendments to Certain Legislative Acts of the Russian Federation in connection with the adoption of the Federal Law "On the Safety of Critical Infrastructure of the Russian Federation". Feder. Law of 07.07.2017. No. 193-FZ]. https://fstec.ru/component/attachments/download/2088. Accessed 19 Jan 2020 (in Russ.)
12. Teorema Kotel'nikova [Kotelnikov's theorem]. Accessed 19 Jan 2020 (in Russ.)
13. Arduino Nano. https://store.arduino.cc/usa/arduino-nano. Accessed 19 Jan 2020 (in Russ.)
14. Sistemy avtomatizirovannogo kontrolya i sbora informatsii (SCADA) [Systems of automated control and information collection (SCADA)]. http://bourabai.ru/dbt/scada.htm. Accessed 19 Jan 2020 (in Russ.)

Neural Network Model with Time Series for the Prediction of the Electric Field in the East Lima Zone, Peru

Juan J. Soria(✉), David A. Sumire, Orlando Poma, and Carlos E. Saavedra

Universidad Peruana Unión, Lima, Peru
jesussoria@upeu.edu.pe

Abstract. Global warming and climate change is a latent problem nowadays because it affects the quality of life of living beings that inhabit an electric planet; therefore, the atmosphere is charged with ions that constantly interact and achieve a continuous balance. Likewise, when the determined value of the electric field is exceeded in one location, this produces an electric discharge, which varies with the time of the day, month and seasons.

The variation of the electric field in the troposphere of the campus of the Universidad Peruana Unión, located in the area of East Lima, Peru, has been evaluated using a EFM-100 Sensor equipment which measures the electric field during the seasons of the year, and this study aims to predict the future measurements using artificial intelligence. The area of East Lima was mapped and the EFM-100 sensor was set for its exact location and to report outputs of the electric field within a radius of 37 km.

A neural network model was found that was supported by the descending gradient algorithm and the Levenberg-Marquardt algorithm of the MatLab libraries in the 2018 version. The neural network model had a mean square error (MSE) of 0.476184, the validation was 0.558515 and the testing was 0.464005. Finally, an electric field of 0.1682 v/m was obtained in the summer season, −0.66 v/m in autumn, −1.62 v/m in winter and −1.43 v/m in spring.

Keywords: Neural networks · Artificial intelligence · Electric field · Time series · Predictive algorithms

1 Introduction

1.1 Context for the Research Study

Atmospheric electric field monitoring is an important area for prediction in the area of East Lima, Peru, which is associated with environmental variables. In this research study, a methodology was used to develop the neural network model with time series considering the electric field measurements made with the EFM-100 atmospheric Electric Field Monitor sensor equipment located on the roof of the Faculty of Engineering and Architecture building of the Universidad Peruana Unión, as shown in Fig. 1.

R. Silhavy (Ed.): CSOC 2020, AISC 1225, pp. 395–410, 2020.
https://doi.org/10.1007/978-3-030-51971-1_33

Fig. 1. EFM-100 atmospheric Electric Field Monitor

According to (Bohari et al. 2014), electromagnetic fields (EMF) have a great impact on any modern society and cause acute effects such as burns and damage to humans. The worrying effect on long-term exposure of weak fields can be a bad influence on the human body; therefore the prediction of electric field values are very important using time series neural networks with high impact algorithms.

(Bohari et al. 2014) mentions that the electromagnetic field (EMF) has many types of radiation, such as microwaves, ultraviolet, gamma rays, X-rays, and it is necessary to determine their occurrence in the atmosphere. The EFM-100 measuring sensor collected the electric field records during the four seasons of the year 2019 as shown in Fig. (2).

Fig. 2. EFM-100 Sensor located in the Universidad Peruana Unión campus, Lima, Peru, In the East Lima Zone

(Nicoll et al. 2019) mentions "in order that truly global signals are considered in understanding the processes within the global circuit, many validating measurements must be made simultaneously at different locations around the world. Beyond thunderstorms, another area of current research in atmospheric electricity is the role that atmospheric electricity plays in modulating cloud properties". Therefore, the research work is a contribution to the study of the measurement of the electric field in the area of East Lima, Peru, as shown in Fig. (3).

(Martinez-Lozano et al. 2014) reports that data from the EFM-100 sensor in the city of Leon, Spain, indicated that the meter collected data from the laboratory at De La Salle Bajio University, country campus, which contains information on clear, partly

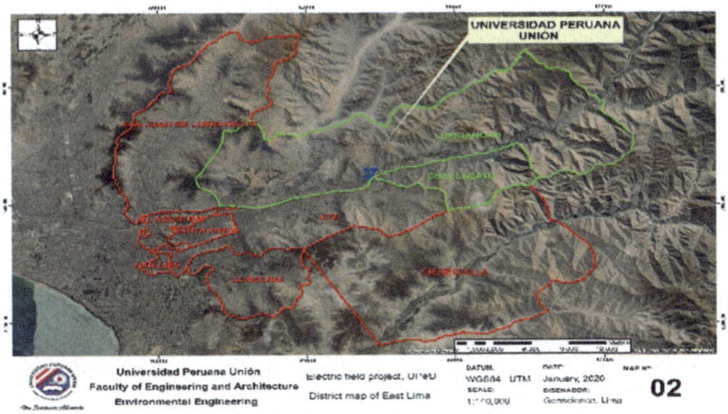

Fig. 3. Zona de Lima Este, Perú

cloudy and overcast days during this period, including thunderstorm days. A similar situation was recorded in the East Lima area, with data during cloudy, partly cloudy and clear days. The determination of air pollution is of great interest here because it is an urban area with a high student population.

According to (Silva et al. 2014) the atmospheric electric field in good weather is dominated by the vertical component, which is generally called the vertical potential gradient (**PG**).

The earth behaves like a giant capacitor, as (Siingh et al. 2007) mentioned, forming a global electric circuit (GEC) that links electricity, field and current flowing in the lower atmosphere, ionosphere and magnetosphere, as shown in Fig. (4), highlighted by (Jánský and Pasko 2015).

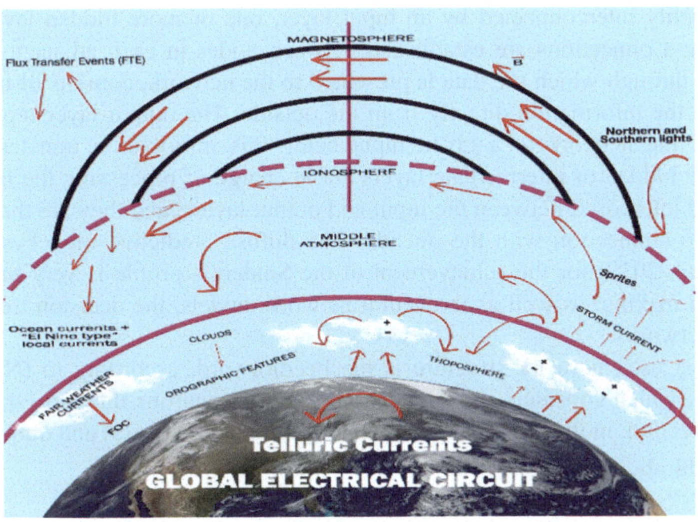

Fig. 4. Schematic diagram of electrical processes in the overall electrical circuit (Roble and Tzur 1986). The vector **B** shows the direction of the Earth's geomagnetic field and the arrows show the direction of current flow in the regions: tropospheric, ionospheric and magnetospheric, according to (Siingh et al. 2007).

This giant capacitor, which constitutes the earth, is charged by electrical storms to a potential of several hundred thousand volts (Roble and Tzur 1986) and conducts vertical current through the atmosphere. The current causes a weak electrification of stratified clouds so it is important to evaluate the role of solar variability in climate; also, to review what is known about the possible relevance of the atmosphere in the planet's electrical circuit associated with climate (Harrison 2004) that produces a vertical potential gradient (PG) in the layers of the atmosphere near the surface. The horizontal current flows freely along the highly conductive earth surface and in the ionosphere, which is closed by the current flowing from the ground into the storm and from the top of the thunderstorm into the ionosphere and back from the ionosphere to the earth through the good overall weather charge resistance ($\sim$100 Ω). The different regions of the atmosphere including the ionosphere and the magnetosphere are electromagnetically linked. (Siingh et al. 2007).

Furthermore, recent studies in this area, point out the need to have a global vision of the flow of currents in different regions of the Earth's environment and its possible linkage of the Global Electric Field, GEF with several other phenomena such as cosmic rays, atmospheric aerosols, climate. In this way, this article shows the electric field values recorded in the area of East Lima, Peru, during the four seasons of the year 2019, taking into account the Neuronal Network Model with time series for the prediction of the electric field.

2 Literature Review

2.1 Neural Networks

A neural network is, as (Aggarwal 2018) mentions, a set of information processing elements highly interconnected by an input layer, one or more hidden layers and an output layer. Connections are established between nodes in each adjacent layer. The input layer, through which the data is presented to the network, consists of input nodes that receive the information directly from the outside. The output layer represents the response of the network to a given input being this information transferred to the outside. The hidden or intermediate layers are in charge of processing the information and they are interposed between the input and output layers, and they are the only ones that have no connection with the outside. The diffuse predictive model achieved by (Saboya et al. 2019) for the achievement of the Student's profile is very important in the decision making; as well as the neural network models, the decision trees and the Bayesian networks.

The most used network structure is the Feedforward, according to (Nunes Silva et al. 2017), which considers the connections between neurons that are established in only one direction, in the order from the input layer, hidden layer(s) and output layer, as shown in Fig. 5.

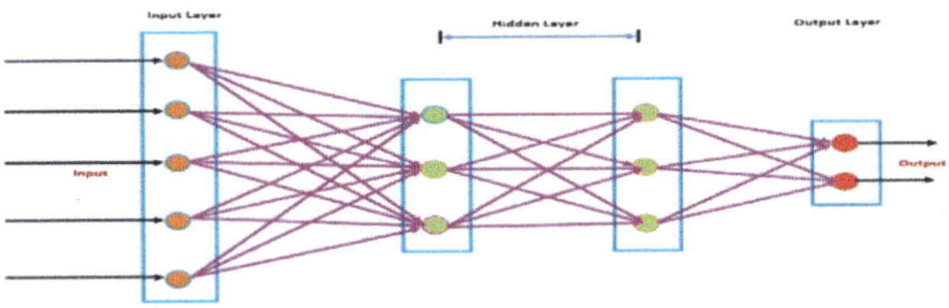

Fig. 5. Artificial Neural Network formed by four layers of neurons

(Martín del Brío and Sanz Molina 2002) argues that a standard artificial neural network model has three aspects:

- A set of inputs $x_j(t)$ and synaptic weights w_{ij}.
- A rule of propagation $h_i(t) = \sigma(w_{ij}, x_j(t))$; $h_i(t) = \sum w_{ij}.x_j$ is the most common.
- An activation function which simultaneously represents the output of the neuron and its activation state as shown in Fig. (6).

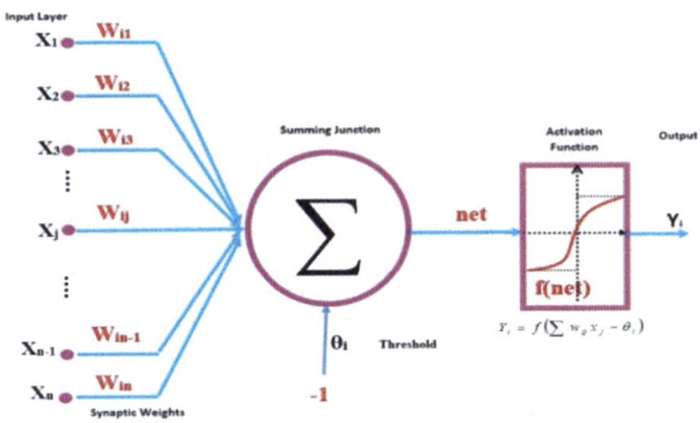

Fig. 6. Standard Artificial Neural Network model.

2.2 Transfer Functions

The activation function, as explained by (Martín del Brío and Sanz Molina 2002), also provides the activation status from the current $a_i(t)$ based on the potential $h_i(t)$ and from its own previously active state $a_i(t-1)$, where the main activation functions are shown in Table 1.

Table 1. Transfer functions

Name	Relation between input/output	Range	Function
Hard limit	$a(n) = \begin{cases} 0;\ n<0 \\ 1;\ n\geq 0 \end{cases}$	$\{0;\ +1\}$	Hardlim
Hard symmetrical limit	$a(n) = \begin{cases} -1;\ n<0 \\ +1;\ n\geq 0 \end{cases}$	$\{-1;\ +1\}$	Hardlims
Lineal	$a(n) = n$	$]-\infty;\ +\infty[$	Purelin
Saturated symmetrical lineal	$a(n) = \begin{cases} -1;\ n<-1 \\ n;\ -1\leq n\leq 1 \\ +1;\ n>1 \end{cases}$	$[-1\ ;\ +1]$	Satlins
Sigmoid logarithmic	$a(n) = \frac{1}{1+e^{-n}}$	$[0\ ;\ +1]$	Logsig
Tangent hyperbolic sigmoid	$a(n) = Tgh(n) = \frac{e^{n}-e^{-n}}{e^{n}+e^{-n}}$	$[-1\ ;\ +1]$	Tansig
Gaussian	$a(n) = A.e^{-B(n^2)}$	$[0\ ;\ +1]$	Gaussian
Sinusoidal	$a(n) = A.Sen(\omega.n+\phi)$	$[-1\ ;\ +1]$	Sinusoidal

2.3 Operation of a Neural Network

The operation of a neural network, as (Nunes Silva et al. 2017) note, should distinguish between the following steps:

Conceptualization of the model for the study of the specific problem. In this model, the inputs, outputs and information available must be indicated.

Adaptation of the available information to the structure of the network to be created. In other words, the learning patterns will be constituted, which start from the information that will be used for the training of the network and the validation patterns, which also, start from the information that will be used as a validation of the network.

Learning phase. The appropriate patterns are presented to the network and the network provides an output, this process is repeated a certain number of stages, also, these outputs are compared with the expected outputs and the various learning algorithms try to minimize the error that exists between the output provided by the network and the expected output.

Validation phase. The set of validation patterns are presented to the network input, and the error made by the network in this set is seen, which is a measure of the goodness of the network.

Generalization phase. A suitable network has been achieved; proceed to use the network as a predictive model, providing a new input, the network will process it and give an output.

2.4 Time Series Analysis Using Neural Networks

A time series, as (Bassis et al. 2016) note, consists of a sequence of several variables that evolve over time. The aim is to predict the future behavior of the phenomenon, which generates these values based on a collection of historical data.

Short-term predictions are achieved using artificial neural networks that are described by a non-linear multivariate function, as shown in Eq. (1).

$$y(t) = F[y(t-1),\, y(t-2), y(t-3), \cdots, y(t-k)] \tag{1}$$

where $y(t)$, $t = m,\, m-1, m-2, \cdots, k$, are the given samples for the time series, F is an unknown non-linear function, and $k < m$.

Multilayer perception is the most widely used model of artificial neural networks for predicting future values. The output unit gives a linear combination of the outputs of all the hidden units as shown in Eq. (2).

$$\hat{y}(t) = w_0 + \sum_{j=1}^{h} w_j \psi_j \left(\sum_{i=1}^{k} w_{ji} y(t-i) + w_{j0} \right) \tag{2}$$

Where ψ_j is the activation function. The synaptic weights w_{ji} y w_j are adjusted during the learning process and they can be positive, negative, or null.

In designing the neural network, the number of hidden neurons to be used must be taken into account. Generally, the number of hidden neurons is proportional to the sample size used for network training. The behavior of the network is evaluated, according to the error function shown in Eq. (3):

$$E(w) = \sum_{k=1}^{p} \left(y(k) - \hat{y}(k) \right)^2 \tag{3}$$

where p is the number of samples used in the training.

The architecture of a neural network for time series prognosis is shown in Fig. (7), where the non-linear autoregressive model is presented in Eq. (4).

$$y_{t+1} = f\left(y_t\, ;\, y_{t-1}; y_{t-2}; \cdots; y_{t-p}\right) + \varepsilon_{t+1} \tag{4}$$

Where y_t is the time observed in the time series for the variable t, ε_{t+1} is the error term in the time $t+1$, in which the model is subject to futures times from the series y_{t+1}.

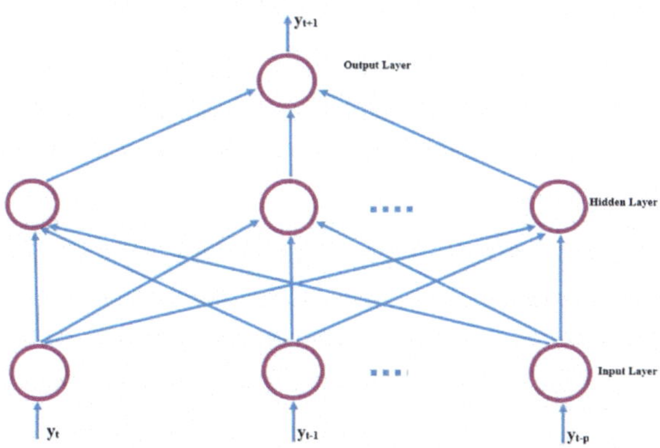

Fig. 7. Typical architecture of a neural network for time series prediction

3 Methodology

3.1 Sensor Data Collection Method for the EFM-100 Atmospheric Electric Field Monitor

In the first phase, as suggested by (Boltek 2014), the EFM-100 Atmospheric Electric Field Monitor was installed on the roof of the Faculty of Engineering and Architecture of the Universidad Peruana Unión by the specialized maintenance personnel and the General Direction of Information Technology.

According to the EFM-100 Atmospheric Electric Field Monitor Installation/ Operators Guide (Boltek Corporation 2015), it measures the electric field within a radius of approximately 37 km measured in (volt/meter).

The ground connection was made through the EFM-100's green well to ground cable as shown in Fig. (8) and the EFM-100's hardware on the building floor, respecting the local electrical code for proper operation and electrical safety. Once in place, the installation cables were connected between the EFM-100 and the computer, which are fiber optic. The computer was located on the fifth floor in a room free of interruptions and interference throughout the day and year, then the software required to start the operation of the EFM-100 was installed.

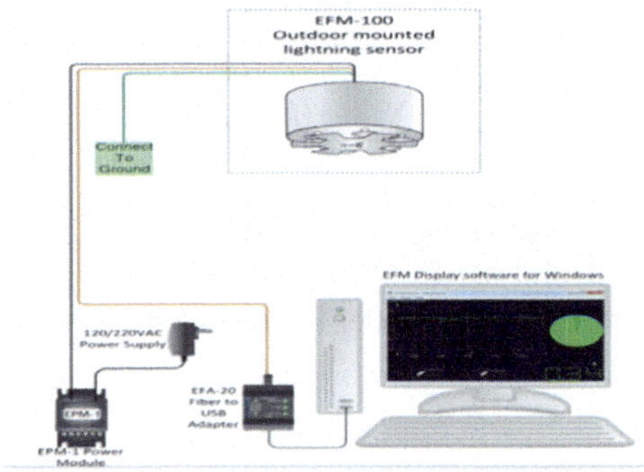

Fig. 8. Connection architecture of the EFM-100

In the second phase (Boltek Corporation 2015), the EFM-100 display software was installed on the computer assigned to the study. The EFA-10 fiber optic adapter was also installed, which converts the EFM-100's optical data into electrical signals compatible with the installed computer and routed to obtain electrical field information. Data is optically transmitted from the EFM-100 at 9600 baud, 8 data bits, 1 stop bit and no parity as shown in Fig. (9).

In the third phase (Boltek Corporation 2015), the data was collected and interpreted from January to December 2019, as shown in Fig. 10 During the summer of 2019 in East Lima, Peru, the solar radiation is intense (Cruz 2009), that is, in January, February and March; the sky is clear and the information recorded by the electric field is positive, but in autumn (March, April and May) and winter (June, July and August) the weather is changing and is generally cloudy with some drizzle. The data obtained contains a history during the 365 days of the year to better understand the disturbance of the electric field in the zone of East Lima.

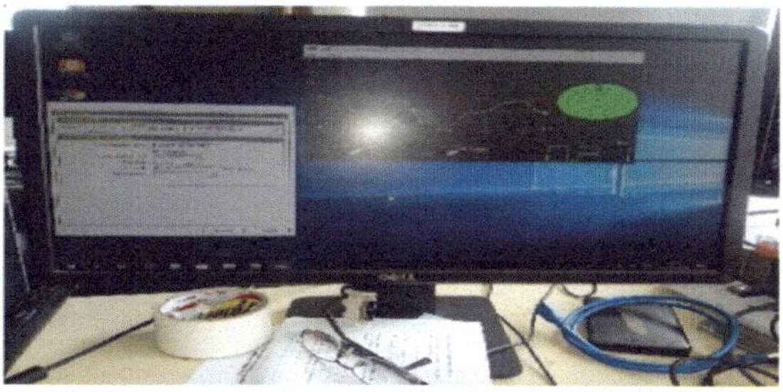

Fig. 9. Installation of the EFM-100 display software

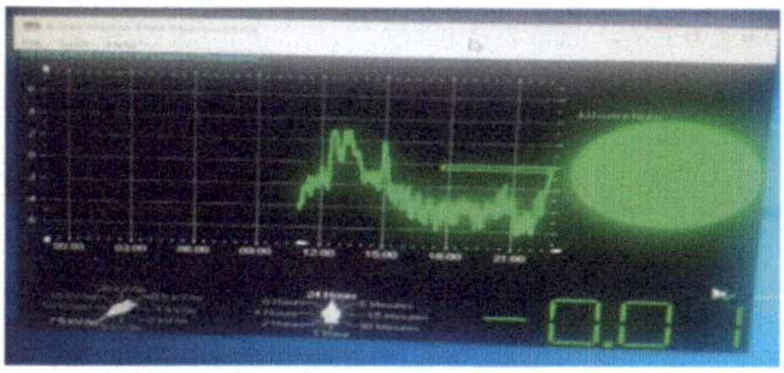

Fig. 10. Electric field measurement January-December 2019

3.2 Formulation Sequence for the Neural Network Model in Time Series with Technology

First, the neural network startup GUI was opened with MatLab version 18 software with the nnstart command, then the time series application was opened to Neural Network Time Series, in which the future values of the time series $y(t)$ were predicted from the past values of the electric field series for the whole year 2019. (Mathworks 2005) argues that the prediction form is the nonlinear autoregressive called NAR, and can be represented in Eq. (5)

$$y(t) = F[y(t-1), y(t-2), y(t-3), \cdots, y(t-d)] \tag{5}$$

Subsequently, the data set with 19564 electric field measurements was loaded into Matlab and iterations were performed. The NARX network is a layered network, a sigmoid transfer function on the hidden layer and a linear transfer function on the output layer. The Feedforward network is precise and of advanced purity in which we worked with 10 hidden layers and 5 numbers of delays.

The algorithm used in the training was the Levenberg-Marquardt (trainlm) mentioned by (Hudson et al. 1992)

3.3 Program Code

The code from the Matlab program that generated our time series neural network is shown in the following sequence:

```
function [y1,xf1] = myNeuralNetworkFunction(x1,xi1)
% Generated by Neural Network Toolbox function genFunc-
tion, 06-Jan-2020 12:33:33.

% [y1,xf1] = myNeuralNetworkFunction(x1,xi1) takes these
arguments:
%   x1 = 1xTS matrix, input #1
%   xi1 = 1x5 matrix, initial 5 delay states for input
#1.
% and returns:
%   y1 = 1xTS matrix, output #1
%   xf1 = 1x5 matrix, final 5 delay states for input #1.
% where TS is the number of timesteps.

% ===== NEURAL NETWORK CONSTANTS =====

% Input 1
x1_step1.xoffset = -20.48;
x1_step1.gain = 0.0618238021638331;
x1_step1.ymin = -1;

% Layer 1
b1 = [2.1772322806255250 96;-1.97247785414382 79245;-
1.5078920227479135 63;-1.0038850798115164231;-
0.20513058820818877437;0.2561905412786299463;-
1.1190092907386639531;-1.2194083620405105073;-
1.6952579373403053875;2.2558141144831767022];
IW1_1 = [-0.769572282857075 6846 -0.79235707332830307426
1.0067100968106219572 -1.3681027098002764841
0.92946358496176650 91;-0.0467116770324393 08244
1.7381826107484801 28 0.42850315815259054641
0.15854987102137613197 -
1.1085156625024257249;1.0467499124557537726 -
1.1839601431287762878 0.2532395228584333 7462
1.2885945199161890073 -
0.065221459237515891361;0.76929638332128824629
0.87020696460853275145 0.86187902332632571056
0.92424185679283199502 -
0.94596573673871930943;1.6547913800324371 98 -
0.15402563699967458666 -0.54287123588919028 45 -
1.2726327688174703 39
0.22107530236244013477;0.24672567942242484 84 -
1.4213484032818666236 1.2367064252751107656
0.36040693310553967299 -1.0822419458314938012;-
```

```
1.4142703782659340472 -0.88948978526801525391
1.097070411568550341 -0.18083678239624406681
0.54662296911944729949;-1.2600967279611210436
0.29836102092443450573 -0.40173942255754913067 -
1.5973967540369522489 0.70582955552834414359;-
1.3855274586648296253 -0.1609307139560552935
0.38806963159303592414 -0.91620325725834550479
1.4244219303298344403;1.1377970002249686576 -
0.51231415224738408032 -1.4030993069264552364 -
1.0822343579691737769 -0.10118128730453158914];
% Layer 2
b2 = 1.0249197970296899385;
LW2_1 = [-0.75998612763886042032 -0.34216271843143264419
0.36006208010673740327 0.21256200559141180673
0.4810811955388696326 -0.18209858940121301241
0.48837981003680741576 -0.94200245109662106291 -
0.28162664541254062156 -0.64364461273588380319];

% Output 1
y1_step1.ymin = -1;
y1_step1.gain = 0.0618238021638331;
y1_step1.xoffset = -20.48;

% ===== SIMULATION ========

% Dimensions
TS = size(x1,2); % timesteps

% Input 1 Delay States
xd1 = mapminmax_apply(xi1,x1_step1);
xd1 = [xd1 zeros(1,1)];

% Allocate Outputs
y1 = zeros(1,TS);

% Time loop
for ts=1:TS

    % Rotating delay state position
    xdts = mod(ts+4,6)+1;

    % Input 1
    xd1(:,xdts) = mapminmax_apply(x1(:,ts),x1_step1);

    % Layer 1
```

```
        tapdelay1 = reshape(xd1(:,mod(xdts-[1 2 3 4 5]-
    1,6)+1),5,1);
        a1 = tansig_apply(b1 + IW1_1*tapdelay1);
        % Layer 2
        a2 = b2 + LW2_1*a1;

        % Output 1
        y1(:,ts) = mapminmax_reverse(a2,y1_step1);
    end
    % Final delay states
    finalxts = TS+(1: 5);
    xits = finalxts(finalxts<=5);
    xts = finalxts(finalxts>5)-5;
    xf1 = [xi1(:,xits) x1(:,xts)];
    end
     % Map Minimum and Maximum Input Processing Function
    function y = mapminmax_apply(x,settings)
    y = bsxfun(@minus,x,settings.xoffset);
    y = bsxfun(@times,y,settings.gain);
    y = bsxfun(@plus,y,settings.ymin);
    end
     % Sigmoid Symmetric Transfer Function
    function a = tansig_apply(n,~)
    a = 2 ./ (1 + exp(-2*n)) - 1;
    end
     % Map Minimum and Maximum Output Reverse-Processing
    Function
    function x = mapminmax_reverse(y,settings)
    x = bsxfun(@minus,y,settings.ymin);
    x = bsxfun(@rdivide,x,settings.gain);
    x = bsxfun(@plus,x,settings.xoffset);
    end
```

4 Results

4.1 Results of the Neural Network Model

In Fig. (11) the performance of the neural network is validated with the regression model against training objectives, validation and test sets. As can be seen, all data are located along a 45° line and this means that the outputs of the network are equal to the targets. In this network, the setting is reasonably good for all data sets with R values, in each case 0.91 or higher. The performance of the network with a linear regression model for the proposal is shown in Eq. (6)

$$Output = 0.87(Target) \pm 0.14 \tag{6}$$

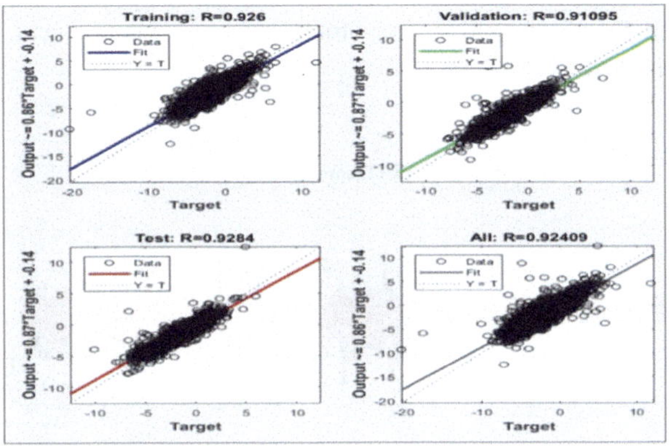

Fig. 11. Performance of the neural network used in this study

Furthermore, the seasons in Peru begin with summer from December 22nd to March 21st, autumn from March 22nd to June 21st, winter from June 22nd to September 22nd and spring from September 23rd to December 21st.

The results obtained from the neural network, measured by season during the year 2019, are shown in Table 2.

Table 2. Electric field values with neural networks by season in 2019.

Seasons of the year 2019	Valid N	(v/m) Mean	Minimum	Maximum	Std. Dev.
Summer	4134	0.1682	−6.6993	3.6323	0.7111
Fall	3179	−0.6550	−6.8502	1.8796	1.0233
Winter	6681	−1.6218	−20.48	11.8697	2.1174
Spring	5570	−1.4309	−8.6725	3.8296	1.9024

Table 2 shows that 4134 records of the electric field were measured at the **Summer** station with intervals of 5 min, from which an arithmetic mean of 0.1682 v/m with a standard deviation of 0.7111 v/m was obtained. This means that in summer there was no cloudiness, since the electric field was positive, by which the solar radiation was intense; damaging the nature and the human beings of the zone.

Likewise, 3179 records of the electric field were measured in the **Autumn** station with intervals of 5 min, from which an arithmetic mean of −0.6550 v/m with a standard deviation of 1.0233 v/m was obtained. This means, that in autumn there was cloudiness, since the electric field was negative, for which the solar radiation was not intense.

In addition, 6681 records of the electric field were measured in the **Winter** station with intervals of 5 min, from which an arithmetic mean of −1.6218 v/m with a standard deviation of 2.1174 v/m was obtained. This means, that in Winter there was cloudiness, since the electric field was negative, for which the solar radiation was not very intense.

Finally, in the **Spring** station 5570 records of the electric field were measured at 5 min intervals, from which an arithmetic mean of −1.4309 v/m with a standard deviation of 1.9024 v/m was obtained. This means, that in **Spring** there was cloudiness, since the electric field was negative, for which the solar radiation was not very intense in the area of East Lima, Peru.

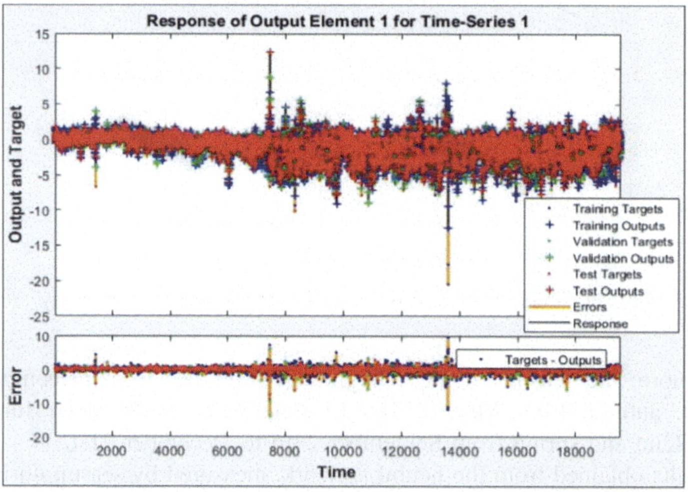

Fig. 12. Response of time series neural network output elements

Figure (12) shows the inputs, targets, and time errors, and indicates which time points were selected for training, testing, and validation, and then creates the MATLAB code by generating the prediction pseudocode.

Table 3. Neural network results

	Target Values	MSE	R
Training	13694	4.76184e−1	9.25997e−1
Validation	2935	5.58515e−1	9.10947e−1
Testing	2935	4.64005e−1	9.28402e−1

In Table 3, it can be seen that the network training using Scaled Conjugate Gradient Training has a mean square error (MSE) of 4.76184e−1 with a regression coefficient of 9.25997e−1 and the validation has an MSE of 5.58515e−1 with a regression coefficient of 9.10947e−1 and finally, the testing has an MSE of 4.64005e−1 with an R value of 9.28402e−1.

5 Conclusions

This research project developed a neural network model with time series that predicts the electric field in the area of East Lima, Peru. The model has incorporated the input values of the electric field records with the EFM-100 sensor and the trained neural network model obtained efficient values in all four seasons. The neural network used the conjugated gradient algorithm and the accuracy of the future prediction of the electric field was obtained with a validation MSE of 0.558515, which makes the performance of the model. A pseudocode was obtained from Matlab software version 18, which predicts the future electric field, which constitutes a support tool for researchers in the area of environmental engineering.

The neural network model is effective because it provides better results of the electric field measurement because it uses powerful algorithms of the artificial intelligence and allows to mitigate the solar radiation accurately in future years.

One recommendation for future prediction of the electric field is to implement a distributed and responsive web application that allows real-time prediction of the electric field for decision making.

References

Aggarwal, C.C.: Neural networks and deep learning. In: Neural Networks and Deep Learning, Yorktown Heights, NY, USA. Springer, Netherlands (2018). https://doi.org/10.1007/978-3-319-94463-0

Bassis, S., Esposito, A., Morabito, F.C., Eros, M., Pasero, E.: Smart innovation, systems and technologies. In: Advances in Neural Networks Computational Intelligence for ICT, vol. 54, Canberra, Australia (2016). https://doi.org/10.1007/978-3-319-33747-0

Bohari, Z.H., Sulaima, M.F., Nasir, M.N., Bukhari, W.M., Jali, M.H., Baharom, M.F.: Int. J. Eng. Sci. (IJES) **3**(6), 59–67 (2014). www.theijes.com

Boltek, N.: EFM-100 Atmospheric Electric Field Monitor Guide. Canada (2014). www.boltek.com

Boltek Corporation. Lightning Detection EFM-100 Atmospheric Electric Field Monitor Installation/Operators Guide EFM-100 Atmospheric Electric Field Monitor. Changes (2015)

Cruz, V.M.: Health risk to non-ionizing radiation by the electricity networks in Peru. Revista Peruana de Medicina Experimental Y Salud Publica. Lima-Perú (2009). https://doi.org/10.17843/rpmesp.2009.261.1341

Martín del Brío, B., Sanz Molina, A.: Redes Neuronales y Sistemas Difusos (Segunda). Alfaomega RaMa (2002). México file:///C:/Users/JUAN SORIA Q/Downloads/Redes_Neuronales_y_Sistemas_Difusos.pdf

Harrison, R.G.: The Global Atmospheric Electrical Circuit and Climate. Surveys in Geophysics. Department of Meteorology, The University of Reading, United Kingdom (2004). https://doi.org/10.1007/s10712-004-5439-8

Hudson, M., Martin, B., Hagan, T., Demuth, H.B.: Deep Learning ToolboxTM User's Guide (1992). www.mathworks.com

Jánský, J., Pasko, V.P.: Effects of conductivity perturbations in time-dependent global electric circuit model. J. Geophys. Res. Space Phys. **120**(12), 10654–10668 (2015). https://doi.org/10.1002/2015JA021604

Martinez-Lozano, M., De, U., Bajío, L.S.: Sizing Optimization of PV Systems Under Commercial Electricity Tariffs Schemes in Mexico View project Effect of Seasonal Variations on the Performance of Distribution Lines View Project (2014). https://doi.org/10.13140/2.1.3635.2323

Mathworks, C.: Deep Learning Toolbox TM Release Notes, vol. 172 (2005). www.mathworks.com

Nicoll, K.A., Harrison, R.G., Barta, V., Bor, J., Brugge, R., Chillingarian, A., Chum, A., Georgoulias, J., Guha, A., Kourtidis, K., Kubicki, M., Yaniv, R.: A global atmospheric electricity monitoring network for climate and geophysical research. J. Atmosph. Solar Terr. Phys. **184**, 18–29 (2019). https://doi.org/10.1016/j.jastp.2019.01.003

Nunes Silva, I., Hernane Spatti, D., Andrade Flauzino, R., Liboni, L.H.B., dos Reis Alves, S.F.: Artificial Neural Networks: A Practical Course. São Paulo, Brazil (2017). https://doi.org/10.1007/978-3-319-43162-8

Roble, R.G., Tzur, I.: The Earth's Electrical Environment. The Earth's Electrical Environment (Primera). National Academies Press, USA (1986). https://doi.org/10.17226/898

Saboya, N., Loaiza, O.L., Soria, J.J., Bustamante, J.: Fuzzy logic model for the selection of applicants to university study programs according to enrollment profile. In: Advances in Intelligent Systems and Computing, vol. 850, pp. 121–133. Springer, Cham (2019). https://doi.org/10.1007/978-3-030-02351-5_16

Siingh, D., Gopalakrishnan, V., Singh, R.P., Kamra, A.K., Singh, S., Pant, V., Singh, R., Singh, A.K.: The atmospheric global electric circuit: an overview. Atmosph. Res. **84**(2), 91–110 (2007). https://doi.org/10.1016/j.atmosres.2006.05.005

Silva, H.G., Conceição, R., Melgão, M., Nicoll, K., Mendes, P.B., Tlemçani, M., Reis, A.H., Harrison, R.G.: Atmospheric electric field measurements in urban environment and the pollutant aerosol weekly dependence. Environ. Res. Lett. **9**(11) (2014). https://doi.org/10.1088/1748-9326/9/11/114025

Theoretical Domains Framework Applied to Cybersecurity Behaviour

Thulani Mashiane(✉) and Elamarie Kritzinger

School of Computing, University of South Africa, UNISA, PO Box 392, Pretoria 0003, South Africa
tmashiane@csir.co.za, kritze@unisa.ac.za

Abstract. The challenge of changing user cybersecurity behaviour is now in the foreground of cybersecurity research. To understand the problem, cybersecurity behaviour researchers have included, into their studies, theories from the Psychology domain. Psychology makes use of several behavioural theories to explain behaviour. This leads to the question, which of these theories are best suited to firstly understand cybersecurity behaviour and secondly to change the behaviour for the better. To answer this question, as a prelude to the current paper, previous publications have 1) established a definition for the different categories of cybersecurity behaviour, 2) identified and applied a framework, the Theoretical Domains Framework, that ties different behavioural theories together into one behaviour change framework. The current study is aimed to show the link between the behavioural constructs discussed in the Theoretical Domains Framework to the different cybersecurity behaviour categories. The contribution of the study is towards the implementors of initiatives that aim to change cybersecurity behaviour.

Keywords: Cybersecurity behaviour · Theoretical Domains Framework · Constructs

1 Introduction

A secure cybersecurity system includes technology and users. Users in the cybersecurity system are the focus of recent research because they are considered to be weakest link in the system [1–3]. Changing user behaviours, for better cybersecurity, was originally approached as a knowledge gain exercise. Users were educated and trained in cybersecurity through cybersecurity awareness programmes and the like. Although this approach did increase the level of cybersecurity awareness of the users it did little for behaviour change. For this reason, cybersecurity awareness now includes behaviour change techniques [1]. The goal of cybersecurity behaviour change initiatives is to curb bad behaviour while promoting good behaviour.

1.1 Cybersecurity and Psychology Theories

Cybersecurity behaviour change is a difficult problem to address because user behaviour in its nature is complicated. The Psychology domain, being a discipline that

R. Silhavy (Ed.): CSOC 2020, AISC 1225, pp. 411–428, 2020.
https://doi.org/10.1007/978-3-030-51971-1_34

focuses on human behaviour, has provided several behavioural theories to support the understanding of user behaviour. Psychology has been used to unpack the difference influences of cybersecurity behaviour. The research presented by cybersecurity researchers makes use psychology theories to explain the influences and barriers of cybersecurity behaviour. This has resulted in many studies being published that employ different psychology theories. There is a need to synthesize these results in order to gain a wider understanding of which psychology theories have the potential of providing effective cybersecurity behaviour change.

1.2 Theoretical Domains Framework (TDF)

Due to the number of different behavioural theories it became a challenge for health care workers to create behaviour change interventions. This is because each theory contained several behaviour constructs. A construct is a characteristic about an item which cannot be directly studied however its impact is visible. For example, attitude towards cybersecurity is not visible. However, the impact of bad attitude can be visible through a users' lack of interest towards learning cybersecurity. A solution to the challenge of dealing with too many behavioural constructs was the development of the Theoretical Domains Framework (TDF).

The TDF is a behaviour change framework. The TDF was developed for intervention implementation by health care workers. The framework is a result of research done by Psychological Theorists, Health Service Researchers, and Health Psychologists to provide an all-encompassing behavioural framework that brings together the existing behavioural theories [4, 5]. As the use of the framework matured, so too did its adoption in different disciplines.

The TDF brings together 33 behavioural theories as well as 128 theoretical constructs [5]. The TDF groups the different constructs from the behavioural theories into domains. In the latest publication of the TDF there are 14 domains with 84 theoretical constructs. Table 1 in the Appendixes provides the TDF. A strength of using the TDF is that it helps to understand behaviour change through the unpacking of the influences on behaviour in the context in which they occur [6].

The paper is presented as follows: Sect. 2 presents the background which is a summary of the previsions work done, Sect. 3 presents the methodology of the paper, Sect. 4 presents the results and discussion of the results of the study and Sect. 5 concludes the paper.

2 Background

The current study concludes a series of conceptual literature studies aimed at better understanding cybersecurity behaviour. Figure 1 presents a summary of the sub-studies that contribute to the main study. The current study is the third sub-study in the series.

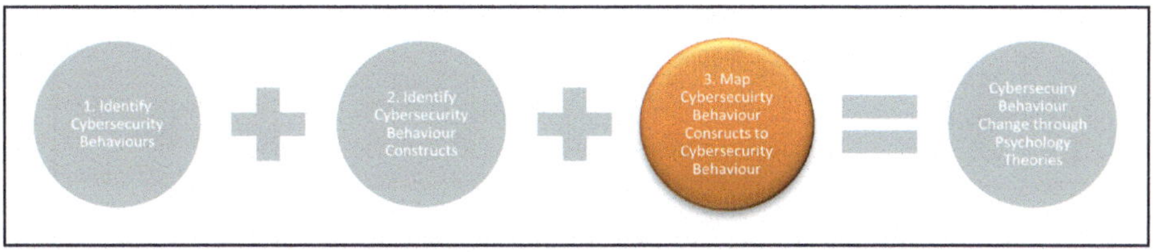

Fig. 1. Context of current study

The first study was accomplished through a literature review study [7], where cybersecurity behaviours were identified and categorised. These behaviours were then categorised and plotted on a graph against *cybersecurity expertise* (Y-axis) and *intention* (X-axis). The results formed a cybersecurity behaviour conceptual taxonomy.

The study separated cybersecurity behaviour in the workplace from cybersecurity behaviour at home. Context is important when considering cybersecurity behaviour because, users behave differently when they are in the workplace compared to at home [8]. The study catered for the difference in environments by changing the X-axis to measure the application of *knowledge and skills* in the home context. The resulting taxonomies are shown in Fig. 2 and Fig. 3.

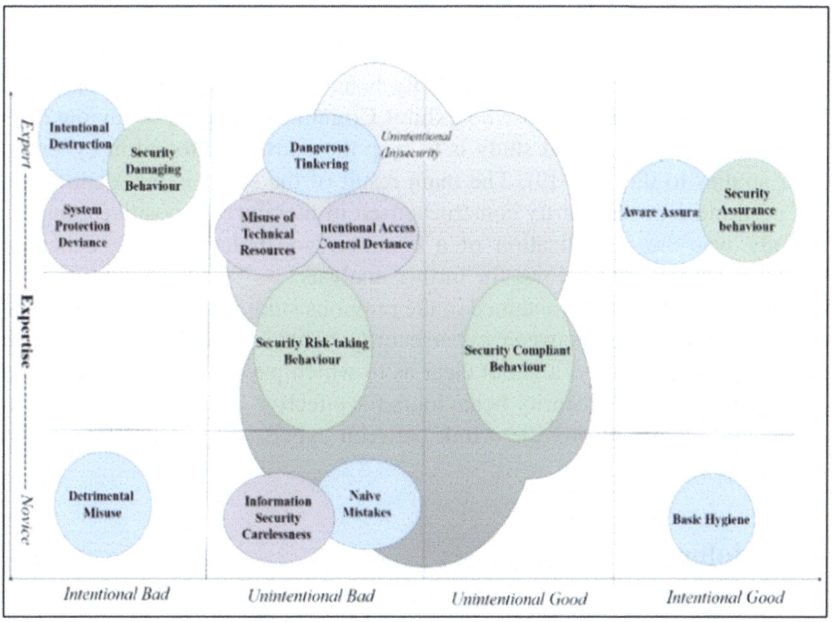

Fig. 2. Cybersecurity behaviour categories at the workplace

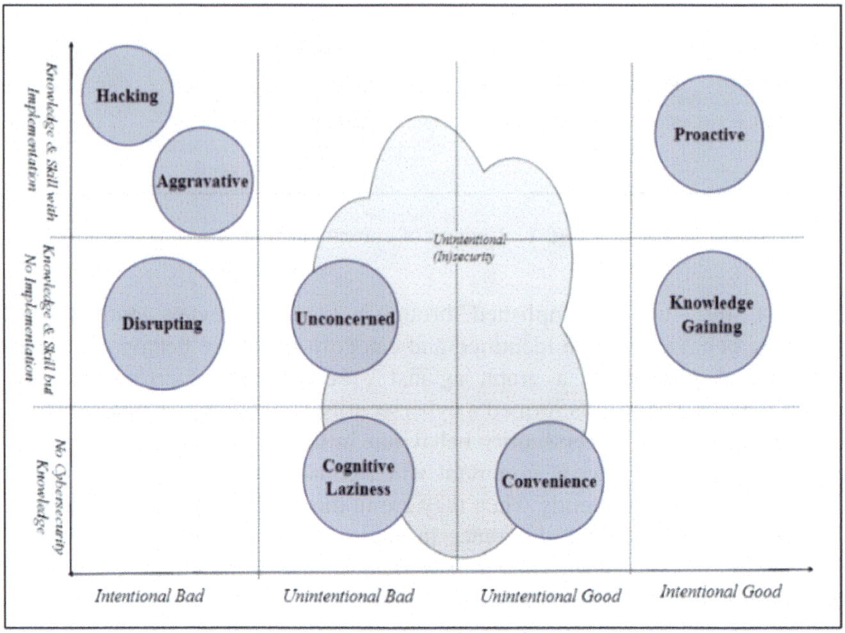

Fig. 3. Cybersecurity behaviours at the home

The results of the study aids in specifying the cybersecurity behaviour change question from "How to change cybersecurity behaviour?" to "How to change cybersecurity behaviour of home users who exhibit Cognitive Laziness behaviour?".

The objective of the second study is to map the constructs found in cybersecurity behaviour studies to the TDF [9]. The main result of the study is the identification of the most prominent cybersecurity constructs used in literature. A positive consequence of the study was the identification of a common vocabulary for the cybersecurity behavioural constructs that allows for further analysis.

With the use of the results obtained in the previous studies, the focus of this study is exploring the relationship between cybersecurity behaviour and cybersecurity behavioural constructs. Currently, it is not clear as to which psychology behaviour theories should be applied to cybersecurity behaviours for effective behaviour change. The aim for the current study is to provide a link between cybersecurity behaviour and behavioural constructs.

3 Methodology

The TDF implementation guideline is used as a tool in this study [6].

Table 1. Theoretical domains framework guideline and implementation

TDF guidance steps	Study implementation
Select and specify the target behaviour	Action: a) Security Assurance Behaviour (Workplace) b) Proactive Cybersecurity Behaviour (Home) Actor: Users of the Internet Target Behaviour: Secure cybersecurity behaviour Context: a) Workplace b) Home
Select your design	Literature review
Develop sampling strategy	Inclusion Criteria: – The publication is in English – The study is an empirical study which makes use of a questionnaire/survey – The study includes hypothesis testing – At least one of the null hypotheses must evaluate the relationship between a construct and the construct "intention to behave" or the behaviour itself – The constructs used were either validated by the study or used previously validated constructs

Develop the study material	foreach TDF Construct:
	{
	foreach Literature Construct:
	{
	compare (TDF Construct Definition, Literature Construct Definition)
	IF (meanings are similar OR the same):
	{
	place Literature Construct under TDF Construct
	}
	ELSE:
	{
	compare (TDF Construct Example Survey Questions, Literature Construct Survey Questions)
	IF (meanings are similar or the same):
	{
	place literature Construct under TDF Construct
	}
	ELSE:
	{
	Add Literature Construct to TDF Construct List
	}
	}
	}
	}
	redo the process starting from the last placed literature construct.

Analyse the data	Questions to answer from the data: What are the different cybersecurity behaviours? Which psychology behavioural theories can be applied to cybersecurity behaviour research? How do we map cybersecurity behaviours to psychology behaviour theories, to encourage good cybersecurity behaviour?
Reporting	A write up of the findings

4 Results

The current section provides the results from synthesising cybersecurity behaviour constructs. The results show constructs that had a direct relationship with the behaviour or the intention to perform the behaviour. Intention to behave is a determent of behaviour [10], for this reason it is included in the results.

4.1 Workplace Cybersecurity Behaviour

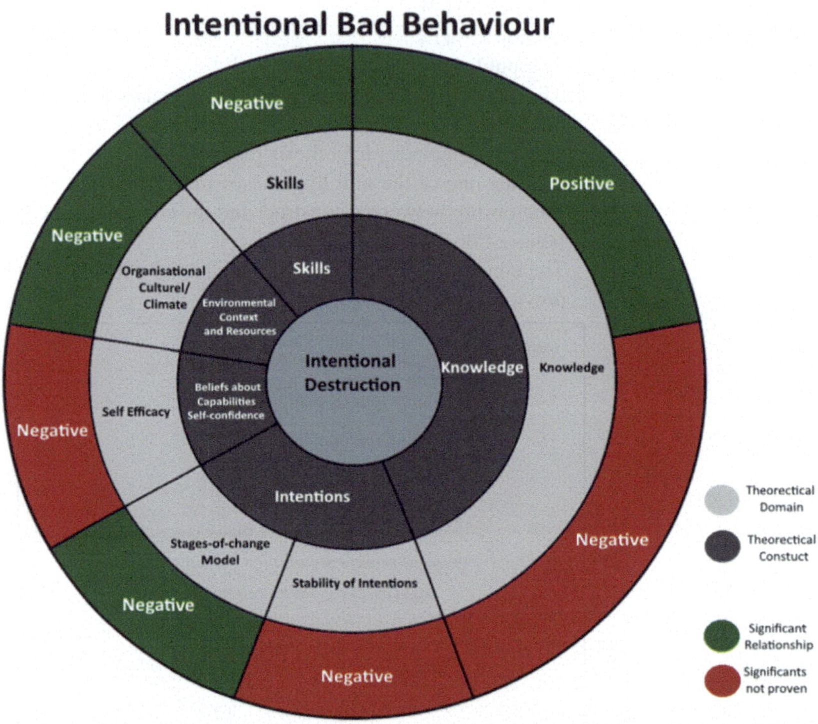

Fig. 4. Workplace intentional bad behaviour

Intentional Bad Behaviour. The results of Intentional Bad Behaviour in a workplace are shown in Fig. 4.

Intentional Destruction. The knowledge theoretical domain is the most evaluated TDF domain (n = 4) with Intentional Destruction. The identification of knowledge theoretical construct acknowledges the need for knowledge in cybersecurity behaviour change. This implies that users that perform intentional destruction understand what they are doing as well as understand the impact that their actions could have. This is appropriate, as insider threats are included in this category.

Equal amounts of researchers (n = 2 each) found knowledge to be an enabler of intentional destruction cybersecurity behaviour.

Skills (n = 1), Organisational Culture (n = 1) and the Stages-of-Change Model (n = 1) are not enablers of intentional destruction cybersecurity behaviour. Self-efficacy (n = 1) and stability of intentions (n = 1) were not found to be enablers of intentional destruction.

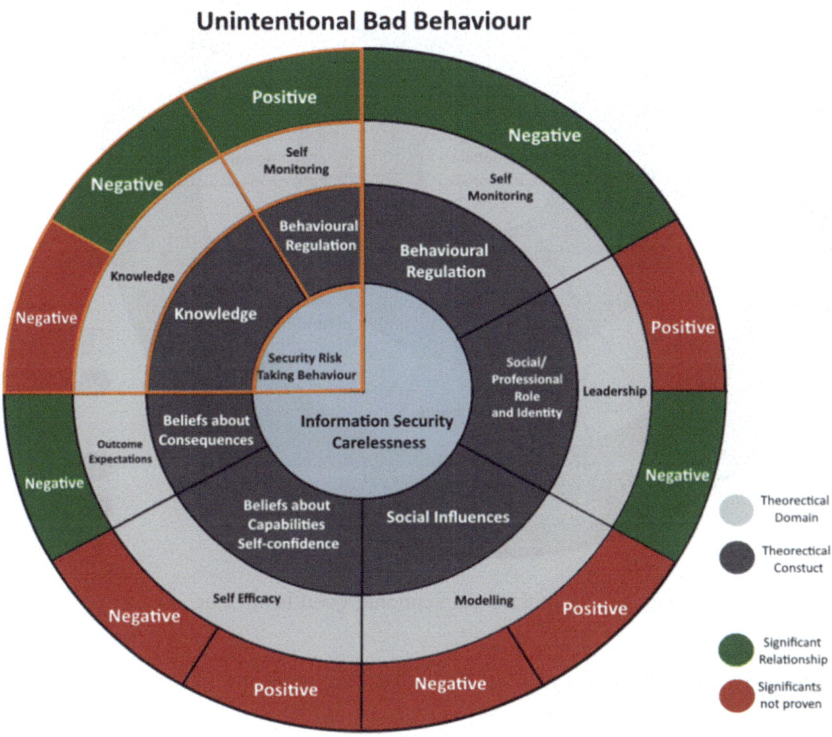

Fig. 5. Workplace unintentional bad behaviour

Unntentional Bad Behaviour. The results of Unintentional Bad Behaviour in an organisation are shown in Fig. 5.

Security Risk Taking Behaviour. Knowledge is the most evaluated TDF domain (n = 2). Knowledge is found to be a barrier by one study and not proven to be a barrier in a different study. *Self-monitoring* was found to be an enabler of Security Risk Taking Behaviour.

Information Security Carelessness. Self-monitoring (n = 2), Outcome Expectations (n = 1) and Leadership (n = 1) are found to be a barrier to Information Security Careless behaviour. The other identified constructs were not proven as barriers or enablers by the included studies.

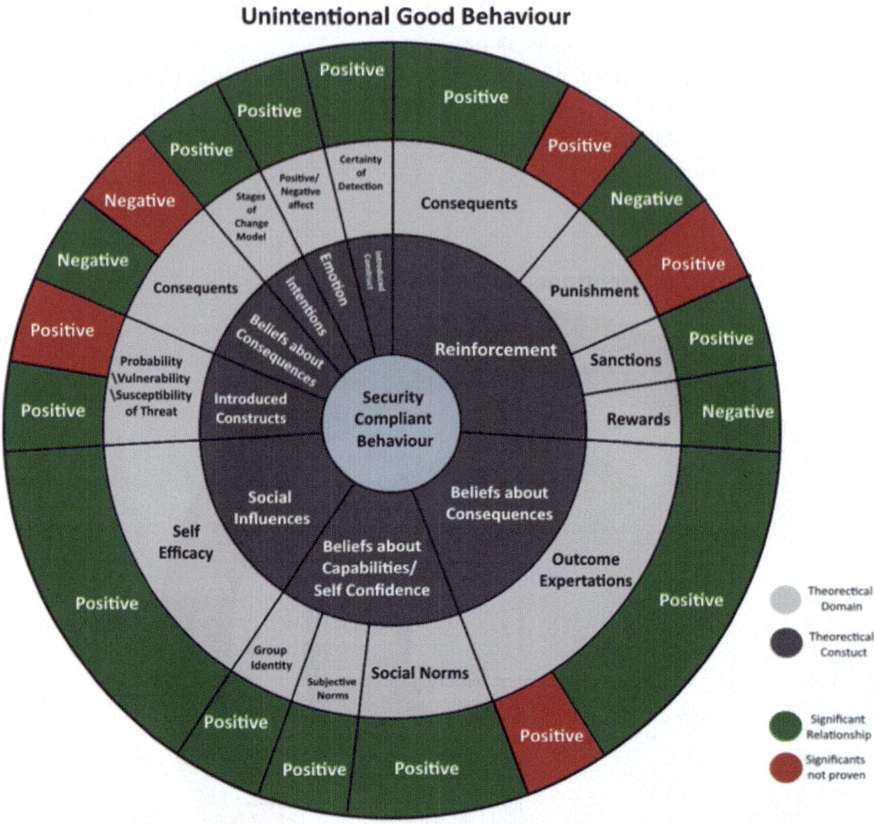

Fig. 6. Workplace unintentional good behaviour

Unintentional Good Behaviour. The results of Unintentional Good Behaviour in an organisation are shown in Fig. 6.

Security Compliant Behaviour. Reinforcements is the most popular TDF domain (n = 7). *Consequents* (n = 3), *Sanctions* (n = 1), *Outcome Expectations* (n = 4), *Social Norms* (n = 2), *Self-efficacy* (n = 4), *Subjective Norms* (n = 2), *Group Identity* (n − 1), *Probability/Vulnerability/Susceptibility of Threat* (n = 1), *Stages-of-Change model* (n = 1), *Positive/Negative Affect* (n = 1) *and Certainly of Detection* (n = 1) are found to be enablers of Security Compliant Behaviour. *Punishment, Rewards, Consequents* are found to be barriers to Security Compliant Behaviour.

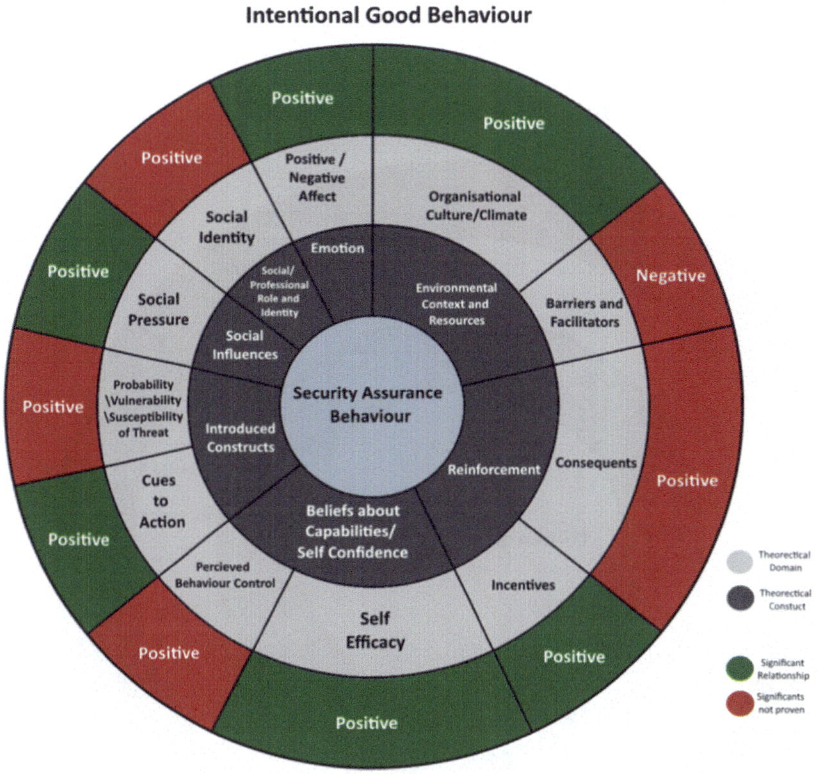

Fig. 7. Workplace intentional good behaviour

Intentional Good Behaviour. The results of Intentional Good Behaviour in an organisation are shown in Fig. 7.

Security Assurance Behaviour. Positive/Negative Affect (n = 1), Organisational Culture (n = 2), *Incentives* (n = 1), *Self-efficacy* (n = 2), *Cues to Action* (n = 1), and *Social Pressure* (n = 1) are found to be enablers of Security Awareness Behaviour. *Probability/Vulnerability/Susceptibility of Threat, Social Pressure, Perceived Behaviour Control*, and *Consequents* are not found to be enablers of Security Awareness Behaviour. *Barriers and Facilitators* is not found to be a barrier to Security Assurance Behaviour.

4.2 Home Cybersecurity Behaviour

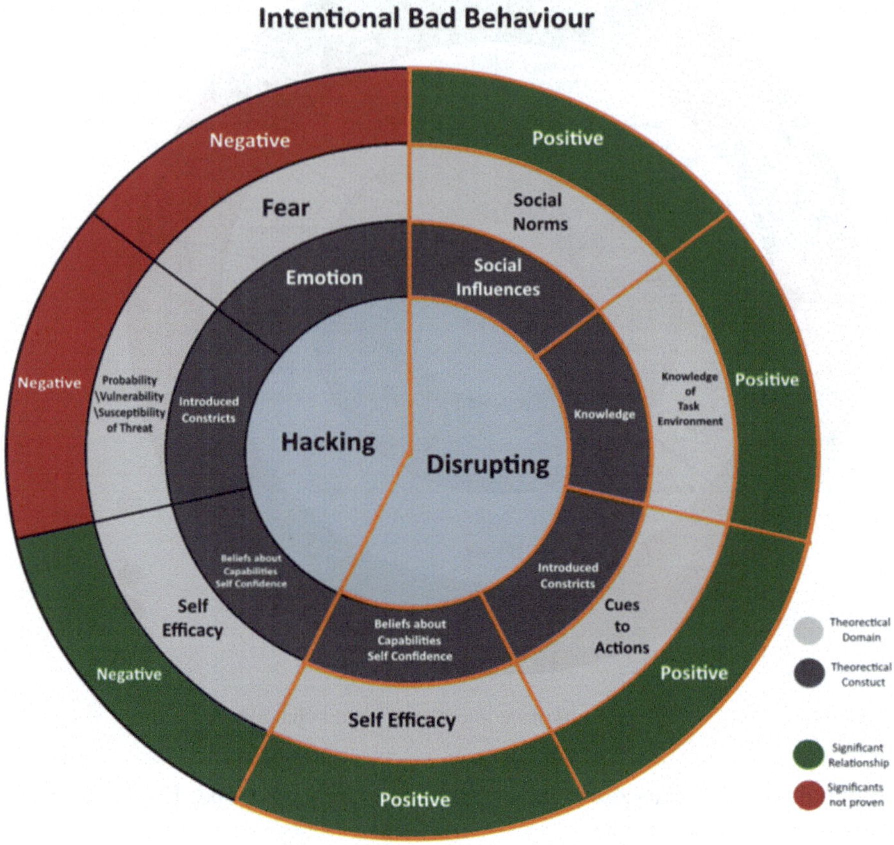

Fig. 8. Home intentional bad behaviour

Intentional Bad Behaviour. The results of Intentional Bad Behaviour at home are shown in Fig. 8.

Hacking. Self-Efficacy (n = 1) is found to be a barrier to hacking. *Fear* (n = 1), *Probability/Vulnerability/Susceptibility of Threat* (n = 1) were not found the be barriers to hacking.

Disrupting. Social Norms, Knowledge of Task Environment, Cues to Action and *Self-efficacy* were found to be enablers of Disrupting behaviour.

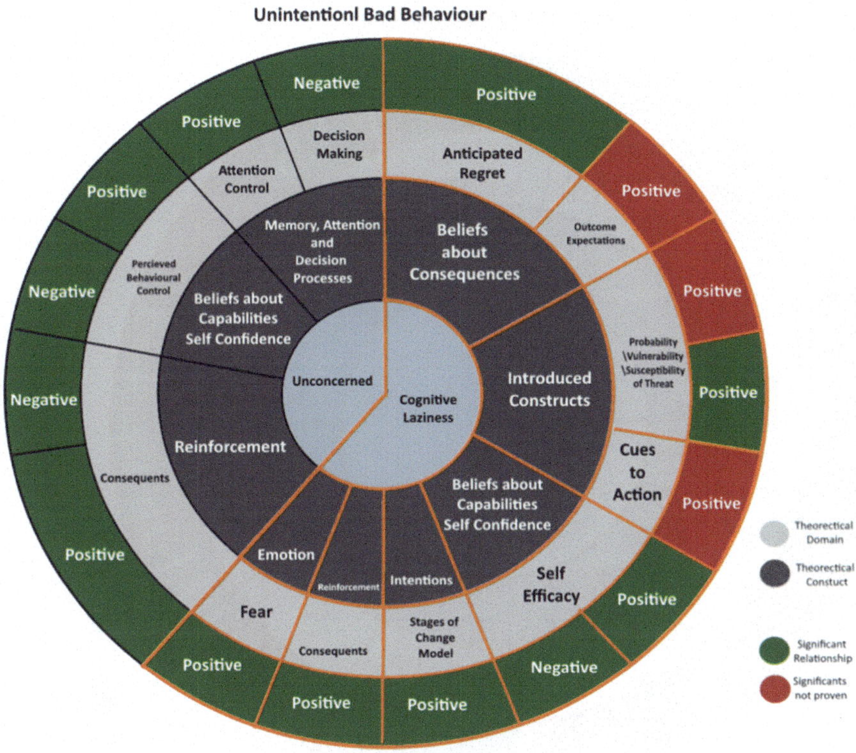

Fig. 9. Home unintentional bad behaviour

Unintentional Bad Behaviour. The results of Unintentional Bad Behaviour at home are shown in Fig. 9.

Cognitive Laziness. Anticipated Regret (n = 2), Probability/Vulnerability/Susceptibility of Threat (n = 1), *Self-efficacy* (n = 1), *Stages-of-Change model* (n = 1), *Consequents* and *Fear* are enablers of Cognitive Laziness. *Self-efficacy* (n = 1) was also found to be a barrier to Cognitive Laziness. *Outcomes Expectations* (n = 1), *Probability/Vulnerability/Susceptibility of Threat* (n = 1), and Outcomes expectations were not found to be an enabler of Cognitive Laziness.

Unconcerned. Attention Control (n = 1), Perceived Behavioural Control (n = 1), Consequents (n = 2) are found to be enablers of unconcerned behaviour. The same constructs are also to be barriers to Unconcerned cybersecurity behaviour.

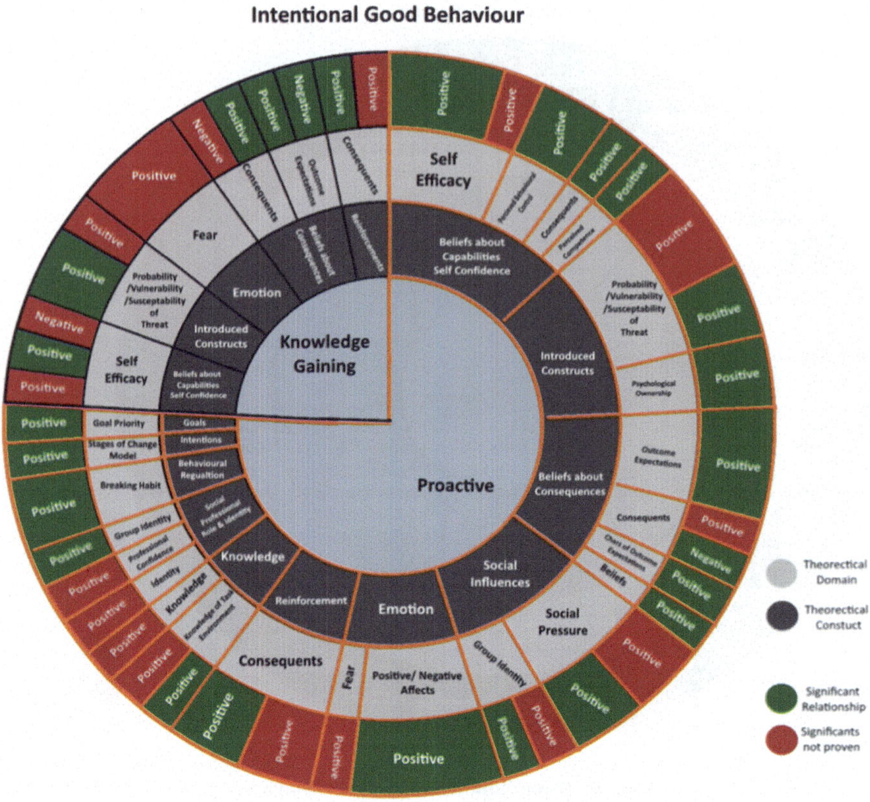

Fig. 10. Home intentional good behaviour

Intentional Good Behaviour. The results of Intentional Good Behaviour at home are shown in Fig. 10.

Knowledge Gaining. Consequents (n = 1), *Outcome Expectations* (n = 1), *Consequents* (n = 1), *Probability/Vulnerability/Susceptibility of Threat* (n = 2) and *Self-efficacy* (n = 2) are enablers of Knowledge Gaining. *Outcome expectations* (n = 1) was also found to be a barrier to Knowledge Gaining behaviour.

Proactive. Self-efficacy (n = 3), Perceived behavioural Control (n = 2), Consequents (n = 1), Perceived Competence (n = 1), Probability/Vulnerability/Susceptibility of Threat (n = 2), *Psychological Ownerships* (n = 2), *Outcome Expectations* (n = 3), *Characterises of Outcome Expectations, Beliefs* (n = 1) *Group Identity* (n = 1), *Positive/Negative* Affect (n = 4), *Consequents* (n = 2), *Knowledge of Environment* (n = 1) *Group Identity* (n = 1) *Breaking Habit* (n = 2) *Stages-of-Change Model* (n = 1) and *Goal Priority* (n = 1) were found to be enablers of Proactive cybersecurity behaviour. The other identified constructs were not found to be enablers of Proactive cybersecurity behaviour at home.

5 Conclusion

The current paper presents the third part of a literature-based study on cybersecurity behaviour and psychological constructs. The TDF was used as a tool to synthesize the constructs found in literature. For each behaviour category (identified in previous study) the associated constructs were identified and evaluated to determine whether they are enablers or barriers of the behaviour. The results of the study can be used in cybersecurity initiative design. The designer can leverage on knowing which constructs enable a desired behaviour. For example, the positive/negative affect was found to be an enabler of proactive cybersecurity behaviour at home. A can include a section in the cybersecurity awareness training that focuses on the positive/negative affect (attitude) that people have towards cybersecurity.

The presented results must be considered under the following limitations. The small sample size of the literature. The analysis done is conceptual, although based on the data found in literature. Future work will consider how to design interventions based on the knowledge gained in this study.

Appendixes

See Tables 2 and 3.

Table 2. Theoretical domains theory [11]

Domain	Constructs
1. Knowledge (An awareness of the existence of something)	Knowledge (including knowledge of condition/scientific rationale) Procedural knowledge Knowledge of task environment
2. Skills (An ability or proficiency acquired through practice)	Skills Skills development Competence Ability Interpersonal skills Practice Skill assessment
3. Social/professional role and identity (A coherent set of behaviours and displayed personal qualities of an individual in a social or work setting)	Professional identity Professional role Social identity Identity Professional boundaries Professional confidence Group identity Leadership Organisational commitment

(*continued*)

Table 2. (*continued*)

Domain	Constructs
4. Beliefs about capabilities self-confidence (Acceptance of the truth, reality, or validity about an ability, talent, or facility that a person can put to constructive use)	Perceived competence Self-efficacy Perceived behavioural control Beliefs Self-esteem Empowerment Professional confidence
5. Optimism (The confidence that things will happen for the best or that Desired goals will be attained)	Optimism Pessimism Unrealistic optimism Identity
6. Beliefs about consequences (Acceptance of the truth, reality, or validity about outcomes of a behaviour in a given situation)	Beliefs Outcome expectancies Characteristics of outcome expectancies Anticipated regret Consequents
7. Reinforcement (Increasing the probability of a response by arranging a dependent relationship, or contingency, between the response and a given stimulus)	Rewards (proximal/distal, valued/not valued, probable/improbable) Incentives Punishment Consequents Reinforcement Contingencies Sanctions
8. Intentions (A conscious decision to perform a behaviour or a resolve to act in a certain way)	Stability of intentions Stages of change model
9. Goals (Mental representations of outcomes or end states that an individual wants to achieve)	Goals (distal/proximal) Goal priority Goal/target setting Goals (autonomous/controlled) Action planning Implementation intention
10. Memory, attention and decision processes (The ability to retain information, focus selectively on aspects of the environment and choose between two or more alternatives)	Memory Attention Attention control Decision making Cognitive overload/tiredness
11. Environmental context and resources (Any circumstance of a person's situation or environment that discourages or encourages the development of skills and abilities, independence, social competence, and adaptive behaviour)	Environmental stressors Resources/material resources Organisational culture/climate Salient events/critical incidents Person x environment Interaction Barriers and facilitators

(*continued*)

Table 2. (*continued*)

Domain	Constructs
12. Social influences (Those interpersonal processes that can cause individuals to change their thoughts, feelings, or behaviours)	Social pressure Social norms Group conformity Social comparisons Group norms Social support Power Intergroup conflict Alienation Group identity Modelling
13. Emotion (A complex reaction pattern, involving experiential, behavioural, and physiological elements, by which the individual attempts to deal with a personally significant matter or event)	Fear Anxiety Affect Stress Depression Positive/negative affect Burn-out
14. Behavioural regulation (Anything aimed at managing or changing objectively observed or measured actions)	Self-monitoring Breaking habit Action planning

Table 3. Papers included in the study

Date	Author name	Title
2019	Jansen, Jurjen, and Paul van Schaik	The Design and Evaluation of a Theory-Based Intervention to Promote Security Behaviour Against Phishing
2018	Vishwanath, Arun, Brynne Harrison, and Yu Jie Ng	Suspicion, Cognition, and Automaticity Model of Phishing Susceptibility
2018	Verkijika, Silas Formunyuy	Understanding Smartphone Security Behaviors: An Extension of the Protection Motivation Theory With Anticipated Regret
2017	Choi, M., Yair Levy, and Anat Hovav	The Role of User Computer Self-Efficacy, Cybersecurity Countermeasures Awareness, and Cybersecurity Skills Influence on Computer Misuse
2017	Matias Dodel and Gustavo Mesch	Cyber-Victimization Preventive Behavior: A Health Belief Model Approach

(*continued*)

Table 3. (*continued*)

Date	Author name	Title
2017	Princely Ifinedo	Effects of Organization Insiders' Self-Control and Relevant Knowledge on Participation in Information Systems Security Deviant Behaviour
2016	Tsai, Hsin-yi Sandy, Mengtian Jiang, Saleem Alhabash, Robert LaRose, Nora J. Rifon, and Shelia R. Cotten	Understanding Online Safety Behaviors: A Protection Motivation Theory Perspective
2016	Ashley N. Doane, Laura G. Boothe, Matthew R. Pearson and Michelle L. Kelley	Risky Electronic Communication Behaviors and Cyberbullying Victimization: An Application of Protection Motivation Theory
2016	Bartlomiej Hanus and Yu Andy Wu	Impact of Users' Security Awareness on Desktop Security Behavior: A Protection Motivation Theory Perspective
2016	Jurjen Jansen and Paul van Schaik	Understanding Precautionary Online Behavioural Intentions: A Comparison of Three Models
2015	Nader Sohrabi Safa, Mehdi Sookhak, Rossouw Von Solms, Steven Furnell, Norjihan Abdul Ghani and Tutut Herawan	Information Security Conscious Care Behaviour Formation in Organizations
2014	Waldo Rocha Flores, Egil Antonsen and Mathias Ekstedt	Information Security Knowledge Sharing in Organizations: Investigating the Effect of Behavioral Information Security Governance and National Culture
2014	Nalin Asanka, Gamagedara Arachchilage and Steve Love	Security Awareness of Computer Users: A Phishing Threat Avoidance Perspective
2014	Justin Cashin and Princely Ifinedo	Using Social Cognitive Theory to Understand Employees' Counterproductive Computer Security Behaviors (CCSB): A Pilot Study
2014	Princely Ifinedo	Social Cognitive Determinants of Non-Malicious, Counterproductive Computer Security Behaviors (CCSB): An Empirical Analysis
2013	Bo Sophia Xiao and Yee Man Wong	Cyber-Bullying Among University Students: An Empirical Investigation from the Social Cognitive Perspective
2013	Sarah Burns and Lynne Diane Roberts	Applying the Theory of Planned Behaviour to Predicting Online Safety Behaviour

(*continued*)

Table 3. (*continued*)

Date	Author name	Title
2012	Anthony Vance, Mikko Siponen and Seppo Pahnila	Motivating IS Security Compliance: Insights from Habit and Protection Motivation Theory
2012	Princely Ifinedo	Understanding Information Systems Security Policy compliance: An Integration of the Theory of Planned Behavior and the Protection Motivation Theory
2010	Anderson, C. L., and Agarwal, R.	Practicing Safe Computing: A Multi method Empirical Examination of Home Computer User Security Behavioral Intentions
2010	Johnston, A. C., and Warkentin, M	Fear Appeals and Information Security Behaviors: An Empirical Study
2009	Tejaswini Herath and H.R. Rao	Encouraging Information Security Behaviors in Organizations: Role of Penalties, Pressures and Perceived Effectiveness
2009	George R. Milne, Lauren I. Labrecque, and Cory Cromer	Toward an Understanding of the Online Consumer's Risky Behavior and Protection Practices
2009	Ng, Boon-Yuen, Atreyi Kankanhalli, and Yunjie Calvin Xu	Studying Users' Computer Security Behavior: A Health Belief Perspective
2009	Hyeun-Suk Rheea, Cheongtag Kimb, Young U. Ryuc	Self-Efficacy in Information Security: Its Influence on End Users' Information Security Practice Behavior
2009	Tim Chenoweth, Robert Minch and Tom Gattiker	Application of Protection Motivation Theory to Adoption of Protective Technologies
2007	Mikko Siponen, Seppo Pahnila, and Adam Mahmood	Employees' Adherence to Information Security Policies: An Empirical Study
2006	S. Chai, S. Bagchi-Sen, C. Morrell, H. R. Rao and S. Upadhyaya	Role of Perceived Importance of Information Security: An Exploratory Study of Middle School Children's Information Security Behavior

References

1. Bada, M., Sasse, A.M., Nurse, J.R.: Cyber security awareness campaigns: why do they fail to change behaviour? arXiv preprint arXiv:1901.02672 (2019)
2. Holmes, M., Ophoff, J.: Online security behaviour: factors influencing intention to adopt two-factor authentication. In: ICCWS 2019 14th International Conference on Cyber Warfare and Security: ICCWS 2019, p. 123. Academic Conferences and Publishing Limited (2019)
3. Jansen, J., van Schaik, P.: The design and evaluation of a theory-based intervention to promote security behaviour against phishing. Int. J. Hum.-Comput. Stud. **123**, 40–55 (2019)
4. Reedman, S.: Theoretical Domains Framework – Behaviour Change (2017)
5. Michie, S., Johnston, M., Abraham, C., Lawton, R., Parker, D., Walker, A.: Making psychological theory useful for implementing evidence based practice: a consensus approach. BMJ Qual. Saf. **14**, 26–33 (2005)
6. Atkins, L., Francis, J., Islam, R., O'Connor, D., Patey, A., Ivers, N., Foy, R., Duncan, E.M., Colquhoun, H., Grimshaw, J.M.: A guide to using the theoretical domains framework of behaviour change to investigate implementation problems. Implement. Sci. **12**, 77 (2017)
7. Mashiane, T., Kritzinger, E.: Cybersecurity behaviour: a conceptual taxonomy. In: IFIP International Conference on Information Security Theory and Practice, pp. 147–156. Springer (2019)
8. Thompson, N., McGill, T.J., Wang, X.: "Security begins at home": determinants of home computer and mobile device security behavior. Comput. Secur. **70**, 376–391 (2017)
9. Mashiane, T., Kritzinger, E.: Theoretical domain framework to identify cybersecurity behaviour constructs. In: International Conference on Innovative Technologies and Learning, pp. 320–329. Springer (2019)
10. Armitage, C.J., Conner, M.: Efficacy of the theory of planned behaviour: a meta-analytic review. Br. J. Soc. Psychol. **40**, 471–499 (2001)
11. Cane, J., Richardson, M., Johnston, M., Ladha, R., Michie, S.: From lists of behaviour change techniques (BCTs) to structured hierarchies: comparison of two methods of developing a hierarchy of BCTs. Br. J. Health. Psychol. **20**, 130–150 (2015)

Method of Recurrent Neural Network Hardware Implementation

Oleg Nepomnyashchiy[1(✉)], Anton Khantimirov[1], Dimitri Galayko[2], and Natalia Sirotinina[1]

[1] Siberian Federal University, Svobodny Prosp. 79, 660041 Krasnoyarsk, Russian Federation
onepomnuashy@sfu-kras.ru
[2] Sorbonne Université, Jussieu Campus, 75006 Paris, France

Abstract. Real-time data processing using recurrent neural networks (NN) is non-trivial task, due to tight timing constraints requirements. It is proposed hardware implementation of recurrent echo state NN (ESN) on the basis of the Cyclone IV FPGA. Advantages of the hardware implementation are high computational parallelism and low power consumption. To solve the problem of neuron weight storage, it is proposed to reduce the space of their values to a set of integers of low capacity. It was determined that the proposed NN model decreases need in hardware resources for the reservoir implementation in 2–3 orders of magnitude in comparison with conventional NN. Modeling results, implementation and testing of the FPGA project confirmed effectiveness of the proposed integer NN in hardware applications #CSOC1120.

Keywords: Recurrent neural networks · Echo state neural networks · Hyperdimensional calculations · Field programmable integrated circuits

1 Introduction

Real-time data processing with the use of neural networks (NN) is non-trivial task, due to tight timing constraints requires. The widespread software approach to NN implementation does not provide necessary performance and requires significant computational power. In this regard, NN implementation as specialized integrated circuits is a promising approach.

Most hardware implementations of NN have the multi-layer perceptron architecture. This NN model is effectively used for classification, pattern recognition, etc. [1]. However, multilayer NNs do not have memory of previous states. This limits their use for solving problems associated with signal sequence processing. Tasks of this kind often arise as a part of task of dynamic system control. In these cases, one of possible solutions is the use of recurrent NN (RNN) architectures, which have dynamic memory [2]. Since embedded control systems must comply with the strict limits of response time and required hardware resources, a hardware implementation is most suitable for these applications.

R. Silhavy (Ed.): CSOC 2020, AISC 1225, pp. 429–437, 2020.
https://doi.org/10.1007/978-3-030-51971-1_35

Advantages of the RNN hardware implementation include high computational parallelism and low power consumption. Due to these advantages, a RNN-based control module can be implemented on a single chip and integrated directly into the control object, for example, in electric motor or energy conversion equipment. However, the conventional RNN model has complex training algorithms and requires a large amount of memory to store weights. A number of works on this subject consider NN of echo states, unstable states, etc. [3]. Effective solutions can be obtained using the hardware implementation of NN, based on hyperdimensional calculations [4].

2 Proposed Solution

The conventional echo state NN (ESN) includes three neuron layers (Fig. 1). The input layer receives the input signal $u(n)$. The hidden layer is called a reservoir. The reservoir state is described by $x(n)$. The output layer elaborates the NN output signal $y(n)$ at the end of the operating phase.

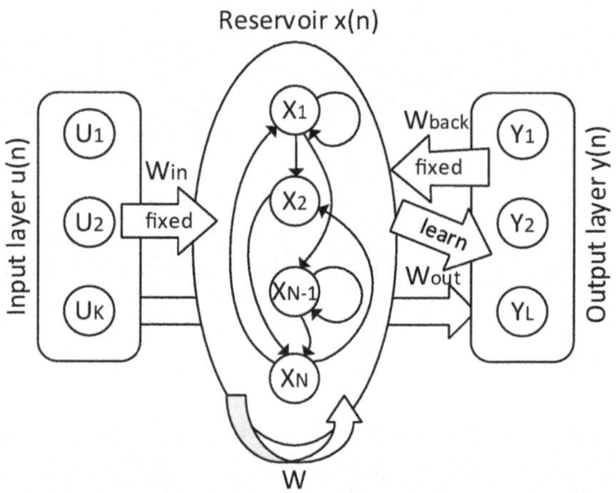

Fig. 1. Conventional echo state neural network architecture

Internal synaptic connections of the considered NN are described by four matrices of weights: input W^{in}, reservoir W, feedback W^{fb} and output W^{out} weights. All matrices, except W^{out}, are set randomly when creating the NN and remain unchanged throughout its functioning. Only matrix W^{out} is trained. Due to this architecture, the task of RNN training is reduced to calculating the linear regression which minimizes the mean square error between predictions and the ground truth.

There are no strict requirements for the projection matrices W^{in} and W^{fb} generating procedure. Usually they are set randomly with a normal or uniform distribution. The reservoir connection matrix W has limitations according to the echo state property. This property is achieved when the spectral radius of the matrix W is less than or equal to 1. For example, it can be generated using the normal distribution and normalized by its maximum eigenvalue. In this paper, an orthogonal matrix will be used to describe reservoir connections. The matrix W is produced by QR decomposition of a random matrix obtained using the standard normal distribution. The matrix W can also be scaled on the feedback parameter according to Eq. (1).

The network reservoir update at time n is described by the following equation:

$$x(n) = tanh\left(\rho W x(n-1) + \beta W^{in} u(n) + \beta W^{fb} y(n-1)\right) \quad (1)$$

where β and ρ denote the projection coefficient and feedback parameter, respectively.

It is assumed that the spectral radius of the reservoir interconnection matrix W is 1. The activation function of the reservoir neuron is hyperbolic tangent. The nonlinearity of the activation function prevents NN degeneration by limiting the possible range of values to the interval [−1; 1]. Output layer signals are calculated as:

$$\widehat{y}(n) = g\left(W^{fb}[x(n); u(n)]\right) \quad (2)$$

where the semicolon means the concatenation of two vectors, g is the activation function of output neurons. This function can be linear or the "winner-takes-all" type. This shows that an efficient hardware implementation of NN requires a transition from a conventional ESN to integer ESN. Integer echo-state NN [5] are the architecture of recurrent echo-state NN implemented based on the paradigm of hyperdimensional computing.

In hyperdimensional calculations, each operand is represented as random vectors of large dimension, reaching thousands of bits. Information is used in a distributed representation in which separate subset of bits cannot be clearly interpreted. Distributed calculations use statistical properties of large-dimensional vector spaces. This allows performing approximate high-parallel calculations that are resistant to errors in individual bits. Several recent works proposed the usage of hyperdimensional computations for energy-efficient realizations of operations of different architectures of artificial neural networks. Amongst the notable examples are integer implementation of unsupervised learning using Self-organized Maps [6] and integer realization of Random Vector Functional Link [7] (RVFL) neural networks, which is a type of Randomly Connected Neural Networks – an attractive alternative to backprop networks for solving practical machine learning problems in edge computing. These results further demonstrate capabilities of the approach for dramatic improvement of resource efficiency of the classical machine learning solutions. In [8] a novel representation of n-gram statistics using hyperdimensional vectors, which in combination with classical machine learning methods lead to the substantial improvement of the memory footprint and runtime performance.

The architecture of the integer ESN (intESN) is depicted in Fig. 2.

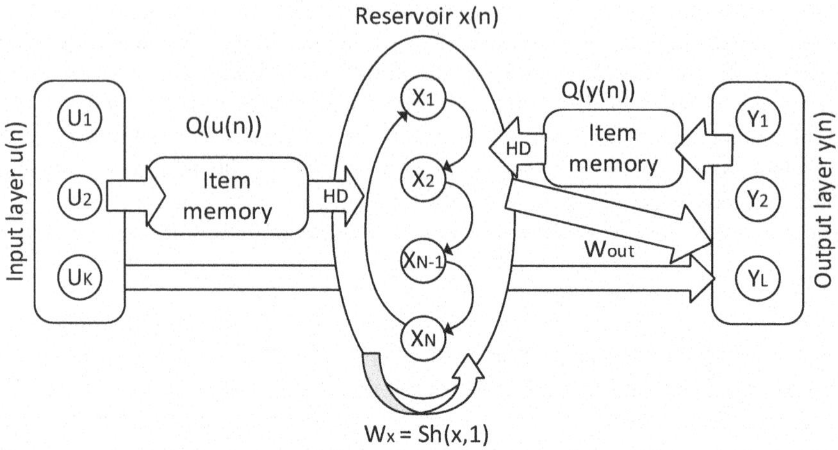

Fig. 2. Integer echo state neural network architecture

The intESN is similar with the conventional ESN which includes three neuron layers (Fig. 1). The input layer $u(n)$ consists of K neurons. The hidden layer $x(n)$ (reservoir) consists K neurons. The output layer $y(n)$ consists of L neurons. Training of the readout matrix W^{out} for the intESN is similar to the conventional ESN.

Other components of the intESN differ from the conventional ESN. First, the activation signals of the input and output layers are projected onto the reservoir in the form of hyperdimensional bipolar vectors [9] of dimension N, denoted by $u^{MM}(n)$ and $y^{MM}(n)$. For tasks where input and output data are described by a finite alphabet, if each symbol can be interpreted independently, their mapping on N-dimensional space is achieved by simply assigning a random hyperdimensional bipolar vector to each symbol in the alphabet and storing them in the item memory [4]. Continuous data (for example, real numbers), are quantized and conversed to the final alphabet. The method and accuracy of quantization depends on the problem being solved. If it is necessary to maintain similarity between quantization levels, distances preserving mapping schemes are used [10, 11]. These schemes allow maintaining linear or nonlinear similarity between levels. Continuous values can also be represented as hyperdimensional vectors by varying their distribution density.

Another feature of the considered integer network of echo states is the method of creating recurrence in the reservoir. Instead of the computationally complex operation of matrix multiplication, recurrence can be realized through permutations of the reservoir vector. Vector permutations can be described in the matrix form that coincides in functionality with the matrix W of a conventional echo states NN (Fig. 1). One of the main conditions for specifying such a matrix is the equality of its spectral radius to unity. An effective implementation of the permutation can be obtained by a cyclic shift marked as *Sh*. Figure 2 shows reservoir neuron connections which implement recurrence through cyclic shift by one position. In this case, the vector-matrix multiplication *Wx* (*n*) is denoted by *Sh* (*x* (*n*), *1*).

As a nonlinearity of the reservoir activation function, the intESN uses constraint (3). For its implementation, it is advisable to store the values of hyperdimensional vector elements in the limited range using a threshold value *k*.

$$f_k(x) = \begin{cases} -k, x \leq -k \\ x, -k < x, < k \\ k, \geq k \end{cases} \quad (3)$$

In the intESN, the reservoir is updated only with integer bipolar vectors and after clipping, neuron output signal values remain integers in the range from −k to k. Thus, each neuron is represented using log2 (2k + 1) bits of memory. For example, if k = 15, then there are 31 different values of the neuron output signal, stored in just 5 bits.

Updating an integer network of echo states can be described as:

$$x(n) = f_k(Sh(x(n-1), 1) + u^{MM}(n) + y^{MM}(n-1) \quad (4)$$

3 Results

To investigate the intESN, software models of the conventional and integer NN have been developed using Matlab environment.

The study objectives are: to identify the ability of the NN to memorize input values and work with time sequences, to compare the reliability of the integer and conventional recurrent NN, to evaluate the possibility of the intESN hardware implementation in the basis of field programmable integrated circuits (FPGAs).

The task solved by the network as the test case, was to implement short-term memory, in other words, NN should reproduce the sequence of characters received by NN during a certain number of steps back. The operating cycle includes two stages: memorization and reproduction. In the memorization step, NN obtains and stores a sequence of characters of the final alphabet. Each character corresponds to a certain hyperdimensional vector consisting of sets 0 and 1, given randomly during network initialization. At the reproducing stage, NN retrieves the character stored a certain number (d) steps back. The variable d defines the delay.

The research results are shown in Fig. 3.

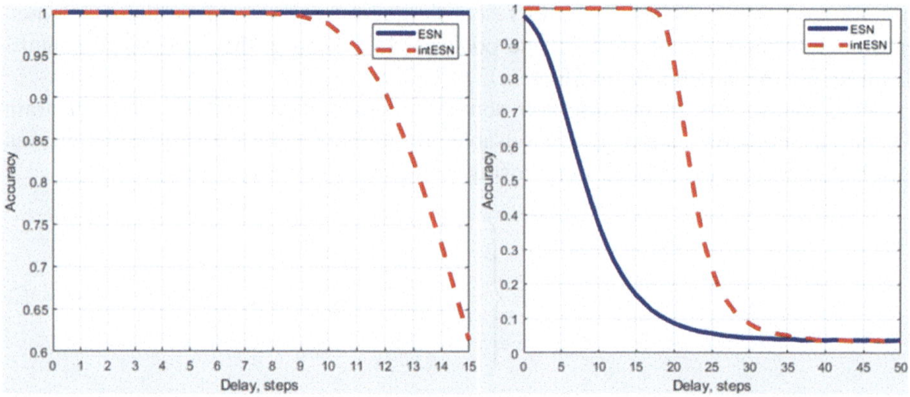

Fig. 3. Dependence of the accuracy of determining an input symbol from the delay duration

Normal and integer recurrent NNs containing 1000 neurons in the reservoir were compared (Fig. 3A). Graphs in the Fig. 3 show, that both networks successfully cope with the task when the delay is up to 9 work cycles, but after that the efficiency of the integer network decreases.

To evaluate the effectiveness of the FPGA-based hardware implementation, the operation of the conventional and intESN was also simulated. In this case study, the number of neurons in the reservoir was chosen taking into account the number of logic gates needs for the reservoir implementation. The graph (Fig. 3B) shows that the intESN significantly exceeds the conventional one in the accuracy of the input character recognition.

Another task used to compare effectiveness of recurrent NN is the prediction of the Mackey-Glass time series. NN was training with the use of one sequence, with the same level of quantization. NN testing results are shown in Fig. 4.

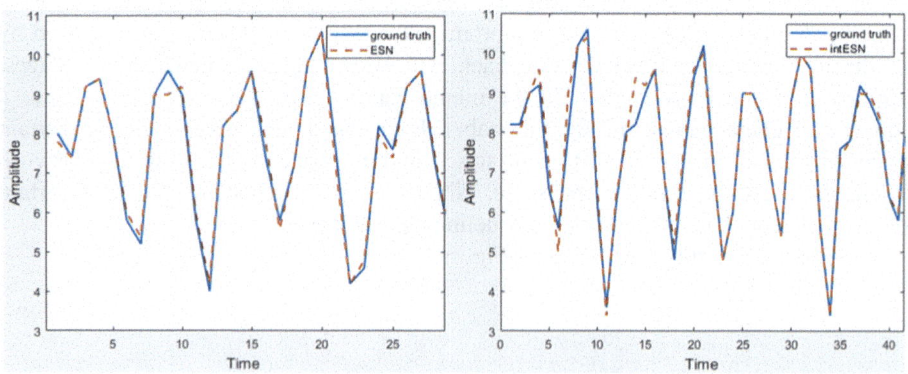

Fig. 4. Mackey-Glass time series prediction

The accuracy of predicting a sequence with quantization of 50 points using a conventional ESN with the reservoir size of 1000 neurons is 51%, and 31% of the predicted points are adjacent to the points of the test sequence. For an intESN with the reservoir size of 40,000 neurons, 53% of the points are predicted accurately, 36% of the predicted points are adjacent. Thus, the experiment study proves that the accuracy of the intESN is no worse than the accuracy of the conventional NN in the case of using more neurons in the reservoir.

Storage of intESN weights needs significantly less hardware resources. Therefore, it is possible to implement a larger number of neurons in the reservoir at the same hardware costs. To confirm this statement, reservoirs for the conventional and the intESN were implemented on the basis of FPGA. Project modules were described in the Verilog hardware description language for the Cyclone IV FPGA. The graph of the logic gates number used for the reservoir implementation is shown in Fig. 5.

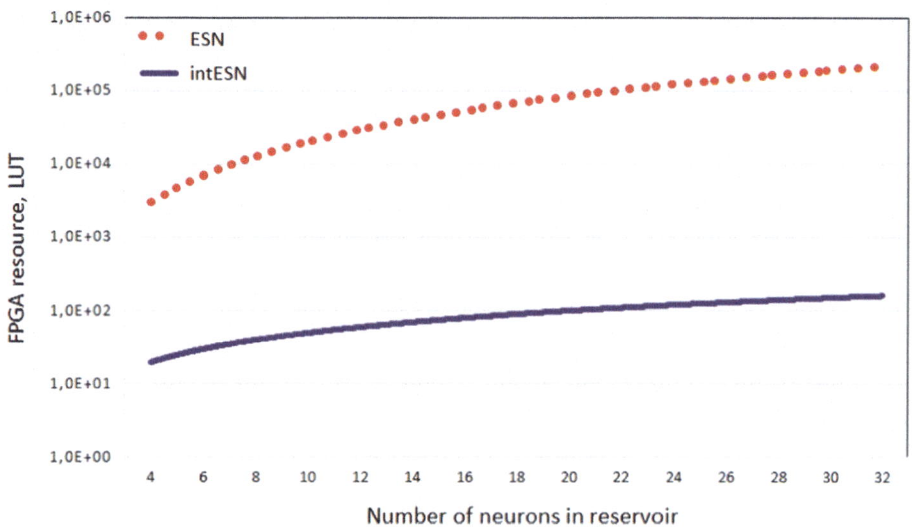

Fig. 5. The graph of the dependence of hardware resources amount (the number of logical gates) for the implementation of the reservoir from the number of neurons

As can be seen from the graph, the reservoir of the intESN takes 2–3 orders of magnitude less hardware resources (logical gates) than the conventional ESN.

4 Discussion

The main advantage of integer neural networks in terms of hardware implementation is the significant reduction in the amount of required hardware resources. The disadvantage of this approach is training time increase. Since the neuron number in the reservoir is up to three order of magnitude more in comparison with the conventional

NN, this significantly increases complexity and computational cost on the linear regression calculation. The number of addition and multiplication operations between the reservoir and the output network layer also significantly increases.

To implement the network retraining, it is supposed to use a specialized coprocessor, which allows to accelerate the calculation of linear regression. It is proposed to implement additions and multiplications between the reservoir and the output layer in a series-parallel circuit using the entire volume of FPGA computing modules. To simplify the developing control circuit, it is necessary to develop tools to synthesize a control automaton based on target platform characteristics.

5 Conclusion

Modeling results, implementation and testing of the FPGA project confirm effectiveness of the use of the proposed model of the intESN in hardware applications. It was determined that the proposed NN model decreases need in hardware resources for the reservoir implementation in 2–3 orders of magnitude in comparison with conventional NN. The further task is to study the effectiveness of applying the integer echo states NN hardware implementation on object control problems in a dynamic environment.

Acknowledgements. Co-funded by the Erasmus + programme of the European Union: Joint project Capacity Building in the field of Higher Education 573545-EPP-1-2016-1-DE-EPPKA2-CBHE-JP "Applied curricula in space exploration and intelligent robotic systems".

The European Commission support for the production of this publication does not constitute an endorsement of the contents which reflects the views only of the authors, and the Commission cannot be held responsible for any use which may be made of the information contained therein.

References

1. Jawandhiya, P.: Hardware design for machine learning. Int. J. Artif. Intell. Appl. **9**(1), 63–84 (2018)
2. Lukosevicius, M., Jaeger, H.: Reservoir computing approaches to RNN training. Comput. Sci. Rev. **3**(3), 127–149 (2009)
3. Maass, W., Natschlager, T., Markram, H.: Real-time computing without stable states: a new framework for neural computation based on perturbations. Neural Comput. **14**(11), 2531–2560 (2002)
4. Kanerva, P.: Hyperdimensional computing: an introduction to computing in distributed representation with high-dimensional random vectors. Cogn. Comput. **1**(2), 139–159 (2009)
5. Kleyko, D., Frady, E.P., Osipov, E.: Integer echo state networks: hyperdimensional reservoir computing. arXiv preprint arXiv:1706.00280 (2017)
6. Kleyko, D., Osipov, E., De Silva, D., Wiklund, U., Alahakoon, D.: Integer self-organizing maps for digital hardware. In: 2019 International Joint Conference on Neural Networks (IJCNN) 2019, pp. 1–8. IEEE, Budapest (2019)
7. Kleyko, D., Kheffache, M., Frady, E.P., Wiklund, U., Osipov, E.: Density encoding enables resource-efficient randomly connected neural networks. arXiv preprint arXiv:1909.09153 (2019)

8. Kleyko, D., Osipov, E., De Silva, D., Wiklund, U., Vyatkin, V., Alahakoon, D.: Distributed representation of n-gram statistics for boosting self-organizing maps with hyperdimensional computing. In: Bjørner, N., Virbitskaite, I., Voronkov, A. (eds.) Perspectives of System Informatics (PSI) 2019. Lecture Notes in Computer Science, vol. 11964, pp. 64–79. Springer, Cham (2019)
9. Gallant, S.I., Culliton, P.: Positional binding with distributed representations. In: International Conference on Image, Vision and Computing (ICIVC), pp. 108–113 (2016)
10. Widdows, D., Cohen, N.: Reasoning with vectors: a continuous model for fast robust inference. Log. J. IGPL **23**(2), 141–173 (2016)
11. Wang, H., Wu, Y., Zhang, B., Du, K.L.: Recurrent neural networks: associative memory and optimization. Inf. Technol. Softw. Eng. **1**(2), 1–15 (2019)

An Experimental Study of the Fog-Computing-Based Systems Reliability

A. B. Klimenko[1(✉)] and E. V. Melnik[2]

[1] Scientific Research Institute of Multiprocessor Computer Systems of Southern Federal University, 2, Chekhov Street, 347928 Taganrog, Russian Federation
anna_klimenko@mail.ru

[2] Federal Research Centre, Southern Scientific Centre of the Russian Academy of Sciences, 41, Chekhov Street, 344006 Rostov-on-Don, Russian Federation

Abstract. The current paper deals with the experimental research of the fog-computing concept application to the information and control systems in terms of the system reliability. The research of publications presented allows to conclude that contemporary research and existing models do not clarify the expediency of the fog-computing concept application to the information systems requiring the reliable functioning. Also some parameters, which should be taken into account, are not incorporated into the known problem models.

In this paper the model of task distribution problem is proposed, which pays attention to such particularity of node workload forming as the transitory data transfer. The transitory data transfer through the particular node generates the additional workload, which is included into the proposed model to estimate the node reliability.

Also the simulation was conducted, whose main result consists in the fact that the expediency of the fog-computing application is quite relative and depends on such parameters as computational complexities of the tasks, transferred data volume and the ratio between the transitory data volume and the additional node workload.

Keywords: Fog computing · Information and control systems · Reliability

1 Introduction

Fog-computing [1, 2] is a computational concept which, in general, was developed to improve latency and scalability of the systems built with the Internet of Things concept usage. Enormous data volumes circulate through the network, so the traditional "cloud" architecture has become insufficient.

Fog computing has been applied to the wide range of subject areas, including:

- Food chains [3];
- Healthcare [4];
- Medical services [5];
- Mobile facilities-based information systems [6];
- Smart cities [7];
- UFV monitoring and control [8], etc.

R. Silhavy (Ed.): CSOC 2020, AISC 1225, pp. 438–449, 2020.
https://doi.org/10.1007/978-3-030-51971-1_36

Also, a theoretical research took place in the area of the fog-computing, including the development of techniques and algorithms of computational workload distribution between the cloud, fog and the edge of the network. Examples of such works can be found in [9, 10].

The key idea of the fog computing is to shift the data processing to the data sources, as near as possible. This idea can be applied to the information and control systems, which already function and have their own network infrastructure. So, locating the data preprocessing on the nodes, which are near to the data sources (for example, sensors), the workload of the central computational cluster can be reduced, as well as the system latency and the network load.

Yet, it must be mentioned, that the question of the fog-computing-based systems reliability is not focused on, while the reliability (and as a consequence - dependability) problems are quite urgent in some application areas. The reliability changes, caused by the fog concept application, are still not clarified, though the reliability of the node itself is connected to the node workload, and, as a consequence, to the way of how the computational tasks are distributed through the heterogeneous network. So, the effect of the fog-computing concept application to the information system reliability is in the scope of this paper.

This research is the continuation of the work, previously done, and published [11–13]. In the current paper the experimental research of the fog-concept in terms of reliability is described. To simulate the computational nodes workload, a particular software was developed, which simulates the tasks distribution and estimates the system reliability, assuming the absence of spare elements.

Besides, some new parameters are used in the problem formal model, which has not been taken into account before.

2 Problem Formalization

Consider G is a graph-based description of the task set, $G = \{<i, x_i, w_i>\}$, where i is a subtask unique identifier, x_i – task computational complexity, w_i– the data size to be transmitted to the communication environment by subtask i.

The subtasks of G are bound to the nodes set P, where P is described by a graph structure $P = \{<j, p_j> , list\}$, where j is a node identifier, p_j – node performance, $list$ – matrix of communication channels.

The tasks distribution through the network is described as matrix A:

$$A = \begin{vmatrix} <t_0^{ij}, u_{ij}> & & \\ & \cdots & \\ & & <t_0^{NM}, u_{NM}> \end{vmatrix} \quad (1)$$

where t_0^{ij} – the time moment of the subtask i computations beginning by the node j, u_{ij} – the fraction of total performance p_j, given by node j for task i accomplishment.

This problem model allows to locate more then one task on one node at the same time, enhancing the classical scheduling problem formulation.

Then, consider the following: full workload of the node consists of the workload, generated by the computational task itself, workload, produced by sending and receiving data of the tasks, located on the node, and the so-called transitory workload, which originates from the need to transfer the data of other computational tasks if the current node is on the route of data transferring.

The latter parameter, the transitory workload, is of a high importance in the model of the workload distribution, though this parameter is not taken into account in a wide range of published models.

So, full node workload L_j contains the following:

$L_t(A)$ – node workload, which is generated by the task, located on the node;

$L_{send}(A)$ – node workload, which is generated by data, which are sent by the tasks, located on the node;

$L_{receive}(A)$ – node workload, which is generated by data, which are received by the tasks, located on the node;

$L_{tr}(A)$ – the transitory workload;

D_{lk} – the list of ribs of graph P, which determines the route between nodes l and k.

Consider the objective function as follows:

$$F = \prod_{j=1}^{M} F_j, \tag{2}$$

$$F_j = e^{-\lambda_j t}, \tag{3}$$

where λ_j - node j failure rate,

t – elapsed time of device operation.

It must be mentioned, that such objective function estimates the lower boundary of the system reliability, because does not take into account the existence of the nodes, which can be spare elements. Also, the objective function describes the system till the first failure.

As $\lambda = \lambda_0 \cdot 2^{\Delta T/10}$, and $\Delta T = kL_j$, where L_j is a device workload, k is a ratio and depends on the device type, the dependency between the reliability function and the workload will be as follows:

$$F_j = e^{\lambda_{j0} \cdot 2^{kL_j/10}} \tag{4}$$

$$L_j(A) = L_{jtask}(A) + L_{jsend}(A) + L_{jreceive}(A) + L_{jtr}(A) \tag{5}$$

We assume that the value of $L_{jtr}(A)$ depends on the volume of transitory data in such a way:

$$L_{jtr}(A) = \xi w_{jtr} \tag{6}$$

as well as

$$L_{jsend}(A) = \mu w_{jsend} \tag{7}$$

$$L_{jreceive}(A) = \mu w_{jreceive} \tag{8}$$

The parameter ξ is quite important in the process of tasks distribution, as will be shown at the simulation stage.

So, to estimate the node workload it is needed to know the way of tasks distribution and, at least, to form the routes of the data transferring (D_{lk}).

The major constraint for this problem is the G completion time T, in other words:

$$\forall i \in G : \frac{x_i}{p_j u_{ij}} + t_{dist}(i) < T \tag{9}$$

where $t_{dist}(i)$ - the maximum time of data delivery from subtask i to the subtasks-receivers of the data.

As the model considers dataflow routes, $t_{dist}(i)$ is calculated with the function:

$$t_{dist}(i) = v(A, G, P) \tag{10}$$

More precisely, the data delivery is calculated on the basis of full information about the subtasks binding to the computational nodes and with the participating of the parameters D_{lk}.

3 The Input Data and the Planning of an Experiment

In the current paper a range of task distribution cases is tested. As is shown in the previous section, the following input data must be determined:

- An experimental network topology, including the node performances;
- An information task graph, including the computational complexities of the tasks and the data values to be transferred through the network;
- Time constraint, task distribution cases and parameter ξ (Eq. 6).

Here the example of experimental data is considered. It must be mentioned that experimental research of the other data gave the similar tendencies.

In the Fig. 1 the network topology is presented.

We can see, that the node performance is presented in the so-called "performance modeling units". Besides, the topology presented models three major layers of the network: the edge (filled with the user devices, sensors, etc.), the fog layer (filled with communication infrastructure objects) and the cloud, represented by computational nodes with high computational performance.

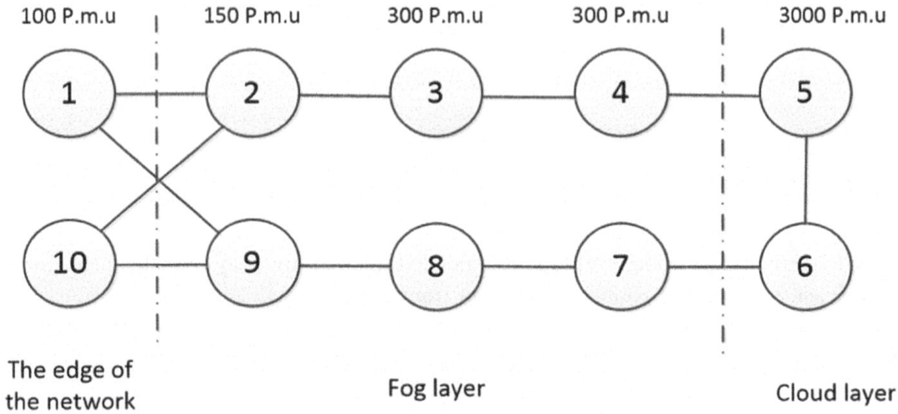

Fig. 1. The network topology

To avoid the model cluttering, we used the bipartite graph, as is shown in the Fig. 2. The computational complexity of the tasks is presented in the "computational complexity units".

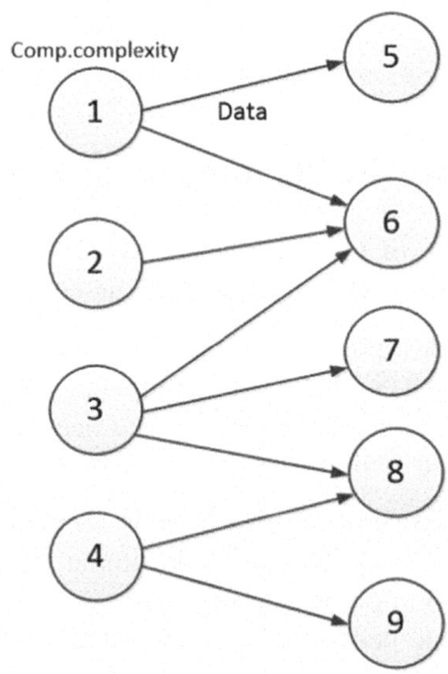

Fig. 2. The task graph

The bipartite graph in our example models the situation, when a part of tasks is located on the edge of the network, while the other tasks process the raw data received from the data sources.

In our experimental research the following parameters are variable: computational complexities of the tasks, the values of the data to be transferred through the network and the ratio between the node load and the data volume, which is transferred through the node, and the task distribution through the network. As is mentioned above, we assume that communicational channels between nodes are of high velocity, so, there is no constraint relating to the time of data transfer.

To conduct the experiment, the plan of the experimental research was developed (Fig. 3).

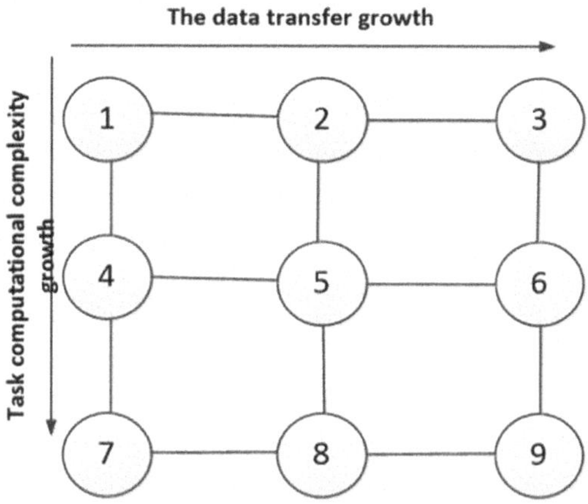

Fig. 3. The plan of the experiment

The parameters for modeling are presented in the Table 1.

Table 1. The simulation parameters

	Small data transfer	Average data transfer	Big data transfer
Low computational complexity	Case 1: Comp. compl.: 50–500 p.m.u. Data value: 500–1000 m.u.	Case 2: Comp. compl.: 50–500 p.m.u. Data value: 5000–10000 m.u.	Case 3: Comp. compl.: 50–500 p.m.u. Data value: 10000–15000 m.u.
Average computational complexity	Case 4: Comp. compl.: 250–2500 p.m.u. Data value: 500–1000 m.u.	Case 5: Comp. compl.: 250–2500 p.m.u. Data value: 5000–10000 m.u.	Case 6: Comp. compl.: 250–2500 p.m.u. Data value: 10000–15000 m.u.
High computational complexity	Case 7: Comp. compl.: 1000–10000 p.m.u. Data value: 500–1000 m.u.	Case 8: Comp. compl.: 1000–10000 p.m.u. Data value: 5000–10000 m.u.	Case 9: Comp. compl.: 1000–10000 p.m.u. Data value: 10000–15000 m.u.

Besides, the following cases of task distribution were considered (Table 2).

Table 2. The task distribution cases

Exp.num./node num.	1	2	3	4	5	6	7	8	9	10
1	1, 2				5, 6	7, 8, 9				3, 4
2	1, 2			5	6	8, 9	7			3, 4
3	1, 2		5		6	8, 9		7		3, 4
4	1, 2	5			6	8, 9			7	3, 4
5	1, 2			5, 6		9	7, 8			3, 4
6	1, 2		5, 6			9		7, 8		3, 4
7	1, 2	5, 6				9			7, 8	3, 4
8	1, 2			5, 6			7, 8, 9			3, 4
9	1, 2		5, 6					7, 8, 9		3, 4
10	1, 2	5, 6							7, 8, 9	3, 4
11	1, 2		5	6		9	8	7		3, 4
12	1, 2	5	6				9	8	7	3, 4
13	1, 2	5		6				9	7, 8	3, 4

You can see that the general strategy of the distribution varying is to shift the data processing tasks from the cloud to the edge of the network.

4 Experimental Results

As we consider the fog computing from the reliability angle, experimental results contain the estimations of the reliability functions of the nodes measured in the particular time moments. All plots are built for the time t = 100 (h).

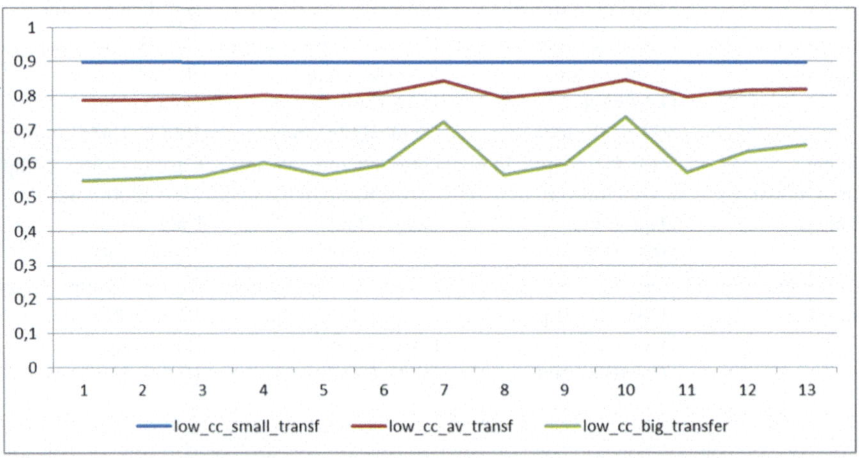

Fig. 4. The reliability function values for tasks with low computational complexity

You can see that for tasks with relatively low computational complexity with the growth of the values of data to be transferred the best distributions are the same: №7 and №10 (see Table 1). The best (in terms of reliability estimations) distributions are when the data processing tasks are located near the data sources. Such result is expected and correlates with the fog computing concept.

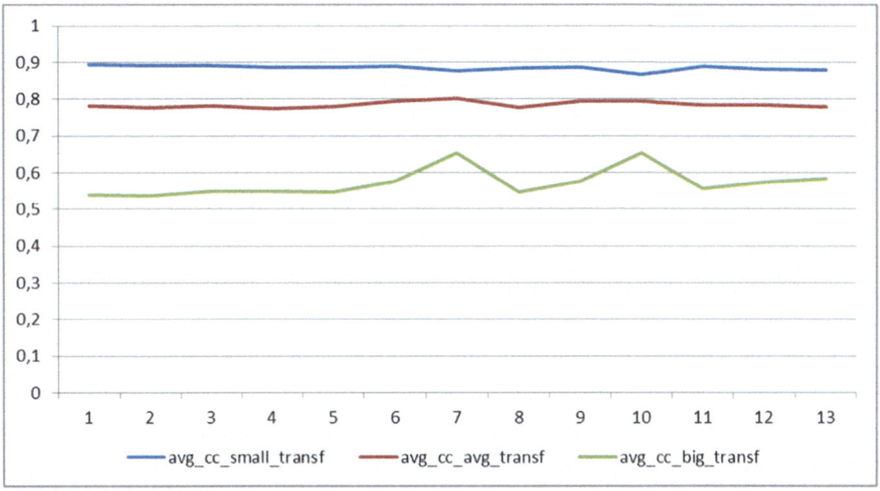

Fig. 5. The reliability function values for tasks with average computational complexity

In the Fig. 5 the plots differs from ones in previous figure. In case of average computational complexity of tasks and low data exchange the best distribution cases are: №1 and № 11 (cloud and distribution through the fog layer), yet, with the growth of data transferring, the picture changes. He best cases are №7 and №10 (the same as for low computational complexity and variable data transfer).

And, finally, for tasks of high computational complexity the following plots were conducted (Fig. 6).

In this figure you can see, that those cases of tasks distribution, which are the best in the previous experiments, are the worst in case of tasks with high computational complexity.

The cases №7 and №10 are the worst (the data processing is near the data sources), while the cases №1 and №11 are the best, when the data processing tasks are in the cloud and are distributed in the fog around the cloud. Also, the Fig. 6 shows that with the growth of the tasks computational complexity the tasks distribution way is rather important: while in Fig. 4 the difference for reliability estimation for all distribution cases is not noticeable, the difference in reliability estimation in Fig. 6 is rather critical.

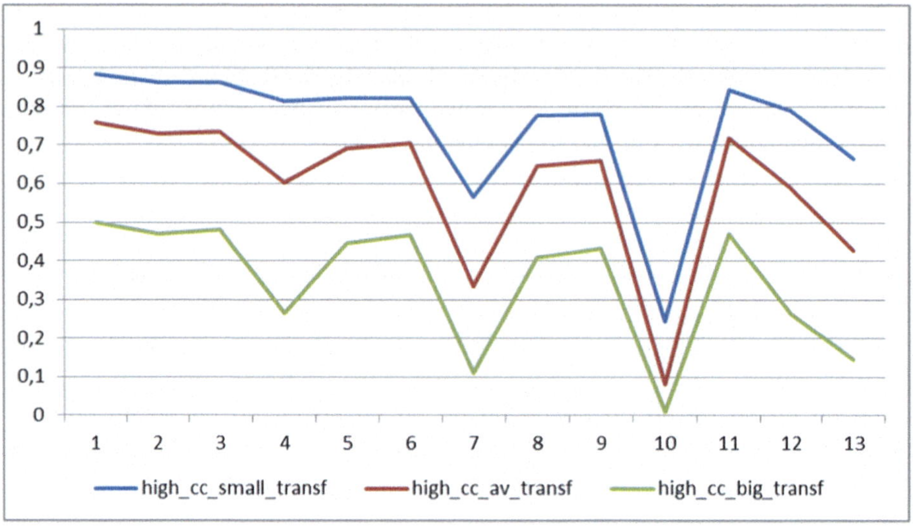

Fig. 6. The reliability function values for tasks with high computational complexity

Also the effect of ratio, which connects the node workload and the value of the transitory data, is rather interesting: the bigger this ration is, the more expedient to distribute the data processing tasks near the data sources at the edge of the network, and vice versa. The ratio effect is shown in Fig. 7, 8 and 9.

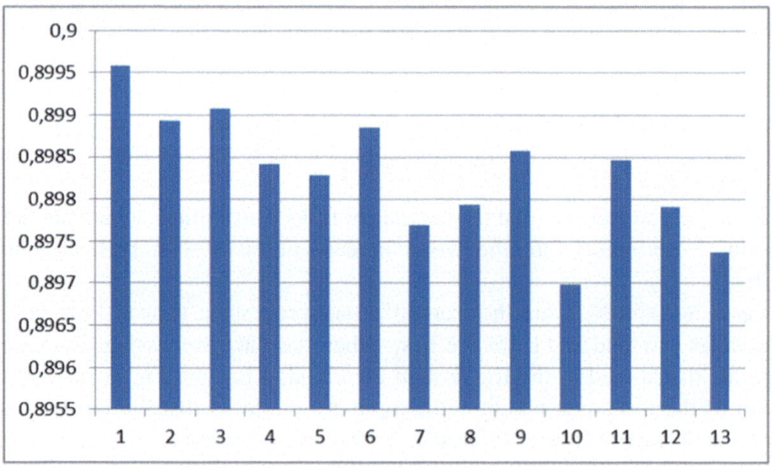

Fig. 7. The reliability function for $\xi = 0, 2$

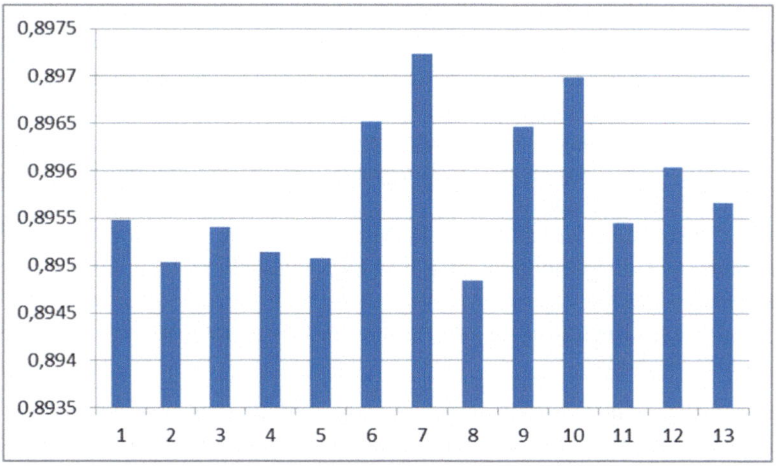

Fig. 8. The reliability function for $\xi = 0, 4$

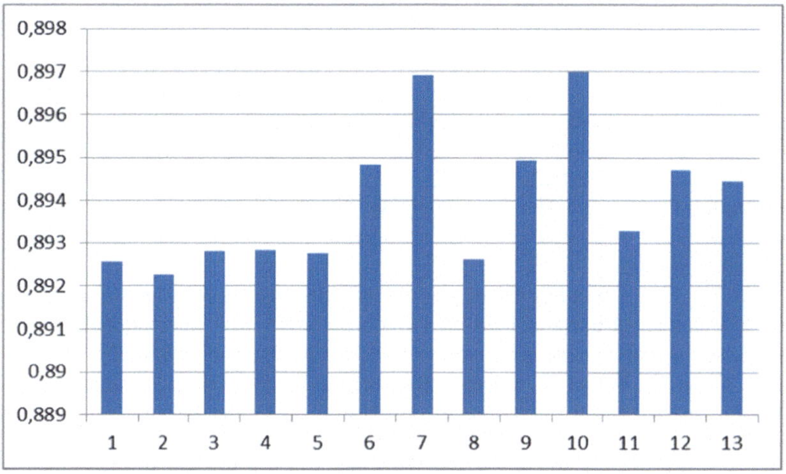

Fig. 9. The reliability function for $\xi = 0, 7$

So, it is seen that the smaller the influence of the transitory data volume is, the more expedient it is to locate the data processing in the cloud and vice versa.

5 Discussion and Conclusion

Summarising the experiments conducted, the following conclusions can be made:

- the preferable task location in the network layers varies according the task parameters, i.e.: task computational complexities, transferred data volume and the ratio between the transitory data and node workload;

- with the task computational complexity growth the task location optimization is needed;
- with relatively small task computational complexity the task locations hardly affects the overall system reliability;
- the usage of the fog-layer as computational tasks locations allows to improve the overall system reliability in case of the low computational complexity of tasks with the varying transfer data volumes;
- the tasks distribution modeling is highly expedient because of the dependency between the task distribution effect and tasks computational complexity and the data to be transferred;
- the ratio between the transitory data volume and the node workload affects the tasks distribution method.

Therefore the main result of our simulation is that some parameter areas can be formed, in which the fog-computing application improves the overall system reliability.

Acknowledgements. This study is supported by the RFBR project 18-05-80092 and the GZ SSC RAS N GR project AAAA-A19-119011190173-6.

References

1. Bonomi, F., Milito, R., Zhu, J., Addepalli, S.: Fog Computing and its role in the internet of things. In: Proceedings of the first edition of the MCC workshop on Mobile cloud computing, pp. 13–16 (2012)
2. Stojmenovic, I., Wen, S.: The fog computing paradigm: scenarios and security issues. In: Proceedings 2014 federated conference on computer science and information systems, vol. 2, pp. 1–8 (2014)
3. Chen, R.Y.: Fog computing-based intelligent inference performance evaluation system integrated internet of thing in food cold chain. In: 2015 12th International Conference on Fuzzy Systems and Knowledge Discovery, FSKD 2015, pp. 879–886 (2016)
4. Shi, Y., Ding, G., Wang, H., Eduardo Roman, H., Lu, S.: The fog computing service for healthcare. In: 2015 2nd International Symposium on Future Information and Communication Technologies for Ubiquitous HealthCare, Ubi-HealthTech 2015, pp. 70–74 (2015)
5. Gia, T.N., Jiang, M., Rahmani, A.M., Westerlund, T., Liljeberg, P., Tenhunen, H.: Fog computing in healthcare internet of things: a case study on ECG feature extraction. In: Proceedings - 15th IEEE International Conference on Computer and Information Technology, CIT 2015, 14th IEEE International Conference on Ubiquitous Computing and Communications, IUCC 2015, 13th IEEE International Conference on Dependable, Autonomic and Se, pp. 356–363 (2015)
6. Sun, X., Ansari, N.: EdgeIoT: mobile edge computing for the internet of things. IEEE Commun. Mag. **54**(12), 22–29 (2016)
7. Perera, C., Qin, Y., Estrella, J.C., Reiff-Marganiec, S., Vasilakos, A.V.: Fog computing for sustainable smart cities: a survey. ACM Comput. Surv. **50**(3), 1–43 (2017)
8. Inaltekin, H., Gorlatova, M., Mung, C.: Virtualized control over fog: interplay between reliability and latency, CoRR, vol. abs/1712.0 (2017)

9. Oueis, J., Strinati, E.C., Barbarossa, S.: The fog balancing: load distribution for small cell cloud computing. In: 2015 IEEE 81stVehicular Technology Conference (VTC Spring), pp. 1–6 (2015)
10. Intharawijitr, K., Iida, K., Koga, H.: Analysis of fog model considering computing and communication latency in 5G cellular networks. In: 2016 IEEE International Conference on Pervasive Computing and Communication Workshops (PerCom Workshops), pp. 1–4 (2016)
11. Melnik, E.V., Klimenko, A.B., Ivanov, D.Y.: Distributed information and control system reliability enhancement by fog-computing concept application. IOP Conf. Ser. Mater. Sci. Eng. **327**(2), 22070 (2018)
12. Melnik, E., Korovin, I., Klimenko, A.: Improving dependability of reconfigurable robotic control system. In: Ronzhin, A., Rigoll, G., Meshcheryakov, R. (eds.) ICR 2017. LNCS (LNAI), vol. 10459, pp. 144–152. Springer, Cham (2017)
13. Melnik, E., Korobkin, V., Klimenko, A.: System reconfiguration using multiagent cooperative principles. In: Abraham, A., Kovalev, S., Tarassov, V., Snášel, V. (eds.) Proceedings of the First International Scientific Conference Intelligent Information Technologies for Industry (IITI 2016). AISC, vol. 451, pp. 385–394. Springer, Cham (2016)

Studies of Big Data Processing at Linear Accelerator Sources Using Machine Learning

Mohammed Bawatna[1,2(✉)] and Bertram Green[3]

[1] Faculty of Computer Science, Technische Universität, 01069 Dresden, Germany
[2] Institute of Radiation Physics, HZDR, 01328 Dresden, Germany
m.bawatna@hzdr.de
[3] Magnet Science and Technology, Tallahassee, FL 32310, USA

Abstract. In linear accelerator sources such as the electron beam of the superconducting linear accelerator at the radiation source Electron Linear accelerator for beams with high Brilliance and low Emittance (ELBE), different kinds of secondary radiation can be produced for various research purposes from materials science up to medicine. A variety of different beam detectors generate a huge amount of data, which take a great deal of computing power to capture and analyse. In this contribution, we will discuss the possibilities of using Machine Learning method to solve the big data challenges. Moreover, we will present a technique that employ the machine learning strategy for the diagnostics of high-field terahertz pulses generated at the ELBE accelerator with extremely flexible parameters such as repetition rate, pulse form and polarization.

Keywords: Cloud computing · Machine learning · Big data · Deep neural networks

1 Introduction

Machine learning (ML) is particularly important in relation to big data, which allows samples to be identified where appropriate, rather than being targeted as in conventional analytical methods. ML deals with methods to automatically learn and develop application approaches for problems that cannot be solved manually at low expense. Big data always has a holistic conception that aims to draw conclusions from an enormous amount of data in order to generate sustainable values. The necessary technologies, which are used to analyze the emerging data volumes, present some challenges for information technology architectures to enable the corresponding analysis of these data. The explanations regarding big data technologies show that there are no uniform technologies and architectures, but that they vary widely depending on the requirements of the analysis options and objectives. The generation and collection of this data must also be considered, such as the storage, analysis, and the presentation possibilities of the analysis carried out. However, unlike traditional analytics tools and legacy databases, the described technologies allow handling a massive amount of data, making it imperative in the field of scientific research. With the increasing availability of data and the technological possibilities that they can use, new ways to create

R. Silhavy (Ed.): CSOC 2020, AISC 1225, pp. 450–460, 2020.
https://doi.org/10.1007/978-3-030-51971-1_37

significant differentiation, opportunities can be found. Decision making can be vastly improved with big data.

This contribution is organized as follows: the second section introduces the challenges of big data generated by most of the linear accelerator sources, as well as big data analysis methods. The third section presents the ML approaches for the big data analysis. The fourth section discuss the current technique used to diagnose the terahertz pulses at the Terahertz source at ELBE (TELBE) user facility. The fifth section present the design and evaluation of using the Artificial Deep Neural Networks (DNNs) method in terahertz spectral analysis. After summarizing the results, foreseeable future development and upgrades are discussed.

2 Challenges of Big Data at Linear Accelerator Sources

Big Data is an enormous amount of data that is collected, generated, stored, and analyzed, is supposed to make the world more predictable. The growing amount of data in recent years poses significant challenges in linear accelerator sources, which require to create a scalable and flexible IT infrastructure. Therefore, understanding of the technological and methodological software requirements and provide an overview of the most current technologies, and analysis methods are playing a significant role in Big data analysis in most of the linear accelerator sources. In this paper, the concept of big data is first defined with a description of the most common big data technologies, as well as the analysis methods.

2.1 Definition and Technologies

There is no standard definition of the term big data in the literature because it is relatively a new term. Big data is the use of large volumes of data from multiple sources at a high processing speed to generate economic benefits, where the storage and analysis of data are no longer sufficient due to the high volume of data. There are four characteristics that describe big data as in [1] and [2]. These include the volume of data, the variety, the velocity, and the analysis. The amount of data will grow to such a high volume, and more demands will be made on its analysis. However, not only the amount of data but also the variety of data structures is increasing. Therefore, a variety of unstructured data resulting from different sources must be considered. The goal here is to achieve real-time processing. The last decisive feature of Big Data is the value of the data. The large amounts of unstructured data that are collected and generated in real-time are only useful if we can make decisive conclusions. Filtering information from a large amount of data has great potential in linear accelerator sources as in [3] and [4]. However, answering questions that cannot be asked before generating the data and drawing the appropriate conclusions from the available data requires analysis.

The data can be stored in databases, where it is possible to be changed, deleted, and called anytime. However, these systems are no longer meet the enormous demands of the big data. The processing capacity necessary to deal and work with big data is a challenging issue in most of linear accelerator sources. Petabytes of raw data are stored in the cloud. Cloud computing makes the handling and processing of these data volumes possible in the first place. Cloud Computing represents a collection of services, applications, and resources that are offered to users in a flexible and scalable way over the Internet without the need for long-term capital commitment and IT experience. The necessary storage space and computing power for the data processing, as well as the software programs for processing these data, are outsourced to the cloud. With this solution, any digital device can access almost unlimited computing and memory performance. In general, the three main functions or service levels of the cloud as in [5] and [6] are Infrastructure as a Service (IaaS), Platform as a Service (PaaS), and Software as a Service (SaaS).

2.2 Analysis Methods

Not only the data itself and the corresponding applied technologies are decisive, whether smart data can be obtained from big data, but also the right choice of the respective analysis model. The focus of the data analysis is on the optimization of the performance, which includes a variety of different statistical models, ML, and data mining to calculate the probabilities from a large amount of data. For all analysis models, the continuous collection of data is essential, as it also runs in parallel to the analysis and at the same time, incorporates all newly generated data into the analysis. Data mining as in [7] and [8] is a generic term for a large number of different methods, procedures, and techniques that are used to promote and exploit valuable data from a wealth of data. The concept originally derived from the field of statistics, which is today equate with the concept of data pattern recognition [9–11]. The data mining process can also be described in five phases. These are the extraction of the relevant data from the respective data sources, then the selection of the data records and attributes, followed by the preprocessing phase. In this phase, the data quality of the selected data is examined in order to avoid errors. The fourth phase is the transformation of the data into a database schema that can be processed by the existing data mining systems. In the fifth phase, the method selection and application for the identification of patterns and relations take place in the analyzed data. Applied methods should then recognize and develop patterns and relations that make it possible to make statements about the analyzed data. There are a variety of analysis methods within the data mining. Cluster analysis can sort a large number of heterogeneous and unstructured data into a homogeneous group, called clusters as in [12]. The data is sorted by similarities based on variables and typologies. It must, therefore, first be determined according to which characteristics this should be done. Therefore, the characteristic of each file must first be checked and evaluated during the analysis process. Classification analysis is another method that divides the data into classes according to a previously known features. A model will then be developed that can predict newly stored data. Possible methods of classification analysis include decision tree techniques and neural networks. Association analysis is used to search for dependencies between the data. Identified patterns can then be translated into if-then rules and clarify exact patterns of action and consequences. In the following section, we will focus on the ML methods.

3 Machine Learning

ML is the technology for the known modern Artificial Intelligent (AI) techniques, which is why, in particular, the terms AI and ML are often used interchangeably. ML and especially the Deep Learning (DL) opened up altogether new possibilities in many research and development fields [13] such as automatic speech processing, image processing, and medical diagnostics.

The publications for DL show a continuing education and research in this field as [14] and [15]. In the future, machines will increasingly generate decisions. For this, it is essential to ensure the robustness and sufficient traceability of automated decision-making processes. Moreover, it must be ensured that ML applications are compatible with legal issues such as liability and accountability for algorithms that make decisions and are technically feasible. Modeling and implementing it is an important and complex issue that requires an interdisciplinary and transdisciplinary approach.

3.1 Concepts, Methods and Existing Models

ML is the key technology for understanding artificial intelligence. AI is a branch of computer science that enables machines to carry out intelligent tasks. An alternative is ML, which is actually the technology of intelligent systems today. ML aims to generate knowledge from experience by learning algorithms from samples to produce intelligent decisions. The ML models as in [16] and [17] are the best choice to automatically generate knowledge for processes that are too complicated to describe analytically from unknown data of the same type, such as sensor readings, pictures, or texts. These models can learn, predict, and make recommendations and decisions without any predetermined rules or calculation.

ML applications are not limited to robots, but can also be implemented in the big data processing. However, the amazing benefits of ML and AI systems do not imply that the machine has any understanding or even awareness of what data it processes, why and in what context it does so. Moreover, the existing ML applications are designed with great effort and trained only for tight tasks and ready for use. Current research is focused primarily on reducing training effort, improving the robustness, safety, and transparency of models, making them easier to adapt to new tasks, and combining the skills of humans and machines.

The learning styles in ML, as shown in Fig. 1, are suitable for different purposes. Depending on which additional information is available, other tasks can be learned. In supervised learning, the correct answers to the raw samples must be provided as labels. Specifying labels usually means more work for data preprocessing, but is necessary when classifying objects or predicting values. In unsupervised learning, the raw samples are sufficient to discover basic patterns in the data. There are now a variety of model types and learning methods or algorithms that are particularly well suited for different tasks.

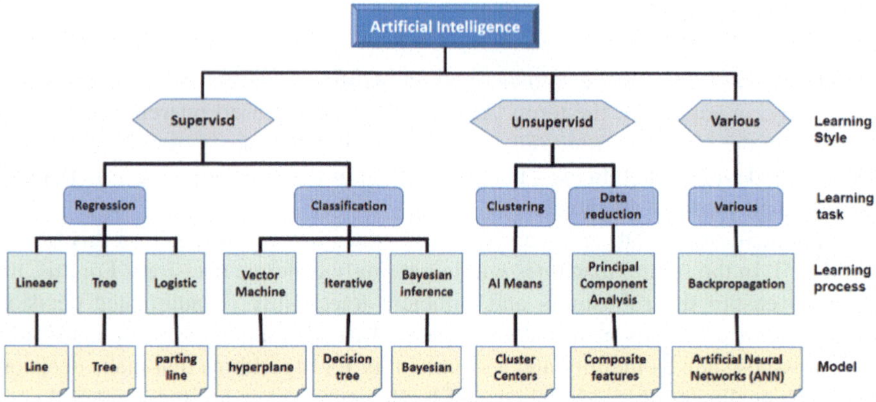

Fig. 1. Common learning methods and their models.

3.2 Deep Neural Networks

Recently, DL or learning with deep Artificial Neural Networks (ANN) has made tremendous progress, especially in the analysis of image, video, voice, and text data as in [18] and [19]. Moreover, ANN applications can identify faces and objects with a lower error rate than humans. In addition, such applications can solve novel complex learning tasks. The deep ANNs consist of many layers of software nodes called artificial neurons. When learning, the weights, which are the connections between the nodes, are changed until the outputs are good enough. Deep ANN form expressive models that can also be efficiently trained in parallel computer systems.

This often only makes sense with big data, because it is not easy for humans to understand what the weighting of an ANN means and how exactly the effort needed for calculations. There is a variety of network architectures that have proven to be useful for different data types and tasks.

4 Machine Learning at Modern Linear Accelerator Sources

Linear accelerators sources are emerging large-scale machines for research in many fields including physics, chemistry, and material science such as in ELBE [20], LCLS [21], and DESY [22]. Their ability to generate light sources such as X-ray and terahertz pulses makes them ideal sources for scientific experiments that requires ultrashort pulses at high intensity. One of the disadvantages of these large-scale facilities is their poor stability that require the electron bunches to travel several hundred meters in length before make these bunches radiate the required coherent light pulses at certain wavelengths. The electron bunches are generated by the electron guns in bunches that contains about half billion electrons, then accelerated in radiofrequency cavities and compressed. The electron bunches then pass through multiple equipment that make them radiate ultra-short coherent light pulse. The fluctuation in the electron gun is translate into fluctuations in the generated pulse properties. There are two main pulses properties that need to be recorded in parallel with the experiments which are the timing jitter and the intensity fluctuation. Depending on the technique used to generate the light source, several sophisticated algorithms are used to process the recorded diagnostics signals to decrease the timing jitter to few femtoseconds, and to minimize the effect of intensity instability.

These sophisticated algorithms are calculated offline because it require several steps of data analysis which are impossible to be processed online. However, the demands on increasing the repetition rates of the generated pulses at these linear accelerator sources produce several big data challenges. This require new approaches in these scientific applications such as the use of machine learning techniques. We could find several proposed solutions that employing machine learning at these large-scale light sources for less than one kilohertz repetition rates as in [23–28], including feedback loop as in [29]. In the next section we will propose and demonstrate the machine learning as a general technique applicable at any terahertz light source facility to obtain full pulses information.

5 Terahertz Pulses Diagnostics at TELBE

The radiation source Electron Linear accelerator for beams with high Brilliance and low Emittance (ELBE), as in [20], delivers various kinds of secondary radiation. As part of an upgrade of the ELBE accelerator, one electron beamline has been modified to allow for the generation and acceleration of ultra-short, less than 150 fs and highly charged up to 1 Nano Coulomb, electron bunches, as in [31]. This upgrade enables the operation of high-field terahertz sources based on super radiant terahertz emission at the ELBE accelerator and thereby opens up the opportunity to generate carrier envelope phase stable high-field THz pulses with extremely flexible parameters with respect to repetition rate, pulse form and polarization. In the high-field terahertz-driven phenomena laboratory, different ultra-fast spectroscopic techniques are performed with few femtosecond time resolution to investigate dynamics in matter driven by high-field THz pulses. At TELBE user facility [31, 32], accurate time resolution is required to utilize the transient THz fields as a novel highly selective excitation for non-linear dynamics. These dynamic processes are typically studied using pump-probe experiments, on few ten fs timescales involving synchronized external laser systems. Figure 2 shows a schematic of the arrival time monitor used at TELBE user facility.

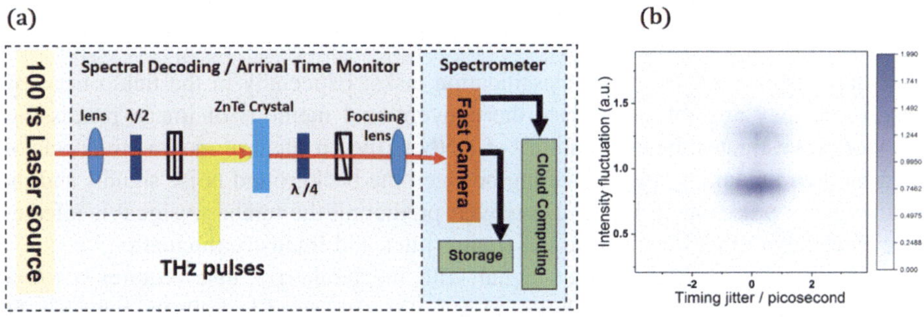

Fig. 2. (a) Schematic of the terahertz pulse-resolved arrival time monitor in TELBE. (b) Terahertz intensity fluctuation versus timing jitter.

Detecting THz radiation can sometimes be problematic because it occupies a position in the electromagnetic spectrum between optical and microwave wave-lengths. There are many different approaches to detecting THz, utilizing a variety of physical phenomena. The Electro-Optic Sampling (EOS) technique as in [30] is implemented at TELBE to detect the terahertz pulses. The basic idea of THz spectroscopy by EOS is to observe refractive index changes which cause an electromagnetic pulse having a frequency bandwidth in the range of several THz in an electro-optical crystal. This refractive index change can be converted into a modulation of its light intensity with an optical probe pulse as shown in Fig. 3.

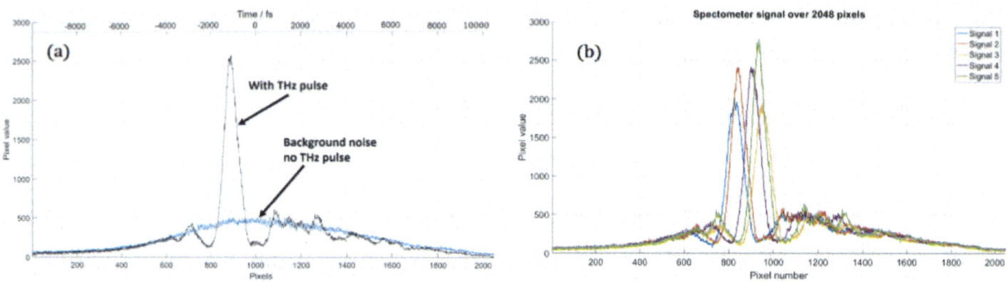

Fig. 3. (a) Modulated laser signal (black) with terahertz pulse and unmodulated laser signal (blue). (b) Five consecutive modulated laser signals with five consecutive terahertz pulses.

The measurement of the modulation allows a direct inference to the electromagnetic field of the THz pulse. The generated volume of raw data depends on the speed of the used imaging detector. At TELBE, 100 kHz frame rate, Basler SPL2048 spectrometer camera is used as a linear array imaging detector which generates a raw data of 358 Megabytes per Second as in [33] and [34]. For each THz pulse, the imaging detector capture 2048 pixels. A large volume of terabytes of raw data is recorded every day at TELBE, and further off-line analysis is required to produce a signal that is readable by the scientists. Based on the status of the terahertz pulse as shown in Fig. 3, a real-time spectral analysis of the individual terahertz pulses, will save the time required to decide whether to record the raw data or to retune the terahertz beamline.

5.1 Terahertz Diagnostics Using Machine Learning

ANNs have been used for many classification tasks. Especially in the field of pattern recognition, they are able to replace the conventional methods of image processing. This work investigates the applicability of ANN to the problem of making the decision to record the raw data by distinguishing between the background noise signals and the valid terahertz pulses as in Fig. 3. Moreover, predicting the terahertz signal is affected by several factors that depends on the timing jitter and intensity fluctuation.

Predicting the modulated laser signal with the terahertz pulse requires a lot of training samples. A total of 200 thousand samples were used to train the network, 50 thousand samples were used in validation, and 15 thousand samples were used for testing. The decision about the architecture of the model and how to train them are made to minimize the prediction error for the validation set. After the validation of the model, the test set is used to calculate the prediction error. The changes in the amplitude and position of the signals is due to the timing jitter and other pulse properties.

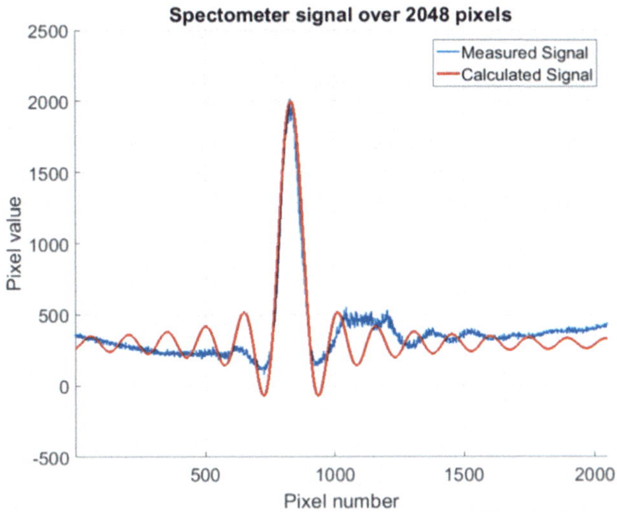

Fig. 4. The measured modulated laser pulse with terahertz (blue), and the calculated modulated laser pulse (red).

Moreover, both of the number of hidden layers and the number of neurons in each of these hidden layers must be carefully considered. Using too few neurons in the hidden layers will result in something called under fitting. Under fitting occurs when there are too few neurons in the hidden layers to adequately detect the signals in a complicated data set. On the other hand, using too many neurons in the hidden layers can result in several problems. First, too many neurons in the hidden layers may result in overfitting. Overfitting occurs when the neural network has so much information processing capacity that the limited amount of information contained in the training set is not enough to train all of the neurons in the hidden layers. A second problem can occur even when the training data is sufficient. An inordinately large number of neurons in the hidden layers can increase the time it takes to train the network. The amount of training time can increase to the point that it is impossible to adequately train the neural network. Obviously, some compromise must be reached between too many and too few neurons in the hidden layers.

5.2 Single-Pulse Shape Prediction Results

We developed a model to predict the shape of the single pulse shown in Fig. 4, by predicting the spectral components that are presented in [30]. The delay prediction errors for the test sets are shown in Fig. 5. The mean error of single pulse detection is 0.24, while the shape agreement of single pulse spectrum is 87%.

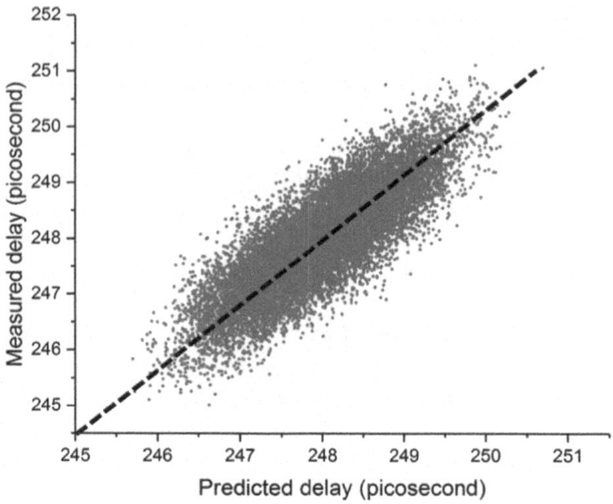

Fig. 5. Delay prediction errors for the test set of 15 thousand samples using the model-based artificial neural network approach.

Moreover, the decision for recording a valid experimental data, as in [34], depends on the result of the neural network. Currently, the implemented ANN can correctly predict the parameters of the signals that contain terahertz pulses. However, more training and evaluation is planned in the future to increase the intelligence and the performance of this neural network using the neuro-fuzzy technique.

6 Conclusion and Future Work

ANNs are adaptive systems that can be used in a variety of ways, particularly in the areas of image processing and analysis. We have shown, using data from TELBE facility, that the timing jitter of the electron bunches in modern linear accelerator sources can be predicted by using machine learning approach. By applying straightforward machine learning procedures, we can accurately predict the spectral shape and time delay of individual terahertz pulses. With these methods, huge amounts of data can be processed on heterogeneous processors such as GPUs and FPGAs in order to train special and deep ANNs. As future work, the presented work shall be implemented on the FPGA architecture as in [33]. This will allow for real-time analysis of terahertz pulses, as well as online decision making that will decrease the cost and time of storing and analyzing invalid recorded raw data.

References

1. Sun, Z., Strang, K., Li, R.: Big data with ten big characteristics. In: Proceedings of the 2nd International Conference on Big Data Research - ICBDR 2018 (2018). https://doi.org/10.1145/3291801.3291822
2. Yambem, N., Nandakumar, A.: Big data: characteristics, issues and clustering techniques. Nciccnda (2018). https://doi.org/10.21467/proceedings.1.55
3. Condron, C., Brown, C., Gozani, T., et al.: Linear accelerator x-ray sources with high duty cycle (2013). https://doi.org/10.1063/1.4802418
4. Adelmann, A., Ryne, R., Shalf, J., Siegerist, C.: From visualisation to data mining with large data sets. In: Proceedings of the 2005 Particle Accelerator Conference (2005). https://doi.org/10.1109/pac.2005.1591735
5. Smoot, S.R., Tan, N.K.: Cloud infrastructure as a service. Private Cloud Comput. 267–297 (2012). https://doi.org/10.1016/b978-0-12-384919-9.00008-8
6. Castro-Leon, E., Harmon, R.: Cloud computing as a service. Cloud Serv. 3–30 (2016). https://doi.org/10.1007/978-1-4842-0103-9_1
7. Talia, D., Trunfio, P., Marozzo, F.: Introduction to data mining. Data Anal. Cloud 1–25 (2016). https://doi.org/10.1016/b978-0-12-802881-0.00001-9
8. Gupta, G., Pathak, D.R.: Role of cloud computing in data mining. Int. J. Eng. Comput. Sci. (2016). https://doi.org/10.18535/ijecs/v5i4.01
9. Petrou, M.: Learning in pattern recognition. In: Machine Learning and Data Mining in Pattern Recognition. Lecture Notes in Computer Science, pp. 1–12 (1999). https://doi.org/10.1007/3-540-48097-8_1
10. Pal, A., Pal, S.K.: Pattern recognition: evolution, mining and big data. Pattern Recogn. Big Data 1–36 (2016). https://doi.org/10.1142/9789813144552_0001
11. Cristianini, N.: Pattern analysis (data mining, intelligent data analysis, pattern discovery, pattern recognition). Dictionary Bioinf. Comput. Biol. (2004). https://doi.org/10.1002/0471650129.dob0521
12. Aggarwal, C.C.: Cluster analysis: advanced concepts. Data Min. 205–236 (2015). https://doi.org/10.1007/978-3-319-14142-8_7
13. Thikshaja, U.K., Paul, A.: A brief review on deep learning and types of implementation for deep learning. Deep Learn. Neural Netw. 30–40 (2020). https://doi.org/10.4018/978-1-7998-0414-7.ch002
14. Karg, M., Scharfenberger, C.: Deep learning-based pedestrian detection for automated driving: achievements and future challenges. In: Development and Analysis of Deep Learning Architectures Studies in Computational Intelligence, pp. 117–143 (2019). https://doi.org/10.1007/978-3-030-31764-5_5
15. Vieira, A., Ribeiro, B.: Deep learning: an overview. In: Introduction to Deep Learning Business Applications for Developers, pp. 9–35 (2018). https://doi.org/10.1007/978-1-4842-3453-2_2
16. Singh, P.: Supervised machine learning. Learn. PySpark 117–159 (2019). https://doi.org/10.1007/978-1-4842-4961-1_6
17. Kung, S.Y.: Unsupervised learning models for cluster analysis. Kernel Methods Mach. Learn. 139–140. https://doi.org/10.1017/cbo9781139176224.008
18. Sejdić, E., Falk, T.H.: Signal Processing and Machine Learning for Biomedical Big Data. CRC Press/Taylor & Francis, Boca Raton (2018)
19. Khan, M., Silva, B.N., Han, K.: Efficiently processing big data in real-time employing deep learning algorithms. In: Deep Learning and Neural Networks, pp. 1344–1357 (2020). https://doi.org/10.4018/978-1-7998-0414-7.ch07

20. Contact. In: Radiation Source at the ELBE Center for High-Power Radiation Sources - Helmholtz-Zentrum Dresden-Rossendorf, HZDR. https://www.hzdr.de/db/Cms?pNid=145. Accessed 24 Jan 2020
21. SLAC Home Page. In: SLAC National Accelerator Laboratory. https://www6.slac.stan-ford.edu/. Accessed 24 Jan 2020
22. FLASH. In: Zur DESY Startseite. http://www.desy.de/forschung/anlagen__pro-jekte/flash/index_ger.html. Accessed 24 Jan 2020
23. Tagliaferri, R., Longo, G., Dargenio, B., Incoronato, A.: Introduction: neural networks for analysis of complex scientific data: astronomy and geosciences. Neural Netw. **16**, 295 (2003). https://doi.org/10.1016/s0893-6080(03)00012-1
24. Baldi, P., Sadowski, P., Whiteson, D.: Searching for exotic particles in high-energy physics with deep learning. Nat. Commun. **5**, 1–9 (2014). https://doi.org/10.1038/ncomms5308
25. Collaboration, T.A.: A neural network clustering algorithm for the ATLAS silicon pixel detector. J. Instrum. **9**, P09009 (2014). https://doi.org/10.1088/1748-0221/9/09/p09009
26. Dieleman, S., Willett, K.W., Dambre, J.: Rotation-invariant convolutional neural networks for galaxy morphology prediction. Mon. Not. R. Astron. Soc. **450**, 1441–1459 (2015). https://doi.org/10.1093/mnras/stv632
27. Aurisano, A., Radovic, A., Rocco, D., et al.: A convolutional neural network neutrino event classifier. J. Instrum. **11**, P09001 (2016). https://doi.org/10.1088/1748-0221/11/09/p09001
28. Kim, E.J., Brunner, R.J.: Star–galaxy classification using deep convolutional neural networks. Mon. Not. R. Astron. Soc. **464**, 4463–4475 (2016). https://doi.org/10.1093/mnras/stw2672
29. Edelen, A.L., Biedron, S.G., Chase, B.E., et al.: Neural networks for modeling and control of particle accelerators. IEEE Trans. Nucl. Sci. **63**, 878–897 (2016). https://doi.org/10.1109/tns.2016.2543203
30. Kueny, E., Calendron, A.-L., Kärtner, F.X.: Electro-optic sampling of terahertz pulses in multilayer crystals. In: Laser Congress 2019 (ASSL, LAC, LS&C) (2019). https://doi.org/10.1364/assl.2019.jtu3a.16
31. Bawatna, M., Green, B., Deinert, J.-C., et al.: Pulse-resolved data acquisition system for THz pump laser probe experiments at TELBE using super-radiant terahertz sources. In: 2019 IEEE MTT-S International Microwave Workshop Series on Advanced Materials and Processes for RF and THz Applications (IMWS-AMP) (2019). https://doi.org/10.1109/imws-amp.2019.8880116
32. Kovalev, S., Green, B., Awari, N., et al.: High-field high-repetition-rate prototype user facility for the coherent THz control of matter. In: 2016 41st International Conference on Infrared, Millimeter, and Terahertz waves (IRMMW-THz) (2016). https://doi.org/10.1109/irmmw-thz.2016.7758880
33. Bawatna, M., Arnold, A., Deinert, J.-C., et al.: Towards real-time data processing using FPGA technology for high-speed data acquisition system at MHz repetition rates. In: Proceedings of the 19th International Conference on RF Superconductivity SRF2019, Germany (2019). https://doi.org/10.18429/JACOW-SRF2019-THP029
34. Bawatna, M., Green, B., Kovalev, S., et al.: Research and implementation of efficient parallel processing of big data at TELBE user facility. In: 2019 International Symposium on Performance Evaluation of Computer and Telecommunication Systems (SPECTS) (2019). https://doi.org/10.23919/spects.2019.8823486

Reducing Digital Geographic Images to Solve Problems of Regional Management Information Support

A. V. Vicentiy[1,2(✉)] and M. G. Shishaev[1,2]

[1] Institute for Informatics and Mathematical Modeling – Subdivision of the Federal Research Centre "Kola Science Centre of the Russian Academy of Science", 24A, Fersmanst., Apatity, Murmansk Region 184209, Russia
alx_2003@mail.ru

[2] Apatity Branch of Murmansk Arctic State University, Lesnayast. 29, Apatity, Murmansk Region 184209, Russia
shishaev@arcticsu.ru

Abstract. Geographic information systems are often used to support decision-making in regional governance. Geoimages synthesized by GIS are an important source of information for decision makers. However, many modern GIS synthesize redundant geoimages from the point of view of the problem being solved. Analysis of redundant geoimages complicates the work of the decision maker. To solve this problem, it is necessary to perform geoimage reduction in accordance with the features of the problem being solved by the user. This paper describes the first step in developing an approach for reducing geoimages synthesized by geographic information systems. A feature of this approach is that during reduction, a change in the level of informativeness of the geoimage is taken into account. Informativeness is evaluated on the basis of a pragmatic measure of information that characterizes the subjective usefulness of a geoimage for solving a problem by a user. The main results of the first step in developing approach to geoimage reduction are the formalized statement of the problem, as well as the user model, the problem model, and the geoimage model. These models represent a methodological basis for further software implementation of geoimage reduction procedures.

Keywords: Decision-making support · Geoimage redundancy · Geoimage reduction · Informativeness · Pragmatic information measure · GIS · Cognitive user interface · Cognitive geovisualization

1 Introduction

Various systems for processing and presenting spatial information have long and successfully been used as one of the decision support tools in various fields of activity. Such systems give a great effect in solving a wide class of problems related to spatial planning and territorial management at different levels. The type and scale of spatially oriented information systems is determined by the problems that can be solved with its help and can differ significantly - from interstate to regional or municipal. Hardware and software systems and information systems that collect, store, process, analyze and

R. Silhavy (Ed.): CSOC 2020, AISC 1225, pp. 461–469, 2020.
https://doi.org/10.1007/978-3-030-51971-1_38

visualize spatio-temporal data (geodata) and associated attribute information are allocated into a separate class of information systems called geographic information systems (GIS). There are various classifications of GIS. In particular, GIS is divided into two types: general-purpose GIS and domain-specific GIS. In our work, we will consider multi-subject GIS, which occupy an intermediate position between general-purpose GIS and subject-oriented GIS.

Another important circumstance is the fact that any geographic information system is also a human-machine system. The main purpose of GIS is to visualize the available information and provide it to the end user for further analysis and decision making. In the process of working with GIS, the user interacts with electronic maps or, more precisely, digital geographic images (geoimages), which the GIS synthesizes upon his request. A geoimage, in its general form, is a scale, spatio-temporal, generalized model of objects or processes of the surrounding world, which is presented in a figurative-symbolic form. At the same time, geoimages synthesized by GIS have a unique heuristic and epistemological potential, which can be revealed only with the direct participation of a person in the process of interpretation and visual analysis of geoimages.

The inclusion of a person (a decision maker (DM)) in the processing circuit of visual cartographic information imposes additional requirements on the visualization of data and causes a number of additional problems related to the peculiarities of the perception of visual information by both individual groups (categories) of users and each user individually. These features can be associated with the cognitive abilities that a particular user has, or the features and conventions of the perception of visual information by certain categories of users, professional knowledge, work experience, both with geographic information systems in general, and with similar problems to be solved or similar subject areas, and many other characteristics.

Therefore, the development of various approaches to improve the efficiency of a person working with synthesized digital geoimages is an important problem, the solution of which will increase the value and effectiveness of the use of geographical information systems in decision making.

2 The Problem of Geoimages Redundancy in Decision Making

A user's interaction with a geographic information system is a sequence of queries to it in a dialogue mode or using a query language (for example, SQL). The result of query processing is a geoimage synthesized by the system, which is built by the visualization subsystem from primitives distributed across thematic layers. This approach to visualization is common in the vast majority of modern GIS and geoservices (ArcGis, QGis, MapInfo, Google Earth, Google Maps, Yandex.Maps and others). The main drawback of this approach is that the effectiveness of data visualization depends both on the problems being solved and on the subjective assessments of users who form search queries and evaluate the result from a pragmatic point of view. As a result, GIS synthesizes, as a rule, redundant visual images, which negatively affects the speed and quality of decisions made with its use.

The negative effects of using redundant geoimages when solving decision-making problems can manifest themselves in different ways. In particular, the use of redundant images leads to additional load on the decision maker in the analysis of geoimages. The decision maker usually conducts a preliminary analysis of the image to determine its most informative fragments, which are the most useful and important for decision making. In the case when the geoimage is redundant, the decision maker at the stage of preliminary analysis is forced to evaluate a larger number of image fragments. Thus, the time of preliminary analysis, and, therefore, the time of decision making increases.

Another negative effect of using redundant geoimages for decision making is an increase in the cognitive load of decision makers [1, 2]. An increase in cognitive load leads to rapid fatigue of decision makers, a decrease in concentration and a potential deterioration in the quality of decisions [3, 4]. The presence of these and other negative effects necessitates to finding ways to solve the problem of reducing the redundancy of geoimages and increasing its informativeness.

Despite the subjectivity of the pragmatic assessment of geoimages for decision support purposes, the problem of the redundancy of synthesized geoimages can be solved to some extent. We propose to solve the problem of redundancy through the development and application of procedures for modifying and reducing geoimages, taking into account the assessment of its informativeness and subjective usefulness for solving specific problems (class of problems).

The problems of estimating the amount of information, developing measures of information and ways to reduce information redundancy are among the fundamental problems of information theory. In this paper, we will rely on a pragmatic measure of information that allows to take into account the usefulness of information to the end user in the context of achieving the goal [5, 6]. In our case, the end user is the decision maker who analyzes the synthesized GIS geoimage. And the goal is to make a quality and informed decision.

We also proceed from the fact that in order to solve the problem of synthesized geoimage redundancy, the most effective way is to reduce the number of primitives to a certain optimal level for decision making. In addition, reducing the number of visual elements of the image will decrease the cognitive load on the decision maker, and thus diminish the risk of making ill-conceived and unreasonable decisions and delay the onset of the fatigue period [7].

3 Problem Statement and Proposed Solution

In view of the foregoing, the statement of the problem in general form can be represented as follows. It is necessary to propose a method for modifying geoimages, which would minimize the redundancy of the image and at the same time maximize its level of informativeness:

$$\begin{cases} \mathrm{E}(g) -> \min \\ \mathrm{V}(g) -> \min \\ \mathrm{I}(g) -> \max \end{cases}$$

where:

E (g) is the redundancy of the geoimage for solving the problem by the user;

V (g) is the volume of the geoimage - a quantitative measure that allows you to give an objective assessment of the "size" of the image (for example, in megabytes);

I (g) is the informativeness (subjective usefulness for the user) of the geoimage for solving a problem by a user;

g is a modifiable geoimage.

In the framework of this formulation of the problem, informativeness of the geoimage can be represented as a function as follows:

$$f : G \times T \times U \rightarrow \mathbb{R}$$

where:

G = {gi}, i = 1..N is the set of geoimages;

T = {tj}, j = 1..M is the set of problems;

U = {uk}, k = 1..K is the set of users.

However, in practice, in such a strict formulation, the solution of the problem is hardly achievable. Therefore, it makes sense to soften the statement of the problem as follows. Let there be some synthesized geoimage from the set of geoimages synthesized by GIS: $g \in G$. Based on the analysis of this geoimage, a user from the set of GIS users $u \in U$ tries to solve a certain problem from the set of problems $t \in T$.

Then the statement of the problem can be represented as follows. It is necessary to propose a method for modifying geoimages, which would reduce the redundancy of geoimages, while not reducing the level of informativeness or not lowering the level of informativeness below a certain minimum level of informativeness of the geoimage necessary to successfully solve this problem:

$$\begin{cases} \mathrm{E}(\mathrm{g}) > \mathrm{E}(\mathrm{g})^* \\ \mathrm{V}(\mathrm{g}) \geq \mathrm{V}(\mathrm{g})^* \\ (\mathrm{I}(\mathrm{g}) \leq \mathrm{I}(\mathrm{g})^*) \vee (\mathrm{I}(\mathrm{g})_{\min} \leq \mathrm{I}(\mathrm{g})^*) \end{cases}$$

where:

E (g)* is the redundancy of the geoimage for solving the problem by the user after the modification procedure;

V (g)* is the volume of the geoimage - a quantitative measure that allows you to give an objective assessment of the "size" of the image after the modification procedure;

I (g)* is the informativeness of the geoimage for solving the problem by the user after the modification procedure;

$I(g)_{min}$ is the minimum level of informativeness of the geoimage required to solve the problem.

In addition, the following assumptions are made:

Assumption 1. The synthesized geoimage initially has some redundancy: E (g) > E (g)*.

Assumption 2. The level of informativeness of the initial geoimage is sufficient to solve the problem by the user: $I(g) \geq I(g)_{min}$.

Given a softer statement of the problem and assumptions, we can hypothesize how to reduce the redundancy of geoimages for a real geographic information system. We believe that the best way to reduce the redundancy of geoimages is to remove primitives from the image that are not relevant to the problem currently being solved by the user. On the one hand, the removal of such primitives will not lead to a decrease in the information content of the geoimage from the point of view of the user, but on the other hand it will decrease the overall information saturation of the image (redundancy) and the volume of the geoimage. Such a way of modifying the geoimage will be called reduction below. We will call such way of geoimage modification as reduction.

4 The Approach to Image Reduction Based on Assessment of Informativeness

From the statement of the problem and the definition of informativeness, it is clear that the most important elements for the successful reduction of geoimages are the user model, the problem model and the geoimage model. Consider these models in more detail.

The User Model. User models are used in the implementation of various information systems and geographic information systems in this sense are no exception. So, in [8] it is proposed to define a user model through a set of user properties. It is proposed to assign users to different groups in order to adapt interfaces, increase the efficiency of interaction with GIS and reduce the likelihood of user errors. It emphasizes the need to take into account the characteristics of the user's psyche and its mental characteristics when creating the model and assigning users to different groups.

In [9], the entire set of GIS users is perceived as a system with natural behavior that can evolve and adapt to changes in the surrounding world. To take into account the characteristic features of the structure and behavior of natural systems, user models are proposed to be implemented on the basis of the mathematical apparatus of the theory of fuzzy sets.

Another approach is proposed in [10]. Its essence is to adapt the capabilities of the user interface to the problems solved by the user. Adaptation is carried out on the basis of providing the user with a basic set of GIS tools and a special mechanism for adapting the capabilities of basic tools for problems that are solved by the user. This technology also supports the division of users into groups based on the commonality of problems solved by users and the tools used.

To assess the level of informativeness of geoimages, one of the most suitable models, in our opinion, is the user model described in [11]. The user model described in this work is based on such concepts as the mental model and perceptual stereotypes. User affiliation to a particular group is determined by the similarity of user perceptions of real-world objects. The similarity of representations is generated by the proximity of mental models of users and the method of ranking attributes of real-world objects by significance. To reduce image redundancy, understanding the significance of a primitive (a potential candidate for reduction) for successfully solving a problem is a key element. Therefore, a user model based on the assessment and consideration of the significance of the attributes identifying the object is well suited for calculating the informativeness of a geoimage.

The Problem Model. The multi-subject geographical information systems considered in this paper are intended primarily to assist users in making more informed decisions through better awareness. Geographic information system allows to increase the efficiency and speed of decision-making by presenting the results of user queries and data analysis in a clear cartographic form, convenient for perception. To build a model of a problem solved by a user using GIS, methods and approaches of system analysis and conceptual modeling, situational conceptual control, a functional-target approach, and others can be applied. In this paper, we propose using a model of the problem based on the formal definition of a multi-subject information resource proposed in. This model allows to take into account the multi-subject nature of the GIS and is in good agreement with the user model.

Let a set of user categories U = {uk}, k = 1..K and a set of resources of the geographic information system R = {rl}, l = 1..L be defined. Then it is possible to determine such a function $z : U \times R \rightarrow \mathbb{R}$ that characterizes the pragmatic value of a certain GIS resource from the user's point of view.

Let also some user uk ∈ U generate a sequence of requests to the GIS of the form: Quk i (qspatial, qtemporal, qsemantic), i = 1..N, which defines the restrictions of the data of interest to the user. Then, the problem model can be represented as a selection of such GIS resources for the synthesis of the visual image g ∈ G that satisfy the spatial, temporal and semantic constraints indicated in the user's queries and for which the function z is not lower than a certain threshold value.

$$\begin{cases} g = \bigcup_{l=1}^{L} r_l \subseteq R \,|\, r_{spatial} \approx q_{spatial}, r_{temporal} \approx q_{temporal}, r_{semantic} \approx q_{semantic} \\ Z \geq Z_{min} \end{cases}$$

where:

$q_{spatial}, q_{temporal}, q_{semantic}, r_{spatial}, r_{temporal}, r_{semantic}$ are spatial, temporal and semantic attributes of user queries and resources of a geographic information system, respectively. The approximate equal sign in this context means that the spatial, temporal, and semantic attributes of the selected resources correspond to the spatial, temporal, and semantic attributes of the user's request.

The Geoimage Model. Currently, there are many varieties of geoimages, depending on the purpose of use, methods of production, basic characteristics and other reasons. The creation and development of a general theory of geoimages is carried out by a separate scientific discipline - geoiconics [12]. In the framework of this work, we propose to use the geoimage model, presented in the form of a set of visualized primitives taking into account spatial, temporal and semantic limitations: g = <P, Spa, Tem, Sem>, where:

P = {p_x}, x = 1..X is the set of primitives included in the geoimage;

Spa = {pol_s}, s = 1..S is a set of polygons that determine the spatial limitations of the geoimage. Each $pol_s = ((x_1, y_1), .. (x_n, y_n))$ is a polygon defined by a set of points with coordinates (x_*, y_*).

Tem = {tem_y}, y = 1..Y is the set of intervals defining the time constraints of the geoimage. Each {tem_y} = {tem_y^{start}, tem_y^{end}} is a pair of values that define the beginning and end of a time interval.

Sem = {sem_w}, w = 1..W is the set of semantic restrictions of the geoimage, where each element of the set Sem is associated with a certain set of semantic values defined in this implementation of the geographic information system.

Specific procedures for reducing geoimages, taking into account the pragmatic assessment of the informativeness of the images synthesized by the geographic information system, are implemented on the basis of the models described above.

We can also try to make a preliminary comparison of the approach proposed in this paper with other similar approaches. It should be noted that at present there are not many works that describe in detail approaches aimed at increasing the informativeness of images. A significant part of these works is related to the processing of medical data, remote sensing data, or to solving the problems of object recognition in photos and videos, as well as for military applications [13–16]. Significantly fewer works are dedicated to improving informativeness and reducing geoimages to solve the problems of decision support. And only in a very small amount of work the authors are trying to give an objective assessment of the informativeness of the image [17].

In most works, informativeness is understood as the change in some parameters of images (contrast, brightness, saturation, entropy, various correlations, etc.) or the definition of some of the most significant, from the point of view of the authors, signs in data sets [18, 19]. In contrast to these approaches, our approach allows us to take into account the semantic correspondence of the geoimage to the problem for which the geoimage was synthesized and the cognitive abilities of the user who solves this problem. Thus, the proposed approach will significantly improve the pertinence and quality of geodata visualization. Besides, it allows to delay the period of occurrence of a cognitive overload of decision maker at the expense of presentation to him more understandable and corresponding to the solved problem of geoimages.

5 Conclusion and Future Research Direction

This paper describes the problem statement and the first step in developing approach to reducing geoimages taking into account its informativeness. The essence of the approach is to reduce the redundancy and objective volume of the geoimage without compromising the level of informativeness. It is proposed to understand the level of informativeness on the basis of evaluating a pragmatic measure of information, which allows taking into account both the features of the problem being solved and the characteristic features of the user of a geographic information system. The paper presents a formalized problem statement of geoimage reduction, as well as the user model, the problem model and the geoimage model, which are necessary for the software implementation of geoimage reduction procedures. The models proposed in this work can be used as a methodological basis for the software implementation of multi-subject geographic information systems. The implementation of multi-subject geographical information systems with the possibility of reducing geoimages is the subject of our future work.

In future research, we plan to study in more detail various measures of information and methods for estimating the amount of information. We will conduct a comparative study of the applicability of existing methods for evaluating informativeness for problems of reducing redundancy and increasing the informativeness of geoimages. We will also evaluate the possibility of using thesauruses and ontologies to describe the characteristic features of users, problems and geoimages. We plan to develop a method for evaluating the usefulness of information contained in the geoimage for a specific user using thesauruses.

References

1. Solso, R., MacLin, M., MacLin, O.: Cognitive Psychology, 8th edn. Allyn & Bacon, Boston (2008). 592 p
2. Vicentiy, A., Shishaev, M., Oleynik, A.: Dynamic cognitive geovisualization for information support of decision-making in the regional system of radiological monitoring, control and forecasting. In: Advances in Intelligent Systems and Computing, vol. 466, pp. 483–495 (2016)
3. Biernat, M., Kobrynowicz, D., Weber, D.: Stereotypes and shifting standards: some paradoxical effects of cognitive load. J. Appl. Soc. Psychol. **33**, 2060–2079 (2003)
4. Vicentiy, A., Shishaev, M., Vicentiy, I.: The development of dynamic cognitive interfaces for multisubject information systems (on the example of geosocial service). In: Advances in Intelligent Systems and Computing, vol. 575, pp. 449–459 (2017)
5. Weinberger, E.: A theory of pragmatic information and its application to the quasispecies model of biological evolution. BioSystems **66**, 105–119 (2002)
6. Andrew, F., Duckham, M., Goodchild, M., Worboys, M.: Pragmatic information content - how to measure the information in a route description. In: Foundations of Geographic Information Science, pp. 47–68 (2003)
7. Toffler, A.: Future Shock. Bantam Books, New York (1990). 561 p
8. Ivanov, S.: Interface design and user model. New Inf. Technol. Autom. Syst. **16**, 227–230 (2013)
9. Didenko, D.: Development of a model for assessing the quality of information in a GIS. In: Izvestiya SFedU. Engineering, vol. 4, pp. 210–216 (2010)
10. Tsvetkov, V., Dyshlenko, S.: Design features of a GIS user based on the basic set of GIS "Karta 2011" In: Land Management, Cadastre and Land Monitoring, vol. 8, pp. 79–84 (2010)
11. Dikovitsky, V., Lomov, P., Shishaev M.: Formalization problem of constructing cognitive user interfaces for multidomain information resources. In: Proceedings of the Kola Science Center RAS. Information Technology, vol. 4, pp. 90–97 (2013)
12. Berlyant, A.: Geoiconic. Astreya, Moscow (1996). 208 p
13. Trenikhin, V., Kobernichenko, V.: Increasing the information content of radar images in remote sensing systems based on fractal processing methods. Ural Radio Eng. J. **3**, 111–131 (2019)
14. Novoselsky, A., Vasiliev, D., Bykov, A., Senitsky, A., Senitsky, I.: A way to increase the information content of the stabilometric study and the hardware complex for its implementation. Invention. Russian Federation Patent Number: 0002665957 (2018). https://edrid.ru/rid/218.016.83fd.html

15. Travina, E.: A way to increase the visual informativeness of digital halftone images. Invention. Russian Federation Patent Number: 0002448367 (2012). https://edrid.ru/rid/219.017.3371.html
16. Bondarenko, M., Drynkin, V.: Image informative increasing for onboard enhanced vision systems. Tech. Vis. **3**, 57–65 (2015)
17. McCamy, M., Otero-Millan, J., Di Stasi, L., et al.: Highly informative natural scene regions increase microsaccade production during visual scanning. Neurosci **34**, 2956–2966 (2014)
18. Mackay, M., Cerf, M., Koch, C.: Evidence for two distinct mechanisms directing gaze in natural scenes. J. Vis. **12**, 1–12 (2012)
19. Nemirovsky, V., Stoyanov, A.: Increasing the information content of full-color images using the neural network algorithm of multi-step segmentation. Mod. High Technol. **3**, 55–60 (2015)

Neural Network Optimization Algorithms for Controlled Switching Systems

Olga V. Druzhinina[1], Olga N. Masina[2], Alexey A. Petrov[2](✉), Evgeny V. Lisovsky[3], and Maria A. Lyudagovskaya[4]

[1] Federal Research Center "Computer Science and Control" of Russian Academy of Sciences, Moscow, Russia
ovdruzh@mail.ru

[2] Bunin Yelets State University, Yelets, Russia
olga121@inbox.ru, xeal91@yandex.ru

[3] Kaluga Branch of Bauman Moscow State Technical University, Moscow, Russia
levgenijv@gmail.com

[4] Russian University of Transport (MIIT), Moscow, Russia
m.ludagovskaya@gmail.com

Abstract. The problems of neural network algorithms designing for modeling of the switching technical systems are considered. The method for designing of dynamic models using polynomial approximation is proposed. The generalized models of switching systems taking into account nonstationary perturbations are constructed. The heuristic optimization algorithms implemented in the form of computer libraries are developed. The problems of neural network algorithm implementation using high-level hybrid computing are considered. The effect of the application of the algorithms proposed in the paper is to reduce the time and energy costs on creating vector thrust. Algorithmic support based on artificial neural networks with training is proposed. The library of high-parallel training algorithms for neural networks in the problems of optimal trajectories constructing for switching technical systems is developed. The obtained results can be used in problems of researching models of controlled technical systems with switching operating modes, in particular, in modeling the dynamics of various classes of aircraft and in modeling intelligent transport systems #CSOC1120.

Keywords: Controlled switching systems · Differential inclusions · Artificial neural networks · Optimization · Algorithms · Machine learning

1 Introduction

The important and intensively developing scientific direction is mathematical modeling of technical systems in the conditions of operating modes switching. The relevant problems in this direction are the construction and the analysis of the specified technical systems models [1–3].

R. Silhavy (Ed.): CSOC 2020, AISC 1225, pp. 470–483, 2020.
https://doi.org/10.1007/978-3-030-51971-1_39

The similar systems find numerous applications in applied problems of mechanical systems control, technological processes in switching power converters, in control of the unmanned aerial vehicles motion, automated transport systems, robotic systems and in other areas [4–7].

The one of the relevant directions at the studying of the technical systems dynamics is adaptive modeling. The adaptive model is understood as a model which is invariant with respect to various input parameters and environmental influences, as well as preserving the adequacy of certain changes in the described system. The principle of constructing models with adaptive properties is of theoretical and applied interest for models with variable structure, autoregressive models, models with polynomial approximation and spline-approximation.

The construction of predictive and control logical components of dynamic systems is connected with the development of artificial intelligence algorithms. In this direction, the use of artificial neural networks (ANN) is effective [8,9]. The problems successfully solved by means of ANN include the following problems: the formation of models and various nonlinear and the systems that are hardly described mathematically, the development forecasting of these systems in time; solution acceptance and diagnostics that exclude the logical conclusion, especially in areas where there are no unfussy mathematical models; control robots, other complex devices. The unique property of ANN is their versatility. Despite the fact that for the problems listed above there are effective mathematical methods of solving, thanks to the universality and promise for solving global problems, for example, the construction of artificial intelligence and modeling complex systems, neural networks are the universal tool for problems solving. The characteristic feature of the neural network as the universal tool is, in particular, that the neural network is the flexible model for the nonlinear approximation of multidimensional functions.

In this paper, the basic model of aircraft dynamics is considered at the condition of minimal fuel consumption. This model is modified and generalized in case of the switching frequency increase, dynamic changes of boundary conditions, the action of nonstationary perturbations. The generalized models can be used in the problems of motion cargo by means of the aircraft in the conditions when the achievement of the final point at the motion along the surface of the earth is impossible.

In Sect. 2 first we consider the base model of the switching technical system along state vector. For this model, the existence of optimal solutions in an analytical form is noted. Then this model is generalized for the multidimensional nonstationary case, continuous polynomial functions of the second order in the right-hand sides are considered, the existence of analytical solutions is shown. The transition is made to the linear nonstationary model with the unknown perturbation, in which nonlinearity is taken into account by linear modes switching of motion. It is noted that to realize this model in the form of a software package, it is necessary to construct the global optimization algorithms and switching generation algorithms. In Sect. 3, the algorithms of optimal parameters search are constructed based on the development of random search method and swarm optimization method, namely, the "Adaptive Search" algorithm and the "Potential Search" algorithm. The comparative analysis of these algorithms performance is also carried out. The switching generation algorithms "Stochastic switching generation algorithm" and "Neural network switching generation algorithm" are developed, the advantages and disadvantages of these algorithms are noted. Moreover, a variant of the neural network algorithm based on high-level hybrid calculations is presented. This paper is the continuation of [10,11], in which we constructed the models and obtained some theoretical results. The main aim of this paper is the elaboration of neural network switching algorithms for generalized models of control systems.

2 Problems Statement, Models Under Consideration and Research Methods

The papers [7,10,11] are devoted to the study of the two-dimensional model of switching technical system. This model is the base model of the aircraft dynamics at the condition of minimal fuel consumption. The two-dimensional base model is defined as follows.

The moving is carried out in Cartesian coordinates x, y with the initial point $(0,0)$ and occurs in two stages. At the first stage, the corresponding interval $(0, t_1)$, the object moves on the xoy plane under the influence of constant vector thrust (f_{x_1}, f_{y_1}) until the height h is reached. At the second stage, corresponding to the interval (t_1, t_2), the object is moved with a constant vector thrust $(-f_{x_2}, f_{y_2})$, reaching the final point $L(l, 0)$. The considered model takes into account the effect of the gravitational force with the magnitude of the acceleration of gravity g and mass m of object. Let

$$\forall x(t), t \in (0, t_2) : \dot{x} > 0;\ \forall y(t), t \in (0, t_1] : \dot{y} \geq 0;\ \forall y(t), t \in (t_1, t_2] : \dot{y} < 0,$$

herewith $x(0) = 0, y(0) = 0, y(t_1) = h, y(t_2) = 0, x(t_2) = l$.

The optimality criterion has the form [11]

$$\int_0^{t_1} (f_{x_1} + f_{y_1})dt + \int_{t_1}^{t_2} (f_{x_2} + f_{y_2})dt \to \min .$$

The physical meaning of the considered optimality criterion is the minimal fuel consumption in the case of jet propulsion.

The differential inclusions describing the system have the form:

$$\begin{aligned} m\ddot{x} &\in f_x, \\ m\ddot{y} &\in f_y - mg. \end{aligned} \tag{1}$$

This differential inclusions describing the system are reduced to a system of differential equations second-order with switching at point t_1. The equations have the form:

$$\begin{aligned} m\ddot{x} &= f_{x_1}, \\ m\ddot{y} &= f_{y_1} - mg, \end{aligned} \quad t \in [0, t_1], \tag{2}$$

$$\begin{aligned} m\ddot{x} &= f_{x_2}, \\ m\ddot{y} &= -f_{y_2} - mg, \end{aligned} \quad t \in (t_1, t_2]. \tag{3}$$

In [7,10,11] the optimal values of the parameters, models satisfying the initial assumptions are found using symbolic calculations of the Mathematica math package. The model (2), (3) is simplified for describing of the controlled object motion, since it does not take into account the possibility for the considering of the final point motion (variability condition). In addition, the model (2), (3) does not take into account air resistance and requires generalization and specification.

The modification of the two-dimensional model is proposed in [12] for the case of final point positioning. Final point positioning refers to moving a point in an additional xoz plane. An algorithm of switches generating for three-dimensional model is proposed. The modeling, based on aerodynamic forces is performed and an algorithm for finding optimal parameters using artificial neural networks is developed in [13].

Further, we will consider the generalization of the model (2), (3) to the multidimensional case taking into account nonstationary perturbations. For this generalization, we propose the numerical interpretation of the optimal trajectories search problem. From the model (2), (3) we turn to the model in which vector thrust is specified using functions of time. We introduce the additional constraint $||\dot{x}(t_2)|| = 0$ for zero speed at the final point.

We will consider the generalized model as a matrix differential equation:

$$m\ddot{x} = PT + mG, \tag{4}$$

where

$$P = \begin{pmatrix} p_{11} & \cdots & p_{1n} \\ \vdots & \ddots & \vdots \\ p_{k1} & \cdots & p_{kn} \end{pmatrix}, \quad T = \begin{pmatrix} t^0 \\ t^1 \\ t^2 \\ \vdots \\ t^k \end{pmatrix}, \tag{5}$$

$x \in \mathbb{R}^n$ – vector of system states, G – vector of potential field, k – maximum degree of parametric curves. Using the Eqs. (4), (5), you can model unallocated mechanical systems without feedback, since polynomial curves allow you to approximate any functions of time in the right sides of the equations for $0 < t < t_2$. The search for the right-hand sides is reduced to finding the matrix of coefficients P.

In [11], we consider the (4) model for matrices of the form:

$$P = \begin{pmatrix} p_{11} & p_{12} & 0 \\ p_{21} & p_{22} & p_{23} \end{pmatrix}, \quad T = \begin{pmatrix} t^0 \\ t^1 \\ t^2 \end{pmatrix}. \tag{6}$$

Taking into account (6) the special case of the system (4) has the form:

$$\begin{aligned} m\ddot{x}_1 &= p_{11} + p_{12}t, \\ m\ddot{x}_2 &= p_{21} + p_{22}t + p_{23}t^2 - mg \end{aligned} \tag{7}$$

The optimality criterion has the form

$$\int_0^{t_2} (p_{11} + p_{12}t)^2 + (p_{21} + p_{22}t + p_{23}t^2)^2 dt \to \min. \qquad (8)$$

the motion will be acceptable under the following conditions:

$$\ddot{x}_1(0) > 0,\ \dot{x}_1(t_1) > 0,\ \ddot{x}_1(t_2) < 0,\ \ddot{x}_2(0) > 0,\ \ddot{x}_2(t_2) > g,$$

where we get

$$p_{11} > 0,\ p_{12} < 0,\ p_{21} > g,\ p_{22} < 0,\ p_{23} > 0. \qquad (9)$$

Given the initial and boundary conditions, the solution has the form:

$$P = \begin{pmatrix} \dfrac{6lm}{t_2^2} & -\dfrac{12lm}{t_2^3} & 0 \\ \dfrac{m(gt_2^2 + 32h)}{t_2^2} & -\dfrac{192hm}{t_2^3} & \dfrac{192hm}{t_2^4} \end{pmatrix}. \qquad (10)$$

It is easy to verify that the (10) values satisfy the conditions (9).

The Eq. (10) implies the uniqueness of the trajectories satisfying the formulation of the problem by the (6) model for each t_2 value. In this case, the fulfilment of the criterion (8) is reduced to choosing the optimal value of t_2 depending on h, l, m.

For the optimality criterion in the form of a function, the analytical form of the extrema is too cumbersome. A Python program is developed to calculate the optimal value of t_2. The single minimum for criterion (8) is obtained in [11].

It should be noted that when increasing the value of k for the model (4), (6), finding the optimal values of the matrices P in the analytical form becomes impossible. In addition, the specified model cannot take into account the effects of disturbances on the simulated object.

In this regard, the use of linear switching models with nonstationary effects is quite effective direction of modeling.

Next, we consider a two-dimensional linear model of the motion of the aircraft under the influence of external disturbances. The equation of the specified model has the form:

$$m\ddot{x} = PT + mG + D, \quad \psi : t \to \Gamma, \quad P \in \Gamma, \qquad (11)$$

where $x \in \mathbb{R}^n$, ψ – the function of the switches for which $|t| = \aleph_1$, $0 \le |\Gamma| < \aleph_0$, $D = d(x,t)$ is an unknown vector function that determines external perturbations. The model (11) is a modification and generalization of the model (4). The vector of initial conditions for the model (11) has the form $I = (a_1, b_1, a_2, b_2)$.

Consider the case then the coefficient matrix has the form

$$P = \begin{pmatrix} p_{11} & p_{12} \\ p_{21} & p_{22} \end{pmatrix}. \qquad (12)$$

Taking into account (12), conditions $x(t_2) = (0, 0)$ and $D = 0$ we obtain

$$P = \begin{pmatrix} \dfrac{2m(-2t_2b_1 + 3l - 3a_1)}{t_2^2} & \dfrac{6m(t_2b_1 - 2l + 2a_1)}{t_2^3} \\ gm - \dfrac{4b_2m}{t_2} - 6ma_2/t_2^2 & \dfrac{6m(t_2b_2 + 2a_2)}{t_2^3} \end{pmatrix}. \tag{13}$$

The switching constructing by the aid of matrix P for the model (11) associates with prediction. However, the constructing of the switching function is the difficult problem that solution requires the use of global optimization methods and neural network methods. The switching process can be divided into following three main stages.

1. Checking the switching condition $\psi(t_n) \neq \psi(t_{n-1})$.
2. Search of matrix P.
3. Constructing the intermediate trajectory of system.

The main purpose of this paper is the elaboration of neural network switching generation algorithm for the model (11). For the indicated algorithms the following conditions must be satisfy.

1. Solvability of the initial-boundary value problem for the Eq. (11) at $D \neq 0$.
2. Minimizing of the vector forces acting in the system, i.e. $\forall P, t : \|PT\| \to \min$.
3. Minimizing of the number of switching, i.e. $|\Gamma| \to \min$.
4. Possibility of high-performance implementation in the form of a computer program (using GPGPU [14]).

The main and the auxiliary algorithms (I, II, III, IV) proposed in this paper, as well as algorithms proposed in [11], are presented in the Table 1 taking into account the main stages of the switching process.

The Potential search algorithm, Adaptive search algorithm, Neural network switching generation algorithm and Neural network learning algorithm are developed in Sect. 3. The Potential search algorithm and Adaptive search algorithm are auxiliary algorithms for the Neural network learning algorithm. Potential search algorithm based on particle swarm optimization, Adaptive search algorithm based on random search method. Neural network switching generation algorithm and Neural network learning algorithm are based on the artificial neural network method.

In general, the finding of the required matrices P is associated to the multicriteria optimization problem. The multi-criteria optimization problem can be reduced to the single-criterion optimization problem by scalar ranking. Despite the fact that the right-hand sides of Eq. (11) everywhere remain smooth, the increase of the ranks of matrices P and T makes the most ineffective methods of local optimization. One way to resolve this issue is to use the global optimization techniques. In [11] finding of the matrix P using the algorithm of differential evolution [15] is considered. However, it is worth noting that the differential evolution algorithm is resource-intensive, and does not always show high performance. Section 3 proposes algorithms that can be used to find optimal

Table 1. Algorithms proposed in this paper.

Stage of the switching	Basic algorithms	Auxiliary algorithms
Checking the switching condition	**Neural network switching generation algorithm (I)** Stochastic algorithm for switching generation Algorithm for switching with increasing frequency	**Neural network learning algorithm (II)**
Search of matrix P	The search of analytical value P The search of numerical value P	Differential Evolution **Potential search (III)** **Adaptive search (IV)**
Constructing the intermediate trajectory of system	Finite–Difference Method (Program) Runge-Kutta, Adams methods etc.	

matrices P for system (11) while ensuring high performance. These algorithms are developed taking into account the possibility of implementation based on the technologies of Python3 with using numpy, scipy, arrayfire libraries.

3 Developed Algorithms and Computer Modeling Results

3.1 Optimization Algorithms

We propose two optimization algorithms based on particle swarm optimization and random search, which are called "Potential search" (Algorithm 1) and "Adaptive search" (Algorithm 2) respectively. This algorithms are developed for the searching of the optimal matrices P for the model (11). The model (11) described in the Sect. 2.

The Potential search algorithm consists of the following steps.

Step 1. Generate the particle swarm.
Step 2. Calculate the mass of each particle.
Step 3. Find the particle with the largest mass.
Step 4. Check the break condition.
Step 5. Move the particle swarm towards the particle with the largest mass.
Step 6. Go to step 2.

The Adaptive search algorithm consists of the following steps.

Step 1. Select the first approximation of the solution ϵ.
Step 2. Calculate n random vectors $\boldsymbol{v}$ taking into account $||\boldsymbol{v}|| \leqslant \delta$.
Step 3. Solve the one-dimensional optimization problem $\boldsymbol{v}, \epsilon + \boldsymbol{v} \rightarrow \min$.
Step 4. Assign $\epsilon := \epsilon + \boldsymbol{v}$.
Step 5. Check the break condition. Go to step 2.

In this algorithm, δ denotes the radius of the hypersphere that defines the search area. These algorithms are implemented in the form of computer libraries by Python. These libraries are included in the module of auxiliary mathematics of the software package.

We perform a comparative performance analysis with the following optimization algorithms [15,16]: 1) differential evolution, 2) Powell's algorithm. These algorithms are part of the scipy mathematical library. The following test optimization functions are used as target functions: Rosenbrock function, Rastrigin function. A series of computer experiments with model (11) showed the consistency of experimental results with theoretical research.

We assume that the algorithm succeeds if the last approximation of the solution lies in a small neighborhood of the global extremum. The productivity of algorithms for these functions is shown in the Table 2.

Table 2. Time of test functions minimization.

	Potential search	Adaptive search	Differential evolution	Powell algorithm
Rastrigin function	1780 ms	–	121 ms	–
Rosenbrock function	1650 ms	532 ms	193 ms	–

The results of the execution speed testing presented in the Table 2 show that the most efficient and universal algorithm for solving the problem under consideration among the tested ones is the differential evolution. The Powell algorithm [16] does not show stable convergence for both the first and the second test function.

It should be noted that the implementation of optimization algorithms efficiency by swarm of particles and random search can be estimated as high enough as they are implemented in an interpreted Python language. The use of other technical means in the resource-intensive parts of the algorithms can increase productivity by 1–2 orders of magnitude.

Since the algorithms discussed in this section are sufficient resource-intensive, you should develop switching algorithms with a small number of optimization procedure calls. Such switching algorithms for the (4) system are proposed in Sect. 3.2.

3.2 Switching Algorithms Based on the Artificial Neural Networks

One of the powerful tools of adaptive modeling is artificial neural networks. The application of artificial neural networks in systems of various types is studied in [8,9] and in other works.

We introduce a controller based on artificial neural networks. We assume that the matrix representation of the neural network has the form of a recursive expression. Consider a simple, fully connected neural network with no hidden layers. Let's denote inputs as $j = 1, \ldots, 4$, outputs as $k = 1, \ldots, 4$, layer number as $I = 1$. Then the weights can be represented by the square adjacency matrix W_i of the elements w_{kj}:

$$W_i = \begin{pmatrix} w_{11} & \cdots & w_{14} \\ & \ddots & \\ w_{41} & \cdots & w_{44} \end{pmatrix}.$$

The values of input nodes are defined by the vector-column $Z_i = (z_1, \cdots, z_4)^T$, which specifies the scalar components of the state vector and the deviation vector from the final point $(l, 0)$ for the system (11). For output neurons the yield will be determined by the formula

$$z_k = \delta\left(\sum_{l=1}^{4} w_{kj} z_j\right),$$

where δ is the neuron activation function. Similarly, you can vectorize the layer activation function. In this case, the layer output is reduced to the matrix expression $Z_{i+1} = a(W_i Z_i)$, where a is vector is the layer activation function. Then the General case the calculation of the outputs of a neural network consisting of n layers, is described by the formula $Z_n = A(W_n A(W_{n-1} A(W_{n-2} \ldots\ A(W_1 Z_1))))$.

The advantage of neural network representation in vector-matrix form is high performance in computer implementation, especially when using hybrid computations based on GPGPU technologies. Construction of networks of direct distribution of any topology is possible due to sparse matrices of weights.

The use of a neural network controller for switching in conjunction with the model (11) enables adaptive regulation of the number of switches depending on the values of external disturbances.

Next, we present the developed Algorithm 4 "Neural network switching generation algorithm". The specified algorithm consists of the following steps.

Step 1. Configure of the neural network controller.
Step 2. Make one iteration of the numerical integration algorithm.
Step 3. Check the shutdown condition. If it is executed, the algorithm is complete. Otherwise, you must proceed to the next step.
Step 4. Calculate outputs of neural network switching generator. In the case that the received vector in the domain of the switches, go to next step. Otherwise, go to step 2.

Step 5. Calculate new matrix P for the current I and L according to the formula (13).
Step 6. Go to step 2.

The above algorithm includes the developed learning algorithm, called "Neural network learning algorithm" (Algorithm 5). The specified algorithm consists of the next step:

Step 1. Create a neural network with random weights.
Step 2. Execute the switching algorithm and get the trajectory.
Step 3. Check the shutdown condition. If it is executed, the algorithm is complete. Otherwise, go to the next step.
Step 4. Optimize the weight coefficients.
Step 5. Go to step 2.

Algorithm "Neural network training algorithm" implements unsupervised learning. As the result, the weights coefficients are calculated that allow the neural network to be invariant with respect to external disturbances, which are not explicitly accounted [17].

The switching algorithms presented in this section are used in the process of finding map Γ for model (11). For this map the power of the set of values is minimal. The problems of the Algorithm 5 implementation using high-level hybrid computing are considered in Sect. 3.3.

3.3 Hybrid Neural Network Computing

Further we consider the implementation variant of the Algorithm 5. In this variant, the target function for optimization at step 5 is the standard deviation of the set of obtained trajectories from those trajectories that are guaranteed to lead to the final point of the system phase space. In this case, the algorithm allows explicit vectorization. Let us use a sample of n trajectories and a neural network without a hidden layer with a matrix of weight coefficients W_γ. We evaluate the increase in the performance of neural network derivation calculation under explicit vectorization using high-level hybrid computations. For this we use the arrayfire library [14] and Python programming language.

We present the weight coefficients of the set of n neural network controllers as a block-diagonal matrix of the form

$$W_\gamma = \begin{pmatrix} W_i \; 0 \; 0 \; \ldots \; 0 \\ 0 \; W_i \; 0 \; \ldots \; 0 \\ \ddots \\ 0 \; 0 \; 0 \; \ldots \; W_i \end{pmatrix}. \tag{14}$$

For matrix (14) the input values are given by $Z_\gamma = (z_1, z_2, \ \ldots, z_\phi)^T$, ϕ defines the size of the matrix W_γ. Then the values of the outputs of the neural networks equal $\delta(W_\gamma Z_\gamma)$, where δ is the activation function in the output layer.

A program for performance testing are developed. Weight matrices are randomly generated, the size of W_i is 16×16 (256 weights). Hyperbolic tangent is used as an activation function. Then the values of the output neurons are determined by the expression

$$Z_1 = \tanh(W_0 Z_0), \tag{15}$$

where Z_0 is the values of inputs of neural networks, W_0 is the block-diagonal matrix of weights of the form (14), Z_1 is the values of neural networks outputs.

The hardware configuration of the computer includes AMD FX-8300 CPU and AMD Radeon RX 580 GPU. The results of the performance test are shown in Fig. 1. On the specified drawing prinyaty the following notations: np.matmul() is the calculation using numpy, af af.matmul() is the calculation using arrayfire library.

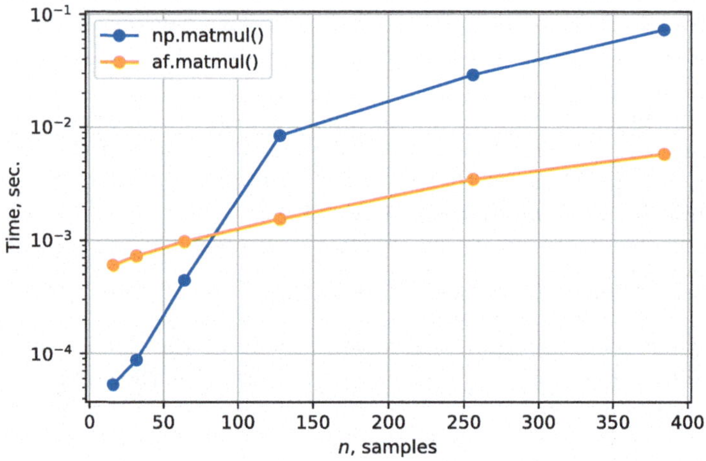

Fig. 1. Computational time of multiple neural networks.

According to the test results, the use of hybrid computations for (15) is reasonable for $n > 100$. It should be noted that the performance measurements are made taking into account the conversion of arrays arrayfire in numpy arrays by calling the method to_array(). This fact is due to the fact that the arrayfire library implements the so-called lazy computing model. In addition, the actual calculation is not performed until the data is loaded from the graphics card memory into the RAM [14]. Thus, the input/output time (I/O) costs are also taken into account in this case. Parallel programming and hybrid computing is also dedicated to the papers [18–22].

The advantages of using Python and the arrayfire library for hybrid computing include high computing performance, high cross-platform portability, and open source libraries. The main disadvantages of this approach include sufficiently large memory consumption when using Python, redundancy approach for highly sparse matrices (14), as well as a limited number of mathematical functions implemented in the library arrayfire.

The program code developed for testing is used to modify the library of artificial neural networks. The results obtained in this direction can be used in the problems of teaching neural networks with a teacher.

The calculation of many trajectories in a single step of Algorithm 4 allows the use of more representative information about the characteristics of motion of the model (11) for training the neural network switching controller.

4 Discussions

The results of [10–13] in the field of mathematical models constructing for technical switching systems are planned to be used in further research. In particular, it is possible to use polynomial curves of the third degree in the system (4) provided that the parameters (5) exist in an analytical form, which makes it possible to consider a greater variety of physical effects affecting the simulated object. In addition, it is possible to search for (5) parameters numerically or symbolically.

The prospects for developing a neural network approach to the modeling of nonlinear systems include the following: the use of recurrent neural networks with a single delay that store information about past switching implementations; the use of deep learning neural networks; using neural networks to search for the coefficients of the matrices included in the right-hand parts of the equations of motion.

The results obtained in the field of hybrid neural network computing are planned to be used in the development of new software libraries for building artificial neural networks with a flexible architecture that combine different methods of artificial intelligence.

Promising areas of research, developing the topic of the paper are: development and modification of models with switching taking into account the complexity of the technical systems classes of transport and robotics; the development of intelligent methods and cognitive technologies of optimal parameters search at technical systems modeling; the expansion of the software libraries taking into account development implementation capabilities of domestic software and hardware platforms; further modification of existing libraries for machine learning and high-parallel GPGPU-computing.

For modern automated control systems in the transport industry, the most urgent tasks are the tasks of data mining and integrated diagnostics of infrastructure facilities [23]. The methods and algorithms proposed in this article can find application in the development of information control systems related to the processing of data obtained using diagnostic cars.

5 Conclusion

The following models of technical systems are considered: basic two-dimensional model, generalized nonstationary nonlinear model, generalized linear nonstationary model with switching. Algorithms of the optimal parameters search on the basis of random search methods development and swarm optimization, namely, proposed Algorithm 1 "Adaptive search" and Algorithm 2 "Potential search". A comparative analysis of the Algorithms 1, 2 is performed, their benefits and drawbacks are identified when solving problems of finding the optimal parameters of the studied models with switching. In addition, in Sect. 3, the algorithms for generating switches are developed in the presence of linear nonstationary modes, namely, Algorithm 3 "Stochastic algorithm for switching generation" Algorithm 4 "Neural network switching generation algorithm", which includes Algorithm 5 "Neural network learning algorithm".

The obtained results can be used in problems of researching models of controlled technical systems with switching operating modes, in particular, in modeling the dynamics of various classes of aircraft and in modeling intelligent transport systems.

References

1. Averchenkov, V.I., Fedorov, V.P., Hejfec, M.L.: Basics of Mathematical Modeling of Technical Systems. Flinta, Moscow (2016)
2. Ayupov, V.V.: Mathematical Modeling of Technical Systems. IPC "Prokrost", Perm (2017)
3. Bahvalov, Y.A., Gorbatenko, N.I., Grechihin, V.V.: Mathematical and computer modeling of complex technical systems. Publ. House J. "News Univ. Electromech." (2014). Novocherkassk
4. Guryanov, A.E.: Quadcopter control simulation. Eng. Gazette (8) (2014). http://ainjournal.ru/doc/723331.html
5. Druzhinina, O.V., Cherkashin, Y.M., Shestakov, A.A.: Study of the safety of the motion of transport systems based on the concept of dynamic rigidity of the trajectories. Theory Saf. Stabil. Syst. (11), 123–136 (2009)
6. Titov, Y.P.: Modifications of the ant colony method for developing software for solving multi-criteria supply management problems. Mod. Inf. Technol. IT-Educ. **13**(2), 64–74 (2017)
7. Masina, O.N.: The problems of the motion control of transport systems. Transp.: Sci. Technol. Control (12), 10–12 (2006)
8. He, W., Chen, Y., Yin, Z.: Adaptive neural network control of an uncertain robot with full-state constraints. IEEE Trans. Cybern. **46**(3), 620–629 (2016)
9. He, W., Dong, Y., Sun, C.: Adaptive neural impedance control of a robotic manipulator with input saturation. IEEE Trans. Syst. Man Cybern.: Syst. **46**(3), 334–344 (2016)
10. Druzhinina, O.V., Masina, O.N., Petrov, A.A.: Models for control of technical systems motion taking into account optimality conditions. In: Proceedings of the VIII International Conference on optimization methods and applications "Optimization and application" (OPTIMA-2017), vol. 1987, pp. 386–391. Petrovac, Montenegro (2017). http://ceur-ws.org/Vol-1987/paper56.pdf
11. Druzhinina, O.V., Masina, O.N., Petrov, A.A.: The synthesis of the switching systems optimal parameters search algorithms. Commun. Comput. Inf. Sci. (CCIS) **974**, 306–320 (2019)
12. Druzhinina, O.V., Masina, O.N., Petrov, A.A.: Model of motion control of transport system taking into account conditions of optimality, multivaluence and variability. Transp.: Sci. Technol. Control (4), 3–9 (2017)

13. Druzhinina, O.V., Masina, O.N., Petrov, A.A.: Approach elaboration to solving of the problems of motion control of technical systems modeled by differential inclusions. Inf.-Meas. Controll. Syst. **15**(4), 64–72 (2017)
14. Chrzeszczyk, A.: Matrix computations on the GPU with ArrayFire for Python and C/C++, January 2012
15. Storn, R., Price, K.: Differential evolution - a simple and efficient adaptive scheme for global optimization over continuous spaces. J. Glob. Optim. **11**(4), 341–359 (1995)
16. Powell, M.J.D.: An efficient method for finding the minimum of a function of several variables without calculating derivatives. Comput. J. **7**(2), 155–162 (1964). https://dx.doi.org/10.1093/comjnl/7.2.155
17. Haykin, S.: Neural Networks. Vilyams, Moscow (2006)
18. Chávez, G., Turkiyyah, G., Zampini, S., Ltaief, H., Keyes, D.: Accelerated cyclic reduction: a distributed-memory fast solver for structured linear systems. Parallel Comput. **74**, 65–83 (2018). Parallel Matrix Algorithms and Applications (PMAA 2016). http://www.sciencedirect.com/science/article/pii/S0167819117302041
19. Deveci, M., Trott, C., Rajamanickam, S.: Multithreaded sparse matrix-matrix multiplication for many-core and gpu architectures. Parallel Comput. **78**, 33–46 (2018). http://www.sciencedirect.com/science/article/pii/S0167819118301923
20. Masliah, I., Abdelfattah, A., Haidar, A., Tomov, S., Baboulin, M., Falcou, J., Dongarra, J.: Algorithms and optimization techniques for high-performance matrix-matrix multiplications of very small matrices. Parallel Comput. **81**, 1–21 (2019). http://www.sciencedirect.com/science/article/pii/S0167819118301091
21. Wolfe, M., Lee, S., Kim, J., Tian, X., Xu, R., Chapman, B., Chandrasekaran, S.: The openacc data model: preliminary study on its major challenges and implementations. Parallel Comput. **78**, 15–27 (2018). http://www.sciencedirect.com/science/article/pii/S0167819118302175
22. Huqqani, A.A., Schikuta, E., Ye, S., Chen, P.: Multicore and gpu parallelization of neural networks for face recognition. Proc. Comput. Sci. **18**, 349–358 (2013). 2013 International Conference on Computational Science. http://www.sciencedirect.com/science/article/pii/S1877050913003414
23. Druzhinina, O.V., Lyudagovskaya, M.A.: Intellectual methods for developement and upgrade of the information control systems at the railways. Transp.: Sci. Technol. Control (8), 3–12 (2019)

A Deep Learning Model with Long Short-Term Memory (DLSTM) for Prediction of Currency Exchange Rates

Thitimanan Damrongsakmethee[1,2(✉)] and Victor-Emil Neagoe[1]

[1] Department of Applied Electronics and Information Engineering, Faculty of Electronics, Telecommunications and Information Technology, Polytechnic University of Bucharest, Bucharest, Romania
manancc@gmail.com, victoremil@gmail.com

[2] Department of Business Information Systems, Suratthani Rajabhat University, Suratthani, Thailand

Abstract. The objective of this research is to implement a Deep Learning model with Long Short-Term Memory (DLSTM) for prediction of the currency exchange rate. The system predicts the currency exchange rate using a fusion of data from the following financial inputs: gross domestic product rate (GDP), interest rate, inflation rate, balance account and trade balance as well as a finite set of previous exchange rates. We have evaluated the model performance by considering the currency exchange rates of the Thai Baht to the US dollar using historical data from the Bank of Thailand for ten years, from April 2009 to April 2019. To evaluate the effectiveness of the DLSTM model, we have considered the mean square error (MSE) and the mean absolute percentage error (MAPE). The best results have shown that the DLSTM model leads to a very low error value for the MSE and the MAPE at 0.0027 and 0.2844, respectively. We have compared the proposed DLSTM model prediction performances with those of the NARX neural network model; our research results show the obvious advantage of the proposed DLSTM model.

Keywords: Deep learning · Long Short-Term Memory · DLSTM · Currency exchange rates prediction · Time series prediction · Financial time series forecasting

1 Introduction

The information necessary for running a business is continuously changing. Therefore, business leaders and organizations need to find tools or develop methods that can be used to make decisions regarding all the effects of future changes [1]. As we know, the techniques of the time series analysis have a strong role in decision making [2]; techniques of predictive models are used to help control current operations and plan future needs. Forecasting is a method with the same goal of using time series data to predict future business events [3]. Time series data is a set of data that is collected continuously over time, such as sales data that are collected continuously for months [4]. Time series data may be in the form of annual, quarterly, or monthly data depending on the suitability of

R. Silhavy (Ed.): CSOC 2020, AISC 1225, pp. 484–498, 2020.
https://doi.org/10.1007/978-3-030-51971-1_40

use [5, 6]. The common techniques in the time series predictions are quantitative forecasting techniques such as regression analysis and time series analysis [7]. These methods rely on the historical data that has been collected over successive intervals and can be used to create equations to find relationships between variables in the forecast [8].

Nowadays, the techniques of computational intelligence have been applied in data analysis, especially to help solve problems that occur in the analysis of financial data [9]. It can be seen that the research involved in applying data mining techniques to solve the problems in the financial analyst sector has many models including: *decision trees, support vector machine (SVM), neural networks (NN), K-nearest neighbor (K-NN),* etc. In addition to the above techniques, there is acknowledged that models using deep learning models can provide extremely accurate results. The deep learning technique has applications for object detection, pattern recognition, handwriting recognition, speech recognition, image processing and time series prediction [10]. A lot of research has been used for deep learning in financial forecasting, especially regarding exchange rate forecasting. Parametric statistical models have also been proposed and used for the prediction of the currency exchange rates [11]. The general mathematical techniques help to describe the patterns and trends in the data. From research reports related to financial forecasting, it is found that the long short-term memory (LSTM) prediction technique is a very popular method. The LSTM model is one of the deep learning architectures and is the most favored technique for bitcoin forecasting [12], stock exchange forecasting [6, 7] and current exchange rate [2, 11, 13–15]. We also found that in financial forecasting, the NARX neural networks technique is still popular [16, 17]. The concept of the NARX neural networks is a *nonlinear autoregressive exogenous* (ARX), representing a simple tool for recurrent nonlinear time series predictions. In general, there are two types of NARX neural network architecture: "open-loop architecture" (called Serial-Parallel architecture (SP)) and "closed-loop architecture" (called Parallel architecture (P)).

The main objective of this research is to implement a predictive technique by using the deep learning long short-term memory (LSTM) technique to forecast the currency exchange rates. We have used the exchange rate dataset of Thailand baht per US dollars from 2009 to 2019 based on the main five economic indicators consisting of the GDP growth rate, interest rate, inflation rate, balance account and trade balance. To measure the effectiveness of our proposed Deep LSTM model, we have selected the best prediction values for the Deep LSTM model to compare with the experimental results of the NARX model based on feed-forward neural networks from the previous research of the same authors [17].

The structure of this research is the following. Section 2 presents the details of the literature reviews. Section 3 describes the proposed model. In Sect. 4 there are presented the financial dataset and the performance measures of our proposed model. In Sect. 5 we present the experimental results. The final conclusions of the research are shown in Sect. 6.

2 Related Work

In this section, we present the details regarding related work on the LSTM model. In research [2], there is proposed a comparison between a forecasting machine learning model based on the technique of long short-term memory (LSTM) versus Super Vector Regression (SVR) to forecast the phone prices in European electronic commerce. The results of this research found that the LSTM technique obtained the best performance results for the next day's price prediction taking into account the error measurement: a root mean squared error (RMSE) of 23.64%. In the research [14], there is presented the forecasting research on a financial time series dataset by comparing the experimental results of three machine learning techniques: Multilayer perceptron (MLP), Convolutional Neural Networks (CNN) and LSTM. The authors found that the LSTM model showed the best hit ratios when compared with MLP and CNN. A novel combination of forecasting models is presented in research [15]; this application implemented the LSTM prediction models by using the dataset of six stock markets. In this research, there are applied Wavelet Transforms (WT) and Stacked Auto Encoders (SAEs) as the WTLSTM model and WSEAs-LSTM model. The final results of this research showed that the WSAEs-LSTM model obtained the best performance when compared to others. For the research of [6], there is a proposed deep learning technique based solution for the stock price forecasting by comparison with linear and non-linear models. The experimental results of the proposed methodology CNNs has been identified as the best model. In the same way, in the research [5], there is a proposed forecasting time series model on the historical monthly financial time series from Yahoo's finance website. This model is compared with the experimental results of the Auto-Regressive Integrated Moving Average (ARIMA) model and LSTM model. The best results of the error measures with Root Mean Squared Error (RMSE) showed that the LSTM method had the best average error rate. In the research [18], the authors have applied a novel *long short-term memory fully convolutional network* (LSTM-FCN) model and *attention mechanism long short-term memory fully convolutional network* (ALSTM-FCN) model for time series classification of the University of California Riverside (UCR) Benchmark datasets. The results showed that the method of LSTM-FCN had improved the accuracy more than the ALSTM-FCN models. In 2015, the authors of the paper [19] proposed the RNN model by applying the concept of the LSTM technique to predict the stock market of China. This research compared the results of stock return prediction with six feature selection categories as a learning feature. The LSTM model was found to improve the performance of stock returns forecasting as well. In the research [20], the forecast model was created to predict the cassava price in Thailand for 72 months. The experimental results were compared on five time series techniques: the Box-Jenkins, simple exponential smooth, Holt's exponential smooth, damped trend exponential method and moving average method. This was done over a

3-month, 6-month and 12-month period, and they compared to the error measures as MAPE and RMSE. This research summarized that the moving average method was the best prediction method when compared to the error measure of MAPE with other models. The research of [12] has introduced a model of machine learning to predict the bitcoin prices by using the bitcoins exchange data with US Dollars from the bit stamp website. This research has focused on the bitcoin price forecasting machine learning algorithms in various ways (Theil-Sen Regression, Huber Regression, LSTM and Gated Recurrent Unit: GRU) to find the model best suited for forecasting the price of bitcoin. There has been found that the results of the regression experiment using different algorithms showed similar effectiveness, but the Huber Regression and Theil-Sen Regression model have the best results with the best accuracy of MSE and R. The research of predictive model [16] has presented a novel approach based on deep learning by using the NARX neural network model to predict Swiss Franc-Romanian Leu exchange rates versus US Dollar-Romanian Leu and there are also improved the general capabilities of the exchange rate with the smallest RMSE error when compared to other models. In 2019, the research [17] has presented the NARX neural network for predicting the exchange rate of Thai Baht per US dollar of the historical dataset from the bank of Thailand from April 2009 to April 2019. The input of the NARX neural network model has been used based on the main five economic indicators, including GDP growth rate, interest rate, inflation rate, account balance and trade balance. The best experimental results have shown that the best performances of forecasting performance measures MAPE and MSE are 3.001 and 0.006, respectively.

3 Methodology

3.1 Recurrent Neural Network (RNN)

Artificial Neural Network algorithms, which are suitable for time series data are called "Recurrent Neural Networks" (RNNs) [6]. This model has status information in a hidden layer so that the previous hidden state is used to calculate the current hidden state and it uses the current hidden state to calculate the information in the next period [21]. This technique is based on Eqs. (1) and (2):

$$h_t = \sigma(x_t W + h_{t-1} U) \tag{1}$$

$$o_t = \sigma(h_t V), \tag{2}$$

where x is the input data, o is the output value, t is the time interval, h is the hidden case, σ is the activation function and W, U, V are the weight matrix for calculating the previous input data, the hidden case and the current hidden state, respectively. The RNNs model has the structure shown in Fig. 1.

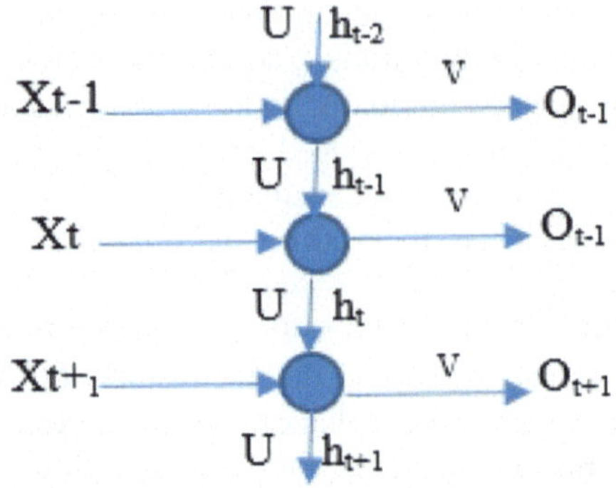

Fig. 1. The structure of Recurrent Neural Networks (RNNs).

Although the RNNs model is suitable for the sequential data and for the time series data, this model has some problems regarding long-term dependencies that are the primary cause of vanishing gradient problems. These problems can be solved with the LSTM model.

3.2 Long Short-Term Memory (LSTM) Model

The LSTM model is a deep learning model developed by solving the problem of vanishing gradients in RNNs [22]. The technique of LSTM classification uses the cell state and hidden state. The cell state is the memory of the LSTM cell. The hidden state is the result of the cell state to store the data and send it to the next time period which relies on various gates for calculation [23]. There are three standard gates in a hidden state: the input gate, output gate and forget gate [22]. Each gate can choose some important data to pass for the next step. In general, the LSTM layer consists of one or more input layers, one or more hidden layers, and an output layer. The number of neurons in the input layer corresponds to the number of variables used to explain the results. Each LSTM model structure can store and record data streams within cells that are connected from one module to another cell [19]. The internal design of the LSTM cell can be described in Fig. 2.

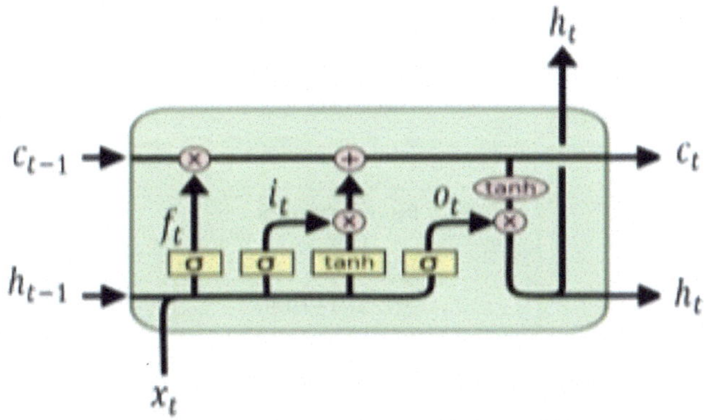

Fig. 2. The structure of the LSTM cell [14].

From Fig. 2, Let i is the input gate, f is the forget gate, o is the output gate, c is cell state, h is hidden state, σ is the activation function, and W, U are the weight matrix and t is the time. The LSTM model helps to avoid vanishing gradient problems at each time step from RNNs method according to the equations given below.

$$i = \sigma(x_t W_i + h_{t-1} U_i) \tag{3}$$

$$f = \sigma\left(x_t W_f + h_{t-1} U_f\right) \tag{4}$$

$$o = \sigma(x_t W_o + h_{t-1} U_o) \tag{5}$$

$$c_t = (c_{t-1} \times f) + (i \times \sigma(x_t W_c + h_{t-1} U_c)) \tag{6}$$

$$h_t = \sigma(c_t) \times o \tag{7}$$

3.3 Our Proposed Model

We proposed the Deep LSTM model for predicting the currency exchange rate. The flow diagram of the proposed DLSTM model architecture is shown in Fig. 3.

From Fig. 3, one can remark that the framework of our DLSTM model has two processing parts. In the first part after data normalization and data preparation, we used the five main economic indicators as the input data $x(t)$: the GDP growth rate, inflation rate, interest rate, balance account and trade balance. The above indicators are applied to the input layer of the Deep LSTM model. By partition phase, the dataset is divided into two sets: training set and testing set. The second processing part consists of exchange rate prediction by the Deep LSTM model using the testing data. We have

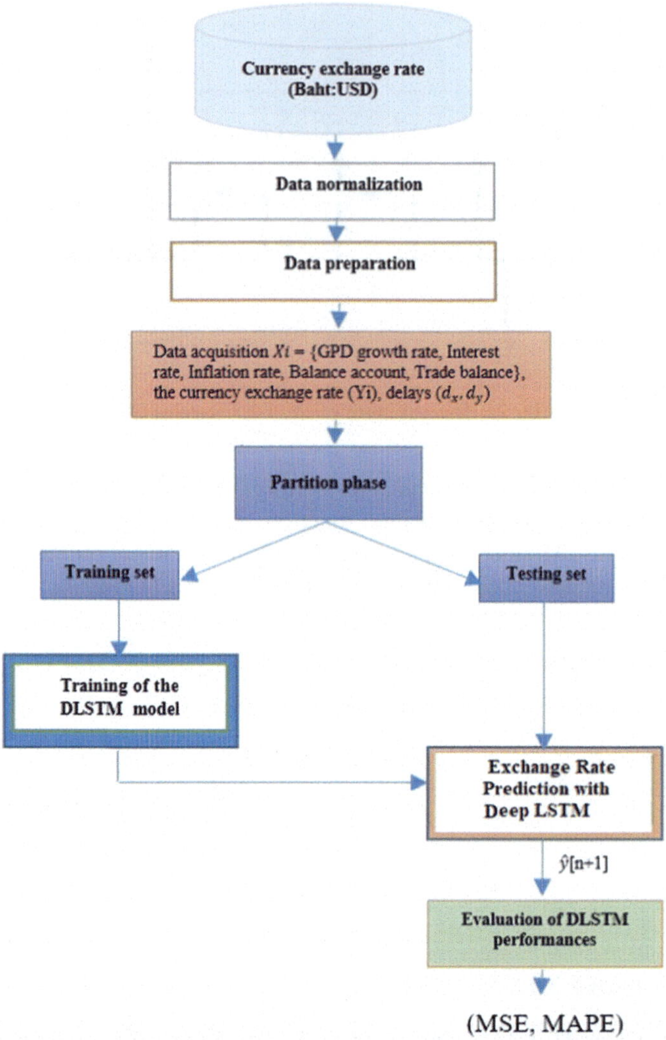

Fig. 3. The framework of the DLSTM model architecture.

denoted a $\hat{y}(t+1)$ currency exchange rate for the Baht:US dollar. The equation of the predicted output of the Deep LSTM model is given by

$$\hat{y}(t+1) = F(y(t), y(t-1), \ldots, y(t-d_y), x(t), x(t-1), \ldots x(t-d_x)) \quad (8)$$

where $F(\cdot)$ is the mapping function of the Deep LSTM model; $\hat{y}(t+1)$ is the predicted value from the present and previous values of $x(t)$ and the true previous values of the time series $y(t)$ at $(t+1)$ time. $y(t), y(t-1), \ldots, y(t-d_y)$ are the true previous values of the time series output; $x(t), x(t-1), \ldots, x(t-d_x)$ are the inputs of the Deep LSTM model. d_x is the maximum delay for the input, and d_y is the maximum delay for the output.

The architecture of our proposed Deep LSTM model is shown in Fig. 4.

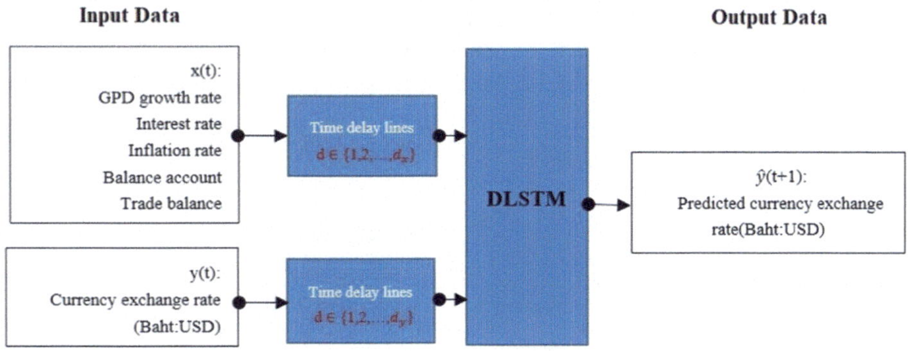

Fig. 4. The architecture of our proposed Deep LSTM model.

We proposed a Deep LSTM model architecture with the following 10 layers:

- Input layer: The sequence input has m neurons corresponding to five main economic indicators, including the GDP growth rate, interest rate, inflation rate, balance account and trade balance $\{m = 5\ d_x + d_y\}$.
- Three LSTM layers: There are three main layers of the LSTM (from layer 2 to layer 4) corresponding to the following numbers of neurons for each of LSTM layers: 50, 100 and 150. Each LSTM layer is linked to a dropout layer.
- Dropout layer: three dropout layers (from layer 5 to layer 7 inclusive) corresponding to a good probability default of 50%.
- Relu layer: the Relu layer performs a threshold operation on each element of the input where any value less than zero is set to zero.
- Fully connected layer: one fully connected layer with one neuron; the fully connected layer acts independently on each time step.
- Regression output layer: one neuron output for prediction of the currency exchange rate (Baht: USD).

We have created a set of options for training the networks by using Adam optimizer. The maximum number of epochs for training are 150, 200 and 250. The maximum iterations are 150, 200, 250 and a mini-batch size with 125 observations at each iteration. The maximum number of input delays (d_x) and the maximum number of output delays (d_y) are from 1 to 3. We specify the learning rate schedule of piecewise and a learning rate of 0.001.

4 Dataset and the Forecast Evaluation Measures

4.1 Dataset

The construction of the Deep LSTM model in this research depends on the historical currency exchange rates of Thai Baht per US dollar from the Bank of Thailand in the 10 years from 10 April 2009 to 10 April 2019 [24]. The currency exchange rates dataset is shown in Fig. 5.

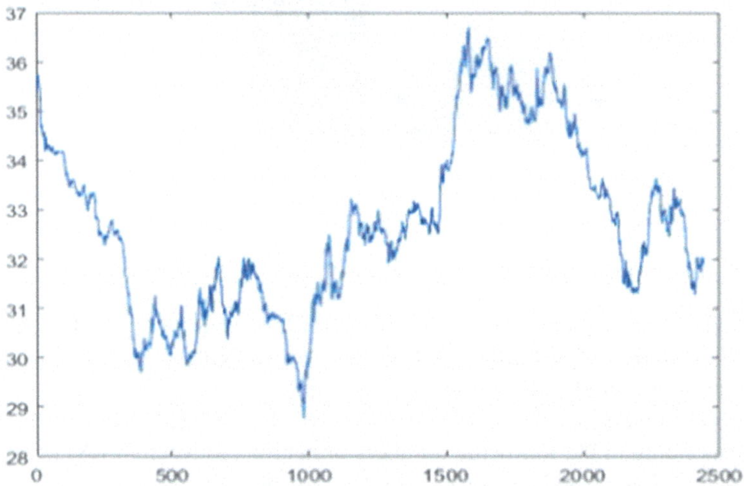

Fig. 5. The currency exchange rates of Thai Baht: USD in the 10 years (between 10 April 2009 and 10 April 2019).

In this research, we have used the main important economic indicators that can effect changes in the currency exchange rates, including the GDP growth rate, the interest rate, the inflation rate, balance account and trade balance. The dataset of currency exchange rates has 2442 rows. We have divided the dataset into a training set and a testing set with the ratio of 80:20 on dataset A and 70:30 on dataset B. The details of the dataset are shown in Table 1.

Table 1. The dataset details of the currency exchange rates of Thai Baht: USD.

Historical data of the currency exchange rates	Ratio (percentage) train: test	Amount of train: test	Total
Dataset A	80: 20	1953: 488	2442
Dataset B	70: 30	1709: 732	2442

4.2 Forecasting Evaluation Measures

The performance of our proposed Deep LSTM model is evaluated using *the mean squared error (MSE)* and *mean absolute percentage error (MAPE).* The defined variables have the following significance: y_t is the actual value, f_t is the forecasted value, $e_t = y_t - f_t$ is the forecast error and n is the size of the test set. The details of the forecasting evaluation measures are as follows:

- *Mean Square Error (MSE)* is the standard deviation measure of the forecast errors. In general, MSE values are used in prediction and regression analysis to verify the results of the experiment. It measures the difference between predictions and actual values. If the MSE value is close to zero, it means that it is a better value of the predictive results. The formula of the MSE is given by:

$$MSE = \frac{1}{n}\sum\nolimits_{t=1}^{n} e_t^2 \tag{9}$$

- *Mean Absolute Percentage Error (MAPE)* is a statistical measure to compute the mean absolute percentage error function for the forecast. This standard measure presents the percentage of the average absolute error that occurred. The concept of MAPE is separated from the measurement level by data conversion. The value of MAPE has very little deviation and does not indicate the direction of the error. The best value of MAPE is closer to zero. The equation to calculate the MAPE is given by:

$$MAPE = \frac{1}{n}\sum\nolimits_{t=1}^{n} \left|\frac{e_t}{y_t}\right| \times 100 \tag{10}$$

5 The Experimental Results

To fit the capabilities of the Deep LSTM model, its parameters have been specified in Table 2.

Table 2. The parameters of the DLSTM model.

Parameters	Parameter Values
Number of input neurons	18 (3 neurons for each of the delayed input variables {GDP growth rate, interest rate, inflation rate, balance account, trade balance} as well as 3 neurons for the delayed output variable)
Number of output neurons	1 (The currency exchange rate)
Epochs	150, 200, 250
Iterations	150, 200, 250
Number of LSTM layers	3
Number of neurons for each LSTM layers	50, 100, 150
Learning rate schedule	Piecewise
Learning rate	0.001
Number of Delays	1:3

We have trained the DLSTM model using two sets of data (A and B); then we have compared the experimental results. The sets of data have the following details:

- Dataset A: partition of the data with 80% for training and 20% for testing. The number of neurons for each LSTM layer has been chosen 50, 100 and 150 and the number of epochs has been chosen 150, 200 and 250. We have used the number of iteration belonging to the set {150, 200, 250}. We have chosen the learning rate of 0.001 and have used a single CPU. The maximum number of input delays (d_x) and the maximum number of output delays (d_y) belong to the set {1, 2, 3}.
- Dataset B: partition of the data with 70% for training and 30% for testing. The number of neurons for each LSTM layer has been chosen 50, 100 and 150 and the number of epochs has been chosen 150, 200 and 250. We have used the number of iteration belonging to the set {150, 200, 250}. We have chosen the learning rate of 0.001 and have used a single CPU. The maximum number of input delays (d_x) and the maximum number of output delays (d_y) belong to the set {1, 2, 3}.

We used the most common error measures to evaluate the accuracy of the variables; MSE and MAPE. Table 3 shows the comparison results of the error measures MSE and MAPE for both of the implemented models on dataset A. Table 4 shows the comparison results of the error measures MSE and MAPE for both of the implemented models on dataset B.

In Table 3, there are given the results for training the model on dataset A (with the divided dataset of 80% for training and 20% for testing). We found that the best structure of the DLSTM model is obtained with three main layers of DLSTM, where the number of neurons for each LSTM layer is 50, 100, 150. The maximum of number of epochs and the maximum number of iterations is 250. The best maximum number of input and output delays is 3. The best performances of MSE and MAPE are 0.0036 and respectively, of 0.3122. The comparison of the observed and forecasted values on the dataset A is shown in Fig. 6.

Table 3. The experimental results on dataset A as a function of the number of iterations (Deep LSTM model has three main LSTM layers with the corresponding numbers of neurons {50, 100, 150}).

Delays	Epochs/Iterations	MSE	MAPE
$d_x = d_y = 1$	150	0.00563	0.3488
	200	0.00455	0.3194
	250	0.00433	0.3231
$d_x = d_y = 2$	150	0.00560	0.3472
	200	0.00414	0.3174
	250	0.00418	0.3150
$d_x = d_y = 3$	150	0.00534	0.4328
	200	0.00497	0.3343
	250	**0.00369**	**0.3122**

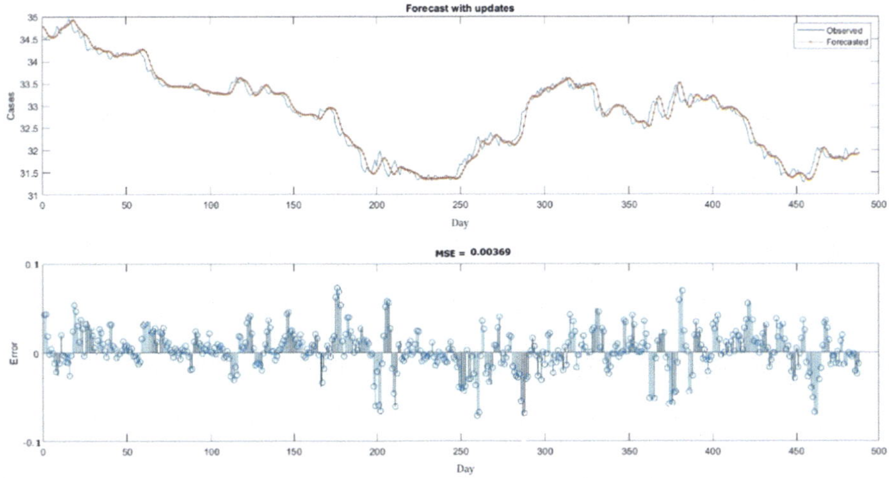

Fig. 6. The comparison of the observed and the forecasted exchange rates and corresponding MSE on test subset of dataset A.

On the other hand, we divided the experimental results on the dataset B in Table 4 into 70% of training data and 30% of testing data and found that the best structure of the Deep LSTM model is obtained with three main layers of Deep LSTM with the number of neurons for each LSTM layer is 50, 100, 150. The maximum number of epochs and the maximum number of iterations is 250. The best maximum number of input and output delays is 2. The best performances of MSE and MAPE are 0.00276 and 0.2844. The comparison of the observed and forecasted values on the dataset B is shown in Fig. 7.

Table 4. The experimental results on dataset B as a function of the number of iterations (Deep LSTM model has three LSTM layers with the corresponding numbers of neurons {50, 100, 150}).

Delays	Epochs/Iterations	MSE	MAPE
$d_x = d_y = 1$	150	0.00532	0.3391
	200	0.00507	0.3405
	250	0.00377	0.3087
$d_x = d_y = 2$	150	0.00792	0.3994
	200	0.00379	0.3032
	250	**0.00276**	**0.2844**
$d_x = d_y = 3$	150	0.00543	0.3418
	200	0.00516	0.3326
	250	0.00333	0.2943

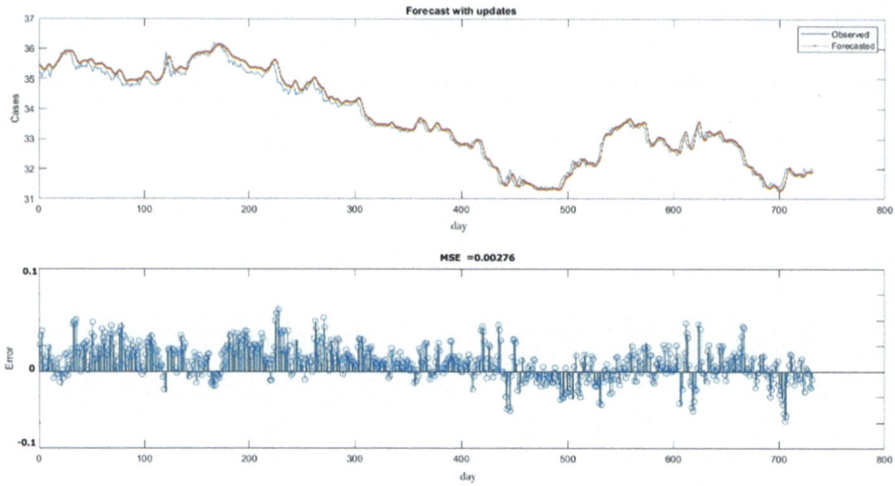

Fig. 7. The comparison of the observed and the forecasted exchange rates and corresponding MSE on test subset of dataset B.

From the experimental results regarding the application of the model on the datasets A and B, we can deduce that by increasing both the number of iterations, and also the number of epochs, and also the maximum number of delays and the number of hidden neurons in the network, significantly improve the results of the DLSTM model.

We have selected the best prediction values for our proposed Deep LSTM model obtained in Table 4 to compare with the experimental results of the NARX model [17], as researches related to forecasting exchange rate data (Baht: USD). We found that the best result of our proposed Deep LSTM model is obtained by using five neurons in the input layer with three main layers of Deep LSTM with the number of neurons for each LSTM layer is 50, 100, 150. The maximum number of epochs and iteration is 250. The best maximum number of input delays d_x and d_y is 2 and the best performance values of MSE and MAPE are 0.0027 and 0.2844. It is obvious that our proposed Deep LSTM model provides the best forecasting results. The details are shown in Table 5.

Table 5. Comparison results of our proposed DLSTM model with the NARX model.

Models	Performance measures		References
	MSE	MAPE	
DLSTM	**0.0027**	**0.2844**	This paper
NARX	0.0060	3.0010	[17][a]

[a] This research used the exchange rate dataset (Baht: USD) from 2009 to 2019, based on the main economic indicators including: GDP growth rate, interest rate, inflation rate, account balance and trade balance.

From Table 5, it is clear that the best experimental results of our proposed Deep LSTM model leads to better results than the considered benchmark NARX model based on feed-forward neural networks for forecasting exchange rate data (Baht: US dollar) by comparing MSE and MAPE values. The our proposed Deep LSTM model has the smallest MSE of 0.0027.

6 Conclusions

In this research, we have presented a deep learning approach for financial time series prediction. We have used a Deep Long Short-Term Memory (DLSTM) model for the issue of predicting the daily currency exchange rates of Thai Baht per US Dollars. The considered dataset is the historical currency exchange rates from the Bank of Thailand over the 10 years from April 2009 to April 2019. The experimental results given in Table 4 have shown that the best performance is obtained when dividing the dataset into a training/test ratio of 70:30 with the maximum number of 250 training epochs; we have used a learning rate of 0.001 and a learning rate schedule of the "Piecewise" variety. The best performance is obtained by choosing the number of delays equal to 3 (both for d_x and also for d_y). The best experimental result has very low error measures (MSE and MAPE) of 0.0027 and 0.2844, respectively. To evaluate the effectiveness of the model, we have considered the best prediction performances for the DLSTM model that was obtained in Table 4. There, the values are compared with the experimental results of already studied and published application of the classical NARX model for the same dataset by the authors of the present paper in [17]. It is obvious the advantage of the present DLSTM model over the NARX model regarding MSE and MAPE performances. We can also conclude that for our DLSTM predictive model, by increasing the number of hidden neurons, by increasing the number of iterations and also by increasing the number of epochs, one can obtain a significant improvement of performances. As future work, we intend to continue our research by evaluating the model performances for other databases. We also intend to compare the performances of our proposed DLSTM model to some benchmark statistical techniques as well as to an evolutionary computation method.

References

1. Damrongsakmethee, T., Neagoe, V.-E.: Data mining and machine learning for financial analysis. Indian J. Sci. Technol. **10**, 1–7 (2017)
2. Bakir, H., Chniti, G., Hedi, Z.: E-commerce price forecasting using LSTM neural networks. Int. J. Mach. Learn. Comput. **8**, 169–174 (2018)
3. Gyamerah, S.: Trend forecasting in financial time series with indicator system. J. Appl. Stat. 0–14 (2019)
4. Adhikari, R., Agrawal, R.K.: An Introductory Study on Time Series Modeling and Forecasting. Lab Lambert Academic Publishing (2013)
5. Siami-Namini, S., Tavakoli, N., Siami-Namin, A.: A comparison of ARIMA and LSTM in forecasting time series. In: 2018 17th IEEE International Conference on Machine Learning and Applications (ICMLA), pp. 1394–1401. IEEE (2018)

6. Selvin, S., Vinayakumar, R., Gopalakrishnan, E.A., Menon, V.K., Soman, K.P.: Stock price prediction using LSTM, RNN and CNN-sliding window model. In: 2017 International Conference on Advances in Computing, Communications and Informatics (ICACCI), pp. 1643–1647 (2017)
7. Roondiwala, M., Patel, H., Varma, S.: Predicting stock prices using LSTM. Int. J. Sci. Res. 1–4 (2018)
8. Sewell, M.V.: Application of machine learning to financial time series analysis (2017)
9. Neagoe, V.-E., Latan, R.-S., Latan, I.-F.: A nonlinear neuro-fuzzy model for prediction of daily exchange rates. In: International Symposium on Soft Computing for Industry held at the 6th Biannual World Automation Congress WAC 2004, pp. 573–578. IEEE, Seville (2004)
10. Neagoe, V.-E., Ciotec, A.-D., Cucu, G.-S.: Deep convolutional neural networks versus multilayer perceptron for financial prediction. In: 12th International Conference on Communications (COMM), pp. 201–206 (2018)
11. Nagpure, A.R.: Prediction of multi-currency exchange rates using deep learning. Int. J. Innov. Technol. Explor. Eng. **8**, 316–322 (2019)
12. Phaladisailoed, T., Numnonda, T.: Machine Learning models comparison for bitcoin price prediction. In: 10th International Conference on Information Technology and Electrical Engineering (ICITEE), pp. 1–6 (2018)
13. Muhammad, G.: Foreign currency exchange rates prediction using CGP and recurrent neural network. Int. Conf. Futur. Inf. Eng. **10**, 239–244 (2014)
14. Kim, S., Kang, M.: Financial series prediction using attention LSTM. arXiv. vol. 1, pp. 1–10 (2019)
15. Bao, W., Yue, J., Rao, Y.: A deep learning framework for financial time series using stacked autoencoders and long- short term memory. PLoS One. 1–24 (2017)
16. Cocianu, C.L., Avramescu, M.-S.: New approaches of NARX-based forecasting model. A case study on CHF-RON exchange rate. Inform. Econ. **22**, 5–13 (2018)
17. Damrongsakmethee, T., Neagoe, V.-E.: A neural narx approach for exchange rate forecasting. In: Proceeding International Conference on 11th Ed. Electronic Electronics, Computers and Artificial Intelligence (ECAI), Pitesti, "Romania", pp. 1–6 (2019)
18. Karim, F., Majumdar, S., Darabi, H.: Insights into LSTM fully convolutional networks for time series classification. IEEE Access **7**, 67718–67725 (2019)
19. Chen, K., Zhou, Y., Dai, F.: A LSTM-based method for stock returns prediction : a case study of china stock market. In: 2015 IEEE International Conference on Big Data (Big Data), pp. 2823–2824. IEEE (2015)
20. Komkul, P.: Forecasting cassava starch price in Thailand by using time series models. J. KMUTNB **27**, 805–820 (2017)
21. Ye, Y.: Study on exchange rate forecasting using recurrent neural networks. Int. J. Econ. Financ. Manag. Sci. **5**, 300–303 (2017)
22. Pulver, A.: LSTM with working memory. In: 2017 International Joint Conference on Neural Networks (IJCNN), pp. 845–851. IEEE (2017)
23. Karim, F., Majumdar, S., Darabi, H., Chen, S.: LSTM fully convolutional networks for time series classification. IEEE Access **6**, 1662–1669 (2018)
24. Historical Foreign Exchange Rates. https://www.bot.or.th/english/_layouts/application/exchangerate/exchangerate.aspx. Accessed 01 Dec 2019

Multi-layer Global Tracing on Base of Bioinspired Method

Boris K. Lebedev, Oleg B. Lebedev(✉), and Ekaterina O. Lebedeva

Southern Federal University, Rostov-on-Don, Russia
lebedev.b.k@gmail.com, lebedev.ob@mail.ru,
ekaterina.o.lebedeva@yandex.ru

Abstract. The composite architecture of a multi-agent bionic search system is proposed to solve the multi-layer global tracing problem based on an ant colony. The method is based on the approach consisting in the decomposition of the connecting network and its representation as a set of two-terminal compounds. At the first stage of the method under consideration, all the circuits are traced directly into a single-layer solution space, and at the second stage the "distribution of compounds into layers" is performed using the same method. The goal is to find such a distribution by region, which satisfies the overflow constraints and the cost of via is minimized. The paper discusses combinatorial approaches to the construction of the global tracing. Both approaches are based on the method of the ant colony. The testing was done on benchmarks. The time complexity of the algorithm is $O(n^2)$. Experiments have shown that the time indicators of the developed algorithm exceed those of the compared algorithms with the best values of the objective function.

Keywords: Solution space · Connection tracing · Multi-layer global tracing · Single-layer tracing · Distribution of compounds in layers · Combinatorial method · Ant colony · Solution search graph

1 Introduction

When performing the design of very large integrated circuits (VLSI), one of the problems that needs to be solved using intelligent computer-aided design (CAD) systems, the problem arises of laying connection routes. In general, when designing VLSI, there is a direct correlation – the higher the result of connection tracing, the overall characteristics of the designed topology are also improved [1, 2]. The connection tracing problem is solved in two stages – global [3, 4] and detailed tracing [5]. Modern technologies can significantly increase the number of components of the designed topology in an area of the same size. With an increase in metallization layers, the number of transitions from layer to layer increases, and as a result, the total area of through transitions between layers increases. These and a number of other factors lead to the problem of trace resource shortage.

The percentage of trace resource can be increased by developing new algorithms that perform global connection tracing. These algorithms should have high efficiency both when finding high quality results, and when the execution time of the algorithm.

R. Silhavy (Ed.): CSOC 2020, AISC 1225, pp. 499–508, 2020.
https://doi.org/10.1007/978-3-030-51971-1_41

Compound tracing is fairly well researched, and there are a large number of methods to solve it. But the emerging modern requirements for VLSI dictate the need to develop new algorithms to solve the trace problem [1, 4]. In the general case, the problem of multi-layer global tracing is solved in two stages. The first stage (it is called preliminary) performs the division of the entire area on which the trace takes place into separate rectangular parts (global cells) of each layer. The second stage carries out the spacing of chains in these areas and forms a list of connections on the border of each area. The main goal of global tracing is to increase the traceability of the switching area (SA) when designing the VLSI topology.

An approach to global tracing was chosen, in which at the first stage all chains are traced directly in a single-layer solution space. At the second stage, the stage of "distribution of compounds by layers" is carried out. This reduces the combinatorial complexity of the global trace algorithm. Many well-known single-layer global tracing algorithms are classified into sequential and parallel methods [1–5]. It is possible to reduce the global trace problem to the combinatorial assignment problem. This procedure is used to trace multiple connections simultaneously. Basically, algorithms incorporating combinatorial procedures apply mechanisms based on blind random search. The main drawback of such algorithms is that the algorithms get into the "local hole", the values of which are not a global optimum. Another disadvantage is that to evaluate the effectiveness of such algorithms, objective functions are used in which the parameters of the circuits (their total length) are considered in priority, which leads away from the main goal of global tracing [6, 7].

To solve these problems, a combined approach is proposed: a new multi-agent method of intellectual optimization, based on collective intelligence modeling, in which all connections are considered at each iteration, and the criterion takes into account the distribution of switching field resources [1–5].

2 Formulation of the Problem

Let $A^l = \{a_i^l | i = 1, 2, \ldots\}$ be the set of regions in l layers. Let $B^l = \left\{b_j^l | j = 1, 2\ldots\right\}$ be the set of boundaries between the regions. Solving the problem of spacing interconnects between regions in l layers, we will use the graph $G^l = (X^l, U^l)$, where $l \geq 2$ – is the number of layers, as a model of a switching area (SA). The vertices of $x_i^l \in X^l$ are the regions of $x_i^l \in X^l$. The set of edges U^l is divided into subsets of the boundary U_b^l and interlayer U_v^l edges. $U_b^l \cap U_v^l = \varnothing$, $U_b^l \cup U_v^l = U^l$. For each edge u_k, the weight β_k is set equal to the throughput of the common boundary between adjacent regions.

There are many $T = \{t_s | s = 1, 2, \ldots, n_s\}$ traces. Any route connects many areas of the SA, where there are terminals connected by this route. On the graph G^l, the set of vertices $t_s \in T$ is associated with the set of regions connected by the path $X_s \in X$. To spread the t_s path into regions is to form a connecting network in the graph G^l on the set of vertices X_s. The task is to build a connecting network in the form of a minimal Steiner tree. The most widespread approaches aimed at improving traceability associated with the distribution of resources in the switching environment. Each route t_s after laying it over regions uses a certain amount of ψ_s resources on each boundary it intersects. Let $E_k \in E$ be the set of connecting networks formed for the set of paths $T_k \in T$ containing the edge $u_k \in U^l$. We introduce the parameter γ_k the total amount of resources. In other words, the total amount of resources required for the traces of the set T_k to go through the border b_k:

$$\gamma_k = \sum \psi_s(s|t_s \in T_k). \quad (1)$$

Each edge $u_k \in U^l$ of the graph G^l corresponds to $\rho_k = \beta_k - \gamma_k$, which is called the resource reserve of the edge u_k [6–8]. In the graph G, we define the minimum value of ρ_k and denote it by ρ_{min}, i.e. $\rho_{min} \to \forall k[\rho_{min} \leq \rho_k]$. The main goal of the global tracer is to find the maximum value of ρ_{min}. A method is chosen in which the first step is to transform the graph G^l into a single-layer graph G^1. The second step is to lay the chains in layers. The article describes the developed general combinatorial method for solving the problems of tracing in one layer and the distribution of compounds between layers, based on the ant colony method. The goal is to find such a spacing of the chains into subdomains and layers, so that restrictions on overflow are fulfilled, and the total cost of interlayer transitions is minimal.

The vertices X^1 of the graph G^1 connected by the path $t_s \in T$ by the Prima algorithm synthesize the minimum connecting tree (MCT) $R = \{R_k|k = 1, 2, ..., n - 1\}$, $R_k = (x_i^1, x_j^1)$ is the edge of the MCT (Fig. 1). Denote by the s-path the trace in the graph G^1 connecting the two vertices (x_i^1, x_j^1). The edge R_k on the graph G^1 has a list $S_k = \Gamma(R_k)$ of s-path variants equal to $S_k = \{s_{kz}|z = 1, 2, ..., n_z\}$. To find S_k, two rules follow: the path s_{kz} must be minimal; other s-paths must match each other.

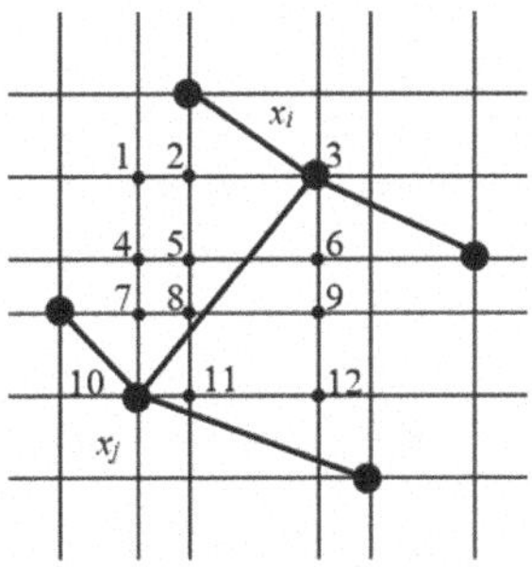

Fig. 1. A set of possible options for s-edges

If the vertex x_j^1 is lower and to the left of the vertex x_i^1, the agent forming the s-route from the vertex x_i^1 to x_j^1 can move down and to the left. The agent forming the s-route from the vertex x_i^1 and x_j^1 can move up and to the right. Moreover, the movement of the agent is limited by the borders of the rectangle. Its diagonal connects the vertices x_i^1 and x_j^1.

Such rules allow all alternatives of the *s*-route connecting two vertices x_i^1 and x_j^1 to have the same distance but differ in the composition of the edges contained in the path. In Fig. 1, for the edge R_k connecting the vertices x_i^1 and x_j^1, there are 10 variants of *s*-routes laid through points 1 through 12:

s_{k1} = (3, 6, 9, 12, 11, 10), s_{k2} = (3, 2, 1, 4, 7, 10), s_{k3} = (3, 2, 5, 4, 7, 10), etc.

Using such a combinatorial method, initially for each edge R_k of each trace, a list $S_k = \Gamma(R_k)$ of *s*-path variants $S_k = \{s_{kz}|z = 1, 2, ..., n_z\}$ is synthesized. A method for synthesizing *s*-routes is presented above.

3 Single Layer Tracing Combinatorial Algorithm

The task of single-layer tracing is reduced to the task of choosing for each edge R_k of each chain a variant of the *s*-route.

In accordance with the methodology presented above, on column *G*, an array of all two-pin interconnects of all routes $D = \{d_\alpha|\alpha = 1, 2, ..., n_d\}$ is synthesized. On the decision search graph (DSG) $H = (V, E)$, a solution search is performed. The number of nodes $n_d \cdot n_v$, where n_d is the number of two-pin interconnects of all circuits; n_v is the number of interconnect alternatives d_α. Graph *H* consists of stages (Fig. 2). The stage models the group of vertices V_α. In fact, the group number is the number of the two-pin interconnect d_α. The vertices of the group V_α are variants (*s*-paths) of the two-terminal interconnect d_α. The number of vertices in the group V_α is n_v – the number of variants of a two-terminal connection. A connecting vertex $o_{\alpha+1}$ is placed between each pair of neighboring groups V_α and $V_{\alpha+1}$. The vertex o_α is connected by the oriented edges emanating from it with each vertex of the group V_α. The peak o_α is the starting peak. The number of vertices o_α is n_d. The number of vertices of the graph *H* is defined as $n_e = n_d \cdot n_v + n_d = (n_d + 1) \cdot n_v$. Each vertex of the group V_α is connected by an outgoing oriented edge with the vertex $o_{\alpha+1}$. The vertex o_α has n_v edges incident, each vertex o_α (for α from 2 to n_d) has $2n_v$ edges incident. The number of edges of the graph *H* is defined as $n_e = (n_d - 1) \cdot 2n_v + n_v = (2n_d - 1) \cdot n_v$.

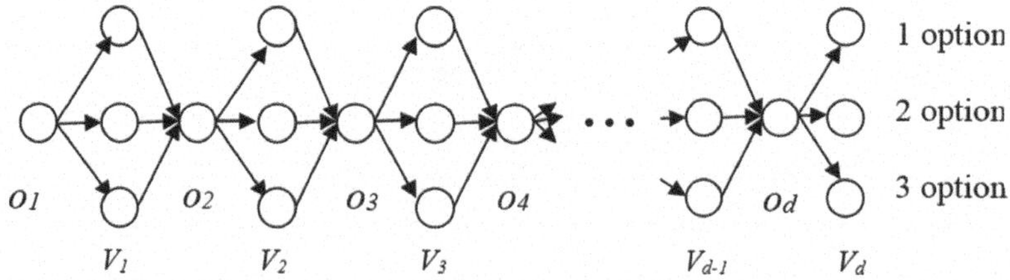

Fig. 2. Decision search graph *H*

The goal – it is required in column *H* to find a path that contains one vertex from each group. This path will contain the selected options (*s*-routes) of all two-pin interconnects of all routes. In the stages, there can be a different number of vertices, which allows one to introduce restrictions on the number of variants of each connection. In Fig. 3 marked lines show the path formed by the ant.

Fig. 3. The path in graph H that determines the selected options for all connections of all chains

Algorithm of Ant Colony Behavior.

1. The conductor $t_i \in T$ on the graph $G^1 = (X^1, U^1)$ is associated with the MCT $R_i = \{r_{ik} | k = 1, 2, \ldots, n_i\}$.
2. For the edge r_{ik} on the graph G, we define the list $S_{ik} = \Gamma(r_{ik})$ of s-path variants, $S_{ik} = \{s_{ikz} | z = 1,2,\ldots,m\}$.
3. We construct a general list of all interconnects of all routes $D = \{d_\alpha | \alpha = 1, 2, \ldots, n_d\}$ and a solution search graph $H = (V, E)$. Synthesized n_d groups of V_α vertices over all interconnects of all routes. A connection is introduced between the vertices of the graph $H \in V_\alpha$ and the variants of s-paths $s_{ikz} \in S_{ik}$.
4. On all edges of the graph H, the same amount of pheromone is applied, equal to Q/n_e, where $n_e = |E|$. The number of iterations n_l is set.
5. $l = 1$. (l is the iteration number).
6. The number of ants n_a that are at the initial vertex o_1 is determined.
7. (Agent Algorithm). Each agent a_σ finds the path M_σ on the graph H, from the vertex o_1, through all the vertices o_2–o_d and containing one vertex from each group (V_1-V_d). The criterion F_σ of the path M_σ is calculated.
8. After all the agents (ants) have constructed the paths, each agent a_σ applies pheromone in the column H on the edges of the obtained path M_σ in the column H in the amount $\tau_\sigma(l) = Q \cdot F_\sigma$, where Q is the initial amount of pheromone applied by the agent a_σ on the edges of the path M_σ.
9. Next, the pheromone on the edges of the graph H evaporates in accordance with the formula: $h_j = h_j(1 - \rho)$, where ρ is the update coefficient, h_j is the total amount of pheromone deposited by the agents on the edge $e_j \in E$.
10. Selecting the best solution found at all iterations performed.
11. In the case of iteration, ($l = n_l$), then the end of the algorithm, otherwise $l = l + 1$ and go to step 6.

Agent Algorithm (clause 7).

1. $\sigma = 1$. (σ is the agent number).
2. $t = 1$, $M_\sigma(t) = \varnothing$. $o(t) = o_t$. *(t is the step number, ($o(t)$ is the connecting vertex that entered the route $M_\sigma(t)$ at step t).*
3. The vertex o_t are included in the path $M_\sigma(t)$. Agent a_σ finds the edges $E_\sigma(t) \in E$ of the candidates for inclusion in their path $M_\sigma(t)$ connecting the vertex $o(t)$ in the graph H with the vertices of the neighboring group V_t.
4. The cost of each edge $e_j \in E_\sigma(t)$, is calculated by the formula $CR_j = (h_j)^\lambda \cdot (\alpha(e_j))^\gamma$, where h_j is the total pheromone volume on the edge e_j;

$\alpha(e_j)$ the minimum weight among the edges of the s-route corresponding to the vertex v_z;
λ and γ are control parameters.

5. For each edge $e_j \in E_\sigma(t)$, the probability P_j of its inclusion in the path under construction $M_\sigma(t)$ is calculated. $P_j = CR_j/\Sigma_j(CR_j)$ for $(j|e_j \in E_\sigma(t))$.
6. Randomly, in accordance with the calculated probabilities, there is an edge $e_j \in E_\sigma(t)$, which is introduced into the path $M_\sigma(t)$.
7. The vertex $v_z \in V_t$ at which the edge e_j enters is remembered. The edge e_j and the vertex v_z are included in the route $M_\sigma(t)$.
8. If the path $M_\sigma(t)$ is found, then go to step 9, otherwise $t = t + 1$, and go to step 3.
9. Routes are placed along the cells of the switching medium based on the path options M_σ. The value of the objective function $F_2 = w_{min}$ is calculated, which is an estimate of F_σ of the path M_σ.
10. If $\sigma < (n_a - 1)$, then $\sigma = \sigma + 1$ and go to step 2, otherwise go to step 11.
11. The end of the agents finding the paths.

Using a combinatorial approach, it must be borne in mind that each individual agent finds one solution.

4 Layering Compounds

We formulate the problem of distribution over layers in the following form. The stated problem is as follows: the l-layer graph $G^l = (X^l, U^l)$, the solution of the single-layer trace problem $E^1 = \{e_s^1|s = 1,2,...,n_s\}$ on (G^1, U^1). Each e_s^1 network is implemented as a Steiner tree. The solution is to lay out fragments of each e_s^1 network into layers. In this case, restrictions limiting the maximum congestion of the routes are taken into account. The main optimization criterion is defined as:

$$F_3 = k_1\Theta - k_2\rho_{min}, \tag{2}$$

where Θ is the total cost of interlayer transitions, k_1, k_2 – control parameters.

The task of distributing the compounds across the layers is to minimize the F_3 index, taking into account the restrictions on maximum congestion [9]. Interlayer transitions can be located at the tops of the Steiner tree. Place a terminal at each vertex of the Steiner tree. In this regard, each Steiner tree is represented as a set of two-term straight sections. Each section is limited by two terminals (contacts) k_i, k_j.

At the stage of splitting the compounds into layers, an array $D = \{d_\alpha|c = 1, 2, ..., n_d\}$ of all two-terminal connections of all chains represented as a Steiner tree on a single-layer graph is formed. Each two-terminal connection d_α on the graph $G^1 = (X^1, U^1)$ corresponds to a pair of terminals connected by it (k_i, k_j). Suppose that the two-terminal connection d_α connects the contacts k_i and k_j, the contact k_i contained in the vertex x_i is located in the layer l_1, the contact k_j contained in the vertex x_j is located in the layer l_2, $l_1 \neq l_2$. Then d_α can be placed in one of the layers l_k in the range from l_1 to l_2 with minimal loading of the edge connecting k_i and k_j. At the vertex x_i of the Steiner tree there is an interlayer transition $v_{i\alpha}$ connecting the terminal k_i with d_α, and at the top x_j of the Steiner tree there is the interlayer transition $v_{j\alpha}$ connecting the terminal k_j with d_α.

We define $\delta_{i\alpha}$ as the cost of interlayer transition $v_{j\alpha}$. Then the total cost Θ of the interlayer transitions is defined as:

$$\Theta = \sum\nolimits_i \sum\nolimits_\alpha (\delta_{i\alpha}). \tag{3}$$

The solution is to find such a separation of the chains into layers, in which the restrictions on overflow are not violated, and the total cost of transitions from layer to layer is minimal. Based on the analysis, for each two-terminal compound d_α, a set of layer options is determined in which it can be placed. The formation of solutions is carried out on the solution search graph $H = (V, E)$, which consists of several stages and whose overall composition is similar to the composition of the decision search graph (see Fig. 2) during single-layer tracing. The number of nodes is defined as n_d•n_v, where n_d, is the number of distributed two-pin interconnects of all circuits; n_v is the number of layers to which interconnects are assigned. Each group of vertices V_α forms a stage. The vertices of stage V_α correspond to the numbers of the layers into which the interconnect d_α can be placed. It is required in column H to select one vertex from each group. These vertices determine the layout options for all traces in layers. Note that in the stages there can be a different number of vertices, which allows you to enter restrictions on the number and number of layers into which connections can be assigned.

The algorithm for the behavior of an ant colony during the distribution of compounds in layers is identical to the algorithm for the behavior of an ant colony during single-layer tracing.

The ant algorithm for the distribution of compounds in layers is slightly different in paragraphs 4 and 9 of the ant algorithms for single-layer tracing.

Points 4, 9 in the ant algorithm when distributing compounds in layers.

4. The value CR_z of each vertex $v_z \in V_\alpha(t + 1)$ is calculated.

Let the vertex $v_z \in V_\alpha$ be the number of the layer i where the two-pin interconnect d_α is located, which has the smallest throughput α_z. By expression:

$CR_z = (h_z)^\lambda \cdot (\alpha_z)^\gamma$ for multiplicative convolution,

$CR_z = \lambda \cdot (h_z) + \gamma \cdot (\alpha_z)$, with additive convolution, the estimated cost CR_z of each vertex is $v_z \in V_\sigma(t + 1)$, where h_z is the amount of pheromone deposited on the peak v_z, λ and γ are control determined by experiment.

9. Interconnections are layered according to the alternatives based on the formed path M_σ. Criterion F_3 is calculated, which is an estimate of F_σ of the path M_σ.

The route defined by the agent should include one vertex from each stage of V_α. The solution is completely determined by the composition of the peaks included in the route and does not depend on the sequence of stages passing by the ant. If violations of the restrictions on maximum congestion occur when the connections are laid out in layers in accordance with the constructed route, then such decisions are discarded.

5 Experimental Research

To determine the effectiveness of the proposed global tracing algorithm, a comparison was made with existing algorithms today [5–16]. For this, standard tests and fixed circuits (the so-called benchmarks) were used. In the analysis, it was determined that the tracers used by the second method to take into account the resources of the switching environment mainly use alternatives of "greedy heuristics", or methods based on linear integer programming, which require a lot of time. The algorithm considered in [14] realizes the solution of the problem as follows. First, the generation of the route posting procedure occurs. This is based on the result of a single layer trace. Next, the solution to the problem of spacing connections into layers is based on dynamic programming procedures. The main disadvantage of this method is the task of choosing the queue of laid traces.

Testing of the proposed GTAC algorithm took place on tests for 6-layer schemes used on ISPD for competitive testing. The results were compared with the well-known results of the MaizeRouter, BoxRouter and FGR tracers. A comparative analysis was performed according to the overflow indices γ on the edges of the model of the switching medium (MSM) (l-layer graph $G^l = (X^l, U^l)$): total - total overflow γ on the edges of the MSM; *max* – maximum overflow among the MSM ribs and the indicator of the total length of the connections. The calculation of the connection length was made according to the rules of the ISPD competition. The connection between neighboring cells on the same layer is one unit of wirelength. The connection between adjacent layers (i.e. end-to-end) is three wirelength units. Table 1 shows the results of comparative experiments.

Table 1. The results of experimental studies

Bench		adate1	adate2	adate3	adate4	adate5	newbl1	newbl2	newbl3
MaizeRouter + COLA	OF	0 (0)	0 (0)	0 (0)	0 (0)	2 (2)	1349 (3)	0 (0)	32573 (207)
	WL	127.49	120.94	259.97	237.00	338.90	118.02	180.50	235.70
BoxRouter + COLA	OF	0 (0)	0 (0)	0 (0)	0 (0)	0 (0)	420 (2)	0 (0)	37895 (364)
	WL	122.07	118.43	262.67	266.70	321.96	115.64	179.73	229.67
FGR + COLA	OF	63 (2)	43 (2)	0 (0)	0 (0)	2266 (2)	2306 (4)	0 (0)	50830 (373)
	WL	119.7	116.8	257.2	234.89	316.91	114.21	172.51	224.22
BoxRouter 2.0 + COLA	OF	0 (0)	0 (0)	0 (0)	0 (0)	0 (0)	389 (2)	0 (0)	39158 (364)
	WL	89.3	94.7	218.5	191.9	273.8	94.7	141.1	172.7
FGR 1.1 + COLA	OF	0 (0)	0 (0)	0 (0)	0 (0)	0 (0)	243 (2)	0 (0)	38482 (399)
	WL	91.9	88.6	191.2	179.5	262.17	91.29	133.59	159.01
GTAC	OF	0 (0)	0 (0)	0 (0)	0 (0)	0 (0)	219 (2)	0 (0)	31994 (206)
	WL	85.23	87.13	187.7	178.1	237.9	82.87	119.8	149.3

* GTAC – global tracing, ant colony.

The contents of the table include: the name of each test task – (benchmarks); OF (overflow) overflow value: total – (total) and in brackets the maximum among MSM ribs– (max): total length of WL connections (wirelength) in arbitrary units.

In each test case, the results were better by 2–4%. Compared with the enhanced versions of BoxRouter 2.0 and FGR 1.1 [9, 10], the results improved by 1.5–2%.

Given by COLA [13], it gets better results than other tracers; nevertheless, the developed algorithm allows one to find results 1–2% better.

The dependence of the running time of the developed algorithm on the number of traced circuits is $O(n^2) - O(n^3)$.

6 Conclusion

The paper considers the combinatorial method of global tracing, based on the consideration of the connecting network and modeling it in the form of associations of two-terminal connections. An approach was taken to perform global tracing, which first translates all traces into a single-layer solution space. Stage two – carries out the separation of interconnects in layers. This reduces the combinatorial complexity of the global trace algorithm. A single combinatorial algorithm based on an ant colony has been developed, used both in the first and second stages of multi-layer global tracing. In contrast to the canonical paradigm of ACO, the decision search graph is presented in the form of a multi-stage structure. The solution is completely determined by the composition of the peaks included in the route and does not depend on the sequence of stages passing by the ant. When solving the problem under consideration, due attention should be paid to the constraints responsible for "chain overflow." The global goal is to search for such separation between layers, in which the restrictions on overflow are not violated and the cost of transitions from layer to layer is minimal. The algorithms developed and presented in the work guarantee layering that does not violate the specified overload restrictions.

Acknowledgements. This research is supported by grants of the Russian Foundation for Basic Research of the Russian Federation, the project №. 18-07-00737.

References

1. Karpenko, A.P.: Modern search engine optimization algorithms. Algorithms inspired by nature: a tutorial, 448 p. Publishing House MSTU. N.E. Bauman, Moscow (2014)
2. Davydov, V.V.: A method of multi-level uniform grids for spatial searching. In: Problems of Developing Promising Micro- and Nanoelectronic Systems, no. I, pp. 32–37. Publishing House IPPM RAS, Moscow (2019)
3. Gavrilov, S.V., Ryzhova, D.I., Vasilyev, N.O., Zhukova, T.D.: Algorithm of structural optimization for digital CMOS circuits. In: Problems of Developing Promising Micro- and Nanoelectronic Systems, no. I, pp. 42–47 (2019)
4. Gao, J.R., Wu, P.C., Wang, T.C.: A new global router for modern designs. In: Proceedings of Asia South Pacific Design Automation Conference, pp. 232–237 (2008)
5. Lebedev, B.K., Lebedev, V.B.: Global tracing based on swarm intelligence, no. 7, pp. 32–39. Izvestia SFedU. Publishing house SFedU (2010)
6. Schemelinin, V.M.: Automation of topological design of LSI, 145 p. MIET, Moscow (2001)

7. Lebedev, O.B.: Channel tracing using the ant colony method, no. 2, pp. 46–52. Izvestia SFedU. Publishing house SFedU (2010)
8. Agasiev, T.A., Karpenko, A.P.: Modern technology of global optimization. In: Information Technologies, no. 6, pp. 370–386. Publishing MSTU. N.E. Bauman, Moscow (2018)
9. Cho, M., Pan, D.Z.: BoxRouter: a new global router based on box expansion and progressive ILP. In: Proceedings of Design Automation Conference, pp. 69–83 (2006)
10. Lebedev, B.K., Lebedev, O.B.: Multi-layer global tracing by the method of collective adaptation. In: V All-Russian Scientific and Technical Conference "Problems of Development of Promising Micro-and Nanoelectronics Systems - 2012". Collection of Papers, pp. 251–257. IPPM RAS, Moscow (2012)
11. Lee, T.-H., Wang, T.-C.: Congestion-constrained layer assignment for via minimization in global routing. IEEE Trans. Comput.-Aided Design Integr. Circ. Syst. **27**(9), 1643–1656 (2008)
12. Pan, M., Chu, C.: FastRoute 2.0: a high-quality and efficient global router. In: Proceedings of Asia South Pacific Design Automation Conference, pp. 250–255 (2007)
13. Cho, M., Pan, D.Z.: BoxRouter 2.0: architecture and implementation of a hybrid and robust global router. In: Proceedings of International Conference on Computer-Aided Design, pp. 503–508 (2007)
14. Moffitt, M.D.: MaizeRouter: engineering an effective global router. In: Proceedings of Asia South Pacific Design Automation Conference, pp. 226–231 (2008)
15. FGR 1.1. http://vlsicad.eecs.umich.edu/BK/FGR
16. Wang, X.: Hybrid nature-inspired computational analysis. Doctoral dissertation. Helsinki University of Technology, TKK Dissertations, 161 p. (2009)

The Impact of the Advanced Technologies over the Cyber Attacks Surface

Willian Dimitrov(✉)

University of Librarian Studies and Information Technologies,
bul. Tsarigradsko Shosse 119, 1574 Sofia, Bulgaria
v.dimitrov@unibit.bg
http://www.unibit.bg

Abstract. The article explores the changes are occurring in the security of the information and communication systems as a result of a state-of-the-art-technologies usage. The research is centered over the impact of new technologies on the cyber attack surface. The focus is on the synergy of a set of advanced technologies with existing digital environment and the consequences for cyber security particularly on the cyber attack surface extension. The study summarizes the evidences for increasing the surface area for attacks due to the advent of the listed advanced technologies. Explores how emerging technologies complicate cyber protection, introducing new risks as consequence from increasing possibilities for malicious threats. Indicate guidelines about future research related to refinement of legislation that is intended to regulate the cyber-physical world.

Keyword: Cyber security attack surface technologies

1 The Impact of Ever-Expanded New Technologies on the Cyber Security

The ICT (Information and Communication Technology) community realizes the impact of ever-expanding emerging technologies on cyber security. When new technology enters into circulation, it turns out that the surface for attacks by malicious actors is increasing [1].

The evolve of security from browsers to virtualization and cloud services has gone that way [2–6]. The article consists a justified summary of the proofs for an attack surface extension because of emerging technology's distribution.

The term surface attack is discussed in detail in [7]. For the purposes of this article, the definition from [8] is used. An attack surface is the total sum of vulnerabilities that can be exploited to carry out a security attack. Adding ICT artefacts into organizations' digital environment increases the attack surface. It leads to a new security measures. Then the need arises to add more controls to compensate for the newly extended attack surface [9]. New exposures are created every day: defects in purchased and internally developed software,

R. Silhavy (Ed.): CSOC 2020, AISC 1225, pp. 509–518, 2020.
https://doi.org/10.1007/978-3-030-51971-1_42

unpatched servers and endpoints, vulnerabilities in network and security devices and mistakes made by overworked administrators. Most are never exploited, but all of them represent potential attack vectors [10].

Methodology of the Research. The summary of evidences of surface extensions for attack is based on review of scientific publications, bulletins from cyber security institutions, empirical experience of implementing cyber protection solutions in ICT systems, examination of technological documentation of manufacturers and conversations with developers and researchers. The method used is systematizing the effects of new technology entry on the attack surface and the security measures that reduce the risks and overall consequences. The list of explored technologies includes Internet of Things (IoT), Artificial Intelligence, Distributed Ledger Technology (DLT), VR (Virtual Reality) and containers.

The surface for attacks is separated on three types of vulnerabilities: **known in the existing technologies**; **unfamiliar in the existing technologies** that are yet to be discovered and **unknown so far** due to the cutting-edge technologies that are entering and positioning themselves in the digital world.

The weaknesses in the old protocols and the vulnerabilities are in huge amounts of devices and applications. Known and unfamiliar vulnerabilities are internal characteristic of the digital space and need to be taken into account in the security assessment of connected cutting-edge technologies.

The interest in the case is concentrated on the third type of vulnerability that forms the increase in attack. These vulnerabilities are characteristic of emerging technologies. The following analyzes prove their weaknesses in terms of cyber security, which is an uncharted terrain for manufacturers and consumers. Finally, brief conclusions about the impact of this trend on society and recommendations to mitigate its effects are presented. The research for this article used information about vulnerabilities from NVD (National Vulnerability Database), cvedetails.com, CVSS (Common Vulnerability Scoring System), CVE (Common Vulnerability Enumeration), CERT-US (Computer Emergency Response Team), and for exploits from exploit-db.com, vuldb.com, attack.mitre.org, SecurityFocus and Rapid7.

2 The Impact of the Emerging Technologies over the Attack Surface

This section consist short review of proofs about impact of listed emerging technologies security weaknesses according application area.

The IoT Security Weaknesses. IoT is already a natural extent of conventional ICT. It comes with BYOD (Bring Your Own device) and many smart devices and the big data. The staff takes care after installation of the smart devices, WSN (Wireless Sensors Network), IoT middleware and the cloud services are not proved for security. This extends the surface for the cyber attack. Malicious actors continue to invent new attack scenarios against cloud technologies and

BYOD. For example, cybercriminals use Google Calendar [11] to steal sensitive information. The technologies listed are synergized with BYOD and cloud based virtualization services. All of them increase in exponential order the surface for threats.

IoT manufacturers for the most part use cheap components and boards which lack the power or ability to run fundamental security features, such as anti-malware, anti-virus, firewalls and encryption. Weaknesses in mobile security multiply the potential for compromising the IoT. The issues are with 5G AKA protocols [12], Stingray, IMSI catchers [13–15]. Big Data is the main part of IoT and the legal analysis have central roles in Big Data projects [16]. The wide area of privacy related to Big Data depend on the security of IoT. An example of a seemingly unusual use case. The project [17] involved running an attack on the EV station's human machine interface (HMI) to communicate with control system to increase the harmonic distortion of the energy flowing through the station.

The Artificial Intelligence Cyber Security. AI technology introduces a new set of issues and application security challenges together with a false sense of security [18]. Many products being rolled out involve "supervised learning", which requires firms to choose and label data sets that algorithms are trained on by tagging code that's malware and code that is clean. Machine learning (ML) algorithms in AI need data to function properly and accurately. The primary method of compromising AI so far has been through data manipulation. If data manipulation goes undetected, any organization could struggle to recover the correct data that feeds its AI system, having potentially disastrous consequences [19]. Classification-based ML algorithms work by finding patterns in its data source. Identifying the algorithm's data source or training method could be a valuable avenue for hackers.

A way for cheating ML algorithms to classify emails as spam is presented in [19]. This is a prerequisite for the emergence of a new type of threat. It can be distributed in aggregated datasets that are sold or marketed in the supply chain. For clarity, an analogy with steganography can be drawn. Hackers who get access to a security firm's systems could corrupt data by switching labels, tagged some malware examples as clean code [20]. Hackers already use AI to create more sophisticated attacks, enhancing traditional hacking techniques like phishing scams or malware attacks. They could use AI and ML to make fake e-mails look authentic and deploy them faster than ever before, apply AI to develop mutating malware that changes its structure to avoid detection. AI could scrub social media for personal data to use in phishing cons. Data poisoning is another danger, in which attackers find out how an algorithm is set up, then introduce false data that misleads on which content or traffic is legitimate and which is not [18].

If found, the bot autonomously produces a "working control flow hijack exploit string" i.e. secures vulnerabilities. Practical AEG (Automatic Exploit Generation) has important applications for defense. For example, automated signature generation algorithms take as input a set of exploits, and output an

Intrusion Detection System (IDS) signature that recognizes subsequent exploits and exploit variants [21]. Rushing to get their products to market, companies use training information that hasn't been thoroughly scrubbed of anomalous data points. That could lead to the algorithm missing some attacks. In this way artificial intelligence opens up new unexplored territories in cyber security for both cyber defenders and attackers.

The Cyber Security Issues with a DLT. Obviously, this technology is susceptible to all known methods by the hacker arsenal: phishing, account takeover, stolen account offering in black market, data leakage. In addition there are described specific type of attack against DLT and a lot of malware tools. Cybersecurity is an unknown territory of the crypto mining society. This is evident from the case the exchange loses its money due to the death of the only owner who knows the password. It is typical exit scam [22]. Malware steals browsers cookies associated with popular cryptocurrency exchanges and wallet service websites. This theft include user names, passwords and credit card information saved in the browser, Cryptocurrency wallet data and keys [23].

The paper [24] highlights the most common methods used by criminal actors, including account takeovers, mining fraud, and scams against initial coin offerings (ICOs). It also includes measures that organizations, consumers, and exchanges can adopt to stay protected. This report has clear references to described cases with breakthroughs in different types of DLT. In the case of successful "SIM swapping", the attackers essentially gained access to the target's mobile phone number using which they can obtain one-time passwords, verification codes, and two-factor authentication in order to reset passwords for and gain access to target's social media, email, bank, and cryptocurrency accounts [25]. SIM swapping is dangerous because the attacker success leads to multiple ATO (Account Take Over).

This highlights a problem with smart contracts. The data that they send around the network is actually money. When it's sent, no bank has to follow up and settle it later. The stakes are high when dealing with security flaws in smart contracts [26,27]. Researchers [28] demonstrated how to use SS7 (Signaling System No 7) to empty a victim's online bitcoin account. They requested a password reset for their target's Gmail account, which meant Gmail sent a token to the linked cellphone number. By accessing the SS7 network, the hackers then just intercepted the text message, and entered the Gmail account themselves [29].

DLT is functioning together with NTP, DNS, SS7 and many more protocols that have famous weaknesses existing in obsolete installations. Internal security weaknesses in DLT are explored in [30,31]. Among the most vulnerable aspects of DLT technology actually comes from outside the chain itself. This technology inherited whole spectrum of environmental weaknesses: viruses, key-loggers, trojans, OS security configuration, compromised crypto keys, lack of security expertise for developing teams.

The VR Security Area. VR and AR (Augmented Reality) tools apply to serious gaming, modelling of real 3D terrain, and the cognitive area. It means that all weaknesses in their security expose vulnerabilities and potential risks to relevant data, personal and corporate interests.

The researchers [32] observe VR/AR security issues analogous to those with IoT and BYOD. Similarity and overlap between those challenges come from technology synergy and that's mean new issues will need to be rethought in this context.

Devices providing immerse feedback can be used by malicious applications to deceive users about the real world. For example, a future malicious application might overlay an incorrect speed limit on top of a real speed limit sign (or place a fake sign where there is none), or intentionally provide an incorrect translation for real-world text in a foreign language. More generally, such an application can trick users into falsely believing that certain objects are or are not present in the real world [33]. The article [34] explains founding that depth perception and balance were temporarily deteriorated in children immediately following 20 min of VR immersion. Furthermore, VR may affect the psychological well being of users given that they feel completely immersed into an environment. Immersion amplifies the consequences of cyber bullying and sexual harassment, where the misconduct "feels all too real" [35]. The ability of VR users to harm one another raises all sorts of complex ethical and legal questions. An attorney who has noticed issues in video game law, explains that society is not yet at the point where virtual assaults should be considered criminal [36].

The study [37] founds that after 30 min of exposure to AltspaceVR or Rec Room, most participants reported feeling unsafe, had difficulty navigating the spaces and struggled with self-expression.

Privacy and data security concerns with AR/VR technologies. According to threat risks and potential damages proved in [32,38,39] the common recommendations about steps organizations must take to address privacy and data security concerns with AR/VR technologies are: an assessment of the need and limit the data they collected, though it is possible that their companies do not collect consumer data in the first place; strengthening data security measures to mitigate the risk of breaches or hacks, updating the policies and disclosures regarding consumer data.

The VR weaknesses leads to future security issues. Malicious software or a DoS attack could temporarily "blind" a user, in real-world environments, such as in medicine and industry, create opportunities for malicious attackers to impact life and safety [39].

Mobile and VR technologies are organically connected, but there are even legitimate spyware apps. In 2018 nearly half mobile applications request location tracking [40], gather phone numbers, email addresses, they use without permissions cameras, microphones, photos, and any personal data stored on the device. The information is far beyond the needs of the application [41].

So far, there are no reported vulnerabilities and no known exploits for the VR/AR. At the same time, there are numerous publications with a proof of use of VR for harmful activities. Should take into account the tendency many VR manufacturers to associate their products with cloud services, which are often provided by third parties.

The Containers Security on the Clouds over The Virtualization. Microservices and containers can introduce hundreds of endpoints and erode the visibility of security risks. A Proof of concept for different weaknesses in the microservices is represented in [42,43].

Providing a deeper and more rigorous analysis to check for things like OS and library vulnerabilities is key to the containers security. A vulnerability in a shared OS kernel can provide a potential way out of a compromised container, active scans can miss most vulnerabilities, containers typically don't include the SSH daemon and credential scans don't work with most containers.

The use of open source code in the software supply chain also introduces standard risks [44]. The risk of attacker-controlled images and the ability to break out beyond the isolation of a container is documented in CVE-2019-5736, MITRE CVE-2018-11757, picked up a well-deserved CVSS v3 9.8 rating. Details include CVE-2018-11756, CVE-2018-8115, CVE-2016-3697, CVE-2018-9862. All they have CVSS v 3.0 ratings over 7.

Organizations are in the midst of a digital transformation (DX), transforming the way they bring products and services to the market. SD-WAN is an example of the paradox of DX: potentially move the business to the next level, but the expanded attack surface it creates can expose the organization to significant risk. They may come from well-known security vendors with well-tested products, but could just as easily be open source components. The last ones can be the weakest security link for many organizations [45–47].

According [48] the reasons for the issues with containers are inadequate cyber security knowledge among teams, visibility into the security of containers and container images is limited, inability to assess risk in container images prior to deployment, lack of tools for detecting of vulnerable containers and experience to handle fundamental differences in securing containers and inability to assess risk in deployed containers. From [48] is visible that 47% have vulnerable containers in production and 46% don't know.

3 Conclusions

The effectiveness of traditional security measures is diminishing. IoT invasion reach critical infrastructure. Building Automation Systems (BMS) and Industrial Control Systems (ICS) exists decades. Together they migrate to IIoT (Industrial Internet of Things). IoT will coexist with such kind of systems in synergy with Smart Properties based on AI embedded already. This multiplies the attack surface. Potentially vulnerable existing BMS, healthcare and different ICS are now prevalent in many buildings and offices, hospitals, airports, sports stadiums, government departments and industrial plants. They coexist with hospital digital devices, utility measured devices, and smart homes. All they are connected to the cloud. The organizational environments are therefore vulnerable to outside control, which has the potential to impact external and internal communications, computer networks, building access, lighting and heating. Downtime on every single one of these systems has a direct link to the well-being of people or the commercial performance of businesses.

An initial estimate of the trend can be obtained from the following model related to the first step of cyber kill chain. The space of elementary independent events contains the probability of penetrating each individual vulnerability. Without taking into account the correlated attacks the probability of penetrating the Existing Attack Surface P(EAS) expands with the probability of penetrating of every one vulnerability of the New Technologies Attack Surface P(NTAS). A result summary of processes behind the risks. **IoT:** *the number of widespread devices, a feverish marketing and non-IP architecture*; **AI:** *unexplored issues, lack of information for vulnerabilities, gives opportunities for extraordinary attacks*; **DLT:** *a distributed architecture*; **VR:** *the extensions of conventional ICT, a unexplored issues*; **container security:** *a background systems integration, a new layer over cloud and the virtualization.* **A common weakness for all of them is the lack of experience of security by design into context of the system integration. In addition to using new technologies**, the attackers change their strategy dynamically, depending on the measures taken by the institutions against them. The cyber-criminals have taken to incorporating a new processes, technologies and communication methods to continue their operations [49]. **The synergy with cutting-edge technologies** puts new demands on the organizations cyber security. The review [50] highlights the need for broad and in-depth regulatory action including limiting the personal information that is collected, shared and used. All this reveals a fuzzy and unclear obstacles around the additional risks after an implementation of the mentioned technologies. The analogy and conclusion: every new technology brings unknown risk expanding attack surface in unexpected way. An in-depth study and exploration of the security aspects and security risks that are hidden in the supply chain is needed. Any introduction of innovative digital technology is accompanied by the need for certain preliminary studies. They are related to avoiding corporate stress and risk management. Security Configuration Management (SCM) is a critical part of maintaining compliance with common regulations, including PCI DSS, HIPPA, SOX, COBIT, NIST 800-53, DASP Top 10 of 2018 [51].

There are still new technologies in areas where the state of threat protection is similar. Outside from the analyses in the article, there remain the areas of cognitive directed to signals from the brain, processing of various forms of physiological information from the human such as external vision and voice, computer vision, natural language processing, and others. The pace of innovation goes beyond the ability of state institutions to respond to cyber security challenges. Weaknesses in the security of new technologies are multiplied in a large number of instances. Their emergence in enterprise ICT is reflected in the digital space. This global impact requires continuity of business regulation and adaptability. Changes are needed in the earliest stages of education and the formation of digital culture. Thus cyber security can become a component of innovative ideas.

Acknowledgements. The research is supported by the KoMEIN Project (Conceptual Modelling and Simulation of Internet of Things Ecosystems) funded by the Bulgarian National Science Foundation, Competition for financial support of fundamental research (2016) under the thematic priority: Mathematical Sciences and Informatics, contract DN02/1/13.12.2016.

References

1. Agrafiotis, I., Nurse, J.R.C., Goldsmith, M., Creese, S., Upton, D.: A taxonomy of cyber-harms: defining the impacts of cyber-attacks and understanding how they propagate. J. Cybersecur. **4**(1), tyy006 (2018)
2. Kadhim, Q.K., Yusof, R., Mahdi, H.S., Al-shami, S.S.A., Selamat, S.R.: A review study on cloud computing issues. J. Phys: Conf. Ser. **1018**, 012006 (2018)
3. Bisong, A., Rahman, S.S.M.: An overview of the security concerns in enterprise cloud computing. Int. J. Netw. Secur. Appl. **3**(1), 30–45 (2011)
4. Noyes, D., Liu, H., Fortier, P.: Security analysis and improvement of USB technology. In: 2016 IEEE Symposium on Technologies for Homeland Security (HST). IEEE, May 2016
5. Prasad, R., Rohokale, V.: Mobile device cyber security. In: Springer Series in Wireless Technology. Springer International Publishing, pp. 217–229 (2019)
6. Almutairy, N.M., Al-Shqeerat, K.H.A., Hamad, H.A.A.: A taxonomy of virtualization security issues in cloud computing environments. Indian J. Sci. Technol. **12**(3), 1–19 (2019)
7. Theisen, C., Munaiah, N., Al-Zyoud, M., Carver, J.C., Meneely, A., Williams, L.: Attack surface definitions: a systematic literature review. Inf. Softw. Technol. **104**, 94–103 (2018)
8. Attack surface, February 2019. https://whatis.techtarget.com/definition/attack-surface
9. Tripware: Unbalanced Security is Increasing Your Attack Surface The State of Security, March 2014. https://www.tripwire.com/state-of-security/featured/unbalanced-security-increasing-attack-surface-2
10. Friedman, J.: Attack your attack surface. How to reduce your exposure to cyberattacks with an attack surface visualization solution, Whitepapper, Skybox security (2016)
11. CISOMAG: Cybercriminals use Google Calendar alerts to steal sensitive information, June 2019. https://www.cisomag.com/cybercriminals-use-google-calendar-alerts-to-steal-sensitive-information
12. Shaik, A., Borgaonkar, R., Park, S., Seifert, J.-P.: On the impact of rogue base stations in 4g/LTE self organizing networks. In: Proceedings of the 11th ACM Conference on Security and Privacy in Wireless and Mobile Networks - WiSec 2018. ACM Press (2018)
13. Sophos: Security weaknesses in 5G, 4G and 3G could expose users' locations, February 2019. https://nakedsecurity.sophos.com/2019/02/04/security-weaknesses-in-5g-4g-and-3g-could-expose-users-locations
14. Chlosta, M., Rupprecht, D., Holz, T., Pöpper, C.: LTE security disabled. In: Proceedings of the 12th Conference on Security and Privacy in Wireless and Mobile Networks - WiSec 2019. ACM Press (2019)

15. 5G Mobile Technology Poses An Espionage Risk, March 2019. https://www.cybersecurityintelligence.com/blog/5g-mobile-technology-poses-an-espionage-risk-4005.html
16. Kemp, R.: Legal aspects of big data: part I legal rights in data. http://www.kempitlaw.com/p97
17. Harris, B., Chin, D., Watson, G.: DOE/DHS/DOT volpe technical meeting on electric vehicle and charging station cybersecurity report. DOT-VNTSC-DOE-18-01, Kevin Harnett (2018)
18. Michael Bruemmer, E.M.B.: Why AI raises your risk of cybercrime and what to do about it, November 2018. https://www.cisoforum.com/why-ai-raises-your-risk-of-cybercrime-and-what-to-do-about-it
19. Zheng, C.: The cybersecurity vulnerabilities to artificial intelligence, June 2019. https://www.cfr.org/blog/cybersecurity-vulnerabilities-artificial-intelligence
20. Giles, M.: AI for cybersecurity is a hot new thing–and a dangerous gamble. MIT Technol. Rev. (2018)
21. Faggella, D.: Artificial intelligence and security: current applications and tomorrow's potentials. https://emerj.com/ai-sector-overviews/artificial-intelligence-and-security-applications
22. Kumar, M.: Cryptocurrency firm loses 145 million after CEO dies with only password, February 2019. https://thehackernews.com/-2019/02/cryptocurrency-exchange-exit-scam.html
23. Kumar, M.: New mac malware targets cookies to steal from cryptocurrency wallets, October 2019. https://thehackernews.com/2019/02/mac-malware-cryptocurrency.htm
24. Digital Shadows, The New Gold Rush: Cryptocurrencies are the New Frontier of Fraud (2018)
25. Khandelwal, S.: First hacker convicted of 'SIM swapping' attack gets 10 years in prison, 04 February 2019. https://thehackernews.com/2019/02/sim-swapping-hack.html
26. Bradbury, D.: Blockchain hustler beats the house with smart contract hack, September 2018. https://nakedsecurity.sophos.com/2018/09/14/-blockchain-hustler-beats-the-house-with-smart-contract-hack
27. Armerding, T.: 300m deleted! How a tiny bug flushed away a fortune, November 2017. https://nakedsecurity.sophos.com/2017/11/09/300m-deleted-how-a-tiny-bug-flushed-away-a-fortune
28. Leyden, J.: Someone checked and, yup, you can still hijack Gmail, Bitcoin wallets etc via dirty SS7 tricks, January 2018. https://www.theregister.co.uk/2017/09/18/ss7-vuln-bitcoin-wallet-hack-risk
29. Cox, J.: You can spy like the NSA for a few thousand bucks, March 2018. https://www.thedailybeast.com/you-can-spy-like-the-nsa-for-a-few-thousand-bucks
30. Prvulovic, M.: Hypothetical blockchain security vulnerabilities and how they can be addressed, June 2018. https://kapitalized.com/blog/hypothetical-blockchain-security-vulnerabilities-and-how-they-can-be
31. Praitheeshan, P., Pan, L., Yu, J., Liu, J., Doss, R.: Security analysis methods on ethereum smart contract vulnerabilities: a survey (2019)
32. Yarramreddy, A., Gromkowski, P., Baggili, I.: Forensic analysis of immersive virtual reality social applications: a primary account. In: 2018 IEEE Security and Privacy Workshops (SPW). IEEE, May 2018
33. Roesner, F., Kohno, T., Molnar, D.: Security and privacy for augmented reality systems. Commun. ACM **57**(4), 88–96 (2014)

34. McKie, R.: Virtual reality headsets could put childrens health at risk (2017)
35. Wong, J.C.: Sexual harassment in virtual reality feels all too real "it's creepy beyond creepy", October 2016. https://www.theguardian.com/technology/2016/oct/26/virtual-reality-sexual-harassment-online-groping-quivr
36. Buchleitner, J.: When virtual reality feels real, so does the sexual harassment, 5 April. https://www.revealnews.org/article/when-virtual-reality-feels-real-so-does-the-sexual-harassment
37. McPherson, R., Jana, S., Shmatikov, V.: No escape from reality. In: Proceedings of the 24th International Conference on World Wide Web - WWW 2015. ACM Press (2015)
38. Hackers Use Ultrasonic Waves To Disrupt VR Headsets - VRScout, July 2017. https://vrscout.com/news/hackers-ultrasonic-waves-disrupt-vr-headset
39. Fineman, B., Lewis, N.: Securing your reality: addressing security and privacy in virtual and augmented reality applications, January 2019. https://er.educause.edu/articles/2018/5/securing-your-reality-addressing-security-and-privacy-in-virtual
40. T. S. Responce. ISTR 24: Symantec's Annual Threat Report Reveals More Ambitious and Destructive Attacks, September 2019
41. de Kerckhove, D., de Almeida, C.M.: What is a digital persona? Tech. Arts **11**(3), 277–287 (2013)
42. Cyber Security Review: Attack Uses Docker Containers To Hide, Persist, Plant Malware Cyber Security Review, February 2019. https://www.csiac.org/digest-article/attack-uses-docker-containers-to-hide-persist-plant-malware/
43. Chelladhurai, J., Chelliah, P.R., Kumar, S.A.: Securing docker containers from denial of service (DoS) attacks. In: 2016 IEEE International Conference on Services Computing (SCC). IEEE, June 2016
44. Bettini, A.: Container security: the foundation of true cybersecurity, February 2019. https://www.itproportal.com/features/container-security-the-foundation-of-true-cybersecurity
45. Fortinet, SD-WAN in the age of digital transformation achieving business agility without complicating network security, December 2018
46. Malley, E.O.: Driving the Convergence of Networking and Security, 3 May 2018. https://www.securityweek.com/driving-convergence-networking-and-security
47. Garson, S.: Warning: security vulnerabilities found in SD-WAN appliances, November 2017. https://www.networkworld.com/article/3238725/warning-security-vulnerabilities-found-in-sd-wan-appliances.html
48. T. B. 1901, Tripwire state of container security report, January 2019
49. Zurkus, K.: Cyber-criminals work around road blocks, June 2018. https://www.infosecurity-magazine.com/news/cybercriminals-work-around-road
50. Kavanagh, C.: New Tech, new threats, and new governance challenges: an opportunity to craft smarter responses? Working Papper, Carnegie Endowment for International Peace, June 2019
51. Decentralized Application Security Project, DASP - TOP 10, November 2018. https://dasp.co

Model of Adaptive System of Neuro-Fuzzy Inference Based on PID- and PID-Fuzzy-Controllers

Ignatyev Vladimir Vladimirovich[1(✉)], Uranchimeg Tudevdagva[2], Andrey Vladimirovich Kovalev[1], Spiridonov Oleg Borisovich[1], Aleksandr Viktorovich Maksimov[1], and Ignatyeva Alexandra Sergeevna[1]

[1] Southern Federal University, Bolshaya Sadovaya Str., 105/42, 344006 Rostov-on-Don, Russia
Vova3286@mail.ru
[2] Chemnitz University of Technology, 09111 Chemnitz, Germany

Abstract. The goal of this work is to develop a model of adaptive system of neuro-fuzzy inference based on PID and PID-FUZZY-controllers, which would allow connecting formalized and informal knowledge in the design of modern automated and automatic control systems for technical processes. To achieve this, we developed an approach to the control of a technical object using an adaptive system of neuro-fuzzy inference. The main control elements of the developed adaptive system of neuro-fuzzy inference are the PID and PID-FUZZY-controllers, as well as the classical and fuzzy control models designed on their basis. The interaction of the two models is provided by the developed hybrid control system. Resulting from the interaction of the two models, the rules base of the fuzzy controller is automatically formed, based on the knowledge about the object obtained when it is controlled by the classical controller. This completely excludes the expert from setting and tuning the parameters of the fuzzy controller. In the developed adaptive system of neuro-fuzzy output, the deviation signals and the deviation and control differential in the classical model are used as data for building a hybrid network. Deviation and control signals in the fuzzy model with automatically generated fuzzy inference rules are used as data to test the hybrid network in order to detect the fact of its overtraining. Thus, the application of the adaptive neuro-fuzzy inference model based on PID and PID-FUZZY controllers allows the effective control of a technical object in the conditions of uncertainty #CSOC1120.

Keywords: Automation · Control · Hybrid systems · Adaptive system · Neuro-fuzzy conclusion · PID-controller · Fuzzy-controller

1 Introduction

Today, an important task in the development of automatic control systems is to improve the control of technical objects in conditions of incomplete data. These topics are widely discussed in works [1–5].

R. Silhavy (Ed.): CSOC 2020, AISC 1225, pp. 519–533, 2020.
https://doi.org/10.1007/978-3-030-51971-1_43

At the same time, three main factors impede the increase in the efficiency of control and the quality of regulation:

- instability of the parameters of the control object in the process of work;
- changing external conditions;
- change of requirements to the quality of regulation in the process of functioning.

At an early stage of development in the field of artificial intelligence, control, engineers discussed about the implementation of artificial intelligence in the control system [6–9].

Today, systems developed on the basis of artificial intelligence and allowing simultaneous use of the advantages of such technologies and traditional controls are actively used to effectively test the impact of the above factors.

Analysis of theoretical research and practical developments in the field of designing intelligent control systems shows that one of the most topical issues in the development of such systems is increasing their adaptability, i.e. ability to reconfigure when the parameters of the control object and external conditions change. At the same time, classical technologies of adaptation are not applicable to intellectual systems, which necessitates further theoretical research and development of new approaches and methods for automatic adjustment of the parameters of intelligent control systems.

Another argument in need of intelligent control for automatic systems is so called human factor [10, 11]. Human factors can be basic for the problem in automatic systems, therefore scholars are working on fully automated systems which are controlled by intelligent controls.

Hybrid control systems connecting, in the opinion of researchers in [12–14], formalized and informal knowledge, are the most promising way to solve the control problems of hard-formalized objects under conditions of uncertainty.

Researchers are trying to combine classical PID and the fuzzy controllers together in control systems. In work [15] scholars tested a pure classical PID controller and a fuzzy controller by simulation. The work concluded that a hybrid system is more effective in comparison with a classical PID controller.

In [16] researcher developed an algorithm for fuzzy PID controller based on hybrid optimization. The main aim of the algorithm is to increase flexibility and capability of the PID controller. Hybrid optimization system integrated genetic algorithm which adopted particle swarm optimization.

In [17], a hybrid fuzzy PID-control approach is proposed to improve the performance of a steam turbine controller, and its efficiency in comparison with the control based on the classical PID controller. The approach in which two variables – system error and error differential are fed to the input of the fuzzy controller is implemented. Then the rules base is built.

In [18] the main attention is paid to the development of a new fuzzy self-adjusting PID-algorithm for temperature control. The goal is to improve the system performance and reduce temperature fluctuations. During operation, the fuzzy controller uses error deviations and error signals. The controller output is three variables (proportional, integral and differential), which are input variables for the PID controller. In this case, there is a constant problem of selecting parameters at the input of the PID controller as the rules for the fuzzy controller are set by the expert.

In [19] hybrid control logic of PID-controller is proposed, the settings of which are provided in real time. The fuzzy controller is configured on the basis of the Mamdani controller, also using the results obtained with the procedure for identifying the values that affect the behavior of the control process.

[20] describes an effective method for determining the optimal parameters of a linear fuzzy PID controller. In this case, training or adapting the controller when changing the parameters of the control object is not provided.

In [21], the stabilization and control of an inverted pendulum on a trolley moving along an inclined surface using PID and fuzzy controllers is considered. To achieve the required results, the use of a complex mathematical apparatus is required.

In [22–28], fuzzy models are used to adjust the PID controller coefficients when changing the parameters of the control object.

[29] shows the organization of interaction between the components of a hybrid controller. The idea is to obtain fuzzy estimates of the efficiency of PID and fuzzy controllers for solving the problem of controlling a technical object with the subsequent introduction of the weighting coefficients of control actions for each controller.

In [30], the classical PID controller is used as the main control device, and the fuzzy controller is used as the compensator of disturbing effects.

An adaptive hybrid controller is described in [31, 32], which uses a fuzzy model of the control object to predict its behavior and adapt the parameters of the control device (PID controller) based on this prediction.

In [33], a hybrid system was developed that uses a neural network to improve the PID controller parameters obtained with the help of a genetic algorithm and to develop a fuzzy model rule base.

A hybrid PID controller is introduced in [34], the tuning of which is performed on the basis of a genetic algorithm.

The key difference between the model of the adaptive neuro-fuzzy inference system developed in this paper based on PID and PID-FUZZY controllers in relation to the studies reviewed above is its ability to learn due to the following factors:

- on the basis of the results obtained in [35], the fuzzy controller rules base is automatically formed in the hybrid system as a result of the interaction of two controllers – classical and fuzzy;
- to adapt the model to changes in the parameters of the control object or external conditions during operation, an adaptive neuro-fuzzy inference system is designed in which the data used for training are data from the classical controller, whereas verifying data are data from the fuzzy controller.

Thus, verification data are further referenced in the operation of the hybrid system; in the event of a change in the parameters of the control object or external conditions and the inability of the classical controller to provide the required control action on the object, the fuzzy controller will provide the desired control.

In [36] PID type fuzzy logic controller is applied for discrete time system with target to track a desired trajectory by using the flatness property. Importance of this work is based on improving the performance of the controller based on genetic algorithm.

2 Proposed Model

In this paper, we propose a new control model based on the use of fuzzy logic apparatus and the classical control theory that allows to simplify and unify the process of creating hybrid control systems and to obtain the desired control of an object by constructing an adaptive system of neuro-fuzzy output based on PID and PID FUZZY controllers. At the same time, the verification data obtained when the two controllers work together are also used for training.

The novelty of the proposed model is the use of training technologies to obtain adaptive qualities in systems of fuzzy inference, which will significantly reduce:

- the size of the rules base;
- delay in the output of control actions;
- the number of computational operations required.

The operation of the hybrid system is divided into two stages.

At the first stage, based on the hybrid algorithm for forming a fuzzy controller base developed in [35], fuzzy rules are automatically generated (obtained as a result of the classical controller of the deviation value of the system θ, the deviation differential $\dot{\theta}$ and the control action U on the object serve as initial values for building the rules base of the model with fuzzy controller).

In this case, the control model based on the hybrid algorithm in [35] is less dependent on expert knowledge and allows to automatically generate fuzzy rules of logical inference and avoid human-induced errors, since this procedure is the most complex in the process of synthesis of fuzzy control systems [37, 38].

The second stage is the construction of an adaptive system of neuro-fuzzy inference. In this case, the deviation values of the system θ, the deflection differential and the control action U are used as test data during training of the obtained (generated) at the first stage fuzzy output system. The values used here are obtained during the operation of the fuzzy controller.

The result of the operations carried out during the two stages above is a full-fledged hybrid control system that makes it possible to obtain the desired control action depending on the degree of complexity of the tasks being solved under uncertainty conditions. The system does not depend on the knowledge of the domain expert and responds quickly to changes in the parameters of the control object.

To implement the proposed model, we use the model of hybrid control developed in [35]. In the example shown in Fig. 1, the hybrid system can be controlled by both PID controller or by PID-FUZZY controller, if its parameters and/or external conditions change.

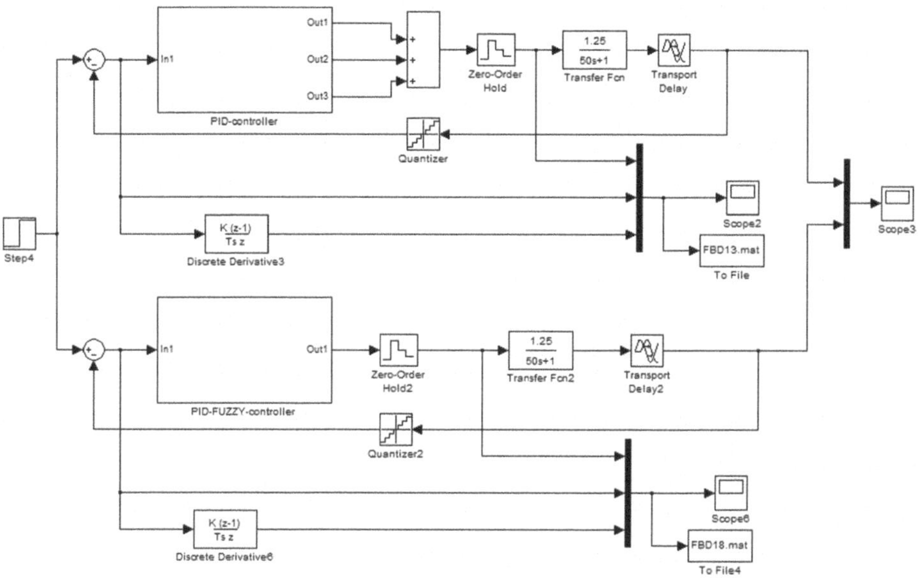

Fig. 1. Hybrid control system in MATLAB

As can be seen from the Fig. 1, the numerical values of the system deviation θ, the deviation differential $\dot{\theta}$ and the control action U are recorded into a special file FBD13.mat for the implementation of the first stage, and FBD18.mat for the implementation of the second stage.

The matrix of system deviation values θ, the deviation differential $\dot{\theta}$ and control action U, obtained as a result of the operation of the classical controller at the first stage and the fuzzy controller at the first stage is shown in Fig. 2.

1.0000	0.6667	17.7333
1.0000	0	8.5500
1.0000	0	9.5000
1.0000	0	10.4500
0.5611	-0.2926	3.4078
0.1462	-0.2766	0.8351
-0.1679	-0.2094	-0.5422
-0.5079	-0.2266	-3.7086
-0.6751	-0.1115	-3.7905
-0.6878	-0.0085	-2.9689
-0.6347	0.0354	-2.5265
-0.4879	0.0978	-0.9959
-0.3049	0.1220	0.3860
-0.1467	0.1055	1.1233
-0.0140	0.0884	1.7952
0.0714	0.0569	1.9933
0.1014	0.0200	1.7425
0.0955	-0.0040	1.4260
0.0641	-0.0209	1.0055
0.0205	-0.0290	0.5912
-0.0180	-0.0257	0.3515
-0.0445	-0.0176	0.2423
-0.0559	-0.0076	0.2607
-0.0515	0.0029	0.4018
-0.0363	0.0101	0.5856
-0.0160	0.0136	0.7677
0.0046	0.0137	0.9216
0.0209	0.0109	1.0139
0.0305	0.0064	1.0424
0.0330	0.0017	1.0203
0.0294	-0.0024	0.9608
0.0217	-0.0051	0.8859
0.0125	-0.0062	0.8154
0.0037	-0.0059	0.7613
-0.0031	-0.0045	0.7304
-0.0071	-0.0027	0.7237
-0.0083	-0.0008	0.7349
-0.0073	0.0007	0.7579
-0.0049	0.0016	0.7846
-0.0020	0.0019	0.8078
0.0006	0.0017	0.8243
0.0024	0.0012	0.8318
0.0033	0.0005	0.8305
0.0031	-0.0001	0.8231
0.0024	-0.0005	0.8133
0.0012	-0.0008	0.8023
0.0000	-0.0008	0.7936
-0.0009	-0.0006	0.7885
-0.0015	-0.0004	0.7866
-0.0017	-0.0001	0.7876
-0.0015	0.0001	0.7915
-0.0011	0.0003	0.7955
-0.0006	0.0003	0.7992
-0.0001	0.0003	0.8029
0.0003	0.0002	0.8043
0.0005	0.0001	0.8044
0.0005	0.0000	0.8042
0.0005	-0.0000	0.8027
0.0003	-0.0001	0.8008
0.0001	-0.0001	0.7997
-0.0000	-0.0001	0.7989
-0.0001	-0.0001	0.7981
-0.0002	-0.0000	0.7983
-0.0002	0	0.7987
-0.0001	0.0000	0.7994
-0.0001	0.0000	0.7996
-0.0000	0.0000	0.8004

1.0000	0.6667	4.5572
1.0000	0	5.3905
1.0000	0	5.3905
1.0000	0	5.3905
0.8872	-0.0752	5.1179
0.7016	-0.1238	3.1075
0.5112	-0.1269	0.8226
0.3265	-0.1231	0.8226
0.1540	-0.1150	0.8226
0.0397	-0.0762	0.7669
0.0099	-0.0199	0.7569
0.0087	-0.0008	0.7963
0.0076	-0.0007	0.7953
0.0080	0.0002	0.7978
0.0092	0.0008	0.8005
0.0096	0.0002	0.7995
0.0094	-0.0001	0.7986
0.0092	-0.0001	0.7983
0.0090	-0.0002	0.7980
0.0087	-0.0002	0.7977
0.0085	-0.0001	0.7976
0.0083	-0.0001	0.7974
0.0082	-0.0001	0.7973
0.0080	-0.0001	0.7971
0.0078	-0.0001	0.7970
0.0077	-0.0001	0.7969
0.0076	-0.0001	0.7968
0.0075	-0.0001	0.7967
0.0073	-0.0001	0.7965
0.0072	-0.0001	0.7965
0.0071	-0.0001	0.7964
0.0070	-0.0001	0.7963
0.0070	-0.0000	0.7963
0.0069	-0.0000	0.7962
0.0068	-0.0000	0.7961
0.0068	-0.0000	0.7961
0.0067	-0.0000	0.7960
0.0066	-0.0000	0.7960
0.0066	-0.0000	0.7959
0.0065	-0.0000	0.7959
0.0065	-0.0000	0.7958
0.0064	-0.0000	0.7958
0.0064	-0.0000	0.7957
0.0064	-0.0000	0.7958
0.0063	-0.0000	0.7957
0.0063	-0.0000	0.7957
0.0063	-0.0000	0.7957
0.0063	-0.0000	0.7956
0.0062	-0.0000	0.7956
0.0062	-0.0000	0.7956
0.0062	-0.0000	0.7956
0.0062	-0.0000	0.7955
0.0061	-0.0000	0.7955
0.0061	-0.0000	0.7955
0.0061	-0.0000	0.7955
0.0061	0	0.7955
0.0061	-0.0000	0.7954
0.0061	0	0.7955
0.0060	-0.0000	0.7954
0.0060	0	0.7955
0.0060	-0.0000	0.7954
0.0060	0	0.7954
0.0060	-0.0000	0.7954
0.0060	0	0.7954
0.0060	0	0.7954
0.0060	-0.0000	0.7953
0.0060	0	0.7954

Fig. 2. Numeric values of signals θ, $\dot{\theta}$, U

The numerical values in the left column (from left to right: θ, $\dot{\theta}$, and U) are the training data used to build the hybrid network, in the right column are the verification data used to test the hybrid network in order to detect the fact of its retraining.

In the hybrid network model (artificial neural network), the artificial neuron is the conceptual basis and a component of the artificial neural network. The structure of the neuron is presented in Fig. 3.

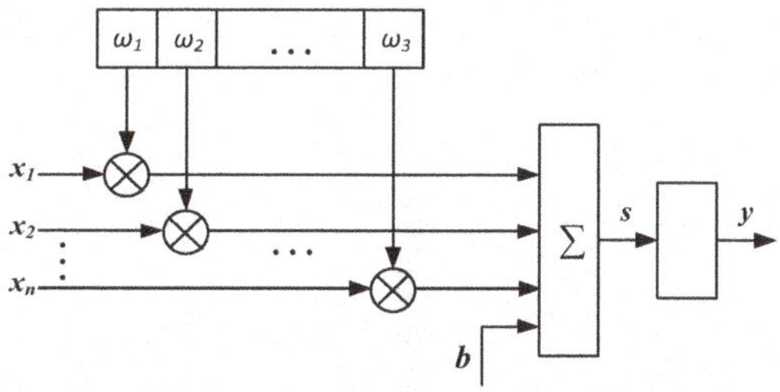

Fig. 3. The structure of the artificial neuron

The mathematical model of the neuron can be written as this analytical expression:

$$s = \sum_{i=1}^{n} \omega_i x_i + b, \tag{1}$$

$$y = f(s), \tag{2}$$

where ω_i is the weight of the multiplier ($i \in \{1, 2, ..., n\}$), b is the offset value, x_i is the component of the input vector or input signal ($i \in \{1, 2, ..., n\}$), s is the summation result, y is the output signal of the neuron, f is the activation function of the neuron (some linear transformation).

The generated structure of the FIS fuzzy inference system after preparation and downloading of training and verification data is shown in Fig. 4, a) and Fig. 4, b), respectively.

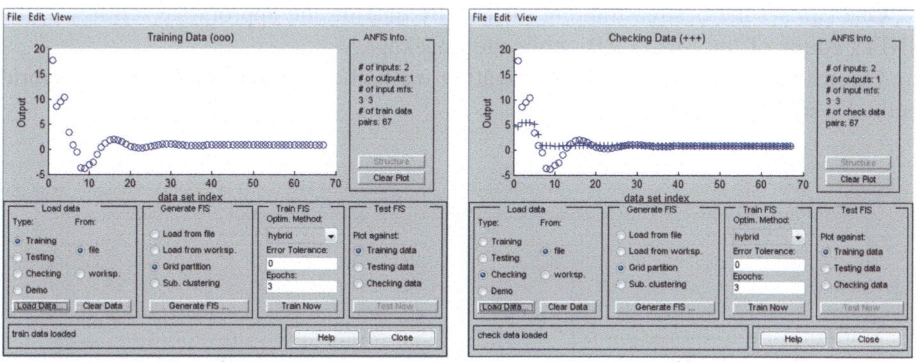

Fig. 4. The structure of training and verification data

Visualization of the generated Sugeno-type FIS structure is shown in Fig. 5.

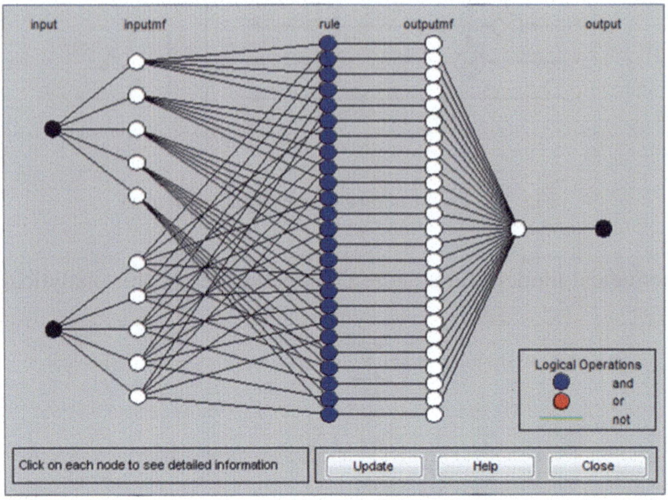

Fig. 5. The structure of the generated fuzzy inference system

Training of the hybrid network is carried out by two methods:

- backward propagation;
- hybrid, which is a combination of the method of least squares and the method of decreasing the inverse gradient [14].

One can set the desired training method by making the appropriate settings in the ANFIS editor of the MATLAB application package.

The results of the backward propagation method are shown in Fig. 6, a); hybrid method – in Fig. 6, b).

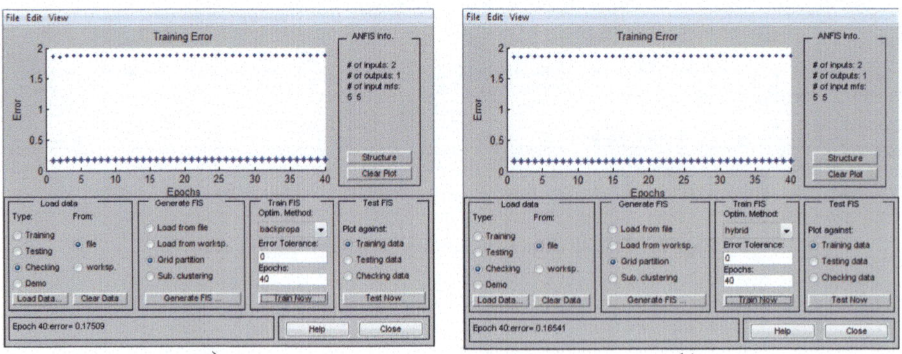

a) b)

Fig. 6. Graphs of dependence of training and verification errors on the number of training cycles

The upper graph in each figure shows the dependence of the verification error on the number of training cycles; the lower graph shows the dependence of the training error on the number of training cycles.

For a more detailed adjustment of the obtained model, one can use the membership function editor of the FIS and the system rules editor of the FIS shown in Fig. 7, a), b).

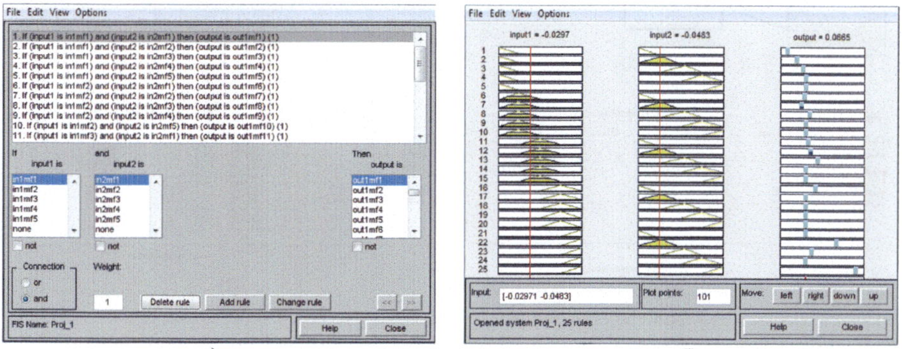

a) б)

Fig. 7. Obtained rules of the created system of fuzzy inference

3 Result Analysis

From the graphs in Fig. 6, a), it is clear that using the back propagation method, training ends in the third step after two cycles, which is confirmed by the data from the command prompt of the MATLAB package:

Start training ANFIS ...

1	**0,175092**	**1,87848**
2	**0,167762**	**1,8593**

Designated epoch number reached --> ANFIS training completed at epoch 2.

The graphs in Fig. 6, b) show that using the hybrid method, training also ends after the second step.

Start training ANFIS ...

1	**0,165412**	**1,86914**
2	**0,165412**	**1,8702**

Designated epoch number reached --> ANFIS training completed at epoch 2.

This confirms that the hybrid algorithm developed in [35] used to form the base of fuzzy controller rules works correctly, because when training, as noted earlier, the verification data are the data obtained as a result of the operation of the fuzzy controller $(\theta, \dot{\theta}, U)$.

Later, one can use the graphic tools of the FUZZY LOGIC TOOLBOX software package to adjust the parameters of the designed and trained network [14]. The form of the membership functions of the input variables and the required values of the output variable of the fuzzy inference system are shown in Fig. 8.

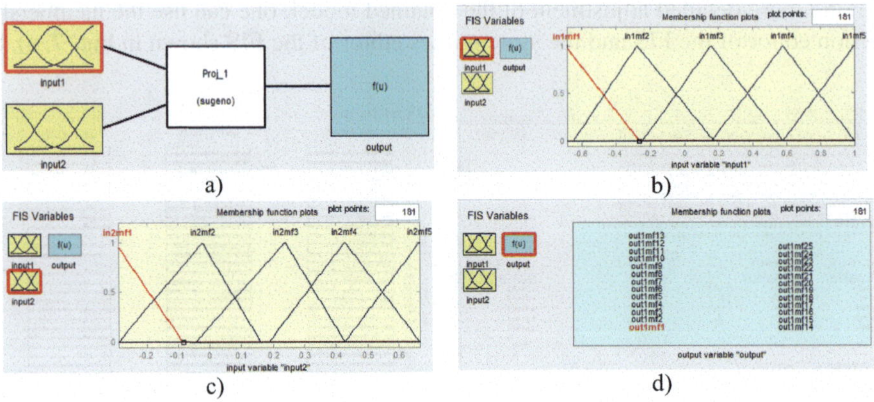

Fig. 8. FIS-editor: a) generated fuzzy inference system, b) terms of the input deviation variable, c) terms of the input variable of the deviation differential, d) the target values of the output variable

After completing all the stages of developing a fuzzy system, an adaptive neuro-fuzzy inference system, training of hybrid network by means of inverse propagation method and combination of least squares method and inverse gradient decreasing method, obtained fuzzy inference model of Sugeno type is integrated into the hybrid control model shown in Fig. 1. After starting the model the transient graphs are obtained, shown in Fig. 9.

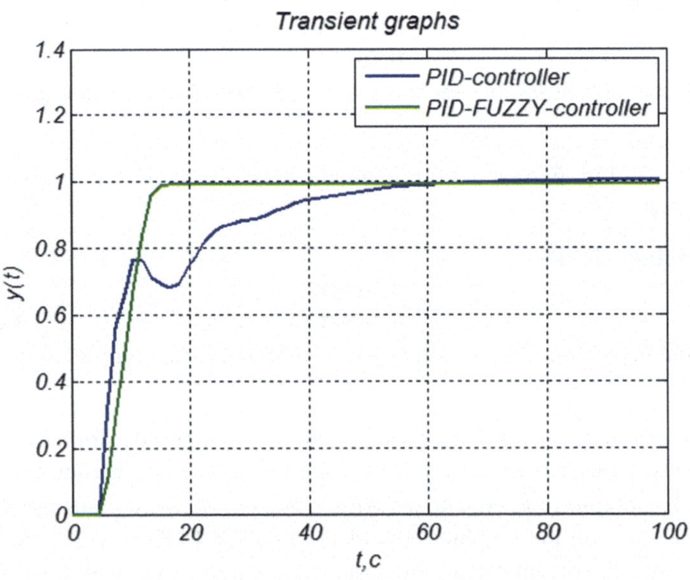

Fig. 9. The transient graphs

The assessment of the quality of transients was carried out on the basis of the following main indicators: the amount of overshoot (σ, %); regulation time (t_{reg}, s); number of oscillations (n), time of reaching the first maximum (t_{1max}, s).

A digital analysis of transient quality for studied hybrid model is given in Table 1.

Table 1. Digital analysis of transient quality.

Main indicators	PID-controller	PID-FUZZY-controller (after training)
σ, %	0	0
t_{reg}, s	32	13
n	3	0
t_{1max}, s	10	13

From the obtained results it can be seen that the developed model of adaptive system of neuro-fuzzy inference based on PID and PID-FUZZY-controllers allows to increase efficiency of technical object control in conditions of incomplete data.

The quality indicators of the transition process obtained by controlling an object using a PID-FUZZY-controller is higher than by controlling an object using a PID controller (classic controller). Thus, at the same amount of overshoot, the regulation time of the PID-fuzzy controller is less than 2 times, and the number of oscillations is less than 3 times. This suggests that using the classic PID controller to automatically form the base of the rules of the fuzzy controller and training technologies to give it adaptive qualities, it is possible to obtain the desired control depending on the degree of complexity of the tasks being solved under conditions of uncertainty, and to not depend on expert knowledge in the subject area and quickly respond to changes in the parameters of the control object.

For a general analysis of the adequacy of the fuzzy model, a fuzzy inference surface is constructed, shown in Fig. 10.

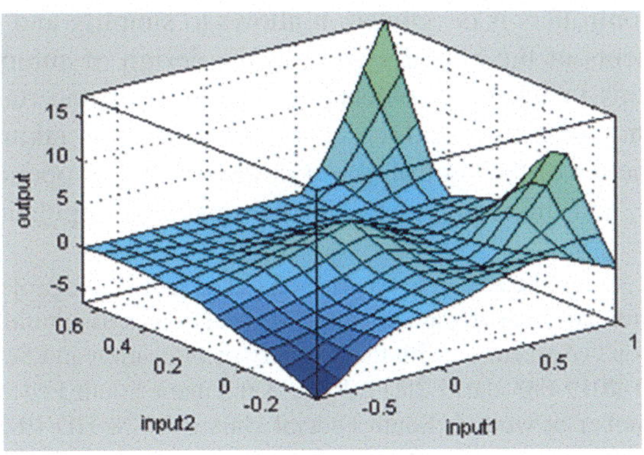

Fig. 10. The surface of the fuzzy output

4 Discussion

The analysis of the quality of the transient processes obtained during the studies in [17–34] makes it possible to state that the required parameters of control of a technical object are achieved. A distinctive feature of the model developed in this paper is the stability of its operation during control of technical systems in the conditions of uncertainty of the parameters, external conditions and changes in the requirements for the quality of regulation in the process of operation.

This will allow:

- to reduce the negative impact of the human on the control process;
- to reduce the setup time and obtain improved quality indicators in comparison with existing methods and approaches.

In addition, in comparison with the results in [17–34], and taking into account the use of PID and PID-FUZZY controllers in them, in this work, training technologies are used to obtain adaptive qualities in control systems.

5 Conclusion

Analysis of recent works in the field of design and application of hybrid control systems using artificial intelligence reveals that methods of automatic adjustment of systems combining classical and fuzzy control technologies in conditions of uncertainty are not sufficiently developed.

The process of developing such systems is characterized by:

- the complexity of parameter setting;
- a high degree of human participation in the design process;
- low automation of the synthesis process of intelligent controllers;
- applied, highly specialized designation of the results obtained.

In this paper, a model of the adaptive system of neuro-fuzzy inference based on PID and PID-FUZZY controllers is developed. It allows to simplify and unify intra-system interaction of elements of the hybrid system in the design of automated and/or automatic control systems for technical objects in conditions of uncertainty, as well as to increase the efficiency of control and the quality of regulation, taking into account the instability of parameters of the control object in the process of operation, the changing external conditions, and the requirements for the quality of regulation.

Acknowledgements. Scientific research was carried out as part of the project "Creating a high-tech production of hardware and software systems for processing agricultural raw materials based on microwave radiation" (Agreement with the Ministry of Education and Science of the Russian Federation № 075-11-2019-083 dated 20.12.2019, Agreement South Federal University № 18 dated 20.09.2019, number of work in South Federal University № HD/19-25-RT.

References

1. Mehta., B.R., Reddy., Y.J.: Industrial Process Automation Systems: Design and Implementation. Butterworth-Heinemann, Oxford (2014)
2. Zhang, P.: Advanced Industrial Control Technology. William Andrew (2010)
3. Dorf., R.C., Bishop., R.H.: Modern Control Systems. Pearson, London (2011)
4. Shu., W., Qian., W., Xie., Y.: Knowledge acquisition approach based on incremental objects from data with missing values. IEEE Access **7**, 54863–54878 (2019)
5. Melvin, I., Grangier, D.: Feature set embedding for incomplete data. U.S. Patent 8,706,668, issued 22April (2014)
6. Tebbutt, C.D.: Artificial intelligence and control system Design. In: Expert Aided Control System Design. Advances in Industrial Control. Springer, London (1994)
7. Mendel., J.M., Zapalac., J.J.: The application of techniques of artificial intelligence to control system design. Adv. Control Syst. **6**, 1–94 (1968)
8. Kumar., V.R., Mani., N.: The application of artificial intelligence techniques for intelligent control of dynamical physical systems. Int. J. Adapt. Control signal process. **8**(4), 379–392 (1994)
9. Nagaraja, G.: Applications of AI in control systems. In: ACE 1990 Proceedings of XVI Annual Convention and Exhibition of the IEEE in India, Bangalore, India, pp. 111–114 (1990)
10. Vassilyev., S.N., Kelina., AYu., Kudinov., Y.I., Pashchenko., F.F.: Intelligent control systems. Proc. Comput. Sci. **103**, 623–628 (2017)
11. Grif, M., Tudevdagva, U., Tsoi, Y.: Man machine systems design methods. In: IEEE Proceedings of the 8th Russian-Korean International Symposium on Science and Technology, KORUS 2004, pp. 42–45 (2004)
12. Kolesnikov, A.V.: Hybrid intelligent systems: theory and technology of development. Publishing house of the St. Petersburg STU, St. Petersburg (2001)
13. Demenkov., N.P.: Fuzzy Control in Technical Systems: A Training Manual. Publishing house of MSTU of N.E. Bauman, Moscow (2005)
14. Leonenkov., A.V.: Fuzzy Modeling in The MATLAB and fuzzyTECH Environment. BHV, Petersburg, St. Petersburg (2005)
15. Erenoglu, I., Eksin, I., Yesil, E., Guzelkaya, M.: An intelligent hybrid fuzzy PID controller. In: Proceedings of ECMS-2006, 20th European Conference on Modelling and Simulation, Bonn, Germany (2006)
16. Ko., C.-N.: A fuzzy PID controller based on hybrid optimization approach for an overhead crane. In: Li., T.-H.S., et al. (eds.) Next Wave in Robotics. FIRA 2011. Communications in Computer and Information Science, vol. 212, pp. 202–209. Springer, Heidelberg (2011)
17. Dettori, S., Iannino, V., Colla, V., Signorini, A.A.: Fuzzy logic-based tuning approach of PID control for steam turbines for solar applications. In: The 8th International Conference on Applied Energy ICAE2016 (2017). Energy Proc. **105**, 480–485
18. Jiang, W., Jiang, X.: Design of an intelligent temperature control system based on the fuzzy self-tuning PID. In: International Symposium on Safety Science and Engineering in China (ISSSE-2012) (2012). Proc. Eng. **43**, 307–311
19. Manenti., F., Rossi., F., Goryunov., A.G., Dyadik., V.F., Kozin., K.A., Nadezhdin., I.S., Mikhalevich., S.S.: Fuzzy adaptive control system of a non-stationary plant with closed-loop passive identifier. Resour.-Efficient Technol. **1**, 10–18 (2015)

20. Kudinov, Y.I., Kolesnikov, V.A., Pashchenko, F.F., Pashchenko, A.F., Papic, L.: Optimization of fuzzy PID controller's parameters. In: XII International Symposium "Intelligent Systems", INTELS 2016, 5–7 October 2016, Moscow, Russia (2017). Proc. Comput. Sci. **103**, 618-622
21. Kharola., A., Patil., P., Raiwani., S., Rajput., D.: Comparison study for control and stabilisation of inverted pendulum on inclined surface (IPIS) using PID and fuzzy controllers. Perspect. Sci. **8**, 187–190 (2016)
22. Nuchkrua., T., Leephakpreeda., T.: Fuzzy self-tuning PID control of hydrogen-driven pneumatic artificial muscle actuator. J. Bionic Eng. **10**(3), 329–340 (2013)
23. Dequan., S., Guili., G., Zhiwei., G., Peng., X.: Application of expert fuzzy PID method for temperature control of heating furnace. Proc. Eng. **29**, 257–261 (2012)
24. Yang., Z., Zhang., J., Chen., Z., Zhang., B.: Semi-active control of high-speed trains based on fuzzy PID control. Proc. Eng. **15**, 521–525 (2011)
25. Mann., G.K., Gosine., R.G.: Three-dimensional min–max-gravity based fuzzy PID inference analysis and tuning. Fuzzy Sets Syst. **156**, 300–323 (2005)
26. Wu., Y., Jiang., H., Zou., M.: The research on fuzzy PID control of the permanent magnet linear synchronous motor. Phys. Proc. **24**, 1311–1318 (2012)
27. Abbasi, E., Mahjoob, M., Yazdanpanah, R.: Controlling of quadrotor UAV using a fuzzy system for tuning the PID gains in hovering mode. In: Fourth International Conference on Advances in Computer Engineering (ACE-2013), Frankfurt, Germany, vol. 10 (2013)
28. Ou., K., Wang., Y., Li., Z., Shen., Y., Xuan., D.: Feedforward fuzzy-PID control for air flow regulation of PEM fuel cell system. Int. J. Hydrogen Energy **40**(35), 11686–11695 (2015)
29. Karli, A., Omurlu, V.E., Buyuksahin, U., Artar, R., Ortak, E.: Self tuning fuzzy PD application on TI TMS320F 28335 for an experimental stationary quadrotor. In: 4th European Education and Research Conference (EDERC 2010), pp. 42–46, IEEE, Nice (2010)
30. Beirami., H., Shabestari., A.Z., Zerafat., M.M.: Optimal PID plus fuzzy controller design for a PEM fuel cell air feed system using the self-adaptive differential evolution algorithm. Int. J. Hydrogen Energy **40**(30), 9422–9434 (2015)
31. Savran., A.A.: Multivariable predictive fuzzy PID control system. Appl. Soft Comput. **13**(5), 2658–2667 (2013)
32. Liem., D.T., Truong., D.O., Ahn., K.K.: A torque estimator using online tuning grey fuzzy PID for applications to torque-sensorless control Of DC motors. Mechatronics **26**, 45–63 (2015)
33. Savran., A., Kahraman., G.: A fuzzy model based adaptive PID controller design for nonlinear and uncertain processes. ISA Trans. **53**(2), 280–288 (2014)
34. Jahedi., G., Ardehali., M.M.: Genetic algorithm-based fuzzy-PID control methodologies for enhancement of energy efficiency of a dynamic energy system. Energy Convers. Manag. **52** (1), 725–732 (2011)
35. Ignatyev, V.V.: Hybrid algorithm of formation of base of rules of the fuzzy controller. Izvestiya SFU, Technical science. Thematic release "Radio-electronic and infocommunication technologies, systems and networks", Publishing house of ETA SFU, vol. 11, no. 172, pp. 177–186 (2015)
36. Gritli, W., Gharsallaoui, H., Benrejeb, M.: A new methodology for tuning pid type fuzzy logic controllers scaling factors using genetic algorithm of a discrete time system. In: Ramakrishnan, S. (ed.) Modern Fuzzy Control Systems and Its Applications, pp. 89–103. IntechOpen (2017)

37. Ignatyev, V.V., Kovalev, A.V., Spiridonov, O.B., Ignatyeva, A.S., Bozhich, V.I., Boldyreff, A.S.: Model of adaptive system of neuro-fuzzy inference based on pi- and pi-fuzzy-controllers. In: Proceedings of SPIE. Emerging Imaging and Sensing Technologies for Security and Defense III; and Unmanned Sensors, Systems, and Countermeasures. SPIE Security + Defense, Berlin, Germany, vol. 10799 (2018)
38. Ignatyev, V.V., Soloviev, V.V., Ignatyeva, A.S., Boldyreff, A.S.: Analysis of the controllers of the vessel course control systems in difficult navigation conditions. In: Proceedings of SPIE, Artificial Intelligence and Machine Learning in Defense Applications. SPIE Security + Defense, Strasbourg, France, vol. 11169 (2019)

Study and Evaluation of Novel Chaotic System Applied to Image Encryption with Security and Statistical Analyses

Hany A. A. Mansour[1] and Mohamed M. Fouad[2](✉)

[1] Department of Electronic Warfare, Military Technical College, Cairo, Egypt
hnhn118799@mtc.edu.eg

[2] Department of Computer Engineering, Military Technical College, Cairo, Egypt
mmafoad@mtc.edu.eg

Abstract. There is no doubt that chaotic sequences attracted much attention in recent decades due to its distinctive properties. These distinct properties give the availability for the chaotic sequences to be applied in different fields. One of the fields that exploit the chaos features is the encryption field, in which chaotic sequences can be used to encrypt texts or images. The randomization attitude of the chaotic sequences can be affected by the initial condition, or by the code length. As these parameters optimized, the performance of the chaotic sequences enhanced and improved from many points of view. This paper presents study, analysis and evaluation for different chaotic sequences from the security viewpoint. The study and analysis include the standard NIST statistical test suite for random and pseudo random number generators for cryptography applications. The presented sequences are the traditional and enhanced version of chaotic sequences based on the logistic map. The new contribution of the paper is to perform the statistical tests over a range of the code lengths, and initial conditions in order to study the effect of each parameter, then optimize its values. Moreover, the proposed sequences are evaluated and applied to image encryption to verify its robustness.

Keywords: Chaotic sequences · Chaotic map · Logistic sequence · Image encryption

1 Introduction

In recent decades, chaos based communication shows high efficiency and reliability. As a result, chaotic sequences (CSs) are applied in various digital communications techniques such as Orthogonal Frequency Division Multiplexing (OFDM) [9], Code Division Multiple Accesses (CDMA) [12], MIMO radar application [6], underwater communication [21], and image processing [26]. Actually, CSs have significant attractive features, especially from the security and correlation point of view. CSs can be efficiently applied in encryption, and has a considerable

R. Silhavy (Ed.): CSOC 2020, AISC 1225, pp. 534–553, 2020.
https://doi.org/10.1007/978-3-030-51971-1_44

resistant to jamming and multi-path fading [18]. Generally, CSs are generated from dynamic nonlinear systems called chaotic maps [16]. It is found that CSs have many privileges over traditional sequences such as the simplicity [25]. Moreover, CSs are wide-band sequences non periodic, which negatively impacts the probability of sequence prediction and reconstruction [25]. In addition, CSs have a unique privilege, which is its sensitivity to the initial conditions. This privilege gives the CSs the excellence to generate large uncorrelated codes [22,23]. The aforementioned features allow the researchers focus on developing new enhanced sequences with improved properties [10].

Recently, the properties and the features of the CSs are investigated and discussed from various points of view. On top of these views are the correlation properties and the security viewpoint [15,17]. As previously mentioned, CSs are generated based on specific chaotic maps controlled by specific parameters. These controlled parameters including the bifurcation parameters of the chaotic map, and the initial conditions. It is found that the choice of the initial value has a serious effect on the performance and the attitude of the chaotic sequence. This means that optimizing the initial condition can significantly improve the properties and the features of the CSs. Regarding to the aforementioned work, it was necessary to apply the NIST suit tests to verify the randomness properties of the applied sequences. The tests are applied on sequences with one initial value and specific code length [15].

This paper aims to study and analyze the effect of the code length and the initial condition individually and simultaneously on the randomization properties of the applied sequences. The study and the analysis are based on the NIST statistical test suite for random and pseudo random number generators for cryptographic applications. The analysis process is performed over chaotic sequences generated from the logistic map. The proposed sequences are the traditional, Zero Mean (ZM), Self-Balanced (SB), and Zero Mean Self-Balanced (ZMSB) sequences. The NIST statistical tests contain 15 different tests this paper discuss just the first 8 tests due to the increase of the result number. After that, the tested sequences are evaluated throughout applying them in encrypted image processing.

The paper is arranged as follows. The mathematical representations of different chaotic sequences are presented in Sect. 2. Whereas, Sect. 3 show the analysis of the proposed chaotic sequences over all underlying tests, including the effect of the code length and the initial value individually, and the simultaneous effect of both on each test. Sect. 4 illustrates the application of the proposed chaotic sequences in image encryption with security and statistical analyses. Finally, conclusions are drawn in Sect. 5.

2 Mathematical Formulation

In this section, the mathematical representation of the applied maps is stated. Starting with the logistic map, it can be considered as one of the simplest and most widely studied nonlinear dynamical systems that are capable of exhibiting chaos [13]. The mathematical representation of the logistic map can be represented as

$$F(x,r) = r\,x\,(1-x)\,, \tag{1}$$

where F is the transformation mapping function, x refers to the output chaotic sequence, and r is the bifurcation parameter of the map. As well, it can be rewritten in its recursive form as,

$$x_{n+1} = r\,x_n(1-x_n)\,,\ \forall\, 0 \le x_n \le 1,\ 0 \le r \le 4\,, \tag{2}$$

where n denotes the iteration number. Since the output of the chaotic map is real values, it is necessary to convert these real values into binary values. The conversion process is performed, such that

$$C_x = g\left(x(n) - E_{t(x(n))}\right), g(x) = 1\ \forall\, x \ge 0, g(x) = -1\ \forall\, x < 0, \tag{3}$$

where $g(.)$ denotes the chaotic function, C_x denotes the output binary value, and $E_{t(x(n))}$ denotes the mean function over continuous time interval t.

The operation of the self-balancing is presented and discussed in [19]. It includes four steps: the inversion step, all upside down step, and radix-S block upside down step, and the shift combination step, where S denotes the length of the short segment. The inversion process depends on multiplying (-1) by the ZM basic sequence X_1, which can be cast as,

$$f(x_i) = -x_i\,,\ \forall\, M \le i \le M+N, \tag{4}$$

where M and N denote the start point and code length, respectively. All up-side down sequence is obtained from the inverted output by rearranging the original sequence in the opposite direction, such that

$$y_i = -x_{(2M+N-1-i)}\,,\ \forall\, M \le i \le M+N. \tag{5}$$

In addition, the radix-S step is operated on the upside down sequence y_i with length N, such that

$$z_i = y_{(2M+(2j-1)S-1-i)}\,, \quad = -x_{(N-(2j-1)S+1)}\,,\ \forall\, M \le i \le M+N\,, \tag{6}$$

where j represents the short segments index number, $j = 1, 2, \ldots, N/S$. The final step is to obtain the balanced sequence by combining the radix-S block upside down sequence z_i and the original sequence x_i. The final output becomes zero mean self-balanced (ZMSB) sequence C_k, such that

$$C_k = \begin{cases} X(k/2) + M,\ \forall\, k \in \{0, 2, 4, \ldots, 2N\}\,, \\ X((k-1)/2) + M,\ \forall\, k \in \{1, 3, 5, \ldots, 2N-1\}\,, \end{cases} \tag{7}$$

where k denotes an integer number representing the index of the final ZMSB sequence.

3 Analysis Results

In this section, the proposed study and analysis are presented. As mention before, the study is focusing on the effect of the control parameters (bifurcation, length, and initial value) on the random and pseudo random properties of the presented sequences. The analysis process is performed by applying the first 8 set of statistical tests described in [5]. The range of the unique bifurcation parameter applied in the logistic map is set to be from 3.57 to 4 with step size 0.01. Regarding to the code length, the applied range is set to be from 500 to 1000 with step size equal to 10. For the initial condition, the applied range is set to be from 0.01 to 0.99 with step size equal to 0.01.

3.1 Frequency (Monobit) Test

This test can be considered as the most important test, in which all other tests depend on the success of this test. The aim of this test is to measure the ratio of the ones and zeros, which is expected to be the same. This ratio is examined under the aforementioned ranges of the bifurcation parameter, sequence length, and initial condition.

Figure 1(a) shows the plot of the code length against the P-value for all the proposed sequences. The figure ensures the results obtained in Fig. 1, in which the SB and ZMSB sequences have perfect attitudes all over the range due to their improvements of perfect balance and zero mean. On the other hand, the traditional has the worst attitude in which it fail at the length 830 and above. The ZM sequence passes the test with attitude lower than the SB and ZMSB sequences.

Figure 1(b) shows the effect of the initial condition on the P-value of the test for all the mentioned sequences. The figure shows that the traditional sequence has a degraded attitude at many initial values before making any improvements or enhancements. As the first improvement performed ZM, the attitude of the resultant ZM sequences become to be acceptable and passes the test for all the initial values. By continuing the improvement throughout the self-balancing operation, the resultant sequences pass the test perfectly.

Figure 1(c) and Fig. 1(d) discuss the simultaneous effect of the both initial values and the code length on the sequences attitude for all the proposed sequences. Figure 1(c) plots the performance of the traditional and ZM sequences against both the code length and the initial value. It can be shown that the ZM sequence almost passes the test all over the ranges of initial and code length. Regarding the traditional sequence, it is found that it fails in the test at certain initial values corresponding to specific lengths. Regarding to the Fig. 1(d), it discusses the attitude of both the SB and the ZMSB sequences. Due to the perfect results of both the SB and ZMSB sequences, the figure ensures the results obtained in both Fig. 1(a) and Fig. 1(b).

3.2 Frequency Test Within a Block

The target of this test is to measure the ratio of the ones and zeros within a block of specific bits. Regarding to the perfect randomness, it is assumed that the ratio is equal for both ones and zeros within the block.

Figure 2(a) illustrates the result of the mentioned test across code length. The figure shows that both the SB and ZMSB sequences still have their perfect attitude over most of the code length range. The figure shows also that both the traditional and ZM sequences pass the test. It can be noted that the ZM process has a negative effect, since the traditional sequence has better performance than the ZM sequence.

In Fig. 2(b), the effect of the initial value is illustrated regarding to the frequency block test for the different mentioned sequences. The figure shows that generally all the proposed sequences pass the test. The figure shows also that the ZM sequence has the worst attitude relatively to the other sequences, including the traditional sequence. This note ensures the fact that the Zero mean process has a negative effect on the randomness feature of the sequence.

Figure 2(c) and Fig. 2(d) show the attitude of the each pair of the sequences under the simultaneous effect of both the initial and the code length. Both the traditional and the ZM are represented in Fig. 2(c), in which the ZM sequences degrades over the traditional sequence due to the effect of the zero mean process. Hence, Fig. 2(c) verifies the results obtained in Fig. 2(a) and Fig. 2(b). Whereas, Fig. 2(d) shows the attitude of both the SB and the ZMSB sequences. The figure shows that the SB sequence has relatively better performance than the ZMSB, due to the zero mean effect.

3.3 The Run Test

The object of this test is to measure the total number of runs along the proposed sequences. In other words, this test measure the speed of the alternation between the ones and the zeros fast or slow.

Figure 3(a) illustrates the effect of changing the length on the attitude of the different sequences. Starting with the traditional sequence, it passes the test

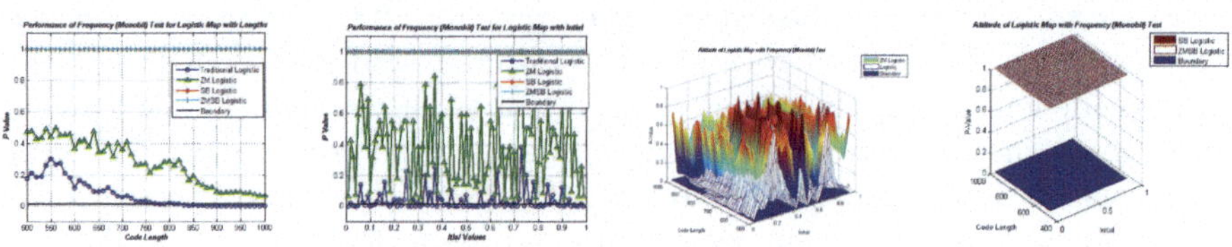

(a) (b) (c) (d)

Fig. 1. Frequency monobit test: (a) P-value against sequence length, (b) P-value against initial value. Frequency monobit test: P-value against both sequence length and initial value for (c) traditional and ZM sequences, and (d) SB and ZMSB sequences.

until length of 840, where it begins to fail. Regarding to the ZM sequence, it clear that it is failed in the test as the length increased over 530. For the SB and the ZMSB, the result shows that they nearly have the same attitude all over the range. In which they pass the test for most of the lengths, and failed at specific points.

The effect of the initial value is discussed in Fig. 3(b), where all the sequences approximately have the same attitudes at the same points. The figure shows that the traditional sequence has comparatively the best attitude compared to the other sequences. This can be obvious as the traditional sequence has the highest peaks over all the other sequences. The figure shows also that all the sequences nearly failed at the same specific initial values.

Figure 3(c) and Fig. 3(d) show the simultaneous effect of both the code length and the initial value on the run test. Figure 3(c) illustrates the attitude of the traditional and ZM sequences. That figure confirms the results obtained at Fig. 3(a) and Fig. 3(b), in which the traditional sequence has a better attitude over the other sequences. Regarding to Fig. 3(d), the attitude of the SB and ZMSB sequences is presented and illustrated. It is clear from The figure that both the zero mean and self-balancing operation has relatively negative effect on the run property.

3.4 The Longest Run Test of Ones in a Block

This test aims to measure the longest run within a block has a specific size. It is desired to have longest run of ones consistent with that would be expected in the perfect random sequence.

Figure 4(a) shows the attitudes of the 4 different sequences under the effect of the different code lengths. It can be noted that the traditional sequence pass the test with accepted attitude for the first half of the range, and begin to degrade for the rest of the code length value. The ZM sequence has the best performance, especially at the first and the last third of the range. SB sequence has also a good attitude, which means that the ZM and the SB operations individually affect positively on the traditional sequence. Finally, the ZMSB sequence has a

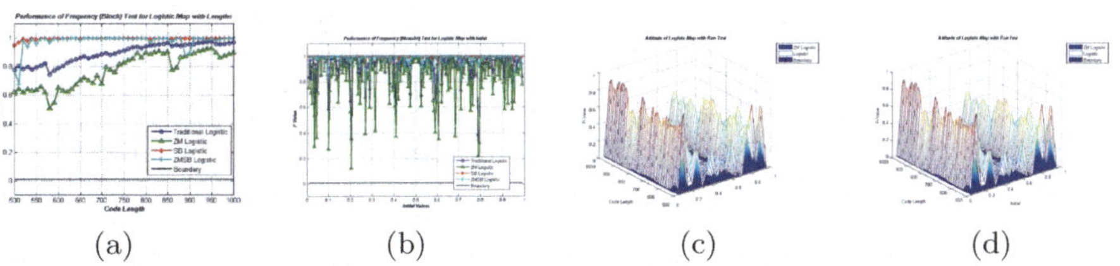

(a) (b) (c) (d)

Fig. 2. Frequency block test: (a) P-value against sequence length, (b) P-value against initial value. Frequency block test: P-value against both sequence length and initial value for (c) traditional and ZM sequences, and (d) SB and ZMSB sequences.

moderate attitude, which means that combining the both operations degrades the attitude of this test.

Figure 4(b) shows the effect of the initial value regarding to the mentioned test. The figure confirms the results shown in Fig. 4(a), in which the traditional, ZM, and SB sequences pass the test for most of the initial values. The figure shows also that the ZM sequences have the best attitude compared to the other sequences. At the same time, the ZMSB sequence has the worst attitude due to the combination of the ZM and the SB operations.

Figure 4(c) and Fig. 4(d) show the effect of both the initial and the code length on different sequences. The traditional and ZM is presented in Fig. 4(c) that confirms the attitude of the ZM sequence shown in Fig. 4(a). Whereas, the SB and the ZMSB sequences are illustrated in Fig. 4(d) that verifies the results obtained in Fig. 16(b), especially at high code lengths.

3.5 The Binary Matrix Rank Test

This test focused on the rank of disjoint sub-matrices of the entire sequence. The binary matrix rank test measures the linear dependence among fixed length sub-strings of the original sequence [G]. Following the same trend, Fig. 5(a) represents the attitude of the proposed sequences through the range of code length. According to result, it can be noted that both the SB and ZMSB sequences have better attitude than the traditional and ZM sequences. This gives indication that the SB and combination between ZM and SB have a positive effect on this test. Concerning the traditional and ZM sequences, the results shows that the attitude of the ZM sequence is better than that of the traditional until length of 770, where the attitudes are reversed. However, all the sequences pass the test.

Figure 5(b) represents the attitudes of the sequences along the range of the initial values. The figure show that the SB and ZMSB also have better performance than the traditional and ZM sequences. The figure shows also that the ZM has relatively bad attitude compared to the other sequences. Generally, all the sequences pass the test except at specific values of the initial.

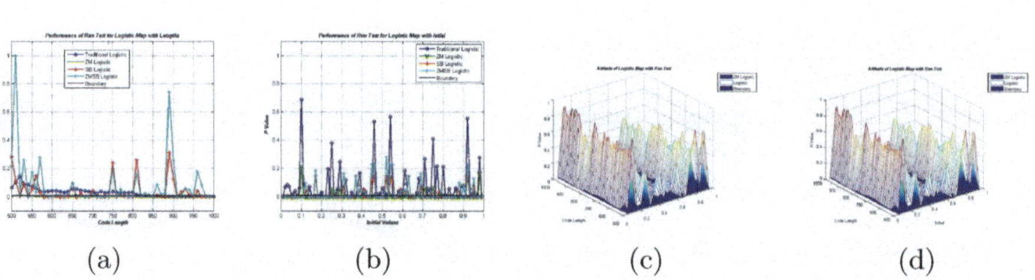

(a) (b) (c) (d)

Fig. 3. The run test: (a) P-value against sequence length, (b) P-value against initial value. The run test: P-value against both sequence length and initial value for (c) traditional and ZM sequences, and (d) ZM and ZMSB sequences.

Figure 5(a) and Fig. 5(b) illustrate the effect of both the code length and the initial values. Figure 5(c) express the attitude of the traditional and ZM sequences, while Fig. 5(d) express the attitude of both the ZM and ZMSB sequences. The figures explained the results illustrated in Fig. 5(a) and Fig. 5(b), where the attitude of the ZM enhanced at high code length and moderate initial values, while the traditional sequence has good attitude high code length and low initial values. Figure 5(d) shows that the attitudes of both the SB and the ZMSB don't have a fixed pattern. This means that the results have alternative values regarding to this test, however the sequences are almost pass the test.

3.6 Discrete Fourier Transform Test

The object of this test is to monitor the periodicity feature of the sequence that can effect on the randomness of the sequence [G]. Regarding to the specified test, Figures Fig. 6(a) and Fig. 6(b) represents the performance of the proposed sequences against the range of the code lengths and the initial values respectively. The plots show that all the sequences have varying attitude along both the code length and the initial ranges. The figures show also that most of the attitude of the sequences has different high P-values.

Figures Fig. 6(c) and Fig. 6(d) ensure the results obtained in figures Fig. 6(a) and Fig. 6(b). The figures show that all the proposed sequences almost have the same attitude with different high p-values along the ranges of the code length and the initial values.

3.7 Non-overlapping Template Matching Test

The purpose of this test is to detect the number of occurrence produced from the generator for a specific pattern [G]. The test is evaluated throughout figures from Fig. 7(a) to Fig. 7(c). Starting with figure Fig. 7(a), the figure plots the p-value of the proposed sequences against the code length. It can be noted that both the traditional and ZM sequences have identical attitudes, and this attitude is decreased as the code length increased. Regarding to the SB and the ZMSB,

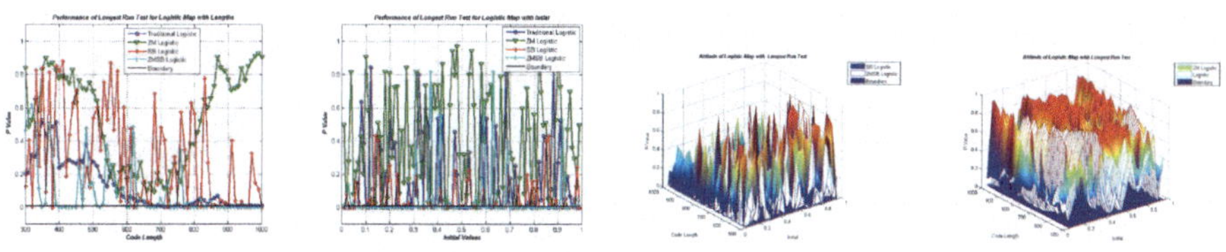

(a) (b) (c) (d)

Fig. 4. The longest run test: (a) P-value against sequence length, (b) P-value against initial value. The longest run test: P-value against both sequence length and initial value for (c) traditional and ZM sequences, and (d) ZM and ZMSB sequences.

the results show that there are decreasing spikes at a specific values of the code length . The results show also that these spikes are increased as the code length increased.

Figure 7(b) illustrates the attitudes of the proposed sequences against the initial values. The results show that also the ZM sequence and these attitudes have fixes values along the initial values. Regarding to the SB, and the ZMSB, it have also spikes at specific initial values, and these spikes nearly has the same values.

Considering the 3D results of each two sequences against both the code length and the initial values shown in Fig. 7(c) and Fig. 7(d). Figure 7(c) plots the result of both the traditional and the ZM sequences. The results become coincident with Figs. 25 and 26. The values of both the two sequences are identical, fixed with the initial values, and decreased as the code length increased. Regarding to the SB and the ZMSB represented in Fig. 7(d), it can be shown that downward spikes appear at the specific values of both the code length and initial values. As usual, this result confirms the results obtained in Fig. 7(a) and Fig. 7(b).

3.8 Overlapping Template Matching Test

This test focused on the number of occurrence of pre-specified target string [G]. Both tests in the aforementioned sections are using a window with m-bits to search for a specific m-bit pattern. Starting with the analysing the attitudes of the mentioned sequences against the code length shown in Fig. 8(a). It can be noted that the performance of all the sequences is improved as the code length increased. In addition, the traditional and ZM sequences have a step pattern during certain intervals of the code length, and they coincident at specific code length intervals. Regarding to the SB and the ZMSB, it is clear that the SB has a spikes pattern, and this spikes are increased as the code length increased. For the ZMSB, it is found that for short code length (up to 570) the sequences has failed at some values, as the code length increased, the sequence performance is increased.

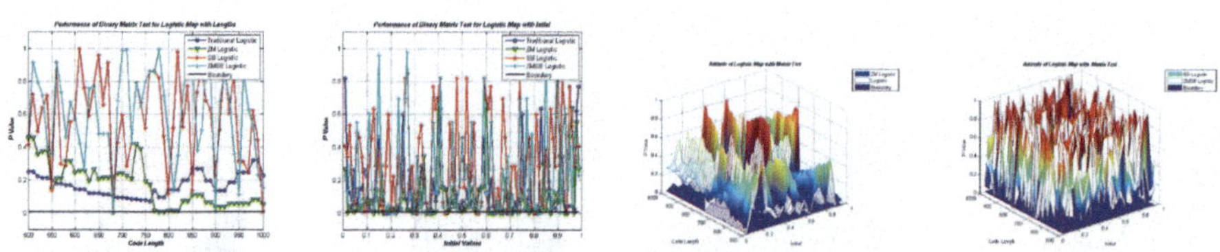

(a) (b) (c) (d)

Fig. 5. The binary matrix rank test: (a) P-value against sequence length, (b) P-value against initial value. The binary matrix rank test: P-value against both sequence length and initial value for (c) traditional and ZM sequences, and (d) ZM and ZMSB sequences.

Regarding to the performance against the initial value, Fig. 8(b) illustrates the attitude of the proposed sequences. It can be notice that almost all the sequences succeeded to pass the test. However the ZMSB has the lowest spikes regarding to the rest of the sequences, which agree the results obtained in the code length case.

Figure 8(c) and Fig. 8(d) represent the 3D plots of the proposed sequences under the mentioned test. The figures ensure the results obtained in Fig. 8(a) and Fig. 8(b). Regarding to the traditional and ZM sequences, it can be shown that along most of the initial values, the performance is increased in a step manner as the code length increased. For the SB and ZMSB, the result shows that the attitude of the mentioned sequences has a spikes pattern, which confirms the attitude with the code length and initial ranges.

3.9 Maurer's "Universal Statistical" Test

This test is concerned with the common bits between matching patterns. In other words, the test is concerning to check whether the sequence is compressed able without losses I information or not. Figure 9(a) discusses the attitude of the proposed sequences against the code length. The figure shows that the traditional and ZM sequences have ascending behaviour as the code length increased. On the other hand, the SB and ZMSB have normal behaviour within the range between the boundary and 0.2. In general all the sequences passed the mentioned test all over the range of the code length.

Regarding to the initial value, Fig. 9(b) represents the performance along the initial range. The figure shows that generally, the traditional and ZM sequences have better performance than the SB and ZMSB sequences. This means that the self-balancing operation has a negative effect under this test. However, generally all the sequence passed the Maurer's "Universal Statistical" Test for all the range of the initial values.

Figure 9(c) and Fig. 9(d) illustrate the 3D plot of each two sequences against both the code length and the initial values. Figure 9(c) represents the traditional

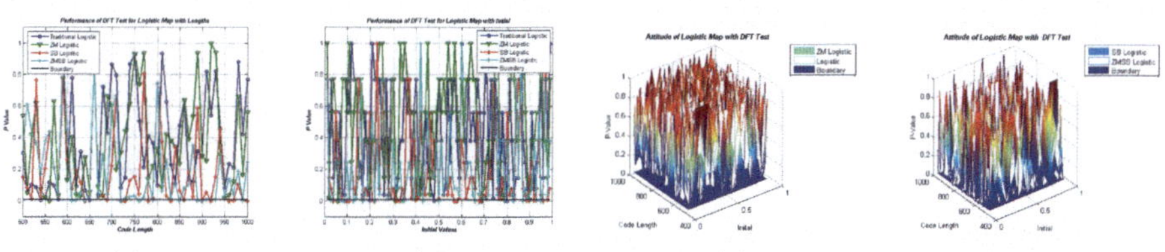

(a) (b) (c) (d)

Fig. 6. Discrete fourier transform test: (a) P-value against sequence length, (b) P-value against initial value. Discrete fourier transform test: P-value against both sequence length and initial value for (c) traditional and ZM sequences, and (d) ZM and ZMSB sequences.

and ZM sequences. The figure shows that same results obtained in Fig. 9(a), in which the attitude increased as the code length increased, especially around the center values of the initial conditions.

3.10 Linear Complexity Test

The purpose of this test is to study the randomness of the generated sequences based on the applied shift registers. Since the proposed sequences are chaotic sequences, which are not generated from the shift register, then this test will be out of scope.

3.11 Serial Test

This test discusses the frequency of all possible overlapping of certain m-bit patterns within the presented sequence. In other words, the test object is to determine whether the number of occurrences of the 2 m m-bit overlapping patterns is approximately the same as would be expected for the presented sequence.

Figure 10(a) represents the attitude of the presented sequences based on different code lengths. The figure shows that the traditional sequences have accepted values for short lengths. As the length increased, the values become to decrease up to code length 950. The figure shows also that the zero mean process has a negative effect, as the attitude of the mentioned sequences degraded. The self-balancing process generally improves the attitude, especially at certain values of the code length. The ZMSB sequence also has better attitude at specific number of code lengths.

Figure 10(b) discusses the attitude based on the initial conditions. Regarding to the traditional and ZM sequences, the figure shows that these sequences pass the test at just limited specific values of initial conditions. The case is improved and enhanced for the SB and ZMSB sequences, as the number of the available initial conditions increased, with relatively high P-values.

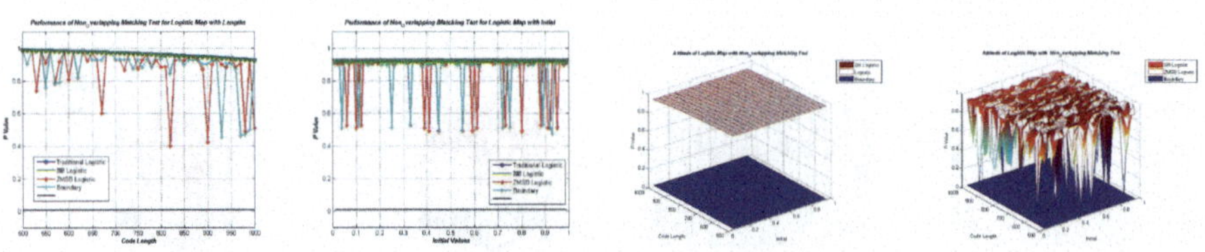

(a) (b) (c) (d)

Fig. 7. Non-overlapping template matching test: (a) P-value against sequence length, (b) P-value against initial value. Non-overlapping template matching test: P-value against both sequence length and initial value for (c) traditional and ZM sequences, and (d) ZM and ZMSB sequences.

Figure 10(c) and Fig. 10(d) are completely coincident and confirm the results obtained from Fig. 10(a) and Fig. 10(b). Figure 10(c) shows that the traditional and ZM sequences pass the test at specific values of initial conditions corresponding to specific code length. Figure 10(d) shows improved attitude of SB and ZMSB than the traditional and ZM sequences, as the number of high spikes are increased.

3.12 Approximate Entropy Test

Similar to the previous Serial test of the previous Section, this test discusses the frequency of all possible overlapping m-bit patterns across the entire sequence. The test is concerning with comparing the frequency of overlapping blocks of two consecutive/adjacent lengths (m and $m+1$) against the expected result for a random sequence.

Figure 11(a) shows that attitudes of the proposed sequences based on different code lengths. The figure shows that generally all the sequences have decreasing attitudes as the code length increased. The figure shows also that the traditional sequence presents the best attitude, as it pass the test all over the code length range. It is clear also that the attitudes of the SB and the ZMSB are degraded as the code length increase over 820.

Figure 11(b) discuss the same attitude against range of the initial conditions. The figure shows that generally the traditional and ZM sequences have better attitudes than the SB and ZMSB. However, it is clear that the attitudes of all the sequences are variable according to the initial values.

Figure 11(c) and Fig. 11(d) represent the 3D plot of the each two sequences against both the code length and the initial conditions. Figure Fig. 11(c) presents the attitude of both the traditional and the ZM sequences. The figure shows that for short code lengths the attitudes are perfect nearly for all the initial conditions, and degrade as the code length increased. Figure 11(d) presents the attitude of the SB and the ZMSB sequences. The figure confirm the results obtained in Fig. 11(a) and Fig. 11(b). The results show that the attitudes of the mentioned sequence are moderate in small code lengths, and degraded as the code length

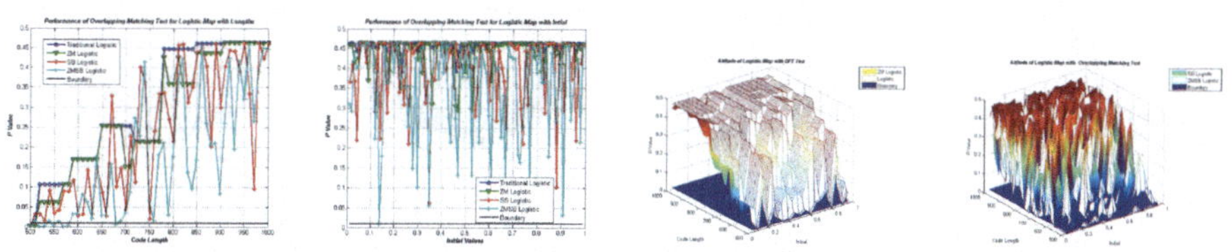

(a) (b) (c) (d)

Fig. 8. Overlapping template matching test: (a) P-value against sequence length, (b) P-value against initial value. Overlapping template matching test: P-value against both sequence length and initial value for (c) traditional and ZM sequences, and (d) ZM and ZMSB sequences.

increased. The figure shows that the obtained results are applied along all the initial conditions.

3.13 Cumulative Sums Test

This test focuses on determine whether the cumulative sum of the partial sequences occurring in the tested sequence is too large or too small relative to the expected behavior of that cumulative sum for random sequences. Figure 12(a) and Fig. 12(d) illustrate the behavior of all the sequences following the same mentioned trend. The results show that all the sequences pass the test with very high *p*-values, nearly reach to 1 for all the code lengths and initial values.

3.14 Random Excursions Test:

This test studies the number of cycles which have K visits in a cumulative sum random walk. It is supposed that the cycle of a random walk consists of a sequence of steps of unit length. These sequences taken at random that begins at and returns to the origin. Generally, the purpose of this test is to determine if the number of visits to a particular state within a cycle deviates from what one would expect for a random sequence.

Figure 13(a) presents the performance against the code length. The figure shows that most the sequences have perfect results all over the code length. The figure shows that the ZMSB has a few lower values at specific code lengths around the length of 650. Figure 13(b) shows the results of the mentioned sequences against the initial values. The figure shows that almost all the sequences pass the test with perfect values. The results show also that the SB sequence has a few lower values at specific code lengths around the initial of 0.2.

Regarding to the 3D plots shown in Fig. 13(c) and Fig. 13(d). As mentioned before, the figures shows that all the sequences have perfect results all over the code lengths and initial values.

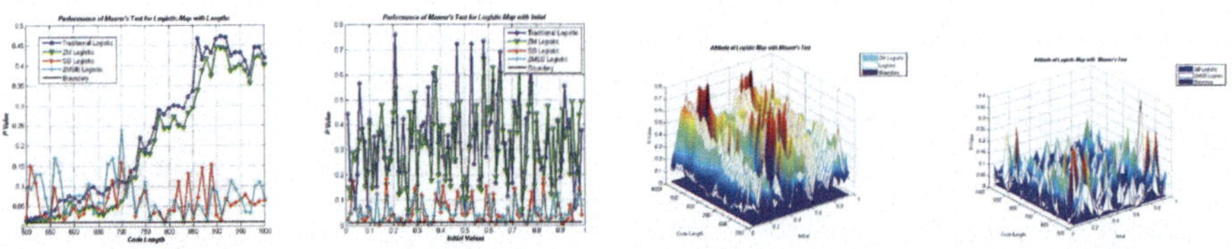

(a) (b) (c) (d)

Fig. 9. Maurer's "universal statistical" test: (a) P-value against sequence length, (b) P-value against initial value. Maurer's "universal statistical" test: P-value against both sequence length and initial value for (c) traditional and ZM sequences, and (d) ZM and ZMSB sequences.

3.15 Random Excursions Variant Test

The focus of this test is the total number of times that a particular state is visited (i.e., occurs) in a cumulative sum random walk. The purpose of this test is to detect deviations from the expected number of visits to various states in the random walk.

Figure 14(a) shows the performance of the presented sequences against the code length. The figure shows that the traditional sequence has the best case. It slightly decreased until reach to length 600, and then gradually increased as the length increased. The rest of the sequences are varying up and down with the different lengths. In general, all the sequences are passed the test.

Regarding to the initial values, Fig. 14(b) shows that almost all the sequences are passed the test. The figure shows that all the sequences have alternative values, however it is clear that the traditional sequences has the best results all over the initial values.

Figure 14(c) and Fig. 14(d) show the 3D results of the mentioned sequences against the code length and initial values under the aforementioned test. Figure 14(c) represents the traditional and the ZM sequences, and confirms the results obtained in Fig. 14(a). The figure shows that there are certain initial values have low p-value for all the lengths. On the other hand, there are certain initial values have high p-value for all the lengths.

Figure 14(d) illustrates the attitudes of the SB and the ZMSB. The figure shows that the results are constructed from spikes, which mean that there is a certain p-value for each pair of code length and initial value.

3.16 Discussion on Tests

This test aims to measure the longest run within a block with a specific size. It is desired to have longest run of ones consistent with that would be expected in the perfect random sequence. shows the attitudes of the 4 different sequences under the effect of the different code lengths. It can be noted that the traditional sequence pass the test with accepted attitude for the first half of the range, and begin to degrade for the rest of the code length value. The ZM sequence has

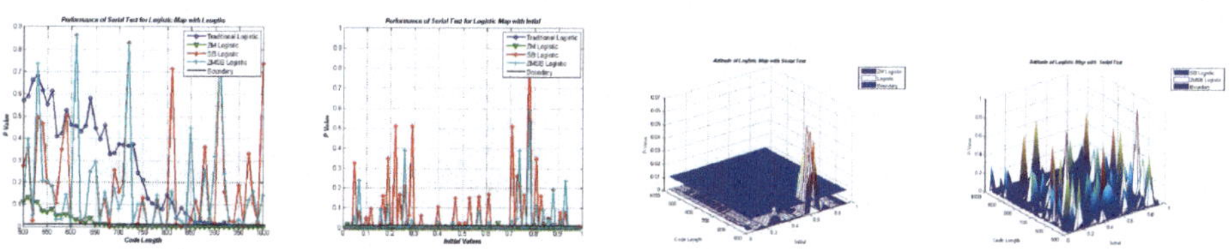

(a) (b) (c) (d)

Fig. 10. Serial test: (a) P-value against sequence length, (b) P-value against initial value. Serial test: P-value against both sequence length and initial value for (c) traditional and ZM sequences, and (d) ZM and ZMSB sequences.

the best performance, especially at the first and the last third of the range. SB sequence has also a good attitude, which means that the ZM and the SB operations individually affect positively on the traditional sequence. Finally, the ZMSB sequence has a moderate attitude, which means that combining the both operations degrades the attitude of this test.

The next section shows applying the proposed chaotic map into the image encryption with security and statistical analyses.

4 Application to Image Encryption

In this section, a set of standard images [1] are used as the input image to be encrypted by the proposed chaotic sequences explained in (1) through (7). Matlab 2017a is utilized to perform the encryption process using a modified version of the standard AES as shown in [2,3]. For example Fig. 15(a) and (b) show the "Peppers" original image and its encrypted (*i.e.,* cyphertext) image, respectively. It can be shown that the encrypted image is a chaotic image and looks like its scrambled image shown in Fig. 15(c). Thus, the encryption process using the proposed chaotic map is subjectively effective.

The security analysis of the proposed chaotic system, in terms of the key space size, is discussed in Sec. 4.1. As well, the statistical sensitivity analysis, in terms of the histogram analysis and chi-square test, is presented in Sect. 4.2.

4.1 Security Analysis

This section depicts the security analysis of the proposed chaotic system shown in Sect. 2 as opposed to the recently competing approaches using the key space size.

It worth noting that the larger the key space, the stronger the ability to resist brute force attacks. In the proposed chaotic system, the initial value key, x_0, is ranging from 0.0 to 1.0 with step-size of 0.01 (*i.e.,* 100 initial values). Whereas, the bifurcation parameter, r, is ranging from 2.5 to 4.0 with step-size of 0.025 (*i.e.,* 60 bifurcation values). Therefore, the key space size becomes $100^{60} \approx 2^{398}$. Generally speaking, in order to prevent brute force security attacks, the key space

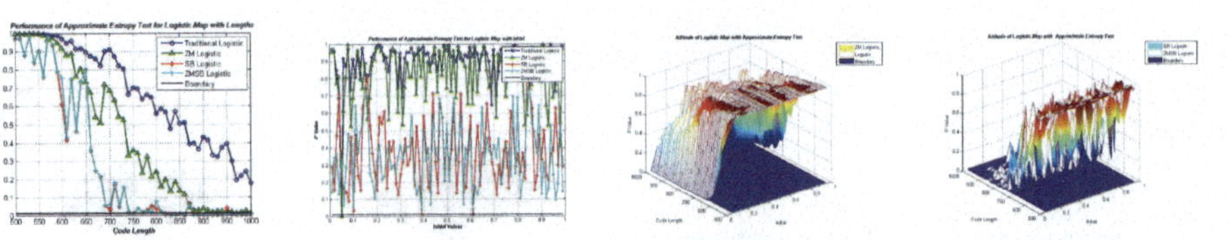

(a) (b) (c) (d)

Fig. 11. Approximate entropy test: (a) P-value against sequence length, (b) P-value against initial value. Approximate entropy test: P-value against both sequence length and initial value for (c) traditional and ZM sequences, and (d) ZM and ZMSB sequences.

of an image encryption system is effective, should its size is greater than 2^{100} [4]. It can shown that the key space of the proposed chaotic-based image encryption approach significantly exceeds not only that point, but also that in [26]. Table 1 shows that the key space of the proposed chaotic system is greater than that of [8, 24, 26]. Recall that the chaotic system, shown in [26], requires determining the Lyapunov exponents, or plotting the phase space trajectories of the chaotic system, in order to avoid weak keys. On the contrary, the proposed chaotic system presents finite number of weak and strong keys, without any additionally computational cost required. Hence, the most strongest key can be easily chosen.

Table 1. Key spaces of the proposed chaotic system compared to the most recent ones, on the basis of the higher, the better; as long as it exceeds 2^{100} as stated in [4].

Reference [8]	Reference [24]	Reference [26]	Proposed
2^{256}	2^{149}	2^{339}	2^{398}

4.2 Statistical Sensitivity

This subsection statistically analyze the image encryption process using the proposed chaotic sequences. This analysis is presented in two perspectives: histogram analysis and chi-square test.

Histogram Analysis. The histogram of a gray image shows the number of pixels of each gray level in that image. This analysis is performed using all standard images shown in [1]. For instance, it can be noticed that the "LivingRoom" image histogram presents many peaks in Fig. 16(b). Whereas, the corresponding encrypted image histogram approaches the uniform distribution as shown in Fig. 16(d). Thus, the encrypted images, using the proposed chaotic sequences, can be considered as a pseudo-random image. Hence, that encrypted image can resist statistical attacks. These statistics can be determined using the Chi-square test as shown in the next subsection.

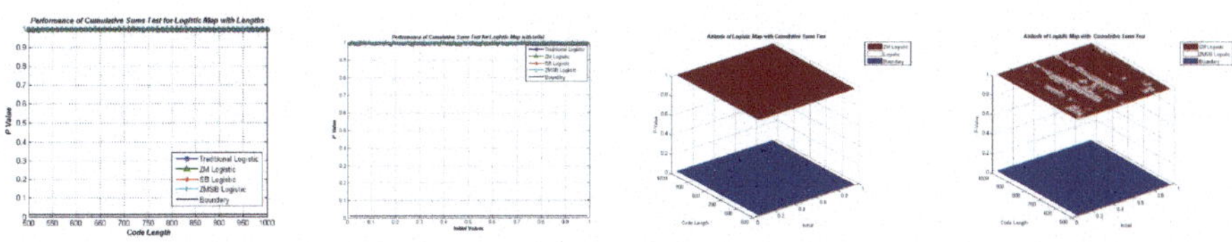

(a) (b) (c) (d)

Fig. 12. Cumulative sums test: (a) P-value against sequence length, (b) P-value against initial value. Cumulative sums test: P-value against both sequence length and initial value for (c) traditional and ZM sequences, and (d) ZM and ZMSB sequences.

Chi-Square Test. The Chi-square test verifies whether an encrypted image can resist statistical attacks or not [11]. The smaller the Chi-square value, the better the uniformity of an encrypted image [7]. The Chi-square function can be cast as,

$$\chi^2 = \sum_{n=1}^{256} \frac{(\Omega_n - \omega_n)^2}{\omega_n}, \tag{8}$$

where Ω_n and ω_n denote the actual frequency and the expected frequency, respectively, of the nth gray level in an encrypted image. For fair comparison with that in [26], the ω_n is set to 256 using the 256×256 standard images in [1]. As well, the encrypted image is generated by changing single bit of the original image. For a confidence level of 0.05, the approach passes the chi-square test, if its χ^2 value does not exceed 295.25. That process is repeated 30 times, then the average value of χ^2 is determined for the whole data set.

The image encryption approach based on the proposed chaotic system is compared to those approaches in [14,20,24,26] using the whole data set in [1]. However, the results of a subset of the standard images are drawn in Table 2 for a fair comparison. It can shown that the χ^2 of the histogram of the encrypted images, using the proposed chaotic system, are less than 295.5. Thus, the histogram of the encrypted images passed the Chi-square test with confidence level = 0.05. Therefore, the proposed chaotic system outperforms those in [14,20,24,26] in terms of χ^2 metric.

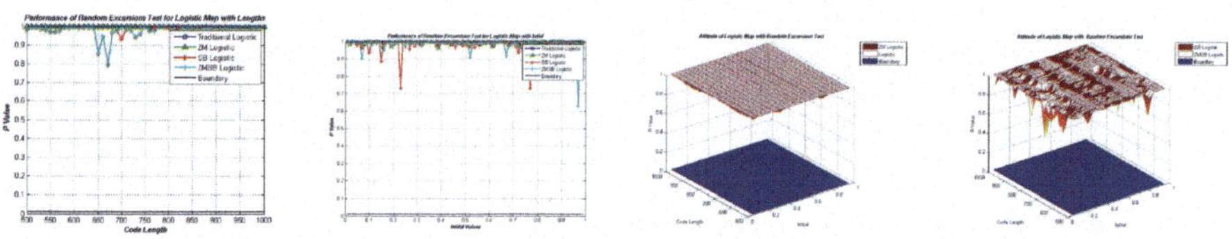

(a) (b) (c) (d)

Fig. 13. Random excursions test: (a) *P*-value against sequence length, (b) *P*-value against initial value. Random excursions test: *P*-value against both sequence length and initial value for (c) traditional and ZM sequences, and (d) ZM and ZMSB sequences.

Table 2. The Chi-square test of the proposed chaotic system compared to the most recent ones in [14,20,24,26] for an encrypted subset of the standard image data set; on the basis of the lower, the better; as long as the χ^2 value does not exceed 295.5.

The image	Reference [14,20]	Reference [24]	Reference [26]	Proposed
Cameraman	288.9823	286.4591	285.3125	**281.6541**
Peppers	269.3387	263.9832	260.3421	**254.7319**
Rice	284.2387	280.0177	278.6172	**272.1128**
Autumn	289.9832	289.4379	288.5792	**283.3692**

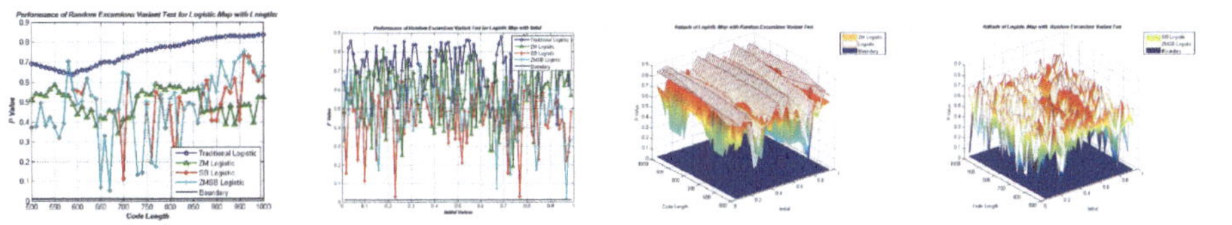

(a) (b) (c) (d)

Fig. 14. Random excursions variant test: (a) P-value against sequence length, (b) P-value against initial value. Random excursions variant test: P-value against both sequence length and initial value for (c) traditional and ZM sequences, and (d) ZM and ZMSB sequences.

(a) (b) (c)

Fig. 15. (a) Peppers original image, (b) its encrypted image, and (c) its scrambled image.

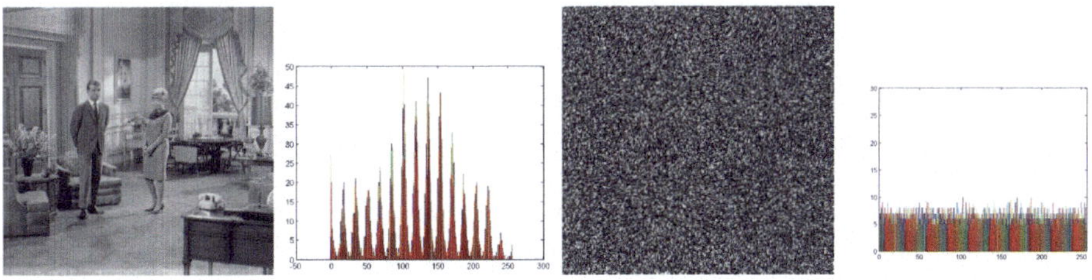

(a) (b) (c) (d)

Fig. 16. Histogram analysis: (a) The 512×512 "Living-Room" original image, (b) the "Living-Room" image histogram; (c) the "Living-Room" encrypted image; (d) the "Living-Room" encrypted image histogram.

5 Conclusions

This paper aims to present a study, an analysis and an evaluation for differently proposed chaotic sequences applied to image encryption using the 15-NIST statistical test suite for random generators. The proposed chaotic system is an improved version of the traditional chaotic sequences based on the logistic map.

The analysis on the proposed chaotic system is performed due to the sensitivity of the chaotic map to the initial condition. The contributions are to study the effect of both initial condition and code lengths variations on the attitude of the chaotic sequences for each test. Nevertheless, a set of optimized values for both initial condition and code lengths are determined based on that analysis. Finally, the proposed chaotic system is applied to image encryption application to verify its robustness against attacks, with security and statistical analyses. The results show that the proposed chaotic system yields significant enhancement in terms of security perspective compared to the conventional systems.

References

1. http://www.imageprocessingplace.com/root_files_V3/image_databases.htm
2. Abdelrahman, A.A., Fouad, M.M., Dahshan, H., Mousa, A.M.: High performance CUDA AES implementation: a quantitative performance analysis approach. In: 2017 Computing Conference, pp. 1077–1085, July 2017. https://doi.org/10.1109/SAI.2017.8252225
3. Abdelrahman, A.A., Fouad, M.M., Dahshan, H.: Analysis on the AES implementation with various granularities on different GPU architectures. Adv. Electr. Electron. Eng. **15**(3), 526–535 (2017). https://doi.org/10.15598/aeee.v15i3.2324
4. Alvarez, G., Li, S.: Some basic cryptographic requirements for chaos-based cryptosystems. Int. J. Bifurcat. Chaos **16**(08), 2129–2151 (2006). https://doi.org/10.1142/S0218127406015970
5. Bassham, L.E., Rukhin, A.L., Nechvatal, J.S.J.R., Smid, M.E., Leigh, S.D., Levenson, M., Vangeland, M., Heckert, N.A., Banks, D.L.: A statistical test suite for random and pseudorandom number generators for cryptographic applications. Technical report 800-22 Rev 1a, National Institute of Standards and Technology (NIST, Gaithersburg, MD, USA (2010)
6. Ben Jemaa, Z., Belghith, S.: Chaotic sequences with good correlation properties for MIMO radar application. In: the 24th International Conference on Software, Telecommunications and Computer Networks, pp. 1–5, September 2016. https://doi.org/10.1109/SOFTCOM.2016.7772127
7. Curiac, D.I., Volosencu, C.: Chaotic trajectory design for monitoring an arbitrary number of specified locations using points of interest. Math. Probl. Eng. **2012**, 1–18 (2012)
8. Guesmi, R., Farah, M., Kachouri, A., Samet, M.: A novel chaos-based image encryption using DNA sequence operation and secure hash algorithm SHA-2. Nonlinear Dyn. **83**, 1123–1136 (2016)
9. Hasan, F.S., Valenzuela, A.A.: Design and analysis of an OFDM-based orthogonal chaotic vector shift keying communication system. IEEE Access **6**, 46322–46333 (2018). https://doi.org/10.1109/ACCESS.2018.2862862
10. Wang, J., Wang, Y.: Analysis performance of MC-CDMA communication system based on improved Chebyshev sequence. In: the 2nd IEEE International Conference on Computer and Communications, pp. 2277–2280, October 2016. https://doi.org/10.1109/CompComm.2016.7925105
11. Khanzadi, H., Eshghi, M., Borujeni, S.: Image encryption using random bit sequence based on chaotic maps. Arab. J. Sci. Eng. **39**, 1039–1047 (2014)

12. Kolosovs, D., Bekeris, E.: Chaos code division multiplexing communication system. In: the 7th International Conference on Computational Intelligence, Communication Systems and Networks, pp. 65–69, June 2015. https://doi.org/10.1109/CICSyN.2015.22
13. Ksheerasagar, T.K., Anuradha, S., Avadhootha, G., Charan, K.S.R., Sri Hari Rao, P: Performance analysis of DS-CDMA using different chaotic sequences. In: International Conference on Wireless Communications, Signal Processing and Networking, pp. 2421–2425, March 2016. https://doi.org/10.1109/WiSPNET.2016.7566577
14. Kulsoom, A., Xiao, D., Aqeel Ur, R., Abbas, S.: An efficient and noise resistive selective image encryption scheme for gray images based on chaotic maps and DNA complementary rules. Multimed. Tools Appl. **75**(1), 1–23 (2016)
15. Litvinenko, A., Aboltins, A.: Use of cross-correlation minimization for performance enhancement of chaotic spreading sequence based asynchronous DS-CDMA system. In: the IEEE 4th Workshop on Advances in Information, Electronic and Electrical Engineering, pp. 1–6, November 2016. https://doi.org/10.1109/AIEEE.2016.7821812
16. Liu, G., Liu, H., Kadir, A.: Hiding message into DNA sequence through DNA coding and chaotic maps. Med. Biol. Eng. Comput. **52**(9), 741–747 (2014). https://doi.org/10.1007/s11517-014-1177-3
17. Makris, G., Antoniou, I.: Cryptography with chaos. In: the 5th Chaotic Modeling and Simulation International Conference, June 2012
18. Manoharan, S., Bhaskar, V.: Pn codes versus chaotic codes: performance comparison in a Gaussian approximated wideband CDMA system over weibull fading channels. J. Franklin Inst. **351**(6), 3378–3404 (2014). https://doi.org/10.1016/j.jfranklin.2014.03.007
19. Mansour, H.A.A., Fu, Y.: A new method for generating a zero mean self-balanced orthogonal chaotic spread spectrum codes. Int. J. Hybrid Inf. Technolo. **7**(3), 345–354 (2014). https://doi.org/10.14257/ijhit.2014.7.3.32
20. Rehman, A., Liao, X., Kulsoom, A., Abbas, S.: Selective encryption for gray images based on chaos and DNA complementary rules. Multimed. Tools Appl. **74**(13), 4655–4677 (2015)
21. Shu, X., Wang, H., Wang, J.: Underwater chaos-based DS-CDMA system. In: the IEEE International Conference on Signal Processing, Communications and Computing, pp. 1–6, September 2015. https://doi.org/10.1109/ICSPCC.2015.7338899
22. Swetha, A., Krishna, B.T.: Generation of biphase sequences using different logistic maps. In: International Conference on Communication and Signal Processing, pp. 2102–2104, April 2016. https://doi.org/10.1109/ICCSP.2016.7754549
23. Tayebi, A., Berber, S., Swain, A.: Performance analysis of chaotic DSSS-CDMA synchronization under jamming attack. Circ. Syst. Signal Process. **35**(12), 4350–4371 (2016). https://doi.org/10.1007/s00034-016-0266-y
24. Wang, X., Zhu, X., Wu, X., Zhang, Y.: Image encryption algorithm based on multiple mixed hash functions and cyclic shift. Opt. Lasers Eng. **107**, 370–379 (2017)
25. Zhou, C., Hu, W., Wang, L., Chen, G.: Turbo trellis-coded differential chaotic modulation. IEEE Trans. Circ. Syst. II Express Brief. **65**(2), 191–195 (2018). https://doi.org/10.1109/TCSII.2017.2709347
26. Zhu, S., Zhu, C., Wang, W.: A new image encryption algorithm based on chaos and secure hash SHA-256. Entropy **20,716**(9), 1–18 (2018). https://doi.org/10.3390/e20090716

Fog Robotics Distributed Computing in a Monitoring Task

Donat Ivanov(✉)

Southern Federal University, 2 Chehova Street, 3479328 Taganrog, Russia
donat.ivanov@gmail.com

Abstract. The paper is aimed at the problem of organizing distributed computing in groups of robots when solving the problem of monitoring natural phenomena using elements of the concept of fog computing. The development of fog computing into fog robotics is shown. A method for organizing foggy robotics is proposed. Particular attention is paid to reducing the load on communications and the redistribution of subtasks between robots in the event of failure of one or more of them.

Keywords: Distributed computing · Fog robotics · Multi-agent technologies

1 Introduction

The idea of combining robots into groups or coalitions to jointly carry out a single task arose a long time ago. Thanks to progress in the field of microelectronics, computer technology, nanotechnology and intelligent control systems, it has become possible to create small-sized robots that are distinguished by a fairly rich set of on-board equipment: sensors, video cameras in various ranges, computing and functional devices. At the same time, in mass production, the cost of such robots is relatively low. All this created the prerequisites for the development of group robotics. In [1–4], the foundations of the theoretical base were laid for the development of robot control systems used in groups to solve a common group problem.

In order to prevent the negative consequences of natural disasters, groups of robots can be used to monitor extended territories, such as the shores of seas and lakes, forests, meadows, etc. On-board sensors can be used to determine the rate of soil erosion, species diversity of the flora, and to collect data from stationary sensors that do not have a permanent connection to a data network. At the same time, the robots of the group need to collect large amounts of data about the study area, pre-process and integrate this information, and transfer data to a remote cloud server.

Transferring all the collected information to the cloud server creates a large load on the data transmission channels. Therefore, the collected information must be partially processed on-board computing devices of the robots of the group. At the same time, solving the problems of moving in space, collecting data, preprocessing data and transferring the results to a cloud server - all these tasks require significant amounts of onboard energy resources. Moreover, in practice, in a group there are almost always robots whose resources are not used to the maximum. In this paper, it is proposed to

R. Silhavy (Ed.): CSOC 2020, AISC 1225, pp. 554–562, 2020.
https://doi.org/10.1007/978-3-030-51971-1_45

use distributed computing and elements of the concept of fog computing in order to equalize the power consumption of the robots of the group and reduce the load on the communication channels.

2 The Fog Computing and Fog Robotics

The concept of fog computing appeared [5, 6], as a further development of the concept of "cloud computing" [7], originally proposed by Cisco experts [8], who showed the technical and economic benefits of equipping network routers and IP cameras with services that allow for initial processing data in the immediate vicinity of data sources or users, which reduced the length of data transmission routes, thereby reducing the burden on the telecommunication environment. Currently, the application of the concept of fog computing is considered promising in the framework of the "Internet of Things" [7, 9, 10], computer networks, global telecommunication networks, projects such as "smart home", wireless sensor networks.

The concept of "fog computing" [11] allows reducing the load on the communication environment by transferring part of the calculations from high-performance servers to local devices ("drops") in order to reduce the computing load on the servers [12]. When distributing the computational load, preference is given to devices located in close proximity to the sources of source data, and/or from the recipients of the calculation results, which can significantly reduce the load on the telecommunication environment. This approach involves obtaining a number of advantages, such as increasing the efficiency of using existing hardware resources and reducing the response time of the system to a user request.

In 2013, Hong et al. proposed the concept of "Mobile Fog" [13] for distributed IoT applications and edge devices. In 2014, a resource allocation model for fog computing was proposed [14]. In the same year, Bonomi et al. [15] examined Internet of Things devices with limited resources as part of fog computing.

In 2017, in [16], the authors propose a basis for minimizing transmission delays in foggy applications by load balancing.

Lee et al. in [17] and Alravis et al. [18] discuss security and privacy issues and propose solutions to reduce the security risks of fog computing.

Thus, the prerequisites were created for the emergence of fog robotic technology.

The pioneering work in the field of fog robotics was considered work [19], which was published in 2017. Just as in the field of distributed computing, fog computing acted as a further development of the cloud technology concept, so in robotics the concept of foggy robotics is presented as a further development of cloudy [20–24] robotics.

The functioning of robots is increasingly associated with data transmission networks, since the volumes of sensory data collected by robots are growing, and the volumes of data required for operation are growing. However, the transmission of such data over public networks leads to the threat of unauthorized access. It was shown in [25] that a number of sensors and actuators of robots using the Robot Operating System (ROS) became vulnerable to unauthorized access via the Internet. The authors of [26] see one of the ways to solve this problem in minimizing the amount of data transmitted by robots via the Internet. And this is achieved through the use of foggy robotics.

The paper considers the computation time in the foggy robotics network [27].

Gudi et al. present the Fog Robotics approach for human-robot interaction [28].

Pop et al. consider the role of fog computing in industrial automation using time-sensitive networks [29].

The work [30] provides a detailed analysis of development trends of approaches to robot control, including a lot of attention paid to the concepts of cloud, edge and fog robotics.

In [31, 32], the use of elements of the concept of fog computing was proposed in relation to groups of robots, where the on-board computing devices of the robots of the group are considered as a distributed computing network.

3 Distributed Computing in Group Robotics

Figure 1 schematically shows the subtasks and information flows of the data collection task by a coalition of intelligent mobile robots (without reference to the devices performing the calculations).

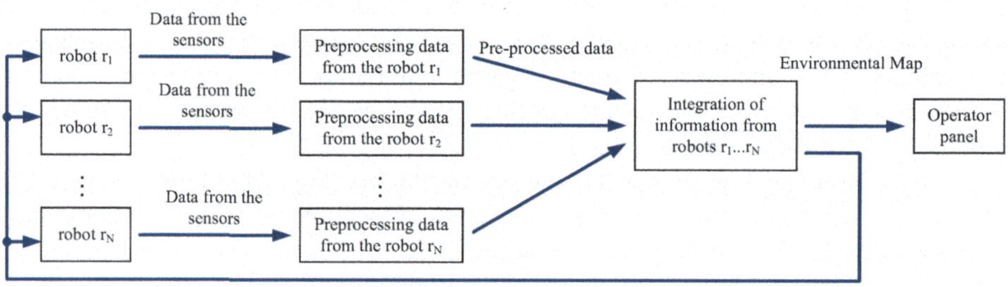

Fig. 1. Sub-tasks and information transfer streams of data collection tasks by a coalition of intelligent mobile robots.

Usually, the information collected by the on-board sensor devices of the robot is processed by the on-board computing device of the same robot, or transmitted to some remote computing device (server or cloud service), which has sufficient computing power to process the received data.

After preliminary processing of data received from individual robots, it is necessary to perform data aggregation to build a map of environmental characteristics common for the group/coalition. That is, the input data for the integration task are pre-processed data from individual robots, and the output is a map of the environmental characteristics in the coalition's working area, which must be sent to the operator's console and, as a rule, to individual coalition robots to ensure high-quality information.

The arising load on the telecommunication and computing devices of robots depends on what data will be transmitted by means of the telecommunication network of a group of robots, and what data is transmitted inside the on-board control system of individual robots.

In the case of combining robots with weak onboard computing devices in a coalition, the task of processing the received data (especially the task of recognizing images on video images) is assigned to a remote powerful computing device, as shown in Fig. 2.

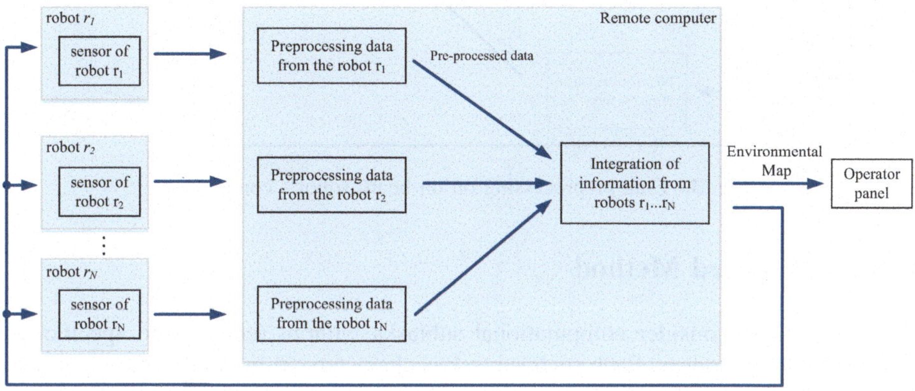

Fig. 2. A common scheme for distributing computational subtasks between computing devices in a data collection task by a coalition of intelligent mobile robots.

In the scheme, shown in Fig. 2, coalition robots are engaged only in collecting environmental information, but all resource-intensive calculations are performed remotely. The disadvantage of such a survey is the high requirements for the bandwidth of the telecommunication network, because in real time it is necessary to transfer large amounts of unprocessed data about the environment and the condition of the robots of the group.

It is possible to propose problems of data preprocessing to be solved on-board computing devices of those robots that received this data. Such a scheme is shown in Fig. 3.

On the one hand, this approach reduces the load on the telecommunication network of a group of robots, but on the other hand, large volumes of data are still required to be transferred between the robots of the group and remote computing devices, which negatively affects the autonomy of the group.

There is a need for such a distribution of computational sub-tasks between participants that would enable the efficient use of on-board computing devices of robots, would not impose excessive requirements on the throughput of the data transmission network and would allow for timely information exchange.

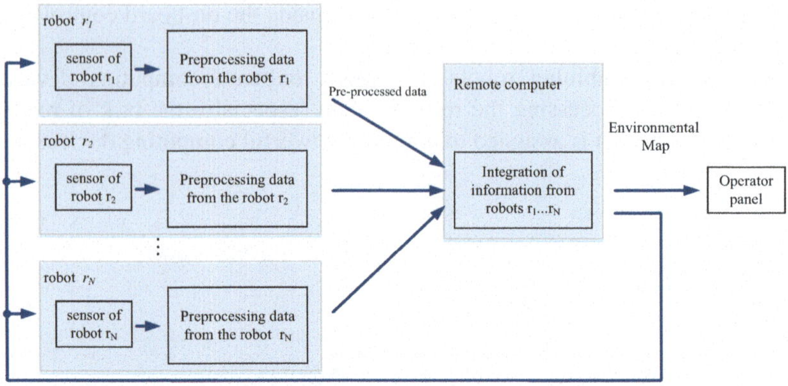

Fig. 3. Data preprocessing on on-board robot's computers

4 The Proposed Method

It is proposed to consider computational subtasks without reference to specific on-board computing devices. Each such subtask is characterized by input and output data. Sources of input for some subtasks are the output of other subtasks. Transfer of output data from one subtask to the input of other subtasks can be carried out directly. And it can be done through a distributed repository of intermediate results (it is advisable if this data is used by several subtasks).

It is proposed to separate the information interaction between robots within the coalition and the information interaction of robots with a dedicated control panel. It is proposed to use the "principle of balancing the computing load" [33, 34] in order to most effectively use the computing resources available in the coalition. And indirectly - to optimize the use of on-board energy resources of coalition robots.

The use of multi-agent interaction is proposed to place computational subtasks on existing computing resources, taking into account the current state of the group robots and the work they perform as part of the overall group task.

The use of elements of the concept of foggy and edge computing is proposed in order to reduce the load on the telecommunication network of the coalition of robots.

With the spread of cloud computing, a number of problems have ripened in telecommunication systems. One of them is the high load on communication channels, caused by the fact that users send large amounts of source data to remote servers, and then significant volumes of data come from these servers to users. And one of the ways to solve this problem was the emergence and development of foggy and edge computing, which consists in shifting the computational load to the "edge" of the network, using both the computing power of communication equipment and controllers, and the computing power of the end (or user) devices.

In this case, the following effect is expected:

- the amount of data transmitted to the cloud is reduced;
- the computational load on the computing environment is reduced, since the data from the end devices are partially or completely processed in the fog layer.

Within the framework of this project, it is proposed the use of elements of the concept of fog computing in relation to coalitions of the MPI. Unlike computing and information management systems, the role of “communication equipment with an available reserve of computing power” can be played by some coalition robots that are geographically located between the coalition’s working area and a remote control device (see Fig. 4).

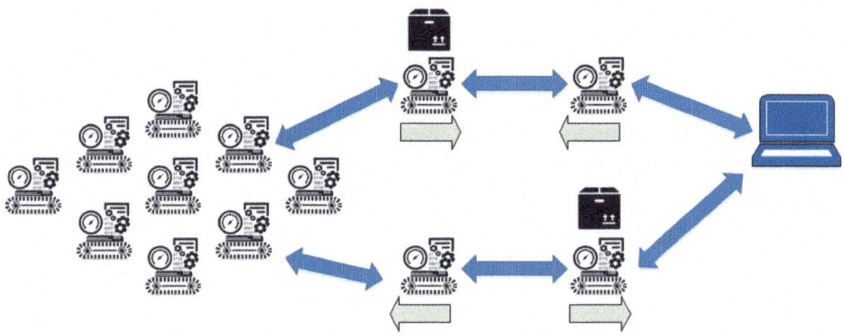

Fig. 4. Schematic illustration of a coalition of intelligent mobile robots and information flows

When performing lengthy work, regardless of the nature of the work itself, some robots are forced to return from the working area to the base due to the exhaustion of the on-board supply of energy for recharging. In some application scenarios, some robots are in reserve, and in some practical tasks, some robots are busy transporting goods (soil samples, etc.) to the base and returning from the base.

In any case, when the robots of the group are advanced to a considerable distance from the base, some robots can be left to perform tasks of maintaining communication with the base (relaying).

It should be noted that the on-board computing facilities of the robots remaining in reserve, en route to the base and returning from it, performing relay tasks are less loaded than the on-board computing power of those robots that perform the main subtasks of the group task.

It is proposed that coalition robots will use multi-agent distributed dispatching methods [33], taking into account the current load of on-board computing devices, as well as the available data on information links between information processing subtasks by a coalition of robots in order to accommodate computational subtasks using available reserves performance, which will reduce the load on the communication network by reducing the distance of data transmission routes between subtasks.

5 Conclusions and Future Work

The paper proposes an approach to the organization of foggy robotics. Particular attention is paid to reducing the communication load and redistributing the computational subtasks between the robots of the group. It is proposed to solve the problems of controlling the distribution of computations using a multi-agent approach.

The proposed approach will find application in systems for monitoring and preventing the negative consequences of natural hazards [35]. As part of such a system, sensor networks, data gateways from meteorological networks, and a group of robots are used to collect data on the state of coastal infrastructure.

Acknowledgement. The reported study was funded by RFBR according to the research project № 18-05-80092, №17-29-07054.

References

1. Dorigo, M.: Swarm-bots and swarmanoid: two experiments in embodied swarm intelligence. In: 2009 IEEE/WIC/ACM International Joint Conference on Web Intelligence and Intelligent Agent Technology, vol. 1 (2009). https://doi.org/10.1109/WI-IAT.2009.370
2. Valentini, G., Ferrante, E., Hamann, H., Dorigo, M.: Collective decision with 100 Kilobots: speed versus accuracy in binary discrimination problems. Auton. Agent. Multi. Agent. Syst. **30**, 553–580 (2016)
3. Cao, Y.U., Fukunaga, A.S., Kahng, A.B., Meng, F.: Cooperative mobile robotics: antecedents and directions. In: Proceedings 1995 IEEE/RSJ International Conference on Intelligent Robots and Systems. Human Robot Interaction and Cooperative Robots, vol. 23, pp. 226–234 (1997). https://doi.org/10.1109/IROS.1995.525801
4. Sahin, E.: Swarm robotics: from sources of inspiration. In: Swarm Robotics Workshop: State-of-the-Art Survey, pp. 10–20 (2005). https://doi.org/10.1007/978-3-540-30552-1_2
5. Bar-Magen, J.: Fog computing: introduction to a new cloud evolution (2013)
6. Bonomi, F.: Connected vehicles, the internet of things, and fog computing. In: The Eighth ACM International Workshop on Vehicular Inter-Networking (VANET), Las Vegas, USA, pp. 13–15 (2011)
7. Бородин, В.А.: Интернет вещей-следующий этап цифровой революции. Образовательные ресурсы и технологии, pp. 178–181 (2014)
8. Bonomi, F., Milito, R., Zhu, J., Addepalli, S.: Fog computing and its role in the internet of things. In: Proceedings of the First Edition of the MCC Workshop on Mobile Cloud Computing, pp. 13–16 (2012). https://doi.org/10.1145/2342509.2342513
9. Atzori, L., Iera, A., Morabito, G.: The internet of things: a survey. Comput. Netw. **54**, 2787–2805 (2010)
10. Familiar, B.: Microservices, IoT, and Azure. Springer, Heidelberg (2015)
11. Yi, S., Hao, Z., Qin, Z., Li, Q.: Fog computing: platform and applications. In: Proceedings - 3rd Workshop on Hot Topics in Web Systems and Technologies, HotWeb 2015, pp. 73–78 (2016). https://doi.org/10.1109/HotWeb.2015.22
12. Stojmenovic, I., Wen, S.: The fog computing paradigm: scenarios and security issues. In: Proceedings of 2014 Federated Conference on Computer Science and Information Systems, vol. 2, pp. 1–8 (2014). https://doi.org/10.15439/2014F503
13. Hong, K., Lillethun, D., Ramachandran, U., Ottenwälder, B., Koldehofe, B.: Mobile fog: a programming model for large-scale applications on the internet of things. In: Proceedings of the Second ACM SIGCOMM Workshop on Mobile Cloud Computing, pp. 15–20 (2013)
14. Aazam, M., Huh, E.N.: Fog computing and smart gateway based communication for cloud of things. In: Proceedings - 2014 International Conference on Future Internet of Things and Cloud, FiCloud 2014, pp. 464–470 (2014). https://doi.org/10.1109/FiCloud.2014.83

15. Bonomi, F., Milito, R., Natarajan, P., Zhu, J.: Fog computing: a platform for internet of things and analytics. In: Big Data and Internet of Things: A Roadmap for Smart Environments, pp. 169–186. Springer, Heidelberg (2014)
16. Yousefpour, A., Ishigaki, G., Jue, J.P.: Fog computing: towards minimizing delay in the internet of things. In: 2017 IEEE International Conference on Edge Computing (EDGE), pp. 17–24 (2017)
17. Lee, K., Kim, D., Ha, D., Rajput, U., Oh, H.: On security and privacy issues of fog computing supported internet of things environment. In: 2015 6th International Conference on the Network of the Future (NOF), pp. 1–3 (2015)
18. Alrawais, A., Alhothaily, A., Hu, C., Cheng, X.: Fog computing for the internet of things: security and privacy issues. IEEE Internet Comput. **21**, 34–42 (2017)
19. Gudi, S., Ojha, S., Clark, J., Johnston, B., Williams, M.-A.: Fog robotics: an introduction. In: IEEE/RSJ International Conference on Intelligent Robots and Systems (2017)
20. Kuffner, J.: Cloud-enabled humanoid robots. In: 2010 10th IEEE-RAS International Conference on Humanoid Robots (Humanoids), Nashville TN, United States, December 2010
21. Hu, G., Tay, W.P., Wen, Y.: Cloud robotics: architecture, challenges and applications. IEEE Netw. **26**, 21–28 (2012)
22. Kehoe, B., Patil, S., Abbeel, P., Goldberg, K.: A survey of research on cloud robotics and automation. IEEE Trans. Autom. Sci. Eng. **12**, 398–409 (2015)
23. Mohanarajah, G., Hunziker, D., D'Andrea, R., Waibel, M.: Rapyuta: a cloud robotics platform. IEEE Trans. Autom. Sci. Eng. **12**, 481–493 (2014)
24. Turnbull, L., Samanta, B.: Cloud robotics: formation control of a multi robot system utilizing cloud infrastructure. In: 2013 Proceedings of IEEE Southeastcon, pp. 1–4 (2013)
25. DeMarinis, N., Tellex, S., Kemerlis, V.P., Konidaris, G., Fonseca, R.: Scanning the internet for ROS: a view of security in robotics research. In: 2019 International Conference on Robotics and Automation (ICRA), pp. 8514–8521 (2019)
26. Tanwani, A.K., Mor, N., Kubiatowicz, J., Gonzalez, J.E., Goldberg, K.: A fog robotics approach to deep robot learning: application to object recognition and grasp planning in surface decluttering. arXiv Preprint arXiv:1903.09589 (2019)
27. Mouradian, C., Naboulsi, D., Yangui, S., Glitho, R.H., Morrow, M.J., Polakos, P.A.: A comprehensive survey on fog computing: state-of-the-art and research challenges. IEEE Commun. Surv. Tutor. **20**, 416–464 (2017)
28. Gudi, S.L.K.C., Ojha, S., Johnston, B., Clark, J., Williams, M.-A.: Fog robotics for efficient, fluent and robust human-robot interaction. In: 2018 IEEE 17th International Symposium on Network Computing and Applications (NCA), pp. 1–5 (2018)
29. Pop, P., Raagaard, M.L., Gutierrez, M., Steiner, W.: Enabling fog computing for industrial automation through time-sensitive networking (TSN). IEEE Commun. Stand. Mag. **2**, 55–61 (2018)
30. Song, D., Tanwani, A.K., Goldberg, K., Siciliano, B.: Networked-, cloud-and fog-robotics. Springer, Cham (2019)
31. Melnik, E.V., Klimenko, A.B., Ivanov D. Y.: The model of device community forming problem for the geographically-distributed information and control systems using fog-computing concept. In: Proceedings of the IV International Research Conference Information Technologies in Science, Management, Social Sphere and Medicine, pp. 132–136 (2017)
32. Korovin, I., Melnik, E., Klimenko, A.: The fog-computing based reliability enhancement in the robot swarm. In: International Conference on Interactive Collaborative Robotics, pp. 161–169 (2019)

33. Melnik, E., Klimenko, A.: A novel approach to the reconfigurable distributed information and control systems load-balancing improvement. In: Application of Information and Communication Technologies, Moscow (2017)
34. Klimenko, A.B., Ivanov, D.I., Melnik, E.V.: The configuration generation problem for the informational and control systems with the performance redundancy. In: International Conference on Industrial Engineering, Applications and Manufacturing (ICIEAM), pp. 1–5 (2016)
35. Orda-Zhigulina, M.V., Melnik, E.V., Ivanov, D.Y., Rodina, A.A., Orda-Zhigulina, D.V.: Combined method of monitoring and predicting of hazardous phenomena. In: Computer Science On-line Conference, pp. 55–61 (2019)

Remote Sensing Image Processing Based on Modified Fuzzy Algorithm

Viktor Mochalov(✉), Olga Grigorieva, Denis Zhukov, Andrei Markov, and Alisher Saidov

Mozhaisky Aerospace Academy, St. Petersburg, Russia
vicavia@yandex.ru

Abstract. The study focused on the problem of natural objects image segmentation using intelligent technology. As the natural objects for this research the authors chose tundra vegetation. Concerned organizations often monitor hard-to-reach remote natural and technological objects using aerospace surveillance tools. Multi- and hyperspectral data provide them with automated image segmentation and assessment of the state of landscape elements. At the same time, the processing of space imagery is a difficult task which often goes under uncertainty. The formation of a training sample for image processing algorithms is a component of the integrated processing technology of heterogeneous data.

The aim of the work is to present a monitoring technology for natural objects that applies the mathematical apparatus of fuzzy logic to automated processing of multi- and hyperspectral aerospace imagery. The technology is likely to be used in regional and industry-specific decision support systems for managing complex natural and technological objects.

Keywords: Fuzzy logic · Spectrometric measurement data · Image segmentation · Identification of the species composition · Entropy · Cluster analysis · Volume requirements · Ground and remote measurements

1 Introduction

Currently, there are a number of practical tasks that require informed management decisions based on real-time automated processing of aerospace imagery. Such tasks, in particular, include monitoring of the dynamics of plant communities in hard-to-reach areas of tundra under conditions of climate change and influence of anthropogenic factors [1].

In the course of the Arctic program, regularly updated by NASA, in November 2018, the team of authors presented the results of assessment of global changes in the state of tundra vegetation. As the input data scientists process satellite imagery since 1982 and assess the state of vegetation according to the NDVI index which in its turn depends on climate conditions. The results of similar studies are given in the article [2].

The presented studies are based mainly on the processing of space imagery from "Landsat" (NASA) and "Sentinel-2" (ESA). This article demonstrates that it is also possible and practically appropriate to use supplementary sources of heterogeneous data, such as: the results of hyperspectral ground-based measurements, as well as a geobotanical description of the territory.

R. Silhavy (Ed.): CSOC 2020, AISC 1225, pp. 563–572, 2020.
https://doi.org/10.1007/978-3-030-51971-1_46

A number of European universities also pay serious attention to the study of the vegetation cover in the Arctic based on Remote sensing (RS) data. He showed an example of the classification of 20 types of vegetation and 5 types of water bodies utilizing the processing of multispectral RS data and ground-based measurements.

In comparison with the outlined approaches, providing a generalized assessment of the state of tundra vegetation cover, in this article the team of authors focused on improving the quality of monitoring by introducing the fuzzy logic apparatus. In the working process, the authors have developed the technological assessment scheme of the state of tundra vegetation [3]. However, it should be noted that identification of the species composition and assessment of the state of tundra vegetation are integral operations of the monitoring technology for the natural objects based on processing of multi- and hyperspectral aerospace imagery.

The significance of the problem is determined by the fact that tundra vegetation in the conditions of the Extreme North changes its properties and areola of growth under the influence of global climate changes. Besides that, vegetation is exposed to human impact in the form of vigorous economic activity and intensive reindeer herding [3].

The technical capabilities of remote sensing data (RSD) from Sentinel-2 in combination with the geobotanical description (GBD) allow us to move from vegetation inventory to monitoring the dynamics of changes in plant communities.

2 Methods

The development of a structural scheme of the technology as a series of cyclic operations. The scheme of image processing based on modified Fuzzy Algorithm is illustrated in Fig. 1.

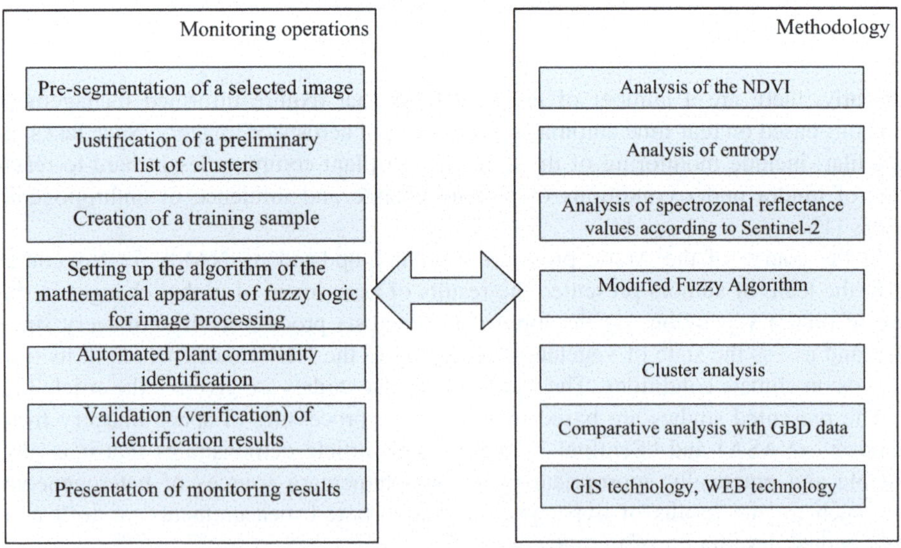

Fig. 1. The image processing scheme based on modified Fuzzy Algorithm

In the diagram, the monitoring operations (on the left) line by line coincide with the conventional name of the methodology (on the right) involved by the investigators. RSD and GBD are used here as the input data.

The monitoring task is an automated image segmentation or transition from the image of multiple dots to the image of multiple classes of objects. In our case, the class of object is a relatively homogeneous plant community. Since the task of recognizing the plant communities on a multispectral image supposes data analysis under conditions of uncertainty, it is valid to use the so-called clustering algorithms under uncertainty.

2.1 Pre-segmentation of the Selected Image

The goal of a preliminary segmentation is to allocate water bodies, as well as the anthropogenic objects (roads) in order to exclude them from consideration when identifying plant communities. During a preliminary segmentation the authors calculated the Normalised Difference Vegetation Index (NDVI) in accordance with the formula (1) [8] and analyzed if the index values are in the given boundaries.

$$NDVI = \frac{NIR - VIR}{NIR + VIR}. \tag{1}$$

2.2 Justification of a Preliminary List of Clusters

In this methodology the aerospace materials and the results of ground-based measurements (geobotanical description of the territory) are used as the initial data.

The methodology involves the following steps:

1. Transformation of the space images by the principal component method [6] in order to switch from the original coordinate system (a set of spectral channels) to a new basis $\rho(x, y, \lambda) \rightarrow \rho(x, y, \lambda')$, which corresponds to the largest spread of spectral reflectance coefficients in a scattering diagram.
2. Selection of the plant communities contours $N = (1, 2, \ldots n)$ with the known properties, previously determined according to the ground measurements, from the transformed image.
3. Calculation of the signal probability (from the image element) to hit a given sequential set of intervals of the reflectance coefficient $P(n, m)$, [0; 1], in the selected contours, where n is a sign of belonging to the contour, and m is a conditional number of the image element (pixel) in the contour.
4. Calculation of the entropy for each contour [9] as a measure of randomness and uncertainty of information in the contour:

$$H_n(x) = -\sum\nolimits_{m=1} P_{n,m} \quad log_2(P_{n,m}). \tag{2}$$

5. Calculation of the entropy for the entire image containing the selected contours and the surrounding territory:

$$H_0(x) = -\sum\nolimits_{m=1} P_{0,m} \quad \log_2(P_{0,m}), \tag{3}$$

where 0 is a sign of the entropy belonging to the whole image, m is a conditional serial number of the image element.

6. Calculation of the combinatorial indicators of entropy $H^k_{n,l}(x)$ for contours' combinations; the number of combinations and calculated entropy indicators (2) is calculated by the classical formula of combinatorics:

$$C^k_n = \frac{n!}{k!(n-k)!}. \tag{4}$$

Formula (4) allows us to find in how many ways we can choose k contours from the considered n contours. The number $C^k_n = l$ determines the total number of the estimated combinatorial indicators of entropy $H^k_{n,l}(x)$ with a sequential increase of k from 1 to n.

So, if at the first step n = 9 contours, then for the combinations of any two contours $(k = 2)$ we calculate $l = 36$ values of $H^k_{n,l}(x)$; and for the combinations of three contours, respectively, it is necessary to calculate $l = 84$ values of $H^k_{n,l}(x)$.

7. Calculation of $k = 1, 2, \ldots, n$ values of the averaged entropy indicators by the formula:

$$\bar{H}_k(x) - \frac{\sum_{l=1} H^k_{n,l}(x)}{C^k_n} \tag{5}$$

8. Analysis of the change in values of $\bar{H}_k(x)$ depending on $k = 2, 3, \ldots, N$, comparison with value of $H_0(x)$ and justification of the number of clusters for the further identification of plant communities.

The methodology assumes that the value of the entropy indicator $H_0(x)$ will be slightly higher, at least for the first few values of the entropy indicator calculated for some combinations of plant communities $\bar{H}_k(x)$.

2.3 Creation of a Training Sample

In the proposed improved algorithm, the cluster centers are not determined randomly, as in the standard fuzzy clustering algorithm [10], but are shaped taking into account preliminary justification of the list of clusters that form training data.

2.4 Setting up the Algorithm of the Mathematical Apparatus of Fuzzy Logic for Image Processing

The problem of image segmentation of plant communities is characterized by the uncertainty; therefore, it is advantageous to apply clustering algorithms based on the fuzzy logic tools to solve it.

The authors analyzed the theory and proposed an improved apparatus of fuzzy logic for identifying the species composition of tundra vegetation [4].

Traditionally, a function calculated by the Euclidean metric can be used as a function of the distance $d(r_i, r_q)$ in the l-dimensional space of spectral characteristics:

$$d(r_{i,}r_q) = \sqrt{\sum\nolimits_{j=1}^{l} (r_i^j - r_q^j)^2}, \tag{6}$$

where R is a set of multispectral image points to be segmented. Each point corresponds to a vector of numerical values $((r_i^1, r_i^2, \ldots r_i^l)$, where r_i^j is the numerical value of the j spectral characteristic for the i point $(i = \overline{1,n}; j = \overline{1,l})$. For Sentinel-2 survey materials, l can be taken equal to 13.

The proposed improved apparatus for image segmentation and further identification of the main types of plant communities is grounded on the detection and further analysis of the membership function matrix. The initialization of the matrix of membership functions U, which determines the degree of belonging of the i-pixel to the k-cluster, is carried out as follows:

1. It is calculated the distance between the center of the class $R_k^{э} = \{r_{kl}{}^{э}\}$ and each pixel brightness vector $R_i = \{r_{il}\}$ where l is the wavelength number.
   In the standard algorithm, the distance from a pixel to a cluster is taken as the Euclidean (3). However, for mixed plant communities, this choice is not always justified. It is valid only if there is no inversion observed between the spectral signatures of individual species of plant communities. In the formula (3), the sum of the squared differences is used as a criterion for a pixel to belong to a cluster. Sometimes it is more efficient to take the Mahalanobis distance [8] for the analyses of vegetation cover, which is a complex unit of objects with different spectral reflective properties. However, in the case of hyperspectral data, there might be problems with inversion of the covariance matrix. Therefore, it was proposed to use the Terebizh metric [9] as a distance measure:

$$\mathrm{d_{ik}} = \Sigma_{\lambda=1}^{\lambda_n} (\mathrm{R_i}\ (\lambda) - \mathrm{R_k^{э}}\ (\lambda))^2 \,/\mathrm{R_k^{э}}(\lambda). \tag{7}$$

   As a result, the n × p matrix D is formed from d_{ik}, where n is a number of clusters, p is a number of pixels of a hyperspectral image.
2. The elements of the membership function matrix U are calculated

$$u_{ik} = t_{ik}/(1_n \sum\nolimits_{k=1}^{n} t_{ik}), \tag{8}$$

   where $t_{ik} = d_{ik}^{-2/(m-1)}$ form the matrix T, m is the fuzzificator.
3. The objective function, which must be minimized, is estimated like this:

$$F = \sum_{i=1}^{p} \left(\sum_{k=1}^{n} d_{ik}^{2} u_{ik}^{m} \right). \tag{9}$$

4. If the number of specified iterations is not completed and the specified classification accuracy is not achieved ε: $|F(j) - F(j - 1)| \leq \varepsilon$, then new class centers are calculated by the formula:

$$r_{kl}{}^{э} = \sum_{i=1}^{p} (u_{ik}{}^{m} r_{il}) \, / \sum_{i=1}^{p} u_{ik}{}^{m} \tag{10}$$

and steps 1–3 are repeated.

2.5 Automated Plant Community Identification

Automated identification is carried out in the special purpose software package. As the initial parameters the authors input the results of tuning of the fuzzy logic algorithm.

2.6 Validation (Verification) of Identification Results

The operation relied on ground-based measurements made in the summer 2019 [7]. The measurements and classification results are the basis for the conditional identification of several (about seven) species of plant communities. For each type of plant community two samples were made, a training sample (the territory with a corresponding cluster) and a control sample (the territory with a similar type of plant community).

2.7 Presentation of Monitoring Results

The results of aerospace imagery processing are presented in the form of a thematic layer of a digital map. In addition, the output is posted on the relevant web resources.

3 Results

To present the methodology of remote sensing image processing, the authors selected Sentinel-2 S2A_MSIL2A_20180801T080711_N9999_R078_T39WXQ_20191107T185442 multispectral survey materials. An atmospheric correction was carried out. Upon that each image pixel contains information about the spectral reflectance coefficients of landscape elements in 13 spectral channels. The spatial resolution of the image is from 10 to 60 meters for various channels. A fragment of the image is shown in Fig. 2.

On the territory of Nenets Autonomous Okrug, Russia, in the region of the Shapkin River and its tributaries, near the highway from 5 to 10 relatively homogeneous species of plant communities were identified during ground work.

The monitoring task is an automated image segmentation or transition from the image of multiple dots to the image of multiple classes of objects. In our case, the class of object is a relatively homogeneous plant community. Since the task of recognizing plant communities on a multispectral image supposes data analysis under conditions of uncertainty, the use of the so-called clustering algorithms in the face of uncertainty is justified.

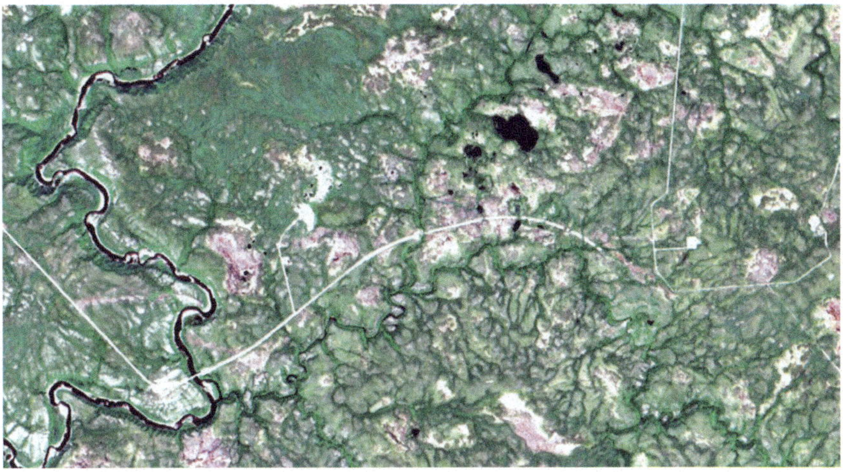

Fig. 2. Fragment of the work area image

The NDVI index was calculated on the basis of the brightness values in the respective Sentinel2 channels

$$NDVI = \frac{Band\,8 - Band\,4}{Band\,8 + Band\,4}. \tag{11}$$

The NDVI index analysis helps to distinguish the territory of anthropogenic objects free from plant communities.

To substantiate the list of clusters for the entropy estimation, the number of them was continuously increased. The authors concluded that, for the chosen example, nine clusters are enough and let us identify the main species of plant communities.

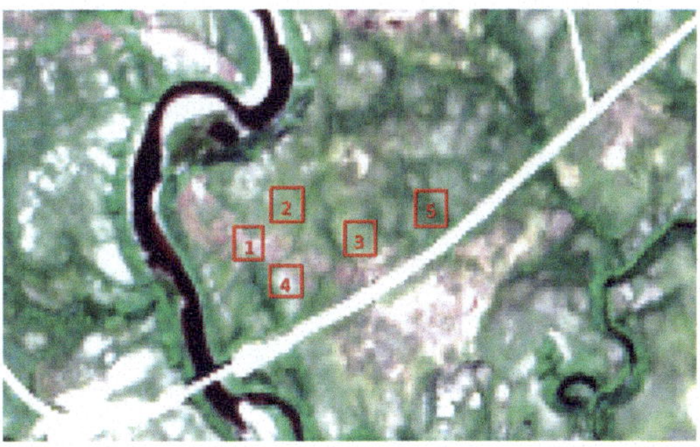

Fig. 3. The spatial layout of the territories selected as training samples

The spatial layout of the territories selected as training samples is shown in Fig. 3. In the image the clusters with conditionally homogeneous species of plant communities are marked with numbers.

The results of substantiating the number of clusters based on entropy analysis are presented in Fig. 4 This dependence is well approximated by the expression:

$$H(x) = \frac{1}{n^{-2,4}} + 0,08\,\mathrm{n} + 1,5, \tag{12}$$

where n - is a number of clusters.

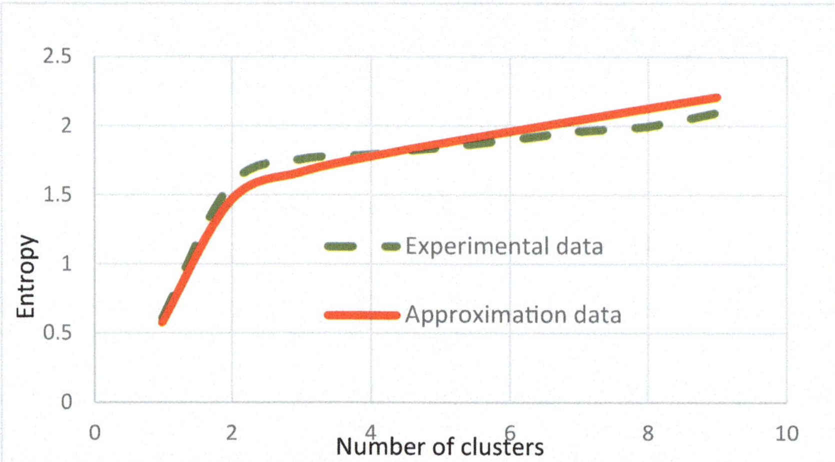

Fig. 4. The spatial layout of the territories selected as training samples

Setting the algorithms of the mathematical apparatus of fuzzy logic made it possible to localize tundra parts in the selected image in accordance with the selected clusters. There were S = 16 iterations when the coordinates of the clusters centers were refined and the values of the membership function were renewed. The results of the automated identification are shown on a digital map in Fig. 5.

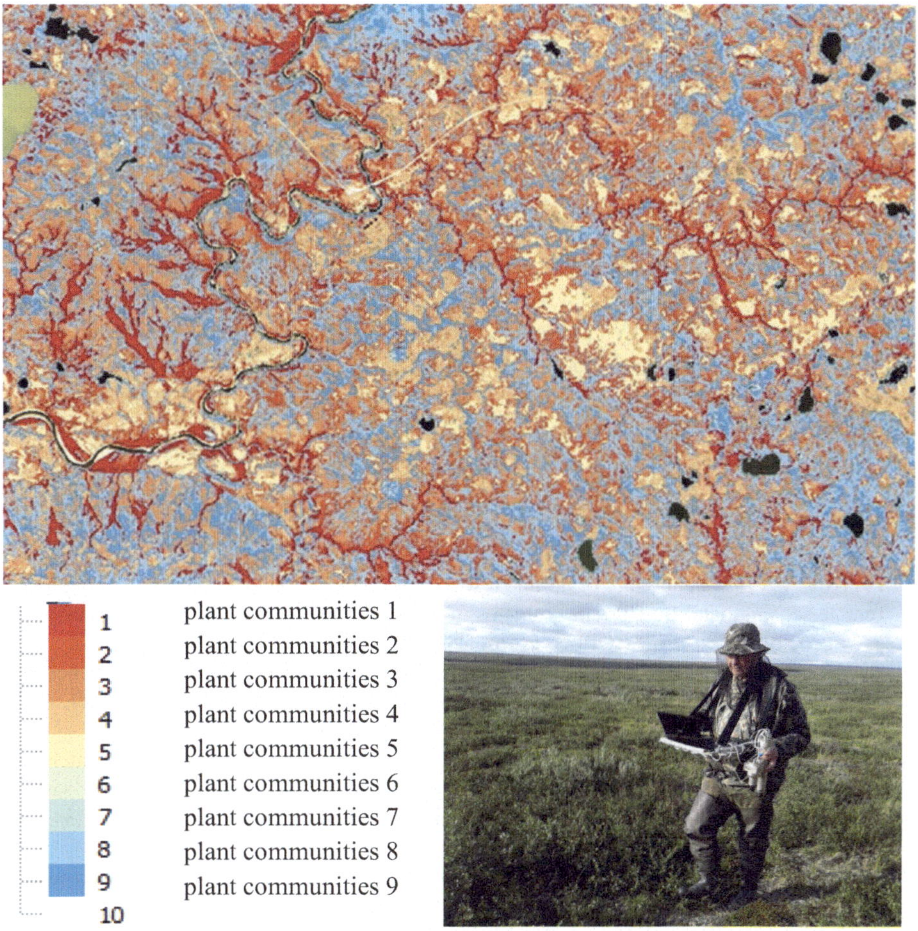

Fig. 5. Results of plant community identification and ground measurements for verification of plant communities identification

4 Discussions

The proposed technology provided processing of aerospace imagery. Tundra plant communities were identified. The verification results confirmed a high quality of the found solution and the validity of the mathematical apparatus of fuzzy logic.

The selected list of initial clusters can be expanded, if necessary. Given the wide variety of tundra plant communities, the list of identifiable surfaces can also be enhanced with the mixed plant communities. The proposed methodology for justification of the number of clusters is original and can be used to assess the information capabilities of multi- and hyperspectral imagery for automated identification of plant communities and other landscape elements.

At the same time, the proposed approach involves operations preceding the direct identification of plant communities. A preliminary analysis of image entropy will allow us to estimate information saturation of images depending on the time of shooting and then to choose the most informative seasonal dates for it and for further processing.

If we consider processing imagery taken over different times it will provide us with the opportunity to analyze the dynamics of spatial changes in the species of plant communities which depends on changes in climatic conditions or come under the influence of anthropogenic factors.

The proposed technology has the following practical significance: it will ensure annual planning of reindeer grazing; provide data on the impact of climatic conditions on the state of tundra vegetation; and will timely identify any possible adverse effects of anthropogenic loading.

The proposed technology can be used not only in Russia, but also in other regions of the world with a hard-to-reach terrain (steppes, deserts, emergency zones, conflicts). Further planning includes justification of the requirements for the volume of ground measurements.

References

1. Elsakov, V.V.: A technology of on-line resource estimation of reindeer pastures from optical remote sensing data. In: Current Problems in Remote Sensing of the Earth from Space, vol. 11, no. 1, pp. 245–255. SCOPUS, Web of Science (2014)
2. Becher, M., Olofsson, J., Berglund, L., Klaminder, J.: Decreased cryogenic disturbance: one of the potential mechanisms behind the vegetation change in the Arctic. Polar Biol. **41**(1), 101–110 (2017). https://doi.org/10.1007/s00300-017-2173-5
3. Mochalov, V., Grigorieva, O., Zelentsov, V., Markov, A., Ivanets, M.: Intelligent technologies and methods of tundra vegetation properties detection using satellite multispectral imagery. In: Advances in Intelligent Systems and Computing, pp. 234–243. Springer, Cham (2019). ISSN 2194-5357
4. Grigoryeva, O.V., Saidov, A.G., Kudro, D.V.: Ensemble algorithm of hyperspectral data processing based on fuzzy set of clusters training in the problem of classification of vegetation. In: Proceedings of the Mozhaisky Military Aerospace Academy. – SPb.: Mozhaisky MAA (2018)
5. Demidova, L.A., Nesterov, N.I., Tishkin, R.V.: Possibilistic-fuzzy segmentation of earth surface images by means of genetic algorithms and artificial neural networks. St. Petersburg State Polytechnical Univ. J. Comput. Sci. Telecommun. Control Syst. **3**, 37–47 (2014). ISSN online 2618-8694
6. Bezdek, J., Ehrlich, R., Full, W.: FCM: fuzzy C-means algorithm. Comput. Geosci. **10**(2), 191–203 (1984)
7. Mochalov, V., Grigorieva, O., Lavrinenko, I.: Initial data for identification of vegetation in southern Tundra based on the processing of multi- and hyperspectral data In: Materials of the 17th All-Russian Open Conference 'Modern problems of Earth Remote Sensing from Space' IKI RAS, Moscow, p. 438 (2019). https://doi.org/10.21046/17dzzconf-2019
8. Schowengerdt, R.: Remote Sensing, Models and Methods for Image Processing, 515 p. Academic Press, Burlington (2007)
9. Terebizh, V.: Introduction to the statistical theory of inverse problems, 376 p. PHYSMATLITIS, Moscow (2005). ISBN 5-9221-0562-0. (in Russian)
10. Karpenko, A.P.: Modern Search Engine Optimization Algorithms. Algorithms Inspired by Nature: Training manual, 446 p. Publishing house of MGTU n N.E. Bauman, Moscow (2014). (in Russian)

Human Pose Estimation Applying ANN While RGB-D Cameras Video Handling

Iakov Korovin and Donat Ivanov(✉)

Southern Federal University, 2 Chehova Street, 3479328 Taganrog, Russia
donat.ivanov@gmail.com

Abstract. In the paper we try to solve the problem of online human pose estimation. All over the world security systems of railway stations, airports and other facilities have control points. Control points are equipped with cameras and other devices for automated data acquisition. One of the tasks within security providing is to recognize the position of body parts of the human, passing the control point. In this paper, a method, based on the use of artificial neural networks. The features of the practical implementation of the proposed approach are described.

Keywords: Human pose estimation · Body parts recognition · Pose recognition · Video recognition

1 Introduction

The task of pose estimation is relevant when solving a large number of practical problems. In the paper we describe a human pose estimation in the safety systems of railway stations, airports and other facilities.

Usually each visitor passes a control point. At the control point, a video camera is installed. One of the tasks of video analysis is to recognize the position of parts of the body of the visitor. In solving this problem, it is necessary to determine the position of the parts of the human body passing the control point, based on data from existing cameras.

The assessment of a human posture is an urgent task [1–7] due to different applications, such as motion capture, manipulation of objects in virtual environments, augmented reality, remote control of robotics means, etc. Evaluation of a person's posture and its change in time also allows to determine the actions performed by a person, or his behavior, which is an important task in the construction of video surveillance and security systems.

The process of assessing a person's posture is related to the search for posture parameters of a human body model that is most suitable for observations on one or more input images. While marker-based systems are already available in many commercial applications, marker-free posture assessment is still a complex research topic.

There are many algorithms that solve this problem with high accuracy from several input images [8, 9] or even one photograph [10, 11]. Often, such systems require manual initialization and cannot process camera images in real time. Promising

R. Silhavy (Ed.): CSOC 2020, AISC 1225, pp. 573–585, 2020.
https://doi.org/10.1007/978-3-030-51971-1_47

methods for assessing human posture in real time use a three-dimensional body model [12–14], time-of-flight depth cameras [13], or RGB-D cameras such as Microsoft Kinect, Intel RealSense, Asus Xonar, etc. [15].

In terms of the type of data used to construct the human skeleton, it can be divided into methods based on the use of only color (RGB) [16–20] and methods combining color and depth data (RGB-D) [21–25]. The definition of human actions can be based either on specified criteria for each type of action, or deep learning algorithms can be used to solve this problem. Regardless of the type of input data (RGB or RGB-D), the main purpose of these methods is to determine either the pose of the person or the actions performed by him.

Methods using only information about the color of the scene may have limited effectiveness due to factors such as camera movement, occlusion, the complex structure of the scene, or changing or low illumination of the scene.

Depth data are stable with respect to changes in the shooting conditions and background, and it makes it fairly easy to segment objects by depth. The use of depth sensors allows you to get a reliable assessment of a person's posture in real time, that is, they demonstrate high recognition accuracy and low time complexity, therefore they are quite popular in works on recognition of human actions. These methods are popular in studies of recognition of human actions [22–24, 26, 27]. However, at present, the accuracy and cost of depth sensors can significantly limit the areas and conditions of use of such systems.

There are three types of depth cameras that are commonly used in computer vision tasks: triangulation (using a pair of RGB cameras or a stereo camera), a time-of-flight camera (TOF, Time-of-Flight camera), and a camera based on structured backlighting.

Depth sensors based on structured illumination and TOF are subject to significant errors and low accuracy of measuring depth when working outdoors due to external illumination of the scene.

Stereo cameras have a lower cost and can give a higher accuracy of measurement, however, the calculation of depth requires more time and computational resources, and such systems cannot be used in poor or low-light scenes.

There is also the possibility of using laser scanners, however, these devices are very expensive and unsuitable for working in real time.

Therefore, when solving the problem of constructing a "skeleton" in order to determine the human posture in real time, the best option is to use RGB-D cameras. In this case, additional markers are not required, which cannot be installed on each person passing the control point. The limiting factors for processing only color streams or depth streams can be offset by a parallel stream. If the scene is poorly lit, you can use the more computationally expensive features for working with scene depth. If it is impossible to obtain a depth map using IR illumination, it is possible to use a pair of RGB-images of a stereo camera to obtain the depth of the scene.

From the approaches described above it follows that to solve the problem of constructing a human "skeleton" the most optimal solution is to use data on the color and depth of the scene. The use of existing methods based on these data streams is ineffective due to the presence of a priori information about the scene being processed (the passage of a control point by a person), since by adding optimizations for a specific task, one can achieve a significant increase in the speed and accuracy of the construction of the "skeleton" human and determination of marker poses for detecting non-standard behavior. For a quick preliminary search for "nodal" points of the skeleton, you can use a neural network, the coordinates and connection of which are already drowned on the depth map.

2 The Proposed Method for Human Pose Estimation from a Video Image

The principle of the proposed method is as follows: using an RGB-D camera, an RGB image and a depth map of the scene are captured. At the first stage, a camera with IR illumination is used, which allows you to quickly and with sufficient accuracy to get a depth map. However, there is no problem switching to using a stereo camera, since the method relies on processing the finished depth map, and where it was received from (ToF camera or stereo camera) does not matter.

After receiving the color image, it goes through the preprocessing process, and a correspondence is established between the pixels in the RGB image and the depth map. A depth map is necessary to obtain information about the three-dimensional coordinates of each part of the human body and the structure of the skeleton as a whole, and that is why the RGB-D camera is used in the method.

With the help of a trained neural network, a color image is used to search for the main parts of the human body (head, neck, trunk, shoulders, elbows, hands, knees, feet), and the correctness of the found parts is preliminary checked based on the conditions described above.

The centers of the elements found are projected onto a depth map, according to which the coordinates of the parts of the body are further refined, a "skeleton" is built, the correctness of its structure is checked, the resulting coordinates of each part are obtained, and a three-dimensional graph of the "skeleton" is constructed.

Thus, the following stages can be distinguished:

- Preparatory stage:
  - Collection and layout of the base of RGB-images with people in various poses
  - Neural network training
  - Calibrate RGB-D cameras
- Main stage:
  - Getting RGB images and depth maps
  - Distortion correction
  - RGB images and depth maps
  - Select a region of interest on a depth map
  - Detecting body parts in an RGB image and obtaining their three-dimensional coordinates
  - Validation of the received skeleton
  - Search for skeleton elements by depth map

The most significant stages of the proposed method are considered on the example of practical implementation.

3 Practical Implementation of the Proposed Method

3.1 Preliminary Stage

The method of preliminary improvement of video sequence images [28] and the general algorithm of a target environment analyzer [29] are used to prepare images from the camera.

To train the neural network used for the preliminary search for body parts in the RGB image, a database of images was collected and marked out (70% of them for training and 30% for verification).

The database of images was collected by shooting people in various poses and performing some standard actions to obtain each part of the body from several angles and under different shooting conditions. Each image contains a set of traceable body parts, not necessarily complete. From frame to frame, changes were made to: scene illumination, angle, scale and rotation, so as not to train the neural network in specific shooting conditions.

To perform the marking of the image database, software was developed to accelerate the marking process. A screenshot of the program used is shown in Fig. 1.

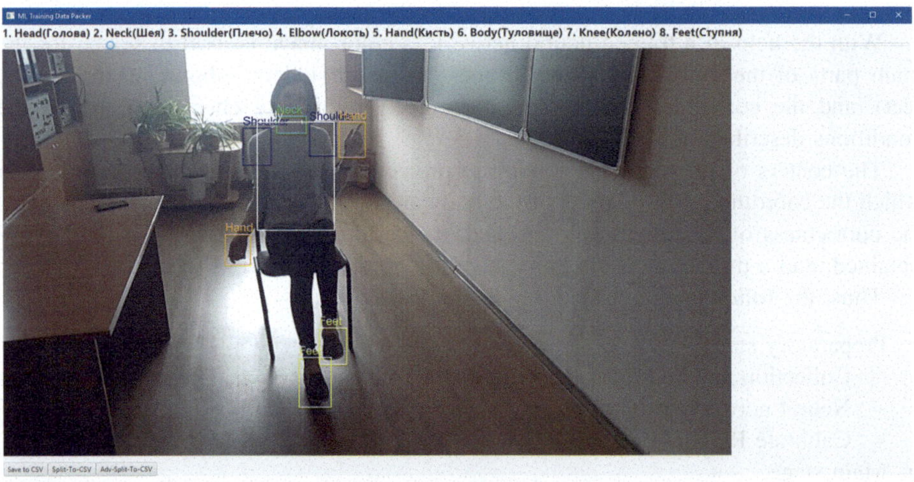

Fig. 1. Image markup software

The developed software allows you to specify a list of labeled classes, switch between them, select the corresponding areas in the image, load and save markup data in an xml file, prepare a batch data set for training a neural network using the Tensorflow framework [30, 31].

To train the neural network, the following set of classes was chosen: head, neck, center of the trunk, shoulder, elbow, hand, pelvis, knee, foot. If one of the classes was absent or was poorly visible on the marked image, then the corresponding part of the body was not marked. Examples of labeled images are shown in the following Fig. 2.

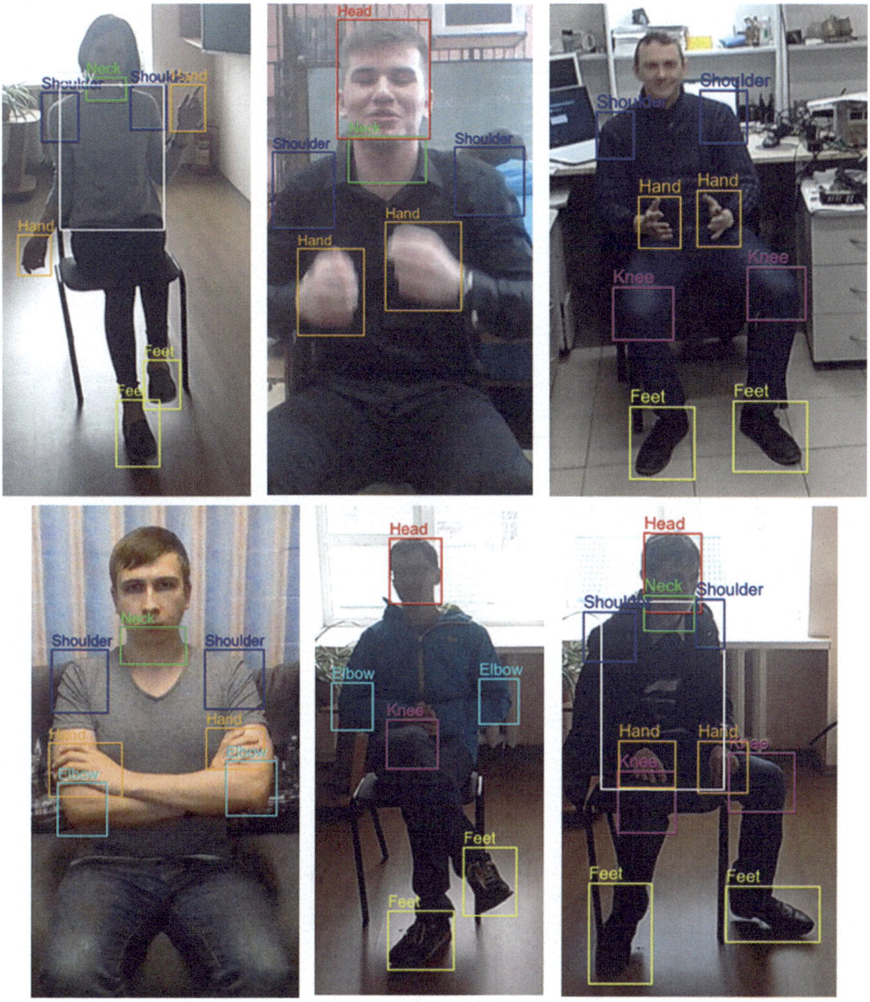

Fig. 2. Tagged image examples

Based on the obtained set of tagged images and xml files with information about the markup, the neural network was trained. To solve the problem of searching and classifying objects on RGB images, we used the Google Object Detection API, which allows you to fine-tune the future network model and select the necessary relationship between accuracy and speed.

Neural network training was conducted using Amazon Web Services, and more than 200,000 steps were taken to prepare the final neural network model. The use of cloud services is due to the high computing and time costs of network training.

When an image is received using an RGB camera due to the distortion effect, significant perspective distortions are observed at the image edges [32]. To correct distortion, it is necessary to calibrate the camera methods [33].

3.2 Main Stage – Building a "Human Skeleton"

Figure 3 shows a block diagram of the algorithm of the method for detecting a human skeleton by the data stream received from an RGB-D camera.

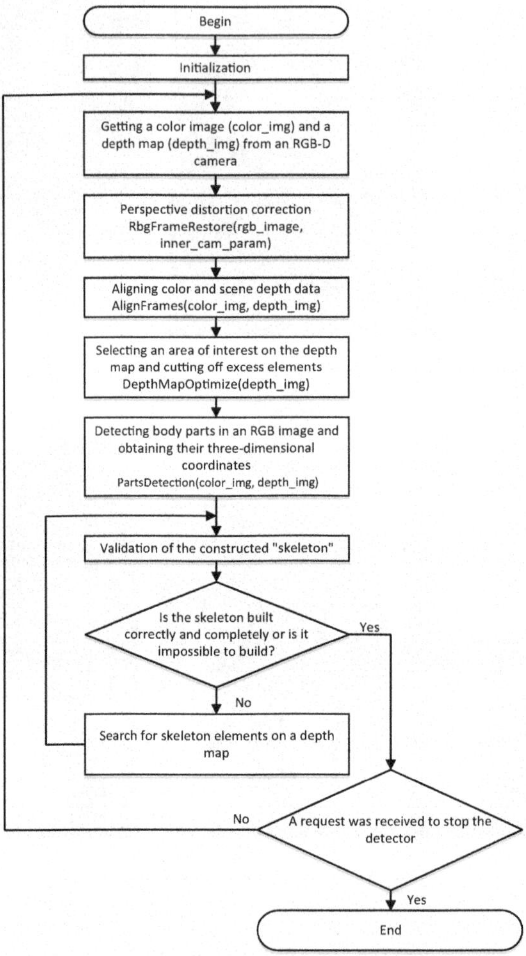

Fig. 3. Human skeleton detection algorithm

Let us consider in more detail each of the stages of the method for detecting a human skeleton.

3.3 Acquiring RGB Images and Depth Maps and Distortion Correction

The RGB-D camera provides a couple of frames: RGB images and depth maps. Frame construction is done by converting the byte stream from the RGB-D camera to the corresponding data structures.

The color image is in RGBA8888 format, the depth map is a 16-bit image in grayscale. The resolution of the RGB image is usually greater than the resolution of the depth map, so at one of the next steps it is necessary to display pixels between this pair of frames.

According to the internal parameters of the camera obtained at the preparatory stage and the calculated distortion coefficients, the perspective distortions of the RGB image are corrected.

3.4 RGB Images and Depth Maps

Knowing the external parameters of a pair of cameras (color and rangefinder), we combine the pair of frames. For each pixel of the depth map, knowing (x, y, z), we calculate the coordinates and size of the region in the color image, the Z coordinate is the value of the depth map in the selected pixel, and it is equal to the distance from the camera plane to the point. The pixel of the depth map corresponds to the region in the color image, and not one dot due to the fact that the resolution of the color image is greater than the resolution of the depth map.

Since the distances to the points in the color image are not known, it is not possible to make directly correspondence of a pixel of the color image to the pixel of the depth map. Therefore, to establish such a correspondence, simply a “back link” is used. The principle of matching by “backlink” is shown in Fig. 4.

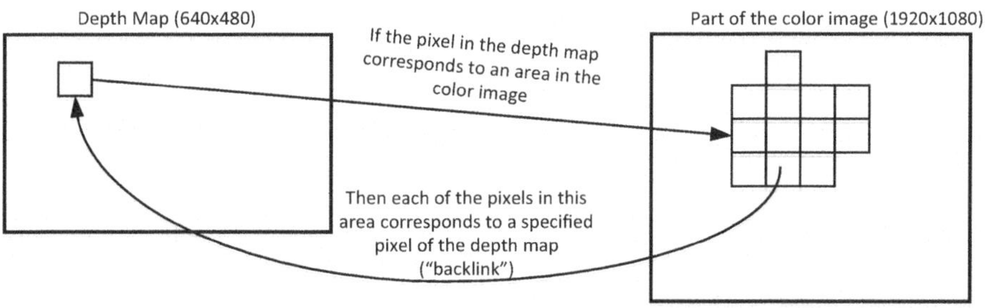

Fig. 4. Отображение пикселей между цветным изображением и картой глубины

The correspondence of pixels is established by storing the indices of the corresponding points.

3.5 Selecting a Region of Interest on a Depth Map

With the known internal parameters of the depth camera, and data on the location of the control point, it is necessary to cut off all unnecessary details on the depth map, except for the surface related to the human body. The floor and objects located outside the control point can be cut off according to the specified threshold values Y and Z of the coordinates of the points in the scene.

Each pixel of the depth map is converted to a vertex in three-dimensional space using expression (1):

$$\begin{cases} w_x = (d_x - c_x) \cdot \frac{d_z}{f_x}, \\ w_y = (d_y - c_y) \cdot \frac{d_z}{f_y}, \\ w_z = d_z \end{cases} \tag{1}$$

where

d_x, d_y, d_z – (x, y) pixel coordinates on the depth map,
(z) distance to the corresponding point,
w_x, w_y, w_z – vertex coordinates in three-dimensional space,
c_x, c_y – optical axis coordinates,
f_x, f_y – focal length along the x and y axes

Depending on the position of the camera and the control point, threshold values are set for cutting off excess vertices and points on the depth map. For example, if the frame of the control point is located directly in front of the camera at a distance of 2 m, has a width of 80 cm, and the camera is at a height of 2 m from the floor, you can specify the following set of restrictions:

$$d_z = \begin{cases} 0, \text{при } w_x(i) < -400 \text{ или } w_x(i) > 400, \\ 0, \text{при } w_y(i) < -2000, \\ 0, \text{при } w_z(i) > 2400 \end{cases} \tag{2}$$

The restriction on z (2.4 m) in (2) was made in order not to capture objects located behind a person passing the control point. Those elements of the depth map that were cut off using expression (2) will not be taken into account when constructing the human "skeleton".

Performing such post-processing of the depth map is aimed at speeding up calculations and reducing the likelihood of detecting "nonexistent" erroneous parts of the body due to foreign objects in the frame. If the method should be applied in the system without the presence of a limited area of human location, the area of interest can be selected dynamically by searching for a person in an RGB image with further translation of the coordinates into the coordinate system of the depth map.

3.6 Detecting Human Body Parts in an RGB Image and Obtaining Their Three-Dimensional Coordinates and Validation of the Received "Skeleton"

Using a trained neural network, a color image searches for parts of the human body. The points found in the color image are transferred to the corresponding points on the depth map, as a result of which, for each found part of the body, real three-dimensional coordinates are calculated using expression (1).

When performing a detection, the main condition is the search for the head. If it was not found in the color image, then an attempt is made to search for it using the

Haar cascade, since the person will most likely look towards the camera. If neither the neural network nor the Haar cascade could perform a search for a person's face, this frame is skipped. This is due to the fact that the head is the simplest element to search, and can act as a starting point for checking the correctness of the construction of the entire skeleton.

After receiving the three-dimensional coordinates of the center of each of the parts of the human body, the correctness of the construction of the skeleton is checked. The following set of criteria is used:

1. The length of the humerus of the left and right hand should differ by no more than 5%;
2. The length of the radius of the left and right hands should differ by no more than 5%;
3. The distance from the center of the neck to the center of the head cannot exceed 1.5 * the height of the head;
4. The center of the body cannot be higher than the neck or head (this test is connected with the assumption of "standard" postures of the person when passing the control point);
5. The lengths of the thighs of the left and right legs should not differ by more than 10% (if the thigh is not visible on the frame, this check is not taken into account);
6. The lengths of the tibia of the left and right leg should not differ by more than 10% (if the tibia is not visible on the frame, the check is not taken into account);
7. The distance from the neck to the shoulders should differ by no more than 15%;
8. The neck and shoulders should be located in a rectangular area responsible for the detection of the human body.

If any of the parts of the body does not satisfy the specified conditions, then it is marked with a marker, according to which at the next stage an attempt will be made to complete the specified part.

On the one hand, the difference in the measurement of the lengths of the left and right arms or legs of 0.5–1.5 cm is a rather large measurement error, however, for the tasks of assessing human postures corresponding to non-standard behavior, there is no need for an absolute measurement error of 1 mm. Even with an error of 1.5 cm, it is possible to evaluate the relative position of body parts relative to each other. If from frame to frame, the resulting lengths will range from −0.5 to +0.5 cm, such vibrations can be smoothed out by applying a filter. This can be either a smoothing filter or a Kalman recursive filter.

If violations are observed during the verification of the criteria described above, then an attempt is made to make corrections to the constructed skeleton; if this fails, then the frame is skipped. In case of successful processing of the next frame, the coordinates of the body parts on the previous "unsuccessful" frame are calculated using interpolation. If the number of iterations of this stage exceeds 5 times and the structure of the skeleton with the available parts does not pass verification, then the current frame is marked as incorrect and is skipped.

Figure 5 shows an example of the operation of the detector of body parts and the distance from the plane of the camera to each point:

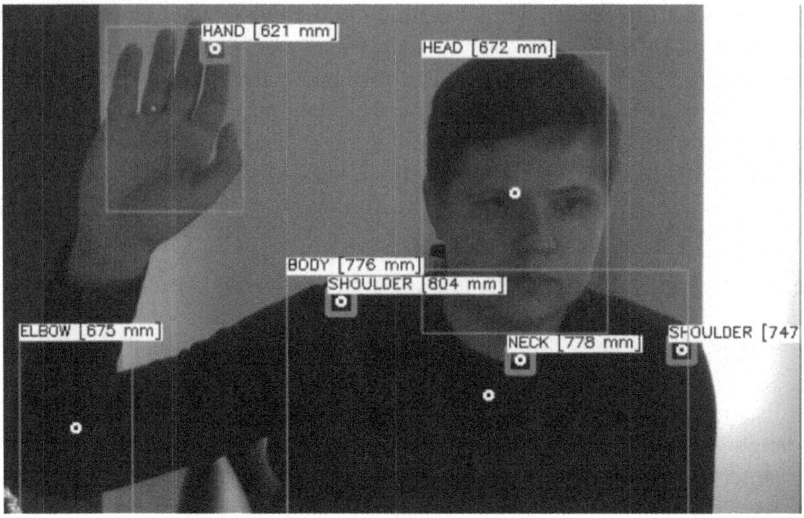

Fig. 5. The result of using a body parts detector

3.7 Search for Skeleton Elements by Depth Map

Following a given set of rules, an attempt is made to complete the missing parts of the skeleton. Such parts are considered elements that were not found when detecting body parts in the RGB image and obtaining their three-dimensional coordinates, as well as elements marked as unreliable when checking the correctness of the skeleton obtained.

The set of rules is based on a priori information about the structure of the skeleton, the relationship of body parts with each other, and the proportions of the human body. Here are a few examples of the applicable rules.

If there is no neck, then the angle of inclination is determined from the known elements of the torso, and the main neck is searched from the center ready in the given direction - the point of intersection of the axis passing through the center of the head and body-axis and the axis passing through the centers of the left and right shoulder.

If there is no information about the coordinates of the elbows, a gradient descent along the visible surface of the arm is made from the shoulder to the side of the hand, and a point is searched between them in accordance with the ratio of the length of the humerus and radius of the person.

A similar approach is used for knee detection, only the two edge points are the feet and lateral parts of the human pelvis.

When obtaining coordinates for each missing part of the skeleton, the skeleton structure is also checked for correctness when adding a new node. If the element is found correctly, then its coordinates on the depth map are converted into three-dimensional world coordinates using the expression (1).

4 Conclusions and Future Work

The obtained results allow real-time recognition of all visible on the image parts of the body of a person passing the control point.

In the future, it is planned to use the information obtained on the coordinates of the parts of the body and the change in these coordinates over time, to analyze the nature of the movements of the person passing the control point. An analysis of the nature of the movements will make it possible to determine some types of deviations in behavior.

Acknowledgment. The reported study was performed within the federal program of the Ministry of Science and Education; the agreement unique ID RFMEFI60819X0281.

References

1. Poppe, R.: Vision-based human motion analysis: an overview. Comput. Vis. Image Underst. **108**, 4–18 (2007)
2. Patrona, F., Chatzitofis, A., Zarpalas, D., Daras, P.: Motion analysis: action detection, recognition and evaluation based on motion capture data. Pattern Recognit. **76**, 612–622 (2018)
3. Zhang, H.-B., Zhang, Y.-X., Zhong, B., Lei, Q., Yang, L., Du, J.-X., Chen, D.-S.: A comprehensive survey of vision-based human action recognition methods. Sensors **19**, 1005 (2019)
4. Aggarwal, J.K., Ryoo, M.S.: Human activity analysis: a review. ACM Comput. Surv. **43**, 16 (2011)
5. Ziaeefard, M., Bergevin, R.: Semantic human activity recognition: a literature review. Pattern Recognit. **48**, 2329–2345 (2015)
6. Papadopoulos, G.T., Axenopoulos, A., Daras, P.: Real-time skeleton-tracking-based human action recognition using kinect data. In: International Conference on Multimedia Modeling, pp. 473–483 (2014)
7. Presti, L.L., La Cascia, M.: 3D skeleton-based human action classification: a survey. Pattern Recognit. **53**, 130–147 (2016)
8. Gall, J., Rosenhahn, B., Brox, T., Seidel, H.-P.: Optimization and filtering for human motion capture. Int. J. Comput. Vis. **87**, 75 (2010)
9. Hofmann, M., Gavrila, D.M.: Multi-view 3D human pose estimation combining single-frame recovery, temporal integration and model adaptation. In: 2009 IEEE Conference on Computer Vision and Pattern Recognition, pp. 2214–2221 (2009)
10. Guan, P., Weiss, A., Balan, A.O., Black, M.J.: Estimating human shape and pose from a single image. In: 2009 IEEE 12th International Conference on Computer Vision, pp. 1381–1388 (2009)
11. Hen, Y.W., Paramesran, R.: Single camera 3D human pose estimation: a review of current techniques. In: 2009 International Conference for Technical Postgraduates (TECHPOS), pp. 1–8 (2009)
12. Hirai, M., Ukita, N., Kidode, M.: Real-time pose regression with fast volume descriptor computation. In: 2010 20th International Conference on Pattern Recognition, pp. 1852–1855 (2010)

13. Luo, X., Berendsen, B., Tan, R.T., Veltkamp, R.C.: Human pose estimation for multiple persons based on volume reconstruction. In: 2010 20th International Conference on Pattern Recognition, pp. 3591–3594 (2010)
14. Tran, C., Trivedi, M.M.: Human body modelling and tracking using volumetric representation: Selected recent studies and possibilities for extensions. In: 2008 Second ACM/IEEE International Conference on Distributed Smart Cameras, pp. 1–9 (2008)
15. Shotton, J., Fitzgibbon, A., Cook, M., Sharp, T., Finocchio, M., Moore, R., Kipman, A., Blake, A.: Real-time human pose recognition in parts from single depth images. In: CVPR 2011, pp. 1297–1304 (2011)
16. Das Dawn, D., Shaikh, S.H.: A comprehensive survey of human action recognition with spatio-temporal interest point (STIP) detector. The Visual Computer **32**(3), 289–306 (2015). https://doi.org/10.1007/s00371-015-1066-2
17. Liu, A.-A., Xu, N., Nie, W.-Z., Su, Y.-T., Wong, Y., Kankanhalli, M.: Benchmarking a multimodal and multiview and interactive dataset for human action recognition. IEEE Trans. Cybern. **47**, 1781–1794 (2016)
18. Liu, A.-A., Su, Y.-T., Nie, W.-Z., Kankanhalli, M.: Hierarchical clustering multi-task learning for joint human action grouping and recognition. IEEE Trans. Pattern Anal. Mach. Intell. **39**, 102–114 (2016)
19. Gao, Z., Zhang, Y., Zhang, H., Xue, Y.B., Xu, G.P.: Multi-dimensional human action recognition model based on image set and group sparisty. Neurocomputing **215**, 138–149 (2016)
20. Fernando, B., Gavves, E., Oramas, J.M., Ghodrati, A., Tuytelaars, T.: Modeling video evolution for action recognition. In: Proceedings of the IEEE Conference on Computer Vision and Pattern Recognition, pp. 5378–5387 (2015)
21. Yang, X., Tian, Y.: Super normal vector for activity recognition using depth sequences. In: Proceedings of the IEEE Conference on Computer Vision and Pattern Recognition, pp. 804–811 (2014)
22. Li, M., Leung, H., Shum, H.P.H.: Human action recognition via skeletal and depth based feature fusion. In: Proceedings of the 9th International Conference on Motion in Games, pp. 123–132 (2016)
23. Chen, C., Liu, K., Kehtarnavaz, N.: Real-time human action recognition based on depth motion maps. Journal of Real-Time Image Process. **12**(1), 155–163 (2013). https://doi.org/10.1007/s11554-013-0370-1
24. Zhang, J., Li, W., Ogunbona, P.O., Wang, P., Tang, C.: RGB-D-based action recognition datasets: a survey. Pattern Recognit. **60**, 86–105 (2016)
25. Liu, J., Shahroudy, A., Xu, D., Wang, G.: Spatio-temporal LSTM with trust gates for 3D human action recognition. In: European Conference on Computer Vision, pp. 816–833 (2016)
26. Oreifej, O., Liu, Z.: HON4D: histogram of oriented 4D normals for activity recognition from depth sequences. In: Proceedings of the IEEE Conference on Computer Vision and Pattern Recognition, pp. 716–723 (2013)
27. Yang, X., Tian, Y.: Effective 3D action recognition using eigenjoints. J. Vis. Commun. Image Represent. **25**, 2–11 (2014)
28. Korovin, I., Khisamutdinov, M., Ivanov, D.: Improvement of a video sequence singular image. In: Proceedings of the 2nd International Conference on Advances in Artificial Intelligence, pp. 12–15 (2018)
29. Korovin, I., Khisamutdinov, M., Ivanov, D.: A basic algorithm of a target environment analyzer. In: Proceedings of the 2nd International Conference on Advances in Artificial Intelligence, pp. 7–11 (2018)

30. Abadi, M., Barham, P., Chen, J., Chen, Z., Davis, A., Dean, J., Devin, M., Ghemawat, S., Irving, G., Isard, M., et al.: Tensorflow: a system for large-scale machine learning. In: 12th Symposium on Operating Systems Design and Implementation 2016, pp. 265–283 (2016)
31. Rampasek, L., Goldenberg, A.: Tensorflow: biology's gateway to deep learning? Cell Syst. **2**, 12–14 (2016)
32. Tsai, R.Y.: An efficient and accurate camera calibration technique for 3D machine vision. Proc. Comp. Vis. Patt. Recog. 364–374 (1986)
33. Zhang, Z.: A flexible new technique for camera calibration. IEEE Trans. Pattern Anal. Mach. Intell. **22**, 1330–1334 (2000)

Framework for Civic Engagement Analysis Based on Open Social Media Data

Igor O. Datyev(✉), Andrey M. Fedorov, and Andrey L. Shchur

Institute for Informatics and Mathematical Modeling - Subdivision of the Federal Research Centre "Kola Science Centre of the Russian Academy of Sciences", 184209 Apatity, Russia
{datyev, fedorov, shchur}@iimm.ru

Abstract. Today, the problem of citizen involvement in governing processes is actively discussed by both officials and scholars around the world. This paper discusses various aspects of creating an information technology framework for civic engagement analysis in social media. The spoken framework will help to search and assess the existing problems and their solutions proposed by citizens, ensure the presentation of the most vocal civic ideas in discussions of various collegial bodies during decision-making processes, and will also contribute to the self-organization of initiative groups and self-government to solve problems at the municipal level. The authors present the general structure of an information technology framework for analyzing civic activity, and also discuss the issues of development and their achievements on the way to its software implementation. The work contains examples of experimental results that can be obtained by using the developed framework.

Keywords: Online social networks · E-participation · Civic engagement analysis · Decision-making support

1 Introduction

The declared global political course aimed at the digitalization of all the possible processes taking place in society implies the need for the existence of information about these processes and the citizens of this society in the corresponding databases or other digital resources. The representation of citizens in the information space can be expressed in several ways: certain information can be downloaded both by direct participants in this process and by a third party. The role of a third party may be: the state, i.e. an official governing institute department, and unofficial sources that have some relation (or information) to the object, problem or question. When third-party information is posted, it becomes impossible to discuss and make a decision taking into account the opinions of all participants in the ongoing processes. Therefore, the representation of direct participants is required, which today is being implemented more and more often by a combination of registered accounts on specialized government resources as part of the implementation of e-participation programs [1, 2].

However, in addition to specialized web resources intended for state, regional and municipal government, there are many other online platforms where citizens are

R. Silhavy (Ed.): CSOC 2020, AISC 1225, pp. 586–597, 2020.
https://doi.org/10.1007/978-3-030-51971-1_48

represented and various issues are discussed in a more open and informal way: news sites, online social networks, forums, blogs, vlogs, etc. Despite various points of view on the term 'social media' [3], we will use it to identify this set of the spoken unofficial resources, not intended for administrative decision and policy making processes. In solving the problems of regional and municipal management, social media data has great potential, which to this day is still not yet fully realized.

In this paper, we will focus on the discussion of issues related to the development of information technology framework for analyzing civic engagement based on open data from social media.

2 Social Media Analysis Application Areas

Let's start with the term 'Social media intelligence' (SOCMINT) - a set of techniques and technologies to monitor social media networking sites (SNSs) such as Facebook, VK, Instagram, Twitter, etc. In paper [4] the authors praise the capabilities of social media: "Measuring and understanding the visage of millions of people digitally arguing, talking, joking, condemning and applauding is of wide and tremendous value to many fields, interests and industries". They also consider the issues of interaction between citizens and the state: "This could help ensure a better flow of information between citizens and the government, especially in times of emergency. With access to social media, passive bystanders can become active citizen journalists, providing and relaying information from the ground". The value of information (for decision-making as well) extracted from social media is underlined by the terms 'crowd-sourced information' and 'wisdom of the crowd'.

Many commercial companies use social media for primarily marketing purposes [5]: brand analysis [6], study of consumer preferences [7] and advertising campaigns [8].

In addition to the brands of commercial companies, there are political brands and relevant political campaigns. Same as with the marketing of commercial products, social media is extensively used in political games, applying similar information technologies to them [9, 10].

Today, many governments highly value the role of social media in monitoring and responding to natural disasters and other emergencies [11, 12].

Social media data is used to counter extremism, terrorism and various types of criminal activity. For example, to counter extremism authors [13] propose "content analysis framework with the focus on classifying tweets into extremist and non-extremist classes… using deep learning-based sentiment analysis techniques to classify the tweets as extremist or non-extremist".

Among other things, social media are also used to research and optimize city transport systems [14, 15].

The use of social media for solving healthcare problems is widely studied. For example, the authors [16] confirm that geotagged photos from the Instagram social network serve to identify "lifestyle" diseases, such as obesity, drinking or smoking. The authors [17] note that "new drug pharmacological properties can be derived patient observations shared in social media forums", "propose to address a challenging problem by applying modern deep neural networks for disease named entity recognition" and "show that it is possible to assess the practicability of using social media data to extract representative medical concepts for pharmacovigilance or drug repurposing".

Based on the open information of social media, the tourist behavior of people, the attractiveness of countries, regions or specific places [18] for local residents or foreigners are studied [19]. For example, in a study [20] based on geotagging of photos and videos (Flickr social network), the authors define the attractiveness of a country or region as the absolute number of media objects published by foreign visitors in relation to the population of that country or region. Such attractiveness, in the long term, can be considered to study migration problems. Another study focuses on the impact of social media on migration processes [21]. One thing in common: researchers study the movement of people for various purposes based on social media data.

Online social networks' data also uses for modeling the propagation and influence of the information [22, 23].

Based on the information extracted from social media, attempts are made to confirm the numerical values published by the official statistics departments or official news channels [24, 25].

The papers [26, 27] discuss the problem of joint decision-making in online social networks as well as methods that can be applied for this.

At first, the main driving force behind the creation of information technologies for social media analysis were the commercial corporations that study and promote their products, lobby their interests. A little later various governments of the world began using social media to disseminate information. The interests of ordinary citizens were studied and taken into account only during user preferences analysis when promoting a product on the market, or during political campaigns. Until recently, the field of application of these technologies rarely reached the level of regional, municipal government, not to mention self-government.

3 Information Technologies Development Issues for Social Media Data Analysis

When developing information technology using data from social media, issues of the following nature can be highlighted: technical, legal, ethical.

Technical issues can be divided into: data extraction difficulties and the difficulties associated with the storage and analysis of data. The first group is more dependent on the features of specific social media: user authorization (passwords, tokens); time limits (for example, limit on the number of requests to a social network site per unit time from one user and the API used); quantitative restrictions (explicit and implicit quantitative restrictions on receiving data set by the site administration); user restrictions on access to profiles (hidden data, differentiation of access rights). Difficulties associated with the storage and analysis of data: big data and the complex of issues it implies (volume, velocity, variety, veracity), a combination of methods from different knowledge fields (computer science and social science). Some of these issues are discussed in more detail in [28].

Discussions about what information on social media is public and what is considered personal have not subsided so far [29]. The problem is complicated not only by the lack of sufficiently developed international legislative norms, but also partly by the "philosophical" nature of the issue itself, which leads to a multipolarity of views on its solution in conjunction with a burden in the form of cultural and ethical differences between nationalities for whom this issue is relevant. The following extreme points of view can serve as examples of two opposing views: 1. Everything that is published by the user or about the user on social media immediately becomes public and all other users get unlimited rights to use the information 2. Everything that is published by the user or a third party about the user (with his consent) is his personal data and cannot be used without the consent of the user.

Since May 2018 a new statutory document General Data Protection (GDPR) [30] has been introduced into European legislation system to regulate the collection and processing of personal data belonging to citizens of Europe, wherever they are territorially located. This document introduces a strict set of rules and establishes liability for violations related to the processing of personal data. The GDPR requires the prior consent of the owner to use the data, it is necessary to specify the purpose of their use and to ensure the principle of "data minimization": personal data are not allowed to be collected in amounts exceeding those required for the declared purpose. The collected data on behavior of the monitored subjects has also become personal.

Along with the legal side of the issue, it is necessary to touch on the ethical component [31]. The following ethical difficulties can be distinguished: maintaining confidentiality, identity, property rights, and reputation. In [32] the authors provide recommendations for researchers using social media data, the main meaning of which is narrowed down to the following: researchers should collect only the basic data necessary to answer the research question and carefully present this data to avoid identifying participants.

In this paper, we will not delve into the legal and ethical aspects of the use of social media data. However, we note that the issue of personal data use in online social networks is very relevant today. Also, it should be noted that the principles of regulation of these issues are currently is a subject of continuous development.

Thus, it is necessary to take these aspects into account when developing information technologies using social media data.

4 The Main Idea for the Civic Engagement Analysis Framework

There are many software tools for evaluating and monitoring numerous aspects of human activity using social media data. However, today, the topic of civic engagement in the process of making socio-political and socio-economic decisions at the regional and municipal levels is very acute, since these levels can provide a tangible "here and now" reaction effect.

Social networking services are considered to be a very typical kind of social media but at the same time they have the most complex structure compared to text or video blogs, news sites, forums and other social media types. Therefore, in this section we will take an in-depth look at our version of an information technology framework for civic engagement analysis based on open data taken specifically from a social networking service.

To analyze and assess civic engagement in online social networks in this paper the following conceptual model (see Fig. 1) is proposed. Objects and processes in a SNS that pose some interest and are accessible for analysis can be divided into three parts. The first part is the multitude of SNS communities (groups). Its analysis allows us to assess the size and structure of communities and also identify the characteristics of their involvement in the studied processes. The second part is formed by content published in communities' public spaces: publication texts (posts) and comment texts. This part also includes other published multimedia content. However, in this paper only textual publications are considered. This group of objects is analyzed for the extraction of syntactic and semantic characteristics (markers). The objects of the third part essentially define the relations of the objects from the first part to the second. Statistical processing and construction of time dependences of such typical SNS counters, such as the number of 'likes', reposts, comments and views, allows identifying the reaction, evaluating the popularity and interest of society in relation to the content.

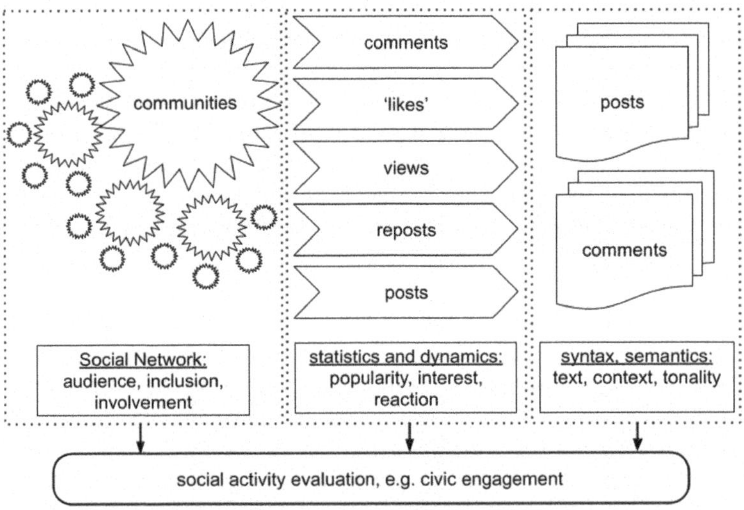

Fig. 1. Civic engagement and online social networks

The resulting set of characteristics makes it possible to evaluate both the general level of social activity, and its subset - civic engagement in a social network. i.e. it is presumed that civic engagement is a type of social activity shown in relation to a particular class of content.

The framework is designed to help evaluate civic engagement in a social network, create a list of issues discussed in user communities and its possible solutions, as well as get a portrait of the audience.

5 Overall Framework Structure

The information technology framework for analyzing civic engagement in a social networking service is a complex of interconnected components. It provides collection, storage, analysis and presentation of data received from SNS. The structure of the components and the relationships between them are shown in Fig. 2. The depicted arrows indicate the direction of data transfer and control, and the lines demonstrate block extensions. The dotted lines show optional connections.

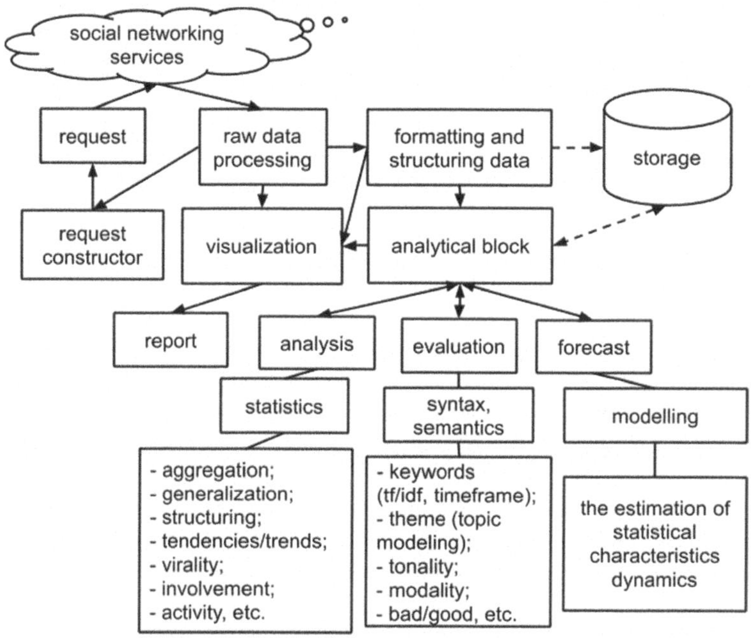

Fig. 2. Framework structure for analysis of civic engagement in online social networks

The main components of the framework are represented by the following groups: *social networking services*, *request constructor* and the subsystem *raw data processing*, data *storage* and the *analytical block*. The interface part of the framework is the block responsible for *visualization* and *report*.

Request constructor and the subsystem *raw data processing*. The main task of these functions is to compile the correct query, taking into account the known restrictions. It is assumed that the framework will be used by an officially registered account of the SNS. This imposes limitations on the amount of open data available for obtaining and analyzing. In this case, the data access is done through API functions.

The second group of functions is designed to provide storage of received data. Recording the received data in the repository (SQL database) requires their preliminary formatting and structuring. The use of data storage isn't required for some of the tasks.

The *Social Networking Services* component is one of the fundamental and requires detailed consideration. A researcher of SNS can relate to it in several forms: as an owner/administrator, a user with a registered account or an unauthorized subject who does not have an account. Thus, access to SNS data can be carried out depending on the type of relationship indicated. Owners and administrators have a free access to the source data stored in the databases, making it possible to get any requested information. Registered users can access SNS data either manually, through the user interface, or by using automation algorithms, which include parsing html-pages of the interface and the mechanism of API functions. Unregistered users of SNS can, in some cases, also gain access to social network data. One of such methods is the use of search engines (Google, Yandex, etc.), which index open data from SNS and include them in search results. Owners control the publicity of data by setting access levels to each of the data elements of the account.

The central component of the framework is the *analytical block*. Here the main functions of data processing are performed: its analysis, evaluation and forecast. The block includes the functions of statistical processing and mathematical modeling, lexical and semantic analysis functions, specialized data processing functions built on the basis of machine learning methods, e.g. with the use of neural networks, as well as other methods of working with big data.

Let's approach the analytical block in more detail. Due to the specifics of modern SNSs, the following data is considered public with respect to each content element: type of content (text, picture, video, etc.), author, number of views, number of likes, number of reposts, number of comments. It should be noted that each comment can be considered as a separate element of content for which the same data is available.

The resulting set sometimes appears to be a rather big array of data. As you can see from the previous description, the structure of this array is simple, but the number of records in it does not allow for high-quality manual processing. Moreover, such processing often needs to be carried out online, on a continuous basis and with subsequent joint processing of all the results obtained.

Based on the available data, it is possible to form generalized characteristics of the studied community that can be of interest to the analyst. Such characteristics are logically grouped by the object that they reflect: a) a number of participants in communication, b) a number of texts of posts/publications and comments, c) a number of indicators specific to SNS.

The first group of characteristics describes in a generalized form the active audience of the community: its gender and age structure, regional binding structure, audience activity. The generalization is carried out on the basis of open data from account profiles that are identified as active audience. If account profiles are closed, then quantitative information about this is also used in the formation of a generalized characteristics of the audience. The percentage division of the audience in terms of profile openness appropriately characterizes the group's communicative context.

The above group of characteristics may raise some questions related to data privacy. However, the use of exclusively open information and the targeted application of structuring and generalization processes to them should remove most of these issues.

The second group of characteristics is focused on (individually posted) texts in SNS communities. These texts include both the content of the community posts themselves and the comments that the active audience writes during the discussion of a post. The following text properties are of interest: a) thematic structure, b) specified key objects, c) tonality, d) modality, etc.

The third group of characteristics is aimed at reflecting the specifics of the social networking service itself. Most of the modern SNSs feature the following 'counters' that measure: a) content views, b) 'likes', c) reposts, d) comments, e) comment responses (as a special kind of comment), e) authorship of posts/publications (if publication is done not on behalf of the group).

Thus, the framework allows you to solve many problems that are classic tasks of any information technology - the formation, aggregation, processing, analysis, transfer and presentation of information. However, when solving these problems, the limitations and complex specificity of the information source - SNS, are taken into account. Task solving of all these problems is an integral part of the solution to the main problem - the assessment and analysis of civic engagement in online social networks.

6 Analysis Results

To demonstrate the capabilities of the framework, an analysis of a typical local VK community was conducted. The selection was made on the criteria of openness and popularity among local residents. The communities formed on such basis are of great research interest, since they reflect the real mood of real groups of people. The analysis of the selected community was conducted in accordance with the concept of civic engagement in online social networks proposed in this work. The results obtained allowed us to evaluate the community in several main areas.

First, community thematic preferences are identified. Two interconnected subsets of content are important here: publications (see Fig. 3) and comments (see Fig. 4). The ratio of volume and tonality of content gives an idea of the popularity of topics within the community. Second, the characteristics of the community's social activity are determined. These characteristics include quantitative indicators: the number of members and active accounts, counters of content views, 'likes', reposts, comments and responses to comments (see Fig. 5).

Third, the evaluation of social activity was obtained, which reflects the frequency of SNS use by the community members (see Fig. 6). The Community activity bar chart shows how often community members check on the SNS. Additionally, generalized characteristics of the community were obtained, reflecting its demographic and spatial structure as well as its social attractiveness.

It should be noted that the presence of data marked "not specified" is common for social networking services. This is due to the fact that not all community members completely fill out their account's profiles. And in some cases, the information in the account profile is obviously false (fake). However, the fraction of such cases is small and insignificant in comparison with the entire community. Therefore, the data used by such studies adequately reflect reality in general.

The results obtained show that the content related to city economy and social issues is popular in the studied community. However, it can be noted that the relative social

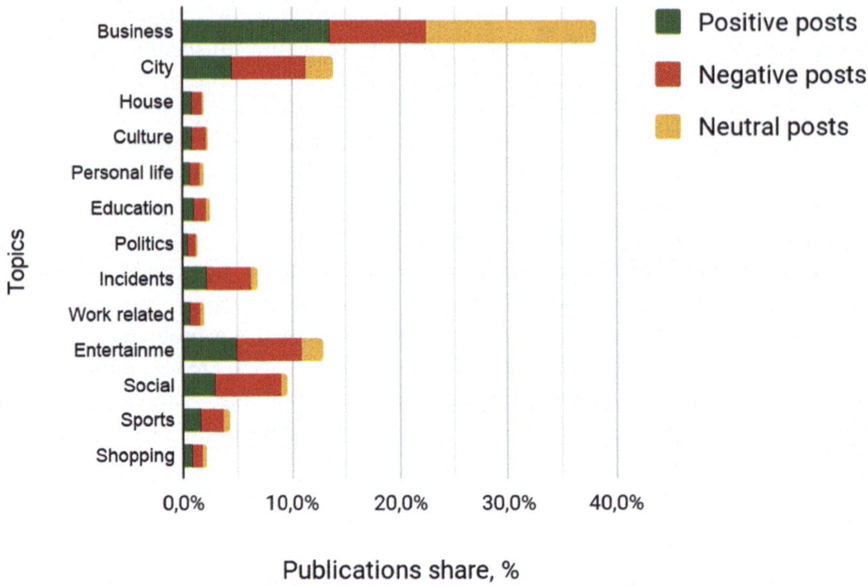

Fig. 3. Community publications: quantity and tonality

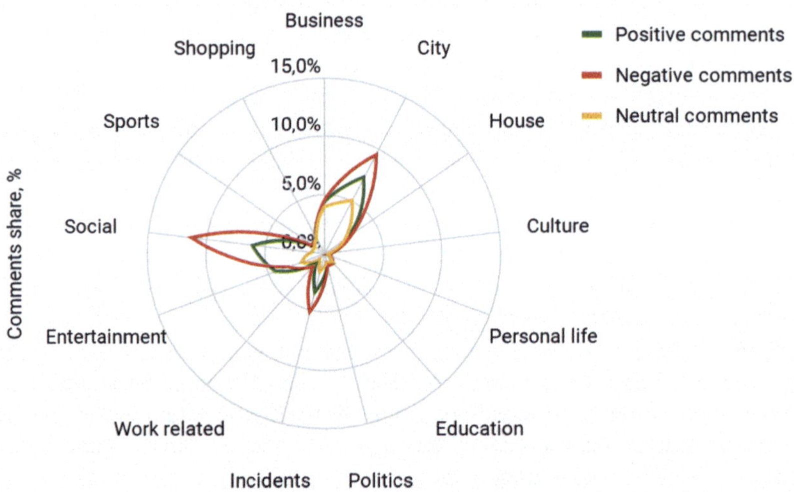

Fig. 4. Community comments: tonality

activity is generally low. As a result of this, it can be concluded that the level of civic engagement of the studied community is at least non-zero.

Further detailed research using the framework functions is required to reveal more possible combinations of social activity and its thematic focus. It is assumed that the results of such studies will introduce more precise measurements of the relative and absolute magnitude of civic engagement in online social networks.

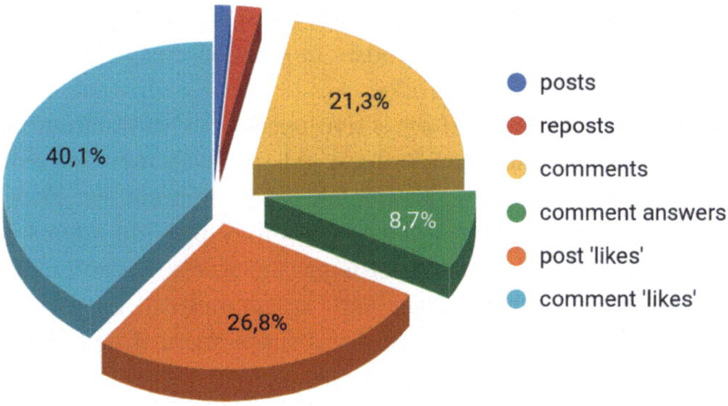

Fig. 5. Community: activity indicators

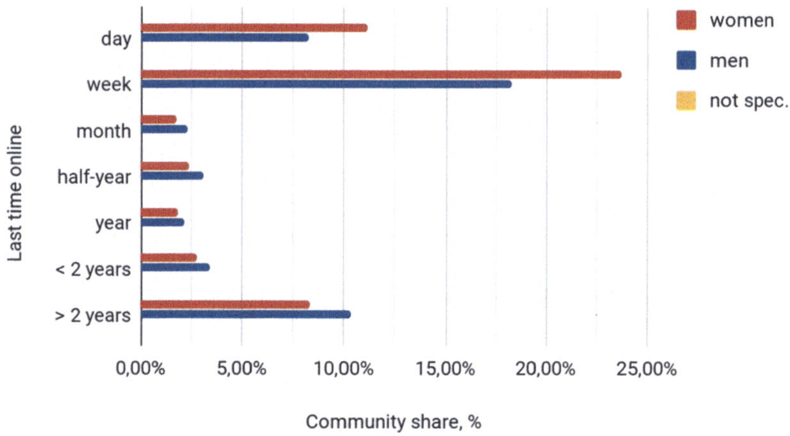

Fig. 6. Community activity: going online

7 Conclusion

Initially, the presented framework was aimed specifically at improving the life of the local population, promoting local initiatives, improving citizens-to-government (C2G) communication and adding a new type of citizens-to-citizens (C2C) communication, which is absent in the classical e-participation concept. However, the scope of the framework turned out to be wider than expected at the initial stage of development.

Using the presented framework, online social network researchers can solve a number of the following high-level tasks. The simplest is the task of statistical express analysis and operational review of the characteristics of various SNS objects and processes. Another interesting task for a researcher is monitoring the time-changing characteristics of objects and processes in social networks. In addition, the lexical and semantic analysis of the publication texts that are distributed and discussed within communities can also present some scholar interest. In this case, the most interesting is the automated identification of the text topics. A separate specific task is the construction of various simulation and prognosis models based on objects and processes data obtained from online social networks.

The framework can also be used by active users of SNS communities to identify content with characteristics that interest them. Also, the framework will be useful to owners and administrators of online social network communities. Depending on the degree of community openness, the opportunities may provide interesting and useful information to outside observers - advertising agencies, government services and representatives of other organizations and communities. Further plans include expanding the block of analytical functions and supplement it with a modeling block, as well as implement these extensions as software.

References

1. Marques, F.: Government and e-participation programs: a study of the challenges faced by institutional projects. First Mon. **15**(8) (2010). https://doi.org/10.5210/fm.v15i8.2858
2. Le Blanc, D.: E-participation: a quick overview of recent qualitative trends. DESA Working Paper No. 163 ST/ESA/2020/DWP/163 (2020)
3. Treem, J.W., Dailey, S.L., Pierce, C.S., Biffl, D.: What we are talking about when we talk about social media: a framework for study. Sociol. Compass **10**, 768–784 (2016). https://doi.org/10.1111/soc4.12404
4. Omand, D., Bartlett, J., Miller, C.: Introducing social media intelligence (SOCMINT). Intell. National Secur. **27**(6), 801–823 (2012)
5. Barger, V., Peltier, J., Schultz, D.: Social media and consumer engagement: a review and research agenda. J. Res. Interact. Mark. **10**(4), 268–287 (2016)
6. Aggrawal, N., Ahluwalia, A., Khurana, P., Arora, A.: Brand analysis framework for online marketing: ranking web pages and analyzing popularity of brands on social media. Soc. Netw. Anal. Mining **7**(1), 1–10 (2017). https://doi.org/10.1007/s13278-017-0442-5
7. Taylor, B.: Understanding consumer preferences from social media data. NIM Mark. Intell. Rev. **11**(2), 48–53 (2019). https://doi.org/10.2478/nimmir-2019-0016
8. Voorveld, H., Noort, G., Muntinga, D., Bronner, F.: Engagement with social media and social media advertising: the differentiating role of platform type. J. Advert. **47**(1), 38–54 (2018). https://doi.org/10.1080/00913367.2017.1405754
9. Papakyriakopoulos, O., Hegelich, S., Shahrezaye, M., Serrano, J.: Social media and microtargeting: political data processing and the consequences for Germany. Big Data Soc. **5**(2) (2018). https://doi.org/10.1177/2053951718811844
10. Stieglitz, S., Dang-Xuan, L.: Social media and political communication: a social media analytics framework. Soc. Netw. Anal. Mining **3**(4), 1277–1291 (2012). https://doi.org/10.1007/s13278-012-0079-3
11. Ghosh, S., Ghosh, K., Ganguly, D., Chakraborty, T., Jones, G.J.F., Moens, M.-F., Imran, M.: Exploitation of social media for emergency relief and preparedness: recent research and trends. Inf. Syst. Front. **20**(5), 901–907 (2018). https://doi.org/10.1007/s10796-018-9878-z
12. Ehnis, C., Bunker, D.: Social media in disaster response: queensland police service - public engagement during the 2011 floods. In: ACIS 2012: Proceedings of the 23rd Australasian Conference on Information Systems, Geelong, Victoria, pp. 1–10, 3–5 December 2012

13. Ahmad, S., Asghar, M.Z., Alotaibi, F.M., et al.: Detection and classification of social media-based extremist affiliations using sentiment analysis techniques. Hum. Cent. Comput. Inf. Sci. **9**, 24 (2019). https://doi.org/10.1186/s13673-019-0185-6
14. Li, D., Zhang, Y., Li, C.: Mining public opinion on transportation systems based on social media data. Sustainability **11**, 4016 (2019)
15. Serna, A., Gerrikagoitia, J.K., Bernabé, U., Ruiz, T.: Sustainability analysis on urban mobility based on social media content. Transport. Res. Procedia **24**, 1–8 (2017)
16. Garimella, V., Alfayad, A., Weber, I.: Social media image analysis for public health. In: CHI 2016: Proceedings of the 2016 CHI Conference on Human Factors in Computing Systems, pp. 5543–5547 (2016). https://doi.org/10.1145/2858036.2858234
17. Miftahutdinov, Z., Tutubalina, E.: End-to-end deep framework for disease named entity recognition using social media data. In: 2017 IEEE 30th Neumann Colloquium, Budapest, Hungary, pp. 47–52 (2017)
18. Mancini, F., Coghill, G.M., Lusseau, D.: Using social media to quantify spatial and temporal dynamics of nature-based recreational activities. PLoS ONE **13**(7), e0200565 (2018). https://doi.org/10.1371/journal.pone.0200565
19. Li, Q., Wu, Y., Wang, S., Lin, M., Feng, X., Wang, H.: VisTravel: visualizing tourism network opinion from the user generated content. J. Vis. **19**(3), 489–502 (2016). https://doi.org/10.1007/s12650-015-0330-x
20. Bojic, I., Belyi, A., Ratt, C., Sobolevsky, S.: Scaling of foreign attractiveness for countries and states. Appl. Geogr. **73**, 47–52 (2016)
21. Dekker, R., Engbersen, G.: How social media transform migrant networks and facilitate migration. Glob. Netw. **14**, 401–418 (2014). https://doi.org/10.1111/glob.12040
22. Atif, Y., Al-Falahi, K., Wangchuk, T., Lindström, B.: A fuzzy logic approach to influence maximization in social networks. J. Ambient Intell. Hum. Comput. (2019)
23. Chen, W., Lakshmanan, L.V., Castillo, C.: Information and influence propagation in social networks. Synth. Lect. Data Manag. **5**(4), 1–177 (2013)
24. Fedorov, A., Datyev, I., Shchur, A., Oleynik, A.: Online social networks analysis for digitalization evaluation. In: Silhavy, R. (eds.) Software Engineering Methods in Intelligent Algorithms. CSOC 2019. Advances in Intelligent Systems and Computing, vol. 984. Springer, Cham (2019)
25. Brakel, J., Söhler, E., Daas, P., Buelens, B.: Social media as a data source for official statistic: the Dutch Consumer Confidence Index. Surv. Methodol. **43**(2), 183–210 (2017)
26. Power, D., Phillips-Wren, G.: Impact of social media and web 2.0 on decision making. J. Decis. Syst. **20**(3), 249–261 (2011)
27. Herrera-Viedma, E., Cabrerizo, F.J., Chiclana, F., Wu, J., Cobo, M.J., Samuylov, K.: Consensus in group decision making and social networks. Stud. Inf. Control. **26**(3), 259–268 (2017)
28. Stieglitz, S., Mirbabaie, M., Ross, B., Neuberger, C.: Social media analytics – challenges in topic discovery, data collection, and data preparation. Int. J. Inf. Manag. **39**, 156–168 (2018)
29. Napoli, P.: User Data as Public Resource: Implications for Social Media Regulation. SSRN, 4 June 2019. https://doi.org/10.2139/ssrn.3399017
30. EU Regulation 2016/679 of April 27, 2016, GDPR, EU. https://publications.europa.eu/en/publication-detail/-/publication/3e485e15-11bd-11e6-ba9a-01aa75ed71a1
31. Lindoo, E.: Ethics in analytics and social media. Lecture Notes in Networks and Systems. LNNS, vol. 69, pp. 970–982 (2019). https://doi.org/10.1007/978-3-030-12388-8_67
32. Moreno, M.A., Goniu, N., Moreno, P.S., Diekema, D.: Ethics of social media research: common concerns and practical considerations. Cyberpsychol. Behav. Soc. Netw. **16**(9), 708–713 (2013). https://doi.org/10.1089/cyber.2012.0334

Reward-to-Variability Ratio as a Key Performance Indicator in Financial Manager Efficiency Assessment

Anna Andreevna Malakhova[1(✉)], Olga Valeryevna Starova[1], Svetlana Anatolyevna Yarkova[2], Albina Sergeevna Danilova[2], Marina Yuryevna Zdanovich[1,2], Dmitry Ivanovitch Kravtsov[1], and Dmitry Valeryevitch Zyablikov[1]

[1] Siberian Federal University, Krasnoyarsk, Russia
amalakhova@sfu-kras.ru, malahovaanna@mail.ru
[2] Krasnoyarsk Institute of Railway Transport - Branch of the Irkutsk State Transport University, Krasnoyarsk, Russia

Abstract. In this paper computational techniques to process financial data and to assess management efficiency are proposed. Personnel evaluation process is formalized on the basis of the proposed key performance indicators based on portfolio efficiency criteria. Personnel efficiency is assessed via the excessive portfolio return over average market performance indicators per unit of risk. Alternative measures to evaluate risk are formulated. The proposed downside risk measures are implemented into portfolio performance evaluation criteria. Comparative analysis of the introduced portfolio performance evaluation criteria is held. Case study via the Trading Organiser 'Moscow Exchange' is performed. The experimental results prove that the introduced portfolio performance evaluation criteria yield better results than the coefficients which do not take into account downside risk measures. It is concluded that the proposed modified 'reward-to-variability' ratio can be incorporated into the system of key performance indicators for assessing financial management efficiency. #CSOC1120.

Keywords: Key performance indicator · Portfolio performance · Sharpe coefficient · Reward-to-variability ratio · Reward-to-volatility ratio · Value at risk

1 Introduction

Quantitative indicators have become widely used in personnel assessment recently due to their recognition as more objective criteria and their convenience in planning personnel management and personnel evaluation processes. For this, various scores, labor contribution factors or labor participation rates, ratings, scales, key performance indicators, and a balanced scorecard are used. One of the conditions to apply such criteria is information processing, e.g. processing personnel statistics, labor indicators, financial and economic indicators of the organization and of the market as a whole. Thus, there is a strong need for techniques to formalize data and evaluation process.

R. Silhavy (Ed.): CSOC 2020, AISC 1225, pp. 598–613, 2020.
https://doi.org/10.1007/978-3-030-51971-1_49

In assessing the results of personnel in the financial sector, one of the key issues remains the evaluation of investment portfolio efficiency. One of the most discussed problems in this respect is the choice of a criterion that most fully and objectively reflects the contribution of the investment analyst or financial manager to the excessive portfolio return over average market performance indicators.

In the present state of the art there are different criteria developed to evaluate financial manager performance, as is demonstrated by many publications on this subject [1, 2]. From the point of view of portfolio diversification the Sharpe coefficient (reward-to-variability ratio) and the Treynor coefficient (reward-to-volatility ratio) most fully reveal the excessive return of risk premium per unit of risk measure. The difference between these approaches is the way to measure risk: the Sharpe coefficient is based on the variance of the portfolio return while the Treynor coefficient takes into account 'beta' of the portfolio, that is its correlation with the market portfolio average return [1, 3].

Despite various modifications of these coefficients [1, 2], the researchers note such their shortcomings as the instability of beta in time, the assumption of the normal or symmetric distribution of financial assets return, which is not fully proved in practice [5, 6]. In particular, in emerging markets economy (to which researchers include, among others, the financial market of Russia and Eastern Europe as well), the financial assets return is closer to the 'downside' distribution, that is, the actual return is shifted to the left, below the average (expected) return, which implies increased investment risk, and therefore, requires taking these deviations into account when constructing Sharpe and Treynor coefficients to assess financial manager performance.

The aim of this paper is to propose new portfolio performance criteria based on the Sharpe coefficient to assess financial manager efficiency. We modify the Sharpe coefficient using the new introduced $(\overline{R} - VaR)$- and $(\overline{R} - R_{low})$-risk measures. The new measures are based on VaR- and *Rlow*-values, which refer to the downside risk measures.

The paper is organized as follows. The techniques of diversification, portfolio performance criteria and modifications of the Sharpe coefficient, based on the new introduced risk measures, are discussed in the second section. An illustrative example is presented in the third section, it is performed via the open trade system of Moscow Exchange and it includes case study of portfolio diversification, portfolio performance evaluation and testing of the selected portfolios. The results of model verification are discussed in the forth section. The main findings of the research are summarized in the conclusion.

2 Methods

2.1 Portfolio Diversification Techniques

First we introduce some concepts to characterize an individual asset.

The monthly moving return at time t of the asset i, $i = \overline{1, n}$ is defined as:

$$R_{it} = \frac{P_{it}}{P_{i(t+1-m)}}, \quad t = \overline{m, N} \tag{1}$$

where $P_{it}, P_{i(t+1-m)}$ are the amounts of money received at the end and invested at the beginning of a period of one 21-day month, $m = 22$, and N is the number of observed values of prices for a period of one 252-day-year, $N = 252$.

The mean value of return on the asset i is:

$$\overline{R}_i = \frac{1}{N}\sum_{t=m}^{N} R_{it} \tag{2}$$

The risk of the asset i is characterized by the variance of the return:

$$\sigma_i^2 = \frac{1}{N-1}\sum_{t=m}^{N} (R_{it} - \overline{R}_i)^2 \tag{3}$$

The covariance summarizes the mutual dependence of two assets i and j:

$$Cov_{ij} = \frac{1}{N-1}\sum_{t=m}^{N} (R_{it} - \overline{R}_i)(R_{jt} - \overline{R}_{jt}) \tag{4}$$

The Market model assumes that an individual asset is correlated with the market portfolio, which can be evaluated by the market index (for example, *S&P*, *DJIA*, *IMOEX*). Thus, the Market model implies a special structural property for the return of the asset. The expected return on the asset i, $i = \overline{1,n}$, is assumed in the form [1]:

$$R_i = \alpha_{iI} + \beta_{iI} \cdot R_I + \varepsilon_{iI}, \tag{5}$$

where α_{iI} is the shift coefficient, β_{iI} is the slope coefficient, ε_{iI} is a random variable, R_I is the return on the market index. One has:

$$\alpha_{iI} = \overline{R}_i - \beta_{iI} \cdot \overline{R}_I, \tag{6}$$

where $\overline{R}_i$ and $\overline{R}_I$ are the mean returns on the asset and on the index, respectively, obtained using (2).

The coefficient β_{iI} represents the sensitivity of asset i to market movement. ‘Beta’ shows how much the asset performance moves when the market moves:

$$\beta_{iI} = \frac{Cov_{iI}}{\sigma_I^2}, \tag{7}$$

where Cov_{iI} is the covariance between the return on the asset and the return on the index; σ_I^2 is the variance of the return on the index.

For the further analysis we consider the Index I_t of the Trading Organiser 'Moscow Exchange' – MOEX Russia Index (*IMOEX*) [8], that is defined as follows:

$$I_t = I_0 \cdot \frac{MC_t}{MC_0}, \tag{8}$$

where MC_t and MC_0 are the total market capitalization of Index component stocks on the current date t and on the initial date *0*, and I_0 is the Index value on the initial date. Thus,

$$MC_t = \sum_{i=1}^{k} P_{ti} \cdot Q_{ti},\ t = \overline{1, N}, \tag{9}$$

where Q_{ti} is the number of stocks i existing on the current date t, P_{ti} is the price of unit stock i on the time t, k is the total number of component stocks used in the Index calculation.

The risk of the asset i is measured by the variance:

$$\sigma_i^2 = \beta_{iI}^2 \cdot \sigma_I^2 + \sigma_{\varepsilon i}^2, \tag{10}$$

where $\sigma_{\varepsilon i}^2$ is the variance of the random variable ε_{iI}.

The mean return of a portfolio of n assets is obtained as:

$$R_p = \sum_{i=1}^{n} w_i \cdot \overline{R}_i, \tag{11}$$

where w_i is the weight of the asset i in the portfolio.

The variance of a portfolio return is defined as [7]:

$$\sigma_p^2 = \sum_{ij} w_i \cdot w_j \cdot Cov_{ij} \tag{12}$$

Assuming that the portfolio consists of n assets and letting the weighting coefficients w_i range over all possible combinations such that $\sum_{i=1}^{n} w_i = 1$, we can plot the mean - standard deviation diagram of the feasible set of portfolios and obtain the efficient frontier, as shown on Fig. 1.

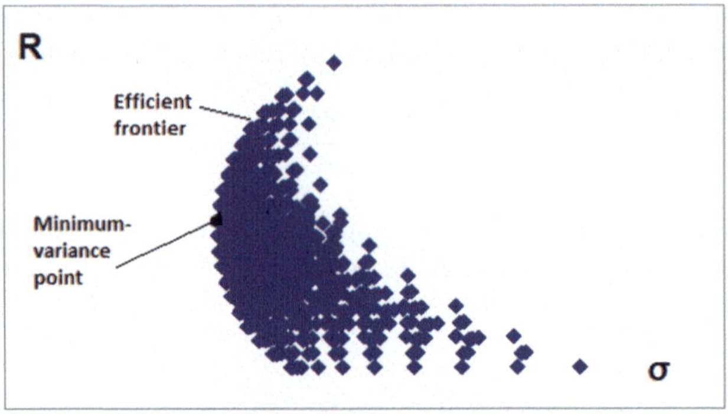

Fig. 1. Feasible set of portfolios

The left part of the boundary is the minimum-variance set and the upper portion of this set forms the efficient frontier of the feasible set.

The points of the efficient frontier are obtained by solving the optimization problem: to minimize the variance of the portfolio under fixed value R of the mean return. That is [7]:

$$J = \sum_{ij} w_i \cdot w_j \cdot Cov_{ij} \to \min \tag{13}$$

subject to

$$\sum_{i} w_i \cdot \overline{R_i} = R \tag{14}$$

$$\sum_{i} w_i = 1$$

$$w_i \geq 0,\ i = \overline{1, n}$$

The solutions of the problems yield the optimal weight coefficients w_i^* for the assets in the portfolio.

2.2 Reward-to-Volatility and Reward-to-Variability Ratio

As a result of solving the optimization problem, the investor obtains a set of portfolios on the efficient frontier. The question is to select a portfolio from the efficient set.

One strand of the literature is based on constructing indifference curves [7]. We consider this approach rather subjective, since it assumes that the investor is able to compare various combinations of portfolio risk and return and to determine which of the combinations are equivalent for him.

Another strand of literature aims at finding the tangent portfolio. Such an approach is implemented in the Tobin model, the Elton – Padberg – Gruber algorithm [3]. The commonly used portfolio performance indicators are the Sharpe and Treynor coefficients, which measure the risk premium per unit of risk. Thus, as a criterion for choosing the optimal portfolio from the efficient set, it is proposed to maximize the risk premium per unit of risk.

The Sharpe coefficient, known as the 'reward-to-variability ratio', is defined to be [1]:

$$RVAR_p = \frac{R_p - R_f}{\sigma_p}, \tag{15}$$

where R_p is the mean-return of the portfolio p, R_f is the risk-free asset return, and σ_p is the standard deviation of the portfolio p.

The Treynor coefficient (the 'reward-to-volatility ratio') is assumed to be [1]:

$$RVOL_p = \frac{R_p - R_f}{\beta_p}, \tag{16}$$

where β_p is the 'beta'-coefficient of the portfolio p, that is defined in the Market Model [1].

The Sharpe coefficient (15) and Treynor coefficient (16) can be equally used for assets evaluation:

$$RVAR_i = \frac{\overline{R}_i - R_f}{\sigma_i} \tag{17}$$

$$RVOL_i = \frac{\overline{R}_i - R_f}{\beta_i}, \tag{18}$$

where $\bar{R}_i$ is the mean return of the asset i, $i = \overline{1, n}$, σ_i is the standard deviation and β_i is the coefficient of sensitivity of the asset i to market movement.

The choice of a security i^* to be added into a portfolio can be based on the maximization of the Sharpe coefficient (17) or the Treynor coefficient (18), i.e. $i^* = \underset{i}{\arg\max}\, RVAR_i$ or $i^* = \underset{i}{\arg\max}\, RVOL_i$, $i = \overline{1, n}$. It means a preference is given to the asset having the largest market prime per risk-unit, measured by the standard deviation (the Sharpe coefficient) or by the 'beta'-value (the Treynor coefficient).

The choice of the coefficient depends on the set of the financial assets in the investor's portfolio. The risk for an investor, possessing other assets that are not included in the portfolio, should be measured by the 'beta'-coefficient since this coefficient evaluates risk relatively to the market. When all instruments are included in the portfolio under consideration, the standard deviation can be seen as a suitable risk-measure, and the Sharpe coefficient can be used as asset evaluation criterion.

2.3 Modifications of the 'Reward-to-Variability' Ratio

Reward-to-volatility and reward-to-variability ratio assume the normal or symmetric distribution of financial assets return. However in emerging markets economy the financial assets return is closer to the 'downside' distribution, that is, the actual return is shifted to the left, below the average (expected) return, which implies increased investment risk, and therefore, requires taking these deviations into account when constructing performance evaluation criteria.

We introduce a new parameter, termed 'low-mean' return of the asset i, defined as:

$$\overline{R}_{ilow} = \sum_{t \in Z^-} p_{it} \cdot R_{it},\ R_{it} < \overline{R}_i, \tag{19}$$

where Z^- is the set of indices t such that $R_{it} < \overline{R}_i$, and p_{it} is the probability of the return R_{it}.

$\overline{R}_{ilow}$ is the mean-return of a left ('bad') part of the return distribution of the asset i, i.e. the mean-value for the returns, which are less than the mean return of the asset $\overline{R}_i$.

We define a new risk-measure, namely the difference between the asset mean return and the 'low-mean' return $(\overline{R}_i - \overline{R}_{ilow})$.

The value-at-risk (VaR) is a measure widely used in financial analysis. For a known asset return distribution, *VaR* defines the return that can be achieved with some probability level [2]:

$$VaR_i = R_{iVaR} : [P\{R_{it} > R_{iVaR}\} = 1 - \alpha], \tag{20}$$

where α is the confidence level, which is usually set equal to 0.01, 0.05, or 0.1.

In the present paper we use method of historical modeling to calculate VaR considering inconsistency of the parametric VaR-models with the Russian stock market. This method is based on empirical distribution for a given period. VaR represents a quantile of an empirically estimated return distribution.

We propose another new risk measure, namely the difference between the asset mean return and the VaR-value for α-confidence level $(\overline{R}_i - VaR_i)$. The choice of the confidence level depends on the investor's attitude to risk. Risk preference allows setting high confidence-level, that increases VaR-value and decreases investor subjective evaluation of risk, measured by $(\overline{R}_i - VaR_i)$-value. And on the contrary, risk aversion implies low confidence-level.

On the base of the risk-measures, we propose the following modifications of the Sharpe coefficient for the asset i:

$$S_{ilow} = \frac{\overline{R}_i - R_f}{\overline{R}_i - \overline{R}_{ilow}} \tag{21}$$

$$S_{iVaR} = \frac{\overline{R}_i - R_f}{\overline{R}_i - VaR_i} \tag{22}$$

The coefficient (21) describes the amount of excessive return (market prime) referred to unit of risk, measured as a deviation of asset mean return from its 'low'-mean return. This coefficient may be recommended to evaluate especially the assets characterized by asymmetric distribution.

The coefficient (22) describes the amount of excessive return per unit of risk, measured as a deviation of asset mean return from its VaR-value. VaR-value can be estimated for different α-confidence levels, which are set regarding the investor's risk preferences.

The proposed modifications of the Sharpe coefficient are based on introduced 'downside' risk measures: $(\overline{R}_i - \overline{R}_{ilow})$- and $(\overline{R}_i - VaR_i)$-measures.

We propose the following modifications of the Sharpe coefficient for the portfolio:

$$S_{plow} = \frac{\overline{R}_p - R_f}{\overline{R}_p - \overline{R}_{plow}} \tag{23}$$

$$S_{pVaR} = \frac{\overline{R}_p - R_f}{\overline{R}_p - VaR_p} \tag{24}$$

The coefficient (23) describes the amount of excessive return (market prime) referred to unit of risk, measured as a deviation of portfolio mean return from its 'low'-mean return.

The coefficient (24) describes the amount of excessive return per unit of risk, measured as a deviation of portfolio mean return from its VaR-value. VaR-value can be estimated for different α-confidence levels, which are set regarding the investor's risk preferences.

The modified coefficients may be recommended to evaluate especially the portfolio characterized by asymmetric distribution, which is typical to emerging markets economy. It is noted, that the financial assets return in developing countries is closer to the 'downside' distribution, that is, the actual return is shifted to the left, below the average (expected) return, which implies increased investment risk, and therefore, requires taking these deviations into account when assessing portfolio efficiency.

3 Results

3.1 Portfolio Optimization

The experimental case study has been performed via open trade system operated by the Trading Organiser 'Moscow Exchange' – Equity and Bond Market (MICEX SE), which lists leading Russian securities that are of great interest to both domestic and foreign portfolio investors [8].

According to a principle of diversification, we consider an investor who distributes the invested amounts among different branches of economy, represented by the following companies: Public Joint Stock Company Gazprom (ordinary share *GAZP*), Public Joint Stock Company "Mining and Metallurgical Company "NORILSK NICKEL" (ordinary share *GMKN*), Public Joint Stock Company "Aeroflot-Russian Airlines" (ordinary share *AFLT*), Public Limited Liability Company Yandex N.V., shares of a foreign issuer (ordinary share *YNDX*), Rosneft Oil Company, ordinary share (ordinary share *ROSN*), Public Joint Stock Company "Magnit", (ordinary share *MGNT*), Mobile TeleSystems Public Joint Stock Company, ordinary share (ordinary share *MTSS*), Sberbank of Russia, ordinary share (ordinary share *SBER*).

We have studied statistic data on selected securities for a one-year period, namely December 2018 – December 2019. We suppose the amounts are invested for a one-month period considering a professional prudent investor who performs market monitoring regularly. Monthly-moving returns have been obtained using (1).

The securities mean returns, standard deviations and 'beta'-coefficients have been calculated using (2), (3), (7). The significance of the 'beta'-coefficients is confirmed by the high values of the coefficient of determination R^2. The results are shown in Table 1.

Table 1. Securities parameters for the period January 2019 – December 2019

Asset	Asset parameters			
	R_i	σ_i	β_{iI}	R^2
GMKN	1,030672	0,049349	1,011474	0,997799
AFLT	1,002309	0,050455	0,983576	0,997429
GAZP	1,041405	0,096502	1,023033	0,993589
MGNT	0,995392	0,063206	0,977365	0,997124
MTSS	1,021866	0,049885	1,003441	0,998921
ROSN	1,005757	0,040527	0,987094	0,998606
YNDX	1,029459	0,083607	1,010474	0,993925
SBER	1,021043	0,058033	1,002838	0,998487

The high values of the R^2-coefficient indicate that the funds' fluctuations are explained by performance fluctuations of the index [5]. Among the selected assets the *GAZP*-asset may be considered to be affected by factors other than the market to the highest degree. The return and the variance of the *IMOEX* Index are as follows: R_I = 1,018437; σ_I^2 = 0,00119. These values have been obtained using (1), (2) and (3).

We have determined the values of the $RVAR_i$-, $RVOL_i$-, and the introduced S_{ilow}- and S_{iVaR}-coefficients, using (17), (18), (21) and (22). The coefficient S_{iVaR} has been calculated for the confidence levels $\alpha = 0.05$ and $\alpha = 0.1$ to assess different risk preferences. The annual expected risk-free rate of return in (17), (18), (21) and (22) is supposed to be 7% as it is estimated for governmental bonds in Russia in 2019 [8]. Thus, annual risk-free return R_f is set equal to 1.07, that is 1.0057 for a one-month period as geometric mean in the considered example. The results are shown in Table 2.

Table 2. Assets performance evaluation criteria

Asset	Asset performance evaluation criteria				
	$RVAR_i$	$RVOL_i$	S_{ilow}	S_{iVaR}, $\alpha = 0.05$	S_{iVaR}, $\alpha = 0.1$
GMKN	0,506954	0,024734	0,750596	0,33616	0,448125
AFLT	−0,06629	−0,0034	−0,12712	−0,04839	−0,05287
GAZP	0,370464	0,034946	0,649321	0,332226	0,385877
MGNT	−0,16236	−0,0105	−0,2247	−0,11704	−0,14175
MTSS	0,324978	0,016156	0,440203	0,225581	0,288132
ROSN	0,002546	0,000105	0,003454	0,001874	0,002143
YNDX	0,284718	0,023558	0,41742	0,199424	0,259973
SBER	0,265184	0,015346	0,346886	0,181994	0,217683

The assets have been ranged according to the $RVAR_i$-, $RVOL_i$-, S_{ilow}- and S_{iVaR}-coefficients, as shown in Table 3.

Table 3. Ranking of the assets according to performance evaluation criteria

Asset performance evaluation criteria				
$RVAR_i$	$RVOL_i$	S_{ilow}	S_{iVaR}, $\alpha = 0.05$	S_{iVaR}, $\alpha = 0.1$
GMKN	GAZP	GMKN	GMKN	GMKN
GAZP	GMKN	GAZP	GAZP	GAZP
MTSS	YNDX	MTSS	MTSS	MTSS
YNDX	MTSS	YNDX	YNDX	YNDX
SBER	SBER	SBER	SBER	SBER
ROSN	ROSN	ROSN	ROSN	ROSN
AFLT	AFLT	AFLT	AFLT	AFLT
MGNT	MGNT	MGNT	MGNT	MGNT

Note, that the applied assets criteria have yielded almost the same results indicating the most efficient assets.

We have composed the portfolio of 6 upper assets, selected from the Table 2 and having the highest $RVAR_i$-, $RVOL_i$- S_{ilow}-, and S_{iVaR}-coefficient values, namely *GMKN*-, *GAZP*-, *MTSS*-, *ROSN*-, *YNDX*-, and *SBER*-assets.

Covariance matrix of the selected assets has been constructed using (4). The results are shown in Table 4.

Table 4. Covariance matrix for the selected assets

Asset	Cov_{ij}					
	GMKN	GAZP	MTSS	ROSN	YNDX	SBER
GMKN	0,002425644	−0,0008194	0,00125529	0,0008496	0,00124137	0,0001964
GAZP	−0,00081943	0,00927567	0,00093926	0,0004933	−0,0005654	0,0013511
MTSS	0,00125529	0,00093926	0,00247863	0,0006858	0,00184348	0,001375
ROSN	0,00084958	0,00049325	0,00068577	0,0016359	0,00067571	0,0010505
YNDX	0,00124137	−0,0005654	0,00184348	0,0006757	0,00696243	0,0010723
SBER	0,000196394	0,00135105	0,00137501	0,0010505	0,00107226	0,0033544

We have solved the variance-minimizing problem to find the efficient frontier of a feasible set. Remind that the efficient frontier is the upper portion of the minimum-variance set that lays upper than a minimum-variance point. The points on the efficient frontiers have been determined by solving the optimization problem (13), (14): minimize the variance of the portfolio under the constraint of a fixed mean return σ. The fixed values R in (14) have been chosen using a 0.1% step.

The obtained results are shown in Fig. 2.

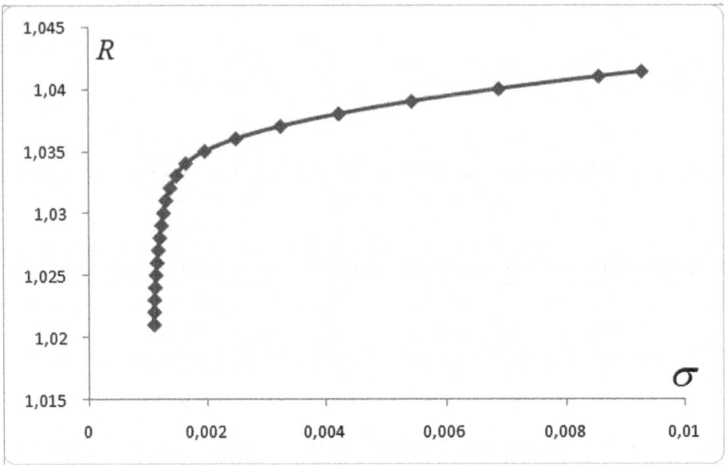

Fig. 2. Efficient frontier of portfolios

The optimal solution yields the weight distribution w_i^*, given in Table 5.

Table 5. Values of R_p, σ_p and the weight distribution w_i^* for efficient portfolios

Portfolio return	Standard deviation	Weights of the assets					
		GMKN	GAZP	MTSS	ROSN	YNDX	SBER
1,041405	0,092685	0,000000	1,000000	0,000000	0,000000	0,000000	0,000000
1,041013	0,083113	0,036501	0,963499	0,000000	0,000000	0,000000	0,000000
1,040012	0,073878	0,129765	0,870235	0,000000	0,000000	0,000000	0,000000
1,039011	0,065123	0,223029	0,776971	0,000000	0,000000	0,000000	0,000000
1,038010	0,057079	0,316294	0,683706	0,000000	0,000000	0,000000	0,000000
1,037010	0,050066	0,409465	0,590535	0,000000	0,000000	0,000000	0,000000
1,036010	0,044407	0,489734	0,498674	0,000000	0,000000	0,011592	0,000000
1,035010	0,040615	0,541632	0,409682	0,000000	0,000000	0,048686	0,000000
1,034010	0,038678	0,593523	0,320703	0,000000	0,000000	0,085774	0,000000
1,033009	0,037245	0,594139	0,269286	0,000000	0,000000	0,090554	0,046021
1,032008	0,036273	0,573069	0,233760	0,000000	0,000000	0,081668	0,111504
1,031008	0,035707	0,552019	0,198269	0,000000	0,000000	0,072791	0,176921
1,030007	0,035226	0,529693	0,179983	0,000000	0,024634	0,068374	0,197316
1,029007	0,034799	0,506621	0,172262	0,000000	0,064310	0,066693	0,190113
1,028006	0,034427	0,483526	0,164533	0,000000	0,104026	0,065012	0,182904
1,027005	0,034114	0,460431	0,156803	0,000000	0,143742	0,063330	0,175694
1,026004	0,03386	0,437336	0,149074	0,000000	0,183458	0,061648	0,168484
1,025004	0,033666	0,414264	0,141353	0,000000	0,223134	0,059968	0,161281
1,024003	0,033529	0,390908	0,133585	0,000891	0,262618	0,058168	0,153830
1,023002	0,033444	0,363136	0,125176	0,016830	0,298175	0,054371	0,142311
1,022001	0,033412	0,335364	0,116766	0,032769	0,333733	0,050575	0,130793

The optimal solution w_i^* has been obtained using the Optimization Toolbox of Excel and the Visual Basic Editor.

3.2 Portfolio Selection Using Performance Evaluation Criteria

We have determined portfolio parameters for the weight distribution w_i^* obtained in Sect. 3.1. We have evaluated portfolio performance using the Sharpe coefficient for a portfolio $RVAR_p$, the Treynor coefficient for a portfolio $RVOL_p$, modified Sharpe coefficients for a portfolio S_{plow}- and S_{pVaR}-coefficients using (15), (16), (23) and (24), respectively. The coefficient S_{pVaR} has been calculated for the confidence levels $\alpha = 0.05$ and $\alpha = 0.1$ to assess different risk preferences. The results are shown in Table 6.

Table 6. Portfolio performance evaluation

Portfolio return	Standard deviation	Portfolio performance evaluation criteria				
		$RVAR_p$	$RVOL_p$	S_{plow}	S_{pVaR}, $\alpha = 0.05$	S_{pVaR}, $\alpha = 0.1$
1,041405	0,096502	0,370464	0,024843	0,558771	0,332226	0,385877
1,041013	0,092685	0,381492	0,025146	0,584157	0,338014	0,397042
1,040012	0,083113	0,413387	0,025987	0,651472	0,359534	0,415895
1,039011	0,073878	0,451514	0,026943	0,704395	0,390774	0,443927
1,038010	0,065123	0,496838	0,028037	0,784864	0,416513	0,474546
1,037010	0,057079	0,549336	0,029303	0,863991	0,442654	0,517757
1,036010	0,050066	0,606315	0,030687	0,881075	0,480951	0,583787
1,035010	0,044407	0,661054	0,032077	0,936465	0,550201	0,622679
1,034010	0,040615	0,698153	**0,033711**	1,008067	0,568637	**0,665058**
1,033009	0,038678	0,707250	0,033038	**1,027810**	**0,575546**	0,648830
1,032008	0,037245	**0,707572**	0,031351	0,983435	0,559560	0,624191
1,031008	0,036273	0,698974	0,029715	0,960907	0,542172	0,606034
1,030007	0,035707	0,682019	0,028597	0,931563	0,524683	0,590323
1,029007	0,035226	0,662941	0,027761	0,899593	0,503408	0,579099
1,028006	0,034799	0,642318	0,026903	0,880509	0,477822	0,560793
1,027005	0,034427	0,620172	0,026024	0,846068	0,468455	0,542368
1,026004	0,034114	0,596530	0,025122	0,809646	0,463934	0,515989
1,025004	0,03386	0,571470	0,024197	0,760013	0,436258	0,491975
1,024003	0,033666	0,545022	0,023240	0,724495	0,421739	0,468593
1,023002	0,033529	0,517397	0,022146	0,673269	0,399149	0,446927
1,022001	0,033444	0,488779	0,021034	0,649726	0,367799	0,423365
1,021000	0,033412	0,459286	0,019904	0,611969	0,343841	0,399553

It is notable, that $RVAR_p$-, $RVOL_p$-, S_{plow}- and S_{pVaR}-coefficients demonstrate a similar trend achieving a maximum value, as shown in Fig. 3. Thus we can assume that introduced portfolio performance criteria enable to select a portfolio with the highest risk premium per unit of risk, measured as deviation of a portfolio return from its 'low-mean' return, or portfolio return from its VaR-return in S_{plow} and S_{pVaR}, respectively. Whereas $RVAR_p$- and $RVOL_p$ coefficients assess risk premium per unit of risk measured by standard deviation and portfolio 'beta', respectively.

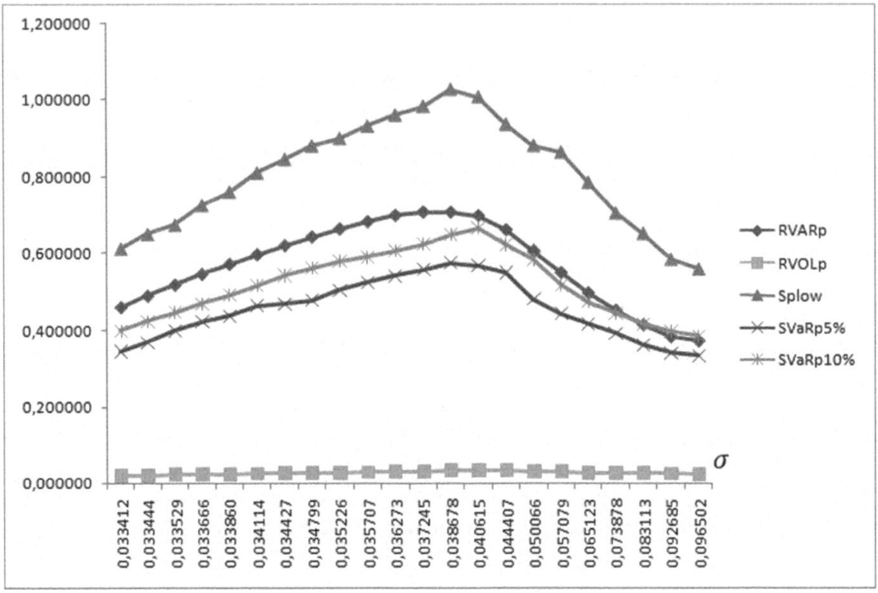

Fig. 3. Portfolio performance evaluation criteria

The portfolios having the maximum values of the $RVAR_p$-, $RVOL_p$-, S_{plow}- and S_{pVaR}-coefficients have been determined. Note, that $RVOL_p$- and S_{pVaR}-criteria for the confidence level $\alpha = 0.1$ indicate the portfolio with the same weight distribution. S_{plow}- and S_{pVaR}-criteria for the confidence level $\alpha = 0.05$ also indicate the portfolio with the same weight distribution. Whereas $RVAR_p$-criterion produces a different weight distribution. Thus, three portfolios, i.e. three weight distributions, have been selected for further approbation.

3.3 Portfolio Testing

In traders' practice it is commonly used to compare the portfolio mean-return with the return of the market to test applicability of the portfolio. An investment portfolio is considered to be efficient if its return is not less than the return of the market.

We consider the return on the *IMOEX* index - the official Moscow Stock Exchange benchmark - to be the market return. The returns on investment portfolios selected in Sect. 3.2 have been compared with the market return. To evaluate their efficiency the monthly-moving portfolio returns and market returns were determined daily for a one-month period, namely January 2020.

Figure 4 shows the returns on the three selected portfolios compared with the returns on the *IMOEX* Index.

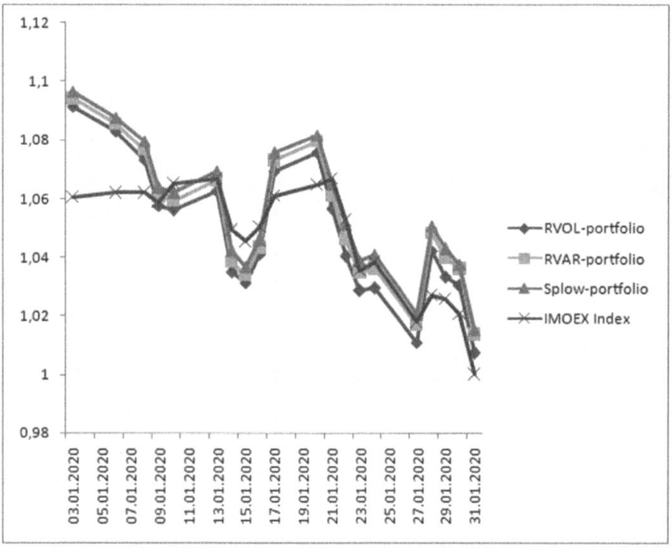

Fig. 4. Returns on the $RVOL_p$-, $RVAR_p$- and S_{plow}-portfolios relatively to the market return (*IMOEX* Index)

To compare the tested portfolio we evaluated portfolio mean return R_p, portfolio standard deviation σ_p, the sum of excessive portfolio return over the market S+, the sum of excessive portfolio return over the market per sum of total deviations from the market (S+/S).

The parameters for the tested portfolios are shown in Table 7.

Table 7. Portfolio and market parameters for a test-period (January 2020)

Portfolio	Portfolio test parameters			
	R_p	σ_p	S+	S+/S
$RVOL_p$	1,047683	0,023272	0,122082	0,553952
$S_{pVaR}(\alpha = 0.1)$	1,047683	0,023272	0,122082	0,553952
$RVAR_p$	1,052124	0,022227	0,164661	0,759838
S_{plow}	1,055105	0,022054	0,199296	0,880431
$S_{pVaR}(\alpha = 0.05)$	1,055105	0,022054	0,199296	0,880431

Note, that the portfolio selected according to $RVOL_p$-criterion (its weight distribution is the same as for S_{pVaR}-criterion ($\alpha = 0.1$), as has been shown in Sect. 3.2) demonstrate lower mean return and higher risk level.

Portfolio selected according to $RVAR_p$- and S_{plow} - criteria produced better results. The highest excessive return over market return was produced by portfolio selected according to S_{plow}-coefficient. Weight distribution for the portfolio, selected according to S_{plow}-criterion, is the same as for S_{pVaR}-criterion for the confidence level $\alpha = 0.05$.

4 Discussions

The applied the $RVAR_i$-, $RVOL_i$-, S_{ilow}- and S_{iVaR}-coefficients for the assets have yielded almost the same results indicating the most efficient assets in their ranking. While the applied $RVAR_p$-, $RVOL_p$-, S_{plow}- and S_{pVaR}-coefficients for portfolio performance produced different results.

In the case study the portfolio selected according to $RVAR_p$- and S_{plow} - criteria produced better results. The highest excessive return over market return was produced by portfolio selected according to S_{plow}-coefficient. Portfolios selected according to $RVOL_p$-coefficient and S_{pVaR}-criterion ($\alpha = 0.1$) demonstrated lower mean return and higher risk level, thus they did not prove their applicability in practice.

The experimental study of S_{pVaR}-criterion for confidence level $\alpha = 0.05$ yields better results than for $\alpha = 0.1$. It can be further recommended to apply other methods for VaR computation. Considering the inconsistency of parametric VaR-methods with the stock market of the developing countries, simulation methods are preferable. In the present paper we have used the method of historical modeling. We suppose that other simulation methods could increase the efficiency of the VaR-approach. For example, Monte Carlo simulation, which is widely used in practice.

Thus we can conclude that the proposed modified S_{plow}-coefficient enables to perform portfolio efficiency evaluation, to select a portfolio from the efficient frontier and to achieve better portfolio parameters relatively to other criteria that do not take into account downside risk measures. There is a strong likelihood that it is more adjusted to unstable markets in order to measure risk of the assets and of the portfolio.

S_{plow}-coefficient can be incorporated into the system of key performance indicators for assessing the personnel efficiency, as it reveals the particular achievements of the financial manager and can serve as the basis for building a differentiated wage system.

5 Conclusion

In this paper new S_{ilow}- and S_{iVaR}-securities selection criteria have been proposed, based on the introduced $(\overline{R}_i - R_{ilow})$- and $(\overline{R}_i - VaR_i)$-risk-measures.

$(\overline{R}_i - R_{ilow})$-value can be a suitable risk-measure especially for the asymmetric distribution of the return of an asset. $(\overline{R}_i - VaR_i)$-value allows the investor to set acceptable deviation of the return from the VaR-value for different confidence levels, considering his risk preferences.

S_{plow}- and S_{pVaR}-coefficients have been introduced for portfolio performance evaluation. The modified coefficients may be recommended to evaluate especially the portfolio characterized by asymmetric distribution, which is typical to emerging markets economy and implies increased investment risk.

The case study for portfolio selected according to $RVAR_p$- and S_{plow} - criteria produced better results. To the best of our knowledge the highest excessive return over market return was produced by portfolio selected according to S_{plow}-coefficient. Portfolios selected according to $RVOL_p$- and S_{pVaR}-criteria did not prove their applicability in practice.

The proposed modified S_{plow}-coefficient enables to perform portfolio efficiency evaluation, to select a portfolio from the efficient frontier and to achieve better portfolio parameters.

The proposed modified Sharpe coefficients and computational techniques enable to process financial and labor data, to formalize personnel evaluation process and therefore to increase management efficiency.

References

1. Sharpe, W.F., Alexander, G.J., Beiley, J.V.: Investments. Prentice Hall International Inc., New Jersey (1995)
2. Fabozzi, F.J.: Investment Management. Prentice Hall International, Inc., New Jersey (1995)
3. Sharpe, W.F.: The Sharpe Ratio, Stanford University (Reprinted from The Journal of Portfolio Management (1994) http://web.stanford.edu/~wfsharpe/art/sr/sr.htm. Accessed 18 Jan 2020
4. Breuer, W., Gürtler, M.: Performance evaluation and preferences beyond mean-variance, swiss society for financial market research. Finan. Mark. Portfolio Manag. **17**(2), 213–233 (2003)
5. Ebner, M., Neumann, T.: Time-varying betas of german stock returns, swiss society for financial market research. Finan. Mark. Portfolio Manag. **19**(1), 29–46 (2005)
6. Jagannathan, R., Wang, Z.: Emperical evaluation of asset-pricing models: a comparison of the SDF and beta methods. J. Finan. **57**(5), 2337–2367 (2002)
7. Luenberger, D.G.: Investment Science. Oxford University Press, New York (1998)
8. Moscow Exchange. https://www.moex.com/. Accessed 21 Feb 2020

Development of Elements of an Intelligent High-Performance Platform of a Distributed Decision Support System for Monitoring and Diagnostics of Technological Objects

Vladimir Bukhtoyarov[1,2(✉)], Vadim Tynchenko[1,2], Eduard Petrovsky[1], Kirill Bashmur[1], and Roman Sergienko[3]

[1] Siberian Federal University, Krasnoyarsk, Russia
vladber@list.ru
[2] Reshetnev Siberian State University of Science and Technology, Krasnoyarsk, Russia
[3] Ulm University, Ulm, Germany

Abstract. In the article the architecture of the intelligent high-performance platform of a distributed decision support system for monitoring and diagnostics of technological objects is developed. The proposed architecture will ensure the implementation of the concept of an integrated cyber-physical system to ensure the safety and sustainability of production in terms of operation of technological equipment. In general version, building a system using such an architecture, on the one hand, ensures openness in the formation of the analytical core of the decision support system, which will ensure a high level of system adequacy in the conditions of production transformation, increasing the intensity of information exchange and the volume of data being analyzed. The approaches based on the methods of automatic construction of artificial neural networks ensemble-distributed solvers (classifiers) are examined. It can be used to create analytical support for decision support systems and their automated initialization in a specific production environment. #CSOC1120.

Keywords: Intelligent platform · Decision support system · Monitoring · Diagnostics · Technological objects

1 Introduction

Nowadays approaches for highly effective decision support systems for monitoring and diagnostics of technological objects are developed in the framework of Industry 4.0 and Smart Production [1–3]. Both of these concepts imply close integration of production processes and computing and telecommunication structures within production cyber-physical systems [4–6]. The main purpose of using such systems is to intensify and intellectualize both data access processes and data processing, which, presumably, will significantly increase the efficiency of the production and business processes implemented in automatic, automated and "manual mode".

R. Silhavy (Ed.): CSOC 2020, AISC 1225, pp. 614–626, 2020.
https://doi.org/10.1007/978-3-030-51971-1_50

Studies show that through the integration of intelligent information and telecommunication systems (cyberphysical systems) into production systems, high flexibility and adaptability of decision support during monitoring and diagnostics of technological objects can be provided. According to some researchers, this allows us to make the transition to a qualitatively new form of implementation of production programs based on the optimal use of resources, technological machines and equipment [7, 8]. A further direction in the development of such an approach on a global of production systems is formed in the postulates of the autonomy of device operation, monitoring, analysis and data processing and forecasting in regimes close to real-time mode, self-diagnosis, self-organization and self-repairing of devices [9].

The basis for the design of cyberphysical systems is the formation of high-precision models (in terms of some studies - "digital twins") of technological equipment and production processes [7, 10–12]. The formation and use of such high-precision models is based on a high degree of integration of the mathematical modeling apparatus and data analysis methods within the framework of the computational analytical core of the cyberphysical system. Another area of research is the development of methods and algorithms that will allow the creation of high-precision software models and simulators that make it possible to partially reject the need to use laboratory research and physical means of diagnostic support of production (the concept of a "virtual laboratory" and its development) [13–15].

The formation of such a system at this level is proposed to be carried out on the basis of the development and implementation of models for the interaction of modern telecommunication devices, including wireless networks, and monitoring and control tools that create a relatively inexpensive highly reliable infrastructure for high-speed remote access to data and devices in an automated production system [16–18].

It is proposed to build such an intelligent platform for decision support systems in the form of distributed, decentralized agents interacting in real time to solve optimization and modeling problems integrated into the cloud computing system with a powerful analytical core that implements big data analysis and visualization algorithms. Thus, decision-making support during monitoring and diagnostics of technological objects as a process, the effectiveness of which is ensured by high-precision digital representation of a physically real production system. Researchers suggest integrating into such systems the capabilities of diagnosing the technical condition of equipment and predicting its life, which can be directly taken into account when determining rational control modes and formulating plans for the operational level. This is achieved through continuous analysis of data obtained from elements of technological equipment and their high-precision models, including a comparison of retrospective, model data and current monitoring data using methods of statistical data processing [19–22]. This, presumably, allows to increase the accuracy of the forecast of the time between failures and to carry out equipment maintenance and repair according to the technical condition on time, providing, on the one hand, minimizing the risks of a sudden stoppage of production and, on the other hand, rationalizing the costs associated with bringing equipment into repair.

The effectiveness of such approaches integrated within a single platform should be ensured by the development of data mining methods and models with the possibility of detecting cause-effect relationships in large volumes of data obtained during the implementation of technological processes [23, 24]. Thus, at the present stage, decision support systems for monitoring and diagnosing technological objects are proposed to be considered as intelligent systems that minimize the influence of the human factor in solving the corresponding problems [25, 26].

2 Development of the Architecture of an Intelligent High-Performance Platform of a Distributed Decision Support System for Monitoring and Diagnostics of Technological Objects

An analysis of approaches developed for creating decision support systems allows us to suggest that it is rational and efficient to form a platform architecture to create such systems, considering several principles, including:

- Focusing on the processing of large amount of data obtained both from retrospective parametric data sources and from real-time data aggregated using high-precision measuring systems.
- Integration of various methods and algorithms for data processing and design of decision support models, including data mining techniques. As one of the approaches, the option of implementing an "open" analytical component within the framework of the decision support system architectures, suggesting the possibility of dynamic integration of components that ensure the functioning of the analytical subsystem, can be proposed.
- Integration of various methods of collecting and aggregating data, ensuring the adaptability of the strategy of forming the information base for decision support models design and application.
- The principle of distributed-centralized information processing, involving the dynamic reconfiguration of the computing structure that ensures the operation of the components of the decision support system in the rational and efficient use of available computing power to ensure high performance of the system as a whole, taking into account the increasing requirements for speed and accuracy of decision making in the aspect of intensification of production processes.

In the framework of the presented study, the main elements that are reflected in the architecture of the high-performance platform for distributed decision support system for monitoring and diagnostics of technological objects (DSS MTO) that is being developed is focus on the implementation of cyber-physical production systems implemented by the multilayer structure of hardware-software and technological solutions that provide control solutions to various levels of the production system.

DSS MTO platform is formed in the form of a set of interconnected specialized subsystems that provide solutions at the software, software, hardware and technical levels of tasks that determine the purpose of the system. The totality of subsystems should include:

1. The subsystem for the integration of hardware and technical objects of the production system with DSS MTO software. The subsystem should be a combination of the following modules, providing the corresponding functionality:
   - Specialized modules for receiving information from technical devices at various levels of an integrated production system.
   - Specialized modules for receiving information generated by hardware elements of an integrated production system.
   - Specialized modules for generating information packets (signals of various kinds) for elements of technical and hardware support for an integrated production system;
   - Module for ensuring adaptation of signal conversion functions for a set of elements of technical and hardware support for DSS MTO.
   - Control and adjustment module for the integration procedures of the hardware and technical components of the production system and the software level of DSS MTO.
2. The subsystem of integration of system-wide and specialized software of the production system with software, hardware and technical support of DSS MTO.
3. The subsystem for the collection, pre-processing and storage of information about technological equipment at the stages of the life cycle.
4. The analytical subsystem that provides support for decision making, in general, the intellectualization of problem-solving process. Basically, it to consist of:
   - Module for the formation of knowledge bases.
   - Module for constructing and adapting models based on data mining technologies.
   - Module for constructing and adapting predictive models based on data mining technologies.
   - Module for constructing and adapting classifiers and cluster analysis models to data mining technologies.
   - Module for the formation of solutions to object-oriented tasks.
   - Design module for intelligent computing and analytical environment.
   - Deployment module for an intelligent computing and analytical environment.
5. The subsystem of the industrial internet of things, ensuring the initialization and operation of a distributed network of industrial internet devices.
6. The subsystem of presentation and data in the context of various levels of management and planning of production activities.
7. A control subsystem that provides functionality in terms of synchronizing the operation of other subsystems and fulfilling special requirements:
   - Module for the basic settings for the functioning of the DSS MTO.
   - Module for setting parameters of subsystems of DSS MTO.
   - Module for ensuring priorities and data synchronization in the distributed structure of DSS MTO.
   - Module for providing a multi-user mode of using software and hardware-software tools of DSS MTO.

- Module for the collection, storage and presentation of specialized management information DSS MTO.
- Control module for initialization of DSS MTO components.

8. The subsystem to ensure the stability of DSS MTO in terms of its technical, hardware and software components. The subsystem should be a combination of the following modules, providing the corresponding functionality:
   - Module for the automated identification of resources, components and assets of the DSS MTO, ensuring the stability of its functioning.
   - Module for automated identification of destabilizing factors, threats to stability of the DSS MTO.
   - Module for quantitative risk assessment of the effects of destabilizing factors.
   - Decision support module in terms of determining the means, methods and measures to ensure the sustainability of the DSS MTO.
   - Module for controlling the implementation of countermeasures to ensure the sustainability of the DSS MTO.

To evaluate the proposed architecture of a high-performance intelligent platform in terms of applicability and adequacy to existing corporate information processing systems at the enterprises, an expert analysis was carried out with the involvement of specialists from a number of oil and gas industry enterprises and related areas, in particular, oil and gas engineering.

The developed architecture of the high-performance platform for DSS MTO being implemented using software and hardware in specific application conditions will ensure the implementation of the concept of an integrated cyber-physical system to ensure the safety and sustainability of production in terms of operation of technological equipment. In a general version, building a system using such an architecture, on the one hand, ensures openness in the formation of the analytical core of the DSS MTO, which will ensure a high level of system adequacy in the conditions of production transformation, increasing the intensity of information exchange and the volume of data being analyzed.

3 Processing of Diagnostic Information to Support Decision-Making in Determining the Technical Condition of Equipment

In the framework of the concept of creating production cyberphysical systems that provide comprehensive data processing methods into the aggregate of production and related tasks, special attention is paid to the creation and application of models for ensuring the reliability, stability, and survivability of such systems. With this it is proposed to build such systems on the basis of the self-x principle, including the implementation of the self-diagnosis functionality, taking into account the formation of an information base for diagnostics and monitoring based on real-time operational information. On the other hand, it seems that the analytical support of the production cyberphysical system is capable of ensuring compliance with the high requirements of the accuracy of construction and application of analytical models based on integrated information processing that reflects aspects of the functioning of the set of elements of the production system. The processing of such volumes of data and the achievement of high-performance indicators of the respective models seems possible within the framework of the functioning of data mining in a high-performance computing environment. Obviously, it is practically impossible to provide the corresponding computing and informational potential for each of the individual elements of the production system within the framework of the self-x concept - the corresponding hardware is economical and energy-efficient with limited computing capabilities and memory capacity.

In this regard, it seems urgent to develop and study a set of solutions for the formation of a distributed-centralized scheme for processing diagnostic information and monitoring the technical condition of production infrastructure facilities in the framework of ensuring the stable and reliable functioning of the cyberphysical production system.

The proposed approach is based on the following information processing model in order to determine the technical condition of technological equipment. Taking into account the fact that one of the most effective methods for solving data mining problems, in particular, recognition of a technical state, is the ensembles of models, it is proposed to build an analytical subsystem of technical diagnostics based on a ensemble distributed approach [27–29]. The essence of such a solution scheme for processing diagnostic information is as follows: at the first stage, the problem is solved by separate, relatively simple models of data mining, distributed on devices that provide hardware self-x for a cyberphysical production system. Corresponding, relatively simple models together form an ensemble of models, but suggest the possibility of developing a solution to appeal to the ensemble, the calculation using which requires significantly more computational costs and can be carried out in particular situations. The totality of such situations is determined either by the inability to provide a given level of confidence in solving a problem of processing diagnostic data, or when it becomes necessary to adapt models and in order to ensure their information integrity.

That is, in the general case, in addition to actually finding the solution, an estimate is calculated that can be described as a "degree of confidence" ξ that the solution is correct. The calculated value ξ is compared with a predefined threshold value Δ. Further, based on the results of this comparison, a decision is made on whether to turn to an ensemble analyzer, formed as a set of relevant data analysis models, for making a collective decision:

- If $\xi \leq \Delta$, then a call is made to the ensemble analyzer (ensemble of models) to develop a solution using the hardware capacities of a high-performance central unit.
- If $\xi > \Delta$, then the solution is calculated using an individual data analysis model.

An important issue is to determine how to calculate the "degree of confidence" ξ. In this work, we used a two-factor estimate ξ, which is based on the following two parameters:

1. ξ_i - "degree of confidence" for *i*-th individual analyzer in its decision ("individual degree of confidence").

2. ρ_i - the degree of confidence for *i*-th individual analyzer, which determines how effective this individual analyzer is.

The final value of the "degree of confidence" was calculated using the following formula:

$$\xi = \xi_i \cdot \rho_i \tag{1}$$

The values ρ_i of the degree of confidence are in the range [0; 1]. The proposed approach for determining the degree of confidence does not imply the need for additional calculations related to obtaining values ρ_i at sample points for individual and ensemble analyzers. The values ρ_i are calculated during the formation of the ensemble of models and are associated with the calculation of the error values of individual analyzers, the values ξ_i when using individual analyzers as artificial algorithms, for example, artificial neural networks, are generated on the output layer of each network. Thus, the costs of additional calculations can be considered insignificant, which is confirmed by the assessment of the operating time of the proposed approach for test problems. When determining the threshold value, requirements to limit the intensity of information exchange can also be taken into account in some form.

4 Experimental Study

4.1 Study of the Efficiency of Methods for Automated Processing of Technical Diagnostic Data

In the framework of the study, the results of which are partially presented in this paper, the task of processing the data of technical diagnostics was considered in the statement of the problem of classification (recognition) of the technical condition.

In this regard, based on the results of previous numerical studies, the following data analysis methods were included and considered in the methods for determining the most effective for the problem being solved, which are intended, inter alia, for solving classification problems: ensemble-distributed artificial neural networks (see Sect. 3), support method machines, decision trees method, multidimensional adaptive splines approach. Adaptive splines, decision trees, and the support vector techniques were considered in the standard versions to assess the basic efficiency and the possibility of further application, taking into account the presence of modifications that also provide higher efficiency in particular classification problems.

It is assumed that the methods indicated above, taking into account the results of a numerical study of their effectiveness in solving the problem of recognizing the technical condition, will be used as elements of analytical nuclei as part of the information and analytical support of the cyber-physical production system. As part of the study, it is further planned to determine the composition of a comprehensive collectively distributed pool of data analysis methods, which will ensure the effectiveness of the implementation of the self-x concept in terms of self-diagnosis.

Research was conducted using proprietary software. The parameters of the classification algorithms were determined during a preliminary study on test problems of diagnosis from a data repository for testing machine learning methods [30]: a set of diagnostic data for ultrasonic flow meters (hereinafter in the table of results - Data Set 1), a set of diagnostic data for pumping units (Data Set 2), as well as data for the diagnostic problem obtained for the Aerzen VMY536M screw compressor. The results were obtained according to a scheme assuming a 10-fold cross-validation. For each test fold, the classification reliability (the ratio of correctly classified examples) was averaged. The main criterion for equalizing computational resources for all algorithms was the time of one test run. The statistical significance of the distinguishability of the results for was tested by the ANOVA method at a significance level of 0.05. The results of the experimental study are shown in Table 1.

Table 1. The results of the experimental study (part 1).

Approach	Machine learning repository data sets		Screw compressor data set
	Data set 1	Data set 2	
Decision trees	0.931	0.910	0.908
Multidimensional adaptive splines	0.950	0.931	0.930
Support vector machines	0.925	0.903	0.923
Artificial neural networks ensembles	0.942	0.928	0.921

The results of the numerical study demonstrate that even in the basic version of the implementation, the selected classification methods are able to provide acceptable recognition efficiency with a reliability of 0.90–0.93, which allows us to consider them as an algorithmic base for a platform solution for the diagnosis of technological equipment. Given the variability of the results, the implementation of several methods in such a platform solution will allow for adaptability and flexibility when using the methods of optimal formation of diagnostic models.

4.2 The Study of the Training Algorithms Methods for Adaptation of Neural Network Models for Determining the Technical Condition

Considering the high efficiency of artificial neural networks ensemble-distributed approach as the data processing technology for DSS MTO platform, the study intends to determine the most effective approaches for design of the structure and training of artificial neural networks. The study examined the following training algorithms for neural networks: the method of back propagation of errors, including modifications of the algorithm with adaptive learning speed, inertial component and multi-start; simulated annealing method; a genetic algorithm method combined with local search; a method based on an optimization method known as ant colony optimization.

The task of determining the state of equipment, namely, "partial" serviceability, is an urgent task of an integrated cyberphysical production system, among which the tasks are to support decision-making in determining rational and safe modes of operation of technological equipment.

The initial data for the task were obtained for the Aerzen VMY536M screw compressor. The locations for installing vibration sensors - vibration velocity transducers with the HART protocol AV02-0.08 were determined. In accordance with a preliminary study of the sensor installation points, areas close to the following installation details were determined: motor bearing on the side opposite to the drive; motor bearing on the drive side; bearing assembly of the compressor unit on the drive side; the bearing assembly of the compressor unit from the side opposite to the drive; subsequent bearing units. Measurement of vibrational parameters (including vibration velocity) was carried out in three planes.

Using the sensors installed at these points, model tests under controlled exposure conditions and expert assessment methods, a sample was obtained that describes various options for changing the generalized health indicator with a volume of 1000 patterns. Sets were formed for two classes: the healthy state of the diagnosed compressor and the faulty state of the diagnosed compressor. The number of observations in the sample for class 1 is 600 patterns; the number of observations for class 2 is 400 patterns.

The studies were carried out using proprietary software that implements neural network training algorithms selected for research. The architecture of the neural networks used is a multilayer perceptron. The parameters of the training algorithms for the study were determined during a preliminary study on test problems of diagnosis from a data repository for testing machine learning methods: a set of diagnostic data for ultrasonic flow meters (Data set 1), a set of diagnostic data for pumping units (Data set 2).

The results were obtained according to a scheme assuming a 10-fold cross-validation. The network structure was initialized in each experimental run and the same for all considered methods was used. As an activation function, a logistic function was used. The main criterion for equalizing computing resources for all algorithms was the time of one test run, which was set to be the same for the computer used. The statistical significance of the distinguishability of the results for was tested by the ANOVA method at a significance level of 0.05. The results of the study are presented in Table 2.

Table 2. The results of the experimental study (part 2).

Learning algorithm	Machine learning repository data sets		Screw compressor data set
	Data set 1	Data set 2	
Modified back propagation	0.861	0.810	0.785
Simulated annealing	0.855	0.907	0.865
Genetic algorithm with local search	0.945	0.918	0.923
Ant colony optimization	0.920	0.920	0.910

In the course of the study, the results of which are given in the paper, results were obtained demonstrating the effectiveness of the neural network approach for determining the technical condition of the compressor equipment. Variants of parametric adaptation of neural network models are determined, the most effective of which for the problem under consideration turned out to be methods based on the global optimization method is a genetic algorithm with local search.

4.3 The Study of the Structural Adaptation Methods of Neural Network Models for Determining the Technical Condition

Along with other methods of data analysis, artificial neural networks are one of the most effective approaches for constructing diagnostic models of various directions, which is proved by the significant number of their successful application for the tasks of medical diagnostics, diagnostics of anomalies in network activity and problems of determining the state of mining equipment. In the framework of the scientific project presented in this paper, comparative studies of various methods of data analysis on the problems of technical diagnostics were carried out, and models based on artificial neural networks turned out to be one of the most effective approaches.

In the aspect of creating appropriate DSS MTO platform, the developed solutions in the field of technical diagnostics should provide a high level of automation in terms of self-organization and adaptation of models and the possibility of distributed information processing. The recognition models of the technical state based on artificial neural networks ensemble-distributed approach fully comply with such requirements, considering the availability of approaches for the automatic generation of structures of neural networks.

In this regard, it seems relevant to study the methods of structural synthesis of neural network models to recognize the technical condition of technological equipment and measuring devices, various types of which are the basis for the implementation of production processes of various kinds. Taking into account the specifics of the task and the need to comply with the requirements of the self-organization of cyberphysical production systems, it seems that the methods of structural synthesis of neural network models should provide automation of the corresponding procedure, minimizing the need for expert participation and, on the other hand, ensuring high efficiency of the solutions obtained in the context of the problem of technical diagnosis.

In the presented study, the following methods for constructing and adapting neural networks were considered: the "optimal surgery" method - trimming the neural network; a method based on an evolutionary genetic algorithm; probabilistic evolutionary method [29]. The architecture of the neural networks used is a multilayer perceptron, a direct distribution network used, according to general estimates, in 80–90% of industrial implementations of the neural network approach.

Research was conducted using proprietary software. The methods were studied on sets of technical diagnostic tasks from the data repository: a set of diagnostic data for ultrasonic flow meters Data Set 1), a set of diagnostic data for pumping units (Data Set 1). Also, the initial data for the study were obtained for the Aerzen VMY536M screw compressor for two classes: the working condition of the diagnosed compressor and the faulty condition of the diagnosed compressor. The number of observations for class 1 is 600 pieces, the number of observations for class 2 is 400 pieces.

The parameters of the algorithms for the formation of structures of artificial neural networks were selected during preliminary research on a set of test tasks from the machine learning repository. Such sets of parameters were used during the numerical study. The main criterion for equalizing computational resources for all algorithms was the time of one test run.

The results were obtained according to a scheme assuming a 10-fold cross-validation. The statistical significance of the results was tested by the ANOVA method at a significance level of 0.05. The results of a numerical study are presented in Table 3.

Table 3. Reliability classification for neural networks, the structure of which is designed by various methods.

Approach	Machine learning repository data sets		Screw compressor data set
	Data set 1	Data set 2	
Optimal surgery	0,913	0,876	0,901
Genetic algorithm based approach	0,945	0,918	0,923
Probabilistic evolutionary method	0,942	0,928	0,921

The results obtained indicate a relatively high efficiency of evolutionary methods (2 and 3). Given fewer settings and automation and self-organization requirements, the latter method is preferred.

5 Conclusion

The developed architecture of the intelligent high-performance platform DSS MTO being implemented using software and hardware in specific application conditions will ensure the implementation of the concept of an integrated cyber-physical system to ensure the safety and sustainability of production in terms of operation of technological equipment. In a general version, building a system using such an architecture, on the one hand, ensures openness in the formation of the analytical core of the DSS MTO, which will ensure a high level of system adequacy in the conditions of production transformation, increasing the intensity of information exchange and the volume of data being analyzed.

The approaches based on the methods of automatic construction of artificial neural networks ensemble-distributed solvers (classifiers) are examined. It can be used to create analytical support for decision support systems and their automated initialization in a specific production environment.

Acknowledgments. The reported study was partially funded Scholarship of the President of the Russian Federation for young scientists and graduate students SP.869.2019.5.

References

1. Lasi, H., et al.: Industry 4.0. Bus. Inf. Syst. Eng. **6**(4), 239–242 (2014)
2. Davis, J., et al.: Smart manufacturing, manufacturing intelligence and demand-dynamic performance. Comput. Chem. Eng. **47**, 145–156 (2012)
3. Coalition, S.M.L.: Smart Manufacturing, Manufacturing Intelligence and Demand-Dynamic Performance, Smart Manufacturing Coalition (2011)
4. Monostori, L.: Cyber-physical production systems: roots, expectations and R&D challenges. Procedia Cirp **17**, 9–13 (2014)
5. Lee, J., Bagheri, B., Kao, H.A.: A cyber-physical systems architecture for industry 4.0-based manufacturing systems. Manuf. Lett. **3**, 18–23 (2015)
6. Wang, L., Törngren, M., Onori, M.: Current status and advancement of cyber-physical systems in manufacturing. J. Manuf. Syst. **37**, 517–527 (2015)
7. Uhlemann, T.H.J., Lehmann, C., Steinhilper, R.: The digital twin: realizing the cyber-physical production system for industry 4.0. Procedia Cirp **61**, 335–340 (2017)
8. Jazdi, N.: Cyber physical systems in the context of Industry 4.0. In: 2014 IEEE International Conference on Automation, Quality and Testing, Robotics, pp. 1–4. IEEE Press, New York (2014)
9. Bagheri, B., et al.: Cyber-physical systems architecture for self-aware machines in industry 4.0 environment. IFAC-PapersOnLine **48**(3), 1622–1627 (2015)
10. Gilchrist, A.: Industry 4.0: the industrial internet of things. Apress (2016)
11. Uhlemann, T.H.J., et al.: The digital twin: demonstrating the potential of real time data acquisition in production systems. Procedia Manuf. **9**, 113–120 (2017)
12. Schroeder, G.N., et al.: Digital twin data modeling with automation and a communication methodology for data exchange. IFAC-PapersOnLine **49**(30), 12–17 (2016)
13. Kelly, J.D., Zyngier, D.: A new and improved MILP formulation to optimize observability, redundancy and precision for sensor network problems. AIChE J. **54**(5), 1282–1291 (2008)
14. Mourtzis, D., et al.: The role of simulation in digital manufacturing: applications and outlook. Int. J. Comput. Integr. Manuf. **28**(1), 3–24 (2015)
15. Grant, T., Eijk, E., Venter, H.S.: Assessing the feasibility of conducting the digital forensic process in real time. In: International Conference on Cyber Warfare and Security-ICCWS 2016, pp. 146–155. Academic Conferences and Publishing International (ACPI), Boston (2016)
16. Ascorti, L., et al.: A wireless cloud network platform for industrial process automation: Critical data publishing and distributed sensing. IEEE Trans. Instrum. Meas. **66**(4), 592–603 (2017)
17. Luo, N., et al.: Cloud computing and virtual reality based virtual factory of chemical processes. Chem. Ind. Eng. Prog. **12**, 171–183 (2012)
18. Yuan, Z., Qin, W., Zhao, J.: Smart manufacturing for the oil refining and petrochemical industry. Engineering **3**(2), 179–182 (2017)
19. Ascorti, L., et al.: A wireless cloud network platform for critical data publishing in industrial process automation. In: 2016 IEEE Sensors Applications Symposium (SAS), pp. 1–6. IEEE Press, New York (2016)

20. Al-Fadhli, M., Zaher, A.: A smart SCADA system for oil refineries. In: 2018 International Conference on Computing Sciences and Engineering (ICCSE), pp. 1–6. IEEE Press, New York (2018)
21. Bey, K.B., Benhammadi, F., Benaissa, R.: Balancing heuristic for independent task scheduling in cloud computing. In: 2015 12th International Symposium on Programming and Systems (ISPS), pp. 1–6. IEEE Press, New York (2015)
22. Xu, L.D., Duan, L.: Big data for cyber physical systems in industry 4.0: a survey. Enterp. Inf. Syst. **13**(2), 148–169 (2019)
23. Qin, S.J.: Process data analytics in the era in big data. AIChE J. **60**(9), 3092–3100 (2014)
24. Zhao, C., et al.: An architecture of knowledge cloud based on manufacturing big data. In: IECON 2018-44th Annual Conference of the IEEE Industrial Electronics Society, pp. 4176–4180. IEEE Press, New York (2018)
25. Joly, M., et al.: Refinery production scheduling toward Industry 4.0. Front. Manag. Eng. **37**, 1877–1882 (2017)
26. Savazzi, S., et al.: Towards a factory-of-things: channel modeling and deployment assessment in PetroEcuador Esmeraldas oil refinery. In: 2016 8th IEEE Latin-American Conference on Communications (LATINCOM), pp. 1–6. IEEE Press, New York (2016)
27. Bukhtoyarov, V., Semenkina, O.: Comprehensive evolutionary approach for neural network ensemble automatic design. In: Proceedings of the IEEE World Congress on Computational Intelligence, pp. 1640–1648. IEEE Press, New York (2010)
28. Bukhtoyarov, V., Zhukov, V.: Ensemble-distributed approach in classification problem solution for intrusion detection systems. In: International Conference on Intelligent Data Engineering and Automated Learning, pp. 255–265. Springer, Cham (2014)
29. Bukhtoyarov, V.V., Tynchenko, V.S., Petrovsky, E.A.: Multi-stage intelligent system for diagnostics of pumping equipment for oil and gas industries. In: IOP Conference Series: Earth and Environmental Science, vol. 272, no. 3, art. 032030. IOP Publishing (2019)
30. Asuncion, A., Newman, D.: UCI machine learning repository. Meta (2007)

Process Automation in the Scenario of Intelligence and Investigation Units: An Experience

Gleidson Sobreira Leite(✉) and Adriano Bessa Albuquerque

Universidade de Fortaleza, Fortaleza, CE, Brazil
gleidson.sleite@gmail.com, adrianoba@unifor.br

Abstract. Being an emerging and very important topic mainly due to the negative economic and social impacts, crime prevention is a worldwide concern. In this context, government institutions in several countries have instituted units or sectors specialized in investigation and intelligence activities to act in different areas and expertise's. Due to the diversity and volume of existing crime practices, concerns about the evolution and expansion of crime help in the rising of an opportunity to adopt concepts of process automation in order to assist the treatment of the high volume of requests directed at these specialized units. Exploring the main characteristics of process automation, this paper presents an overview of different application trends and performs an experience in a real scenario of a specialized unity. Results point out the benefits achieved, and discussions about challenges and needs inherent to the application of process automation in the context of units that perform or support intelligence and investigation activities are presented.

Keywords: Intelligence units · Investigation units · Process automation

1 Introduction

In recent years, issues related to the high crime rate and the performance of criminal organizations have been constantly making headlines around the world, where, being an emerging and very important topic mainly due to the negative economic and social impacts, crime prevention is a worldwide concern.

Many of the benefits of globalization and the rise of technology in society, such as, for example, agility and ease of communication, financial movement and mobility, also created opportunities for the flourishing, diversification, expansion and organization of criminal groups [1].

A survey conducted in March 2018 by [2], attended by 2,373 senior managers from large global organizations across 19 countries, found about $ 1.45 trillion of total lost turnover estimated as a result of financial crimes, and statistical studies by [3] pointed out that in 2018 there were an estimated 1,206,836 violent crimes in the United States alone.

R. Silhavy (Ed.): CSOC 2020, AISC 1225, pp. 627–641, 2020.
https://doi.org/10.1007/978-3-030-51971-1_51

Figure 1 presents the total violent crime reported in United States from 2013 to 2018 extracted from a statistical survey presented by Statistita with data reported by the Federal Bureau of Investigation (FBI).

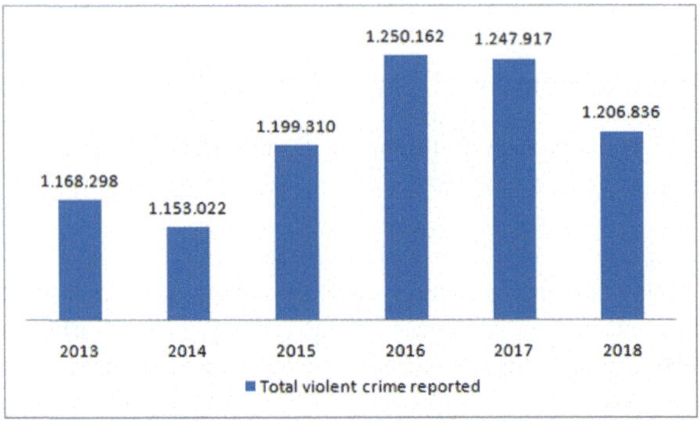

Fig. 1. Total violent crime reported in the United States from 2013 to 2018. Survey period: 2013–2018. Publication date: September 2019. Source: statista.com

These and several other studies point to concerns about the evolution and expansion of crime, where, due to the diversity and volume of existing crime practices, preventive, repressive (impediment of continuity), control or punitive action by government institutions is essential in order to minimize the damage caused to society.

To this end, several government institutions in various countries, which work in the fight against crime, have established units or sectors specialized in investigation and intelligence activities to act in different areas and expertise, such as, for example, FBI or EUROPOL, which have units specialized in combating crimes such as corruption, criminal organizations, violent crimes, white-collar crime, financial crimes, among others [4, 5].

However, due to the high crime rate, even with the excellent results obtained with the help of the solutions currently adopted, over time, the high volume of requests directed at these specialized units ended up resulting in limitations and challenges regarding to their treatment.

One of the reasons is due to the fact that a good part of the technological resources and team with the appropriate expertise is centralized in specialized units, in which stakeholders end up forwarding a high volume of requests due to limitations of knowledge, technological or specialized expertise's, among others. Thus, among the disadvantages normally found in centralized approaches [6], they can also lead to needs for prioritizing requests or even defining acceptance criteria.

Therefore, an opportunity arises to adopt solutions that contribute to agile results, simplification or optimization of processes, as well as reducing the use of resources necessary to carry out operational activities. Being a practice increasingly adopted in the market, concepts of process automation can also be used in order to contribute to these guidelines when using technology and the integration of systems and data to improve the control and progress of workflows and, when possible, replacing manual activities with automated ones [7].

Motivated by this scenario and exploring the main characteristics of process automation, this paper presents an overview of different application trends, and performs an experience in a real scenario of a specialized unity. Results point out the benefits achieved, and discussions about challenges and needs inherent to the application of process automation in the context of units that perform or support intelligence and investigation activities are presented.

This work also intends to be a contribution to the body of knowledge related to studies regarding the use of information technology in the fight against crime, and can be adopted to help both practitioners and researchers in order to identify possible applications trends and improvements.

This paper is organized as follow: the methodology and procedures performed in the work (Sect. 2); background and related work (Sect. 3); the experience and discussions (Sect. 4), followed by the final considerations (Sect. 5).

2 Research Methodology

To accomplish the objective of this work, three specific actions were carried out. They were:

1) Research and exploration of the main characteristics and benefits of process automation presented in academic studies and application trends in different domains.
2) Bibliographic research and selection of approaches proposed in academic research and related to applications of process automation.
3) The conduction of an experience in the context of units that perform or support intelligence and investigation activities, and present discussions about challenges and needs.

3 Background and Related Work

According to [8], intelligence is a product created through the process of collecting, grouping and analyzing data for dissemination as useful information to inform future interventions or support decision making. With regard to the investigation, [16] adds that it covers the collection of information and evidence to identify, apprehend and convict suspected criminals.

Although they appear similar, there are fundamental differences in several areas of their components, such as: final product objective, time orientation, data collection and analytical techniques, skill set requirements, nature of the conclusion, dissemination of information, among others [8–10]. However, in both contexts there are similar activities such as carrying out operational activities aimed at collecting, grouping and analyzing information, and producing/presenting results in order to achieve a certain objective.

According to [11–13], processes are composed of sequences of tasks ordered and organized with the participation of agents, resources and tools, and are usually executed to achieve a certain objective.

Automation, which means moving by itself, is the application of specific techniques, software and/or equipment in a given machine or industrial process, with the objective of increasing its efficiency, maximizing production with the lowest consumption energy and/or raw materials, or to reduce human effort or interference with this process or machine [14–16].

In this context, the concept of process automation can be understood as a factor that allows the integration of technology into the routine of organizations, in various areas, from the simplest to the most complex tasks, assisting managers, leaders and other employees in the most efficient performance of their activities.

At this point, automation shows its importance by strengthening and creating elementary values for the organization and advantages such as, for example [11–17]:

- Efficiency and greater productivity: it is possible to reduce the number of manual and bureaucratic processes (as well as reducing the occurrence of failures) in the business environment, leaving this task to specific software for this purpose.
- Innovation: in this context, automation enters as a factor that enhances innovation, as it employs concepts that are more aligned with the dynamics of the current market, such as information technology, among others.
- Quality of operations: automating processes is also linked to the issue of quality, including the fact that, depending on the task, work performed by machines tends to be less prone to errors and to have greater work capacity.
- Reduction of costs and time: the reduction of operational costs and time is a common consequence with the application of automation (although it is sometimes costly in the beginning, in the medium and long term it exceeds investments). Application of technology, for example, makes it possible to achieve results more quickly than work done manually, also allowing for less investment.
- Standardization of services and ease of follow-up: When automating a process, it is usually necessary to design the flow of existing activities with a view to further automation. After its automation, it is easier to monitor the flow of standardized activities

Over the years, several studies and practices focused on the application of process automation concepts have been applied in the market and in the academy motivated by different objectives, such as, for example, contributing to greater agility of results, reduction of time, resources and costs for carrying out operational activities, among others, as indicated in [18].

[19] for example, analyzed the historical perspectives on management automation identifying the impact of automation in sectors like farming, manufacturing, shipping, home care, food preparation and transportation, which showed progress over time. Office automation, agriculture and automation in travel and tourism was also studied in [20–22].

A survey performed by [23], aimed to provide information about the automation and control applications in developing regions analyzing an industry perspective of emerging technologies and challenges focused on technology projects regarding e-citizen services and smart city approach. Real industry illustrations in regards to the barriers and challenges faced during implementation of information systems in different context were provided.

[17] also pointed that in the highly industrialized countries, process automation serves to enhance product quality, master the whole range of products, improve process safety and plant availability, efficiently utilize resources and lower emissions. In the rapidly developing countries, mass production is the main motivation for applying process automation. The greatest demand for process automation identified was in the chemical industry, power generating industry, and petrochemical industry. The fastest growing demand for hardware, standard software and services of process automation was in the pharmaceutical industry.

Regarding the application of process automation in the government and public sector, [24] presented a literature review examining the impact of, and lessons from, automating government process in middle-income countries and fragile and conflict affected environments. It was pointed that automation of aspects of public sector staff recruitment, performance review, management, and monitoring can address nepotistic practices, can lead to efficiency savings on the salary bill, can improve staff and institutional performance, and can increase transparency and trust in institutions, among other impacts.

In concern of office automation, [25] designed a synthesized rural E-government system for use at the municipal to village levels. Daily information management functions are included in the system, such as social and public affairs, resources, financial transactions and so on. Because of the integration of office automation and geographical information system, the system could query the information in detail and provide an easy-to-use management platform for data management and external publicity.

In public sector, [26] mentioned that Robotic Process Automation (RPA) can reduce the amount of time staff spend on repetitive and routine activities, allowing more time to be spent on interaction with the public and jobs requiring a greater degree of complex problem solving or human judgment. Data improvements driven by RPA could also improve the quality of information available for management decision-making. RPA can therefore contribute to cost reduction targets, drive productivity and allow organizations to refocus on delivering critical public services. Opportunities in the frontline like in policing, health and education, as well as in the support services like finance, information technology and human resource were also mentioned.

According to [27–30], RPA is an emerging approach that uses software-based robots to perform structured tasks that require manual labor, for example, to capture and interpret existing applications to process a transaction, manipulate data, trigger responses and communicate with other digital systems. [27] added that, although RPA does not have as main focus the optimization, modeling or simplification of the process (traditional process automation), the approach allows the virtual workforce to perform tedious and repetitive work.

Against the background of current activities towards administrative modernization based on the digitalization of processes, according to [31] the usage and integration of RPA software into public administration work processes can significantly improve their efficiency, reduce process costs and provide better services for citizens. Discussions on the particular potential and challenges of RPA in the public administration context are also presented. Furthermore, an application example of a new cognitive RPA approach for automated data extraction and processing that was used in a trade tax assessment scenario using deep convolution neural networks (CNN) was demonstrated. Based on the findings it can be concluded that RPA has considerable potential for the improvement of the efficiency of administrative work processes and for administrative modernization in general.

Application of RPA in administrative process of public administration was also analyzed in [32], where discussions on the advantages of using RPA technology were made, and its use in administrative processes was investigated.

In concern of industry automation investments, a research performed by [33], for example, pointed out that the market for automation and artificial intelligence business operations spend will hit $15.4 billion in 2021. With the growing interest in automation technologies, one can foresee the future of process automation as one of the progressing technologies of current era. As presented in Fig. 2, a huge increase in business operations spend has been observed in the intelligent process automation segment and robot process automation (RPA).

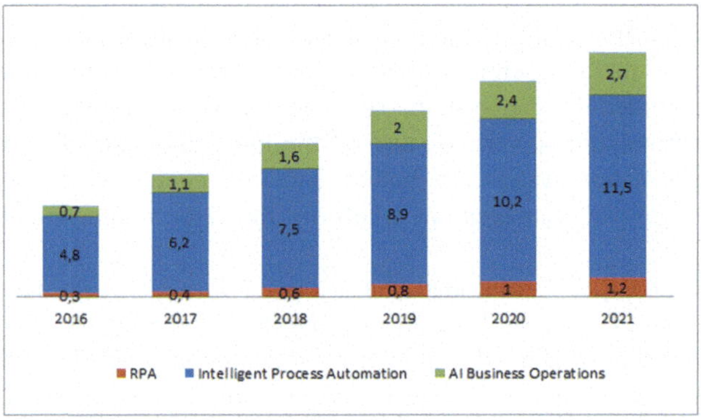

Fig. 2. Automation and AI business operations spend 2016–2021

Regarding existing approaches aimed at the application of process automation, the adoption of a strategy for the implementation of process automation can be an important factor to contribute to a higher success rate. In this context, in his work [18] mentions that the approach presented by [34], named as the USA (Understand, Simplify and Automate) principle, is common sense, and can have its strategy applied to different types of projects aimed at automation of processes.

In his approach [34] points out that the first step of the process automation application project (Understand) is to understand the current process in all its details, being able to be executed through different methods such as walking through the process, conducting an analysis of the structure work breakdown (WBS) or fishbone analysis.

With the knowledge obtained after the execution of the first step, the next step is to look for ways to simplify the process, checking, for example, the possibility of excluding unnecessary steps, unnecessary or inefficient use of resources, among other improvements. Finally, once the process has been reduced to its simplest form, automation can be considered [18, 34].

According to [34], applying the USA principle to a system implementation leads to an increase in the involvement and participation of stakeholders affected by technology, and also leads to higher levels of productivity and acceptance.

In addition to the strategy for application in projects aimed at process automation, it is worth mentioning the importance of an adequate process representation where, according to [13], information expected to be extracted from a process model must be integrated to adequately describe software processes. Among the forms of information that people ordinarily want to extract from a process model are what is going to be done, who is going to do it, when and where will it be done, how and why will it be done, and who is dependent on its being done.

For this, [13] presented four of the most commonly represented perspectives (see Sect. 4.3): functional, behavioral, organizational and informational.

It is also worth mentioning the importance of identifying criteria for applying automation in processes or workflows in order to contribute to the identification of which processes to automate. In this regard, [35] pointed out nine characteristics that can be used as criteria references for the application of automation, resulting from literature reviews and interviews with some information technology professionals.

The characteristics cited were high transaction volume, high transaction value, frequent access to multiple systems, stable environments, limited human intervention, limited exception handling, manual IT processes prone to errors or rework, ease of decomposition of the process into sub - well-defined processes and a clear understanding of current manual costs [35–37].

Thus, despite the existence of several studies that present different approaches or applications of process automation in industrial or academic environments, to the best of our knowledge, no records of application-oriented works or experiences in the context of intelligence and investigation units were found.

4 The Experience

4.1 Scenario

Considering a case of action in the fight against crimes such as, for example, corruption and money laundering, we will consider a scenario of specialized units that act or support complex criminal investigations whose involved ones deal with high and complex volumes of data [38].

Established in 2003 by Brazil's Ministry of Justice, the Department of Assets Recovery and International Legal Cooperation, at the National Secretariat of Justice, created the Technology Laboratory Against Money Laundering (LAB-LD) in 2007 to support complex investigations into corruption and money laundering [39, 40].

Being part of the Federal Laboratory Network against Money Laundering (REDELAB), in 2019 there were 52 laboratories in several investigative bodies, police and public prosecutors across Brazil. In these laboratories, a vast amount of data from a variety of sources (ex: bank account activities, email exchanges, phone records, company registrations, data from people and companies investigated) is analyzed to, for example, uncover and freeze illicit assets (among other activities), using a methodology developed by specialists and replicated throughout all laboratory units [39, 40].

An example of a general flow of requests forwarding to these laboratories is presented in Fig. 3.

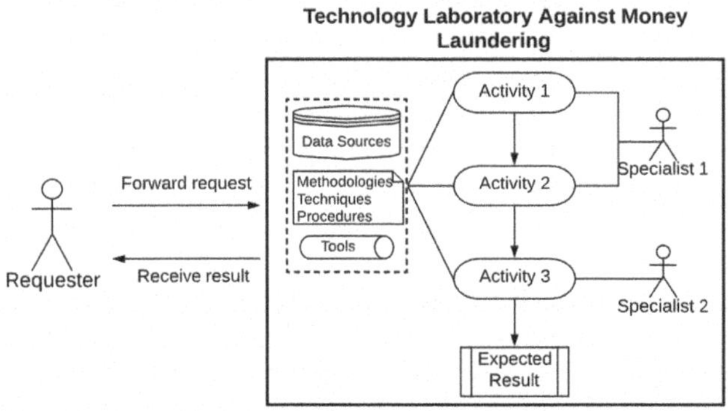

Fig. 3. Example of request flow to a technology laboratory against money laundering

In the example presented, stakeholders submit requests to laboratories, where specialists use tools, data available from various sources, methodologies, techniques, or operational procedures to perform certain activities to achieve the expected results. In order to fulfill the request, one or more specialists may be involved during the flow of activities performed to reach the expected result.

For this experience, a Technology Laboratory against Money Laundering located in a State Prosecutor's Office in Brazil was selected.

Some factors contributed to the institution's adherence to the experience of use, among them the fact that the institution is carrying out investments and research in technological solutions that contribute to actions in the fight against criminal organizations, is facing difficulties in meeting the current high volume of specialized demands (creating bottlenecks and service limitations), and having, as well as several other public institutions, limitations on financial and personnel resources.

Another important factor would be the need and importance that the institution gives on the agile treatment of requests aimed at combating crime, motivated by the impact that the performance of malicious agents has on society.

4.2 Experience Planning

To carry out the experiment, the following actions were performed:

- Selection of a process to apply automation and identification of the specialists involved.
- Interviews and information collection inherent to the process activities, aiming at its modeling and simplification before the start of automation
- Implementation of an information system developed with the objective of automating the selected process

4.3 Experience Execution

The criterion adopted by the managers and specialists team of the selected unit for the selection of the process in the experiment (after specialists identification), motivated by [35–37], was a process that had a high volume of transactions or that take a considerable time to be completed, and have frequent access to multiple systems/data sources, ease of decomposition into sub-process and pre-defined business rules.

For the extraction of the information inherent to the process activities aiming at process modeling and simplification before the start of automation, interviews were conducted with the specialists in order to identify the four of the most commonly represented perspectives according to [13]. Table 1 present the descriptions of the perspectives and examples of questions adopted in the interviews.

Table 1. Perspectives adopted for information extraction

Perspectives	Descriptions	Questions (e.g.)
Functional	Represents the activities being performed and the dependencies between them, and what flows of informational entities (e.g., data, artifacts, products), are relevant to these process elements	- What are the activities performed? - What are the expected input and output parameters for each activity? - What are the dependencies between activities? - What flows of informational entities (e.g., data, artifacts, systems, products), are relevant?
Behavioral	Represents when process elements are performed (e.g., sequencing), as well as aspects of how they are performed and control dependency information between the activities	- What is the order of execution of the identified activities? - How does the execution take place (details of operating procedures)? - How does the information transition between activities (input/output) occur?
Organizational	Describes who performs each process activity, where in the organization, and the communication mechanisms	- Who are the people responsible for carrying out each activity? - How does the iteration/communication take place between the specialists involved?
Informational	Includes the description of the entities that are manipulated by the process (e.g., artifacts, products, etc.)	- What are the entities that are manipulated by the process? - How is the interaction with systems/data sources (if any)? - What are the research standards/methodologies/techniques adopted? - What and how is the information used?

The selected process involves procedures of collecting, crossing and analyzing information, applying operational procedures based on investigative techniques and methodologies, and, finally, generation of reports resulted from analysis of individuals (natural and legal persons) within investigative contexts in the fight of crimes.

In the collection and crossing phase, specialists carry out data and information searched from various sources of organizations in which the selected institution has cooperation agreements, as well as public data bases. Then cross-checking and grouping of these data are performed in order to obtain the maximum amount of information available from investigated individuals.

Then, investigative techniques and methodologies are carried out on the information collected in order to obtain evidence of suspicious behavior normally linked to certain types of crimes. After that, the results are organized and structured into reports.

Finally, after the conclusion of the extraction and modeling of the process, the implementation of a system with the objective of automating the selected process was performed. It should be added that investments were not necessary in the implementation of the project since the solution was developed with the adoption of free tools (e.g. programming languages: PHP, Javascript and CSS; application server: Apache), and other resources already existing in the institution (a web application was developed).

4.4 Results and Discussions

In order to analyze the benefits reached with the execution of the experiment, an analysis of data obtained before and after using the developed tool was performed.

The criteria used in the analysis were indications of contributions to saving resources (e.g. time) for carrying out operational activities, obtaining results in a more agile way (quantity of requests answered) and the number of specialists involved in the process.

Table 2 shows the comparative analysis before and after the implementation of the developed tool, comparing the three criteria mentioned before its implementation, with data obtained after.

Table 2. Comparison before and after the use of the system

	Before	After
Quantity of requests answered (average 12 months)	60 per month	664 per month
Average time per request	6 to 18 h	1 to 5 min
Number of specialists involved	3	*

It should be added that the information contained in the “Before” column was provided by the specialized sectors of the institution, reporting a monthly average of the last 12 months before the implementation of the project. The information in the “After” column resulted from the verification of LOG history records of obtained from use of the system in the period of 12 (twelve) months after the beginning of its use.

The average time presented for the fulfillment of each request after the adoption of the system is only an average of the time that the tool took to attend each request and make available the final result, time that can vary depending on the speed of the internet and the location of the request.

Regarding the participation of specialists after using the system, in the initial period of use, specialists still received requisitions and used the tool to attend to them. However, from the second month of use, it was decided to make the system available directly to interested parties, where it only caused the use by specialists in a sporadic way.

Due to the possibility of using the tool directly by interested parties who previously requested the specialized unit, the number of requests handled increased exponentially.

It should be added that, from the product developed as a result of the experience performed, it was used by other specialized units that also carry out similar activities. It should also be added that the tool and the approach used for its development were awarded by evaluation committees of organizations at the state [41] and national [42] level.

Finally, despite the indications of positive benefits obtained with the execution of the experience resulting from the application of process automation in a real scenario of an intelligence and investigation unit, some considerations must be pointed out.

Before the beginning of the extraction of information to enable the modeling, simplification and subsequent automation of a process, there is a need not only to have a guide or method to select and validate processes that can be automated, but also to obtain means to enable forms to prioritize processes, as well as methods to simplify them before automation.

Despite the existence of several studies that present different approaches to process automation in industrial or academic environments, to the best of our knowledge, there is a lack of approaches aimed at guiding the application of process automation in environments of intelligence and investigation units, mainly due to the complexity and specificity of their activities, as well as the fact that they also work with classified information [43].

These issues also generate a high concern with security and confidentiality [44] during all internal operational processes including the generation, storage and/or sharing of information or generated digital assets between members of the same unit or even between different units or institutions.

For these situations, in case of exposures or leaks of information or even of digital assets such as electronic documents containing standards, methodologies, techniques or strategies adopted, it can harm the performance not only of a specialized unit, but also of the organization and others involved [9].

For the case of adopting approaches currently used in the market or even those involving the hiring of companies that provide specialized consultancy services for process optimization and improvement, or automation of operational activities with the use of specialized systems, for example, there are limited or partial applications due to restricted access to existing operating procedures, workflows and information.

Consequently, there are also challenges regarding the location and adoption of market solutions to meet many of the needs of these specialized units. For the case of specialized systems that aim to automate processes or operational activities, for example, specialized units tend to develop their own solutions and underutilize market tools previously acquired, as indicated [38] in a survey applied to the main specialized agencies in fighting crime organized in Brazil.

Therefore, in the scenario of units that perform or support intelligence or investigation activities, there is a need to conduct studies aimed at building approaches aimed at applying process automation (e.g. that assists in the analysis, validation, potential analysis, prioritization and simplification) in a way that also includes issues inherent to the management, storage or sharing of information or generated digital assets.

5 Conclusion

Due to the evolution and expansion of crime, as well as the diversity and volume of existing crime practices, researches aimed at helping or strengthening the performance of institutions that focus on fighting crime are essential.

Motivated by this scenario, and exploring the main characteristics and benefits of process automation, this paper presented an overview of different application trends and performed an experience in a real scenario of a specialized unity.

Despite the indications of positive benefits obtained with the execution of the experience, a need to conduct studies intended to the construction of approaches aimed at applying process automation to assist and guide researchers and professionals interested in studies and practices aimed at simplification and/or automation of processes in the context of intelligence and investigation units was pointed out.

Therefore, future work recommends expanding the scope of this work by building and performing an approach that assists in the analysis, validation, potential analysis, prioritization and simplification of process that considers characteristics inherent to the context of intelligence and investigation units.

References

1. UNODC: Organized Crime. United Nations Office on Drugs and Crime. https://www.unodc.org/unodc/en/organized-crime/intro.html. Accessed 18 Jan 2020
2. REFINITIV: Revealing the true cost of financial crime - 2018 Survey Report. https://www.refinitiv.com/content/dam/marketing/en_us/documents/reports/true-cost-of-financial-crime-global-focus.pdf. Accessed 20 Jan 2020
3. FBI: FBI Releases 2018 Crime Statistics. Federal Bureau of Investigation National Press Office, Department of Justice, United States, 30 September 2019. https://www.fbi.gov/news/pressrel/press-releases/fbi-releases-2018-crime-statistics. Accessed 16 Jan 2020
4. FBI: What We Investigate. Federal Bureau of Investigation Department of Justice, United States. https://www.fbi.gov/investigate. Accessed 16 Jan 2020
5. EUROPOL: Crime Areas – Fighting Crime on a Number of Fronts. European Union Agency for Law Enforcement Cooperation. https://www.europol.europa.eu/crime-areas-and-trends/crime-areas. Accessed 16 Jan 2020
6. Karjalainen, K.: Estimating the cost effects of purchasing centralization – empirical evidence from framework agreements in public sector. J. Purchasing Supply Manag. **17**, 87–97 (2011)

7. Sharma, K.L.S.: Overview of Industrial Process Automation, 2nd edn, pp. 1–14. Elsevier (2017). https://doi.org/10.1016/b978-0-12-805354-6.00001-3
8. Metscher, R., Gilbride, B.: Intelligence as an Investigative Function. International Foundation for Protection Officers, New York (2005). https://www.ifpo.org/wp-content/uploads/2013/08/intelligence.pdf. Accessed 13 Jan 2020
9. UNODC: Criminal Intelligence – Manual for Analysts. United Nations Office on Drugs and Crime. Vienna, United Nations (2011). https://www.unodc.org/documents/organized-crime/Law-Enforcement/Criminal_Intelligence_for_Analysts.pdf. Accessed 18 Jan 2020
10. Osterburg, J.W., Ward, R.H.: Criminal Investigation: A Method for Reconstructing the Past. Lexis Nexis, New Providence (2010). ISBN 9781422463284
11. Weske, M.: Business Process Management: Concepts, Languages, Architectures, 2nd edn, pp. 1–24. Springer, Heidelberg (2012). https://doi.org/10.1007/978-3-642-28616-2
12. Scheel, V.H., Rosing, V.M., Scheer, A.W.: The Complete Business Process Handbook, pp. 1–9 (2015). ISBN 978-0-12-799959-3
13. Curtis, B., Kellner, M.I., Over, J.: Process modeling. Commun. ACM **35**, 75–90 (1992). https://doi.org/10.1145/130994.130998. Special issue on analysis and modeling in software development
14. Love, J.: Process Automation Handbook: A Guide to Theory and Practice (2007). https://doi.org/10.1007/978-1-84628-282-9
15. Goldberg, K.: What is automation? IEEE Trans. Autom. Sci. Eng. **9**(1), 1–2 (2012). https://doi.org/10.1109/tase.2011.2178910
16. Nof, S.Y.: Springer Handbook of Automation. Springer, Heidelberg (2009). https://doi.org/10.1007/978-3-540-78831-7
17. Jämsä-Jounela, S.-L.: Future trends in process automation. Ann. Rev. Control **31**(2), 211–220 (2007). https://doi.org/10.1016/j.arcontrol.2007.08.003
18. Groover, M.P.: Automation, Production Systems and Computer-Integrated Manufacturing, 2nd edn. Prentice Hall, Upper Saddle River (2001)
19. Brown, A., Safford, H., Sperling, D.: Historical perspectives on managing automation and other disruptions in transportation. In: Empowering the New Mobility Workforce, pp. 3–30 (2019). https://doi.org/10.1016/b978-0-12-816088-6.00001-8
20. Azma, F., Izanlou, A., Mostafapour, M.A.: The survey relationship between office automation and employees performance in the yield tax affairs office. Procedia Technol. **1**, 153–157 (2012). https://doi.org/10.1016/j.protcy.2012.02.029
21. Tussyadiah, I.: A review of research into automation in tourism: launching the annals of tourism research curated collection on artificial intelligence and robotics in tourism. Ann. Tourism Res. **81**, 102883 (2020). https://doi.org/10.1016/j.annals.2020.102883
22. Jha, K., Doshi, A., Patel, P., Shah, M.: A comprehensive review on automation in agriculture using artificial intelligence. Artif. Intell. Agric. **2**, 1–12 (2019). https://doi.org/10.1016/j.aiia.2019.05.004
23. Aliu, A.: Automation and control applications in developing regions: An industry perspective of emerging technologies and challenges. IFAC-PapersOnLine **52**(25), 568–572 (2019). https://doi.org/10.1016/j.ifacol.2019.12.607
24. Herbert, S.: Automation of government processes. K4D Helpdesk report. Institute of Development Studies, Brighton (2019). https://www.gov.uk/dfid-research-outputs/automation-of-government-processes. Accessed 12 Jan 2020
25. Wang, J., Dong, J., Li, L., Wang, Y.: Design and implementation of an integrated office automation/geographic information system rural E-government system. In: 2010 World Automation Congress, Kobe, Japan, 19–23 September 2010, pp. 377–384 (2010)

26. Willmer, A., Duhan, J., Gibson, L.: The new machinery of government - Robotic Process Automation in the Public Sector. Deloitte Development LLC (2017). https://www2.deloitte.com/content/dam/Deloitte/uk/Documents/Innovation/deloitte-uk-innovation-the-new-machinery-of-govt.pdf. Accessed 2 Jan 2020
27. Leshob, A., Bourgouin, A., Renard, L.: Towards a process analysis approach to adopt robotic process automation. In: 2018 IEEE 15th International Conference on e-Business Engineering (ICEBE), Xi'an, China, 12–14 October 2018, pp. 46–53 (2018). https://doi.org/10.1109/icebe.2018.00018
28. Jovanović, S.Z., Đurić, J.S., Šibalija, T.V.: Robotic process automation: overview and opportunities. Int. J. Adv. Qual. **46**(3–4), 34–39 (2018). https://doi.org/10.25137//ijaq.n3-4.v46.y2018
29. Slaby, J.: Robotic automation emerges as a threat to traditional low-cost outsourcing, pp. 1–18. HfS Research, Ltd., 24 October 2012. https://www.hfsresearch.com/pointsofview/robotic-automation-emerges-threat-traditional-low-cost-outsourcing. Accessed 6 Jan 2020
30. IRPA: RPA - Definition and Benefits. Institute for Robotic Process Automation. https://irpaai.com/definition-and-benefits/. Accessed 4 Jan 2020
31. Houy, C., Hamberg, M., Fettke, P.: Robotic Process Automation in Public Administrations. Digitalisierung von Staat und Verwaltung. Lecture Notes in Informatics (LNI). Gesellschaft für Informatik, pp. 67–74 (2019)
32. Uskenbayeva, R., Kalpeyeva, Z., Satybaldiyeva, R., Moldagulova, A., Kassymova, A.: Applying of RPA in administrative processes of public administration. In: 2019 IEEE 21st Conference on Business Informatics (CBI), Moscow, Russia, pp. 9–12 (2019). https://doi.org/10.1109/cbi.2019.10089
33. Fersht, P.: Automation and AI Business Operations Spend 2016–2021. HFS Research report, November 2017
34. Kapp, K.M.: A framework for successful e-technology implementation: understand, simplify, automate. J. Organ. Excell. **21**(1), 57–64 (2001). https://doi.org/10.1002/npr.1119
35. Fung, H.P.: Criteria, use cases and effects of information technology process automation (ITPA). Adv. Robot. Autom. **3**(3), 1–11 (2014). https://doi.org/10.4172/2168-9695.1000124
36. Slaby, J.R., Fersht, P.: Robotic Automation Emerges as a Threat to Traditional Low Cost Outsourcing. HfS Research Ltd., October 2012. https://www.hfsresearch.com/pointsofview/robotic-automation-emerges-threat-traditional-low-cost-outsourcing. Accessed 8 Jan 2020
37. Sutherland, C.: Framing a Constitution for Robotistan – Racing with the Machine for Robotic Automation. HfS Research, Ltd., October 2013. https://www.hfsresearch.com/pointsofview/framing-constitution-robotistan. Accessed 2 Jan 2020
38. Santos, R., Nunes, F., Oliveira, M., Júnior, M.: A survey on the use of data mining and data analytics techniques by Brazilian criminal investigation agencies. In: Brazilian Symposium on Information Systems (SBSI), Lavras. Proceedings of the 13th Brazilian Symposium on Information Systems, pp. 593–600. Sociedade Brasileira de Computação, Porto Alegre (2017)
39. Ministry of Justice and Public Safety. Technology Laboratory Against Money Laundering - LAB-LD. Federal Government. https://www.justica.gov.br/sua-protecao/lavagem-de-dinheiro/LAB-LD. Accessed 3 Jan 2020
40. UNODC: Organized Crime. United Nations Office on Drugs and Crime. Collection of Information Prior to the Sixth Intersessional Meeting of the Open-Ended Intergovernmental Working Group on Prevention Established by the Conference of States Parties to the UN Convention against Corruption - Response to Brazil. https://www.unodc.org/documents/treaties/UNCAC/WorkingGroups/workinggroup4/2015-August-31-to-September-2/Contributions_NV/Contribution_-_Brazil.pdf. Accessed 4 Jan 2020

41. DOECE: Homologation Act and Medal Praise and Functional Merit Award. Official Gazette [of the State of Ceará], Fortaleza, Ceará, Brazil. Series 3, Year IX, no. 241, pp. 151–152, 27 December 2017
42. DIÁRIO DO NORDESTE. Public prosecutor's office of Ceará receives three national awards (2018). https://diariodonordeste.verdesmares.com.br/editorias/metro/online/ministerio-publico-do-ceara-recebe-tres-premios-nacionais-1.1999466. Accessed 8 Jan 2020
43. Drezewski, R., Sepielak, J., Filipkowski, W.: The application of social network analysis algorithms in a system supporting money laundering detection. Inf. Sci. **95**, 18–32 (2015). https://doi.org/10.1016/j.ins.2014.10.015
44. UNDP: Investigation Guidelines. United Nations Development Programme. https://www.undp.org/content/dam/undp/library/corporate/Transparency/Investigation_Guidelines_ENG_August_2019.pdf. Accessed 4 Jan 2020

Analysis, Study and Optimization of Chaotic Bifurcation Parameters Based on Logistic/Tent Chaotic Maps

Hany A. A. Mansour(✉)

Department of Electronic Warfare, Military Technical College, Cairo, Egypt
hnhn118799@mtc.edu.eg

Abstract. In recent years, chaotic sequences attracted much attention due to its attractive features. It is found that chaotic sequences are very sensitive to the initial conditions. This feature gives the ability to generate large number of sequences with special correlation properties. Chaotic sequences are generally generated from specific chaotic maps with specific control parameters. These control parameters are the initial conditions in addition to the bifurcation parameters. Actually, it's found that the bifurcation parameters have a series effect on the attitude of the output sequences and its correlation properties, especially in the multiple access applications. The contribution of this paper is to study, analyze and optimize the effect of the bifurcation parameters of the chaotic map. The applied chaotic maps are the logistic and tent map as a model of the 1D chaotic maps. The study, analysis and optimization process are performed based on the correlation properties in presence of different number of chaotic sequences.

Keywords: Chaotic Sequences · Chaotic map · Multiple accesses · Cross correlation

1 Introduction

Recently, the focus on applying the Chaotic Sequences (CSs) in various fields is increased [1]. The reason of attracting much attention of the chaotic sequences is due to its desired properties. CSs can be found in different applications of digital communications such as Code Division Multiple Accesses (CDMA), Orthogonal Frequency Division Multiplexing (OFDM), radar, in addition to image processing. CSs also can be efficiently applied in encryption, and has a significant resistant to jamming, multipath, and fading [2]. As a result, a lot of researches are focused on developing new types of sequences with enhanced properties and improved features [3]. Generally, CSs are found in dynamic nonlinear systems, or generated from chaotic maps, which gives more privileges over the traditional sequences such as the PN sequences [4]. Unlike the traditional sequences, increasing the code length of the CSs has not a serious effect on the complexity [5], meanwhile in PN sequences, the complexity increased as the code length increased, since it depends on the number of the shift register. In addition, CSs are wideband sequences have the advantage of lack of reparability; this feature

R. Silhavy (Ed.): CSOC 2020, AISC 1225, pp. 642–652, 2020.
https://doi.org/10.1007/978-3-030-51971-1_52

increases the difficulty of the prediction and reconstruction process, especially in the brute force attacks [6]. Moreover, CSs has high sensitivity to the initial values, which enable the generators to create large different spreading sequences [7, 8]. All of these presented features motivate the researches to focus on improving its properties and widely apply the CSs in various applications.

Actually, the properties of the CSs are studied, analyzed and investigated from different points of views. One of the most dominant points of view is to improve the correlation properties [9]. The correlation point of view is considered, as it has a significant effect on the Multiple Access (MA) techniques, especially the cross correlation properties. As previously mentioned, the CSs are generated from chaotic maps; these maps are constructed from bifurcation parameter(s) with specific ranges of value. The pattern of the chaotic output is strongly dependent on the value of the bifurcation parameter(s). In other words, it is found that the values of these bifurcation parameters have a significant effect on the properties of the generated sequences; consequently it is very important to study this effect on the attitude of the generated sequence.

The idea of this paper is to study, analyze, and optimize the values of the bifurcation parameters. The optimization process is manipulated based on the minimum cross correlation value between the generated sequences. The minimum cross correlation value are averaged over 10 independent sequences, to generalize the results. The study and the analysis are performed based on the logistic map and the tent map, since these maps are the most popular 1D maps. Nevertheless, the study, analysis, and the optimization process can be applied over any other map. The optimization process is performed over a small scale (short length for each code), and a large scale (long length for each code) to ensure the obtained results. Regarding to the logistic map, the study and the analysis focused on its unique bifurcation parameter, and determine its optimized value. On the other hand, the tent map has 3 different bifurcation parameters with different ranges; therefore the analysis studies the effect of each parameter individually. After that, the effect of each two parameters on each other, and on the cross correlation are studied and analyzed.

The paper is arranged as follows; Sect. 2 discusses the mathematical representation of the proposed systems. Section 3 explains the simulation results for the different cases. Finally, Sect. 4 represents the conclusion of the work.

2 Mathematical Formulation

In this section, the mathematical representation of the applied maps in addition to the binary transformation is stated. Starting with the logistic map, this can be considered as one of the simplest and most widely studied nonlinear dynamical systems that are capable of exhibiting chaotic output [10]. The mathematical representation of the logistic map can be represented as

$$F(x, r) = rx(1 - x) \tag{1}$$

Or can be represented in the recursive form as

$$x_{(n+1)} = rx_n(1 - x_n), \qquad 0 \leq x_n \leq 1, \qquad 0 \leq r \leq 4 \tag{2}$$

Where, F is the function of the transformation mapping, r is the bifurcation parameter of the map.

Considering the tent map, it can be also considered as one of the main chaotic maps that used to generate the chaotic sequences. The state space description of the first-order generalized tent map can be represented as

$$x(n+1) = A - B|x(n) - C| \tag{3}$$

Where A, B, and C represent the bifurcation parameters have specific ranges, such that $A \geq 1$, $1.5 \leq B \leq 2$, and $C \leq 1$. It is found that these parameters have serious effect on the chaotic property.

Since the output of the chaotic maps is real values, it is necessary to convert these real values into binary values. Actually there are many methods to obtain the binary values, however the conversion process is performed according to the following formula

$$C_x = g\{x(n) - E_t(x(n))\} \tag{4}$$

Where $g(x) = 1$ for $x \geq 0$, and $g(x) = -1$, for $x < 0$. $E_t(x(n))$ represents the mean function over the continuous time and C_x represents the binary output value.

As previously mentioned, the optimization process is performed based on the average cross correlation cost function, which is averaged over the total number of the generated sequences. The cross correlation cost function can be represented as:

$$C_c(\tau_i) = \sum_{i=1}^{N} x_i . y_{i+\tau_i} \tag{5}$$

Where N is the code length, $x_i \cdot y_{i+\tau_i}$ are different generated chaotic codes shifted by τ.

3 Simulation Results

This section represents the simulation of the proposed study and analysis. The analysis is started with studying the variation of the bifurcation parameter regarding to the logistic map against the average cross correlation value.

Figure 1 represents the orbit diagram, which is an attempt to track the dynamics of the function F for different values of the bifurcation parameter r [11, 12]. The orbit of x is represented against the scaling parameter r for an initial condition $x_0 = 0.1$. In this figure it is clear that the dynamics of the system can change attractively depending on the value of r and exhibiting periodicity or chaos. It is clear also that the first bifurcation occurs a $r = 3$, leading to a stable period with 2 cycles, which eventually lose stability, as $r \approx 3.45$, it gives rise to a stable with 4 cycle period. As r increases further, the scenario repeats itself over and over again. Each time a period of $2k$ cycle of the map F loses stability through a bifurcation of the map F.

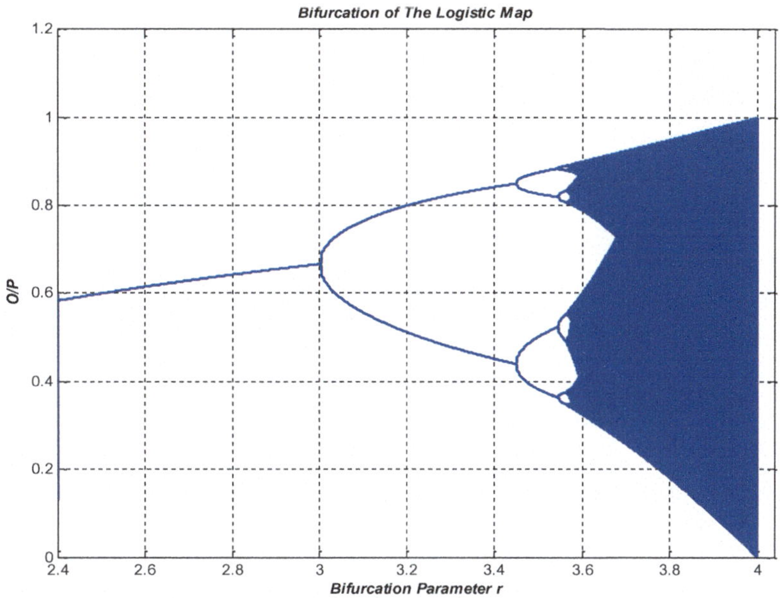

Fig. 1. Bifurcation diagram of logistic map

For $0 < r < r_c = 3.57$, the sequence $\{x_n\}$ of values of r at which cycles of period 2k appear has a finite accumulation at point $r \approx 3.57$. For $r_c < r < 4$, the sequence is, for all practical purposes, non-periodic and non-converging, and the resultant sequence will be chaotic sequence.

Moreover, a very interesting and useful feature of chaotic maps is their sensitivity to the initial value x_o, any small disturbance in the value of x_o results in completely different output sequence. For the case of the multiple accesses, the bifurcation parameter has a serious effect on the cross correlation coefficient between the spreading sequences.

Figure 2 plots the average cross correlation coefficient against the bifurcation parameter values. The cross correlation is averaged over the different sequences generated by the logistic map for the large scale and small scale cases. Regarding to the large scale case, the code length is set to be 10000 chips for each user, while in the case of the small scale; the code length is set to be 100 chips for each user. In both cases, the simulation is performed for 10 different sequences, and the results are averaged over a 50 independent trials. It is clear from the figure that generally the two plots have similar attitudes, which ensure that the results are correct. It is found that the cross correlation coefficient has large values for $r < r_c = 3.57$. As $r = r_c = 3.57$, the cross correlation coefficient begins to decrease rapidly as r increased, until reach to the minimum value at $r = r_{opt} = 3.9$.

Regarding to the Tent map, as previously mentioned, it consists of 3 different bifurcation parameters A, B, and C. These parameters have serious effect on the chaotic property, which can be clear from the bifurcation plot of the tent map shown in Fig. 3.

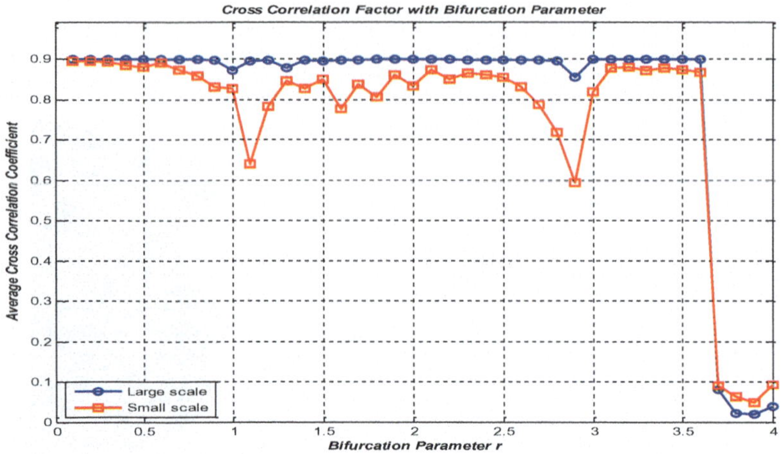

Fig. 2. Effect of bifurcation parameter on the average cross correlation factor

In this figure, the output value is plotted against the bifurcation parameter B which is ranging between 0.8 and 2 and the chaotic sequences value is confined in the interval range $\{-1, 1\}$. The figure shows that the tent map exhibits the chaotic behavior as the value of B being larger than 1. However, as the number of both ones and negative ones would be judge balanced, the parameter B should have to assign to be in the range of $1.8 \leq B \leq 2$.

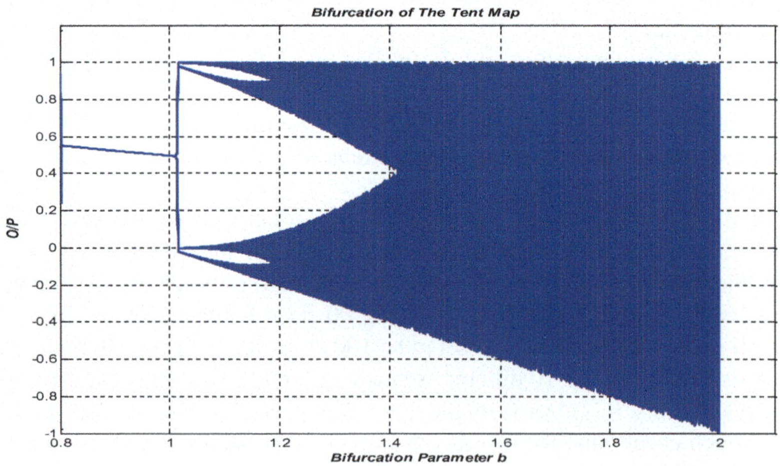

Fig. 3. The bifurcation diagram of tent map

Concerning the study of the effect of the bifurcation parameter of the tent map, the effect of each parameter will be studied and analyzed individually, and then the effect of each pair of the parameters will be studied and analyzed simultaneously. Similarly as stated in the logistic map, the simulation is performed on both the small scale and the large scale. In the small scale, the applied code length is set to be 100 chips for different 10 different users, which represents the MAI. While in the large scale, the applied code length increased to be 1000 chips for the same 10 MAI. The obtained cross correlation is averaged over 50 independent trials with different random initial condition for each user all over each trial.

Figure 4 inspects the first bifurcation parameter A for both the large scale and small scale case. The inspection range is set to be from 1 to 2 with step size 0.01, and the values of the other bifurcation parameters B and C are set initially to be 2 and 0.5 respectively. The figure plots the value of the average cross correlation (ACC) against the range of parameter A. It shows that generally the ACC for the large scale is lower than that of the small scale. It is noted also that the general ACC is accepted all over the range except for a certain values which are [1, 1.25, 1.5, 1.75, and 2]. Referred to the small scale, the minimum value of the ACC is 0.0664 obtained at value of A equal to 1.02. While for the large scale, the minimum value of the ACC is 0.0205 obtained at value of A equal to 1.11.

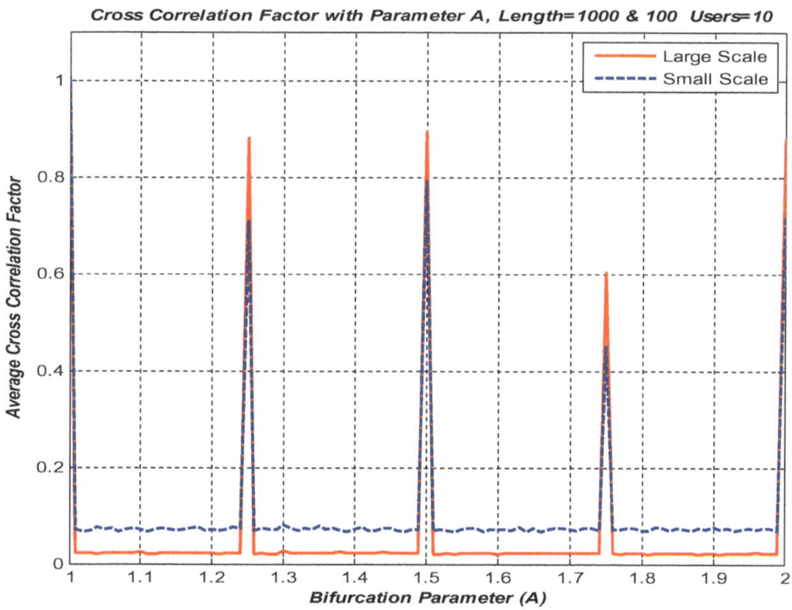

Fig. 4. Average cross correlation value with parameter A, length = 1000 & 100 users = 10

Figure 5 discusses the attitude of the second bifurcation parameter B against the ACC. The range of parameter B is set to be from 1.5 to 2 with step size 0.01, while the values of the other parameters A and C are set to be 1.91, and 1 respectively. Similarly as mentioned in Fig. 4, it can be concluded from the Fig. 5 that generally the small scale has ACC value higher than that of the Large scale. In addition, in case of the small scale, the attitude of the parameter B with the ACC is slightly improved as the value of B is increased. While in case of the large scale, the attitude of the parameter B is nearly fixed. From the results, it is found that the optimum value of B is 1.97 which is correspondent to the minimum ACC equal to 0.067 related to the small scale. Related to the large scale, the optimum value of B is 1.91, which gives the minimum value of ACC equal to 0.0236. The figure also shows that the accurate value of B at which the ACC begin to be accepted is 1.51.

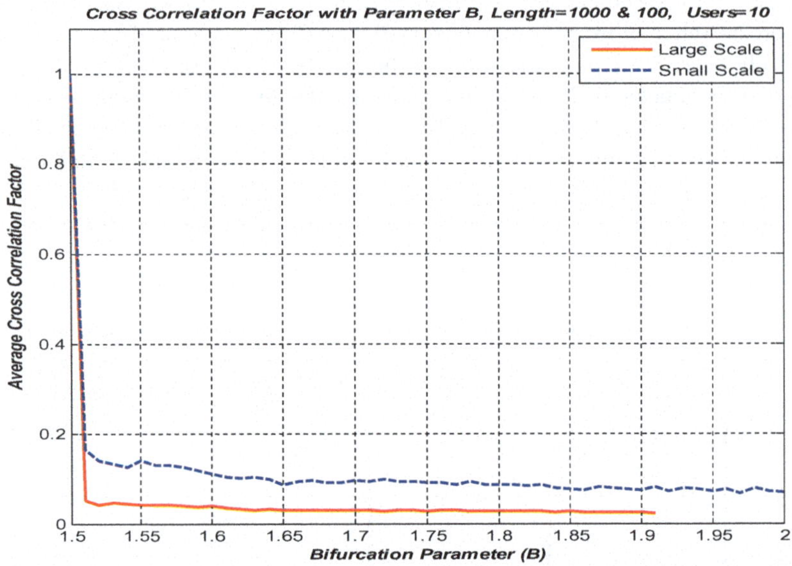

Fig. 5. Average cross correlation value with parameter B, length = 1000 & 100 users = 10

Following the same trend, Fig. 6 plots the performance of parameter C against the ACC on both cases of large and small scale. The range of the parameter C is set to be from 0 to 1 with step size 0.01, while the values of parameters A and B are set to be 1.91, and 2 respectively. The figure shows that both the large scale and small scale have the same attitude, in which the minimum value of C at which the ACC is being accepted is 0.05. The figure shows also that the large scale has ACC values lower than that of the small scale. Related to the small scale the optimum value of C is 0.7 correspondent to the minimum ACC equal to 0.0676. While for the large scale, the optimum value of C is 0.4 correspondent to the minimum ACC equal to 0.0204.

Regarding to the variation of the ACC with each two parameter, the effect of each pair of the parameters on each other and on the ACC is studied. Figure 7 presents the plot of both parameters A and B across the ACC. The ranges of parameters A and B are set to be from 1 to 2 and from 1.5 to 2 respectively with step size equal to 0.05 for each. The fixed value of the third parameter C is set to be 0.5. The figure ensures the results obtained in Fig. 3, in which the ACC is high at certain values of parameter A as mentioned before. It is found that the minimum value of the ACC is obtained at the optimum value of A equal to 1.11, and optimum value of B equal to 1.89.

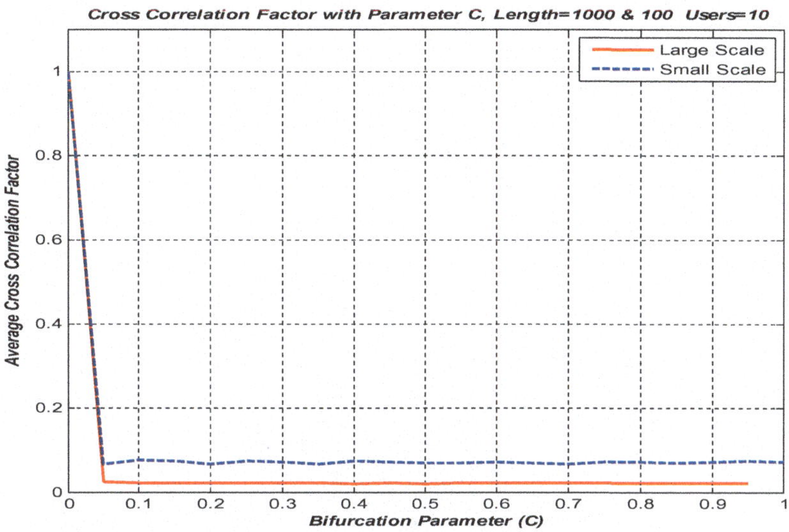

Fig. 6. Average cross correlation value with parameter C, length = 1000 & 100 users = 10

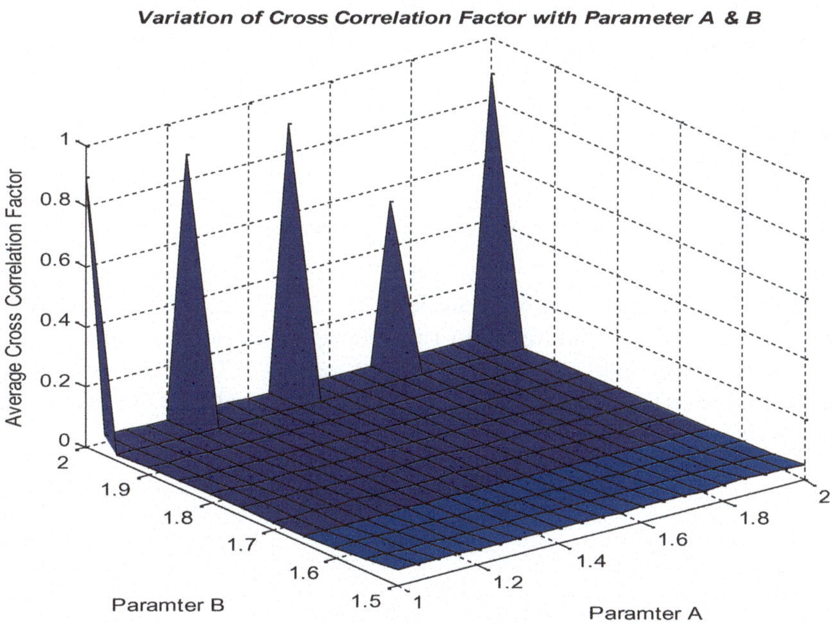

Fig. 7. Average cross correlation value with parameters A, B length = 1000, users = 10

Figure 8 discusses the effect of parameters A, C across the ACC. The rages of A, C are set to be from 1 to 2 and from 0 to 1 respectively with step size equal to 0.05 for each, while the fixed value of parameter B is set to be 1.9. The results show that the minimum ACC is equal to 0.0222, and obtained at the optimum value of A equal to 1.55, and optimum value of B equal to 0.55.

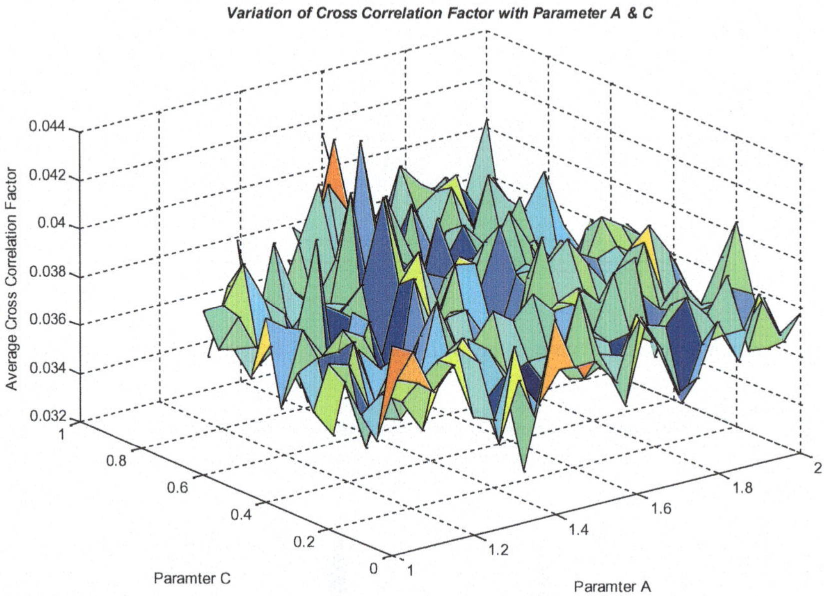

Fig. 8. Average cross correlation value with parameters A, C length = 1000, users = 10

Figure 9 illustrates the effect of parameters B, C across the ACC. The rages of B, C are set to be from 1.5 to 2 and from 0 to 1 respectively with step size equal to 0.05 for each, while the fixed value of parameter A is set to be 1.1. The results show that the minimum ACC is 0.0.0218 and obtained at the optimum value of B equal to 2, and optimum value of C equal to 0.7500.

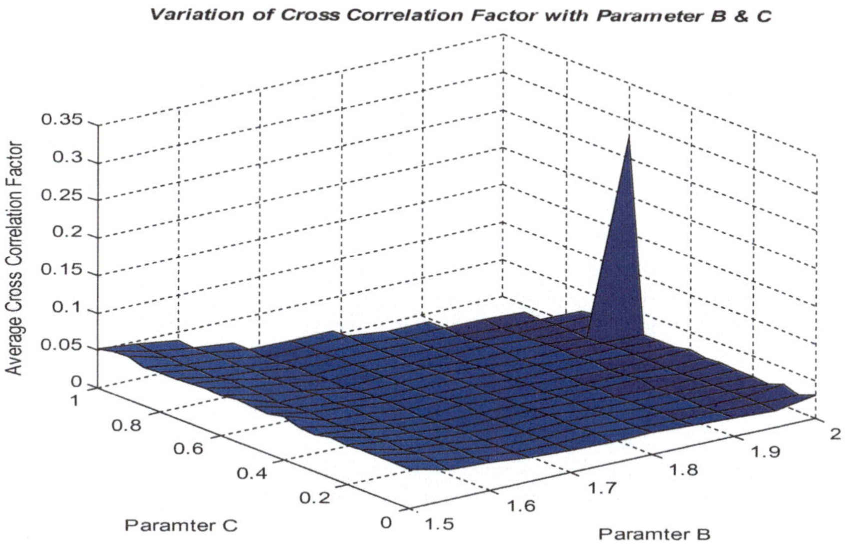

Fig. 9. Average cross correlation value with parameters B, C length = 1000, users = 10

4 Conclusions

The paper studies and analyses the effect of the bifurcation parameter on the cross correlation properties. Based on the study and the analysis, the value of the bifurcation parameter is optimized according to the minimum average cross correlation. The study, analysis, and optimization can be applied and generalized over any chaotic maps, however the logistic and the tent maps are considered due to their simplicity and popularity. The optimization process is performed over a small scale (code length equal 100), and over large scale (code length equal 10000) in presence of 10 different sequences. The considered cross correlation is averaged over 10 independent sequences. Regarding to the logistic map, it is found that its unique parameter has an optimum value equal to 3.9, corresponding to the minimum cross correlation. Considering the tent map, the 3 bifurcation parameters A, B, and C are studied, analyzed, and optimized individually over both the small and the large scales. The optimum value of each parameter is determined corresponding to the minimum average cross correlation. In addition, non-recommended values are determined which gives relatively high values of cross correlation. Moreover, the study is also including the effect of each two parameter across the average cross correlation. The results determine the recommended value of each parameter corresponding to the minimum average cross correlation.

References

1. Li, S., Zhao, Y., Wu, Z.: Design and analysis of an OFDM-based differential chaos shift keying communication system. J. Commun. **10**(3), 199–205 (2015)
2. Manoharan, S., Bhaskar, V.: PN codes versus chaotic codes: Performance comparison in a Gaussian approximated wideband CDMA system over Weibull fading channels. J. Franklin Inst. **351**, 3378–3404 (2014)
3. Wang, J., Wang, Y.: Analysis performance of MC-CDMA communication system based on improved Chebyshev sequence. In: 2nd IEEE International Conference on Computer and Communications (2016)
4. Liu, G., Liu, H., Kadir, A.: Hiding message into DNA sequence through DNA coding and chaotic maps. Med. Biol. Eng. Comput. **52**, 741–747 (2014). https://doi.org/10.1007/s11517-014-1177-3
5. Zhou, C.: Turbo trellis-coded differential chaotic modulation. IEEE Trans. Circ. Syst. II: Exp. Briefs **65**(2), 191–195 (2018)
6. Sharifi, M., Jafarpour Jalali, M.: Using chaotic sequence in direct sequence spread spectrum based on code division multiple access (DS-CDMA). ARPN J. Eng. Appl. Sci. **12**(20) (2017)
7. Tayebi, A., Berber, S., Swain, A.: Performance analysis of chaotic DSSS-CDMA synchronization under jamming attack. Circ. Syst. Signal Process. **35**(12), 4350–4371 (2016). https://doi.org/10.1007/s00034-016-0266-y
8. Swetha, A., Krishna, B.T.: Generation of biphase sequences using different logistic maps. In: International Conference on Communication and Signal Processing, 6–8 April 2016
9. Litvinenko, A., Aboltins, A.: Use of cross-correlation minimization for performance enhancement of chaotic spreading sequence based asynchronous DS-CDMA system. IEEE Trans (2016)
10. Ksheerasagar, T.K., Anuradha, S., Avadhootha, G., Charan, K.S.R., Sri Hari Rao, P.: Performance Analysis of DS-CDMA using different Chaotic Sequences. 978-1-4673-9338-6/16/$31.00@ IEEE (2016)
11. Reddy, G.V.: Performance evaluation of different DS-CDMA receivers using chaotic sequences. Master thesis (2007)
12. Gizem, A., Ayese, O.: Communications & Networks. Network Books, ABC Publishers (2009)

Correction to: Development of an Educational Device Based on a Legacy Blood Centrifuge

Mohamed Abdelkader Aboamer

Correction to: Chapter "Development of an Educational Device Based on a Legacy Blood Centrifuge" in: R. Silhavy (Ed.): *Artificial Intelligence and Bioinspired Computational Methods*, AISC 1225, https://doi.org/10.1007/978-3-030-51971-1_11

In the original version of the book, the following belated correction has been incorporated:
In Chapter 11, the affiliation of author Mohamed Abdelkader Aboamer has been changed from "Department of Medical Equipment, Faculty of Applied Science, Majmaah University, Abha, Saudi Arabia" to "Department of Medical Equipment Technology, College of Applied Medical Sciences, Majmaah University, Al Majmaah 11952, Saudi Arabia".
The book and the chapter have been updated with the change.

The updated version of this chapter can be found at
https://doi.org/10.1007/978-3-030-51971-1_11

R. Silhavy (Ed.): CSOC 2020, AISC 1225, p. C1, 2021.
https://doi.org/10.1007/978-3-030-51971-1_53

Author Index

R. Silhavy (Ed.): CSOC 2020, AISC 1225, pp. 653–655, 2020.
https://doi.org/10.1007/978-3-030-51971-1

The manufacturer's authorised representative in the EU is Springer Nature Customer Service Centre GmbH, Europaplatz 3, 69115 Heidelberg, Germany. If you have any concerns regarding our products, please contact ProductSafety@springernature.com

Printed and bound by CPI Group (UK) Ltd, Croydon, CR0 4YY
22/09/2026
02220979-0014